Advances in Intelligent Systems and Computing

Volume 1171

AF892984

The series “Advances in Intelligent Systems and Computing” contains publications on theory, applications, and design methods of Intelligent Systems and Intelligent Computing. Virtually all disciplines such as engineering, natural sciences, computer and information science, ICT, economics, business, e-commerce, environment, healthcare, life science are covered. The list of topics spans all the areas of modern intelligent systems and computing such as: computational intelligence, soft computing including neural networks, fuzzy systems, evolutionary computing and the fusion of these paradigms, social intelligence, ambient intelligence, computational neuroscience, artificial life, virtual worlds and society, cognitive science and systems, Perception and Vision, DNA and immune based systems, self-organizing and adaptive systems, e-Learning and teaching, human-centered and human-centric computing, recommender systems, intelligent control, robotics and mechatronics including human-machine teaming, knowledge-based paradigms, learning paradigms, machine ethics, intelligent data analysis, knowledge management, intelligent agents, intelligent decision making and support, intelligent network security, trust management, interactive entertainment, Web intelligence and multimedia.

The publications within “Advances in Intelligent Systems and Computing” are primarily proceedings of important conferences, symposia and congresses. They cover significant recent developments in the field, both of a foundational and applicable character. An important characteristic feature of the series is the short publication time and world-wide distribution. This permits a rapid and broad dissemination of research results.

**** Indexing: The books of this series are submitted to ISI Proceedings, EI-Compendex, DBLP, SCOPUS, Google Scholar and Springerlink ****

More information about this series at http://www.springer.com/series/11156

Suresh Chandra Satapathy · Vikrant Bhateja ·
B. Janakiramaiah · Yen-Wei Chen
Editors

Intelligent System Design

Proceedings of Intelligent System Design: INDIA 2019

Springer

Editors
Suresh Chandra Satapathy
School of Computer Engineering
KIIT Demmed to be University
Bhubaneswar, Odisha, India

B. Janakiramaiah
Department of Computer Science
and Engineering
PVP Siddhartha Institute of Technology
Vijayawada, Andhra Pradesh, India

Vikrant Bhateja
Department of Electronics
and Communication Engineering
Shri Ramswaroop Memorial Group
of Professional Colleges (SRMGPC)
Lucknow, Uttar Pradesh, India

Dr. A.P.J. Abdul Kalam Technical
University
Lucknow, Uttar Pradesh, India

Yen-Wei Chen
College of Information Science
and Engineering
Ritsumeikan University
Kyoto, Japan

ISSN 2194-5357 ISSN 2194-5365 (electronic)
Advances in Intelligent Systems and Computing
ISBN 978-981-15-5399-8 ISBN 978-981-15-5400-1 (eBook)
https://doi.org/10.1007/978-981-15-5400-1

This Springer imprint is published by the registered company Springer Nature Singapore Pte Ltd.
The registered company address is: 152 Beach Road, #21-01/04 Gateway East, Singapore 189721, Singapore

Preface

This volume contains the papers that were presented at the 6th International Conference on Information System Design and Intelligent Applications (INDIA) organized by the Department of Computer Science and Engineering, LIET, Visakhapatnam, India, during November 1–2, 2019. It provided a great platform for researchers from across the world to report, deliberate, and review the latest progress in the cutting-edge research pertaining to smart computing and its applications to various engineering fields.

The response to INDIA was overwhelming with a good number of submissions from different areas relating to machine learning, intelligent system designing, deep learning, evolutionary computation, etc. After a rigorous peer review process, with the help of program committee members and external reviewers, only quality papers were accepted for publication in this volume of Springer.

Our thanks are due to Dr Neeraj Gupta, India, and Dr Naeem Hannon, Malaysia, for keynote address during these two days. We are thankful to Dr PVGD Prasad Reddy, Hon. VC, Andhra University, Visakhapatnam, for his inaugural address being the Chief Guest. We would like to express our appreciation to the members of the Program Committee for their support and cooperation in this publication. We are also thankful to the Team from Springer for providing a meticulous service for the timely production of this volume.

Our heartfelt thanks to management committee members of LIET for their support for hosting the conference. We place our appreciation to the Principal of LIET and HOD of the Department of CSE for continuous support and motivation. Special thanks to all Guests who have honored us in their presence in the inaugural day of the conference. Our thanks are due to all special session chairs, track managers, and reviewers for their excellent support. Support of all faculty members and student volunteers of LIET is praise worthy. Last but certainly not least, our special thanks go to all the authors who submitted papers and all the attendees for their contributions and fruitful discussions that made this conference a great success.

Bhubaneswar, India — Suresh Chandra Satapathy
Lucknow, India — Vikrant Bhateja
Vijayawada, India — B. Janakiramaiah
Kyoto, Japan — Yen-Wei Chen

Editorial Board of INDIA

Contents

About the Editors

Suresh Chandra Satapathy is a Professor, School of Computer Engg, KIIT Deemed to be University, Bhubaneswar, India. His research interest includes machine learning, data mining, swarm intelligence studies and their applications to engineering. He has more than 140 publications to his credit in various reputed international journals and conference Proceedings. He has edited many volumes from Springer AISC, LNEE, SIST etc. He is a senior member of IEEE and Life Member of Computer society of India.

Vikrant Bhateja is Associate Professor, Department of ECE in SRMGPC, Lucknow. His areas of research include digital image and video processing, computer vision, medical imaging, machine learning, pattern analysis and recognition. He has around 150 quality publications in various international journals and conference proceedings. He is associate editor of IJSE and IJACI. He has edited more than 22 volumes of conference proceedings with Springer Nature and is presently EiC of IGI Global: IJNCR journal.

B. Janakiramaiah is a Prof in the Dept of CSE, PVPSIT, Vijayawada, India. He obtained his PhD in Computer Science from JNTU Hyderabad. He has ample experience being in Editorial board of Spinger AISC and LNNs of IC3T 2016 and also acted as reviewer for various international journals from Elsevier, springer, IOS press etc. His research domain is Intelligent system with Soft Computing. He has quality publications in international journals and also many publications in conf proceedings. He is a member of CSI, IEEE etc.

Yen-Wei Chen received the B.E. degree in 1985 from Kobe Univ., Kobe, Japan, the M.E. degree in 1987, and the D.E. degree in 1990, both from Osaka Univ., Osaka, Japan. He was a research fellow with the Institute for Laser Technology, Osaka, from 1991 to 1994. From Oct. 1994 to Mar. 2004, he was an associate Professor and a professor with the Department of Electrical and Electronic Engineering, Univ. of the Ryukyus, Okinawa, Japan. He is currently a professor with the college of Information Science and Engineering, Ritsumeikan University, Japan. He is also a visiting professor with the

College of Computer Science, Zhejiang University, China. He was a visiting professor with the Oxford University, Oxford, UK in 2003 and a visiting professor with Pennsylvania State University, USA in 2010.

His research interests include medical image analysis, computer vision and computational intelligence. He has published more than 300 research papers in a number of leading journals and leading conferences including IEEE Trans. Image Processing, IEEE Trans. SMC, Pattern Recognition. He has received many distinguished awards including ICPR2012 Best Scientific Paper Award, 2014 JAMIT Best Paper Award, Outstanding Chinese Oversea Scholar Fund of Chinese Academy of Science. He is/was a leader of numerous national and industrial research projects.

Acceptance of Technology in the Classroom: A Qualitative Analysis of Mathematics Teachers' Perceptions

Perienen Appavoo

Abstract The degree of integration of ICT in education varies within contexts. Accordingly, teachers have different opinions and beliefs on the practicability and educational worth of the integration. This research was carried out to collect the views of Mathematics teachers after a blended model, combining the traditional approach with ICT-based lessons, was used to teach the topic fractions to junior secondary school students. These class teachers were thus able to give an informed opinion of the process. Data collected led to the construction of a Technology Implementation Model. The four emerging themes of the interviews were 'learner empowerment', 'effective teaching', 'inhibiting factors' and 'teacher support'. Generally, teachers were positive about the pedagogical worth of ICT and expressed their willingness to see technology as part of the teaching/learning process. However, apprehension and concerns were also voiced out and one key element highlighted was the systemic and systematic professional development of teachers.

Keywords Teacher professional development · Integration of technology · Mathematics · ICT-based lessons

1 Introduction

Today, we are witnessing a major shift in the way we conduct business and do our activities because of the tremendous influence of technological affordances. The education sector has not escaped this wave of technology integration and teaching and learning are taking new turns to bring learning content in innovative ways. Many of those who are called to embrace this new paradigm have never used any technology in their learning. What is required of them is a novel way of teaching, to which they have scantly been exposed. It is, therefore, appropriate to investigate to what extent teachers are ready to accept computing tools in the classroom. Closely linked to that is the readiness of teachers to operate these tools. As mentioned by Tondeur et al. [1]

P. Appavoo (✉)
Open University of Mauritius, Moka, Mauritius
e-mail: p.appavoo@open.ac.mu

S. C. Satapathy et al. (eds.), *Intelligent System Design*, Advances in Intelligent Systems and Computing 1171, https://doi.org/10.1007/978-981-15-5400-1_1

'merely providing ICT does not inevitably improve learning, but beyond access, it is how teachers use ICT that makes a difference, and Teacher professional development (TPD) is critical to achieving valued outcomes'. Moreover, teachers' opinions and beliefs count, and helping them to develop their knowledge and attitudes can promote a culture that supports ICT as an integral part of the learning and teaching process. By now, every teacher in Mauritius has been exposed to educational technology to some extent and every school is equipped with computers [2]. But there is scant information about what is happening in the classroom and what teachers as key players feel about this integrative process. This paper seeks to analyze the opinions and beliefs of practicing Mathematics teachers who have been exposed to a real classroom situation whereby both traditional practices and ICT-based lessons were used to teach the Mathematics topic, fractions.

2 Literature Review

Today, there is mounting pressure from all quarters to use innovative tools to meet the emerging learning styles of students. Great challenges are thus awaiting teachers in this new era of technological transformation. Teachers in Israel have positive perceptions of their competence in technology and have embraced technology in the teaching of Mathematics [3]. It was found that teachers with routine access to computers tend to employ teaching practices that put students at the centre of learning [4, p. 1]. Teachers, being an integral part of the teaching and learning process, will play a major role in the adoption and implementation of ICT in education [5]. Lim et al. [6] purported that teachers' personal ICT conceptions affected how they used ICT in their teaching, and teachers' belief in the potential of ICT was an important factor that contributed to the high frequencies of ICT usage. The overall conclusion from the works of Uluyol and Sahin [7] is that more concrete encouragement, support and opportunities must be developed to increase teachers' motivation, and thus improve the level and quality of ICT use in classrooms. Gul [8] investigated technology integration and reported that teachers' attitude to technology and willingness to use technology were significant factors. Teachers need ICT-pedagogical skills to be able to integrate technology-enhanced lessons in their teaching.

Unfortunately, many teachers are grappling with ICT tools to channel them towards sound pedagogical gains and meet the emerging demand for education [9]. In an educational process still constrained by the traditions and practices of the past, the integration of technology is not an automatic process and encompasses a number of factors, with the teacher as the lead agent of change. It is known that in the history of new technologies in education, many teachers have in varying ways resisted, through fear and anxiety, lack of competence, poor leadership from senior staff and inadequate technical support [10]. How are teachers supposed to integrate what they know with technology? Koehler et al. [11] contend that there is no 'one best way' to integrate technology with curriculum. Following a survey carried out by Blackwell et al. [12] with early childhood educators, it was found that more than anything else, attitudes toward the value of technology to aid children's learning have the strongest effect on technology use, followed by confidence and support in using technology. It was argued that apart from the knowledge, skills and attitudes, teachers' need, beliefs about teaching and learning with technology also mattered for adequate teaching in the knowledge society [13]. Ertmer [14] purported that any new knowledge base will remain unused unless teachers make sense of its application within their prevalent pedagogical understandings. Whether through choice or necessity, there is an increasing number of teachers using both ICT-mediated and face-to-face teaching and learning [15].

When it comes to the teaching of Mathematics, teachers have to be trained in the innovative use of technological tools [16]. As highlighted by Tondeur et al. [1], systemic (stakeholders, local factors) and systematic (gradual and evolving) teacher professional development (TPD) is one of the five challenges to effective teacher empowerment. Despite the panoply of academic works already documented in this field, Aris and Orcos [17] believe that it is crucial to continue research on the teacher educational experiment for future implementation of ICT. This study will investigate the teachers' perception/beliefs regarding the pedagogical worth of technology in the teaching of Mathematics.

2.1 TAM and UTAUT

Successful uptake of technology in the teaching/learning process starts with technology acceptance and one framework that has been widely adopted and researched is the technology acceptance model (TAM), based on the theory of reasoned action (TRA) by Fishbein and Ajzen [18]. TRA examines the relationship between beliefs, intentions, attitudes and the behaviour of individuals. According to this model, a person's behaviour is determined by its behavioural intention to perform it. Bandura [19] also highlighted the importance of perceived usefulness and perceived ease of use in predicting behaviour. Perceived Usefulness (PU) measures the efficacy identified by the user while perceived ease of use (PEOU) identifies the difficulty level of the technology perceived by the user [20]. PEOU and PU form the two major components of TAM which is a theoretical framework that provides a systematic way to make predictions about technology acceptance and computer usage behaviours. This technology acceptance model rests upon the fact that the way a person perceives the usefulness (PU) of technology and its ease of use (PEOU) will determine the way that person makes use of that particular technology. The first proponent of TAM [21] based the model on the assumption that user motivation can be explained by three factors: (1) perceived ease of use, (2) perceived usefulness and (3) attitude toward usage. Davis [21] hypothesized that the attitude of a person towards a system was a major factor that influenced whether he/she would use or reject the system. In turn, the person's attitude seems to be influenced by two major beliefs: perceived usefulness and perceived ease of use, where perceived ease of use has a direct influence on perceived usefulness.

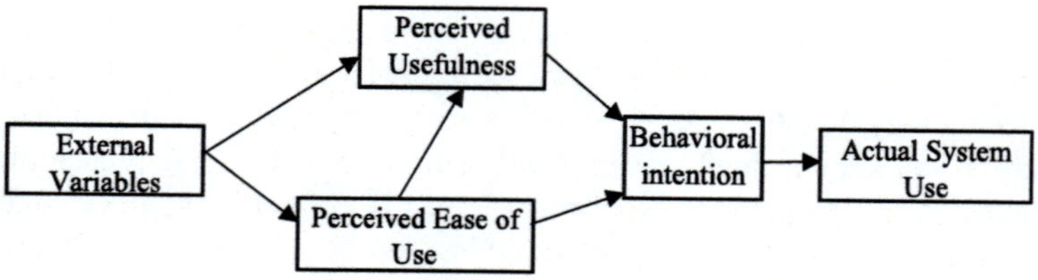

Fig. 1 Final version of TAM by Venkatesh and Davis [23] (*Source* Lala [20])

Davis et al. [22] later found that perceived usefulness was a major significant determinant of peoples' intention to use computers while perceived ease of use was only a significant secondary determinant. In 1996, Venkatesh and Davis [23] proposed a new refined model of TAM (Fig. 1).

However, according to Lala [20], there is a need for future research that can focus on developing new models that exploit the strengths of TAM because though it is a very popular model for explaining and predicting the use of a system, it does have some uncertainty among some researchers about its application and accuracy. This model has been largely reviewed by many researchers and additional factors and variables identified to include extrinsic and intrinsic motivation, self-efficacy [19], behavioural intention [21] and opinion of others [23], all of which were found to influence the adoption and usage of new technologies. Ensminger [24] proposed that examining teachers' perceptions of these variables can help those in charge of planning for technology gain a deeper insight into what might affect the final use of technology in the classroom.

2.2 *Aim and Objectives of the Study*

The aim of this study is to draw a theoretical framework that depicts teachers' opinions and beliefs about the integrative process of ICT in the classroom. This was achieved through the following research questions:

1. Which elements of ICT-based lessons did teachers find beneficial?
2. Which elements of ICT-based lessons were challenging to teachers?
3. What kind of support must be given to teachers for the integration of technology?

3 Methodology

The topic fractions was taught in five different schools during a period of two weeks in each school. Teaching was done using a digital learning software comprising interactive powerpoints, instructional videos and apps, e-exercises and worksheets, all loaded on tablets. Students worked in pairs using the tablets and class teachers were present throughout the experiment to witness how technology, blended with the traditional practices, was used to teach Mathematics. The researcher conducted the experiment in each school, hence guaranteeing consistency. Once the experiment was over, semi-structured interviews, guided by questions based on literature review, were conducted with the class teachers. Subsequent interviews were slightly modified based on emerging findings in view of seeking further clarification. Five teachers were interviewed, two female and three males. Each interview lasted around one hour. Beliefs, attitudes and opinions of teachers regarding the experiment were recorded and codified around central themes addressing the pedagogy, learning conditions, performance gains and apprehensions.

4 Data Analysis

Once the five interviews had been transcribed verbatim, the Word documents were then uploaded on the computer-assisted qualitative data analysis software Atlas.ti. During the experiment, informal chats were held with the class teachers to collect views and opinions of the process of technology integration in the teaching/learning of fractions.

This qualitative method resulted in a vast amount of richly detailed data that was contextually laden and subjective, hence revealing the perception and beliefs of those teachers exposed to the experiment. The process of thematic content analysis was adopted to identify themes and categories that 'emerge from the data'. Particular care was given to deviant or contrary cases—i.e. findings that were different or contradictory to the main ones, or were simply unique to some or even just one respondent.

One hundred and thirteen quotations were highlighted from the five Word documents to generate 32 codes. These codes were later reduced to 14 with appropriate renaming, and then these were finally grouped under four main themes namely, *learner empowerment*, *effective teaching*, *inhibiting factors* and *teacher support* (Fig. 2).

Teachers noted four key elements with regard to *learner empowerment*. They observed that students were motivated and showed great interest in their Mathematics lessons, factors which have been vastly researched and reported upon by Jewitt et al. [25] and Penuel [26]. They discussed the benefits of students taking ownership of their learning, giving them in the same breath independence to learn and progress at their own pace. They also appreciated that peer tutoring evolved as a natural practice of this new learning environment. Tsuei [27] reported that peer tutoring is one of the most well-studied strategies in Mathematics instruction and proved that technology was effective in enhancing the learning of Mathematics for students with learning disabilities.

The second emerging theme related to *effective teaching*. Teachers mentioned that learning content, enhanced with video clips, slides presentation, mathematical games and interactive exercises, could make them more effective in their teaching. They

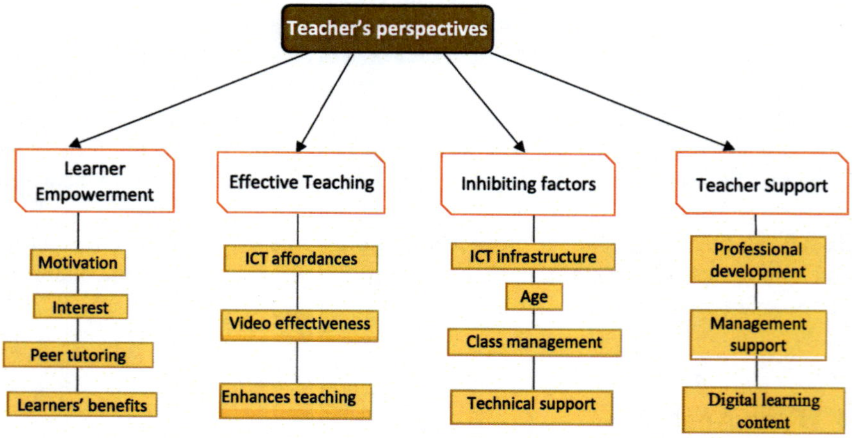

Fig. 2 Technology implementation framework—a teacher's perspective

appreciated the pedagogical worth of the videos which offered greater and enhanced learning experiences to students [28], but were also concerned with the narrator's language and pronunciation.

Teachers valued the interactivity of the learning content and its multimedia components as driving forces that rendered the Mathematics lessons more appealing and engaging. Moreover, they could see technology facilitating effective lesson planning, where future lessons can be easily enhanced, edited and updated. These interviews demonstrated that teachers were able to appreciate the worth of ICT as a teaching tool that could ease the teaching of abstract and difficult concepts.

One teacher considered the computer as a convenient tool for managing student's personal data, marks and grades. In general, teachers opined that this mode of instruction provided more class time to attend to needy students while high performers proceeded with further work. This was seen as convenient to work with mixed-ability students, offering teachers the ability to transform the quality of instruction and promote a more student-centered learning environment as proposed by Penuel [26].

The third theme related to factors that they considered as inhibiting the successful uptake of technology in their teaching. If teachers were positive about the benefits of technology integration, yet they drew attention to the poor and inappropriate technological infrastructure and restricted access in schools. They voiced out the unpreparedness in terms of skills required to channel the affordances of technology in their teaching, especially by the elder teachers who were the most resistant to change. As described by Archambault et al. [29], there is a need to provide teachers training in progressing technologies that can help them transform their pedagogy to leverage the affordances provided by ICT integration. For two teachers, managing digitally enhanced classrooms can have its own challenges, in terms of discipline, class management and ensuring that computer activities were really geared towards learning, and attending to technical failures especially where technical support was scarce. Penuel [26] reported that in addition to teacher professional development and positive teacher attitudes, access to technical support is a key factor directing the effective implementation of ICT in schools. One teacher also proposed to reduce the class size to make the class more manageable.

A fourth theme emerged from the interviews and focused on teacher's support. Professional development to master the skills of working with digital content was commonly mentioned by the teachers as a non-negotiable prerequisite for the successful uptake of ICT by teachers. Such findings have been reported by Penuel [26] and Minshew and Anderson [30] where attention was drawn to the fact that the lack of adequate professional development can become a source of frustration.

Teachers said it would be appropriate if the digital learning content could be made readily available, but also wished they could have the expertise to amend and contextualize the latter to fit the level of their students. Lim [31] proposed to develop a framework for teachers within the same department to collaboratively design ICT-mediated lessons, and share ICT resources, and lesson plans. Teachers also requested the sustained support of management as reported by Solar et al. [32]. One teacher suggested that school's management should encourage collaboration and discussions among colleagues to foster confidence and beliefs in ICT integration. Moreover, the role of parents was raised and comments were geared towards making them aware of the potential of ICT to impact learning and to encourage them take responsibility for the proper use of technology at home. Zhang and Zhu [33] found that in order to improve students' digital media literacy, cooperation between school and home is necessary.

5 Limitations

This experiment was very time consuming, lasting two weeks in each of the five schools. Hence data was collected from only five class teachers. Moreover, only one topic, namely fractions was taught using this approach. In the future, more teachers could be exposed to technology-enhanced teaching. For example, three to four teachers could be present while the experiment is carried out in one class. Other topics could be taught using this approach. Opinions and views would then be collected from a greater number of teachers, and the findings would then be more generalizable.

6 Conclusion and Recommendations

Despite the restricted number of participants in this study, there was a good representation of male and female teachers, working in both high and low performing schools. Interviews were intense and a rich array of data collected to form an opinion of what teachers perceived of the integration of technology in schools and hence their acceptance thereof. Teachers' feedback focused on four themes namely, *learner empowerment*, *effective teaching*, *inhibiting factors* and *teacher support*. They saw both sides of the coin, one side showing all the benefits teachers and students could derive from ICT-enhanced lessons and the other one showing the hindrances and hence measures to be taken to facilitate the integration of technology in schools.

The major concerns of teachers evolving from observations, discussions and interviews can be summarized as follows:

- Fear of losing control of the class as students might demonstrate greater mastery of the tool than the teacher.
- Inability to attend to hardware malfunctioning during classes.
- Lack of specific teaching skills and strategies to integrate ICT in the curriculum.
- Restricted access to the latest technology and appropriate logistics.
- Belief that planning and conducting ICT-based lessons is more time consuming, hence the fear of not completing the syllabus on time.
- Challenges of managing a digital classroom, with IT equipment and students working more independently.
- Some were apprehensive that the uptake of ICT will discard prevailing teaching methods completely and recommended rather a blended approach that would support and enhance existing teaching practices. Teachers maintained the importance of the explanation copybook for revision.
- Lack of traceability of work done by students. Contrary to exercise books, the tablet left no trace of work accomplished by the student.

This study did reveal though a significant acceptance of technology in the classroom by teachers. However, readiness to maximize on the affordability of ICT to revamp teaching and learning remains a grey area. Teachers need to be reassured through ongoing professional development, and they must also be accompanied in the integrative process. More ICT-enhanced model lessons should be made available to teachers and they must be provided with the appropriate guidelines. Most teachers have studied in the traditional way while they were students, and today shifting to the use of technology poses problems. The lessons learnt from this study are numerous and should add academic discourse to the uptake of technology in education.

References

1. Tondeur, J., Forkosh-Baruch, A., Prestridge, S., Albion, P., & Edirisinghe, S. (2016). Responding to challenges in teacher professional development for ICT integration in education. *Educational Technology & Society, 19*(3), 110–120.
2. Central Statistics Office. (2016). http://statsmauritius.govmu.org/English/Publications/Pages/all_esi.aspx#2017.
3. Baya'a, N., & Daher, W. (2012). Mathematics teachers' readiness to integrate ICT in the classroom: The case of elementary and middle school Arab teachers in Israel. In *Interactive Mobile and Computer Aided Learning (IMCL) International, Conference*, IEEE. https://doi.org/10.1109/imcl.2012.6396470.

4. Office of technology assessment: Teachers and technology: Making the connection. U.S. Government Printing Office (1995).
5. Liu, R. (2010). Psychological research in educational technology in China. *British Journal of Educational Technology, 41*(4), 593–606.
6. Lim, L. Y. T. S. K., Lim, C. P., & Koh, J. H. L. (2012). Pedagogical approaches for ICT integration into primary school English and Mathematics: A Singapore case study. *Australian Journal of Educational Technology, 28*(4), 740–754.
7. Uluyol, C., & Sahin, S. (2016). Elementary school teachers' ICT use in the classroom and their motivators for using ICT. *British Journal of Educational Technology, 47*(1), 65–75.
8. Gul, K. Y. (2015). The views of Mathematics teachers on the factors affecting the integration of technology in Mathematics courses. *Australian Journal of Teacher Education, 40*(8), 32–148.
9. Virginia, E. (2016). Transforming the classroom. Technology counts. *Education Week, 35*, 35. https://eric.ed.gov/?id=ED566602.
10. Nagel, D. 6 technology challenges facing education. Ed Tech Trends, https://thejournal.com/articles/2013/06/04/6-technology-challenges-facing-education.aspx.
11. Koehler, M. J., Mishra, P., & Cain, W. (2013). What is technological pedagogical content knowledge (TPACK)? *Journal of Education, 193*(3), 13–21.
12. Blackwell, C. K., Lauricella, A. R., & Wartella, E. (2014). Factors influencing digital technology use in early childhood education. *Computers & Education, 77*, 82–90.
13. Sim, J., & Theng, L. B. (2007). Teachers' perceptions of the use of ICT as an instructional tool in Mathematics and Science.
14. Ertmer, P. A. (2005). Teacher pedagogical beliefs: The final frontier in our quest for technology integration. *Educational Technology Research and Development, 53*(4), 25–39.
15. Latchem, C. (2017). Using ICTs and blended learning in transforming TVET. United Nations Educational, Scientific and Cultural Organization and Commonwealth of Learning.
16. Clarke, T., Ayres, P., & Sweller, J. (2005). the impact of sequencing and prior knowledge on learning Mathematics through spreadsheet applications. *Educational Technology Research and Development, 53*(3), 15–24.
17. Aris, N., & Orcos, L. (2015). ICTs and school education. Special issue on teaching Mathematics using new and classic tools. *International Journal of Interactive Multimedia and Artificial Intelligence, 3*(4), 13–18.
18. Fishbein, M., & Ajzen, I. (1975). Belief, attitude, intention and behaviour: An introduction to theory and research, Reading, MA.
19. Bandura, A. (1982). Self-efficacy mechanism in human agency. *American Psychologist, 37*(2), 122–147.
20. Lala, G. (2014). The emergence and development of the technology acceptance model (TAM). *Proceedings of the International Conference Marketing- from Information to decision, 7*, 149–160.
21. Davis, F. D., Bagozzi, R. P., & Warshaw, P. R. (1989). User acceptance of computer technology: A comparison of two theoretical models. *Management Science, 35*, 8.
22. Venkatesh, V., & Davis, F. D. (1996). A model of antecedents of perceived ease of use: Development and test. *Decision Science, 27*(3), 451—481.
23. Svendsen, G. B., Johnsen, J. K., Almas-Sorensen, L., & Vitterso, J. (2013). Personality and technology acceptance: The influence of personality factors on the core constructs of the technology acceptance model. *Behaviour & Information Technology, 32*(4), 323–334.
24. Ensminger, D.(2016). Technology planning in schools. In N. Rushby & D. Surry (Eds.), *The Wiley handbook of learning technology* (p. 461). Wiley: New Jersey.
25. Jewitt, C., Hadjithoma-Garstka, C., Clark, W., Banaji, S., & Selwyn, N. (2010). *School use of learning platforms and associated technologies*. Becta: University of London.
26. Penuel, W. C. (2006). Implementation and effects of one-to-one computing initiatives: A research synthesis. *Journal of Research on Technology in Education, 38*(3), 329–348.
27. Tsuei, M. (2014). Mathematics synchronous peer tutoring system for students with learning disabilities. *Educational Technology & Society, 17*(1), 115–127.

28. Willmot, P., Bramhall, M., & Radley, K. (2012). Using digital video reporting to inspire and engage students. http://www.raeng.org.uk/education.
29. Archambault, L. M., Wetzel, K., Foulger, T. S., & Williams, M. K. (2010). Professional development 2.0: Transforming teacher education pedagogy with 21st century tools. *Journal of Digital Learning in Teacher Education, 27*(1), 4–11.
30. Minshew, L., & Anderson, J. Teacher self-efficacy in 1:1 iPad integration in middle school science and math classrooms. *Contemporary Issues in Technology and Teacher Education, 15*(3). http://www.citejournal.org/volume-15/issue-3–15/science/teacher-self-efficacy-in-11-ipad-integration-in-middle-school-science-and-math-classrooms.
31. Lim, C. (2007). Effective integration of ICT in Singapore schools: Pedagogical and policy implications. *Educational Technology Research and Development, 55*(1), 83–116.
32. Solar, M., Sabattin, J., & Parada, V. (2013). A maturity model for assessing the use of ICT in school education. *Journal of Educational Technology & Society, 16*(1), 206–218.
33. Zhang, H., & Zhu, C. (2016). A study of digital media literacy of the 5th and 6th grade primary students in Beijing. *The Asia-Pacific Education Researcher, 25*(4), 579–592.

Smart Agriculture Using IOT

Bammidi Deepa, Chukka Anusha, and P. Chaya Devi

Abstract An automated agriculture system is developed to monitor and maintain the important aspects of farming like temperature, humidity, soil moisture content and sunlight using IoT technology. The sensors must be placed at appropriate places and positions to sense and communicate the details using cloud computing to the mobile phones of farmers, to optimize the agriculture yield by automating the field maintenance system. Improved water supply process, brightness maintenance, temperature conditions adjustments can be achieved in the automated system using the proposed idea. Single board Node MCU microcontroller is used as the decision making and controlling device between various sensors and the farm maintenance equipment. The proposed system is expected to be helpful to the farmers in controlling an irrigation system in a better and accurate way.

Keywords Farm automation · Node MCU · Sensors · Cloud computing · Smart agriculture · Monitoring

1 Introduction

Agriculture and cultivating plants is a science and art. The food and livestock development through agriculture have facilitated the human population to progress many times larger than that could be achieved by hunting and collecting food. Due to the enormous increase in population and changes in climatic conditions, the application of IoT has become necessary for the sustainable growth of food for living beings. IoT is a shared network that can interact with objects through an internet connection. IoT systems help farmers in getting information and making decisions throughout

B. Deepa (✉) · C. Anusha · P. Chaya Devi
Anil Neerukonda Institute of Technology and Sciences, Visakhapatnam, India
e-mail: deepa.ece@anits.edu.in

C. Anusha
e-mail: anoosha.ece@anits.edu.in

P. Chaya Devi
e-mail: chayadevi.ece@anits.edu.in

S. C. Satapathy et al. (eds.), *Intelligent System Design*, Advances in Intelligent Systems and Computing 1171, https://doi.org/10.1007/978-981-15-5400-1_2

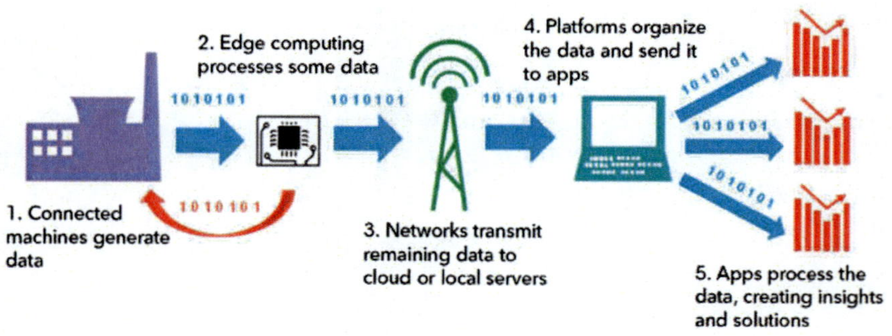

Fig. 1 IOT architecture (*Source* http://www.appletoninnovations.com)

the farming cycle [1–3]. Smart farming consists of IoT technologies such as cloud servers, different sensors and automated irrigation system, to increase the yield of crops and minimize the manual and monetary costs [1–4].

The architecture of IoT is shown in Fig. 1 that explains that the Autonomy, Monitoring, Controlling and Optimization of processes can be achieved by sensing, communicating and computation [5, 6].

As in Fig. 2, a Node is an End Device, which interacts with the Physical Environment. The node may consist of microcontroller, Sensors/Actuators and Connectivity. It can be a sensor, actuator and a processor, which can sense its surroundings, process data locally and transmit data. A Network transfers data from one device to another device. The network can connect multiple nodes to each other and the data can be exchanged among nodes. These networks can be established using cables or wireless media like Wi-Fi. The network may use Mesh network topologies (e.g. Zigbee, Bluetooth), Star network topologies (e.g. Wi-fi-Lan, LoRaWan), Ring network topologies or Tree network topologies. The gateway is the coordinating device that coordinates the communication between nodes or between nodes and cloud. The processing at the gateway is called edge computing/fog computing. The cloud Processes the data from the gateway and is called "cloud computing". The IoT cloud examples are Blynk, ThingSpeak, AWS IoT, etc. Thus, Multiple nodes (sensors) can communicate to single gateway and processing can be done at gateway and cloud.

1.1 *The Microcontroller—Node MCU*

Node MCU, shown in Fig. 3, is a low-cost 32-bit microcontroller that can work on

Fig. 2 IOT dataflow

either analog or digital data inputs and outputs. The basic features of Node MCU are GPIO, ADC, UART, SPI, Built-in Wi-Fi and Low-power operated.

In the pin description shown in Fig. 4, General Purpose Input Output (GPIO) pins can be either set to act as input or output, individually, and values can be either logically low (0) or high (1) for digital signaling. The IDE (Integrated Development Environment) facilitates programming the board. Enter a program on the computer, with the help of IDE and then upload the program to the MCU where it is stored in the program memory (flash) and executed in RAM. ADC is used to read analog values from sensors, such as temperature sensors, potentiometers, light sensors, etc. ADC discretizes the input signal based on different voltage levels (i.e. states) by 1024 states (10 bits) with the smallest measurable voltage change is 3.3 V/1024 or 3.22 mV.

1.2 Sensors

DTH11 shown in Fig. 5 is digital sensor for measuring the weather in the field. It gives a digital output value, and therefore these outputs can be directly transferred to the microcontroller based on the conditional coding, the MCU decides to achieve suitable temperature values by controlling air fans. DHT11 has a capacitive sensor that measures humidity. Based on the capacitor voltage, humidity value can be known. This sensor can get the updated data from it after every 2 s. It is highly reliable and provides excellent stability.

The amount of water content in the soil is measured by the soil moisture sensor with the help of its electrical resistance properties. Based on different environmental conditions, the relationship between the soil moisture and the measured property varies according to the soil type, the extent of its connection to the electrification and the atmospheric temperature. These parameters and the relationship are analyzed. The moisture in the soil of the field is measured and transferred to microcontroller to decide to control the operation and extinguishing of the water pump (Fig. 6).

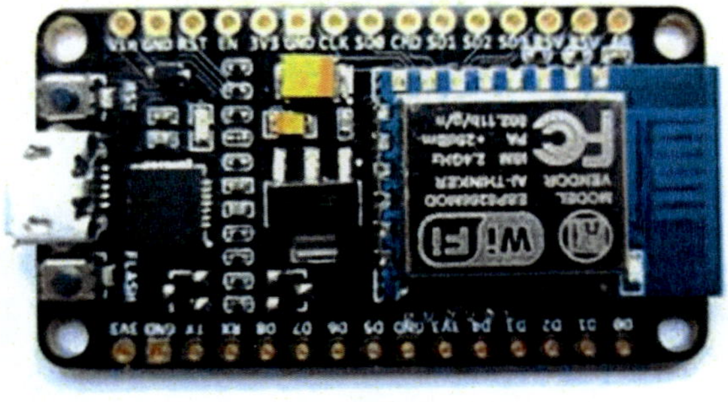

Fig. 3 NODEMCU

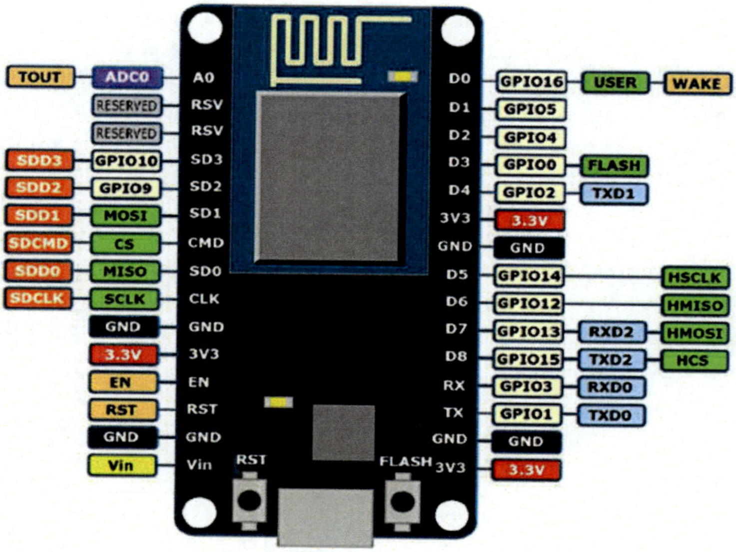

Fig. 4 Pin description of Node MCU ESP-12 development kit V1.0

Fig. 5 DHT11 sensor

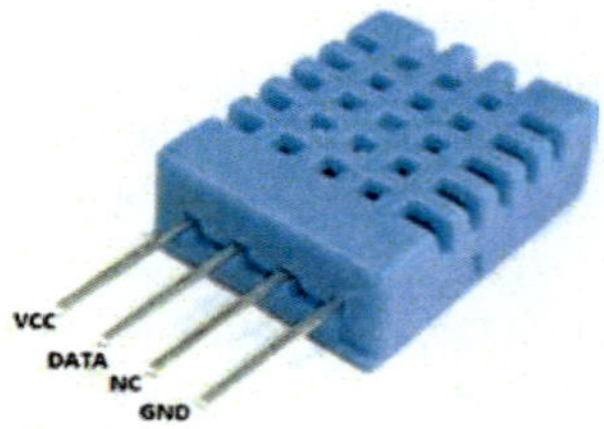

Fig. 6 Soil moisture sensor

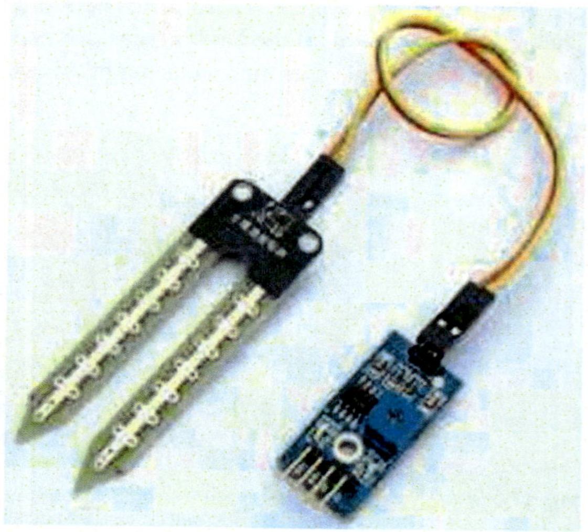

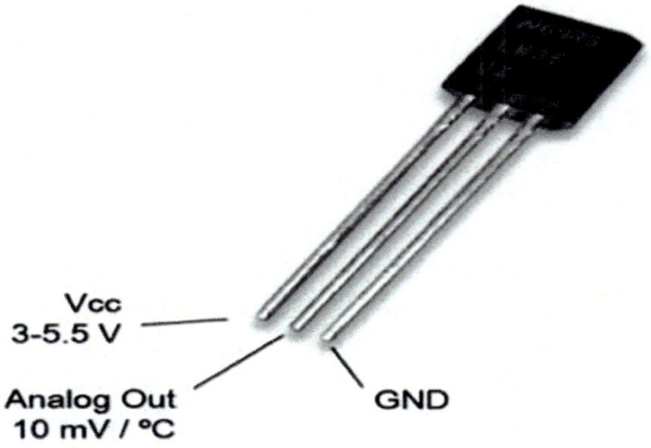

Fig. 7 LM35 temperature sensor

Fig. 8 Photoresistor (LDR)

LM35 shown in Fig. 7 is a temperature sensor and gives an output voltage proportional to the Celsius temperature with a scale factor of 0.01 V/°C. The LM35 does not require any trimming or external calibration and maintains an accuracy of ±0.4 °C at room temperature and ±0.8 °C over a range of 0 to +100 °C. The sensor has a sensitivity of 10 mV/°C. The output voltage can be converted into temperature by a simple conversion factor which is the reciprocal that is 100 °C/V. Thus Temperature = 3.3 * ADC Value * 100/1024.

A photoresistor or LDR (light dependent resistor) shown in Fig. 8 is an analog sensor with any orientation and a resistor whose resistance depends on the exposed light intensity.

1.3 Wi-Fi Connectivity

STA stands for stations, consists of connected devices in Wi-Fi network. Access Point (AP) provides Connection to Wi-Fi, and acts as a hub for one or more stations. Wi-Fi network to other devices Connectivity and further to wired network interconnections are provided by an access point (AP). SSID (Service Set Identifier) recognizes each access point and SSID is the network selected when connecting a device (station) to the Wi-Fi. Node MCU acts as a station and the Wi-Fi network can be connected to it. It can also operate as a soft access point (soft-AP), to establish its own Wi-Fi network.

2 Agriculture Automation System Design

Figure 9 shows the prototype for the agriculture automation system. NodeMCU is connected to input sensors like soil moisture sensor, LDR, LM35 temperature sensor and DHT 11 sensor. The MCU is also connected to a DC motor (through a relay) and an LED. When the moisture of the soil is below the acceptable level of the sensor, the Node MCU turns on the relay and the DC motor runs. In real-time, the relay output can be connected to a water motor. Therefore, the system automates the process of the water irrigation system of a farm and supplies sufficient water to the need of soil for the better yield of the crop. The LDR senses the light intensity and if there is less brightness, the Node MCU turns on the LED. LEDs may be replaced by an electric bulb in real-time applications. Based on the temperature sensor data, the fan can be turned on by the microcontroller. And the humidity sensor data can be used to switch ON/OFF a humidifier or cool mist vaporizer.

Integrated Development Environment is used to code the Node MCU device to process the data of sensors and communicate to the output pins of the MCU for the required action to be taken. Figure 10 shows the coding done in the IDE for the farm automation system.

Fig. 9 Agriculture automation setup prototype

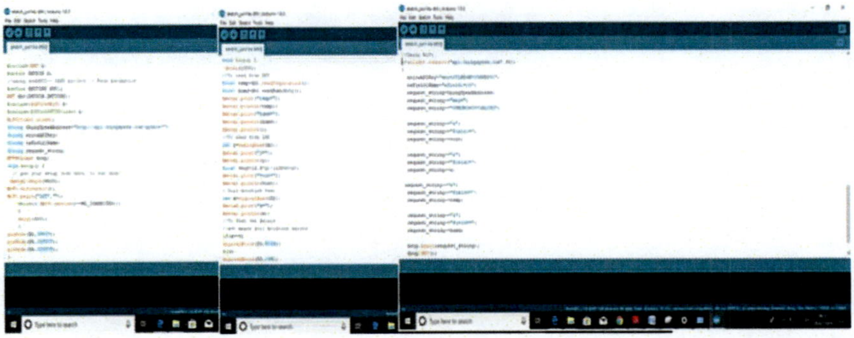

Fig. 10 Node-MCU coding

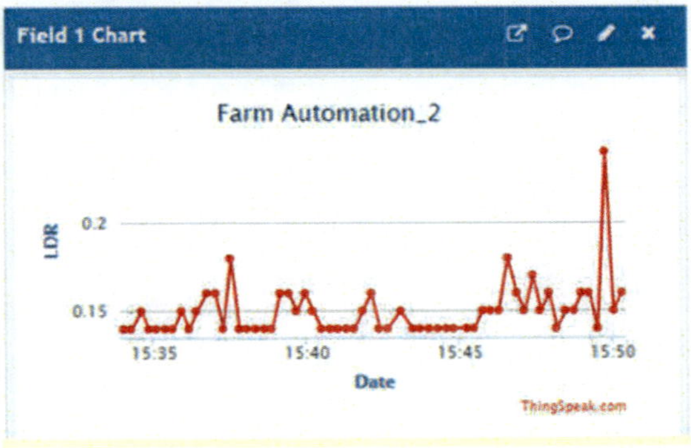

Fig. 11 LDR analog output response

3 Results

Figure 11 shows the LDR analog output response over a time duration. ThingSpeak is the open IoT platform with MATLAB analytics. The MCU communicates to ThingSpeak platform.

Figures 12, 13 and 14 show the responses of soil moisture sensor digital output, temperature sensor analog output over a period of observation and the humidity sensor analog outputs.

4 Conclusion

A prototype system is developed to sense the important parameters of healthy crop maintenance. Using IDE, the sensor data if taken into Node MCU, and then it's processed to the need, like conversion of parameters. Then based on the decision taken by the logic system, the output pins of the MCU operates the relay and the relay

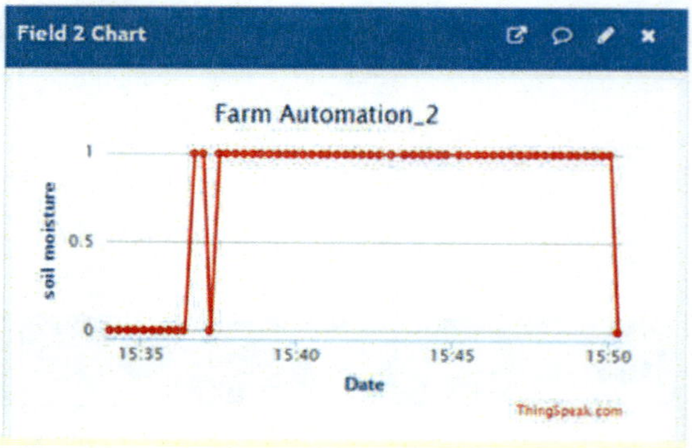

Fig. 12 Soil moisture sensor digital output response

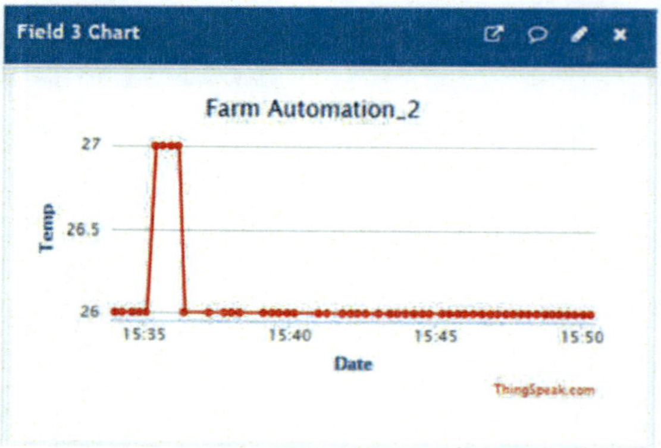

Fig. 13 Temperature sensor analog output response

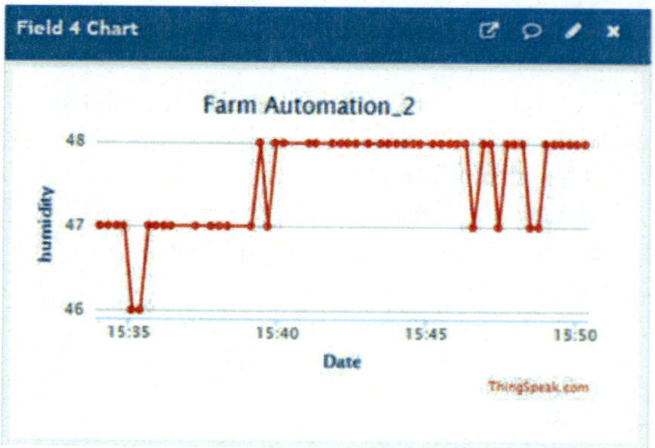

Fig. 14 Humidity sensor analog output

further turns ON/OFF the output devices like DC motor, LED, etc., automatically for better crop production with more precision.

References

1. Sivakumar, S. A., Mohanapriya, G., Rashini, A., & Vignesh, R. (2018, February). Agriculture automation using internet of things. *International Journal of Advance Engineering and Research Development, 5*(02); (2016 November) *The International Conference on Communication and Computing Systems (ICCCS-2016).*
2. Muthunpandian, S., Vigneshwaran, S., Ranjitsabarinath, R. C., & Manoj Kumar Reddy, Y. (2017, April). *IOT Based Crop-Field Monitoring And Irrigation Automation, 4*(19).
3. Mohanraj, I., Kirthika, A., & Naren, J. (2015, June). Field monitoring and automation using IOT in agriculture domain. *IJCSNS, 15*(6).
4. Nageswara Rao, R. IOT based smart crop-field monitoring and automation irrigation system. In *Proceedings of the Second International Conference on Inventive Systems and Control (ICISC 2018)*, IEEE Xplore Compliant Part Number: CFP18J06-ART, ISBN: 978-1-5386-0807-4; ISBN: 978-1-5386-0806-7.
5. Lee, M., Hwang, J., & Yoe, H. (2013). Agricultural protection system based on IoT. In *IEEE 16th International Conference on Computational Science and Engineering.*
6. Mirabella, O., & Brischetto, M. (2011). A hybrid wired/wireless networking infrastructure for greenhouse management. *IEEE Transactions on Instrumentation and Measurement, 60*(2), 398–407.
7. Gutiérrez, J., Villa-Medina, J. F., Nieto-Garibay, A., & Porta-Gándara, M. Á. (2017). Automated irrigation system using a wireless sensor network and GPRS module. *IEEE Transactions on Instrumentation and Measurement, 17.*
8. Mohanraj, I., Ashokumar, K., & Naren, J. (2016). Field monitoring and automation using IOT in agriculture domain. In *6th International Conference on Advances in Computing & Communications, ICACC,* 6–8 September 2016, Cochin. Elsevier: India.
9. Liu, C., Ren, W., Zhang, B., & Lv, C. (2011). The application of soil temperature measurement by lm35 temperature sensors. In *International Conference on Electronic and Mechanical Engineering and Information Technology* (Vol. 88, No. 1, pp. 1825–1828).
10. Wang, Q., Terzis, A., & Szalay, A. (2010). A novel soil measuring wireless sensor network. *IEEE Transactions on Instrumentation and Measurement*, 412–415.

CSII-TSBCC: Comparative Study of Identifying Issues of Task Scheduling of Big data in Cloud Computing

Chetana Tukkoji, K. Seetharam, T. Srinivas Rao, and G. Sandhya

Abstract The world moves day by day toward cloud computing and large data, where it includes huge information sets from many sources such as community medium, company, astronomic, healthcare, economics, depository and many additional field causes. All this information generates processing and use complexities. Due to improved hardware possessions and software algorithms, system throughput, information accuracy, locality, management, examination, storage space, data privacy and task scheduling, many variables can influence system efficiency. In big data and clouds, all these problems generate difficulties. This section describes the issues of scheduling caused by cloud computing.

Keywords Big data · Cloud computing · MapReduce

1 Introduction

Big data (BD) is produced and composed from multiple sources via electronic operations. It needs adequate processing power and high analytical capacity. The importance of BD is the analytical use that can contribute to a better and faster service delivery decision. The term big data means the huge quantity of high-speed big data of various kinds where the data cannot be processed or stored on ordinary computers.

Big data faces difficulties related to mainly "3 V", and these are discussed below [1].

C. Tukkoji (✉)
GITAM Deemed to be University, Bengaluru Campus, Bengaluru, India
e-mail: chetana20.kle@gmail.com

K. Seetharam
Sambhram Institute of Technology Bengaluru, Bengaluru, India

T. Srinivas Rao
GITAM Deemed to be University, Visakhapatnam Campus, Visakhapatnam, India

G. Sandhya
KNS Institute of Technology Bengaluru, Bengaluru, India

S. C. Satapathy et al. (eds.), *Intelligent System Design*, Advances in Intelligent Systems and Computing 1171, https://doi.org/10.1007/978-981-15-5400-1_3

- Volume: The issue of dealing with data handling and storing.
- Variety: Different sources with various configurations.
- Velocity: Much information needs speed handling.

Big data has several problems, was posed in a range of fields, storage, computing and data transfers [2, 3]. The following topics are addressed:

Storage Issues

A database is an organized file. In the delayed 1960s, 70s relational databases were created for these purposes. Structured Query Language (SQL) for storing, editing and recovering information is used in relational database management systems (RDBMS). The missing of assistance for unstructured information contributed to fresh techniques, such as BLOB in the 2000s. Multimedia information can be used as unstructured data. NOSQL is a scalable database that can spread information across a number of computers. NOSQL is used for cloud computing, as an information storage space server may be included or eliminated at any time in the cloud.

Computing Issues

We need to gather, evaluate and alter big data when we store it. The significant aspect of data collection is to analyze big data and convert pure data into useful data that might enhance a business process or decision-making operation. Using a number of CPUs and RAMs in cloud computing software can address this problem.

Transfer Issues

Big data transmission is another problem. For instance, the transfer of DNA from China to the USA, which is a sort of big data, has some pause in the core of the Internet, which creates an issue when they obtain information in the USA. BGI (one of Beijing Genomics Institute's biggest manufacturers of genomic information) might move 50 DNA through median volume of 0.4 terabyte over Internet in 20 days, that is not suitable achievement.

- Traffic jam: Big information transition may take place among two local sites, towns otherwise around the world via the Internet, other than this transition would lead in an extremely great traffic jam between any locations.
- Precision and privacy: Big information is often transferred via unsecured networks, for example, the Internet. Internet transactions must be held safe from unlawful entry [3–5].

2 Cloud Computing

Cloud computing has full-grown quickly then gained significant attention as provides organizations with flexibility as well as scalability. NIST recognized significant elements of the cloud, shortening the cloud computing idea as follows in five features:

- **On-demand Self-service**: Cloud services provide, as required, software resources such as storage and handling without human interference.
- **Broad network connectivity**: Network accessible computing services are portable devices, even cloud computing services can be accessed by detectors.
- **Resource pooling**: Cloud stage clients share a wide scope of registering resources; clients can decide the asset nature and geographic position.
- **Rapid elasticity**: Handling devices and apps was always accessible and can be enhanced or reduced almost instantly.
- **Measured service**: Cloud schemes can fully transparently access resource procedures and consumption as well as monitoring, regulate and reporting [6].

2.1 Services of Cloud Computing

The distribution system for cloud computing is provided as facilities, essentially at three stages: software, platform and infrastructure.

A. **Software as a service (SaaS)**: Here, clients who can navigate it through the network are supplied with an implementation as a provider. Cloud Information Centers host the request. Since the request is not stored on the client page, the client does not have to worry about the application's servicing and help.
Ex salesforce.com: to purchase on-demand software [7, 8].

B. **Platform as a service (PaaS)**: As a business model, platform offers all the funds needed to construct Internet apps and facilities. The software does not need to be installed or downloaded. PaaS includes designing, developing, monitoring, deploying and installing applications.

C. **Infrastructure as a service (IaaS)**: It merely provides the hardware to maintain anything on it from the client. IaaS enables the client to carry on lease such as server room, CPU periods and memory space and network facilities.

2.2 Scheduling in Cloud Computing

The goal of planning methods in cloud setting is to enhance device performance with load balance, enhance resource usage, save energy, decrease expenses and minimize overall handling cost. The scheduler should therefore recognize the virtualized assets and the necessary limitations of customers in order to achieve an effective match between employment and resources [5]. There are two levels of resources in cloud computing: VM level and host level.

- A scheduler which is based on VM type is used at host level to assign VMs to physical hardware is called as VM scheduling.
- Using a task/job scheduler, assignments are linked to the assigned VMs at the VM level. It is called the scheduling of tasks. Task scheduling relies on effectively

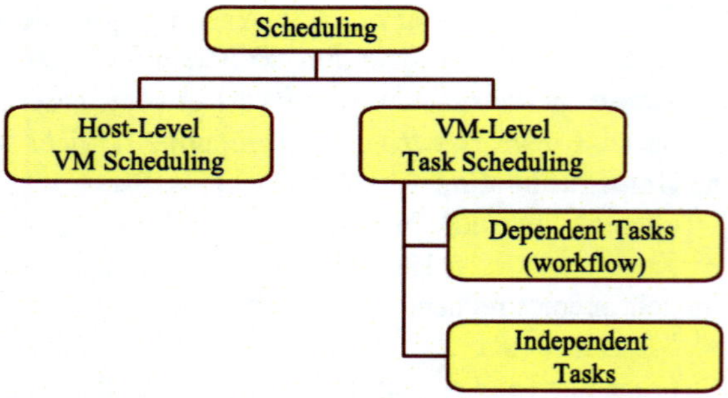

Fig. 1 Levels of scheduling

routing duties to suitable VMs depending on assignment dependence; assignments can be categorized as autonomous or contingent tasks [7]. Figure 1 demonstrates the workflow planning stages.

The primary goal of planning the duties is to minimize the make pan. If tasks are dependent, it can be possible to minimize the make pan by reducing the cost of the computation that takes the time to implement every node and the price of communication that takes the time period to handover data among the two nodes [9].

2.3 Scheduling Types in Cloud Environment

A main method for clouds is to schedule the asset. Resource allocation is used to economically allocate accessible funds. Scheduling is a technique of giving entry to machine assets (e.g., CPU time) to threads, procedures or information movements[10–12]. Timing can normally be performed for load handling in a scheme to efficiently attain the goal performance of delivery. Different cloud computing system scheduling types and categories are shown in Fig. 2.

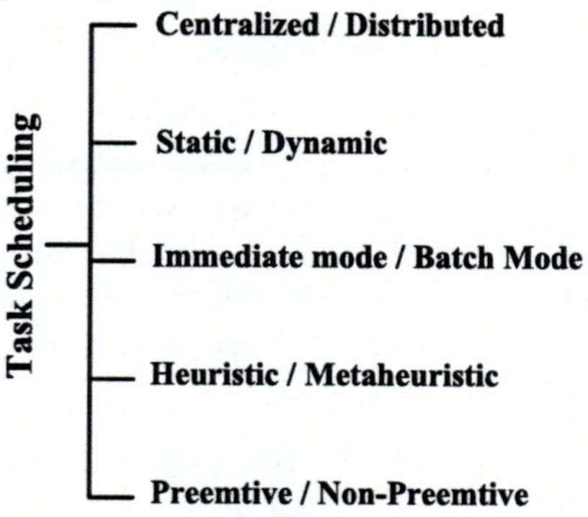

Fig. 2 Scheduling types

- **Centralized Scheduling/Distributed Scheduling**

It is the responsibility of the centralized scheduler to make worldwide decisions. By using distributed planning, the advantages obtained are ease of execution, effectiveness and more resource command and tracking. When the distributed scheduling is more feasible for actual grids due to its poor effectiveness as opposed to distributing planning, and where local schedulers are required to handle and retain the status of the job queue, and hence key command agency no longer exists.

- **Static/Dynamic Scheduling**

Each timing information on tasks is available in static scheduling before execution schedule of every task is calculated before some task is executed. The timing data on the assignments is not understood at runtime in vibrant planning. Thus, the job plan of execution may alter as required by the customer. Compared to dynamic planning, dynamic planning results in overhead runtime.

- **Preemptive/Non-Preemptive Scheduling**

Preventive planning empowers each job activity to be intruded during execution and the task to be relocated to another benefit. The virtual machine cannot be removed in non-preemptive planning until the job operating on it is complete. It does not allow interruption of the assignment while performing.

2.4 Task Scheduling Forms in Cloud

Variety of facilities delivered (implementation, transport, CPU, etc.) and the need to render them accessible to customers. The task scheduling is illustrated in Fig. 3.

- **Job Scheduling**: Usually, a number of autonomous duties are called to work. Most methods start by assessing the weight of the assignments in their reservoir (queue moment, date, arrival time, execution moment, etc.) in attempt to determine the precedence of each assignment to assess finest mapping.
- **Workflow Scheduling**: Having set of autonomous assignments, depicted by Directed Acyclic Graph (DAG) anywhere, every node shows assignments, and

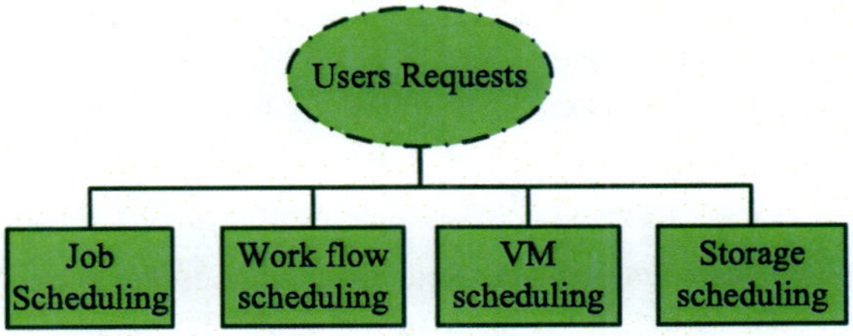

Fig. 3 Task scheduling in cloud environment

every border constitutes interdependence functions message. The assignment can only be accomplished once all the sibling assignments are calculated.

- **VM Scheduling**: In a storage base, cloud suppliers generate organizations of VMs, and then give each VM to a client. It is one of the primary problems when applying IaaS as there are numerous limitations identified in Service Level Agreement (SLA) paper shows how VMs associated with each client.
- **Storage Scheduling**: Implying collection of information blocks, regularly with big ability and heterogeneous form that are placed in groups in separate geographical locations. Its magnitude also continuously improves as is the situation with large information. In addition, we need a strong handling technique for saving and retrieving information to achieve a successful optimization, so planning is regarded as a helpful alternative to cope with this sort of problem.

2.5 Schematic Procedures for Scheduling Problem

Scheduling issues are related to wide range of combinational optimization issues are aimed at discovering ideal match from finite set of items, so collection of viable alternatives is generally discreet rather than constant. An issue relates to one with a tiny number of instances, so polynomial algorithms or enumerations can simply work it out.

a. **Enumeration Method**

If all feasible alternatives are enumerated and contrasted one by one, their ideal answer can be chosen for an optimization problem. In the worst case, exact enumerative algorithms have the complexity of exponential time. However, for some weak-sense NP-hard problems, when the number in one instance is quite small, a pseudo-polynomial algorithm can solve it, time complexity which is bounded by a polynomial expression of input size and maximum number of problems. In addition, there is another type of enumeration is called explicit enumeration that assesses all feasible alternatives without clearly mentioning them all.

b. **Heuristic Method**

Exhaustive enumeration is not possible for planning issues because in polynomial time, only a few special cases of NP-hard issues have precisely solvable algorithms. We strive to discover suboptimal alternatives for the sake of exercise that are great enough to match precision and moment. Heuristic is a suboptimal method for finding fairly quick, fairly useful alternatives. In terms of a specified performance criterion, iteratively increases a candidate solution but does not ensure the greatest alternative.

c. **Scheduling Hierarchy**

We implemented associated concept about planning issues and their schematic techniques in the last chapter of the Cloud Datacenter. We indicate planning issues in cloud settings from this chapter. Service planning allows cloud computing distinct from other computing paradigms as a main feature of resource management. Centralized scheduler in the cluster scheme seeks to improve the general efficiency of the scheme, while generalized scheduler in the grid system seeks to improve the efficiency of particular end-users.

d. **Real-Time Scheduling**

There are evolving categories of apps for cloud computing that can profit from enhanced timing guarantees for cloud facilities. Typically, these critical apps require timetable demands, and any wait for the entire implementation is deemed a mistake.

3 MapReduce

The MapReduce is a programming technique in which the problem execution on dispersed systems can be easily parallelized to render big handling tasks tractable. The MapReduce works and reduces with two fundamental features titled map. Based on issue requirements, the customer should identify map and decrease features. The chart and reduction features both bring information entry and transfer result in the <key, value> tuple format. The map function receives all the information and produces some tuples <key, value>, whereas the reduction function discovers the tuples with the same buttons and combines their characteristics to produce the ultimate outcome. The reduction procedure can be carried out in three subsections, i.e., shuffling, routing and reduction. The MapReduce models are shown in Fig. 4.

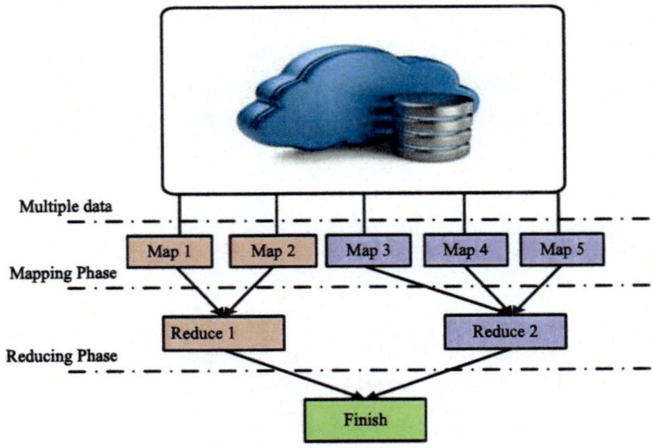

Fig. 4 Basic MapReduce architecture by handling big data

3.1 MapReduce Implementations

- **Google MapReduce**: In 2004, the initial MapReduce structure at Google was suggested and enacted by Jeff Dean and Sanjay Ghemawat. It was created primarily for Google's inner use. Together with functions in Java and Python, its software was published in C++ and licensed by Google. According to Jeff Dean, Google uses MapReduce to a greater magnitude.
- **Hadoop**: Hadoop is an accessible source MapReduce system application. Doug Cutting created it to promote the Nutch Search engine in 2005. Many organizations such as Yahoo, Facebook, Amazon, IBM use Hadoop extensively. Yahoo is using a Hadoop node with 10,000 nodes to generate search index. Hadoop put a target in the month of April 2008 by arranging 1 terabyte of records in 209s with 910-node array.
- **Disco**: Nokia Research Centre has led to the development of Map Reduces technique as another open-source version called Disco, which can be used to solve problems in real-time that manage huge amounts of information.
- **Skynet**: Sky net is another open source version of MapReduce programming structure produced at Geni using Ruby Language in which apps for MapReduce are also published in Ruby.
- **Dryad**: Another MapReduce system flavor is Microsoft-developed Dryad. Dryad's structure promotes huge scaling of thousands of nodes from a tiny group of multi-core pcs to large-scale data center.

4 Conclusion

The fast adoption of the various applications generates large data that requires a flexible, effective as well as cost-effective platform for the data managements from viewpoint of storage, accessing and processing the data. Cloud computing is the paradigm which qualifies these requirements. But the task scheduling is the major task in cloud.

References

1. Anuradha, J. (2015). A brief introduction on big data 5Vs characteristics and hadoop technology. *Procedia Computer Science, 48,* 319–324.
2. Emani, C. K., Cullot, N., & Nicolle, C. (2015). Understandable big data: A survey. *Computer Science Review, 17,* 70–81.
3. Philip, C. C. L., & Zhang, C.-Y. (2014). Data-intensive applications, challenges, techniques and technologies: A survey on big data. *Information Sciences, 275,* 314–347.
4. Jin, X., Wah, B. W., Cheng, X., & Wang, Y. (2015). Significance and challenges of big data research. *Big Data Research, 2*(2), 59–64.
5. Zhang, Q., Cheng, L., & Boutaba, R. (2010). Cloud computing: state-of-the-art and research challenges. *Journal of Internet Services and Applications, 1*(1), 7–18.

6. Mell, P., & Grance, T. (2011). The NIST definition of cloud computing.
7. Beloglazov, A., Abawajy, J., & Buyya, R. (2012). Energy-aware resource allocation heuristics for efficient management of data centers for cloud computing. *Future Generation Computer Systems, 28*(5):755–768.
8. Subashini, S., & Kavitha, V. (2011). A survey on security issues in service delivery models of cloud computing. *Journal of Network and Computer Applications, 34*(1), 1–11.
9. Ramakrishna Murty, M., Murthy, J. V. R., Prasad Reddy, P. V. G. D, Sapathy, S. C. (2013, November) Performance of teaching learning based optimization algorithm with various teaching factor values for solving optimization problems. In *International Conference FICTA-13 at Bhuveneswar* (Vol. 247, pp. 207–216).
10. Carlin, S., & Curran, K. (2012). Cloud computing technologies. *International Journal of Cloud Computing and Services Science, 1*(2), 59.
11. Marinescu, D. C. (2013). *Cloud computing: Theory and practice.* Newnes.
12. Qian, L., Luo, Z., Du, Y., & Guo, L. Cloud computing: An overview. In *IEEE International Conference on Cloud Computing* (pp. 626–631). Springer: Berlin, Heidelberg.

De-Centralized Cloud Data Storage for Privacy-Preserving of Data Using Fog

Gadu Srinivasa Rao, G. Himaja, and V. S. V. S. Murthy

Abstract The cloud storage technology gets a massive growth in development and attraction with respect to the growth of unstructured data (Fu et al., Trans Inf Forensics Sec 12(8):1874–1884, 2017 [11]). There are chances of privacy leakage risks for face detection and data control rights can be lost due to this schema (Dinh et al., Wirel Commun Mob Comput 13(18):1587–1611, 2013 [2]). A Three-Layer Approach is designed in order to store and access the data in a secure manner from the cloud server. The fog server concept is integrated for the current cloud in which the data can be stored on multiple nodes rather than in a single storage medium. The data gets partitioned into a number of blocks where each block's encryption standard is monitored by the data owner (Feng, A data privacy protection scheme of cloud storage 14(12):174–176, 2015 [5, 14]). Once if any data user tries to access the file, he needs to request the file access from the cloud server, where the users can view the file in a decrypted manner and for the remaining, the data cannot be viewed in a plain text manner.

Keywords Fog server · Cloud server

1 Introduction

The administration over a network system can be conveyed by the utilization of processing assets done by cloud computing. Gradual growth of cloud computing is projected behind the effort of many people [4]. Local machine capacity is not sufficient anymore to satisfy the requirements of the user. The massive growth of

G. Srinivasa Rao (✉) · G. Himaja · V. S. V. S. Murthy
Department of Computer Science & Engineering, Gitam Institute of Technology, Gitam University, Visakhapatnam, India
e-mail: gadusrinivasarao5@gmail.com

G. Himaja
e-mail: himaja.himajagadi@gmail.com

V. S. V. S. Murthy
e-mail: murthy522@gmail.com

S. C. Satapathy et al. (eds.), *Intelligent System Design*, Advances in Intelligent Systems and Computing 1171, https://doi.org/10.1007/978-981-15-5400-1_4

user requirements leads to the selection of cloud storage. The process of making different storage devices work together in a coordinated way is pursued by a group of applications such as network technology and distributed file system technology, cloud storage [12, 18]. In the present scenario still, there are some famous events representing the data privacy leakage in cloud storage. For example, the WikiLeaks where the storage is getting leaked. Therefore, we propose an Advanced Encryption Standard and design using HMAC representing Hash-based message authentication code and also the access control policy [11, 13, 16]. Computing using fog is an extension to a computing model based on computation using cloud which is consisting of a cluster of fog nodes. In our schema, the user's data is partitioned into three parts and they are separately stored in the cloud server, the fog server and the user's local machine [4]. The Advanced Encryption Standard is used to provide security to the data by encrypting it and whereas the Hash-based message authentication code is used to generate hash keys using which the data can be retrieved to access and for any operation, the access control policy describes whether the access can be given at same without leading to conflicts [1, 4]. Thus, we take advantage of CI to do some calculating works in the fog layer.

2 Related Works

In the Industrial field and academics as well as in other fields the feature of cloud storage, i.e. security has attracted a lot of attention. There are numerous researches going on about the architecture of cloud storage and its security. In order to solve the privacy issue in cloud computing, [12, 14] (AES) Advanced Encryption Standard encryption algorithm is used for generating secret key for the conversion of plain text to cipher text. The AES is the best cryptography algorithm in network security domain as of now because it can be done on three different key sizes like 128, 192, 256 key sizes. Even if the key in transmission is hacked [15]. Hence it is proved to be best. The experiment results represent that this is an efficient schema [6]. HMAC uses two phases for message authentication code generation [2, 7, 16]. The National Institute of Standards and Technology in its document relating to Computer security specifies various aspects of cloud computing such as services, deployment methods and utilization for people relating to technology as well as consumers [12]. The address virtualization approach where numerous virtual machine instances can be hoisted on a server without forgoing functionality [2]. Wang addresses the importance of privacy in cloud servers through encryption and proposed a multi-keyword ranked search scheme for encryption of information before outsourcing the data to cloud servers [20]. This paper proposes a method of using local machine and fog servers along with cloud servers for data privacy to protect from attacks inside cloud server [5, 18].

3 Role of Fog Computing in Providing Security for Cloud Storage

The cloud storage system quality can be measured by the degree of security which plays an important role [8]. The security of cloud storage server depends on the security provided to the data and it comprises of three aspects: data availability, data security and data integrity. The concept of data integrity and data security plays a key role in most of the research studies [8, 20].

3.1 The Computation Using Fog

The fog computing is defined to be as an extension to cloud computing. Ciscos Bonomi in the year 2011 firstly came up with fog computing [3]. Fog computing is represented as an architecture of three-level where the cloud computing layer is the top-most layer having powerful storage capacity and a powerful computing capability as represented in Fig. 1 [18]. The role of this layer is to collect the data and upload the data to the fog server. The work efficiency of the cloud computing layer can be improved by the introduction of fog computing.

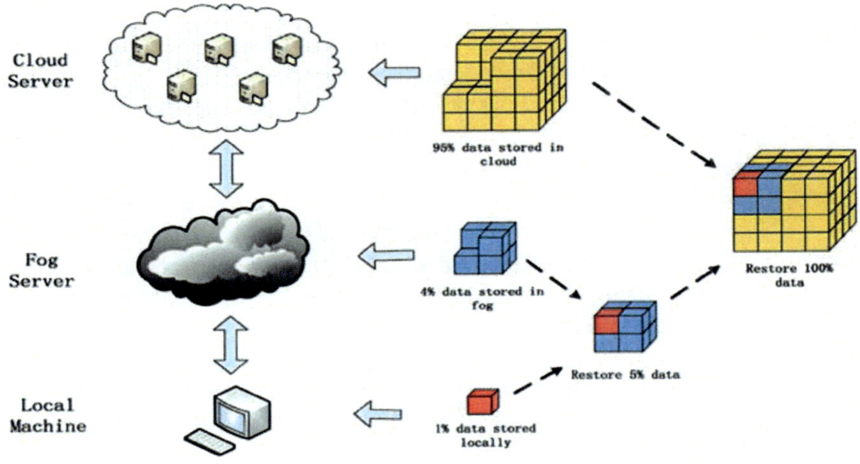

Fig. 1 Representation of fog computing based three-layer storage architecture

3.2 Three-Layer Cloud Data Storage Scheme for Privacy-Preserving of Data Using Fog Computing

The data gets partitioned and is stored in the three data modules namely:

- Data owner: Is the person who has the facility to upload the files into the cloud server. For example, he is the one who can access directly with the cloud.
- Data user: Is the one who tries to access the files which are uploaded into the cloud server and they will not have all rights especially the upload access is not given. They only have access to retrieve, download and update the files.
- Cloud server: This module is the storage area the access type files are stored, been uploaded by the data owner. This is the only medium to store and access the files.
- Fog server: Is the one that divides the data into multiple blocks. Here the data gets stored safely inside the cloud and then the data can be stored in the fog nodes in order to give privacy for the data [9].

It is not possible to recover the original data even if they get the data from any particular server [9]. The following are the step by step procedure for this proposed application. This is as follows:

Step 1: Initially the data user or data owner needs to register into the application with all his/her basic details in order to get login.

Step 2: After registering, the data owner tries to login with his valid credentials and then try to upload a file into the cloud server.

Step 3: Now the input file is divided into multiple parts in which each and every part is encrypted and divided into four fragments.

Step 4: The cloud server will occupy 60–70% of data content blocks from the owner and the fog nodes receiver 20% of data and the least amount like 10% of data is received by the local server.

Step 5: Once the data is stored as per step 4, now the data owner gets a confirmation like data is uploaded into the de-centralized cloud server.

Step 6: The user tries to login to the account with his valid login credentials. Once the user is login to the system, the user can able to search for the files which are uploaded into the cloud server.

Step 7: Now the multiple attributes in the cloud need to give authorized permission for accessing the file blocks and once every individual key access is received by that user. The user can able to download the file by restoring all the blocks into a single file.

Step 8: The system will try to recombine all the individual blocks into a single file block and try to generate the file in a plain text manner.

The architecture consists of three layers namely the local machine, the cloud server and the fog server [18].

3.3 *HMAC (*Hashed Message Authentication Code*)*

The *Hash-Based or Hashed Message Authentication Code* is purely used for generating message authentication codes [21]. In other words, if the data is divided into multiple parts by the fog server then in order to identify easily which block is contained in which storage medium, we need to use the HMAC algorithm. This is used for identifying the data block easily [21].

3.4 *AES (*Advanced Encryption Standard*)*

The *Advanced Encryption Standard* encryption algorithm is used for generating secret key for the conversion of text in plain format to cipher format [11]. The AES is the best cryptography algorithm in network security domain as of now because it can be done on three different key sizes like 128, 192, 256 key sizes [4]. Even if the key got hacked in transmission mode [15]. Hence it is proved to be best.

3.5 *Access Control Policy*

An access control policies, also known as ABE algorithm, in cloud domain, where ABE is abbreviated to be as attribute-based encryption [10]. The attribute-based encryption means the data which is uploaded by data owner cannot be accessed by everyone [17, 19]. This can be restricted for unauthorized users by the cloud server by providing some access rules [10]. This type of process is known as access-based policy.

3.6 *Working Procedure of Its Implementation*

Procedure for Storing: When the file is ready to get stored by the user to the cloud server. Firstly, the file uploaded by the user gets encrypted by the Advanced Encryption Standard algorithm. And then, the file gets partitioned into data in the block form and the system will also feedback encrypted information in a simultaneous way [12]. After then receiving the blocks consisting of 99% data from local machine, these data blocks can be identified easily by which block is contained in which storage medium and for this, the HMAC algorithm is considered. Access control policies in the cloud domain is applicable to provide restriction for unauthorized users by the cloud server by providing some access rules [1]. This type of process is known as access-based policy [12]. Then after, the cloud server receives the data blocks from

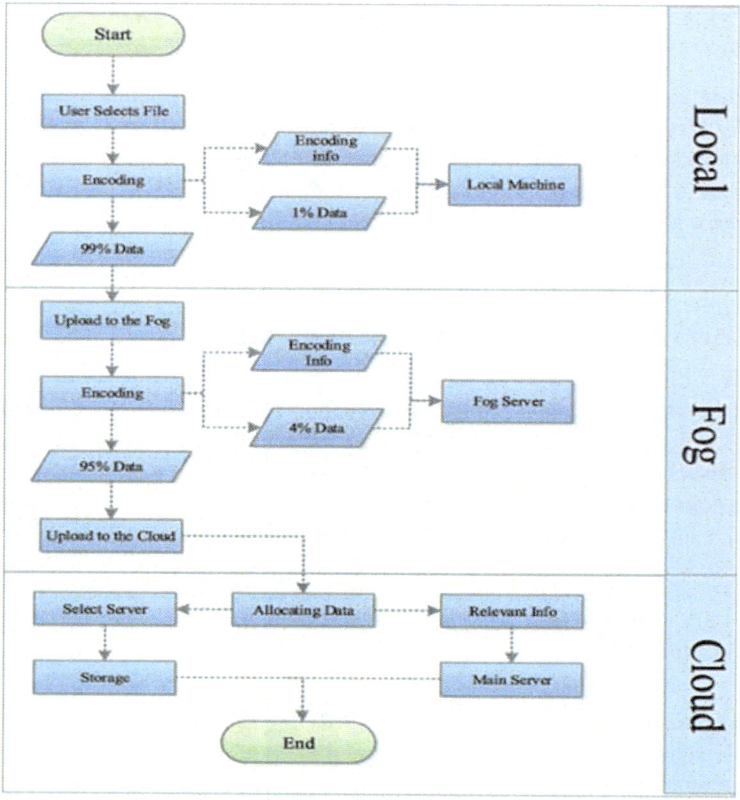

Fig. 2 Diagrammatic representation of procedure for storing

the fog server, then the data blocks get distributed by the cloud manager system as represented in Fig. 2 [18].

Procedure to Download: When the user needs to download the required file present in the server from cloud. First the request from the user is received by the cloud server and then based on the request the data stored in different distributed servers gets integrated. After the integration process, 95% of data from cloud server is sent to the server in fog [4]. Second, the data sent from cloud server gets received by the fog server. The server from fog sends 4% of data and while integrating all the received data, now we have completely 99% of data present [4, 17]. Third, the data from the fog server gets received by the user based on the request. And now the users can get the data completely by the repetition of the process (Fig. 3) [4].

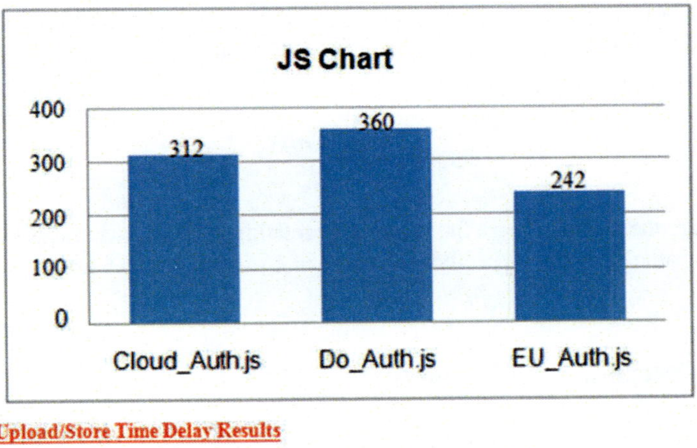

Fig. 3 Representing the results of time delay to store/upload the file

3.7 *The Analysis of Its Efficiency*

In this scheme, the concept of storage efficiency and coding efficiency is been discussed [7]. Storage efficiency can be discussed by the Storage Industry Networking Association as:

$$\text{Storage Efficiency} = \frac{\text{Data Space}}{\text{Data Space} + \text{Check Space}} \tag{1}$$

Here the storage efficiency can be represented as $E_s = \frac{k}{k+m}$. Here the following formulas (2), (3) can be derived. We can see that the storage efficiency and the ratio of k and m are said to be proportional to each other.

$$E_s = \frac{k}{k+m} = \frac{\frac{k}{m}}{\frac{k}{m}+1} \tag{2}$$

$$\lim_{\frac{k}{m}\to\infty} = \frac{\frac{k}{m}}{\frac{k}{m}+1} = 1 \tag{3}$$

The equation $2^{\omega} > k + m$ gets satisfied by the relationship between ω, k and m. Where ω and the RAM consumption are proportional to each other [12]. Therefore, the coding efficiency can be represented by the reciprocal of ω and this can be expressed as

$$E_c = \frac{\ln(k+m)}{\ln 2} \tag{4}$$

The m value is said to be set to 2. Hence to take into consideration the storage and coding efficiency, a new index is to be designed. The scheme's comprehensive efficiency can be expressed as

$$E_w = C_1 \frac{\ln(k+m)}{\ln 2} + C_2 \frac{k}{k+m} \tag{5}$$

Here the C_1 and C_2 are the parameters relating to the storage ratio. For example, if the value of m is set to 2, then the value of C_1 and C_2 are 0.6 and 0.4 respectively.

4 Experiment and Analysis

4.1 Environment Adaptable for Experiment

In the experimental system we have used three types of files:

- Picture (Format: .nef, Size: 24 MB)
- Audio (Format: .MP3, Size: 84.2 MB)
- Video (Format: .RMVB, Size: 615 MB).

Saving $m + 1$ data block ensures reduction of storage pressure in lower servers [18].

4.2 Resultant Representation of Experiment

During experimentation, we ensure the following [9]:

- Consideration of degree of delay.
- Adjusting the value of k according to the machine performance of the operating user.
- Value of m and removed data are set as 2.

Decoding time increases when:

- The number of data blocks increase.
- Number of removed data increases.

Therefore decoding efficiency can be increased by downloading much data from upper server. When the k cost value of Vandermonde is said to be very large and Cauchy increases.

5 Conclusion and Future Scope

A Three-Layer Approach has been proposed in an order to store and access the data in a secure manner from the cloud server [4]. Once if any data user tries to access the file, then they need to request the file access from the server in cloud in which the server from cloud will try to give access permission for multiple blocks and once if the cloud server provides access then those users can view the file in a decrypted manner and the remaining ones who are not having permission from multi-level, the data cannot be viewed in a plain text manner [4, 17]. By conducting various experiments on this, the comparison results clearly tell that the proposed approach is best suitable for representing that the security can be provided for the sensitive data which is stored inside the server space. The future scope can be best achieved by consideration of the Replica concept.

References

1. Ayache, M., Err, M. (2015) Access control policies enforcement in a cloud environment: Openstack, International conference, IEEE.
2. Barham, P., et al. (2003). Xen and the art of virtualization. *ACM SIGOPS Operating Systems Review, 37*(5), 164–177.
3. Bonomi, F., Milito, R., Zhu, J., Addepalli, S. (2012). Fog computing and its role in the internet of things. In *Proceedings 1st Edition MCC Workshop Mobile Cloud Computing* (pp. 13–16).
4. Dinh, H. T., Lee, C., Niyato, D., & Wang, P. (2013). A survey of mobile cloud computing: Architecture, applications, and approaches. *Wireless Communications and Mobile Computing, 13*(18), 1587–1611.
5. Feng, G. (2015). A data privacy protection scheme of cloud storage *14*(2), 174–176.
6. Fu, Z., Huang, F., Ren, K., Weng, J., & Wang, C. (2017). Privacy-preserving smart semantic search based on conceptual graphs over encrypted outsourced data. *IEEE Transactions and Information Forensics Security, 12*(8), 1874–1884.
7. Hou, Q., Wu, Y., Zheng, W., & Yang, G. (2011). A method on protection of user data privacy in cloud storage platform. *Journal of Computational Research and Development, 48*(7), 1146–1154.
8. Kaewpuang, C. R., Yonggang, W., & Niyato, D. (2014). Joint virtual machine and bandwidth allocation in software defined network (sdn) and cloud computing environments. In *Proceedings in IEEE International Conference on Commuications* (pp. 2969–2974).
9. Kulkarni, R., Forster, A., Venayagamoorthy, G. Computational intelligence in wireless sensor networks: A survey. *IEEE Communications Surveys & Tutorials, 13*(1), 68–96, First Quarter (2011).
10. Li, H., Sun, W., Li, F., & Wang, B. (2014). Secure and privacy-preserving data storage service in public cloud. *Journal of Computational Research and Development, 51*(7), 1397–1409.
11. McEliece, R. J., & Sarwate, D. V. (1981). On sharing secrets and reed-solomon codes. *Communications of the ACM, 24*(9), 583–584.
12. Mell, P., & Grance, T. (2009). The NIST definition of cloud computing. *National Institute of Standards and Technology, 53*(6), 50.
13. Plank, J. S. (2005). T1: Erasure codes for storage applications. In *Proceedings in 4th USENIX Conference File Storage Technology* (pp. 1–74).
14. Rewagad, P., Pawar, Y. (2013). Use of digital signature with Diffie Hellman key exchange and AES encryption algorithm to enhance data security in cloud computing. In *International Conference on Communication Systems and Network Technologies.*

15. Shen, J., Liu, D., Shen, J., Liu, Q., & Sun, X. (2017). A secure cloud-assisted urban data sharing framework for ubiquitous-cities. *Pervasive Mobile Comput., 41,* 219–230.
16. Vinay Chopra, S. (2015). Data security approach based on HMAC algorithm for cloud environment. In *Proceedings of International Conference on Networking and Computer Application,* July 15–16. ISBN: 9788193137314.
17. Wang, C., Chow, S. S., Wang, Q., Ren, K., & Lou, W. (2013). Privacy-preserving public auditing for secure cloud storage. *IEEE Transactions on Computers, 62*(2), 362–375.
18. Wang, T., Zhou, J., Chen, X., Wang, G., Liu, A., Liu, Y. (2018). A three-layer privacy preserving cloud storage scheme based on computational intelligence in fog computing. *IEEE Transactions on Emerging Topics in Computational Intelligence.*
19. Wei, L., et al. (2014). Security and privacy for storage and computation in cloud computing. *Information Sciences, 258,* 371–386.
20. Xia, Z., Wang, X., Sun, X., & Wang, Q. (2016). A secure and dynamic multi-keyword ranked search scheme over encrypted cloud data. *IEEE Transactions on Parallel and Distributed Systems, 27*(2), 340–352.
21. Xiao, L., Li, Q., & Liu, J. (2016). Survey on secure cloud storage. *Journal of Data Acquisition Process, 31*(3), 464–472.

Multilayer Perceptron Back propagation Algorithm for Predicting Breast Cancer

K. Satish Kumar, V. V. S. Sasank, K. S. Raghu Praveen, and Y. Krishna Rao

Abstract Machine learning applications are growing rapidly in the present world due to its learning capabilities and improved performance. Supervised learning is a concept of machine learning. In supervised learning, target values are known. Classification problem is addressed by various techniques, and in this paper, we are focusing on perceptron-based learning. In perceptron-based learning, they are single layer perceptron, multilayer perceptron, and RBF network. The work is based on the prediction of breast cancer in women. For prediction task, we designed an artificial neural network (ANN) and the trained the model with back propagation algorithm. The weight optimization in back propagation algorithm is done using stochastic gradient decent algorithm. The proposed model predicts the accuracy of the classifier using performance metric. The proposed algorithm is compared with decision tree algorithm.

Keywords Supervised learning · Classification · Multilayer perceptron · Single layer perceptron · RBF network · Back propagation algorithm · Stochastic gradient decent algorithm

K. Satish Kumar (✉)
CSE, Teegala Krishna Reddy Engineering College, Hyderabad, Telengana, India
e-mail: satish7433@gmail.com

V. V. S. Sasank
CSE, Koneru Lakshmaiah University, Vijayawada, A.P, India
e-mail: sasank.2712@gmail.com

K. S. Raghu Praveen
CSE, Malla Reddy Engineering College, Hyderabad, Telengana, India
e-mail: raghupraveen06@gmail.com

Y. Krishna Rao
Koneru Lakshmaiah University, Vijayawada, A.P, India
e-mail: me.krishnayellanki@gmail.com

S. C. Satapathy et al. (eds.), *Intelligent System Design*, Advances in Intelligent Systems and Computing 1171, https://doi.org/10.1007/978-981-15-5400-1_5

1 Introduction

Breast cancer in women is a serious concern that needs to be addressed, and early detection of breast cancer results in better treatment. The traditional prediction algorithms have not shown significant improvement in performance when new sample of data is given. But these algorithms failed to predict the breast cancer accurately. To address this issue, we proposed an ANN with back propagation model.

The proposed model predicts the breast cancer accurately for test samples. The ANN with back propagation has been integrated with stochastic gradient decent for weight optimization, and the learning rate is also fixed for good convergence. The hidden layer of ANN gets activated based on the logistic sigmoid function.

It have been noted that roughly ten percent of the women across the globe are suffering with breast cancer disease at some stage of their lives. This motivated us in working on Wisconsin breast cancer dataset for early prediction to help the patient for better cure.

This paper discusses about back propagation neural network with stochastic gradient as a weight optimizer. The rest of the paper follows literature survey, problem, objectives, multilayer perceptron algorithm with back propagation, architecture, training algorithm, testing algorithm, dataset, k-fold, performance metric, results, and conclusions.

2 Literature Survey

[1] Design an ANN using particle swarm optimization for improved classification accuracy and low classification error rate. [2] Develop a decision tree classifier—CART for improved accuracy for particular dataset. [3] Provided a comprehensive review on different classification techniques in machine learning. [4] Described radial basis function network is a type of feedforward network used for EEG data classification task. [5] developed particle swarm optimization using Naive Bayes algorithm for attribute selection to improve Naive Bayes classifer and compared its performance with decision tree (C4.5), K-neighbor(KNN), Naive Bayes, and Cfs-Best first algorithms. [6] proposed an improved particle swarm optimization and discrete particle swarm optimization for joint optimization of three layer feedforward artificial neural network structure and parameters (weights and bias) on real-world problems for better generalization capability of classifier. [7] constructed a neural net from decision tree, and its accuracy is comparable to the nearest neighbor algorithm. [8] proposed a decision tree classifier using gain ratio and gini index for classification task. [9] proposed a particle swarm optimization technique for improving classifier performance. [10] Designed a support vector machine and probabilistic neural network for classification task. [11] Proposed an ant colony optimization with genetic algorithm for classification task.

3 Problem

Design an artificial neural network to predict the class label of Wisconsin breast cancer dataset. Train the artificial neural network using back propagation algorithm and optimize its weights by stochastic gradient descent algorithm. Determine the prediction accuracy using performance metrics.

4 Objectives

1. Design and implement ANN using back propagation algorithm with stochastic gradient decent as a weight optimizer.
2. Compare the accuracy of the model with standard algorithms.

5 A Multilayer Perceptron Algorithm with Back Propagation

Back propagation algorithm [12] has gained its significance in neural networks. The algorithm is applied to multiple layers in feedforward neural networks; in each layer, it has neurons with activation function that are differentiable. Back propagation networks (BPNs) are neural networks with back propagation algorithm.

For any given training instance, the algorithm updates the weights in the network by differentiating the activation units with respect to the error obtained in forward neural network. The proposed algorithm aims at developing optimized BPNs that give near prediction to the input pattern.

In back propagation algorithm, first the input is fed through the input layer, and each input is mapped to the hidden layer neurons through synapse. The synapse calculates the product of input value with its associate weight value. Each neuron in layer 2 calculates the sum of synapse of layer 1 with bias of layer 2 added to it. The activation function we used is logistic sigmoid function emits output value at the node. The sum of synapse of layer 2 and bias of layer 3 is fed as input to the output layer. The activation function in the output layer gives the predicted value. This completes the feedforward phase. Now, the cost is calculated in which the sum of squared error of actual minus predicted. In back propagation phase, the cost function is partially differentiated with respect to the weight and updates the weights at each hidden layer and propagates toward input neurons. The larger partial differential cost to weight value the change at weight is also more. For each input instance, we find the gradient, so we use stochastic gradient decent. We repeat this process for the training set until the cost function is minimized. During the training, the network is slow, but at testing, the network gives the best results.

In stochastic gradient decent (SGD), we update the weights based on each training example but not as a batch as whole as done in gradient decent. SGD minimizes the

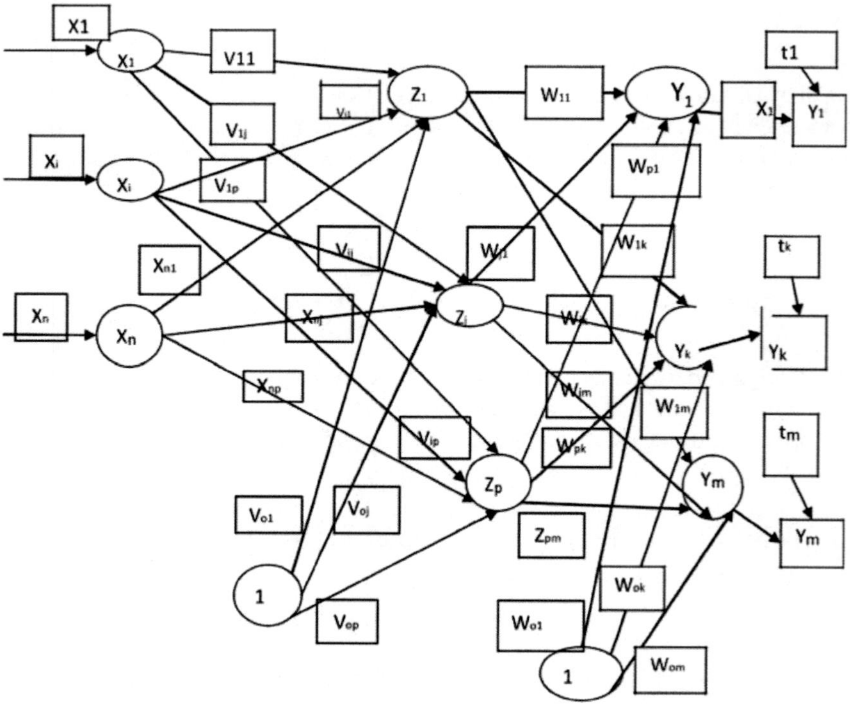

Fig. 1 Back propagation network architecture

loss function faster. We integrated SGD as a weight optimizer in multilayer perceptron back propagation network algorithm (MLPBPN) (Fig. 1).

6 Architecture

Terminologies presented in the proposed algorithm are

X denotes input instance $(X_j, \ldots, X_i, \ldots, X_n)$
t denotes target output $(t_1, \ldots, t_k, \ldots, t_m)$
α = it represents learning rate
$X_i = i$ is input unit (the input and output signals are similar because the input layer has identity activation function.)
V_{oj} = it represents jth hidden neuron bias
W_{ok} = represents kth neuron output bias
Z_j = hidden neuron j.

Z_j final input computed is

$$Z_{inj} = V_{oj} + \sum_{i=1}^{n} X_i V_{ij} \tag{1}$$

And the output is

$$Z_j = f(z_{inj}) \tag{2}$$

Y_k = output unit *k*. Y_k final input is given by

$$Y_{ink} = W_{ok} + \sum_{j=1}^{p} Z_j W_{jk} \tag{3}$$

and the output is

$$Y_k = f(Y_{ink}) \tag{4}$$

∂_k = weight adjustment for W_{jk} with respect to error correction.
∂_j = weight adjustment for V_{ij} with respect to error correction (Fig. 2).

Logistic sigmoid activation function

$$\phi(Z) = 1/1 + e^{-z} \tag{5}$$

This function predicts the probability as an output. For any given pair of points on curve, we can find the slope, and therefore, the function is said to be differentiable.

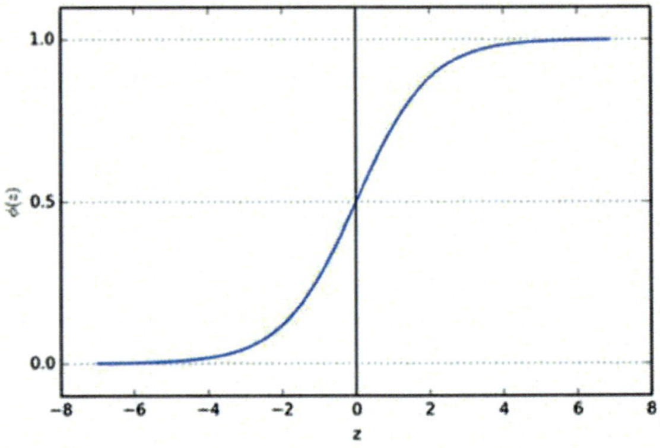

Fig. 2 Logistic sigmoid activation function graph varying between 0 and 1

7 Back Propagation Network Training Algorithm

Step 0: Random number generator function is used to assign values to weights and learning rate.

Step 1: If the stopping condition is flagged as false, then go through steps 2–9.

Step 2: For individual input pattern, go through steps 3–8.
Feedforward phase-I

Step 3: Input signal X_i is received from each input unit and forwarded to hidden neuron ($i = 1$ to n).

Step 4: Z_j ($j = 1$ to p) represents each hidden unit with weighted input signals is added to obtain final input:

$$Z_{inj} = V_{oj} + \sum_{i=1}^{n} X_i V_{ij} \tag{6}$$

Activation functions on Z_{inj} are applied to compute hidden unit response

$$Z_j = f(z_{inj}) \tag{7}$$

the obtained response is given as input to neurons in output layer

Step 5: Individual output unit Y_k ($k = 1$ to m), compute the final input:

$$Y_{ink} = W_{ok} + \sum_{j=1}^{p} Z_j W_{jk} \tag{8}$$

And activation function is applied to estimate output response

$$Y_k = f(Y_{ink}) \tag{9}$$

Back propagation of error (Phase II).

Step 6: Individual output unit Y_k ($k = 1$ to m) gets the target value for the given input vector and estimates the error correction term

$$\partial_{k-} = (t_k - Y_k) f_1(Y_{ink}) \tag{10}$$

Update the change in weights and bias based on the computed error correction term:

$$\Delta w_{jk} = \alpha \partial_k Z_j \tag{11}$$

$$\Delta w_{ok} = \alpha \partial_k \tag{12}$$

∂_k is propagated backwards through the hidden layers.

Step 7: Individual hidden neuron ($Z_j, j = 1$ to p) adds its delta inputs obtained from the output units:

$$\partial_{inj} = \sum_{k=1} \partial_k W_{jk}^m \tag{13}$$

Error term is computed by multiplying ∂_{inj} with the derivative of $f(Z_{inj})$

$$\partial_{j-} = \partial_{inj} f_1(Z_{inj}) \tag{14}$$

the derivative of $f^1(Z_{inj})$ is calculated on the sigmoidal function used. Renew the weights and bias on the basis of computed ∂_j:

$$\begin{aligned} \Delta V_{ij} &= \alpha \partial_j X_i \\ \Delta V_{oj} &= \alpha \partial_j \end{aligned} \tag{15}$$

Weight and bias updation (Phase III):

Step 8: Individual output unit (Y_k, $k = 1$ to m) renews the weights and bias:

$$W_{jk}(\text{new}) = W_{jk}(\text{old}) + \Delta W_{jk} \tag{16}$$

$$W_{ok}(\text{new}) = W_{ok}(\text{old}) + \Delta W_{ok} \tag{17}$$

Individual hidden unit ($Z_j, j = 1$ to p) renews its weights and bias:

$$V_{ij}(\text{new}) = V_{ij}(\text{old}) + \Delta V_{ij} \tag{18}$$

$$V_{oj}(\text{new}) = V_{oj}(\text{old}) + \Delta V_{oj} \tag{19}$$

Step 9: Activate stopping condition after certain number of epochs attained or target output equals actual output.

8 Testing Algorithm

Step 0: Weights are assigned with obtained weights after learning from input patterns.
Step 1: For a given input, steps 2–4 are executed.
Step 2: Input neuron X_i ($i = 1$ to n) is enabled with activation function.
Step 3: Compute the final input to hidden neuron X and its response. For $j = 1$ to p,

$$Z_{inj} = V_{oj} + \sum_{i=1}^{n} X_i V_{ij} \tag{20}$$

$$Z_i = f(z_{inj}) \tag{21}$$

Step 4: Output layer neuron response is calculated. For $k = 1$ to m,

$$Y_{ink} = W_{ok} + \sum_{j=1}^{p} Z_j W_{jk} \tag{22}$$

$$Y_k = f(Y_{ink}) \tag{23}$$

Use sigmoid activation functions for calculating the output.

9 Dataset

The multilayer perceptron back propagation neural network with stochastic gradient descent as weight optimizer is applied on Wisconsin breast cancer dataset. The dataset is obtained from UCI machine learning repository [13]. The dataset consists of 699 samples with nine attributes and one target field. After removing the noise in the data, we obtain 683 samples out of which 444 belong to benign class and 239 belong to malignant class. All attribute values are discrete in nature and vary between 1 and 10 (Tables 1 and 2).

10 K-Fold Cross Validation

K-fold cross validation is used to assess the classifier. In K-fold, split the data into K equal folds. Let i running from 1 to k, in each iteration, $k - i$ is the training set, and i is the test set. Here, i represents the ith fold. For example, k is 10 means, for $i = 1$,

Table 1 Dataset description

Attribute	Domain	Description
Clump thickness	1–10	Normal cells belong to mono layers, and cancerous cells belong to multilayers
Uniformity of cell size	1–10	Cancerous cells vary in size
Uniformity of cell shape	1–10	Cancerous cells vary in shape
Marginal adhesion	1–10	Normal cells are surrounded, and cancer cells tend to lose their ability
Single epithelial	1–10	Normal epithelial cell has uniform cell size, whereas cancer cell is significantly enlarged
Bare nuclei	1–10	These are seen in benign tumors and are not bounded by cytoplasm
Bland chromatin	1–10	They generally say about uniform texture of cells, and in cancerous cells, they tend to be coarser
Normal nuclei	1–10	In cancerous cell, the nucleus is very prominent compared to normal cells
Mitosis	1–10	In cancerous cells, the cells divide and replicate
Target field		Class 1: benign
		Class 2: malignant

Table 2 Specifications to train MLPBPN in orange tool

Neurons in hidden layers	10
Activation	Logistic sigmoid function
Solver for weight optimization	Stochastic gradient descent
Regularization	$\alpha = 0.0009$
Maximum number of iterations	100

training has nine folds except first fold. First fold is test set. For $i = 2$, training has nine folds except second fold. Second fold is test set. So on up to $i = 10$; for each i, training has nine folds except the ith fold. Ten folds are used for evaluation of an proposed algorithm.

11 Performance Metrics

Performance metrics are used to evaluate how reliable the classifier is. We have predefined performance metrics to assess the reliability of a classifier (Table 3).

TN: Negative samples are labeled as negative.

TP: Positive samples are labeled as positive.

FP: Negative samples are incorrectly labeled as positive.

FN: Positive samples are incorrectly labeled as negatives.

Table 3 Performance metrics formulae

Measure	Formula
Classifier accuracy	*TP* + *TN*
	TP + *TN* + *FP* + *FN* (24)
Classifier precision	*TP*
	TP + *FP* (25)
Classifier recall	*TP*
	TP + *FN* (26)
Classifier F-measure	2 * Precision * Recall
	Precision+Recall (27)

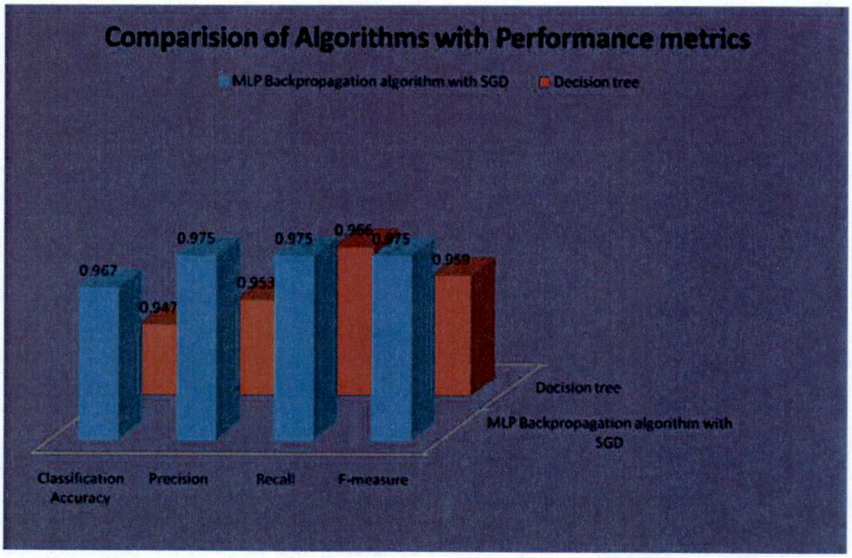

Fig. 3 Comparison of algorithms with performance metrics

F-measure is the harmonic mean of precision and recall. Accuracy represents the performance of the classifier (Figs. 3, 4 and 5).

12 Results

All the experiments were carried out in Orange open source tool. The results in terms of numerical values and bar graphs are presented (Table 4).

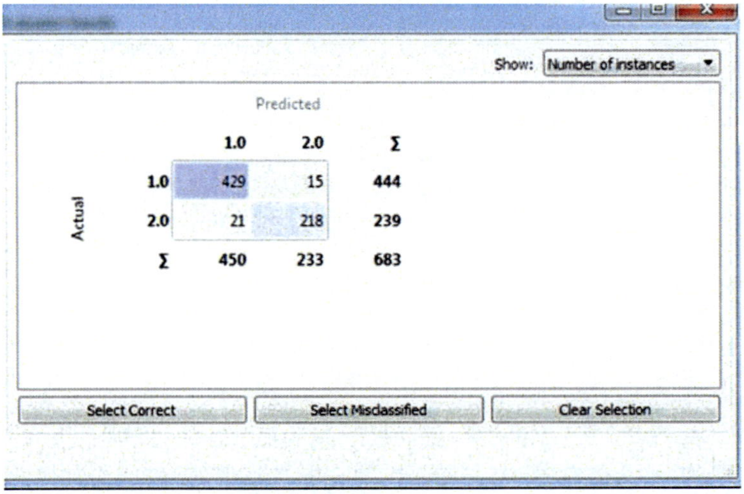

Fig. 4 Confusion matrix of decision tree

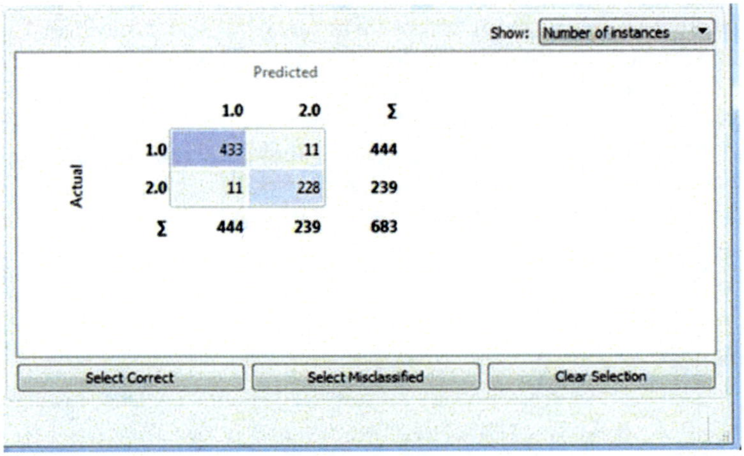

Fig. 5 Confusion matrix of MLP back propagation neural network

Table 4 Performance metric values of different algorithms on Wisconsin breast cancer data

	Classification			
	Accuracy	Precision	Recall	F-measure
Performance metrics algorithms				
MLP Back propagation algorithm with SGD	0.967	0.975	0.975	0.975
Decision tree	0.947	0.953	0.966	0.959

13 Conclusions

The proposed multilayer perceptron back propagation neural network (MLPBPN) with stochastic gradient decent as a weight optimizer when applied on Wisconsin breast cancer data showed improved performance. An improvement of 0.20 is observed with MLPBPNs when compared with decision tree. MLPBPNs have shown an improvement in 0.022 in precision, 0.009 with recall, 0.016 with F-measure when compared with decision tree. The proposed algorithm says with 0.975 precision that the patient sample is benign, and with 0.975 recall rate, the algorithm justifies the test sample as benign. Since the false positive and false negatives of decision tree are different, the proposed algorithm is compared in terms of F-measure also. The F-measure of proposed MLPBPN algorithm is 0.975, which shows that our algorithm is reliable than decision tree classifier. The false positives and false negatives of MLPBPNs are same, so the algorithm performance is compared in terms of accuracy also, in classifier accuracy, MLPBPN performed better than decision tree. The overall improved performance of MLPBPNs over decision tree in terms of performance metrics is presented. MLPBPNs justify a new patient sample with 0.967 accuracy, and this helps the patient to know apirori that he has benign or malignant. If he has malignant, then he has better chance of cure with good medication. Thus, our algorithm predicts the target class by learning from past patient records. This saves time for the doctors by restricting them to perform fewer experiments to ensure the patient is cancerous or not. The algorithm optimizes doctor's work and helps patient for better treatment in time. This algorithm is useful in medical field in terms of reduced cost and time. Early prediction of the patient disease helps in better cure for patient. Further, the proposed model is enhanced with evolutionary algorithms for better prediction with less computational complexity and execution time.

References

1. Garro, B. A., & Vazquez, R. A. (2015). Designing artificial neural networks using particle swarm optimization algorithms. *Computational Intelligence and Neuroscience*, 1–21.
2. Lavanya, D., & Usha Rani, K. (2011). Analysis of feature selection with classfication: Breast cancer datasets. *Indian Journal of Computer Science and Engineering*, 756–763.
3. Soofi, A. A., & Awan, A. (2017). Classification techniques in machine learning: Applications and issues. *Journal of Basic & Applied Sciences*, 459–465.
4. Girish, C., & Ferat, S. (2014). A survey on feature selection methods. *Computers and Electrical Engineering*, 16–28.
5. Jun, L., Lixin, D., & Bo, L. (2014). A novel naive bayes classification algorithm based on particle swarm optimization. *The Open Automation and Control Systems Journal*, 747–753.
6. Jianbo, Y., Shijin, W., & Lifeng, X. (2007). Evolving artificial neural networks using an improved PSO and DPSO. *Neuro Computing*, 1–7.
7. Brent, R. P. (1991). Fast training algorithms for multi layer neural nets. *IEEE Transactions on Neural Networks*, 346–354.
8. Kalagotla, S. K., Sita, Mahalakshmi, T., & Kamadi, V.S.R.P. (2013). Optimal classification rule discovery using hybrid model: A case study on bank note authentication using gain ratio and gini index. In *National Conference on Advance Computing and Networking*.

9. Kalagotla, S. K., Sita, Mahalakshmi, T., & Vedavati, K. (2016). Computational intelligence approach for prediction of breast cancer using particle swarm optimization: A comparative study of the results with reduced set of attributes. s.l.: Springer. *Computational Intelligence Techniques in Health Care*, 31–44.
10. Kalagotla, S. K., & Sita Mahalakshmi, T. (2016). Performance Variation of support vector machine and probabilistic neural network in classification of cancer datasets. *International Journal of Applied Engineering Research*, 2224–2234.
11. Kalagotla, S. K., & Sita Mahalakshmi, T. (2015). Computational intelligence techniques for classification of cancer data. *International Journal of Computer Applications*, 0975–8887.
12. Sivanandam, S. N., & Deepa, S. N. (2011, 13 October). *Principles of soft computing*. s.l., 2nd edn. Wiley.
13. Dua, D., & Graff, C. {UCI} Machine learning repository. https://archive.ics.uci.edu/ml/index.php. University of California, Irvine, School of Information and Computer Sciences, may 27, 2017. (Cited: August 27, 2019) http://archive.ics.uci.edu/ml.

IOT-Based Borewell Water-Level Detection and Auto-Control of Submersible Pumps

Sujatha Karimisetty, Vaikunta Rao Rugada, and Dadi Harshitha

Abstract Water crisis has been a major problem in day-to-day life. So, people depend on underground water to perform daily activities. Submersible pumps are used in utilizing underground water. Due to the technological advancement in household usage of water is increasing drastically resulting in a decrease in groundwater level. In this scenario, submersible pumps cannot reach water and the pump still works leading to the failure in the pump functioning. So, users are unable to identify the exact problem behind the misfunctioning of the pump. IOT-based Borewell Water-Level Detection and Auto-Control of Submersible pumps identify the underground water level by using water detecting sensors and sends the information to the server for Auto-Control of submersible pumps. If the sensor does not reach the water level, it will send an alert message to the user's mobile and shut down the pump automatically. The pump can be controlled over a mobile app which allows the user to control the pump remotely from anywhere. Hence, this reduces the manpower intervention and users need not worry about the functioning of the pump.

Keywords Water crisis · Submersible pumps · IOT-based borewell water-level detection and auto-control of submersible pumps (IBAS)

1 Introduction

In India, whether it is urban or rural the drinking water major source is groundwater. This is used for industry and agriculture too. The availability usually depends on recharge done automatically by ground resources and on rainfall. Due to the

S. Karimisetty (✉) · D. Harshitha
Dadi Institute of Engineering & Technology, Visakhapatnam, India
e-mail: sujathakota29@gmail.com

D. Harshitha
e-mail: harshithad@diet.edu.in

V. R. Rugada
Raghu Engineering College, Visakhapatnam, India
e-mail: vrugada@gmail.com

S. C. Satapathy et al. (eds.), *Intelligent System Design*, Advances in Intelligent Systems and Computing 1171, https://doi.org/10.1007/978-981-15-5400-1_6

rapid usage of water over the years, water scarcity has been generated. Other causes are misutilization of water resources and environmental degradation. This crisis is observed in many parts in India varying in level based on usage and season.

The water requirement and utilization due to rapid rising population and advancement in lifestyle of human along with rapid growth of utilization of groundwater in agriculture and industry resulting in reducing the levels of groundwater. Many states in India are unable to reach water levels. Though it is raining heavily there is a lack of attention of conserving water levels, efficient use of water and reusability of water.

Due to the specified issues, many times submersible pumps are failing to reach water levels. However, the consumer being unaware of the situation that pump is not reaching the groundwater level is continuously running idle. This eventually results in the failure of submersible pump. The need is to verify if the submersible pump reaches water level then only switch on motor and alert consumer on the failure of reaching groundwater levels periodically [1].

2 Literature Review

Jonthan in Water Tank Depth Sensor revealed concept on switching on automatically the pumps. This paper creates acquaintance with the preciousness of water resource and about tank control and makes [1]. Akinlalu addressed the problem related to find the depth of freshwater boreholes and the recommendation of areas to trace the location during installation time [2]. Getu proposed a system for finding water levels automatically and controlling them [3]. Poovizhi proposed a system for water level monitoring in trains for filling the water tanks in trains optimally [4].

3 IOT-Based Borewell Water-Level Detection and Auto-control of Submersible Pumps (IBAS)

The groundwater level decreases with an increase in its usage [2]. In such a scenario, the pump is activated with no water in the ground. The pump siphons air and pumps stops working and forcing the user to fix it. Due to this loophole, users cannot understand the actual problem behind the misfunctioning of the pump. The IOT-based Borewell Water-Level Detection and Auto-Control of Submersible pumps are build up with sensors to monitor the underground water level. This sends notifications to the user's mobile about underwater level.

IOT-based borewell water-level detection process identifies the underground water level by using water detecting sensors. When the ground level water decreases, the sensor couldn't reach it, it will give an alert message or notification to the user's mobile. If the pump is working in this scenario, then it will be automatically turned off without causing any further damage to pump and pumps. Hence the user need not worry about the functioning of the pump. The sensors used are cheap and best with high capacity and performance.

3.1 Objectives of the IBAS

- Consumer can operate pump remotely from anywhere and anytime using an internet connection.
- Autodetection of low underground water level.
- Avoid pump failure due to idle running.
- Send notification to users to make alternate arrangements for water if the pump cannot switch on after trying for specified number of attempts.
- Give auto SMS to plumber specified in database.

The IBAS architecture diagram is as shown in Fig. 1.

There are four basic modules of IBAS. Server Module, IBAS App Module, IOT Kit, Water-level Sensors are fixed with submersible pumps. Every module has their own importance and all together work as an integrated module.

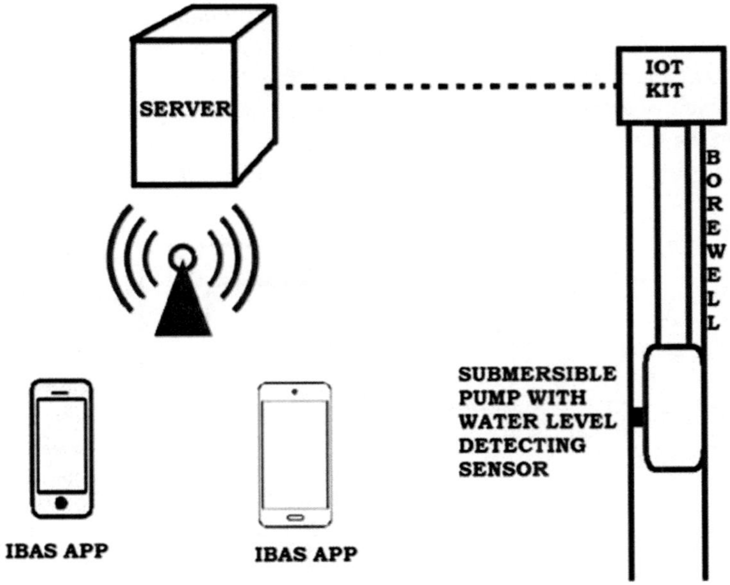

Fig. 1 Architecture of IBAS

3.2 Advantages

The notifications and alerts help the consumer in taking good decision before running in scarcity of water. This helps in maintaining good health of pump. This also saves time and money of the consumer.

4 Results

The IBAS is a device with low cost and is designed based on live requirements of the user. The mobile application is user friendly and helps novice smart mobile users operate with minimal features. The sensors used are reliable and run for years. Two types of designs are proposed where one is at low cost and other available at moderate range. This helps middle class and high class to make a choice in choosing the model. This saves time in measuring water depth and switches on pumps.

4.1 IBAS Server Module

IBAS server module is installed in any server and data obtained is periodically posted on server is used by the IBAS App for deriving solutions for queries on water level for monitoring the pump and for analysis purpose. The data can be used for Analyzing water levels of a particular resource seasonally and eventually this is used for regularizing the water supply. The cautious measures can be taken to handle situations such as going into scarce situations. The workflow is shown in Fig. 2.

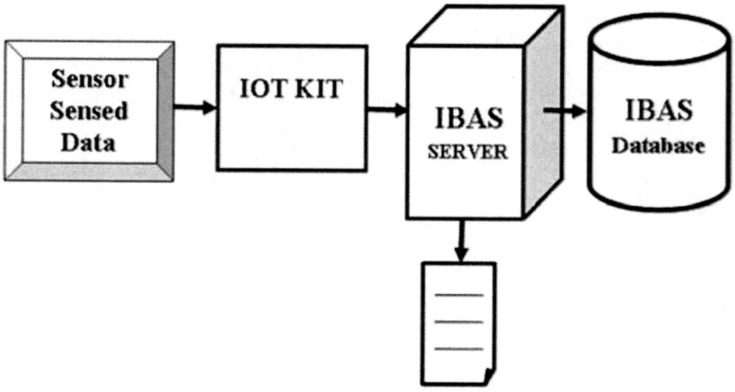

Fig. 2 IBAS server workflow

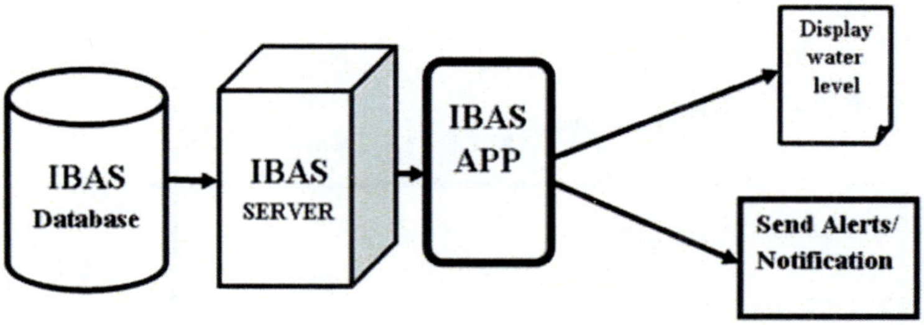

Fig. 3 IBAS mobile app workflow

4.2 IBAS App Module

IBAS App module serves as the instant query answering machine which takes data from the latest updated information on the server. Whenever the user opens mobile app it displays the information such as water level in graphical format. This also sends notifications and alerts when required such as situations arising due to low water level. This is shown in Fig. 3.

4.3 IOT Kit

Internet of Things is used to connect the things and transmit data to server. In IBAS, IOT collects the data from sensors and transmit data to server periodically [5]. This plays an important role in updating information online.

4.4 Water-Level Sensors

Different sensors such as Ultrasonic Sensors, Pressure Sensors, Radar Sensors, Open Channel Sensors, Capacitance Sensor, Submersible Hydrostatic Sensor, Magnetostrictive Sensor, Hydrostatic Sensor, Magnetic Float Level Sensor, Piezo Level Sensor, Piezoresistive Sensor, Leak Detection Sensor are being tested for accurate water-level detection. After working with all sensors Capacitance and Hydrostatic Sensors are found to be accurate in tracing the water levels and cost-effective.

4.5 Test Results

The IBAS system is developed and tested and found to be accurate compared to manual systems. The confusion matrix is as shown in Fig. 4.

n=200	Yes (Predicted)	No (Predicted)	
Yes (Actual)	130(TP)	7(FN)	137
No (Actual)	5(FP)	58(TN)	63
	135	65	

Fig. 4 IBAS confusion matrix

The confusion matrix shown in Fig. 4 is obtained by noting the values after running the IBAS system for 200 times denoted by n. The accuracy is calculated by using the formula shown in Eq. (1)

$$\text{Acc} = (\text{TP} + \text{TN})/(\text{TP} + \text{TN} + \text{FP} + \text{FN}). \tag{1}$$

where Acc is the accuracy, TP is True Positive, TN is True Negative, FP is False Positive and FN is False Negative Value. Here the Acc is calculated as

$$\text{Acc} = (130 + 58)/(130 + 58 + 5 + 7) = 0.94.$$

5 Conclusion

The proposed system is implemented for detecting electronic water-level indicator including borewell starter. The automated system helps in eradicating wastage of water and informs borewell water levels periodically to consumer. This starts pump when water-level drops below a certain level and stops when it reaches maximum level. Easy to install and operate, it gives trouble-free service for longer periods. The system keeps the entire system safe. This can be used for geological survey for checking well depths for checking well safety. This can be used in new construction projects.

Acknowledgements The authors would like to express their deepest gratitude to Sri Dadi Ratnakar, Chairman of Dadi Institute of Engineering and Technology for his encouragement and support in executing this work successfully.

References

1. Jonathan, O., Hugh, B. (2011). Water tank depth sensor. In *Practical Arduino: Cool Projects for Open Source Hardware* (pp 211–239).
2. Akinlalu, A. A., Afolabi, D. O. (2018). Borehole depth determination to freshwater and well design using geophysical logs in coastal regions of Lagos, southwestern Nigeria. Applied Water Science, *8*, 152, 1–17.
3. Getu, B. N., Attia, H. A. (2016). Automatic water level sensor and controller system. In *2016 5th International Conference on Electronic Devices, Systems and Applications (ICEDSA)*, Ras Al Khaimah (pp. 1–4).
4. Poovizhi, S., Premalatha, M., Nivetha, C. (2017). Automatic water level monitoring and seat availability details in train using wireless sensor network. In *2017 International Conference on Computation of Power, Energy Information and Commuincation (ICCPEIC)*, Melmaruvathur (pp. 321–324).
5. Poornisha, K., Keerthana, M. R., Sumathi, S. (2018). Borewell water quality and motor monitoring based on IoT gateway. In *2018 International Conference on Communication, Computing and Internet of Things (IC3IoT)*, Chennai, India (pp. 514–518).

Institute Examcell Automation with Mobile Application Interface

Sujatha Karimisetty, Sujatha Thulam, and Surendra Talari

Abstract Nowadays examcell runs in every college which maintains the student lists and results. However, activities are mostly done manually and this involves a lot of paperwork and delays in searching for relevant data. The manual process of student registration and result generation is always bulky and tedious. All these problems can be eliminated if the college examcell system is automated. Institute Examcell Automation with Mobile Application Interface will make examcell activities more efficient by covering the important drawbacks of manual system, namely speed, precision and simplicity. By this system, the examination coordinators can easily conduct the registration of the students and generation of instant results systematically. This also needs 'less manpower' to execute the system and is more efficient in producing Graphical Output. Organizations can easily check the performance of the student that they give in examinations. As a result of this Organizations can view this result easily. This will also help the students and parents in knowing their percentage and backlogs. Using machine learning the results are predicted to check the students will pass/fail. This system can be used by any colleges who need to automate examcell in the college.

Keywords Institute examcell automation with mobile application interface (IEAM) · Instant results · Graphical output · Machine learning algorithms

S. Karimisetty (✉) · S. Thulam
Dadi Institute of Engineering & Technology, Visakhapatnam, India
e-mail: sujathakota29@gmail.com

S. Thulam
e-mail: sujathat@diet.edu.in

S. Talari
GITAM Institute of Science, Visakhapatnam, India
e-mail: Surendrat.bw@gmail.com

S. C. Satapathy et al. (eds.), *Intelligent System Design*, Advances in Intelligent Systems and Computing 1171, https://doi.org/10.1007/978-981-15-5400-1_7

1 Introduction

Assessment with proper evaluation methodology plays a key role in all educational institutions [1]. Assessment is generally done in the examination cell. The examination cell usually runs with a head by the Chief Superintendent of Examinations with supporting staff. The Prime responsibility of examination cell is conducting all examinations in a fair and systematic manner under the directions of the Chief Superintendent of Examinations and maintaining all records related to marks and examinations. The examination cell of the Institute is the most active and vibrant body which remains active throughout the year, This wing works full-fledged with a good number of well trained, dedicated and work minded people. The examination cell needs to conduct the exams on stipulated dates specified by university and need to look after other related works. Practical Assessment and Internal exams are also conducted by the Institute. The overall departments rely on the examcell for the marks and analysis of reports regarding results [2].

However many examination cells are running manually and hence they are unable to retrieve data instantly when a need arises such as a parent enquiring for his son, University asking to register students for regular and supplementary, listing students for placements as per eligibility criteria and many more queries. Though some examination cells are atomized, updating data is becoming cumbersome as there will be various types of marks to be updated such as internals, externals and assignments. Many times faculty, student, parent would like to know the status of backlogs or percentages, but eventually, have to wait for getting data from concerned department. Hence an Institute Examination Cell Automation with Mobile Application Interface is required for getting information instantly.

Exams are held at regular intervals as per predefined schedules that play an important role in any educational institute. Examcell cell staff need to take care of many things otherwise questions papers or answer sheet shortage may arise or sometimes staff substitution may be required. There are huge manual calculations which are paper based. The requirement of centralized system is required that manages the activities defined by the examination cell. A centralize system is required for making the conduction of examinations easy and automate the complete life cycle of managing examinations.

Examcell serves with a large number of duties due to which the process becomes very slow and tedious. Lot of documents are still maintained in examcell leading to scarcity of physical space and making searching crucial. Retrieval of documents or answering enquiries will be delayed due to this issue.

2 Literature Review

Fatima discussed the concept of computerized system for E-examcell which involves forms filled online with interfaces related to student and examcell. E-examcell is faster and required for applications and hall ticket generations. E-Examcell explains the problems in manual examcell and stresses the need for automation. This deals with the application process and downloading hallticket online [3]. However, the need of the day is multiple queries required by the student and parents such as updated backlogs status and so on.

Kaiiali in Secure Exam Management System for M-Learning Environments proposed e-learning techniques which allow to make exam system that focuses on the issues which violate the security in conducting examinations. This focuses on random distribution of questions and good assessment preventing the unattended exam issue which usually happens if the exam is conducted in a dedicated classroom [4]. Ebel highlighted the purpose of marks and their effects. Institutional marking systems is to be enforced for a clearly defined marking system [5].

3 Institute Examcell Automation with Mobile Application Interface (IEMI)

Institute Examcell Automation with Mobile Application Interface aims to ensure examcell activities make the examcell activities to be managed effectively. IEMI aims is to introduce a centralized system to ensure in relation to examination, evaluation and documentation all the activities are effectively managed. The coordinators of examcell are struggling with present system and with IEMI their work becomes enjoyable. Automation of examination system allows to conduct examinations without any errors in a systematic manner.

3.1 The Objectives of IEMI are Listed as Follows

- Maintain Complete Student and Marks Database and other records pertaining to examcell
- Serves all Examination notices to all Concerned Individuals/department
- Provision to update results released in pdf to database by automatic pdf to text conversion
- SMS results to students and parents
- Support to different level users (a) administrator (b) Faculty (c) Students (d) Parents
- Backlog reports semester wise and aggregate
- Support for placement drive by segregating eligible students as per proposed criteria
- Restrict Wrong exam application registration by eliminating detained, passed students in a particular subject to avoid fines by university for false registrations
- Accept Question Bank from Concerned Faculty
- Generate Question paper from Question Bank basing on predefined Criteria

- Invigilation duty charts and Seating plans
- Result Analysis Graphs Generation
- Cumbersome can operate pump remotely from anywhere and anytime using an internet connection.

3.2 IEMI Working Methodology

The working methodology of the IEMI system is shown in the following Fig. 1.

There are two main modules installed on webserver and as Mobile App. The web server is used for maintaining the student and marks database. This requires to retrieve data from pdf and post them into database. Mobile App is installed on Staff, Students and Parent mobiles.

There are mainly four users of the system: Examcell admin who updates information to the webserver and then IEMI app is used by Staff, Student and Parent. The functionality of the users are unique and are as defined in the below sections.

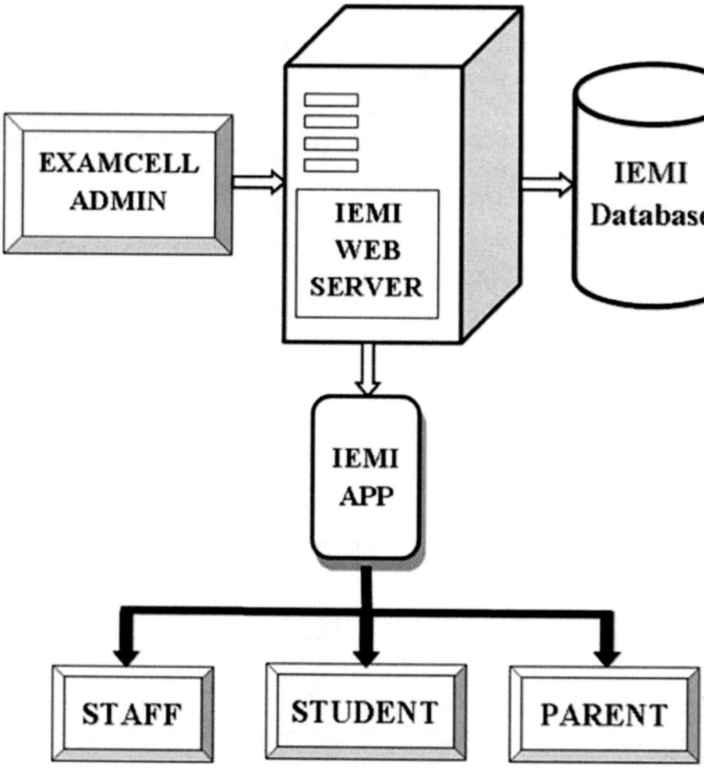

Fig. 1 Working methodology of IEMI

3.2.1 Examcell Admin

The admin responsibility is to periodically update the database without any issues. This is updated to the server. The admin maintains Student data along with all marks semesterwise. The IEMS has a feasibility to convert the pdf files of results delivered by university automatically.

3.2.2 Staff

Staff installs the IEMI app onto his mobile and he periodically checks for assigned examcell duties and notifications. This user has a provision to check results and access previous question papers.

3.2.3 Student

The student has provision only to verify his personal and academic data. He can check results instantly and can also view semester wise and aggregate reports. This tremendously motivates students to clear their backlogs by viewing the count that is piling up.

3.2.4 Parent

Parent has a provision to view the details of his son provided he registers his mobile number in the college. The results can be viewed anytime along with backlogs list.

4 Results

This system allows only registered students or faculty or staff to login to the system. Hence it avoids unauthorized access. Students can view their results and backlogs and even give their feedback or complain about any issues. This makes immediate availability and improved accuracy of student data. Better convenience for students with less human effort. This is easy to handle, operate and update involving zero paper. Students and parents get instant updates on results. Faculty get notifications to upload question banks.

The comparative analysis of manual system along with IEMI is as shown in below Table 1.

The student database is used to predict the students would fail in upcoming exams and basing on this the False Acceptance Rate (FAR) and False Rejection Rate (FRR) which are tabulated in Table 2. FAR is the Rate of identifying students as fail, who

Table 1 Comparative table of manual and IEMI systems

Category	Manual existing system	IEMI system
Speed of accessing individual results	Slow	Fast
Data maintenance	Paperbased	Database
Notifications and SMS alerts	Not available	Staff, students, parents get SMS alerts and notifications
Backlog reports	Cumbersome	Staff, students, parents get SMS alerts and notifications
Result analysis reports and graphs	Manually done	Automatically generated

Table 2 Result analysis table with FAR and FRR

Data set	FAR	FRR
Set 1	1.3	0.8
Set 2	0	1.5
Set 3	0.5	2
Set 4	1.12	1.2
Set 5	0.75	1.3

are reliable to pass by manual calculations. Similarly, FRR is Rate of identifying students as pass, who are reliable to fail by manual calculations.

The sets indicate the values taken for a batch of students and predicted accordingly. The tabulated values of FAR and FRR show that the system is successful and is accurate than manual system.

5 Conclusion

The need for this automation is required because of a large amount of data is being on paper and it made analysis of results a tedious task, apart from consolidating the amount of data that is generated in an institution from various departments. The automation system is like an intermediary between staff and students, thus easing the activities of each regarding examination results and many more. This makes examcell organized basing on crucial data where other systems are dependent on this data acquisition. The solution, however, will manage a great deal of menial work. The system under consideration is prepared in order to replace the current paper record system with an automated online management system. College staff are able to directly access all aspects of a student's academic progress through a secure, online interface through a website.

6 Future Scope

Future scope requires to put the data in cloud so that the data can be accessed more securely and quickly. Add additional modules like conducting online exams and online exam application formats and downloading hall tickets online.

Acknowledgements The authors would like to express their deepest gratitude to Sri Dadi Ratnakar, Chairman of Dadi Institute of Engineering & Technology for his encouragement and support in executing this work successfully.

References

1. Mojtaba, A. (2018). How to evaluate the TEFL students' translations: Through analytic, holistic or combined method? *Language Testing in Asia, 8,* 1–8.
2. Sujatha, K., Nageswara Rao, P. V., & Arjuna Rao, A. (2016). Encrypted examination paper distribution system for preventing paper leakage. *International Journal of Advance Research in Science and Engineering, 5*(1), 394–400.
3. Ansari, F., Ahmad, U. M., Saqlain, K., Sohail, S. M. (2017). E-exam cell. In *2017 International Conference on Innovations in Information, Embedded and Communication Systems (ICIIECS)* (pp. 1–4). Coimbatore: IEEE.
4. Kaiiali, M., Ozkaya, A., Altun, H., Haddad, H., & Alier, M. (2016). Designing a secure exam management system (SEMS) for m-learning environments. *IEEE Transactions on Learning Technologies, 9*(3), 258–271.
5. Ebel, R. L. (1974). Marks and marking systems. *IEEE Transactions on Education, 17*(2), 76–92.

Plasmonic Square Ring Resonator Based Band-Stop Filter Using MIM Waveguide

P. Osman, P. V. Sridevi, and K. V. S. N. Raju

Abstract This paper presents a plasmonic square ring resonator based band-stop filter using MIM (metal-insulator-metal) waveguide. The waveguide operates at optical wavelength $\lambda = 1296$ nm. The plasmonic square ring resonator is formed with square ring cascaded with two feed-lines. The input feed-line is given at 180^0 and output feed-line is given at 270° of the square ring resonator. A full-wave simulation has been performed using a full-wave simulation CAD software CST microwave studio. The plasmonic square ring resonator based band-stop filter is very attractive and useful in the applications of photonic integrated circuit (PICs).

Keywords Band-stop filter · MIM waveguide · Plasmonics · Photonic integrated circuits

1 Introduction

Nanophotonics offers extensive awareness as a new technical innovation to overcome the diffraction limit for reducing the size of the photonic integrated circuits into nanoscale [1]. The limitations of the light wave in subwavelength scale lead through the localization and surface plasmons propagation along with the interface of MIM waveguide recently, several research on nano-plasmonic metal-insulator-metal (MIM) waveguides [2, 3] are confirmed to be more effective technique for guiding the light with nanoscale mode. Different nanophotonic waveguide devices have been

P. Osman (✉)
Department of Electronics and Communication Engineering, Dr. Samuel George Institute of Engineering and Technology, Prakasam-Dt, Darimadugu, Markapur, Andhra Pradesh, India
e-mail: pathanosman@gmail.com

P. V. Sridevi
Department of Electronics and Communication Engineering, Andhra University College of Engineering (A), Andhra University, Visakhapatnam, Andhra Pradesh, India

K. V. S. N. Raju
Department of Electronics and Communication Engineering, SRKR Engineering College, Bhimavaram, Andhra Pradesh, India

S. C. Satapathy et al. (eds.), *Intelligent System Design*, Advances in Intelligent Systems and Computing 1171, https://doi.org/10.1007/978-981-15-5400-1_8

proposed to achieve the photonic integrated circuits like MIM waveguide, nanoparticles and nanowires. Because of its subwavelength nature and maximum degree of the light internments, MIM guiding structure has been identified as perfectly suitable such systems.

Surface Plasmonic Polaritons (SPPs) are the basic EM-waves that travel along the metal-dielectric interface [4], which are laterally limited to the diffraction limit subwavelength metallic designs. Several SPPs based MIM waveguide devices have been demonstrated numerically and experimentally, such as tooth-shaped filters [5], bends [6], biosensors [7], optical modulators [8], Mach–Zehnder interferometers [9], Bragg reflectors [10] and Plasmonic switches [11]. However, each one of these circuit elements have been designed for single-band frequency operation at a time.

This paper demonstrates the plasmonic square ring resonator based band-stop filter using MIM waveguide. Plasmonic square ring resonator based band-stop filter has been proposed for the first time which is operated at optical O band. The reflection and transmission coefficients have been observed for the resonator. The proposed design has been carried out using the full-wave simulation using a commercially available tool CST microwave studio. The analysis of a plasmonic MIM waveguide based band-stop filter have been analyzed with perfectly matched layer (PML) boundary conditions.

Recently, several plasmonic devices like plasmonic antenna [12], plasmonic directional coupler [13], step impedance resonator based square ring resonator [14] and split-mode ring resonator [15] has been proposed and numerically analyzed. The above plasmonic devices are also designed at optical wavelengths which are further used in PICs.

2 Geometry of the Plasmonic Square Ring Resonator Based Band-Stop

The proposed square ring resonator based band-stop filter using MIM waveguide has been designed and analyzed using CST microwave studio suite. The plasmonic square ring resonator based band-stop filter using MIM waveguide has been analyzed and the results show the band-stop characteristics in the optical spectrum. The square ring resonator refractive transmission spectrum is observed when the insulator refractive index and the feed-line width are fixed. We carry out the full-wave analysis with fixed values of the parameters, time step, PLM, boundary conditions. In this dielectric as silica and the metal is assumed as silver, respectively. The drude model of the silver is fixed to be Epsilon infinity = 3.7, $\omega_p = 1.38 \times 10^{16}$ rad/s, $\Upsilon = 2.73 \times 10^{13}$ rad/s and silica as $\varepsilon i = 2.50$ that is obtained through experimental results [16].

In the past, a reconfigurable band-pass to band-stop filter using pin diodes based on the square ring resonator has been proposed by Salman et al. [17] at microwave frequencies with millimeter lengths. We had proposed a plasmonic square ring resonator based band-stop filter using MIM waveguide has been proposed and analyzed at THz frequency. The input feed-line is given at 1800 and output feed-line is given at 2700 of the plasmonic square ring resonator.

L_2 is the common length and w_2 is the common width of both feed-lines. To avoid multiple mode generation the width and length of the feed-lines are kept the same. The circumference of the square ring resonator is given by $L_k = P * \lambda$, where P is the number of mode, λ is the guiding wavelength and L_k is the circumference of the square ring resonator.

$$\lambda = \frac{c}{f\sqrt{\varepsilon_r}} \tag{1}$$

where$f$ is the operating frequency, c is the light speed and ε_r is the dielectric constant. The dimensions of the square ring resonator are $l_1 = 990$ nm, $l_2 = 500$ nm, $w_1 =$ 148 nm and $w_2 = 140$ nm. The square ring resonator band-stop filter rejects the filter from wavelength 1210 to 1380 nm. Figure 1 shows the geometry proposed square ring resonator based band-stop filter using MIM waveguide. Figure 2 shows the variation in reflection and transmission coefficients has been observed with respect to wavelength as a function of width w_1. Figure 3 shows the field distribution at wavelengths 1296 (nm).

3 Conclusion

A plasmonic band-stop filter using MIM waveguide has been implemented using a square ring resonator. Full-wave EM simulation results using CST microwave studio show the presence of stop-bands at 1296-nm, i.e. at O band. The variation in reflection and transmission coefficients has been observed with respect to wavelength and width w_1 as function. The field distribution at wavelengths 1296 (nm) has been observed at optical bands. The proposed band-stop filter geometry is compact with low insertion loss; hence it can be used for high-density multi band photonic integrated circuits (PICs).

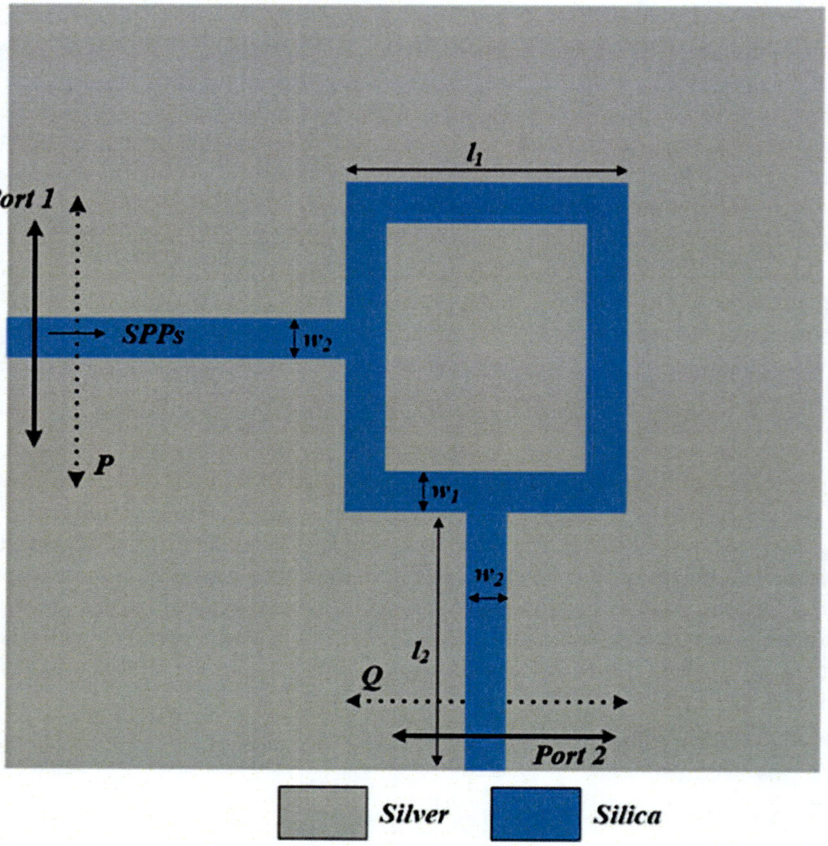

Fig. 1 Geometry of the Plasmonic square ring resonator based band-stop filter using MIM waveguide

Fig. 2 Variation in reflection and transmission coefficient with wavelength as a function of width (w_1)

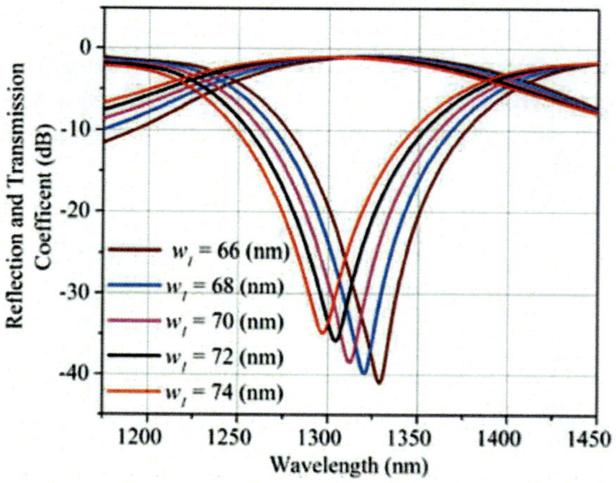

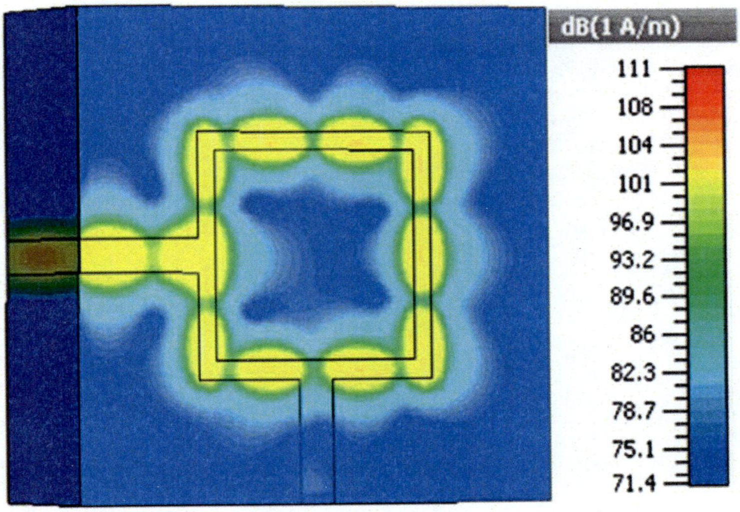

Fig. 3 Field distribution of Plasmonic square ring resonator based band-stop filter using MIM waveguide at wavelength $\lambda = 1296$ (nm)

References

1. Zhu, J. H., Wang, Q. J., Shum, P., & Huang, X. G. (2011). A simple nanometeric plasmonic narrow-band filter structure based on metal-insulator-metal waveguide. *IEEE Transactions on Nanotechnology, 10*(6), 1371–1376.
2. Veronis, G., Fan, S., & Member, S. (2007). Modes of subwavelength plasmonic slot waveguides. *Journal of Lightwave Technology, 25*(9), 2511–2521.
3. Taylor, P., Li, C., Qi, D., Xin, J., & Hao, F. (2014). Metal-insulator-metal plasmonic waveguide for low- distortion slow light at telecom frequencies. *Journal of Modern Optics, 61*(8), 37–41.
4. Mu, W., & Ketterson, J. B. (2011). Long-range surface plasmon polaritons propagating on a dielectric waveguide support. *Optics Letters, 36*(23), 4713–4715.
5. Lin, X., & Huang, X. (2009). Numerical modeling of a teeth-shaped nanoplasmonic waveguide filter. *Journal of the Optical Society of America B, 26,* 1263–1268.
6. Pile, D. F. P., & Gramotnev, D. K. (2018). Plasmonic subwavelength waveguides: Next to zero losses at sharp bends. *Optics Communications, 414,* 16–21.
7. Podoliak, N., Horak, P., Prangsma, J. C., & Pinkse, P. W. H. (2015). Subwavelength line imaging using plasmonic waveguides. *IEEE Journal of Quantum Electronics, 51*(2), 7200107.
8. Melikyan, A., Alloatti, L., et al. (2014). High-speed plasmonic phase modulators. *Nature Photos, 8*(2), 229–233.
9. Lu, H., Gan, & X., et.al. (2018). Flexibly tunable high-quality-factor induced transparency in plasmonic systems. *Scientific Reports, 8*(1), 1558, 1–9.
10. Han, Z., Forsberg, E., He, S. (2007). Surface plasmon bragg gratings formed in metal-insulator-metal waveguides, *19*(2), 91–93.
11. Zheng, Y. B., et al. (2011). Incident-angle-modulated molecular plasmonic switches: A case of weak exciton-plasmon coupling. *Nano Letters, 11*(5), 2061–2065.
12. Chityala, R. K. (2019). Nanoplasmonic concurrent dual-band antennas using metal-insulator-metal step impedance resonators. *Microwave and Optical Technology Letters.*
13. Chityala, R., & Chandubatla, V. K. (2019). Concurrent dualband nanoplasmonic MIM slot waveguide based directional Coupler. *International Journal of Electrical and Electronics Research, 7*(1), 217–219.

14. Osman, P. et al. (2019). A novel dual-band band-pass filter using plasmonic square ring resonator. IJSSST.
15. Osman, P., et al. (2019). Dual band band-pass filters using plasmonic split-mode ring resonator. *International Journal of Innovative Technology and Exploring Engineering, 8*(4), 563–565.
16. Jhonson, P. B., & Christy, R. W. (1972). Optical constants of noble metals. *Physical Review Letters, 6*(12), 4370–4379.
17. Salman, A. et al. (2015). A reconfigurable bandpass to bandstop filter using pin diodes based on thesquare ring resonator. *Progress in Electromagnetics Research Symposium, 45*(20), 1415, 1–5.

Interactive and Assistive Gloves for Post-stroke Hand Rehabilitation

Riya Vatsa and Suresh Chandra Satapathy

Abstract The inability to fold fingers and move the wrist due to stroke, cardiovascular injuries or emotional shock is one of the most common illnesses wherein conventional rehabilitation therapies are propitious in functional recovery. However, implementation of these methods is laborious, costly and resource-intensive. The structure of the prevailing healthcare system challenges us to design innovative rehabilitation techniques. A desktop-based interactive hand rehabilitation system is, therefore, developed to ensure a more feasible and cost- effective approach. It will encourage a higher number of participation as it is designed to be interesting and interactive than the traditional physiotherapy sessions. The system uses sensor data from Arduino microcontroller and is programmed in Processing IDE allowing user interaction with a virtual environment. The data is further received in an Android application from where it is stored using ThingSpeak Cloud.

Keywords Smart glove · Hand rehabilitation · Arduino · Processing · App inventor · ThingSpeak

1 Introduction

Stroke is a condition wherein adequate blood supply from the heart is discontinued due to heart failure. Hand disability is the most common consequence of a stroke attack. It is the inability to fold palm, move the fingers and bend the arm. The patient often finds it difficult to either bend the fingers or rotate the wrist. Sometimes the fingers shake vigorously. He or she feels anxious to hold an object. This inability to control one's hand movement is called a hand disability.

R. Vatsa (✉)
Department of Information Technology, Kalinga Institute of Industrial Technology, Deemed to be University, Bhubaneswar, Odisha, India
e-mail: riya31047@gmail.com

S. C. Satapathy
School of Computer Engineering, Kalinga Institute of Industrial Technology, Deemed to be University, Bhubaneswar, Odisha, India
e-mail: sureshsatapathy@ieee.org

S. C. Satapathy et al. (eds.), *Intelligent System Design*, Advances in Intelligent Systems and Computing 1171, https://doi.org/10.1007/978-981-15-5400-1_9

Physiotherapy is the best exercise given to every victim who suffers from hand disability. In some cases, the fingers tremble and any movement often renders pain. The most common reason behind this disability is the stroke, which disables a hemisphere of the brain causing paralysis of the opposite part of the human body. Apart from hand immobility, people who have met with shocking accidents suffered from depression, bipolar syndrome and anger are also prescribed with physiotherapy as a relief cure. However, with the introduction of Virtual Reality (VR), physiotherapy sessions can become effective as well as engaging. In the case of stroke, the frequency of physiotherapy sessions, improvement in finger movements and folding capabilities need to be mapped periodically and assessed by doctors.

There are several gloves designed commercially as well as for research purposes for achieving the same goal. They are: Post-stroke assistive rehabilitation robotic gloves [1], Portable Assistive Glove [2], Supervised Care and Rehabilitation Involving Personal Tele-robotics (SCRIPT) [3], and RAPAEL Rehabilitation Solutions [4].

However, these gloves aren't user friendly because they appear extremely complicated with electronic equipment and have significant weight making it quite difficult to maintain. They are costly since they use multiple sensors and highly accurate microcontrollers, but at the same time, the measurements are precise.

The basic motivation of the project comes with the fact that it aims to assist a large group of people with a wide age group. Therefore, cost-effectiveness, feasibility, comfort, light weightiness, portability and reproducibility are some of the properties that the glove must satisfy. In the paper titled "SMART HAND GLOVES FOR DISABLE PEOPLE" [5], gloves for the disabled people is suggested. Textile sensor can be one such type of sensor which may account for the aforementioned properties. In the paper titled "Sensorized Glove for Measuring Hand Finger Flexion for Rehabilitation Purposes" [6], viewers can read step by step procedure to build various types of sensors from some easily available textile sensors.

Android Lilypad is the most common microcontroller used with textile sensors. In the paper "Managing Wearable Sensor Data through Cloud Computing" [7], the use of Arduino Lilypad as a wearable microcontroller has been explained explicitly. A comparison table between Arduino Lilypad and Arduino Uno has been shown in Table 1.

The change in data is reflected in the form of a virtual environment through an interactive visual art platform. The data is also sent to cloud so that the doctors can regularly inspect the performance. An Android application is also developed which helps users to track daily exercises, set up reminders read motivational quotes, etc.

Table 1 Comparison between Arduino Lilypad and Uno

Properties	Arduino Lilypad	Arduino Uno
Comfort	Wearable directly	Needs support
OTG	No port available	Port available
Cost	INR 2000/-	INR 250/-
Availability	Uncommon	Easy available

Thus, this VR-based rehabilitation method is engaging, interesting and motivating from the patient's point of view. This aspect is crucially important for an accelerated and improved recovery.

1.1 System Design

Block diagram of the entire process
This entire project is divided into two parts (Fig. 1).

(1) Internet of Things part(IoT)—In this part, real-time data is sent from the assistive gloves to Arduino where wireless communication is established with a customized Android application through a Bluetooth (HC-05) module. The app is useful for the patient in terms of portability and tracking his/her performance. The user can set a reminder accordingly for his/her physiotherapy sessions. The sensor data from the app is transferred to ThingSpeak Cloud where it is stored permanently. The doctor can virtually monitor the progress of the patient through the cloud and prescribe changes in the session as per requirement.

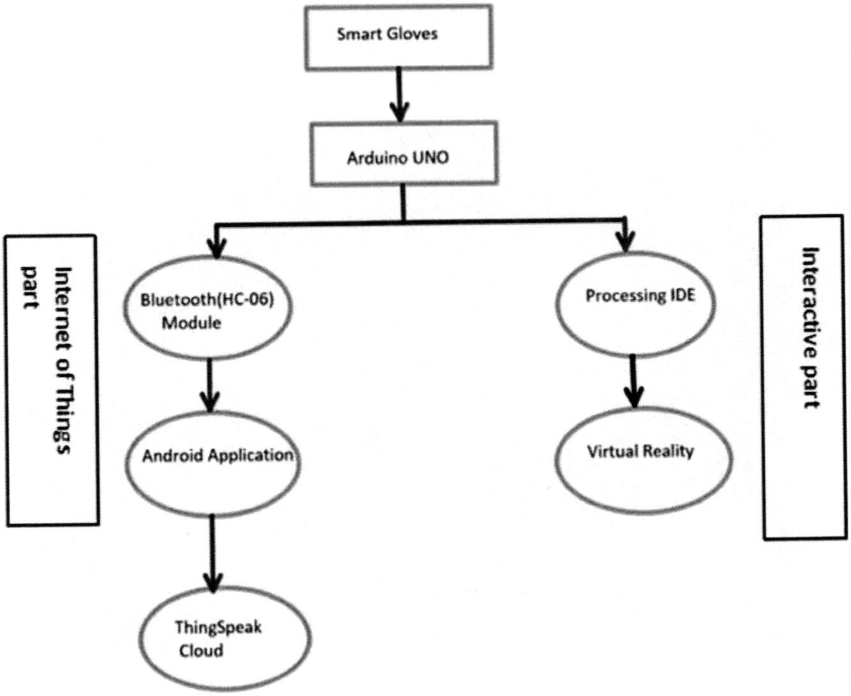

Fig. 1 Block diagram of the implemented System

(2) Interactive part—The real-time data is made to serially communicate with open-source software for creating graphics known as processing. As the Arduino data is received in processing, a virtual environment is created which allows the patient to interact with the virtual world. Through this process, the patient will not feel like a patient which in turn will have a motivating effect on the progress of the patient. He/She would be engrossed in the virtual world due to its interactive approach. Thus, this is an important aspect of the emotional as well as functional recovery of stroke patients.

Hardware

(1) Arduino Uno: It is the most common microcontroller board used in electronics. It has 14 pins that work as digital input or digital output and 6 pins that work as analog pins, a USB connection, as 16 MHz quartz clock, a reset button and a power jack. It works just like the ATmega328P microcontroller. Also, it can be easily powered by an AC to DC battery (Fig. 2).
(2) Bluetooth Module: The HC-05 is a type of Bluetooth Serial Port Protocol V2.0 + EDR (Enhanced Data Rate) module. In this project, the HC-05 module is used as a mode for wireless communication between the Arduino IDE and the Android application (Fig. 3).
(3) Connecting Wires: Male to Female jumper wires are used for connecting HC-05 Bluetooth module with the Arduino Uno board. They are roughly about 20 cm or 8 inches long.

Fig. 2 Arduino Uno

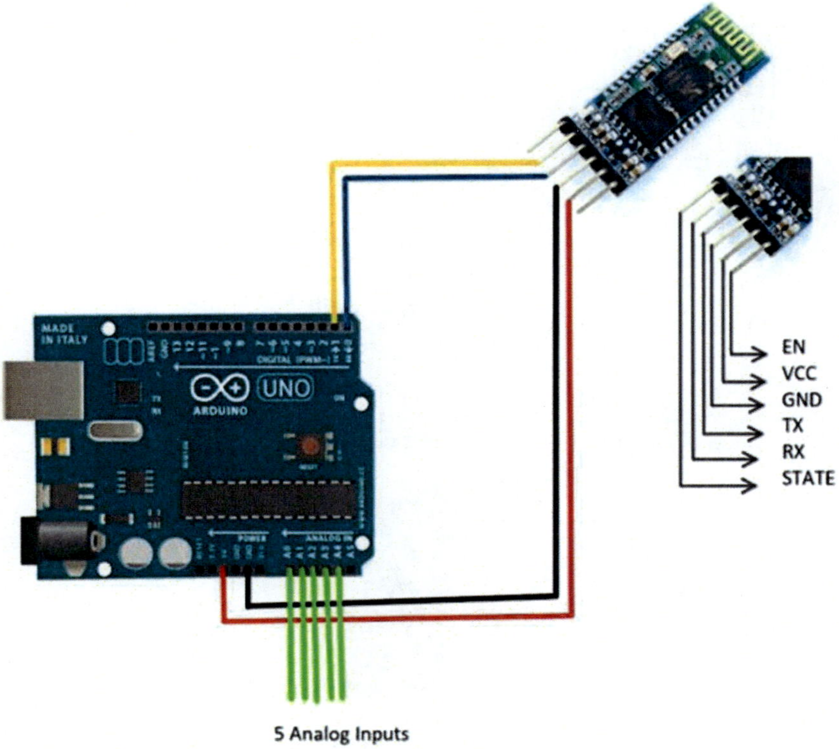

Fig. 3 Connection of the Bluetooth (HC-05) module with Arduino

2 Experimental Prototype

The most easily available gloves are the summer cotton gloves which are easy to wear, washable and comfortable. The length of these gloves reach up to the wrist or till the mid-portion of the arm. The glove must fit exactly so that it is neither too tight nor too lose. However, the size of the hands or fingers of every human being differs. Since the project addresses people of varying age groups, deciding an average palm or limb size and thereby designing a glove is impossible. Velcro straps can be used to tighten or loosen the gloves according to the user's size of the palm and the arm (Fig. 4).

The readings from the five fingers are mapped into a percentage to make it more comprehensive. When the fingers are fully outstretched, the reading would be 0% and when they are fully bent it would be 100%. They are observed by the serial monitor. The observations attained can be summarized in a tabular format for the following five instances:

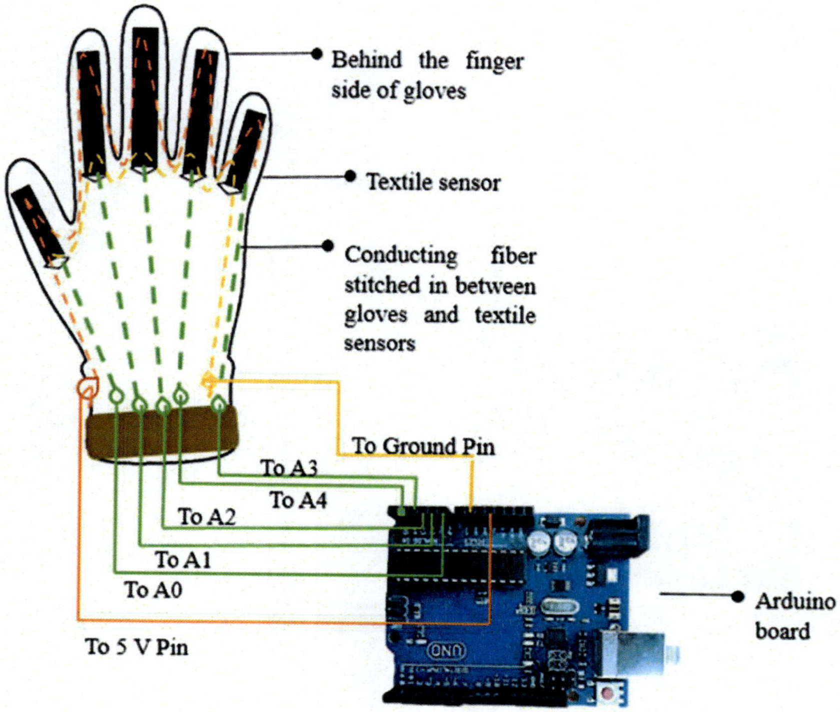

Fig. 4 Schematic diagram of the gloves with Velcro straps

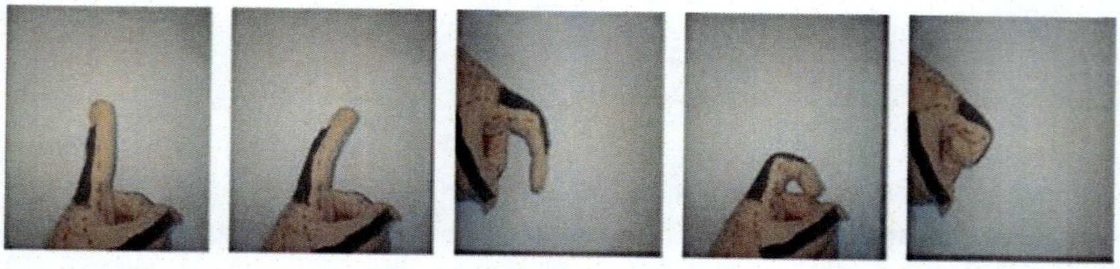

Fig. 5 Different bending positions of the fingers

(1) When the finger is straight
(2) When the finger is slightly bent
(3) When the finger is half bent
(4) When the finger is almost bent
(5) When the finger is full bent (Fig. 5, Table 2).

Rea-time sensor data from Arduino is sent to processing IDE over the serial port and a virtual environment is created for the patient to see and practice in the front-end.

Table 2 Observation as per seen in the serial monitor

Position of fingers	% Bending observed
Position 1	0–20
Position 2	20–40
Position 3	40–60
Position 4	60–80
Position 5	80–100

Fig. 6 Image of hand and smileys created in processing

Five smileys are designed bearing five different expressions that work in accordance with the bending position of the five fingers. An illusion of a palm is added in the background to give it a realistic touch.

In Fig. 6, smileys 1–5 corresponds to bending of figures. Smiley 1 denotes position 4, smiley 2 denotes position 5, smiley 3 denotes position 2, smiley 4 denotes position 3 and smiley 5 denotes position 1.

An Android application named Smart Gloves is developed keeping in mind the feasibility aspect of the project. With this feature, the patient can practice any number of times irrespective of location and time. To transfer real-time sensor data of the assistive gloves from Arduino to the app, a wireless transmission is always preferred. Hence, a Bluetooth (HC-05) module is chosen (Fig. 7).

In the cloud, data is plotted in % graphically with respect to time. The doctor can analyze the progress of the patient according to the % bend in the five fingers. Data is also stored so he will have the provision to check the history of the patient's earlier physiotherapy sessions for future reference (Fig. 8).

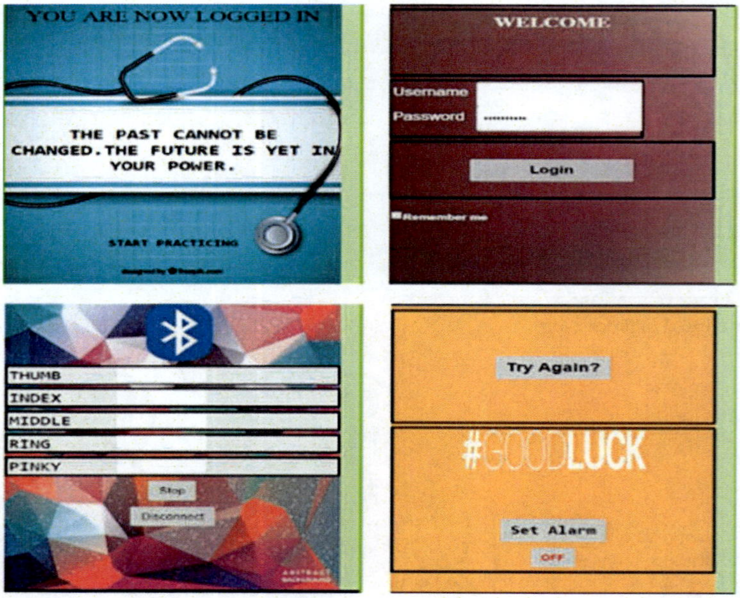

Fig. 7 Instances of the Android application

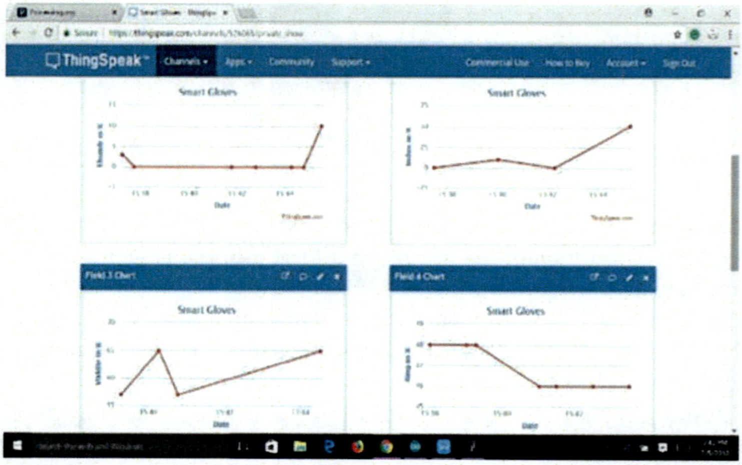

Fig. 8 Graphical visualization of the real-time sensor data from the app to ThingSpeak Cloud

References

1. Popescu, D., Ivanescu, M., & Popescu, R. (2016). Post-stroke assistive rehabilitation robotic gloves. In *2016 International Conference and Exposition on Electrical and Power Engineering (EPE), IEEE Explore*, December 12, 2016.
2. Fischer, H. C., Triandafilou, K. M., Thielbar, K. O., Ochoa, J. M., Lazzaro, E. D. C., Pacholski, K. A., et al. (2015). Use of a portable assistive glove to facilitate rehabilitation in stroke survivors with severe hand impairment. *IEEE Transactions on Neural Systems and Rehabilitation Engineering, 24*(3), 344–351.
3. Prange, G. B., Hermens, H. J., Schäfer, J., Nasr, N., Mountain, G., Stienen, A. H. A., & Amirabdollahian, F. SCRIPT: TELE-ROBOTICS AT HOME Functional architecture and clinical application. *Community Research and Development Information Service (CORDIS)*, Nov 01, 2011–Dec 31, 2014.
4. Shin, J.-H., Kim, M.-Y. Ji-Yeong, Lee, Jeon, Suyoung Kim, Y.-J., Lee, S., Seo, B., et al. (2016). Effects of virtual reality-based rehabilitation on distal upper extremity function and health-related quality of life: a single-blinded, randomized controlled trial. *Journal of Neuro Engineering and Rehabilitation, 13,* 17.
5. Patel, D. L., Tapase, H. S., Landge, P. A., More, P. P., & Bagade, A. P. (2008). SMART HAND GLOVES FOR DISABLE PEOPLE. *International Research Journal of Engineering and Technology (IRJET), 05*(04).
6. Borghetti, M., Sardini, E., & Serpelloni, M. (2013). Sensorized glove for measuring hand finger flexion for rehabilitation purposes. *IEEE Transactions on Instrumentation and Measurement, 62*(12), 3308–3314.
7. Doukas, C., Maglogiannis, I. (2011). Managing wearable sensor data through cloud computing. In *2011 Third IEEE International Conference on Cloud Computing Technology and Science.*

An Approach for Collaborative Data Publishing Using Self-adaptive Genetic Grey Wolf Optimizer

T. Senthil Murugan and Yogesh R. Kulkarni

Abstract This paper introduces an algorithm, termed self-adaptive genetic grey wolf optimizer (self-adaptive genetic GWO), for privacy preservation using a C-mixture factor. The C-mixture factor improves the privacy of data, in which the data does not satisfy the privacy constraints, such as *l*-diversity, *m*-privacy, and *k*-anonymity. Experimentation is carried out using the adult dataset, and the effectiveness of the proposed self-adaptive genetic GWO is checked depending on the information loss and the average equivalence class metric values and is evaluated to be the best when compared to other existing techniques with low information loss value as 0.1724 and average equivalence class value as 0.71, respectively.

Keywords Data publishing · Privacy preservation · *m*-privacy · Self-adaptive genetic GWO · Information loss

1 Introduction

Privacy-preserving data publishing and analysis has become a hot research topic over the past few years in various applications, which have been paid more attention to privacy problems and in sharing the data [1]. Due to the linking attacks, identifying attributes, like user ID and names, is not distributed via sensitive information, but leakage is remaining [2]. Several organizations bring out microdata-tables contain unaggregated information about persons [3–5]. Privacy-sensitive is adopted in data mining techniques, but it faces several issues. Thus, there is a requirement for developing data mining techniques, which can deal with privacy problems [6]. The main aim of data anonymization is protecting the privacy of the users [7, 8]. Due to the advancements in online services and big data, privacy-preserving data publishing

T. Senthil Murugan
Department of Computer Science and Engineering, Vel Tech Rangarajan Dr. Sagunthala R&D Institute of Science and Technology, Chennai, India
e-mail: senthilmuruganme@gmail.com

Yogesh R. Kulkarni (✉)
Vel Tech University, Avadi, Chennai, Tamil Nadu 600062, India
e-mail: kulkarni.yr@gmail.com

S. C. Satapathy et al. (eds.), *Intelligent System Design*, Advances in Intelligent Systems and Computing 1171, https://doi.org/10.1007/978-981-15-5400-1_10

(PPDP) has received considerable attention [9]. The generalization and the other processing are performed using the original data for enhancing the security that results in information loss [10].

Several mechanisms and definitions are developed for the privacy preservation in the existing methods, among which, one of the most significant rigorous methods of the privacy definition [11] is the differential privacy [12]. Zakerzadeh et al. [13] proposed a fragmentation-based anonymization technique for huge-sized data. Hua et al. [14] introduced a technique of privacy-preserving utility verification for Diff-Part. An effective algorithm for *k*-anonymization was developed by Kabir et al. [15] to reduce the information loss that occurs while anonymizing the data, assuring data quality. Zhang et al. [16] introduced a privacy-aware set-valued data publishing system, named Cocktail. On data publishing phase, an approach, called Extended Quasi-Identifier-partitioning (EQI-partitioning), was developed.

The contribution of the paper is the self-adaptive genetic GWO algorithm, which is proposed by introducing the self-adaptive concept in genetic GWO algorithm such that the update rule of genetic GWO can be preceded without any user interaction. The proposed algorithm predicts the best solution so that the privacy of data is maintained during data publishing.

2 Proposed Method of Secure Data Publishing Using Self-adaptive Genetic GWO Algorithm

In this paper, three privacy constraints and the C-mixture are utilized for enhancing the privacy measures. Also, an algorithm, named self-adaptive genetic GWO, is developed for determining the better solution for publishing the privacy ensured

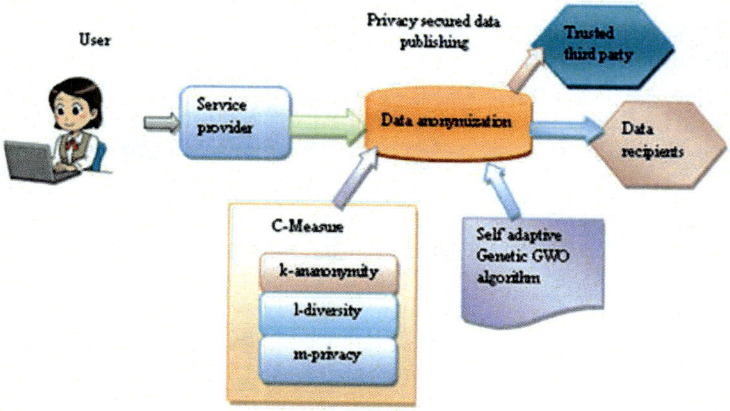

Fig. 1 Proposed privacy preserved collaborative data publishing

data. Figure 1 depicts the privacy preserved data publishing strategy of the proposed method.

The set of reports or data given by the service providers is expressed as

$$D = \{D_i \in G_i; 1 \leq i \leq M\} \quad (1)$$

where M denotes the total number of providers, D_i represents the report given by the ith service provider, and G_i is the ith service provider. The data published by the service provider consists of attributes and quasi-identifier and is represented as

$$D = \left\{G, J_i, J_2, J_{0,} l_i, l_2, l_T, K_i, K_2, K_m\right\} \quad (2)$$

where D indicates the data report of the service provider, G is the name of the service provider, J represents the common attribute, l denotes the quasi-identifier, and K is the sensitive attribute.

2.1 *Data Privacy Enhancement Based on the C-Mixture Model*

C-mixture measures the privacy of the published data for modifying the privacy measures, namely k-anonymity, l-diversity, and m-privacy. Assume that the report D contains quasi-identifiers K, and the report D must fulfil the C-mixture. The privacy constraint based on C-mixture is depicted below:

$$[j = X * E][p = h * E][s = G * E] \quad (3)$$

where X indicates the total records, whose value is 8, E represents the C-mixture, h denotes the class attributes, and the value of class attributes is 2, and G refers to the number of service providers, and it is 2. The privacy constraints, l-diversity, k-anonymity, and m-privacy are improved depending upon the C-mixture. The range of E is denoted as $0 \leq E \leq 1$.

2.2 *Proposed Self-adaptive Genetic GWO Algorithm for Collaborative Data Publishing*

This section presents the proposed self-adaptive genetic GWO developed for privacy preservation.

2.2.1 Solution Encoding

The objective of the solution encoding is to encode all the data of the table into a single vector to preserve the privacy of the data during the process of anonymization. The generalization procedure of the solution encoding has b quasi-identifiers with the anonymization level that ranges in the interval $1 < x < K$. It determines the level of zip code, gender, age, and education. After converting all the attributes into the general format, the fitness is evaluated, and the privacy data has been published. Thus, the report R is converted into a generalized report R^*.

2.2.2 Fitness Evaluation

After solution encoding, the fitness function is evaluated for the privacy preservation model. The fitness function depends upon the generalized information loss $\left(G^{\mathrm{IL}}\right)$ and the average equivalence size of the class, which is denoted as $\left(E_{\mathrm{avg}}\right)$. The value of $\left(\mathrm{Gen}^{\mathrm{IL}}\right)$ and $\left(E_{\mathrm{avg}}\right)$ is minimized to handle the privacy of data. Therefore, the fitness constraint equation is given below,

$$W(S) = \lambda * \mathrm{Gen}^{\mathrm{IL}}(S) + \chi * E_{\mathrm{avg}}(S) \tag{4}$$

The Eq. (4) can be represented into three cases such as

$$\left.\begin{array}{r} k \geq H_{\mathrm{ano}}(J, D) \\ l \geq H_{\mathrm{div}}(J, D) \\ m \geq H_{\mathrm{pri}}(J, D) \end{array}\right\} \tag{5}$$

where

$$\mathrm{Gen}^{\mathrm{IL}}(S) = \frac{1}{M \times T} \times \sum_{s=1}^{T} \sum_{m=1}^{M} \frac{V_{sm} - L_{sm}}{V_s - J_s} \tag{6}$$

$$E_{\mathrm{avg}}(S) = \frac{M}{|Kl_T| * j} \tag{7}$$

The sth quasi-identifier of the upper and lower bounds is denoted as V_s and J_s,$H_{\mathrm{ano}}(J, D)$ indicates the duplicate records,V_{sm} and L_{sm} are the upper and the lower bounds found in the generalized interval, T represents the total number of quasi-identifiers, and the total number of records is represented as M. The number of service providers is calculated based on the function $H_{\mathrm{pri}}(J, D)$, and $H_{\mathrm{div}}(J, D)$ is the number of defined sensitive attributes.

2.2.3 Self-adaptive Genetic GWO Algorithm

The proposed self-adaptive genetic GWO algorithm is designed by modifying the genetic GWO, which is the integration of GA [17] and GWO [18], based on the self-adaptive concept. GA applied in GWO decreases the convergences time and addresses the problem of converging to the nearly the best solution instead of the best solution. The proposed algorithm makes genetic GWO self-adaptive by adjusting the control parameters such that the algorithm significantly improves the convergence speed, offering the best solution.

The steps involved in self-adaptive genetic GWO are illustrated below:

1. **Initialization**

The first step involves the initialization of the grey wolf population to select the best position.

$$B = \{B_1, B_2, \ldots B_j \ldots B_w\} \tag{8}$$

Assume the position of the best solution as B_α, while B_β and B_δ are the second best solution and the third best solutions, and w represents the total number of wolves present in the population

2. ***Encircling phase***

To make the algorithm self-adaptive, these random numbers are selected based on the distance between the positions of the best search agents, B_α, B_β and B_δ as given below,

$$h_1 = \frac{\text{dist}(B_\alpha, B_1)}{\text{dist}(B_\alpha, B_1) + \text{dist}(B_\beta, B_2) + \text{dist}(B_\delta, B_3)} \tag{9}$$

$$h_2 = \frac{\text{dist}(B_\beta, B_1)}{\text{dist}(B_\alpha, B_1) + \text{dist}(B_\beta, B_2) + \text{dist}(B_\delta, B_3)} \tag{10}$$

whereh_1 and h_2 indicate the random vectors range from [0, 1] and a is a parameter that has the value from 2 to 0. The random vectors h_1 and h_2 are used to reach in any positions inside the search area. Equation (10) forms the position update equation of the proposed self-adaptive genetic GWO.

3. ***Hunting process based on GA***

Consider α, β, δ having knowledge of the location, in which prey resides. According to the best search agents, the positions are updated. Hence, the distance measures are expressed as

$$\overrightarrow{t_\alpha(v)} = \left|\overrightarrow{W_1}.\overrightarrow{B_\alpha} - \overrightarrow{B}\right| \tag{11}$$

$$\overrightarrow{t_\beta(v)} = \left|\overrightarrow{W_2}.\overrightarrow{B_\beta} - \overrightarrow{B}\right| \tag{12}$$

$$\overrightarrow{t_\delta(v)} = \left|\overrightarrow{W_3}.\overrightarrow{B_\delta} - \overrightarrow{B}\right| \tag{13}$$

In self-adaptive genetic GWO, the update is performed using a genetic algorithm by adding a new term B_4 in GWO. Hence, the position update of the proposed self-adaptive genetic GWO is mathematically represented as

$$\overrightarrow{B(v+1)} = \frac{\overrightarrow{B_1} + \overrightarrow{B_2} + \overrightarrow{B_3} + \overrightarrow{B_4}}{4} \tag{14}$$

where $\overrightarrow{B(v+1)}$ indicates the position of the next iteration, and the position of the best search agents is calculated based on the below equations,

$$\overrightarrow{B_1} = \overrightarrow{B_\alpha} - \overrightarrow{Y_1}.\overrightarrow{t_\alpha} \tag{15}$$

$$\overrightarrow{B_2} = \overrightarrow{B_\beta} - \overrightarrow{Y_2}.\overrightarrow{t_\beta} \tag{16}$$

$$\overrightarrow{B_3} = \overrightarrow{B_\delta} - \overrightarrow{Y_3}.\overrightarrow{t_\delta} \tag{17}$$

The value of $\overrightarrow{B_4}$ is calculated using GA [19].

1. *Crossover*: The chromosomes chosen from the preceding step are utilized to determine the crossover points. The child chromosomes obtained from the parent positions are depicted below.

$$[\gamma_1, \gamma_2] = B_\alpha \otimes B_\beta \tag{18}$$

where B_α and B_β are the positions of the chromosomes α and β.γ_1 and γ_2 are the child chromosomes.

2. *Mutation*: In the mutation step, again, two children chromosomes via a mutational operator with the random number are generated. The mutated chromosomes attained are expressed as

$$[\tau_1, \tau_2] = [\gamma_1 \circ \gamma_2] \tag{19}$$

After obtaining the new mutated chromosomes, the value of B_4 is found based on the optimal position of the mutated chromosomes, as below,

$$\overrightarrow{B_4} = \text{Best}[\tau_1, \tau_2] \tag{20}$$

4. ***Finding the best solution***

The fitness of the solution is computed using Eq. (4), based on metrics, such as $\text{Gen}^{\text{IL}}(S)C_{\text{avg}}(S)$. The minimum value assures the maximum utility and the privacy of the data. If the fitness is not fulfilled, then the privacy metrics are enhanced using the C-parameter, and then, the same process is repeated.

5. ***Termination***

The steps from 1 to 4 are repeated for identifying the fittest search agent that can be published. The data to be published assures the higher protection to the secrecy of the user data.

3 Results and Discussion

This section presents the results and discussion of the proposed method.

3.1 Experimental Analysis

The proposed technique is implemented on 4 GB RAM, Windows 8 OS with Intel core i-3 processor and is executed on JAVA. The evaluation of the proposed technique is done using two metrics, namely generalized information loss (Gen^{IL}) and average equivalence class size metric, (E_{avg}). "Adult data set, 1996" or "Census Income" dataset [18] is the dataset utilized for the experimentation. This data is extracted from the census bureau database. The methods, such as GA [20], Grey Wolf Optimization algorithm (GWO) [19], Genetic + GWO algorithm, and m-privacy method [21] are used for the comparison with the proposed self-adaptive genetic GWO algorithm to prove the effectiveness of the proposed method.

3.2 Comparative Analysis

The comparative analysis of the existing methods and the superiority of the proposed method are explained in the following subsections by varying the values of λ, χ, and population size.

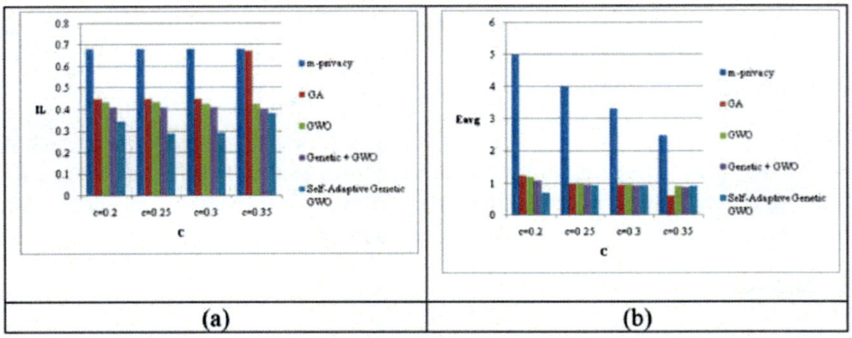

Fig. 2 Comparative analysis of the proposed method in terms of **a** IL, **b** E_{avg}

3.2.1 For λ = 0.4, χ = 0.6 and the Population Size = 10

The analysis of the comparative techniques for $\lambda = 0.4$ and $\chi = 0.6$ with the population size = 10 is depicted in Fig. 2. The analysis based on IL with varying C-mixture value is shown in Fig. 2a. When the C-mixture value is 0.2, the values of IL for existing techniques, like *m*-privacy, GA, GWO, Genetic + GWO, and proposed self-adaptive genetic GWO are 0.6804, 0.4529, 0.436, 0.411, and 0.3441. The analysis based on E_{avg} metric is depicted in Fig. 2b. When the C-mixture value is 0.25, the corresponding E_{avg} values obtained by *m*-privacy, GA, GWO, Genetic + GWO, and the proposed self-adaptive genetic GWO are 4.008, 1.002, 0.99, 0.98, and 0.9494.

3.2.2 For λ = 0.6, χ = 0.4 and Population Size = 20

The comparative analysis of the proposed self-adaptive genetic GWO algorithm on the basis of IL and E_{avg} metric for $\lambda = 0.6$, $\chi = 0.4$, and population size = 20 is depicted in Fig. 3. The analysis based on IL parameter with C-mixture value is

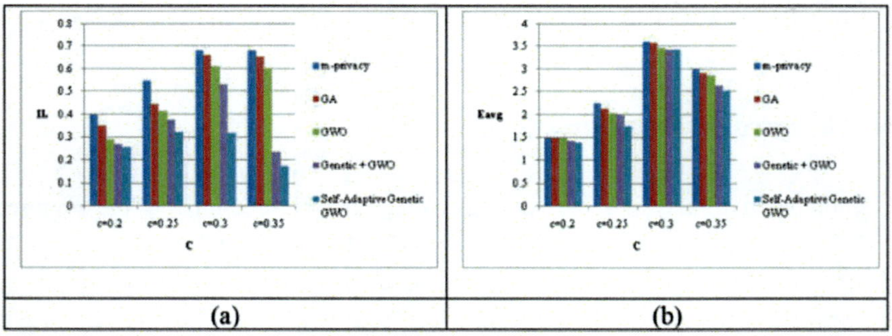

Fig. 3 Comparative analysis of the proposed method in terms of **a** IL, **b** E_{avg}

Table 1 Comparative discussion

Methods	IL	E_{avg}
m-privacy	0.4003	1.5
GA	0.3529	0.98
GWO	0.2904	0.91
Genetic + GWO	0.2335	0.9
Self-adaptive genetic GWO	**0.1724**	**0.71**

depicted in Fig. 3a. When the C-mixture value is 0.2, the IL values measured by *m*-privacy, GA, GWO, Genetic + GWO, and the proposed self-adaptive genetic GWO are 0.4003, 0.3529, 0.2904, 0.2715, and 0.2550. The analysis with E_{avg} metric for varying C-mixture value is depicted in Fig. 3b. When the C-mixture value is 0.25, the E_{avg} values measured by *m*-privacy, GA, GWO, Genetic + GWO, and, the proposed self-adaptive genetic GWO are 2.25, 2.1416, 2.0334, 1.9940, and 1.7531.

3.3 Discussion

Table 1 describes the discussion regarding the minimum values attained by the existing techniques with the proposed technique. The comparative results prove that the proposed self-adaptive genetic GWO algorithm attained the minimum IL value as 0.1724 and the minimum E_{avg} value as 0.71 that increases the utility and preserves the privacy of the data.

4 Conclusion

In this paper, an algorithm, named self-adaptive genetic GWO, is proposed to enhance the privacy of the data via three privacy constraints, such as *l*-diversity, *m*-privacy, and *k*-anonymity. These three privacy constraints are based on the value of the C-mixture. The proposed technique is experimented using adult database. From the analysis, it is noted that the proposed technique provides minimum information loss of 0.1724 with a minimum E_{avg} of 0.71, respectively.

References

1. Luo, Y., Jiang, Y., & Le, J. (2011). A self-adaptation data publishing algorithm framework. In *Proceedings of International Conference on Mechatronic Science, Electric Engineering and Computer*.

2. Hasan, A. S. M., & Jiang, Q. (2017). A general framework for privacy preserving sequential data publishing. In *Proceedings of 31st International Conference on Advanced Information Networking and Applications Workshops.*
3. Gao, A., & Diao L. (2009). Privacy preservation for attribute order sensitive workload in medical data publishing. In *Proceedings of IEEE International Symposium on IT in Medicine & Education*, August 2009.
4. Ragit, S. M., & Badhiye, S. S. (2016). Preserving privacy in collaborative data publishing from heterogeneity attack. In *Proceedings of World Conference on Futuristic Trends in Research and Innovation for Social Welfare (Startup Conclave)*, March 2016.
5. Yaseen, S., Abbas, S. M. A., Anjum, A., Saba, T., Khan, A., Malik, S. U. R., et al. (2018). Improved generalization for secure data publishing. *IEEE Access, 6*, 27156–27165.
6. Kargupta, H., Datta, S., Wang, Q., & Sivakumar, K. (2003) On the privacy preserving properties of random data perturbation techniques. In *Proceedings of Third IEEE International Conference on Data Mining*, November 2003.
7. Goswami, P., Madan, S. (2017). Privacy preserving data publishing and data anonymization approaches: A review. In *Proceedings of International Conference on Computing, Communication and Automation (ICCCA).*
8. Karle, T., & Vora, D. (2017). Privacy preservation in big data using anonymization techniques. In *Proceedings of International Conference on Data Management, Analytics and Innovation (ICDMAI) Zeal Education Society*, Pune, India, February 2017.
9. Zhu, T., Li, G., Zhou, W., & Yu, P. S. (2017). differentially private data publishing and analysis: A survey. *IEEE Transactions on Knowledge and Data Engineering, 29*(8), 1619–1638.
10. Loukides, G., & Gkoulalas-Divanis, A. (2011). COAT: Constraint-based anonymization of transactions. *Knowledge and Information Systems, 28*(2), 251–282.
11. Zhu, T., Xiong, P., Li, G., Zhou, W., & Yu, P. S. (2017). Differentially private query learning: From data publishing to model publishing. In *Proceedings of IEEE International Conference* (pp. 1117–1122).
12. Dwork, C. (2011). A firm foundation for private data analysis. *Communications of the ACM CACM Homepage Archive, 54*(1), 86–95.
13. Zakerzadeh, H., Aggarwal, C. C., & Barker, K. (2016). Managing dimensionality in data privacy anonymization. *Knowledge and Information Systems, 49*(1), 341–373.
14. Hua, J., Tang, A., Fang, Y., Shen, Z., & Zhong, S. (2016). Privacy-preserving utility verification of the data published by non-interactive differentially private mechanisms. *IEEE Transactions on Information Forensics and Security, 11*(10), 2298–2311.
15. Kabir, M. E., Wang, H., & Bertino, E. (2011). Efficient systematic clustering method for k-anonymization. *Acta Informatica, 48*(1), 51–66.
16. Zhang, H., Zhou, Z., Ye, L., & Du, X. (2018). Towards privacy preserving publishing of set-valued data on hybrid cloud. *IEEE Transactions on Cloud Computing, 6*(2), 316–329.
17. McCall, J. (2005). Genetic algorithms for modelling and optimisation. *Journal of Computational and Applied Mathematics, 184*(1), 205–222.
18. Adult Data Set. (1996). From https://archive.ics.uci.edu/ml/datasets/Adult.
19. Mirjalilia, S., Mirjalilib, S. M., & Lewisa, A. (2014). Grey wolf optimizer. *Advances in Engineering Software, 69*, 46–61.
20. Kulkarni, Y. R., & Senthil Murugan, T. (2016). C-mixture and multi-constraints based genetic algorithm for collaborative data publishing. *Computer and Information Sciences.*
21. Goryczka, S., Xiong, L., & Fung, B. C. M. (2011). m-privacy for collaborative data publishing. In *Proceedings of 7th International Conference on Collaborative Computing: Networking, Applications and Worksharing*, October 2011.

Review of Optical Devanagari Character Recognition Techniques

Sukhjinder Singh and Naresh Kumar Garg

Abstract Optical character recognition techniques are capable of automatic translating of document images into equivalent character codes, so it helps in saving human energy as well as cost. These techniques can play a key role to improve or enhance the interaction between human and machine in many applications such as postal automation, signature verification, recognition of city names and automatic bank cheque processing/reading. This paper gives a review of various techniques explored for Devanagari word/text and isolated character recognition in the past few years. Different challenges to optical character recognition are also presented in this work. In the end, practical aspects towards the development of a robust optical character recognition system has been discussed along with directions for future research.

Keywords Optical character recognition · Feature extraction · Classification · Devanagari script

1 Introduction

With the rapid increase of technologies, computational potential, novel sensing and rendering instruments, computers are becoming more and more intelligent. Ability of computers to interface or interact with humans have been proven by various researchers in their research projects and many commercial products are also available for the same. One such ability is Optical Character Recognition (OCR) which is concerned with the automatic converting of scanned documents/images of machine-printed/handwritten characters into corresponding machine-readable digital form. OCR systems help to process large amount of textual data automatically. These systems minimize the exhaustive labor of manual processing, improve the speed

S. Singh (✉)
ECED, Giani Zail Singh Campus CET, MRSPTU, Bathinda, Punjab, India
e-mail: er.ssrakhra@gmail.com

N. K. Garg
CSED, Giani Zail Singh Campus CET, MRSPTU, Bathinda, Punjab, India
e-mail: naresh2834@rediffmail.com

S. C. Satapathy et al. (eds.), *Intelligent System Design*, Advances in Intelligent Systems and Computing 1171, https://doi.org/10.1007/978-981-15-5400-1_11

of operation, reduce errors or noise in the documents and decrease storage space needed for paper documents. These systems can be employed in various applications including postal addresses reading for automatic sorting, cheque verification, digitalization of ancient documents, text extraction and recognition from handwritten structured form documents, forensic and medical analysis, reading registration numbers from vehicle license plate images, automatic data entry and city name recognition [1, 2]. As many OCR based techniques can read/recognize characters of one script only and hence are script dependent. India is a multilingual country where 23 languages and 13 scripts (including English/Roman) exist and hence automation recognition of both printed as well as handwritten characters/script leads to many applications. Most available OCR techniques are specific to script and therefore, recognition/processing of documents with multi-script is not an easy task.

There has been always a great need for research work in the field of optical character recognition for Indian languages/scripts even though there are many challenges and lack of commercial market [3]. Research towards handwritten character recognition of Indian scripts has gained much attention in recent years even though first research work for the recognition of Devanagari script was reported in 1977 [4]. Many methods for recognition of word/text and isolated characters written in Devanagari scripts have been developed in last few decades and are briefly outlined in this work. The remaining paper is outlined as follows. Section 2 presents challenges to optical character recognition. The optical character recognition techniques are overviewed in Sect. 3. Section 4 reviews the various techniques proposed by different researchers for optical character recognition. Conclusions and future directions had been presented in Sect. 5.

2 Challenges to Optical Character Recognition

Challenges to handwritten character recognition are due to change in writing styles from one writer to another and even change in a single person's writing from time to time [5]. Due to these factors, handwritten character recognition becomes a challenging and harder task. As the recognition accuracies of OCR systems are directly concerned with input images. Therefore, higher accuracies of the same can be obtained with high quality and good resolution scanned documents/images. Better recognition results of OCR techniques can also be obtained by overcoming various errors that affect image quality significantly. Different sources of error are due to aspect ratios, blurring and degradation, complex backgrounds, presence of irregular illumination, low-resolution, multilingual content, scene complexity, skewness, tilting or perspective distortion, change in text layout/fonts and warping which will make it difficult for OCR systems to distinguish text from non-text [6–8]. In contrast with printed character recognition, handwritten character recognition is a more challenging task because of various factors such as individual styles of inscription, speed of writing, size of letters, physical and or mental situation of the writer, overlap of letters, large symbolic set, character complexity and shape similarity. These factors may affect and cause problems in correct character recognition by computer systems. Thus, there is a need for technique(s) that overcomes these challenges or factors so that OCR systems produce correct machine-readable digital form, automatically.

3 Optical Character Recognition Techniques

Researchers used several techniques for the handwritten character recognition with Devanagari script. Generally, these techniques/approaches can be categorized into three categories as mentioned below [9].

3.1 Segmentation-Based or Analytical Techniques

In this technique, each word is segmented into individual equivalent components (or character parts or even character subparts) and each individual equivalent component (or character) is recognized and allotted a unique symbol and the resultant symbols are reconstructed (or reassembled) for the identification of a word. Recognition is considered as a matching process which is usually carried out with the help of dictionary/library words [10]. The segmentation for obtaining each individual character is usually based on heuristics and done before the recognition of each character.

3.2 Segmentation Free or Holistic Techniques

It is a word-based technique, where each word is not segmented into individual equivalent components (or character parts or even character subparts). The word is processed as a single unit (or a whole) and there is no attempt to identify characters individually. In this, the word as a whole is recognized with the help of a trained classifier whose training is done after extracting the properties of whole set words. Challenges to this technique include complexity as the whole word is processed as a single indivisible entity/unit and lesser discrimination abilities. The applications of holistic techniques are usually limited to areas where a small number of word groups exist (small size vocabulary) and constrained to a predetermined lexicon such as bank cheque number. Holistic techniques have been less studied than analytical techniques and therefore recently have gained significant attention.

3.3 Hybrid

In hybrid, the combination of segmentation-based and holistic techniques is used.

4 Related Work

There has been always a great need for research work in the area of OCR for Indian languages/scripts despite many challenging factors involved and lack of commercial market [11]. Many methods for recognition of word/text and isolated characters written in Devanagari script have been developed by various researchers during the past few years and are briefly reported in the following section.

4.1 Word/Text Recognition

Parui et al. presented a method for Devanagari word recognition by considering stroke-based features and Hidden Markov Model (HMM) classifier [12]. Shaw et al. proposed an offline Devanagari word recognition system based on directional chain code and HMM [13]. They carried out their experimentation on dataset of 22,500 (Training) and 17,200 (Test) words obtained 80.2% accuracy. Furthermore in [14], authors worked with stroke-based features and HMM classifier using the same dataset and achieved 84.31% accuracy. Singh et al. presented a recognition system for handwritten words based on curvelet transform, SVM and kNN using dataset of 28,500 words and achieved promising results [15]. Ramachandrul et al. developed a handwritten Hindi words recognition system, using directional element and dynamic programming based approaches [16]. Shaw et al. proposed an offline handwritten word (Devanagari) recognition technique based on Directional Distance Distribution (DDD) and Gradient-Structural Concavity (GSC) by considering 100 Indian town names and improved recognition accuracy significantly [17]. Kumar proposed a segmentation-based approach/technique for recognition of isolated hand-printed Devanagari words taking more than 3500 words as database [9] and classification were carried out using Multi-Layer perceptron (MLP). Bhunia et al. proposed a novel technique for cross-language handwritten text recognition and word spotting, considering Bangla, Devanagari and Gurumukhi scripts [18].

In Table 1, recognition results of word/text by different researchers for Devanagari script with various features and classifiers have been presented.

4.2 Isolated Character Recognition

Hanmandlu et al. carried out the handwritten Hindi character recognition based on fuzzy models and obtained overall recognition rate of 90.65% for 4750 samples [24]. Pal et al. proposed a recognition system for offline handwritten Devanagari characters by considering combination of MQDF (Modified Quadratic Discriminant Function) and SVM (Support Vector Machine) based on gradient and curvature features and achieved accuracy 95.13% on 36,172 samples [22]. Agrawal et al.

Table 1 Recognition results of word/text

Authors	Script (language)	Test data size	Feature extraction technique	Classifier	Accuracy (%)
Parui et al. [12]	Devanagri (offline)	7000 (training) 3000 (test)	Stroke based	HMM	87.71 (training set) 82.89 (test set)
Shaw et al. [13]	Devanagri (offline)	22,500 (training) 17,200 (test)	Directional chain code	HMM	80.2
Shaw et al. [14]	Devanagri (offline)	22,500 (training) 17,200 (test)	Stroke based	HMM	84.31
Shaw et al. [31]	Devanagri (offline)	7000 (training) 3000 (test) 3000 (validation)	Stroke based (stage-1); wavelet (stage-2)	HMM (stage-1); modified byes (stage-2)	85.57 (test) 91.25 (training)
Singh B et al. [15]	Devanagri (offline)	28,500	Curvelet transform	SVM and kNN	85.6 (SVM) 93.21 (kNN)
Ramachandrula et al. [16]	Hindi	39,600	Directional element	Dynamic programming	79.94 (30 vocabulary words); 91.23 (10 vocabulary words)
Shaw et al. [32]	Devanagri (offline)	22,500 (training) 17,200 (test)	Combination of skeleton and contour based	SVM	79.01
Shaw et al. [17]	Devanagri (offline)	22,500 (training) 17,200 (test)	DDD and GSC	Multi-class SVM	88.75
Kumar [9]	Isolated hand-printed Devanagari	More than 3500 words	Neighbor pixel weight and gradient feature	MLP	80.8 (for 2 character words); 72.0 for 6 character words
Bhunia et al. [18]	Bangla, Devanagari and Gurumukhi	3856, 3589 and 3142	PHOG feature	HMM (for middle-zone) SVM (for upper/lower zone)	Above 60

proposed algorithm for identification of existence and position of the vertical bar in offline handwritten Hindi characters and achieved classification rate of 97.25% [25]. Kubatur et al. developed a NN approach for recognition of Devanagari handwritten characters and obtained recognition rates 97.2% (maximum) for 2760 characters [20]. In [21], Yadav et al. developed an OCR based system for Hindi text (printed) by considering Artificial Neural Network (ANN), histogram of projection based on mean distance, pixel value and vertical zero crossing, so that recognition rate can be improved. The testing was carried out at individual letter (1000 individual characters) level and paragraph (15 paragraphs consisting of 650 words) level and obtained recognition rates of 98.5% and 90%, respectively. Furthermore, Kale et al. in [19], achieved overall recognition rate of 98.25% and 98.36% for 3750 basic Devanagari and 11,250 compound characters, respectively, considering Legendre moment and ANN. Dixit et al. proposed a handwritten Devanagari character recognition system using wavelet-based feature extraction, Back Propagation NN (BPNN)and obtained 70% accuracy on 2000 samples [26]. Furthermore, Jangid and Srivastava [23] explored GLAC (Gradient Local Auto-Correlation) algorithm for recognition of Devanagari handwritten characters and achieved accuracies of 93.21% and 95.21% for ISIDCHAR and V2DMDCHAR datasets, respectively.

In Acharya et al. [27] introduced a dataset of Devanagari handwritten characters (92,000 samples) and achieved 98.47% accuracy using Deep Convolutional Neural Network (Deep CNN). Furthermore, Dongre and Mankar [28] developed a Devanagari numeral and character recognition system using MLP-NN, structural and geometric features and achieved 82.7% accuracy on 5375 samples. Shelke and Apte optimized the performance of neural networks or handwritten Devanagari character recognition. They also carried out the comparative analysis of neural networks for the same, using Structural features, Feed-Forward Back Propagation Network (FF-BPN), Cascade-Forward BPN (CF-BPN) and Elman BPN (EB-PN) based classifiers [29]. Jangid et al. developed a handwritten Devanagari character (ISIDCHAR and V2DMDCHAR) recognition technique using layer-wise technique of Deep CNN and achieved higher recognition accuracy and faster convergence rate as compared as shallow technique of handcrafted features and standard Deep CNN [30]. In Table 2, recognition results of isolated characters by various researchers for Devanagari script with different features and classifiers have been presented.

It has gathered from the tables that recognition rate/accuracy of OCR systems highly depends upon the size of dataset taken for experiment work and feature extraction and classifiers (i.e. classification techniques) considered for the same.

5 Conclusions and Future Work

In this work, efforts have been carried out to review marvelous advancements in the area of optical character recognition with Devanagari script specifically to recognition of word/text and isolated characters. The literature has been reviewed systematically and presented in the form of tables so as to provide a guide and update for researchers

Table 2 Recognition results of isolated characters

Authors	Script (language)	Test data size	Feature extraction technique	Classifier	Accuracy (%)
Hanmandlu et al. [24]	Hindi	4750	Box approach	Coarse	90.65
Pal et al. [22]	Devanagari	36,172	Gradient and curvature	MQDF and SVM	95.13
Agrawal et al. [25]	Hindi	100 samples of each character	–	Coarse	97.25
Kubatur et al. [20]	Devanagari	2760	DCT	NN based	97.2
Yadav et al. [21]	Hindi	1000 characters and 15 paragraphs consisting of 650 words	Histogram of projection	ANN	98.5 and 90, respectively
Kale et al. [19]	Devanagari	3750 basic and 11,250 compound characters	Legendre moment	ANN	98.25 and 98.36, respectively
Dixit et al. [26]	Devanagari	2000	Wavelet-based	BPNN	70
Jangid and Srivastava [23]	Devanagari	ISIDCHAR (36,172 characters) and V2DMDCHAR (20,305 characters)	GLAC	SVM	93.21 and 95.21
Acharya et al. [27]	Devanagari	DHCD	–	Deep CNN	98.47
Dongre and Mankar [28]	Devanagari	5375	Structural and geometric	MLP-NN	82.7
Shelke and Apte [29]	Devanagari	40,000	Structural	FFBPN, CFBPN and (EBPN	97.20, 97.46 and 98.10 respectively
Jangid and Srivastava [30]	Devanagari	ISIDCHAR and V2DMDCHAR	RMSProp adaptive gradient	DCNN	98

working in this area. At present scenario, optical character recognition has achieved considerable attention from various researchers because of huge application areas. It has been gathered that OCR techniques be increasingly used in real-world applications and products. Although numerous research work for Roman, Japanese/Chinese and Arabic scripts exists but only limited work had been carried out towards recognition of Indian scripts specifically with non-isolated handwritten words. Segmentation of characters as well as their recognition is a tedious job for Indian scripts because of their complex nature due to various factors viz. compound characters, presence of modifiers, overlapping/touching characters. Although, various efforts have been done by various researchers in the field of OCR for Indian scripts, but still there is the need for developing more efficient methods of optical character recognition with different scripts.

Due to assorted writing styles of peoples, paper/material's quality, sometimes unusual ligatures adjacent characters written in Devanagari frequently add misperception to recognize the words viz. "झ", "ड" and "इ".There is still a need for reseach in the field/area of recognition of word, complete sentence and full document written in Devanagri script along with its semantics/lexicon. The future work also needs more effective segmentation techniques that can properly segment the overlapping and light characters. Further, recognition rate can be increased/improved by proposing/developing a novel feature extraction and classification techniques that can differentiate between most confusing/similar characters.

References

1. Pal, U., Roy, K., & Kimura, F. (2008). Bangla handwritten pin code string recognition for Indian postal automation. In *11th International Conference on Frontiers in Handwriting Recognition* (pp. 290–295). Canada: L. Lam Publications.
2. Pal, U., Roy, K., & Kimura, F. (2009). A lexicon-driven handwritten city-name recognition scheme for Indian postal automation. *IEICE Transactions on Information and Systems, 92*(5), 1146–1158.
3. Jayadevan, R., Kolhe, S. R., Patil, P. M., & Pal, U. (2011). Offline recognition of devanagari script: A survey. *IEEE Transactions on Systems, Man, and Cybernetics, Part C Applications and Reviews, 41*(6), 782–786.
4. Sethi, I. K., & Chatterjee, B. (1977). Machine recognition of constrained hand printed Devanagari. *Pattern Recognition, 9*(2), 69–75.
5. Sarkhel, R., Das, N., Das, A., Kundu, M., & Nasipuri, M. (2017). A multi-scale deep quad tree based feature extraction method for the recognition of isolated handwritten characters of popular indic scripts. *Pattern Recognition, 71*, 78–93.
6. Chaudhuri, B. B., & Pal, U. (1997). Skew angle detection of digitized indian script documents. *IEEE Transactions on Pattern Analysis and Machine Intelligence, 19*(2), 182–186.
7. Liang, J., DeMenthon, D., & Doermann, D. (2008). Geometric rectification of camera-captured document images. *IEEE Transactions on Pattern Analysis and Machine Intelligence, 30*(4), 591–605.
8. Singh, P. K., Sarkar, R., & Nasipuri, M. (2015). Offline script identification from multilingual indic-script documents: A state-of-the-art. *Computer Science Review, 15*(16), 1–28.
9. Kumar, S. (2016). A study for handwritten devanagari word recognition. In *IEEE International Conference on Communication and Signal Processing* (pp. 1009–1014). Melmaruvathur: IEEE Press.
10. Garg, N. K., Kaur, L., & Jindal, M. K. (2015). Recognition of handwritten hindi text using middle region of the words. *International Journal of Software Innovation, 3*(4), 62–71.
11. Jayadevan, R., Kolhe, S. R., Patil, P. M., & Pal, U. (2011). Automatic processing of handwritten bank cheque images: A survey. *International Journal on Document Analysis and Recognition, 15*(4), 267–296.

12. Parui, S. K., & Shaw, B. (2007). Offline handwritten Devanagri word recognition: An HMM based approach. In A. Ghose, R. K. De, & S. K. Pal (Eds.), *PReMI* (Vol. 4815, pp. 528–535). Berlin: Springer-verlag, LNCS.
13. Shaw, B., Parui, S. K., & Shridhar, M. (2008). Offline handwritten Devanagari word recognition: A holistic approach based on directional Chain code feature and HMM. In *IEEE International Conference of Information Technology* (pp. 203–208). Bhubaneswar: IEEE Press.
14. Shaw, B., Parui, S. K., Shridhar, M. (2008). A segmentation based approach to offline handwritten Devanagri word recognition. In *IEEE International Conference on Information Technology* (pp. 256–257). Bhubaneswar: IEEE Press.
15. Singh, B., Mittal, A., Ansari, M. A., & Ghosh, D. (2011). Handwritten word recognition: A curvelet transform based approach. *International Journal of Computer Science and Engineering Survey (IJCSES), 3*(4), 1658–1665.
16. Ramachandrula, S., Jain, S., & Ravishankar, H. (2012). Offline handwritten word recognition in Hindi. In *Workshop on Document Analysis and Recognition* (pp. 49–54). New York: ACM.
17. Shaw, B., Bhattacharya, U., & Parui, S. K. (2015). Offline handwritten Devanagari word recognition: Information fusion at feature and classifier level. In *IEEE 3rd IAPR Asian Conference on Pattern Recognition* (pp. 720–724). Malaysia: IEEE Press.
18. Bhunia, A. K., Roy, P. P., Mohta, A., & Pal, U. (2018). Cross-language framework for word recognition and spotting of indic scripts. *Pattern Recognition, 79,* 12–31.
19. Kale, K. V., Chavan, S. V., Kazi, M. M., & Rode, Y. S. (2013). Handwritten Devanagari compound character recognition using Legendre moment: An artificial neural network approach. In *IEEE International Symposium on Computational and Business Intelligence* (pp. 274–278). New Delhi: IEEE Press.
20. Kubatur, S., Sid-Ahmed, M., & Ahmadi, M. (2012). A neural network approach to online devanagari handwritten character recognition. In *IEEE International Conference on High Performance Computing and Simulation* (pp. 209–214). Madrid: IEEE Press.
21. Yadav, D., Sanchez-Cuadrado, S., & Morato, J. (2013). Optical character recognition for Hindi language using a neural-network approach. *Journal of Information Processing Systems (JIPS), 9*(1), 117–140.
22. Pal, U., Chanda, S., Wakabayashi, T., & Kimura, F. (2008). Accuracy Improvement of Devnagari character recognition combining SVM and MQDF. In *11th International Conference on Frontiers in Handwriting Recognition* (pp. 367–372). Canada: L. Lam Publications.
23. Jangid, M., & Srivastava, S. (2014). Gradient local auto-correlation for handwritten Devanagari character recognition. In *IEEE International Conference on High Performance Computing and Applications* (pp. 1–5). Bhubaneswar: IEEE Press.
24. Hanmandlu, M., Murthy, O. V. R., & Madasu, V. K. (2007). Fuzzy model based recognition of handwritten hindi characters. In *9th IEEE Biennial Conference of the Australian Pattern Recognition Society on Digital Image Computing Techniques and Applications* (pp. 454–461). Glenelg: IEEE Press.
25. Agrawal, P., Hanmandlu, M., & Lall, B. (2009). Coarse classification of handwritten hindi characters. *International Journal of Advanced Science and Technology, 10,* 43–54.
26. Dixit, A., Navghane, A., & Dandawate, Y. (2014). Handwritten devanagari character recognition using wavelet based feature extraction and classification scheme. In: *Annual IEEE India Conference* (pp. 1–4). Pune: IEEE Press.
27. Acharya, S., Pant, A. K., & Gyawali, P. K. (2015). Deep learing based large scale handwritten devanagari character recognition. *9th International Conference on Software, Knowledge, Information Management and Applications* (pp. 1–6). Nepal: IEEE Press.
28. Dongre, V. J., & Mankar, V. H. (2015). Devanagari offline handwritten numeral and character recognition using multiple features and neural network classifier. In *2nd International Conference on Computing for Sustainable Global Development* (pp. 425–431). New Delhi: IEEE Press.
29. Shelke, S., & Apte, S. (2016). Performance optimization and comparative analysis of neural networks for handwritten Devanagari character recognition. In *IEEE International Conference on Signal and Information Processing* (pp. 1–5). Vishnupuri: IEEE Press.

30. Jangid, M., & Srivastava, S. (2018). Handwritten Devanagari character recognition using layer-wise training of deep convolutional neural networks and adaptive gradient methods. *Journal of Imaging, 4*(14), 1–14.
31. Shaw, B., & Parui, S. K. (2010). A two stage recognition scheme for offline handwritten Devanagri words. In A. Ghosh, R. K. De, & S. K. Pal (Eds.), *Research machine interpretation of patterns 2010. SSIR* (Vol. 12, pp. 145–165). Singapore: World Scientific.
32. Shaw, B., Bhattacharya, U., & Parui, S. K. (2014). Combination of features for efficient recognition of offline handwritten Devanagri words. In *IEEE 14th International Conference on Frontier in Handwritten Recognition* (pp. 240–245). Heraklion: IEEE Press.

A Comprehensive Review on Deep Learning Based Lung Nodule Detection in Computed Tomography Images

Mahender G. Nakrani, Ganesh S. Sable, and Ulhas B. Shinde

Abstract Lung Nodules detection plays an important role to detect early stage lung cancer. Early stage lung cancer detection can considerably increases the surviving rate of patients. Radiologist diagnosis the Computerized Tomography (CT) images by detecting lung nodules. This task of locating lung nodules from CT images is rigorous and becomes even more challenging due to the structure of lung parenchyma region and also due to the size of lung nodules is small even less that 3 cm. Many Computer Aided Diagnosis CAD systems were proposed to detect lung nodules to assist radiologists. Recently, Deep learning neural network has found its way into lung nodule detection system. Deep learning neural network has shown better results and performance than traditional feature extraction based lung nodule detection techniques. This paper will focus on different deep learning neural network proposed for lung nodule detection and also we will analyze the result and performance of this detection network.

Keywords Lung nodules · Convolutional neural network · Deep learning · Nodule detection · Classification · False positive reduction

1 Introduction

Cancer is among the top leading causes of disease deaths accounted globally in present scenario. It has become a big threat to our society. It is the major cause of disease deaths in India. In 2018 only around 67,795 new cases of Lung cancer have been registered in India out of which 63,435 has resulted in death [1]. Lung cancer

M. G. Nakrani (✉) · U. B. Shinde
CSMSS's Chh. Shahu College of Engineering, Aurangabad, Maharastra, India
e-mail: nakrani.mahender@gmail.com

U. B. Shinde
e-mail: drshindeulhas@gmail.com

G. S. Sable
GS Mandal's Maharashtra Institute of Technology, Aurangabad, Maharastra, India
e-mail: sable.eesa@gmail.com

S. C. Satapathy et al. (eds.), *Intelligent System Design*, Advances in Intelligent Systems and Computing 1171, https://doi.org/10.1007/978-981-15-5400-1_12

is among top five types of cancer in India. The survival rate of early detected cancer patients is much higher compare to the cancer detected in advanced stage when it has metastasis to other parts of body like lymph nodes, liver, bone, brain etc. For detecting early stage lung cancer, lung nodule detection plays an important role. Lung nodules are round shaped opacity generally surrounded by lung parenchyma and smaller than 3 cm in diameter. Lung nodules are of two types benign and malignant. Benign lung nodules are lesion coin which are less than 2 cm in diameter and are noncancerous whereas a malignant lung nodule are larger than 2 cm and grows constantly and are cancerous. The detection of early stage malignant nodule can help in early treatment of lung cancer. Apart from these pulmonary nodules can also be classified into four classes based on the position of the nodule and surrounding structures which are well-circumscribed, Juxtavascular, Pleural tail and Juxtapleural shown in Fig. 1. Radiologist receives the CT scans of the patients and diagnoses it by finding the malignant nodules. This task is affected by many factors such as professional experience of radiologist, distractions and dose used in CT scans. This may lead to misinterpretation of the data available. To assist radiologist, many computer aided diagnosis system for detection and classification of lung nodules have been proposed. In traditional CAD system, the detection of nodule candidates from CT scans is done by using template matching, region growing, and thresholding. Different features

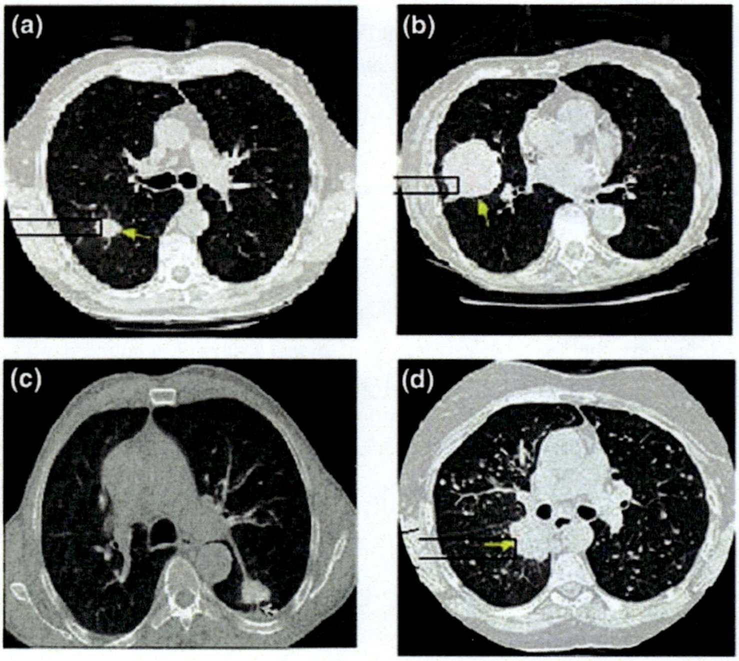

Fig. 1 Four classes of pulmonary lung nodules. **a** Well-circumscribed nodule. **b** Juxtavascular nodule. **c** Pleural tail nodule. **d** Juxtapleural nodule

like shape, texture, size, contrast, entropy of nodule candidates are then extracted. These features are used to reduce false positive by using support vector machine, genetic programming based classifier, Bayesian supervised classifier and rule based classifier. Recently, deep learning had showed its power in extracting various features from an image in various fields such as object detection, character recognition; this attracted many researchers in medical image processing and exploited deep learning in their research. This paper will identify the deep learning architecture that are implemented for lung nodule detection and will compare the result and performance of these techniques. It will also discuss about deep learning architecture that can improve the lung nodule detection system.

2 Literature Review

Many CAD systems for detection and classification of lung nodules have been proposed over years. The traditional CAD system generally has four important stages as shown in Fig. 2. The system begins with segmentation of lung lobe region by removing other regions surrounding lung such as rib cage, blood vessels, air gaps etc., which is called as lung segmentation. Then the nodule candidates are detected and its location is identified in second stage. From these nodule candidates features like shape, texture, size etc., are extracted in the third stage and finally in fourth stage these features are used by different clustering techniques like fuzzy clustering, support vector machine (SVM), Genetic programming based classifier (GPC) etc., for classification of nodules as true positive and to decrease number of false positives. This traditional system gave comparable results of accuracy around 90%. The breakthrough result of AlexNet a convolutional neural network in image classification attracted interest of many medical image researchers. Since then CAD system adopted different Deep learning architecture for lung nodule detection. This technique starts with lung segmentation and then uses deep learning architecture for lung nodule candidate detection purpose and finally another deep learning architecture for false positive reduction as shown in Fig. 3.

2.1 *Datasets*

As with all deep learning architecture, huge data is required to train and validating the network before testing it. To facilitate this, many public archive data base are available which are used by researchers to train and also for validation and testing purpose. In [2–8] used LUNA16 dataset which is the collection form the largest public archive database for pulmonary nodule, Lung Image Database Consortium's Image Database Resource Initiative (LIDC-IDRI). References [9–13] adopted complete LIDC-IDRI database of 1018 cases. In [14], a collection of 43,292 chest radiographs of patients from January 2010 to December 2015 at Seoul National University Hospital was

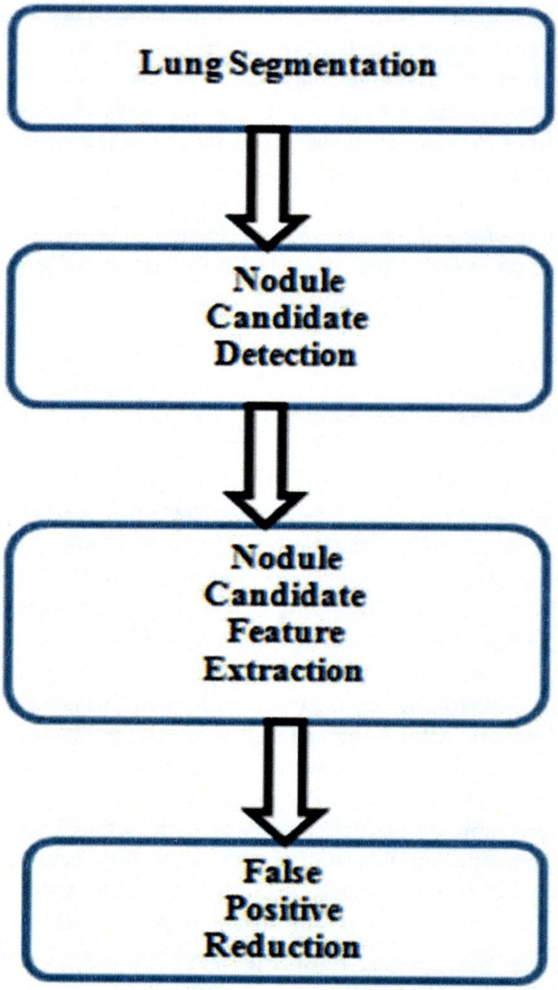

Fig. 2 Traditional CAD system for lung nodule detection

used for algorithm development. The validation was done by dataset of 693 patient chest scans collected from Boramae Hospital, Seoul National University Hospital, University of California San Francisco Medical Center and National Cancer Center. In [15], dataset from Tian Chi competition containing CT scans from hospitals in Chain with images of 1000 patients was used. They divided dataset in three groups, 600 images for training, validation set of 200, and the remaining 200 was used as the test set. Winkels and Cohen [16] used LIDC-IDRI as well as NLST National Lung Screening Trial dataset. The NLST dataset was used for validation and LIDC-IDRI dataset was used for testing. In [17], 233 scans containing nodules with diameter bigger than 3 cm from LIDC-IDRI dataset were used. In [18], LUNA16 dataset for fully supervised nodule detection and for weekly supervised nodule detection, NLST

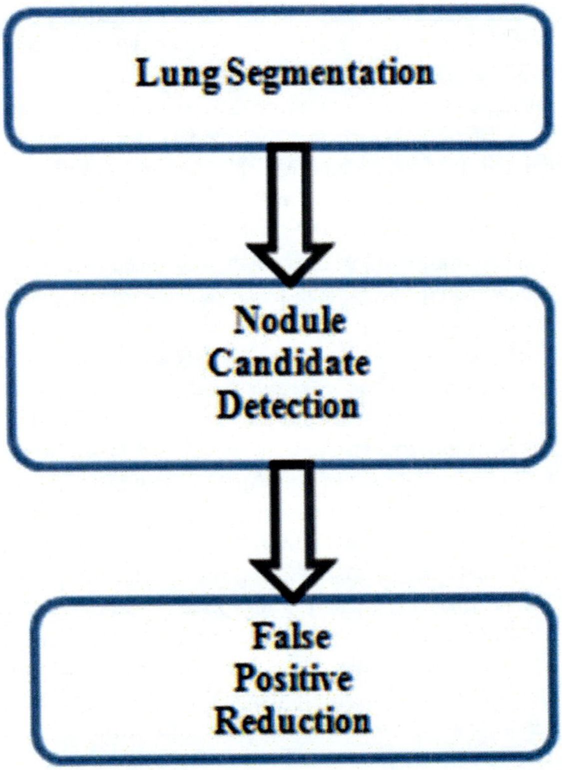

Fig. 3 Deep learning based lung nodule detection CAD system

dataset was used. For testing they used Tianchi Lung Nodule Detection dataset. In [19], LIDC-IDRI dataset was utilized for training and validation purpose, then ANODE09 challenge dataset was used for testing purpose. The system was evaluated on Danish Lung Cancer Screening Trial dataset. From the dataset of Lung CT scans, lung parenchyma region is segmented before using it for training, validation and testing purpose.

2.2 *Lung Nodule Candidate Detection*

Lung segmented images from the above stage are used to train different deep learning architecture developed to detect lung nodule candidates. In [2], a faster Residual-2D Convolutional Neural Network which consists of three sub-networks was proposed for nodule candidate detection. These three sub-networks were region proposal network, feature extraction network and region-of-interest classifier. The Feature extraction network was implemented using a VGG16 convolutional network with five group convolutions and a deconvolution layer to extract a 148 × 148 feature map. The system used two networks for region proposal which were concatenated to middle convolutional layer and deconvolution layer. Anchors of size 12 × 12, 18 × 18, 27 × 27, 36 × 36, 51 × 51, 75 × 75 and 120 × 120 are designed to predict multiple region-of-interest of different size termed as nodule candidates.

A MU-Net which is a modified U-Net was used in [20], for nodule candidate detection. Multi-layered MU-Net was employed to detect nodules of different size as single MU-Net layer was able to detect large nodules but not smaller one. The non-detected small nodule (false positive) from previous layer which detected large nodules compared to next layer were used as the training set for the next layer in concatenated layers. The network was able to detect nodules which were of size greater than 10 mm.

A deep residual 3D Faster R-CNN architecture consists of more than 30 convolutional layers and had extensive residual shortcut connections with transposed convolutional layers was proposed in [15] for nodule candidate screening. The input image has been split into an overlapping 128 × 128 × 128 input volumes and the output was a 32 × 32 × 32 map of coordinates, diameter and nodule probability which resulted in five features with three for size 5, 10 and 30 mm as per nodule size distribution in the dataset and the final output size was 32 × 32 × 32 × 5 × 3.

GoogLeNet consist of 22 layers was used in [12]. This GoogLeNet also consists of 3 softmax layers with 9 inception modules. Each inception module was combination of four layers comprised of a 1 × 1 convolution, a 3 × 3 convolution, a 5 × 5 convolution, and a 3 × 3 maxpooling layer. Data from the previous layer was divided into four different batches are applied to inception module which are again concatenated to form single batch. The softmax layers at the end of network were used to classify lung nodules.

A 3D convolutional neural network was proposed in [4], for nodule candidate detection. The non-linear activation function for every convolution layer was Rectified Linear Unit ReLU, except the last convolution layer activation was done by sigmoid function to output the probability of predicted nodules. The stride of (1, 1, 1) was used in all the convolution layers and the inputs are padded with zeros to make the output and input feature map of same size.

Feature vector extraction was carried out by AlexNet in [21]. The AlexNet network comprised of eight layers consisting of five convolutional layers and three fully connected layers. The convolutional layer consists of 11 × 11 convolution template in three channels which was followed by a ReLU, Norm Transformation and pooling. The final output is given by softmax classifier.

2.3 False Positive Reduction

The lung nodule detection technique detects all nodules like masses. It also deems the masses which are not nodules as nodules this are called as false positive. The next step in the CAD system is to decrease the number of false positive candidates. Sun et al. [2] used a 2D Convolutional Neural Network based boosting classifier to remove false positive candidates. In boosting, voting is used to obtain final result from several CNNs which are trained independently. The dataset was divided into 5 subsets out of which 3 subsets were used for training, 1 subset each for validation and testing. The training subset was again divided into 3 different parts. Each part is used to train the classification models independently. The week classification model 1 was trained by first subset of training dataset. The new second model was trained with the misclassified samples from the model1 and second subset independently. Similarly, model3 is trained using misclassified samples from model1 to model2 and third subset independently. In the training process, the misclassified sample of previous round was used to train the next model.

3D Deep convolutional neural network classifier was used for false positive reduction. This 3D DCNN was trained on difficult samples extracted during candidate screening in [15]. This 3D DCNN contains several residual blocks of Convolution with Batch Normalization and activated by ReLU. The 3D DCNN was terminated to a fully connected layer to produce final classifications.

SVM classifier was employed in [17] for false positive reduction. The VGG-16 having ReLU activation with fully connected layers was used to extract 4096-dimensional output. This output features were used to train the linear SVM classifier. A deep neural network of two-staged stacked autoencoder was proposed in [13] for false positive reduction. The scaled conjugate gradient Equations with the updated rules and activation functions was used for the training of network.

A deep 3D residual convolutional neural network was constructed in [6] to reduce false positive nodules from candidate nodules. The proposed 3D-RCNN with 27 layers was used for reducing false positives. To extract multi-level contextual information of CT data, the spatial pooling and cropping (SPC) layer was designed. This 27-layer RCNN includes three residual groups with four residual units each. The kernel size of $5 \times 5 \times 3$ was used in each residual unit which consists of two convolutional layers.

3 Comparisons

The comparison of CAD systems for detection and classification of lung nodules is done in Table 1. It givens the details about the dataset used for the system, CNN architecture used for the detection and classification of lung nodules and reduce false positive with its performance metrics. The performance metrics are Area under the ROC curve (AUC) which distinguishes between two diagnosis groups, CPM is average sensitivity and the accuracy.

Table 1 Comparison of deep learning based CAD systems for detection of lung nodules

Nodule candidate detection architecture	Dataset	Performance metrics	Scores	False positive reduction architecture	Performance metrics	Scores
Faster R-CNN	LUNA16	Area under ROC curve (AUC) CPM	0.954 0.775	Boosting 2D CNN	CPM	0.790
MU-Net	LIDC-IDRI	Area under ROC curve (AUC) Accuracy	0.99 97.4%	D-CNN	CPM	0.815
3D Faster R-CNN	TianChi competition dataset	CPM	0.815	SVM classifier	Accuracy	87.2%
GoogLeNet	LIDC-IDRI	Accuracy	81%	Two staged autoencoder	Accuracy	96.9%
3D CNN	LUNA16	CPM	0.795	3D residual CNN	Sensitivity	94.4%

4 Conclusion

This paper has presented a comparative study on latest trend in detection and classification of lung nodules. It also presented a generalized structure for detection and classification of lung nodules CAD system based on deep learning. The algorithms and methods adopted for realization of various stages of existing system were described in details. The review of existing system demonstrates that wide range of deep learning convolutional neural network has been adopted for detection of lung nodules and reduction of false positives. It also demonstrates that the sensitivity, accuracy and reliability of the system majorly depend on detection of nodule candidates and reduction of false positive components. The performance of the system can be further improved by improving by implementing more advance deep learning architecture like ResNet, ResNeXt etc., for detection of nodule candidates and reduction of false positive components as they have performed well compared to other architecture in ImageNet competition. In future, we will try to implement advance CNN architecture for both detection of nodule candidates and reduction of false positive components.

References

1. National Institute of Cancer Prevention and Research. http://cancerindia.org.in/lung-cancer/.
2. Sun, N., Yang, D., Fang, S., & Xie, H. (2018). Deep convolutional nets for pulmonary nodule detection and classification. In W. Liu, F. Giunchiglia, & B. Yang (Eds.), *Knowledge science, engineering and management, KSEM 2018, lecture notes in computer science* (Vol. 11062). Cham: Springer.
3. Pezeshk, A., Hamidian, S., Petrick, N., & Sahiner, B. (2018) 3D convolutional neural networks for automatic detection of pulmonary nodules in chest CT. *IEEE Journal of Biomedical and Health Informatics.*
4. Gu, Y., Lu, X., Yang, L., Zhang, B., Yu, D., & Zhao, Y. (2018). Automatic lung nodule detection using a 3D deep convolutional neural network combined with a multi-scale prediction strategy in chest CTs. *Computers in Biology and Medicine, 103,* 220–231.
5. Liu, M., Dong, J., Dong, X., Yu, H., & Qi, L. (2018). Segmentation of lung nodule in CT images based on mask R-CNN. In *9th International Conference on Awareness Science and Technology (iCAST)*, Fukuoka (pp. 1–6).
6. Jin, H., Li, Z., Tong, R., & Lin, L. (2018). A deep 3D residual CNN for false positive reduction in pulmonary nodule detection. *Medical Physics, 45,* 2097–2107.
7. Tran, G. S., Nghiem, T. P., Nguyen, V. T., Luong, C. M., & Burie, J.-C. (2019). Improving accuracy of lung nodule classification using deep learning with focal loss. *Journal of Healthcare Engineering, 5156416,* 9.
8. Sajjanar D., Rekha, B. S., & Srinivasan, G. N. (2018). Lung cancer detection and classification using convolutional neural network. *Jasc Journal of Applied Science and Computations, 5*(6).
9. Srivenkatalakshmi, R., & Balambigai, S. (2018). Lung nodule classification using deep learning algorithm. *Asian Journal of Applied Science and Technology (AJAST), 2*(2), 692–699.
10. Nóbrega, R. V. M. D., & Peixoto, S. A., Silva, S. P. P. D., & Filho, P. P. R. (2018). Lung nodule classification via deep transfer learning in CT lung images. In *IEEE 31st International Symposium on Computer-Based Medical Systems (CBMS)*, Karlstad (pp. 244–249).
11. Xie, Y., Xia, Y., Zhang, J., Song, Y., Feng, D., Fulham, M., & Cai, W. (2018). Knowledge-based collaborative deep learning for benign-malignant lung nodule classification on chest CT. *IEEE Transactions on Medical Imaging.*
12. Fang, T. (2018) A novel computer-aided lung cancer detection method based on transfer learning from GoogLeNet and median intensity projections. In *IEEE International Conference on Computer and Communication Engineering Technology (CCET)*, Beijing (pp. 286–290).
13. Naqi, S. M., Sharif, M., & Jaffar, A. (2018). Lung nodule detection and classification based on geometric fit in parametric form and deep learning. *A Neural Computing and Applications.*
14. Nam, J. G., Park, S., Hwang, E. J., Lee, J. H., Jin, K.-N., Lim, K. Y., et al. (2019). Development and validation of deep learning–based automatic detection algorithm for malignant pulmonary nodules on chest radiographs. *Radiology, 290*(1), 218–228.
15. Tang, H., Kim, D. R., & Xie, X. (2018) Automated pulmonary nodule detection using 3D deep convolutional neural networks. In IEEE *15th International Symposium on Biomedical Imaging (ISBI 2018)*, Washington, DC (pp. 523–526).
16. Winkels, M., & Cohen, T. S. (2018). 3D G-CNNs for pulmonary nodule detection. arXiv:1804.04656.
17. Shi, Z., Hao, H., Zhao, M., Feng, Y., He, L., Wang, Y., et al. (2018). A deep CNN based transfer learning method for false positive reduction. *Multimedia Tools and Applications, 78*(1), 1017.
18. Zhu, W., Vang, Y. S., Huang, Y., & Xie, X. (2018) Deepem: Deep 3d convnets with em for weakly supervised pulmonary nodule detection. In *Medical Image Computing and Computer Assisted Intervention MICCAI.*
19. Setio, A. A. A., Ciompi, F., Litjens, G., Gerke, P., Jacobs, C., van Riel, S. J., et al. (2016). Pulmonary nodule detection in CT images: False positive reduction using multi-view convolutional networks. *IEEE Transactions on Medical Imaging, 35*(5), 1160–1169.

20. Hu, Z., Muhammad, A., Zhu, M. (2018). Pulmonary nodule detection in CT images via deep neural network: Nodule candidate detection. In *ICGSP'18, Proceedings of the 2nd International Conference on Graphics and Signal Processing* (pp. 79–83).
21. Wang, Z., Xu, H., & Sun, M. (2017). Deep learning based nodule detection from pulmonary CT images. In *10th International Symposium on Computational Intelligence and Design (ISCID)* (pp. 370–373), Hangzhou.

ROS-Based Pedestrian Detection and Distance Estimation Algorithm Using Stereo Vision, Leddar and CNN

Anjali Mukherjee, S. Adarsh, and K. I. Ramachandran

Abstract Pedestrian detection systems are of paramount importance in today's cars, as we depart from conventional ADAS toward higher levels of integrated autonomous driving capability. Legacy systems use a variety of image processing methodologies to detect objects and people in front of the vehicle. However, more robust solutions have been proposed in recent years with the advent of better sensors (lidars, stereo cameras and radars) along with new deep learning algorithms. This paper details a ROS-based pedestrian detection and localization algorithm utilizing ZED stereo vision camera and Leddar M16, employing darknet YOLOv2 for localization, to yield faster and credible results in object detection. Distance data is obtained using a stereo camera point cloud and the Leddar M16, which is later fused using ANFIS. YOLOv2 was trained using the Caltech Pedestrian dataset on a NVIDIA 940MX GPU, with the sensor interfacing done via ROS.

Keywords Leddar · Stereo vision · YOLOv2 · Pedestrian detection · Adaptive neuro-fuzzy inference system (ANFIS) · Robotic operating system (ROS)

1 Introduction

Driver assistance systems in vehicles have been around since the 90s in the form of cruise control, power steering, ABS, ESC, etc. To transcend the capability of the driver assistance systems to true autonomy, the system should be able to sense

A. Mukherjee (✉) · S. Adarsh
Department of Electronics and Communication Engineering, Amrita School of Engineering, Amrita Vishwavidyapeetham, Coimbatore, India
e-mail: sonuanjali2005@gmail.com

S. Adarsh
e-mail: s_adarsh@cb.amrita.edu

K. I. Ramachandran
Centre for Computational Engineering and Networking, Amrita School of Engineering, Amrita Vishwavidyapeetham, Coimbatore, India
e-mail: ki_ram@cb.amrita.edu

S. C. Satapathy et al. (eds.), *Intelligent System Design*, Advances in Intelligent Systems and Computing 1171, https://doi.org/10.1007/978-981-15-5400-1_13

the environment and navigate without human input. However, the road to autonomy poses numerous safety issues, prominent among them being pedestrian safety.

Pedestrians are among the most unpredictable elements of traffic in a road scene. Hence, for ensuring safety, the major requirement of a pedestrian detection system must be fulfilled: real-time speed and accuracy. This will enable the system to predict hazardous events and make decisions that ensure the safety of the pedestrians themselves, along with the driver and passengers. Early pedestrian detection systems used conventional image processing techniques like segmentation and histogram of oriented gradients, which exhibit their share of drawbacks.

Self-driving cars use sensors like radars, lidars and stereo vision cameras for the perception of the environment. For making sense of these streams of data, they are combined using various algorithms. This greatly increases the accuracy and helps the system overcome the limitations of individual sensors, improving the reliability of the entire system.

This paper proposes an approach which combines convolutional neural networks (CNN) and sensor fusion algorithms to create a robust pedestrian detection and alerting system. YOLOv2 by Redmon et al. [1] is the CNN used to detect the regions of interest (ROI) in the incoming video stream from ZED stereo camera. This is followed by distance estimation using the fusion of Leddar distance data with ZED point cloud distance data using adaptive network-based fuzzy inference system (ANFIS) [2]. The system uses Robotic Operating System (ROS), which simplifies the implementation of multi-sensor systems making it more robust in real time.

Section 2 elaborates the ongoing related research in the field. Section 3 describes the methodology followed by this work. Section 4 gives the experimental results and analysis. Section 5 gives conclusions.

2 Related Work

Obstacle detection and tracking methods can be broadly classified into the following based on the data and sensors in use (1) methods using range sensors like RPLiDAR 360, Leddar™ M16, Velodyne LiDAR, etc. (2) methods using vision sensors like monocular cameras and stereo vision cameras and (3) methods that use a combination of both vision-based and range sensors for detection and distance estimation. OpenCV has a pedestrian detection built in function that uses histogram of oriented gradients (HOG) along with support vector machine (SVM) as the classifier [3]. The studies [4, 5] examine various pedestrian detectors in use.

CNN-based methods are computationally complex, but they give better accuracy compared to other machine learning-based methods. The study [6] compares various CNN-based methods using mean average precision(mAP) as the evaluation metric. Out of all the methods, YOLOv2 [1] performs well with respect to speed as well as accuracy with 76.8% mAP. The works [7, 8] use only vision sensors like stereo vision camera and Kinect camera for detecting pedestrians and estimating the distance from the camera. [9] implemented an unmarked speed hump detection algorithm using

ZED stereo camera with SSD. In [10], a fuzzy UV-disparity based approach for obstacle detection has been proposed by Irki et al. The inputs to the fuzzy system are the depth in millimeter and difference between center of acquired image and center of obstacle. Becsi et al. [11] have implemented a low cost 3D LiDAR-like system by mounting Leddar™ M16 on a rotating platform interfaced with ECU using CAN.

The algorithm proposed by Wang et al. [12] detects the obstacle using lidar and generates an ROI which is projected on the camera image as prospective candidate window for object detection. Chen et al. in [13] combine the information from an IR radar sensor with the video information from camera. Jose et al. in [14] implemented ROS-based system on an embedded platform for performing autonomous vehicle tasks like navigation and vision.

3 Methodology

The pedestrian detection system proposed in this work consists of two sub-systems: visual perception and distance estimation. The visual perception module uses ZED stereo vision camera and a convolutional neural network (CNN) for detection of pedestrians in the road scene. The ZED camera captures the scene as video and passes it on to the CNN, YOLOv2 trained for detecting pedestrians. YOLOv2 predicts multiple bounding boxes and class probabilities for those boxes. Features from entire image are used to predict bounding boxes, simultaneously. YOLOv2 architecture has 24 fully convolutional layers in total. This work uses the transfer learning approach, by using the pre-trained weights of the first 23 layers and fine-tuning the final layers as per the required classes, in this case, only one 'person.' Bounding boxes are drawn around the detected pedestrians, and the centroid of this box is used for further distance calculation.

The distance estimation part takes the centroid coordinates and extracts corresponding point cloud 3D coordinate. This is then used to find the distance and the corresponding segment number from Leddar M16 as explained in subsequent sections. Distance values from Leddar are then obtained from the corresponding segments where the pedestrians lie. The system uses ANFIS for fusion of distance data from Leddar M16 and depth information from ZED stereo camera to increase the accuracy and range of the existing system. The platform for interfacing is ROS, which can handle multiple sensors at a time, and the user has to worry only about the processing of data, while ROS takes care of data acquisition.

3.1 Dataset Description

The dataset used for training the CNN is Caltech Pedestrian benchmark [15]. Caltech dataset has 10 h of 30 Hz video taken from a vehicle driving through regular traffic in an urban environment (680 × 480 resolution) divided into training and testing sets. They are available in sequence animation (seq) format. The annotations are in video bounding box (vbb) format. The videos were converted into JPEG images, and annotations were generated as per requirements of YOLOv2. During the experimentation, we have used our own custom data as all standard datasets available do not have Leddar distance data which can be used for testing. Dataset was created by positioning pedestrians in the line of sight of the sensors at different positions and recording the corresponding distance data.

3.2 Experimental Setup

The experimental setup consists of the Leddar™ M16 evaluation kit, ZED stereo camera and laptop. Light Emitting Diode Detection and Ranging (Leddar™) is a patented technology by LeddarTech which uses time of flight of light principle to measure distances from obstacles. The Leddar™ M16 sensor has different modes of communication like USB, Modbus, CAN and serial communication using RS485. The mode of communication chosen was USB.

The ZED is stereo vision camera from Stereolabs which tries to reproduce human vision using two cameras at a given baseline separation. It produces a 3D map of the scene by comparing the displacement of pixels in left and right images. The ZED camera has other functionalities like odometry, positional tracking, depth mapping, etc., and can work at a high frame rate of 100fps. Specifications of Leddar M16 [16] and ZED camera [17] used in this work are furnished in Fig. 1.

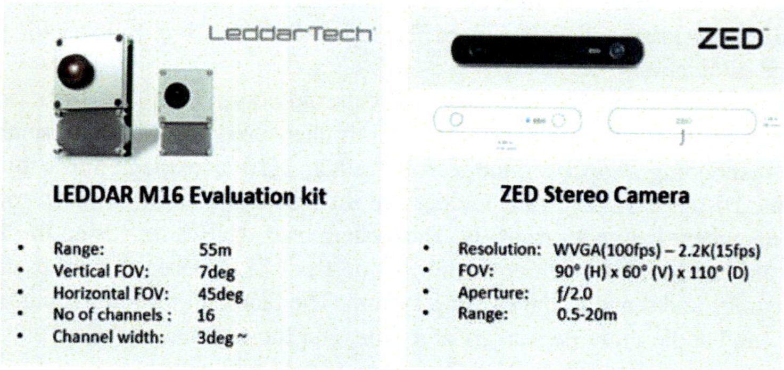

Fig. 1 Specifications of the Leddar™ M16 evaluation kit and ZED stereo camera used

3.3 ROS Framework

Robot Operating System (ROS) is an open-source framework designed for code reuse in robotics. The package consists of the implementation of basic functionalities provided by a normal operating system like hardware abstraction, package management, inter-process communication (IPC), etc. ROS can manage multiple processes on number of hosts connected to each other at run time. It is a language neutral framework that currently supports C++, Python, Octave and Lisp. Basic architecture of ROS is given in Fig. 2. ROS master controls and keeps track of all the nodes in system. Nodes send data by publishing to a topic and receive data by subscribing to a topic. Nodes communicate to each other using a peer-to-peer connection. Using ROS, outputs of sensors can be synchronized in real time and implementation becomes effortless.

For the experiment, the ROS Kinetic package, ZED ROS wrapper and Leddar ROS driver were set up on Ubuntu 16.04. The Leddar™ M16 was used in its default 24 V configuration. Leddar™ M16 and ZED camera were set up on a golf cart as shown in Fig. 3. YOLOv2 model trained on Caltech Pedestrian dataset was used with

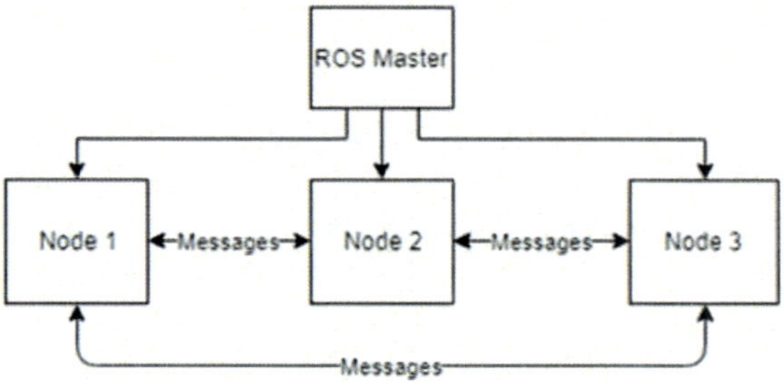

Fig. 2 Basic ROS architecture

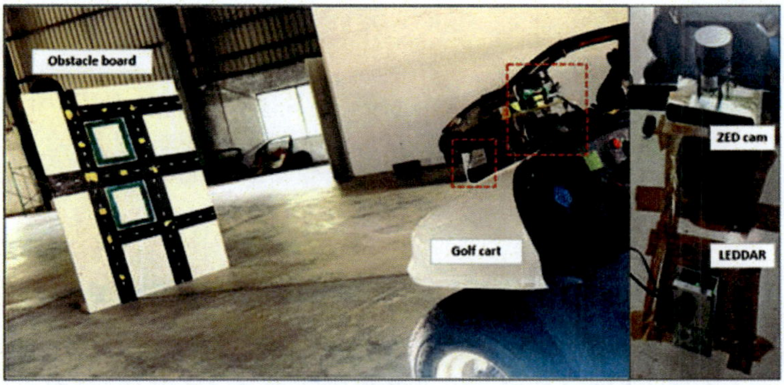

Fig. 3 Experimental setup for sensor data collection with golf cart and obstacle board

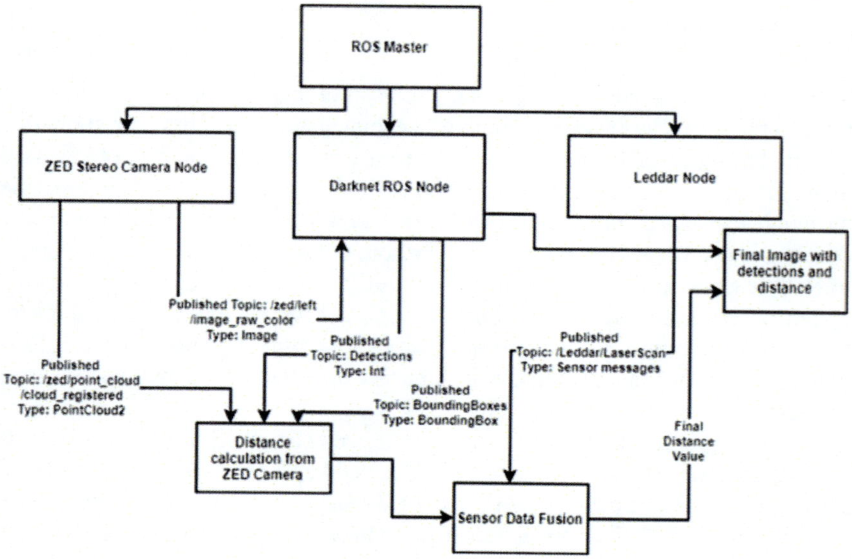

Fig. 4 ROS architecture of the proposed system showing the topics from each node and the communication between them

the darknet ROS wrapper that implements YOLOv2 in ROS. The ROS architecture of the proposed system is shown in Fig. 4. For communication between nodes, message filters were used to ensure that the sensor messages are synchronized with each other. Message filters ensured that the sensor messages with same time stamp are processed together.

3.4 Proposed System for Pedestrian Detection

Flow diagram of the proposed algorithm is shown in Fig. 5. The left image from ZED stereo camera is given as input for the pedestrian detector, YOLOv2. The number of detections and corresponding bounding box coordinates generated by YOLOv2 are then used to determine the centroids of the ROIs obtained. From the point cloud of the ZED camera, corresponding x-, y- and z-coordinates of the centroids are extracted and used to calculate distance value as given below,

$$d = \sqrt{x^2 + y^2 + z^2}. \quad (1)$$

This gives the distance estimate from ZED camera.

The ZED camera has a horizontal field of vision (FOV) of 87° in VGA mode and Leddar^{TM} M16 has 48°. The FOV of Leddar is further subdivided into 16 segments of 3° each. If the left camera and Leddar are aligned, the sensor data fusion area is

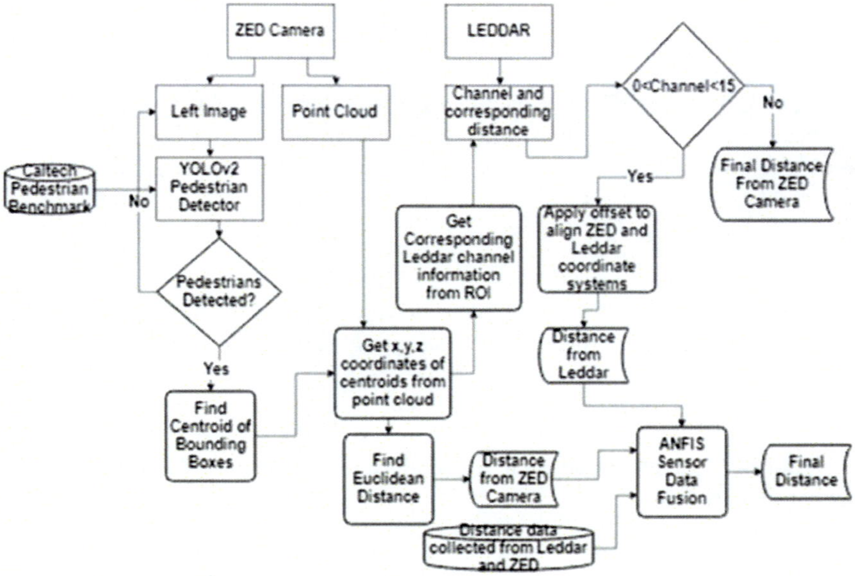

Fig. 5 Process flow diagram of the proposed system

as shown in Fig. 6. ANFIS sensor fusion is done when the pedestrian is in the FOV of Leddar. In all other cases, only ZED distance data is used.

Mapping between Leddar and ZED camera outputs. Distance data comes in 16 channels in the Leddar™ M16. So, the total horizontal distance covered by the Leddar with FOV of α radians at distance $\dot{d}$ from Leddar would be $\alpha\dot{d}$, where $\dot{d}$ is d from (1) with offset applied with respect to Leddar. Thus, each channel of Leddar with n channels will cover the horizontal distance, w, given by Eq. (2).

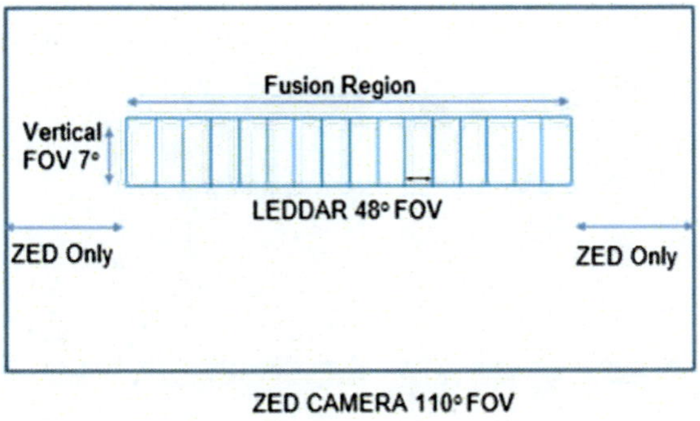

Fig. 6 Sensor fields of vision and fusion area

$$w = \frac{\alpha \dot{d}}{n} \tag{2}$$

From the ZED camera, the x-coordinate of the centroids is obtained. If the Leddar channel 7 at the center is aligned with the left camera of ZED, the following relation is obtained to find the corresponding channel S in which the object lies.

$$S = 7 + \frac{x}{\psi} \tag{3}$$

Vertical FOV of η radians is also considered to ensure that object is in the fusion region. Vertical distance covered by each segment is given by

$$h = \eta \dot{d} \tag{4}$$

From the channel S, distance value is obtained which is given to the ANFIS module for sensor data fusion.

3.5 Sensor Data Fusion

Python implementation of Sugeno-type neuro-fuzzy system which uses Gaussian membership functions has been trained using distance data collected from sensors. The data for training ANFIS was collected using the setup shown in Fig. 7. Ground truth measurements were marked on the floor, for a total distance of 15 m at an interval of 0.5 m. An obstacle board was moved over the markings, and distance values were recorded using each sensor. The data was then given to ANFIS to generate the sensor fusion model. The final distance value is the output from the ANFIS model, which is displayed on the detection image. RMSE obtained was 0.069.

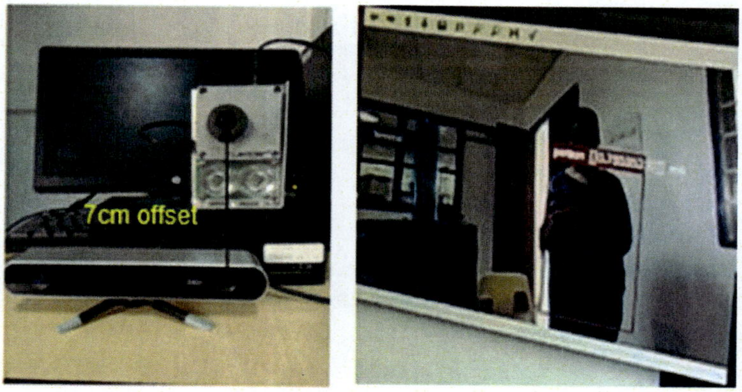

Fig. 7 Setup for single pedestrian. Sensor alignment and sample detection image

4 Experiment Results and Discussion

The Leddar and ZED camera sensors were mounted on the golf cart as shown in Fig. 3, and coordinate systems were aligned using extrinsic calibration. The origin of both the sensors was aligned together. YOLOv2 pedestrian detector was trained for 30,000 iterations, and a mAP of 80% was obtained with a log average miss rate of 40%. The trained model was given to the darknet ROS node. The Leddar and ZED camera nodes were powered up and connected to ROS master. Distances 2, 5, 7 and 10 m were marked on the ground from the sensors. The FOV of Leddar was also marked on the ground with the help of RVIZ tool in ROS.

Single Pedestrian Case. Pedestrian was positioned at random designated locations with reference to the ground truth markings. The distance measurement using Leddar and ZED camera was performed first, followed by distance calculation using the proposed fusion algorithm. The results are summarized in Table 1.

Multiple pedestrian Case. Multiple pedestrians were made to stand in the designated locations, and results were recorded in the same way as the single pedestrian case. Results of one trial with three pedestrians have been summarized in Table 2. Figure 8 gives the pictorial representation of the experimental setup.

Table 1 Summary of results single pedestrian case

Actual distance (m)	Actual channel number	ZED distance (m)	Leddar distance (m)	Calculated channel number	ANFIS result (m)
2.5	8	2.44	2.5	8	2.47
2.8	7	2.61	2.83	8	2.81
3.1	6	2.94	3.16	7	3.17
3.4	5	3.23	3.56	5	3.5
3.7	10	3.41	3.78	10	3.72
4.0	12	3.86	4.05	12	4.01
4.3	8	4.1	4.52	8	4.48
4.6	7	4.42	4.71	8	4.66

RMSE (Leddar Only) = 0.10083
RMSE (ZED Only) = 0.157909
RMSE (Sensor Fusion) = 0.0694

Table 2 Summary of results multiple pedestrian case

Pedestrian number	Actual distance (m)	Actual channel number	ZED distance (m)	Leddar distance (m)	Calculated channel number	ANFIS result (m)
1.	2.5	8	2.44	2.5	8	2.47
2.	4	3	4.25	4.29	4	4.15
3.	7	12	7.31	7.4	12	7.19

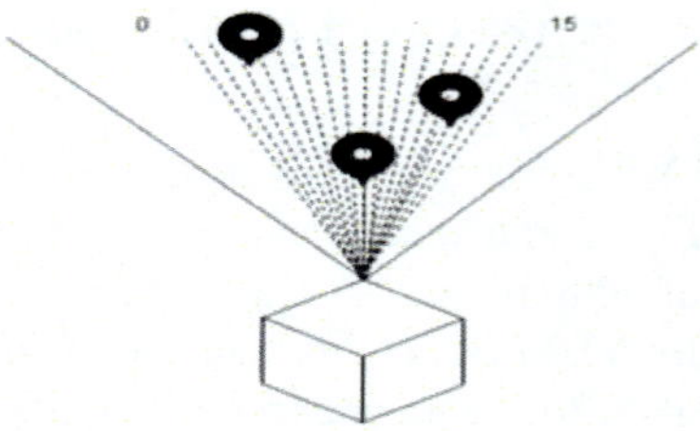

Fig. 8 Pictorial representation of experimental setup for multiple pedestrians. Black markers represent pedestrians

Compared to the case where single sensor value is considered, the sensor fusion approach gives more accurate distance values. Pedestrian detection on the real-time video was obtained at 10 fps using NVIDIA 940 MX 2 GB GPU.

5 Conclusions and Future Work

This work proposes and implements a pedestrian detection and distance estimation system that uses YOLOv2 for detection and Leddar and ZED stereo camera for distance perception. The RMSE value indicates that the distance values after sensor fusion are more accurate compared to the single sensor data. The detections are also faster as high frame rates are achievable. The system can be implemented on a faster GPU or an embedded platform for a standalone implementation. This work currently includes scenarios concerning vehicle moving in straight path. In future, scenarios like maneuvers, turns and overtaking can also be added. Higher performance can be achieved by using more powerful GPUs.

Declaration We have taken permission from competent authorities to use the images/data as given in the paper. In case of any dispute in the future, we shall be wholly responsible.

References

1. Redmon, J., & Farhadi, A. (2017). YOLO9000: Better, faster, stronger. In *Proceedings - 30th IEEE Conference on Computer Vision and Pattern Recognition, CVPR 2017*. https://doi.org/10.1109/CVPR.2017.690.
2. Jang, J. S. R. (1993). ANFIS: Adaptive-network-based fuzzy inference system. *IEEE Transactions on Systems, Man, and Cybernetics*. https://doi.org/10.1109/21.256541.
3. Dalal, N., Triggs, B. (2005). Histograms of oriented gradients for human detection. In *Proceedings—2005 IEEE Computer Society Conference on Computer Vision and Pattern Recognition, CVPR 2005*. https://doi.org/10.1109/CVPR.2005.177.
4. Brunetti, A., Buongiorno, D., Trotta, G. F., & Bevilacqua, V. (2018). Computer vision and deep learning techniques for pedestrian detection and tracking: A survey. *Neurocomputing, 300*, 17–33. https://doi.org/10.1016/j.neucom.2018.01.092.
5. Benenson, R., Omran, M., Hosang, J., & Schiele, B. (2015). Ten years of pedestrian detection, what have we learned? In *Lecture Notes in Computer Science (including subseries Lecture*

Notes in Artificial Intelligence and Lecture Notes in Bioinformatics). https://doi.org/10.1007/978-3-319-16181-5_47.
6. Kuang, P., Ma, T., Li, F., & Chen, Z. (2018) Real-time pedestrian detection using convolutional neural networks. *International Journal of Pattern Recognition and Artificial Intelligence*. https://doi.org/10.1142/s0218001418560141.
7. Wang, B., Florez, S. A. R., & Fremont, V. (2014). Multiple obstacle detection and tracking using stereo vision: Application and analysis. In *2014 13th International Conference on Control Automation Robotics and Vision, ICARCV 2014* (pp. 1074–1079). https://doi.org/10.1109/ICARCV.2014.7064455.
8. Mane, S. B., Vhanale, S. (2017). Real time obstacle detection for mobile robot navigation using stereo vision. In *International Conference on Computing, Analytics and Security Trends, CAST 2016* (pp. 637–642). https://doi.org/10.1109/CAST.2016.7915045.
9. Varma, V. S. K. P., Adarsh, S., Ramachandran, K. I., & Nair, B. B. (2018). Real time detection of speed hump/bump and distance estimation with deep learning using GPU and ZED stereo camera. *Procedia Computer Science, 143*, 988–997. https://doi.org/10.1016/j.procs.2018.10.335.
10. Irki, Z., Oussar, A., Hamdi, M., Seddi, F., & Numeriques, S. (2014). A Fuzzy UV -disparity based approach for obstacles avoidance. In *International Conference on Systems Signals, Image Processing* (pp. 12–15).
11. Bécsi, T., Aradi, S., Fehér, Á., & Gáldi, G. (2017). Autonomous vehicle function experiments with low-cost environment sensors. *Transportation Research Procedia*, 333–340. https://doi.org/10.1016/j.trpro.2017.12.143.
12. Jun, W., & Wu, T. (2016). Camera and lidar fusion for pedestrian detection. In *Proceedings–3rd IAPR Asian Conference on Pattern Recognition, ACPR 2015*. https://doi.org/10.1109/ACPR.2015.7486528.
13. Chen, X., Ren, W., Liu, M., Jin, L., & Bai, Y. (2015). An obstacle detection system for a mobile robot based on radar-vision fusion. https://doi.org/10.1007/978-3-319-11104-9_79.
14. Jose, S., Sajith Variyar, V. V., & Soman, K. P. (2017). Effective utilization and analysis of ros on embedded platform for implementing autonomous car vision and navigation modules. In *2017 International Conference on Advances in Computing, Communications and Informatics, ICACCI 2017*. https://doi.org/10.1109/ICACCI.2017.8125952.
15. Dollár, P., Wojek, C., Schiele, B., & Perona, P. (2009). Pedestrian detection: A benchmark. In *2009 IEEE Computer Society Conference on Computer Vision and Pattern Recognition Workshops, CVPR Workshops 2009*. https://doi.org/10.1109/CVPRW.2009.5206631.
16. Solid-State LiDAR. Leddar M16 Multi-segment sensor module. http://leddartech.com/lidar/m16-multi-segment-sensor-module/.
17. Stereolabs: ZED Stereo Camera—Stereolabs. https://www.stereolabs.com/zed/.

An Enhanced Prospective Jaccard Similarity Measure (PJSM) to Calculate the User Similarity Score Set for E-Commerce Recommender System

H. Mohana and M. Suriakala

Abstract E-commerce is an exemplary representation for online commercial transactions which permits business applications in dealing with different organizations. A business intelligent (BI) technique implicated in predicting the user's buying preference in advance such as Flipkart and Amazon is said to be recommender system (RS). It builds a product–customer relationship in an efficient manner. Thus, collaborative filtering(CF) is the most promising approach followed by the E-commerce firm in generating product recommendations to the end user.

Collaborative filtering approach searches for a "like - minder " people with similar buying preferences. The conventional machine learning algorithms used to calculate the user similarity index (USI) value are identified as Pearson correlation coefficient (PCC), cosine (COS) and Jaccard similarity measure. These algorithmic similarity measures are not successful in several situations particularly in cold start and data sparsity problem. From the literature survey, it is inferred that data sparsity is the most important primary issue that needs to be resolved first in CFRS in order to generate a reliable product recommendation listing to the targeted user. So that a novel method called prospective Jaccard similarity measure (PJSM) was proposed. Its novelty is to overcome the data sparsity issues in CFRS method as well as the conventional algorithmic method limitations. MovieLense dataset is extracted and applied for an implementation. Research objective involves in showing rapid improvements in users' rating prediction accuracy. Compare conventional algorithms results with the newly proposed PJSM algorithm results by finding the MAE and RMSE. Accordingly, the proposed PJSM algorithm results in showing superiority for the predicted rated data and reduces the error ratio as well. The PJSM algorithm is also effective in managing the 10 lakh data too.

Keywords Collaborative filtering (CF) · Recommender system (RS) · E-commerce · User–item matrix · PJSM · Pearson · Cosine · Jaccard

H. Mohana (✉)
Dr. Ambedkar Government Arts College (Autonomous), Chennai, India
e-mail: vkmohana85@gmail.com

M. Suriakala
Government Arts College for Men (Autonomous), Nandanam, Chennai, India
e-mail: suryasubash@gmail.com

S. C. Satapathy et al. (eds.), *Intelligent System Design*, Advances in Intelligent Systems and Computing 1171, https://doi.org/10.1007/978-981-15-5400-1_14

1 Introduction

Electronic commerce (e-commerce) is an icon of communication, data and security management that deals business application in different organizations. Its functionality is to exchange information to the sales of goods and services. In Internet, the e-commerce industry plays a paramount role in the WWW. Enterprises mainly focus on rebuilding the relation with their old customer and simultaneously focusing on new customer [1]. The effectual technology focused by the e-commerce industry is said to be data mining and Web mining. The methodology used to extract the immersing patterns from huge database is called data mining. Similar way, Web mining is the usage of data mining techniques to extract interesting information from the Web data. The business intelligent technique (BI) handled by the e-commerce industry is found to be e-commerce recommendation system (RS). Predicting the user's preferences in advance by analyzing the user's past records is called recommender system. E-Commerce firm such as Amazon, Alibaba and Snapdeal enlarges the customer support system using recommender system [2]. "The main goal of recommender system is to generate meaningful recommendations to a collection of users for items/products that might interest them" [3]. Therefore, recommender system plays an imperative role in information overloading issues in several e-commerce application areas like movies, music, news, books, research articles, etc. Recommender system is broadly classified into three different characteristics, namely collaborative filtering, content and hybrid-based recommender system. CF is again classified into memory-based CF and model-based CF. The methodology that adheres both user-based and item-based is done by memory-based CF type. The research work targets the memory-based collaborative filtering recommender system (CFRS) for further analysis due to the consideration of user-based and item-based CF method. The large and the commercial e-commerce Web sites make benefit on it. There are several predefined user similarity conventional algorithms such as Pearson correlation coefficient, cosine, Jaccard and MSD. These conventional algorithmic similarity measures are not successful in several situations particularly in cold start and data sparsity problem. Thus, a new similarity measure called prospective Jaccard similarity measure (PJSM) was proposed. This improves the recommendation performance when only few ratings are available.

The limitations which reduce the recommendation performance in CFRS are identified to be data sparsity, cold start problem, poor security and scalability. Rating parameter describes whether the product purchased by the user is liked or not, user feelings on purchased products as well as services provided by the e-commerce industry are expressed by the scale of rating parameter, whereas scale of rating ranges from 1 to 5. Data sparsity is defined as users' unrated rated data, i.e., number of rated data by the total number of actual data. The item purchased by them will not be rated by users due to laziness; not interested in giving feedback leads to increase in the sparsity level of the data, i.e., users' unrated data for the purchased item. For example, the sparsity level for the extracted data is said to be 20:100, i.e., only 20% of data has been rated by the user, and balance data leads to null, i.e., data sparsity instead of 100% rated data. Therefore, this issue leads to poor or bad recommendation to the end user. It will moderately reduce the e-commerce profit ratio. Similarly, when new user enters into the Web site and searches for a particular product/item, at that time, recommendation is difficult for the new user because no past history record is available. There is a possibility to recommend a product/item to the new user if 100% rated data is available. Thus, converting unrated data into rated data helps to generate an exact recommendation list to the end user. This list will increase the e-commerce turnover and helps to acquire a net profit. To do so, the rating prediction plays a paramount role in e-commerce recommender system.

Methodology used in identifying the solution for data sparsity issue is found to be Pearson correlation coefficient similarity measure [4], cosine similarity measure [5] and Jaccard similarity measure. The conventional similarity measures were not much effective in resolving the data sparsity issue and its prediction accuracy is relatively low. The reason is , PCC does not consider the impact of absolute values as well as the proportion of size of the set of common ratings given by the user. In cosine, the existing cosine similarity measure does not consider the preference of the user's ratings, and it provides high similarity index value instead of low similarity index value. In Jaccard similarity measure, it does not consider the absolute ratings; rather it considers number of items rated. Thus, if these limitations are overcome effectively, then it will be an adequate platform for the e-commerce giant market to perform a trustworthy recommendation to active user. Two different methods are followed in collaborative filtering recommender system called user-based and item-based CF method. The majority of the research work was carried forward in focusing the IBCF method only. Proving efficiency in both the methods is truly a challenging one. This paper targets to prove efficiency into both the methodologies, and it is implemented in algorithm and proposed a new framework for e-commerce recommender system for future extent. R-programming, LenseKit, GroupLense, Python, NetBeansIDE, Java are the platforms used to implement recommender algorithms. Most of the researchers used to extract Epinions, Jester and MovieLense dataset for further enhancement works in CFRS. "It can be easily identified that most students will prefer to the fantasy movies, and the popularity of comedy movies far surpasses drama" [6].

The need of improving the data sparsity in CFRS is to upgrade precision of predicting users' unrated rating data. So that higher precision leads to well-grounded top N recommendation to the end user. Thus, online commercial e-commerce industries acquire a huge net profit. Therefore, the foremost motivation in improving the recommendation algorithm for e-trade industry is to enhance the marketing strategy in a booming way, improve the sales frequency range, well-equipped customer support system, target the e-commerce bottom line and increase the profitability.

2 Related Works

Sun et al. [7] developed new similarity measure (NSM) called integrating triangle and Jaccard similarity measures for e-commerce recommender system. His work considers the size of the length of rating vectors with its co-rated items and its vector angle involving them. Haifeng et al. [8] deal with data sparsity issue and generate a new model which targets global context as well the local context of user's behavior in buying the item and prove the enhancement of the proposed work using three datasets. Zhe et al. [9] planned a new CFRS framework in order to find the user's interest on mobile and how to build suitable recommendations for the mobile users depends on users' behavior and ratings were analyzed. Zhipeng et al. [10] target in finding a nearest neighbor using UBCF method. The covering-based rough set helps in reducing the redundant users from all the other users. Zhang et al. [11] developed personalize book recommendation algorithm for an e-commerce Web site which depends on time sequence method for digital libraries. How frequently the books were hired by the customer and its return cycle time were analyzed. His proposed algorithm creates a successive impact between college students, and it shows the demand in professional learning method. Weimin et al. [12] projected a modified fitting recommendation model in order to calculate the user missed rated data which depends on user similarity score set available using SVM regression technique. Mohana et al. [13] developed a new heuristic similarity method called prospective business intelligent (BI) technique in collaborative filtering recommender system by integrating modified Pearson correlation coefficient (MPCC) and exponential function-based Pearson correlation coefficient (EPCC) algorithm to alleviate the data sparsity issues in collaborative filtering recommender system. The accuracy of the algorithm is evaluated for both conventional and proposed algorithms. Hence, the proposed one shows successive improvement on it by reducing the rating prediction error ratio. Alexander et al. [14] target on how to increase the e-commerce profit ratio. The authors developed a recommender system framework along with big data, and it is executed on real time to one of the e-commerce companies. The output of the proposed work is said to be delivering a personalized recommendation to the end user through email services. Several similarity measures have been taken into account, and finally, Tanimoto coefficient produces better performance in prediction part by showing the least precision, recall, MAE and RMSE error ratio.

Jaccard similarity coefficient does not consider the absolute ratings; rather it considers number of items rated [8, 15]. When used with rating dataset, it does not yield accurate results because it ignores the rating value [16]. Jaccard similarity measure takes number of preferences common between two users into account [17]. To overcome the limitation existing in the Jaccard similarity coefficient, the mean square differences (MSD) have been combined to the conventional Jaccard algorithm. Various types of similarity measures which have been adopted and designed for these issues are Jaccard mean squared differences(JMSD) [18] and Jaccard proximity, significance and singularity (JPSS) [8] similarity measure. Most recently, the highlighted term used in proposing a new measure for recommendation is integrated triangle and Jaccard similarities. The triangle similarity considers both the length and angle of rating vectors between them [7]. Two users will be more similar, when two users have more common rated items.

3 Prospective Jaccard Similarity Measure

3.1 Conventional Collaborative Filtering Similarity Measures

The conventional similarity measure like Pearson correlation coefficient, cosine and Jaccard describes the closeness of the user's behavior in buying the items.

The mathematical formula for Pearson, cosine and Jaccard is as follows.

$$\mathrm{sim}(x, y)^{PCC} = \frac{\sum_{P\in I}\left(r_{x,p} - \bar{r}_x\right)\left(r_{y,p} - \bar{r}_y\right)}{\sqrt{\sum_{P\in I}\left(r_{x,p} - \bar{r}_x\right)^2 \cdot \sum_{P\in I}\left((r_{y,p} - \bar{r}_y)^2\right)}}. \tag{1}$$

Let us consider $X = \{x_1, x_2, x_3 \ldots x_n)$ and $Y = \{y_1, y_2, y_3 \ldots y_n\}$ are number of users. Similarly, $P = \{p_1, p_2, p_3 \ldots p_m\}$ are number of items. The UI (user, item) rating matrix is defined as $R = (r_{i,j})\ N * M$, where $i = 1, 2, 3, \ldots N$ and $j = 1, 2, 3, \ldots M$. I is number of common ratings given by the user for the purchased item p.$r_{x,p}, r_{y,p}$ is scale of rating. $\bar{r}_x, \bar{r}_y$ is average mean rating for user x and y.

$$\mathrm{sim}(x, y)^{\cos} = \frac{\vec{r}_x \cdot \vec{r}_y}{\|\vec{r}_x\| \cdot \|\vec{r}_y\|} \tag{2}$$

where $\vec{r}_x$ and $\vec{r}_y$ are rating scales of rating vector for user x and user y. $\|\vec{r}_x\|, \|\vec{r}_y\|$ represents the magnitude representation of $\vec{r}_x$ and $\vec{r}_y$, respectively.

$$\mathrm{sim}(x, y)^{\mathrm{Jaccard}} = \frac{|I_x| \cap |I_y|}{|I_x| \cup |I_y|} \tag{3}$$

where I_x and I_y indicate number of items rated by both users. Actually, this conventional similarity measure does not produce reliable results for cold user condition and data sparsity problem. Therefore, new similarity measures called prospective Jaccard similarity measure were proposed, and it is discussed briefly in the upcoming section.

Table 1 An experimental dataset using UI rating matrix

User/item	I1	I2	I3	I4
U1	4	3	5	4
U2	5	3	–	–
U3	4	3	3	4
U4	2	1	–	–
U5	4	2	–	–

3.2 User–Item Rating Matrix—An Experimental Dataset

An experimental sample data has been created stated 5 × 4 UI (user, item) rating matrix. This example is the platform used to prove the efficiency of enhanced PJSM similarity measure.

In Table 1, symbol "-" represents the user unrated data. A 5 by 4 matrix table was generated. Totally, 20 rated data has to be available, but 14 data only existed. Remaining place is addressed as data sparsity. It is mandatory to predict the unrated data into rated data by means of handling the flawed data in a proper manner.

The user similarities have to be computed for the values in Table 1, by finding the relationship between two users (U1, U2), (U1, U3), (U1, U4), (U1, U5) using conventional and new PJSM similarity algorithm. The conventional formula 1 for PCC, 2 for Cosine, and 3 for Jaccard is applied in to Table 1 and thus the resultant USI value obtained is shown in Fig. 1.

3.3 Limitations

The predicted drawback for the conventional algorithm is as follows.

(i) The foremost limitation commonly found in all the algorithms results in high predictive error ratio, i.e., MAE and RMSE.
(ii) Suggest bad recommendation to the end user based on absolute ratings.
(iii) Pearson does not consider the proportion of size of the set of common ratings given by the users.
(iv) COS does not consider the user's rating preference in buying the item.
(v) Jaccard similarity measure produces absolute values as resultant similarity score set, i.e., either 0.5 or 1.0. So that it is difficult to find out the closest user with similar buying preferences.

3.4 Motivation of Prospective Jaccard Similarity Measure (PJSM)

To overwhelm the drawbacks found in the conventional algorithm in addition to increase the rating prediction accuracy for the unrated data, a modified Jaccard similarity measure has been proposed, namely prospective Jaccard similarity measure (PJSM). The mathematical formalization of prospective Jaccard similarity measure (PJSM) is as below

$$\mathrm{sim}(x, y)^{\mathrm{MJaccard}} = \frac{|I_x| \cup |I_y|}{|I_x| \cap |I_y|} \tag{4}$$

where I_x, I_y sets of users, U—union (sets of all), n—intersection (common sets), $|r_u - r_v|$—magnitude representation of difference between r_x and r_y, respectively.

$$\mathrm{significance} = \frac{1}{1 + |r_x - r_y|} \tag{5}$$

$$\mathrm{sim}(x, y)^{\mathrm{PJSM}} = \frac{\mathrm{significance}}{\mathrm{Jaccard}} \tag{6}$$

Modified Jaccard (MJaccard) similarity measure does not result in the scale limit as −1 to +1. It exceeds the similarity scale limit. So that a significant difference between two user rating vectors is calculated, and it helps to bring back the newly proposed PJSM algorithm within the similarity scale limit. Still, the limitation involved in the conventional Jaccard method persists in both modified Jaccard and significance measure so that a prospective measure has been derived, and thus, it overcomes the limitations involved in getting back the absolute values. This means the newly proposed PJSM algorithm results in different similarity measure which is shown in Fig. 2.

Prediction methodology:

$$\hat{r}_{x,t} = \frac{\sum_{j \in p_{t(x)}} \mathrm{sim}(y, t)^{\mathrm{Proposed}} \cdot S_{y,j}}{\sum_{j \in p_{x(x)}} |\mathrm{sim}(j, t)^{\mathrm{Proposed}}|}$$

The above methodology is used to convert the user similarity index value into actual rating value, i.e., weighted sum for prediction purpose.

3.5 Discussion on PJSM Method

The pitfalls that occur in the conventional algorithms are discussed earlier. The resultant USI value obtained for the proposed PJSM method is shown in Fig. 2.

Figure 2 illustrates the PJSM USI value for Table 1 experimental setup. Therefore, resultant figure is compared to conventional Jaccard method with PJSM to show how come the existing similarity measure limitations are overcome using PJSM method.

Comparison between Jaccard with PJSM:

(a) Conventional jaccard USI value results in absolute values. In Fig. 1, jaccard shows 0.5 and 0.1 as USI value for all the users. So it is difficult to find the correlated users, which means most of the users were closest each other, whereas, in Fig. 2, the PJSM results in four different similarity measures as 0.166, 0.25, 0.33 and 1.0. So that it overcomes the limitations of getting back the absolute values

(b) Conventional Jaccard method results in showing high similarity index value instead of low similarity index value, which means the conventional jaccard method predicts the user with dissimilar buying preference, i.e., in Table 1 example, users U2 and U4 who do not similar with each other, and they are the least pair, but in Fig. 1, Jaccard shows 1.0, i.e., high similarity score set. In Fig. 2, the PJSM results in 0.1666, i.e., least pair; thus, the proposed PJSM predicts the low similarity users accurately than the predefined one. The user's pair which actually does not correlate with each other will be identified.

Fig. 1 Conventional algorithm USI value

	u2	u3	u4	u5
u1	0.707	0.0	0.707	0.707
u2		1.0	1.0	1.0
u3			1.0	1.0
u4				1.0

(a) PCC

	u2	u3	u4	u5
u1	0.612	0.975	0.606	0.605
u2		0.703	0.997	0.997
u3			0.696	0.695
u4				1.0

(b) COS

	u 2	u 3	u 4	u 5
u 1	0 .5	1 .0	0 .5	0 .5
u 2		0 .5	1 .0	1 .0
u 3			0 .5	0 .5
u 4				1 .0

(c) Jaccard

Fig. 2 PJSM user similarity index score set

	u2	u3	u4	u5
u1	0.25	0.33	0.1	0.25
u2		0.25	0.166	0.33
u3			0.1	0.25
u4				0.25

(f) PJSM

(c) The error ratio in predicting the user's missed out rating data is reduced in PJSM method as much as possible compared to the existing Jaccard method. The predictive error ratio for the existing Jaccard method is 0.824, whereas in PJSM 0.808. Result shows that PJSM algorithm outperforms well in terms of MAE and RMSE.

3.6 Prediction Accuracy

Two types of evaluation metrics are there, online and offline evaluation. Online evaluation is done by e-commerce research experts by handling real-time data whose data is available in their own laboratory. The second type of evaluation is done by research scholars. Online evaluation is cost effective and needs high influence in running the algorithm as well. So, therefore, offline evaluation method is executed and calculated the error ratio in terms of mean absolute error (MAE) and root mean square error (RMSE). Lower error ratio results lead to high prediction accuracy.

$$\text{MAE} = \frac{1}{n}\sum_{i=1}^{n} |p_i - r_i| \tag{8}$$

$$\text{RMSE} = \sqrt{\frac{1}{n}\sum_{i=1}^{n} |(p_i - r_i)^2|} \tag{9}$$

where p_i = predicted rating, r_i = original rating, n = number of ratings predicted.

3.7 Experimental Dataset

MovieLense (ML), Epinion and Jester are the most famous open-source datasets used in collaborative filtering recommender system. "Extensive experiments were done on real-time MovieLense datasets" [19]. The data collected from the source is (https://www.movielense.umn.edu). In ML-100k dataset, there are 100,000 data available.

Table 2 Conventional Jaccard resultant table

Similarity method	MovieLense dataset	UBCF		IBCF	
		MAE	RMSE	MAE	RMSE
Jaccard	ML-100K random 1	0.824	1.031	0.814	1.027
	ML-100K random 2	0.814	1.024	0.801	1.008
	ML-1M	0.777	0.974	0.788	0.990

It contains anonymous rating data from 943 active users and 1682 movies. Another one is ML-1M datasets; there are 1,000,000 data available. It contains anonymous rating data from 6040 active users and 3952 movies. 80% data was taken as training data, and 20% data was taken as test data. "The sparsity level of dataset is analyzed by calculating the sparseness of the data" [20]. Therefore, data sparsity is defined as $1 - \frac{\text{non zero entries}}{\text{total entries}}$. For ML-100K, the sparsity level of data is calculated as, $1 - \frac{100{,}000}{943*1682}$, i.e., 0.9369. Similar way, for ML-1M dataset, $1 - \frac{1{,}000{,}000}{6040*3952}$, i.e., 0.9581. So, therefore, the sparseness of data is high in both ML-100 K and ML-1 M. Java platform be used to implement the proposed work. The input files taken up for implementation purpose is named as ratings1.cvs and toBeRated1.csv files. The ultimate goal is to predict the users unrated rating data in to 100% rated data. Thus, results are stored in the result.csv file. At the end, result.csv file is compared to actually rated data csv file to check how exact the predicted rating is evaluated. Implementation is done for conventional PCC, Cosine, Jaccard method and PJSM individually for MovieLense dataset, and its predicted error ratio results will be discussed in the next section.

4 Result Analysis

4.1 Conventional Jaccard Similarity Measure Resultant Table

Table 2 shows the resultant prediction accuracy for MovieLense 100 k and 1 M dataset for both user-based (UBCF) and item-based (IBCF) collaborative filtering methods. Two random datasets are considered for evaluation and checked the efficiency. Double-blind peer-reviewed resultant table was produced.

4.2 Proposed Prospective Jaccard Similarity Measure (PJSM) Resultant Table

Table 3 shows the resultant prediction accuracy for MovieLense 100 k and MovieLense 1 M dataset for both user-based (UBCF) and item-based (IBCF) collaborative filtering methods. Two random datasets are considered for evaluation and checked the

Table 3 Proposed PJSM result

Similarity method	MovieLense dataset	UBCF		IBCF	
		MAE	RMSE	MAE	RMSE
PJSM	ML-100K random 1	0.808	1.017	0.815	1.022
	ML-100K random 2	0.801	1.011	0.802	1.002
	ML-1M	0.764	0.962	0.789	0.983

efficiency too. Two-blind peer reviewed resultant table was produced in Table 2 and Table 3 respectively to compare the efficiency of the proposed PJSM method with conventional jaccard method. In traditional Jaccard algorithm, it results in 0.824, 0.814, 0.777 and 1.031, 1.024, 0.974, respectively. Similarly in Table 3, the new prospective Jaccard similarity measure (PJSM) results in 0.808, 0.801, 0.764 and 1.017, 1.011, 0.962, respectively. Thus, the reduced error ratio will create a big impact in prediction part. Here, in user-based method, the new Jaccard algorithm works out well in handling the big data too. Similarly, the MAE and RMSE for item-based method are described next. In Table 2, the traditional Jaccard algorithm results in 0.814, 0.801, 0.788 and 1.027, 1.008, 0.990. In Table 3, the new prospective Jaccard (PJSM) similarity measure results in 0.815, 0.802, 0.789 and 1.022, 1.002, 0.983, respectively. Thus, the reduced error ratio definitely enables a big impact in prediction part. Therefore, item-based method has the ability to handle big data.

4.3 *Jaccard with PJSM Resultant Chart*

Figures 3 and 4 illustrate the user-based collaborative filtering (UBCF) presentation chart for the MAE and RMSE values obtained by both conventional algorithm

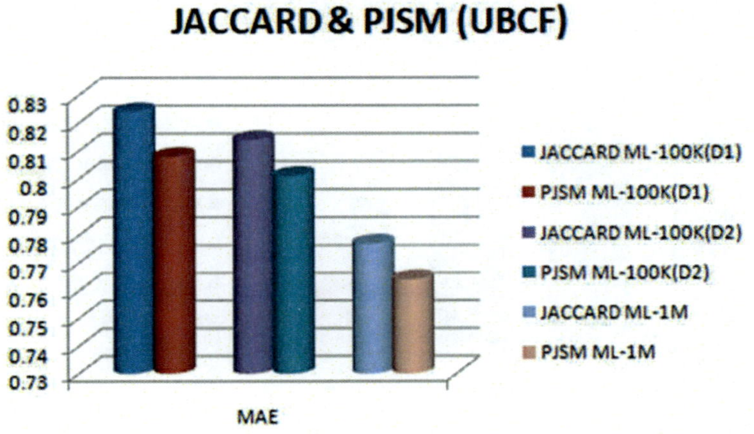

Fig. 3 MAE resultant chart using MovieLense dataset for Jaccard and PJSM UBCF method

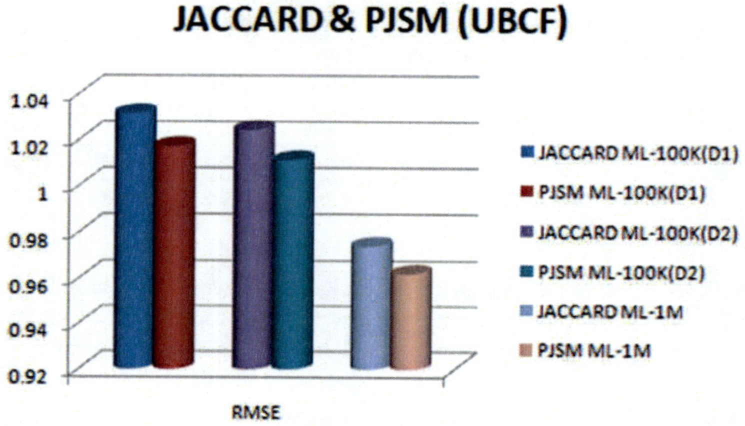

Fig. 4 RMSE resultant chart using MovieLense dataset for Jaccard and PJSM UBCF method

and new PJSM for MovieLense dataset. Our improved new similarity measure can obtain the best mean absolute error ratio as the traditional one reaches the range of 0.82, 0.81 and 0.77 and above, whereas the PJSM similarity one retains at 0.8, 0.79 and 0.76, respectively. Similarly, the root mean squared error (RMSE) ratio for the traditional one reaches the range of 1.02, 0.96 and above, whereas the prospective Jaccard (PJSM) new similarity one retains at 1.01 and 0.95, respectively. Thus, the presentation chart clearly declares that the new prospective Jaccard (PJSM) similarity measure outperforms well compared to the traditional Jaccard method.

Figure 5 illustrates the item-based collaborative filtering (IBCF) presentation chart for the MovieLense dataset. Our improved new similarity measure (PJSM) can obtain the best root mean squared error (RMSE) ratio as the traditional one reaches the

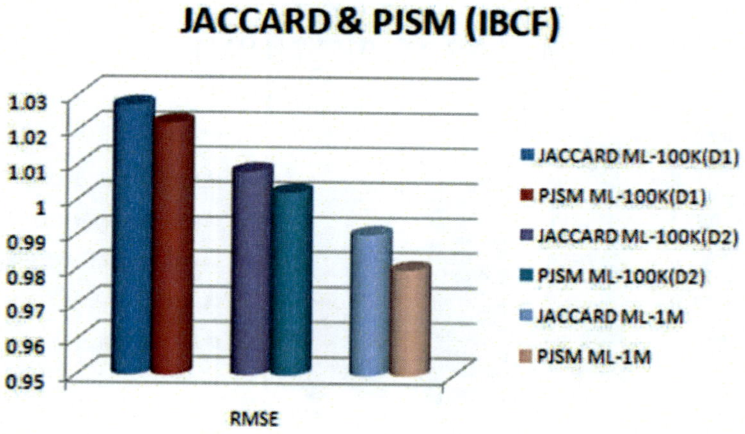

Fig. 5 RMSE resultant chart using MovieLense dataset for Jaccard and PJSM IBCF method

range of 1.02, 1.0, 0.98 and above, whereas PJSM similarity one retains at 1.01, 1.0 and 0.97, respectively. Thus, the presentation chart clearly declares that the new prospective Jaccard (PJSM) similarity measure outperforms well compared to the traditional cosine one.

5 Conclusion

The concept of e-commerce is all about surfing the Internet to do business better and faster. The conventional algorithm and the proposed PJSM algorithm in CF recommendation algorithm in calculating the user similarity index value were addressed. The tailored recommendation framework in collaborative filtering method describes the systematic approach in predicting the user's unrated data and also discussed on how to show the effectiveness for the predicted rated data. The determination of the predicted data is exposed by integrating the characteristics of the user/item similarity score set. Almost most of the existing similarity measure only considers the actual original rating data in the user–item rating matrix. The drawback addressed in traditional method provides results as absolute values. So it is difficult in finding the nearest neighbor since it shows that all are nearest one. However, the PJSM results in different USI values. So that it helps to find the nearest neighbor without any confusion. The resultant output proves that PJSM similarity method illustrates the improvements in recommendation performance by showing the satisfactory accuracy results. Future work is to carry forward the implementation in real time as an online evaluation, and its successive ratio in predicting the user's missed out rated data is evaluated. According to the proposed recommender framework, a data sparsity issue is overcome with high accuracy rate.

References

1. Mohana, H., & Suriakala, M. (2017). An overview study on web mining in ecommerce. *International Journal of Scientific Research (IJSR), 6*(8), ISSN No 2277–8179, Aug 2017.
2. Cho, Y.S., & Moon, S.C. (2016). Frequent pattern to promote sale for selling associated items for recommender in E-Commerce. *Indian Journal of Science and Technology, 9*(38). https://doi.org/10.17485/ijst/2016/v9i38/102552, oct 2016.
3. Melville, P., & Sindhwani, V. (2010). Recommender systems. In *Encyclopedia of Machine Learning*. Springer Science & Business Media.
4. Resnick, P., Iacovou, N., Suchak, M., Bergstrom, P., Riedl, J. (1994). GroupLens: An open architecture for collaborative filtering of netnews. In *Proceeding of the ACM Conference on Computer Supported Cooperative Work* (pp. 175–186).
5. Adomavicius, G., & Tuzhilin, A. (2005). Toward the next generation of recommender systems: a survey of the state-of-the-art and possible extensions. *IEEE Transactions on Knowledge and Data Engineering* (pp. 734–749).
6. Li, W., Li, X., Yao, M., Jiang, J., & Jin, Q. (2015). Personalized fitting recommendation based on support vector regression. *Human-Centric Computing and Information Sciences.*

7. Sun, S. B., Zhang, Z. H., Dong, X. L, Zhang, H. R, Li, T. J, Zhang, L., & Min, F. (2017). Integrating triangle and Jaccard similarities for recommendation. *PLoS One, 12*(8), e0183570. https://doi.org/10.1371/journal.pone.0183570.eCollection 2017.
8. Liu, H., Hu, Z., Mian, A., Tian, H., Zhu, X. (2014). A new user similarity model to improve the accuracy of collaborative filtering. *Knowledge-Based Systems, 56*, 156–166.
9. Yang, Z., Wu, B., Zheng, K., Wang, X., Lei, L. (2016). A survey of collaborative filtering based recommender system for mobile internet applications. *IEEE Access*.
10. Zhang, Z., Kudo, Y., & Murai, T. (2015). Neighbor selection for user-based collaborative filtering using covering based rough sets. SpringerLink.com. IUKM.
11. Zhang, Z. (2016). A personalized time sequence based book recommendation algorithm for digital libraries, IEEE. https://doi.org/10.1109/access.2016.2564997.
12. Li, W., Li, X., Yao, M., Jiang, J., Jin, Q. (2015). Personalized fitting recommendation based on support vector regression. *Human-centric Computing and Information Sciences, 5*, 21.
13. Mohana, H, & Suriakala, M. (2018). A prospective BI technique using INSM model to alleviate data sparsity issues in recommender system for an e-trade industry. *International Journal of Pure and Applied Mathematics, 119*(16), 305–314, ISSN: 1314-3395.
14. Gunawan, A. A. A. S., & Suhartono, D. (2016). Developing recommender systems for personalized email with big data, IWBIS. *IEEE Access*.
15. Agarwal, A., & Chauhan, M. (2017). Similarity measures used in recommender systems: A study. *International Journal of Engineering Technology Science and Research, 4*(6), ISSN 2394–3386.
16. Suganeshwari, G., & Syed Ibrahim S. P. (2018). A comparison study on similarity measures in collaborative filtering algorithms for movie recommendation. *IJPAM, 119*(15), 1495–1505, ISSN: 1314–3395.
17. Candillier, L., Meyer, F., & Fessant, F. (2008). Designing specific weighted similarity measures to improve collaborative filtering systems. *ICDM*, 242–255.
18. Bobadilla, J., Ortega, F., Hernando, A., & Bernal, J. (2011). A collaborative filtering approach to mitigate the new user cold start problem. *Knowledge-Based System, 26*, 225–238.
19. https://www.movielense.umn.edu.
20. Sarwar, B., & George, K., Joseph, K., & John, R. (2001). Item-based collaborative filtering recommendation algorithms. In *Proceedings of the 10th international conference on the World Wide Web*. ACM (pp. 285–295).

Spliced Image Detection in 3D Lighting Environments Using Neural Networks

V. Vinolin and M. Sucharitha

Abstract Digital image forensics is a trending research domain that validates the authenticity of the digital images. The traditional methods invested time on detecting the device used for capturing and identifying the traces. Nowadays, it is interesting to note that the illumination deviations in the image provide an effective trace for detecting the forgeries in the image. Accordingly, a method is developed for detecting the spliced image based on the illumination features. Initially, the human faces in the composite image are detected, and the three-dimensional model of all the faces is derived using the landmark-based 3D morphable model (L3DMM). The light coefficients are determined using the 3D shape model for extracting the features. To identify the spliced/pristine images in the input composite image, neural network (NN) is used, which is trained using the standard back-propagation algorithm. The experiments were conducted on DSO-1 and DSI-1 datasets. Performance metrics such as accuracy, true positive rate (TPR), true negative rate (TNR) and ROC are used to prove the efficiency of the proposed method.

Keywords Image forgery detection · Spliced images · Neural network · Image forensics · Three-dimensional model

1 Introduction

Images are excellent information carriers in the era of digital technology. Experts in image processing get easy access over the contents in the image without any visual traces of identification [1]. The availability of low-cost and user-friendly editing tools is never time restricted to the experts to perform tampering and counterfeit of visual content. Due to the aforementioned reasons, the modification of the images is prevailing in common than found before. Thus, digital image forensics ensures the

V. Vinolin (✉)
Research Scholar, Noorul Islam Centre for Higher Education, Kumaracoil, Tamil Nadu, India
e-mail: v.vinolin@gmail.com

M. Sucharitha
Malla Reddy College of Engineering and Technology, Maisammaguda, Telangana, India

S. C. Satapathy et al. (eds.), *Intelligent System Design*, Advances in Intelligent Systems and Computing 1171, https://doi.org/10.1007/978-981-15-5400-1_15

security of the multimedia along with digital watermarking, rendering contrast and explosive malicious manipulation in the image [2].

The prime goal of this paper is to design and develop an automatic forgery detection method using the neural networks, which is duly based on the light variations of the individual faces in the input composite image.

Neural Networks for Image forgery detection: The automatic forgery detection is performed using the standard back-propagation algorithm-based neural networks (NNs). The input composite images are subjected to the image forgery detection using the features based on the lightning coefficient variations such that the accuracy of classification is assured. The rest of the paper is structured as: The review of the forensic detection approaches based on physical detection is concentrated in Sect. 2. The proposed method of forgery detection is demonstrated in Sect. 3, and the results of the methods are depicted in Sect. 4. Finally, Sect. 5 concludes the paper.

2 Literature Review and Challenges

Image forgery detection using the transformed spaces was progressed in [3], and there were some inconsistencies when there was forged reflection in the image. The research in [4] dealt with the basic ideas related to the reflective geometry in addition to the linear perspective projection that depended truly on the concept for the occurrence of the geometric inconsistency if there is any forged reflection in the image. Since there required an extra effort for understanding the photometric inconsistencies prevailing in the shadow, the detection accuracy was poor. However, there was a need for detecting the composites in the image. The method in [5] found the physical inconsistency from the image shadings and shadows through which there was a possibility for detecting the lighting conditions in the image.

There are numerous challenges that exist in the literature in the field of image forgery detection. The main challenge in the image forgery detection was about the lack of methods to differentiate similar light sources. The second challenge was regarding the detection of the 3D model of the human face with higher accuracy. Additionally, there existed a rare persistence of the wrong detection when images with near light situations prevailed, and at last, the conventional methods based on physical approaches suffered as a result of the high computational complexity when there was a requirement of extra images. Apart from everything, the existing methods insisted that the applicability of the automated methods would be an effective solution in case of the online applications and promoted us to establish an automatic method based on machine learning methods to find the inconsistencies in the lighting conditions [6].

3 Proposed Method of Forgery Detection in Spliced Images

Figure 1 shows the block diagram of the forgery detection strategy using the neural networks. Initially, the input composite image is subjected to face detection using the Viola–Jones algorithm [7], and the faces are subjected to the generation of the 3D model using the standard L3DMM [8]. The 3D model of the face images enables the effective evaluation in order to perform the forgery detection for which initially the lightning coefficients of the images are determined. Using the light coefficient matrix, the features, such as the Euclidean distance, Seuclidean distance, Bhattacharyya distance, Hamming distance, Chebyshev distance and correlation coefficient are determined. Using the features, the forgery images are determined using the neural networks that are the simple and easy means of detection.

3.1 *Detection of Face and Feature Extraction*

Consider the input composite image as *M*, to which the Viola–Jones algorithm is applied for detecting the faces. Once the faces are detected, the 3D model is generated using the L3DMM, which facilitates the extraction of the effective features for forgery detection. Thus, the light coefficient matrix is determined for all the faces in the input composite image, and the features are extracted. The feature vector is represented as,

$$f = \{f_1, f_2, \ldots, f_6\} \tag{1}$$

where $f_1, f_2, \ldots, f_6$ refer to the distances, such as Euclidean, Seuclidean, Bhattacharyya, Hamming, Chebyshev, and correlation coefficients. f forms the input to the NNs, and using the features of the individual faces, the spliced image is detected to mark the effective forgery detection.

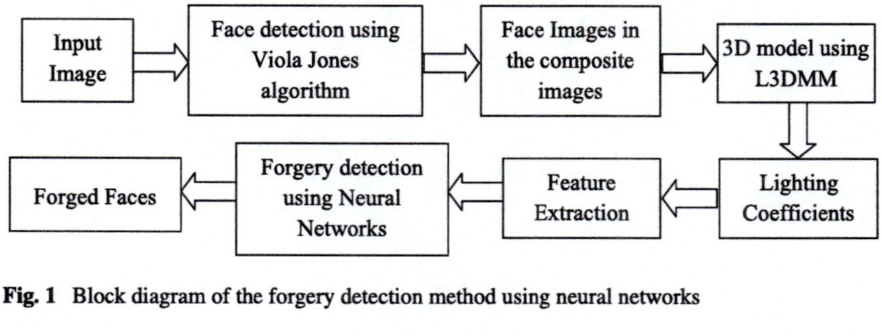

Fig. 1 Block diagram of the forgery detection method using neural networks

3.2 Image Forgery Detection Using Neural Networks

The image forgeries are detected using the NN [9] that is trained using the standard Levenberg–Marquardt (LM) back-propagation algorithm [10]. The input to the NN is the feature vector of the individual faces present in the input composite image. Figure 2 shows the LM-based neural network that possesses three functional layers, such as an input layer, hidden layers and the output layer. The input features are processed in the input layer, where the input features are normalized that restricts the features within the predefined limits of the activation function, which contributes to the precision of the entire network. Hidden layers extract the features for analysis and are responsible for the internal functioning of NN, and the output layer generates the final output.

Let us assume that there are k input neurons represented as,

$$I = \{I_1, I_2, \ldots I_i, \ldots I_k\} \tag{2}$$

where k refers to the total input neurons in NN. The hidden layers are represented as,

$$H = \{H_1, H_2, \ldots H_i, \ldots H_z\} \tag{3}$$

where z is the total hidden neuron present in NN and ith hidden layer is given as,

$$H_i = \frac{1}{k}\sum_{i=1}^{k} W_i I_i \tag{4}$$

where W_i is the weight in the ith input neuron. The output layer is denoted as,

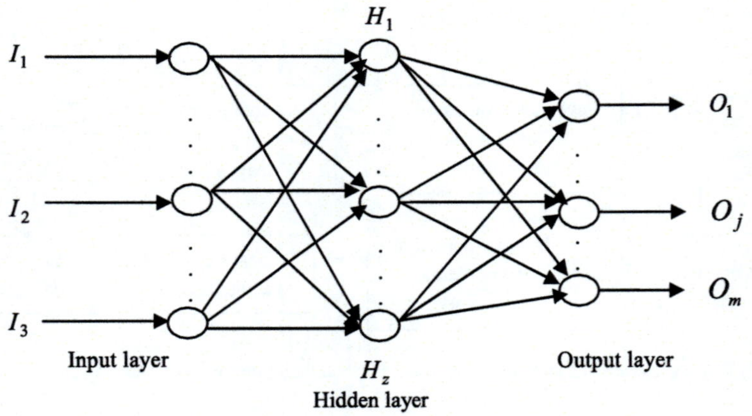

Fig. 2 Architecture of LM-based NN

$$O = \{O_1, O_2, \ldots, O_j, \ldots, O_m\} \quad (5)$$

where m is the total output neurons and O_j is the jth output neuron in NN. The weights of NN are given as,

$$W = \{W_1, W_2, \ldots, W_N\} \quad (6)$$

where N is the total weights in NN.

The output of NN is calculated as,

$$O_i = \sum_{i=1}^{z} H_i \times W_i \quad (7)$$

where O_i is the output neuron and H_i is the ith hidden layer. Thus, the output of the layer is represented as $O_i = \mathrm{Fn}(W_t, I)$. Accordingly, the output layer is represented as the function of weights and the input neurons, and W_t denotes the weight of the current iteration.

If the output of NN corresponding to an image acquires the value '1', the presence of the spliced image is confirmed or else the image is the pristine image. It is worth interesting to note that if any of the individual images in the input composite image is found to be spliced, the input composite image is marked as the forged image.

(a) **Training Algorithm for NN**

Step 1 *Initialization*: Initialization as the first step, initializes all the weights in NN, which is denoted as, $W = \{W_1, W_2, \ldots, W_N\}$ and in addition, set the learning rate and delay rate as, $\gamma_t = 0.1$ and $\delta_t = 0.1$, respectively.

Step 2 *Compute the output of NN*: To acquire the output from NN, the weights are provided corresponding to the input of NN. Thus, the output $f(\mathrm{net})$ is determined to find the value of O_i^t.

$$f(\mathrm{net}) = \begin{cases} 1 \text{ if } W^T I \geq \theta \\ 0 \text{ otherwise} \end{cases} \quad (8)$$

where $f(\mathrm{net})$ refers to the scalar product of the input features with the corresponding weights of NN and $f(\mathrm{net})$ acquires the value as one when the scalar product exceeds the activation function and $f(\mathrm{net})$ is zero whenever the scalar product lies below one. Thus, the output of NN is $O_i = Fn\,(W_t,\ I)$.

Step 3 *Evaluate the sum of squares for all the input neurons*: The sum of squares otherwise corresponds to the error, and the weight is decided based on the minimal value of the error.

$$\varepsilon = \sum_{j=1}^{i} \left(O_i^t - G_i\right)^2 \quad (9)$$

where O_i^t is the output of the ith neuron and G_i is the targeted output of NN.

Step 4 *Compute the incremental weights using LM:* The incremental weight is calculated for efficient recognition that is computed based on the Jacobian matrix and the learning rate.

$$\Delta W = \left[J^T J + \gamma_t I\right]^{-1} \cdot J^T \varepsilon \tag{10}$$

where J is the Jacobian matrix,γ_t is the learning rate and ε is the error at the current iteration.

Step 5 *Update the new weights*: The new weights are updated using the LM algorithm, which is the sum of the current iteration and the incremental weight. The weights of NN are determined as,

$$W_{t+1} = W_t + \Delta W \tag{11}$$

where W_{t+1} is the new weight determined using the incremental weight and W_t is the weight of the previous iteration.

Step 6 *Determine the output of NN using the updated weight*: Once the new weights are updated, the error is determined and the output is computed as,

$$O_i^{t+1} = F(W_t,\ I) \tag{12}$$

Step 7 *Recompute the error*: The error between O_i^{t+1} and the output of the previous iteration O_i^t is calculated using the sum of the square errors as,

$$\varepsilon_{t+1} = \sum_{j=1}^{i} \left(O_i^{t+1} - O_i^t\right) \tag{13}$$

where ε_{t+1} is the error computed using the output of O_i^t and O_{i+1}^t.

Step 8 *Update the learning rate*: The weights of NN determine the output, and the learning rate gets updated at the end of every iteration as given by,

$$\gamma_{t+1} = (\gamma_t * \delta) \tag{14}$$

where γ is the learning rate, ρ is the adjustment factor and δ is the delay rate.

Step 9 *Terminate*: The above steps are repeated for the maximal number of iterations, and the weights and bias are determined for finding the forged image.

4 Results and Discussion

4.1 Experimental Set up

The implementation is made in MATLAB 2018a, and the system utilized for the implementation consists of i5 processor with 4GM RAM. The datasets used for the analysis include DSO-1 and DSI-1 [11]. DSO-1 consists of 200 indoor–outdoor images with the image resolution of 2048 × 1536 pixels. Out of the 200 images in the database, there are about original 100 images and 100 forged images. DSI-1 possesses a total of 50 images with 25 original images and 25 doctored images. The metrics used for comparing the works include accuracy, TPR and TNR.

$$\text{Accuracy} = \frac{\text{TP} + \text{TN}}{\text{TP} + \text{TN} + \text{FP} + \text{FN}} \tag{15}$$

$$\text{TPR} = \frac{\text{TP}}{\text{TP} + \text{FN}} \tag{16}$$

$$\text{TNR} = \frac{\text{TN}}{\text{TN} + \text{FP}} \tag{17}$$

TP, FP, TN, FN signify true positive, false positive, true negative, false negative, respectively.

4.2 Performance Analysis

In this section, the analysis of the methods based on the two datasets is deliberated, and the analysis is progressed depending on the performance metrics. The effectiveness of the proposed method is established through the comparison of the proposed method with the existing methods, such as Kee and Farid's [12], shape-from-shading (SFS) algorithm [13], Random Guess [6] and Peng et al. [6]. Figure 3 shows the comparative analysis for DSO-1 dataset, and Fig. 4 shows the comparative analysis for DSI-1 dataset. The accuracy, TPR, TNR and ROC values of DSO-1 dataset are 84.5%, 92.8%, 87.3% and 91%, respectively, and that of DSI-1 dataset is 76.9%, 87.7%, 54.3% and 87.7%, respectively. The performance analysis is shown in Table 1, which reveals that the proposed neural network method outperforms the existing methods in terms of all evaluation metrics.

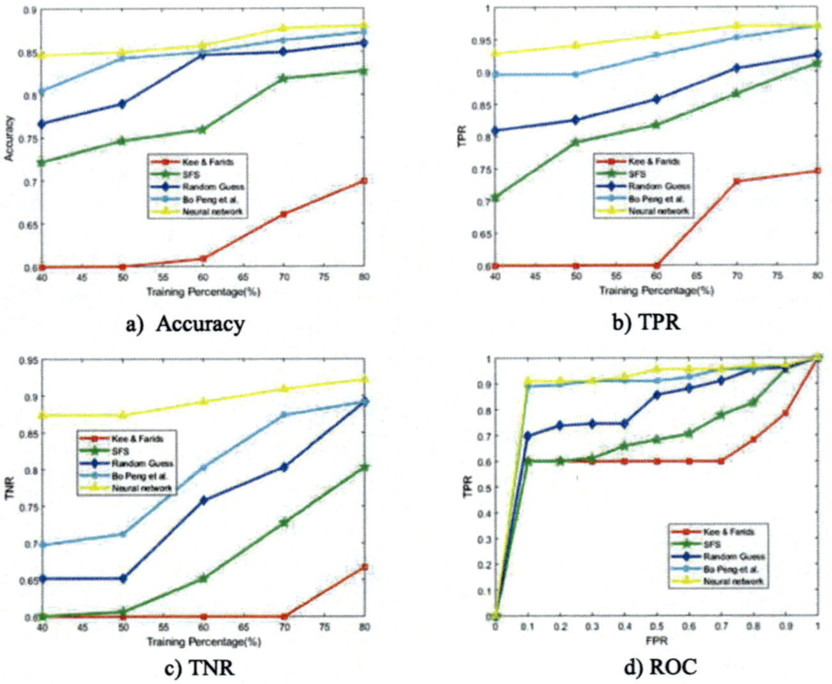

Fig. 3 Comparative analysis of the methods using DSO-1 dataset

5 Conclusion

In this paper, image forgery detection is progressed using neural networks. The variation in the illumination of the face images in the input composite image marks the presence of the spliced images for which the light coefficients are extracted for the individual faces in the image. The features are extracted for all the face images, which include six distance measures, and any deviation in these features is investigated effectively using the neural networks to acquire the surety regarding the presence of spliced or pristine images. The neural networks are trained based on the back-propagation algorithm, and the accurate detection is assured using the neural networks. The analysis of the methods is carried out using the DSO-1 and DSI-1 to prove the efficiency of the proposed approach. The future dimension can be based on any of the hybrid optimization-based deep learning methods.

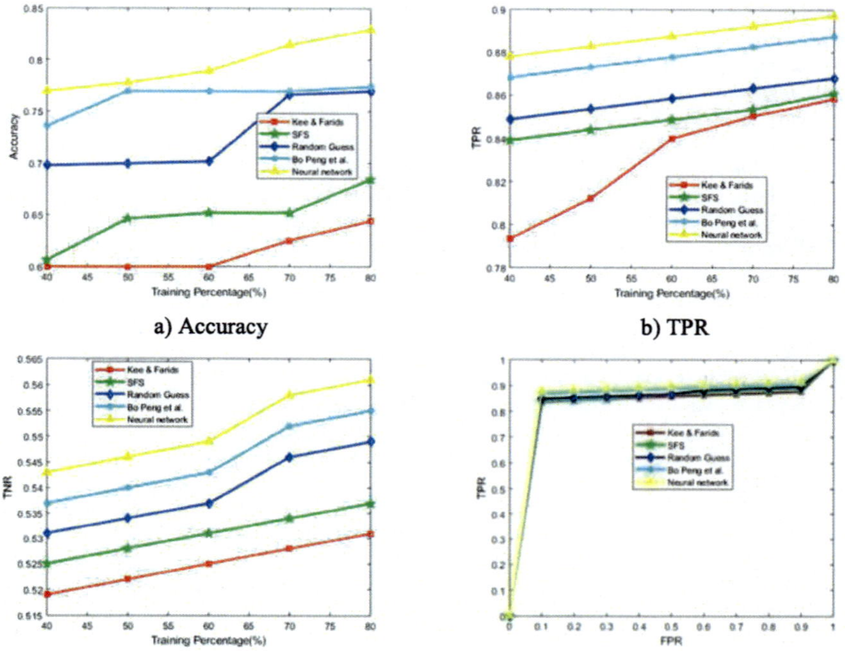

a) Accuracy

b) TPR

c) TNR

d) ROC

Fig. 4 Comparative analysis of the methods using DSI-1 dataset

Table 1 Performance analysis

	Accuracy		TPR		TNR		ROC	
	DSO-1	DSI-1	DSO-1	DSI-1	DSO-1	DSI-1	DSO-1	DSI-1
Kee and Farid's [12]	0.600	0.600	0.600	0.793	0.600	0.519	0.600	0.836
SFS [13]	0.721	0.606	0.700	0.839	0.600	0.525	0.600	0.839
Random Guess [6]	0.766	0.697	0.809	0.848	0.651	0.531	0.698	0.848
Peng et al. [6].	0.804	0.735	0.895	0.868	0.697	0.537	0.888	0.868
Neural network	0.845	0.769	0.928	0.877	0.873	0.543	0.910	0.877

References

1. Zhou, J., Ni, J., & Rao, Y. (2017). Block-based convolutional neural network for image forgery detection. In *International Workshop on Digital Watermarking* (pp. 65–76). Springer, Berlin.
2. Redi, J. A., Taktak, W., & Dugelay, J.-L. (2011). Digital image forensics: A booklet for beginners. *Multimedia Tools and Application, 51*(1), 133–162.
3. Carvalho, T., Faria, F. A., Pedrini, H., Da Torres, R. S., & Rocha, A. (2016). Illuminant-based transformed spaces for image forensics. *IEEE Transactions on Information Forensics and Security, 11*(4), 720–733.

4. Liu, Q., Cao, X., Deng, C., & Guo, X. (2011). Identifying image composites through shadow matte consistency. *IEEE Transactions on Information Forensics and Security, 6*(3), 1111–1122.
5. Bermano, A. H., et al. (2014). Exposing photo manipulation from shading and shadows. *Siggraph, 1*(212), 1–12.
6. Peng, B., Wang, W., Dong, J., & Tan, T. (2017). Optimized 3D lighting environment estimation for image forgery detection. *IEEE Transactions on Information Forensics and Security, 12*(2), 479–494.
7. Viola, P., Jones, M. (2001). Rapid object detection using a boosted cascade of simple features. In *Proceedings of the IEEE Computer Society Conference on Computer Vision and Pattern Recognition. CVPR*, December 8–14, 2001.
8. Peng, B., Wang, W., Dong, J., Tan, T. (2016). Automatic detection of 3d lighting inconsistencies via a facial landmark based morphable model. In *IEEE International Conference on Image Processing (ICIP)*, September 2016.
9. Tahmasebi, P., & Hezarkhani, A. (2012). A hybrid neural networks-fuzzy logic-genetic algorithm for grade estimation. *Computers & Geosciences, 42,* 18–27.
10. Sapna, S., Tamilarasi, A., Pravin Kumar, M. (2012). Backpropagation learning algorithm based on Levenberg Marquardt algorithm. *CS & IT-CSCP*, pp. 393–398.
11. DSO-1 and DSI-1 dataset, https://recodbr.wordpress.com/code-n-data/#dso1_dsi1. Accessed on December 2018.
12. Kee, E., Farid, H. (2010). Exposing digital forgeries from 3-D lighting environments. In 2010 *IEEE International Workshop on Information Forensics and Security (WIFS), IEEE, Conference Proceedings* (pp. 1–6).
13. Wei, F., Kai, W., Cayre, F., Zhang, X. (2012). 3D lighting-based image forgery detection using shape-from-shading. In *Signal Processing Conference (EUSIPCO), 2012 Proceedings of the 20th European, Conference Proceedings* (pp. 1777–1781).

Two-Level Text Summarization Using Topic Modeling

Dhannuri Saikumar and P. Subathra

Abstract Since the dawn of the Internet, the size of textual data has been steadily growing, every single day, due to frequent usage of digital libraries, social media, and online search engines, concomitant with storage of a mountainous amount of raw text. Gleaning useful content toward generation of a credible summary is a challenging task. This work reports on the implementation of a two-level summarization document, using latent Dirichlet allocation (LDA). The proposed method is constituted of two steps: the first step involves usage of maximal marginal relevance (MMR) and (b) text rank (TR) summarization procedures. These techniques generate summaries of a document, for a given corpus with multiple topics. In the second step, LDA modeling algorithm is applied to the summaries generated by MMR and TR, in the previous step. This process generates more shortened summaries, with differentiated topics. We used customers opinion reviews on products, hotels. [Section 4.1] as input corpus. The performance of this two-level document summarization (DS) using LDA is compared with MMR and TR. The comparison results show the two-level document summarization using LDA generates better summaries Sect. 5.

Keywords Text mining · Text summarization · Topic modeling

1 Introduction

Text mining (TM) [1] is undertaken to find relevant information, and it is a way toward examining extensive accumulations of composed assets to produce new data. Text mining's main aim is to change the unstructured text into organized or structured information for use in further investigation. The increasing availability of text requires new techniques to convert raw data into useful information. Text mining includes a large number of tasks that can be distinguished into four stages: (a). information retrieval (IR), (b). natural language processing (NLP), (c). information extraction (IE), and(d). data mining (DM).

D. Saikumar · P. Subathra (✉)
Department of Computer Science and Engineering, Amrita School of Engineering, Coimbatore, Amrita Vishwa Vidyapeetham, Coimbatore, India
e-mail: p_subathra@cb.amrita.edu

S. C. Satapathy et al. (eds.), *Intelligent System Design*, Advances in Intelligent Systems and Computing 1171, https://doi.org/10.1007/978-981-15-5400-1_16

The first stage, IR, is to find the required document from the collection of online text documents. For example, web search engine Google, Bing, etc.

The second stage, NLP, is one of the components of artificial intelligence; it can understand the natural language like humans not exactly but near to it. NLP makes us classify the words into the grammatical way and removes the uncertainty of meaning from words among the multiple meaning, etc.

The third stage, IE, is for changing unstructured into structured data; this level is called extraction of information. The structured information it is going to store in the database in a structured way. To identify the specific terms in a document, we are using this IE. IE maintains the relationship between names and entities.

The fourth stage, DM, uses the extracted information from the previous stage to convert into useful knowledge. The major tasks of DM are mining association, classification, prediction, and clustering. There are different text mining techniques that are available, to bring useful information from the collection of raw texts.

The following are text mining techniques used for various applications, text normalization, text classification, text clustering, topic modeling, and document summarization.

Text normalization [2] is used to convert the text into the standard form. This is normally used to convert the text to speech and used for automatic speech recognition. Text classification is a fundamental task of NLP; based on the content it will assign the tags to the text it will use in an application like sentiment analysis, spam detection, etc. Text clustering uses the NLP and machine learning to understand the unstructured data, and also, it used to categorize the text data. Topic modeling is used to find the abstract or hidden themes from the particular document; LDA is one example of topic modeling. Document summarization is used to summarize the documents or texts; in the following sections, we gave more details about it. In this work, we proposed two-level document summarization by using topic modeling algorithm called latent Dirichlet allocation (LDA), to improve the summarization of documents. The opinion dataset is used for summarization. The results of LDA with summarization techniques are compared with the results of summarization algorithms maximal marginal relevance (MMR) and text rank (TR). The results show that our proposed algorithm performance better than the other summarization techniques.

2 Related Work

2.1 *Document Summarization*

Document summarization [3, 4] is one of the text mining techniques. The objective of automatic text summarization is introducing the source text into a shorter variant with semantics. The most essential preferred standpoint of utilizing a synopsis is that it diminishes the perusing time. There are five types of summarization methods that are available; those are: (a) abstractive and extractive, (b) indicative and informative,

(c) generic and query based, (d) single document and multi-document, and (e) mono-lingual and multi-lingual.

A. **Abstractive and Extractive**: These summarization methods are called approach-based summarization techniques. An abstractive summarization gives the summary by changing the content in the input documents. Extractive summarization strategy comprises of choosing essential sentences, sections, and so on from the first record and linking them into a shorter frame.
B. **Indicative and Informative**: These are based on the type of detail summarization techniques. The objective of indicative summarization provides only the main idea of the, and based on that, it gives the fundamental thought of the original content. Informative summary fills in as a substitution to the original record. It gives brief data about the original document to the user.
C. **Generic and Query Based**: These are content-based summarization techniques. Generic summarization approach can be utilized by the user and the outline of document or content independent upon the subject of the report. All the data is at the same level of significance which is not user particular. Query-based summarization is like a question and answers compose where the outline is the consequence of query.
D. **Single Document and Multi-document**: These summarizations are based on the number of input documents. In the single document, we are passing single document as input and we will get summary for that document only. In multi-document summarization, we are passing multiple documents means corpus as input and we will get summary for the whole corpus.
E. **Mono-lingual and Multi-lingual**: These summarizations are language-specific summarization techniques. Mono-lingual summarization accepts the document with specific language, and it will generate an output based on that language only. Multi-lingual summarization accepts the document with several languages. It is difficult to implement because the input document contains a different type of languages.

2.2 Maximal Marginal Relevance (MMR)

MMR [5] is an automatic summarization technique which is frequently used to summarize the documents. We can use the MMR either in between word or in between words. By using a unit step function of maximum marginal relevance is decided about the sentences which are required in the summary. This unit step function will provide the importance of the sentences. The system likewise had a database of words which are futile in the report and regardless of whether they wipe out cannot affect the significance of primary file. The MMR algorithm contains the following steps.

a. Give input document which we need to summarize.
b. Now the algorithm will start, and it will go through the entire document to compare the word in the document with database words after that it will remove the unnecessary words from the document.

c. It starts with the first sentence in the document and it will run until the document ends.
d. To summarize the most extreme number of words in document, we are utilizing the unit step work, and also it will compute the useful pertinent data.

The following notations are used in MMR:
I: input document
A: individual arrangement of sentences in doc d
N: most extreme group of individual data permitted
R: a collection of repeated sentences which are to be removed in the doc d.

The MMR procedure:

i. Initialize k= 0; AI, k is constant.
ii. Repeat Sd = Sd + R(I), the unnecessary words to be removed from the document.
iii. While (Sd) k size is less than N.
iv. Select the next unit u(k+ 1) according to the above equation.
v. A(k+ 1) = Ak U A(k+ 1).
vi. k= k+1.
vii. End.

2.3 *Text Rank*

Text rank [6, 7] is one of the graphs based ranking algorithm. It is mainly working based on page rank algorithm which is founded by Google. The page rank algorithm is used to find out the ranking of the web pages based on the ranking of web pages. In the text rank instead of web pages, we will find out the ranking of sentences. The basic idea of implementing the graph-based ranking algorithm is for voting or recommendations; it is basically casting a vote for another vertex. The one vertex which got the highest number of votes that will be the most important vertex.

This algorithm is mainly used to identify the most relevant sentences and relevant keywords. With the end goal to locate the most important sentences in the content, a graph is developed where the vertices of the diagram display to each sentence in a document. The edges between sentences depend on the substance cover, to be specific by ascertaining the number of words that two sentences share for all intents and purpose. Based on the votes or ranking, it will generate the summary.

By observing Fig. 1, we can understand the work flow of text rank algorithm.

Text rank algorithm contains the following steps:

i. Give the collection of text documents as input.
ii. Now the algorithm goes to the entire corpus, and it will divide the text into sentences.
iii. In the third stage, all the sentences convert into vector form so that it is easy to compare sentences.

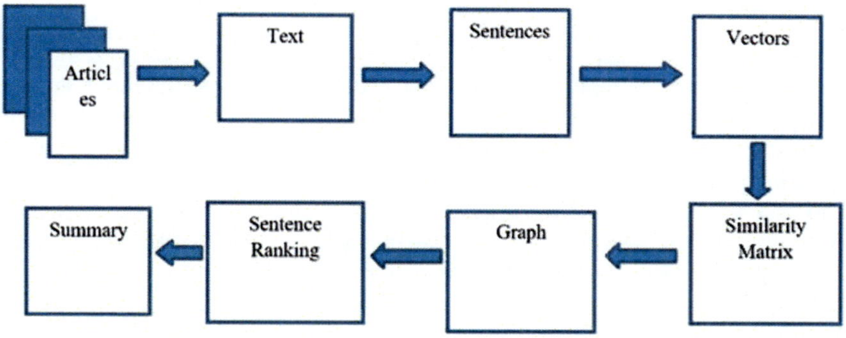

Fig. 1 Text rank process flow [8]

iv. In the next step, it will compare all sentences by using the cosine similarity, and it will find out the similarity between each sentence, and it generates similarity matrix.
v. In the fifth stage, based on the similarity matrix, we can find the connection between the two sentences so that it will generate a connected graph. Here, each sentence links to another [in the graph, each vertex is a sentence].
vi. Based on the number of links or connection between the sentences, it provides the ranking, means the vertex which is having more number of connections or links can have higher importance.
vii. Based on the rankings or score of a vertex, it will sort the sentences and make a summary. By using the following equation, we can assign ranking or score of a vertex.

$$S(V_i) = (1 - d) + d * \Sigma(jsI_n(V_i))\frac{1}{|\mathrm{Out}(V_i)|}S(V_i)$$

V_i = Vertex which is given.
In (V_i) = Incoming vertices to *Vi* (predecessors).
Out (V_i) = Outgoing vertices from *Vi* (successors).
D = Damping factor [value can be set between 0 or 1].

In our work, we followed multi-document and extractive summarization approaches, and we were using the MMR and text rank algorithms for summarization. To improve these summaries, we are applying second-level summarization with topic modeling. For this second-level summary, we are using topic modeling algorithm called latent Dirichlet allocation (LDA).

2.4 Topic Modeling

The primary goal of topic modeling [9, 10] is to discover the hidden themes in a corpus. We can define topic modeling in different ways like in dimensionality reduction, unsupervised learning, and form of tagging. There are different techniques available for archiving the topic modeling, in that we are focusing on LDA. Latent Dirichlet allocation (LDA) [8, 11] is one of topic modeling technique. It is also a generative probabilistic topic model. LDA, first presented by Blei, Ng, and Jordan in 2003 [12], is a standout among the most well-known techniques in point displaying. LDA speaks to themes by the probabilities of word. In each topic, the words with the most outstanding probabilities give more often than not give a smart thought of what the topic is can offer LDA word likelihoods. Consider a corpus D, and it contains M documents, and each document d having the total number of words *N* (d 1,…, M); LDA models the corpus (D) according to the following way.

LDA is a one of the distinguished tools for distribution of a latent topic for a big corpus. Therefore, LDA has the excellent power to find the sub-topics either in document or in the whole corpus. LDA area is composed of many patents, and each parent represents with an array of topic distributions. By using the LDA, the words in the corpus produce a vocabulary, and again, this is used to generate the latent topics for collection of documents. Generally, the LDA consider each document as a mixture of topics, where a topic is a probability distribution over this set of terms. Each document then looks like as a probability distribution over the collection of topics. We can think that generative process is providing the data and then that is defined by the joint probability distribution over what is observed and what is hidden. The LDA displays above forms an inactive subject layer. The topic quantity is much lesser than words. In the content of the document, topics have the profound association. So, it is match for being the establishment of the sentence articulation. In the space comprised of a point, we can elaborate the word, sentence, archive, corpus as a uniform articulation (Fig. 2).

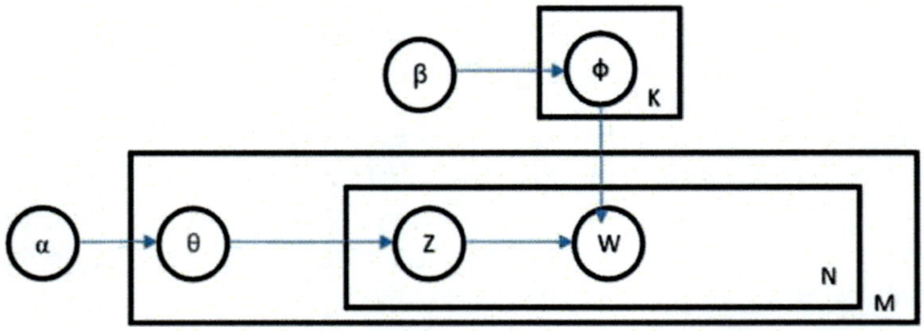

Fig. 2 Notations used in LDA [13, 14]

Notations used in LDA:

K = number of topics in the collection.
Z = topic index for the word w_i.
α = parameter of the per-document-topic distributions.
β = parameter of the per-topic-word distribution.
φ_k = word distribution for topic k.
θ_m = topic distribution for document m.
N = number of words.
W = words per document.

Steps followed to implementing the LDA:

i. Give the MMR or text rank output as input to the LDA.
ii. Divide the whole text into tokens.
iii. LDA generates document term matrix; here, it will form document to topic and topic-word distribution.
iv. It will iterate through whole text and provides the probability to each word by using probability of word given in topic P (word/topic) and to each topic by probability of topic given in a document P (topic/document).
v. Based on these probabilities, LDA generates the topics. We must set the number of topics we need based on that it will generate the topics.

3 Proposed Work

Figure 3 explains the proposed work.

Here, we are taking opinion data set as corpus; in this, we have fifty different topics and we are doing the prepossessing for this corpus. At this stage, we are deleting

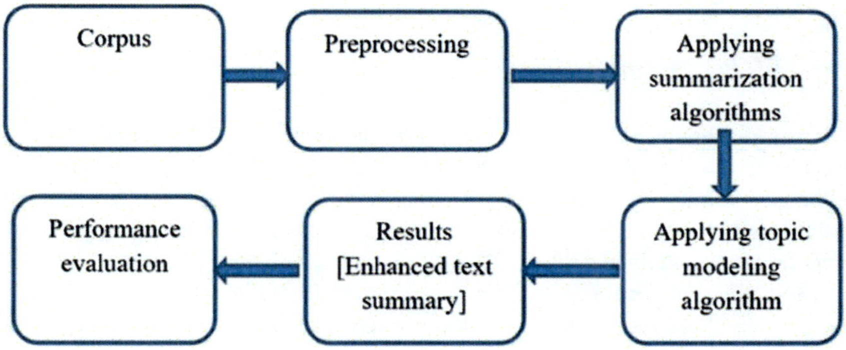

Fig. 3 Architecture of proposed work

the unnecessary symbols, white spaces, etc. Next stage is applying summarization algorithms, To convert the lager documents text into shorter and condensed form, here we are applying the two automatic document summarization techniques called MMR and text rank; after cleaning the documents, we are passing it as input to these summarization algorithms; at this stage, we will get the summary of whole corpus, By applying the summarization techniques, we are getting the short summaries, and this summary may be having different topics, so user may not able to get what does that summary up to. At this stage, we are completing one level of our work. Next level is applying topic modeling algorithm. So here. we are trying to improve the summaries by applying the topic modeling technique called LDA. For LDA, we are passing output of the summarization algorithms as input, means the summaries of MMR and text rank. LDA divides text into different topics; it will generate the frequently used words in that topics based on the document-topic and topic-word probability distributions. By this, the user may get proper idea about the summary. At the end, we are getting enhanced summary as final output; in the final stage, we are doing performance evaluation to check the performance of the summary.

4 Implementation

4.1 Data Set Description

We are using the customer's opinion reviews as data set. It contains fifty documents; each document contains individual topics, and each topic explains about reviews of the products. For example, review of a hotel, kindle, gps, etc. These reviews are generally taken from online shopping websites like Amazon, Flipkart, etc. Each topic contains nearly three pages of content.

4.2 Preprocessing

Data preprocessing [15] is one of the text mining techniques. The preprocessing is mainly used to convert the raw data into simple and understandable format, and it eliminates the stop word unnecessary symbols. Simply saying, it will clean the raw data. Here, we mainly do stop words removal, lemmatization, and it will remove unnecessary symbols like (,(,:¡).

Main steps in preprocessing are tokenization, stemming, and parts of speech tagging. In tokenization phase, we will convert text into sentences and sentences into words, and stemming will be reduced; the word length means consider words like consist, consisting, consists; if we do stemming on the above word, it will give only one word apart from three words like consist. After completing the stemming step, we will assign pats of speech tagging to each stemmed word.

4.3 *Tool Description*

We are using python, Natural Language Toolkit (NLTK); it is one of powerful toolkit which contains libraries of NLP, and these libraries are helpful to machine for understanding the human language; here for the prepossessing techniques, we are mainly using this tool and also NumPy and SciPy; these are mainly used for faster manipulation of array dimensionality. Python is coding language for implementing the summarization and topic modeling algorithms.

5 Results

In this work, we are using 50 different topics of corpus; each topic tells about the reviews of different items; it is hard to go through whole document for reading the reviews of those items.

5.1 *MMR Results*

Each document contains more than one page of content; we are passing these documents as input to the MMR, and finally, we are getting the summary, but it is combination of different topics, so it somewhat difficult to read and understand and it is long too. [In our case, we are considering more than 20 lines summary because here we are getting the summary of different topics, so if we take 5- or 10-lines summary, it is difficult to read, and it will not give proper result]. For our understanding in Table 1, we are showing MMR results of three documents only.

5.2 *Text Rank Results*

Table 2 explains about text rank results. Here also, we are passing corpus as input to the text rank algorithm, and we are getting single summary as output.

5.3 *Two-Level Document Summarizations Using LDA*

Table 3 showing the output of MMR and results of LDA for MMR output. For MMR, we are passing the fifty different topics as input to the MMR for summarization, and we are doing a summary for the whole corpus. By observing the output of MMR Check in Table 3, we can say it still not specifiable because it is still not condensed,

Table 1 Results of MMR

Input document	MMR results
Reviews about GPS. kindle and hotel rooms	In closing, this is a fantastic GPS with some very nice features and is very accurate in directions and like the easy to read graphics, the voice used to tell you the name of the street you are to turn on it is a nice adjunct to a travel trip and the directions are accurate and usually the quickest, but not always. Accuracy is determined by the maps. The rooms were nice, very comfy bed and very clean bathroom. There is a wall mounted hair dryer in the bathroom. The only possible complaint was that the sink in the bathroom had a crack in it a king bed. It's also easy to charge the I charged this thing for a couple of hours, hours ago it's still running and just at of full power plus the Kindle Battery life is days longer per charge the original Kindle had a removable battery brought long ago and you could power source for an extended time Battery life is very good

Table 2 Results of text rank

Input documents	Text rank results
Reviews about GPS, kindle and hotel rooms	The room was not overly big, but clean and very comfortable beds, a great shower and very clean bathrooms. Large comfortable room, wonderful bathroom with a king bed, and a truly claustrophobic bathroom well-appointed with nice refrigerator and toiletries, and is very, very accurate. In closing, this is a fantastic GPS with some very nice features and is very accurate in directions. But, it's always very accurate. The map is pretty accurate and the Point of interest database also is good very easy and quite accurate to use. estimates of mileage and time of arrival at your destination. I can't believe how accurate and detailed the information estimated time of arrival. It's also easy to charge Kindle in the car if you have a battery charger with a USB port Short battery life. I charged this thing for a couple of hours, hours ago it's still running and just at of full power plus the Kindle Battery life is days longer per charge. The original Kindle had a removable battery and you could buy extra batteries in case you'd be away from a power source for an extended time Battery life is very good, even with the wireless on constantly

Table 3 Comparison of MMR results and summary after applying LDA to text rank summary

Input	MMR results	Summary after applying LDA to MMR suminarv
Reviews about GPS, kindle, and hotel rooms	In closing, this is a fantastic ***GPS*** with some ***very*** nice features and is very ***accurate*** in **directions** and like the **easy** to read ***graphics*** and ***operate***, the voice used to tell you the name of the street you are to turn on it is a ***nice*** adjunct to a ***travel*** trip and the directions are accurate and usually the quickest, but always ***Accuracy*** is determined by the maps The ***rooms*** were nice in ***hotel***, very comfy bed and very ***clean bathroom*** There is a wall ***mounted*** hair ***dryer*** in the bathroom The only possible complaint was that the ***sink*** in the bathroom and truly ***claustrophobic*** bathroom had a ***crack*** in it a ***king bed***. It's also ***easy*** to ***charge*** the I charged this thing for a couple of hours, hours ago it's still running and just at of full power Plus the ***Kindle Battery life*** is days longer per charge original Kindle had a removable battery I ***brought*** this kindle ***long ago*** and you could power source for an extended time Battery life is ***very good***	Kindle battery hours good kindle life power good brought long ago easy charge nice graphics GPS maps very accurate determined gives accurate directions nice travel with good accuracy easy operate hotel clean bathroom good king bed sink had crack mount dryer truly enough space claustrophobic room

and it contains the mixture of topics, so we came up with an idea; i.e., by applying topic modeling techniques, we can reduce the summary. So, we chosen LDA to reduce the summary of MMR; it is one of the topic modeling technique. By applying the LDA, we are reducing the summary and differentiating the topics in the summary. We cannot apply LDA directly to the primary documents.

By applying the LDA directly to the primary document, we will not get a proper result; it will give repeated words, and mainly, we need a summary not a collection of a mixture of words. Apart from this, LDA is not a summarization algorithm. So, for this reason, we are doing the summary first and applying the LDA to that summary; after that, we are framing sentence; it is more like filtering. For text rank also, we are following the same approach. For the results, see Table 4. [Here, highlighted italic words are retrieved by LDA from the summary of MMR].

Table 4 Comparison of text rank results and summary after applying LDA to text rank summary

Input	Text rank results	Summary after applying LDA to text rank summary
Reviews about GPS, kindle, and hotel rooms	The ***hotel room*** was overly big, but ***clean*** and ***very comfortable bed***s, a great shower and very clean bathrooms. Large comfortable room, wonderful ***bathroom*** with a king bed, and a truly ***claustrophobic*** well-appointed with ***nice refrigerator*** and ***toiletries***, nice ***towels*** and is very, very accurate. In closing, this is a fantastic ***GPS*** with some very nice features and is very ***accurate*** in directions. But, it's always very accurate. The ***map*** is pretty accurate and ***great*** the Point of interest ***database*** also is ***good*** very easy and quite accurate to use ***estimates*** of ***mileage*** and gives ***detailed*** map time of arrival at your destination. I can't believe how accurate and detailed the information estimated time of arrival, it's also easy to charge Kindle in the car if you have a battery charger with a USB port Short battery life I charged this thing for a couple of hours, hours ago it's still running and just at of full power Plus the ***Kindle Battery life*** is days longer per ***charge*** The original Kindle had a ***removable*** battery and you could buy extra batteries in case you'd be away from a power source for an extended time Battery life is ***very good***, even with the wireless on constantly ***brought*** it ***long*** time ago	Hotel decorated claustrophobic bathroom clean very comfortable bed big toiletry shower bath refrigerator towels nice, map accurate GPS easy use for direction friendly estimates mileage database good great detailed map kindle battery life long charge time easy charge brought long ago removable battery very good

Here, our main intention is to compare the results of MMR and text rank with the LDA output for the MMR and text rank result. So, we haveshown in the tables. In Table 3, we have comparison between the MMR and LDA; in Table 4, we have comparison between the text rank and LDA.

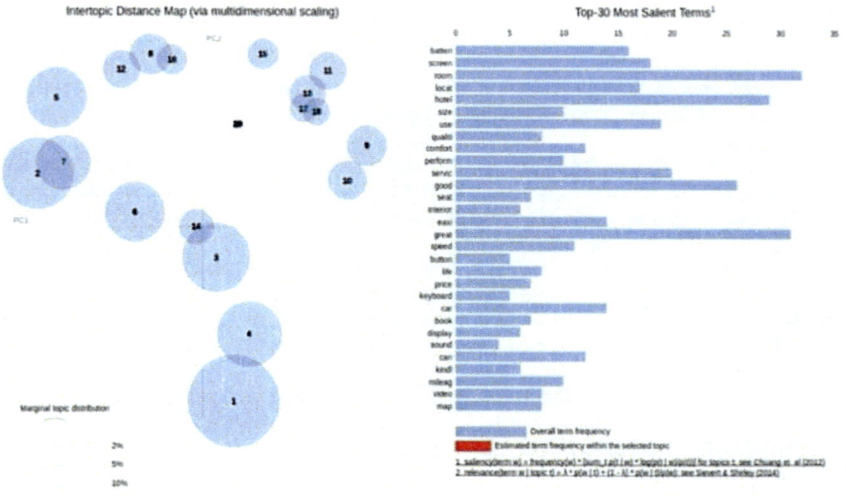

Fig. 4 Visualization of LDA

5.4 Visualization for LDA Output

Figure 4 showing the visualization of LDA for Table 3.

Each bubble on the left-hand side plot represents a topic. Larger the bubble, the more prevalent is that topic. Right-side chart tells about words which are used frequently used; by this only, we can say that the output which we got in the LDA is correct because the collisions of topics are very less.

6 Conclusion

In this paper, we established that the two-level document summarization, and using LDA, generates better summaries. In the first level, we applied MMR and TR to the opinion reviews corpus, which generated the summaries of the corpus, with different topics. In the second level, we improved the summary by applying the LDA to the summaries that were generated by the MMR and TR. At the end, we compared the results two-level document summarization using LDA with results of MMR and TR, and the results of comparison showed that two-level document summarization using LDA improved the summaries. In the future, by combining the summarization and topic modeling algorithms, we can improve the performance of summarization, and we can get more condensed summaries.

References

1. Naveen Gopal, K. R., & Nedungadi, P. (2014). Query-based multi-document summarization by clustering of documents. In *International Conference on Interdisciplinary Advances in Applied Computing*, ACM.
2. https://medium.com/lingvo-masino/do-you-know-about-text-normalization-a19fe3090694.
3. Rajasundari, T., Subathra P., & Kumar, P. N. (2017). Performance analysis of topic modeling algorithms for news articles. *Journal of Advanced Research in Dynamical and Control Systems,* 175–183.
4. Munot, N., & Govilkar, S. S. (2014). Comparative study of text summarization methods. *International Journal of Computer Applications (0975 8887), 102*(12).
5. Kurmi, R., & Jain, P. (2014). Text summarization using enhanced MMR technique. In *2014 International Conference on Computer Communication and Informatics (ICCCI-2014),* Coimbatore, INDIA.
6. Mihalcea, R., & Tarau, P. (2004). *Taxtrank: Bringing order into texts.* Stroudsburg, Pennsylvania: Association for Computational Linguistics.
7. https://medium.com/the-artificial-impostor/use-textrank-to-extract-most-important-sentences-in-article-b8efc7e70b4.
8. https://www.analyticsvidhya.com/blog/2018/11/introduction-textsummarization-textrank-python/.
9. Radev, D. R., Hovy, E., & McKeown, K. (2002).. Introduction to the special issue on summarization. *Journal Computational Linguistics Summarization, 28*(4), 399–408.
10. Rohani, V. A., Shayaa, S., & Babanejaddehaki, G. (2016). Topic modeling for social media content: A practical approach. In *3rd International Conference on Computer and Information Sciences (ICCOINS).*
11. Foster, I., Kesselman, C., Nick, J., & Tuecke, S. (2002). *The physiology of the grid: An open grid services architecture for distributed systems integration.* Global Grid Forum: Technical report.
12. Blei, D. M., Ng, A. Y., & Jordan, M. I. (2003) Latent Dirichlet allocation. *Journal of machine Learning research,* 993–1022.
13. https://en.wikipedia.org/wiki/LatentDirichletallocation.
14. Kumar, A. Sharma, A., Sharma, S., & Kashyap, S. (2017). Performance analysis of keyword extraction algorithms assessing extractive text summarization. In *International Conference on Computer, Communications and Electronics (Comptelix),* Manipal University Jaipur, Malaviya National Institute o/Technology Jaipur IRISWORLD.
15. Gaikwad, S. V., Chaugule, A., & Patil, P. (2014). Text mining methods and techniques. *International Journal of Computer Applications (0975–8887), 85*(17), 42–45.
16. Goldstein, J., Kantrowitz, M., Mittal, V., & Carbonell, J. (1999). Summarizing text documents Sentence selection and evaluation metrics. In *Proceedings of ACMSIGIR99*, pp. 121–128.
17. Ajij, M., Pratihar, S., & Ganguly, K. (2016). Detection and retargeting of emphasized text for content summarization. In *2016 International Conference on Advances in Computing, Communications and Informatics (ICACCI),* Jaipur, India.
18. Moratanch, N., & Chitrakala, S. (2017). A survey on extractive text summarization. In *IEEE International Conference on Computer, Communication, and Signal Processing (ICCCSP-2017).*
19. Kumar, A., Sharma, A., Sharma, S., & Kashyap, S. (2017). Performance analysis of keyword extraction algorithms assessing extractive text summarization. In *2017 International Conference on Computer, Communications and Electronics (Comptelix),* Manipal University Jaipur, Malaviya National Institute o/Technology Jaipur IRISWORLD.
20. Mirani, T. B., & Sasi, S. (2017). Two-level text summarization from online news sources with sentiment analysis. In *2017 International Conference on Networks Advances in Computational Technologies (NetACT),* 22 July 2017, Trivandrum.
21. Sajid, A., Jan, S., & Shah, I. A. (2017). Automatic topic modeling for single document short texts. In *2017 International Conference on Frontiers of Information Technology.*

22. Tong, Z., & Zhang, H. (2016) A text mining research based on LDA topic modeling, Jordery School of Computer Science, Acadia University, Canada, pp. 201–210.
23. Lalithamani, N. (2018). Text summarization. *Journal of Advanced Research in Dynamical and Control Systems, 10*(3), 1368–1372.
24. Allahyari, M., Pouriyeh, S., Assef, M., & Safaei, S. (2017). A brief survey of text mining: Classification, clustering and extraction techniques. arXiv:1707.02919v2 [cs.CL].
25. Murty, M. R., Murthy, J. V. R., Pradas Reddy, P. V. G. D., & Sapathy, S. C. (2012) A survey of cross-domain text categorization techniques. In *International conference on Recent Advances in Information Technology RAIT-2012*, ISM-Dhanabad, 978-1-4577-0697-4/12, IEEE Xplorer Proceedings.

A Robust Blind Oblivious Video Watermarking Scheme Using Undecimated Discrete Wavelet Transform

K. Meenakshi, K. Swaraja, Padmavathi Kora, and G. Karuna

Abstract Intellectual property rights and claiming ownership are two prime requirements of the digital video watermarking. The issues which researchers are interested in digital video watermarking applications lie in the creation of new algorithms to cater four requirements of making oblivious, robust, high-capacity, and secured watermarking. This work presents an improved video watermarking scheme based on the undecimated discrete wavelet transform (UDWT). The frames of cover video are divided into 8 × 8 blocks. Two AC coefficients are selected in each 8 × 8 block to insert the watermark bit. The process is applied on four bands of UDWT and the redundancy in this transform allowed to produce a video watermarking with large capacity. Due to the masking properties of human visual system of UDWT, the watermarking scheme is made oblivious. The experimental results prove that proposed video watermarking scheme is providing all the four requirements of watermarking that is security, oblivious, robustness, and capacity.

Keywords UDWT · Quantization index modulation · Normalized cross correlation · Spread spectrum

1 Introduction

With sophisticated mobile technology and Internet of Things (IOT), there is a large-scale spurt in usage of video chatting, videoconferencing, video-on-demand, consumer video, and medical videos for tele-medicine, tele-surgery, tele-diagnosis, etc. [1, 2]. The consequence of such large usage of video has compelled to safeguard copyrighted multimedia data from malicious tampering and undesired distribution [3]. Video watermarking is a scheme to hide owners' authentication information into the frames of cover video by slightly altering its content transparently. Video watermarking algorithms are categorized into three types namely blind [3], semi-blind, and non-blind [1] based on information needed at extraction. In blind algorithms,

K. Meenakshi (✉) · K. Swaraja · P. Kora · G. Karuna
GRIET, Bachupally, Hyderabad, India
e-mail: mkollati@gmail.com

S. C. Satapathy et al. (eds.), *Intelligent System Design*, Advances in Intelligent Systems and Computing 1171, https://doi.org/10.1007/978-981-15-5400-1_17

it neither requires the frames of cover media nor the concealed watermark at the extraction stage. It requires only the frames of watermarked video for watermark detection [4]. To detect watermark in semi-blind manner, knowledge of both watermark sequence and secret key are needed. And finally, in non-blind method, frames of both host and signed watermark video are required [1] for watermark extraction.

Though watermarking techniques are initially used for images [5], now few works are reported using watermarking in video. Watermarking in video is more complex than image because in the former technique, temporal dimension has to be taken into account in addition to spatial dimensions. Further, it must also deals with attacks such as H.264 compression, frame dropping, frame swapping, and collusion attacks which are unique to video. A video watermarking resistant to rotation and collusion is proposed by [6], using discrete cosine transform and zernike moments. The advantage of the scheme is that they exploited rotational invariance of zernike moments. However, zernike moments are computationally complex. Further, it requires two transforms, whereas the proposed algorithm utilizes only one transform. Another drawback is that the capacity of this watermarking scheme is one-eighth of the capacity of the proposed algorithm.

The rest of this paper is organized as follows: Sect. 2 presents the background material for UDWT. Section 3 provides the methodology used for watermarking with UDWT. Extensive simulation results are conducted and are presented in Sect. 4. Finally, conclusion is drawn in Sect. 5.

2 Background Material

In this work, the transform UDWT [7, 8] is used for watermark concealing.

2.1 *Undecimated Discrete Wavelet Transform*

DWT [9] has attracted researchers due to its multi-resolution properties. However, it has the limitation of shift variance due to down-sampling at different wavelet levels which may introduce blocking artifacts in the image. In order to rectify this problem, the UDWT [10] is used in its place by introducing redundancy by eliminating the step of the down-sampling used in classical DWT. Similar to DWT, UDWT has four sub-bands (LL_C, LH_C, HL_C, and HH_C) with LL_C band as same size as the size of frame of host video [8]. Figure 1a is the frame of Claire video resolution 144×176. The resolution of LL_C band is 72×88 and 144×176 in DWT and UDWT, respectively, as shown in Fig. 1b, c. This feature can be utilized for enhancing watermark embedding capacity. Therefore, if same number of frames are utilized in DWT and UDWT, the number of pixels available for embedding in UDWT is four times than that of DWT.

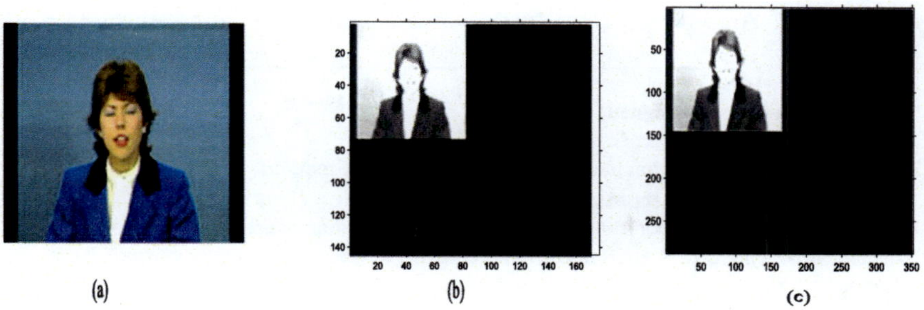

Fig. 1 **a** Frame of claire video **b** 1-level decomposition of luminance component of claire with one-dimensional DWT **c** 1-level decomposition of luminance component of claire with one-dimensional UDWT

3 Proposed Watermark Concealing and Extraction Algorithm

In this algorithm, a blind watermarking is developed based on UDWT.

3.1 Watermark Embedding

The flow diagram of watermark embedding with UDWT is shown in Fig. 2. The algorithmic steps of watermarking are given in Algorithm 1. The high motion frames are selected based on the histogram difference of the present and next frame. If this is more than the specified motion threshold, then the frame is considered having large motion. The watermark embedded in highly motion frames is difficult to perceive. Suppose the resolution of video is 144 × 176. After application of UDWT, the resolu-

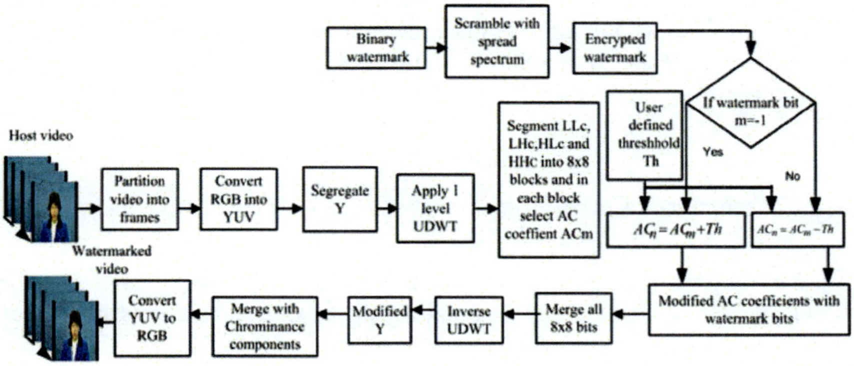

Fig. 2 Flow diagram of watermark concealing algorithm

Algorithm 1 Algorithmic steps for watermark embedding in cover video using UDWT

Input: **Cover video, Logos**
Output: **Watermarked(Signed) video**
Partition cover video into frames.
Embed watermark information in frames where there is more motion.
For each motion frame **perform the following steps**
1. The frame is in RGB color standard. To isolate achromatic luminance component (Y) from chromatic components U & V, RGB is converted into YUV format . To insert watermark, luminance component Y is used leaving U & V unmodified.
2. Segment the selected fast motion frames into non-overlapping blocks of 8 × 8.
3. Apply forward UDWT which segments frame into four bands of LL_C, LH_C, HL_C, and HH_C.

4. Watermark taken is binary. It is converted from unipolar (0,1) to bipolar (-1,1), and later, spread spectrum(SS) scheme is employed. In spread spectrum, -1 is transmitted as [-1 1 -1], and 1 is transmitted as [1 -1 1]. SS ensures the security of watermark, and the capacity of proposed watermarking scheme is therefore increased to 144 × 176 × 4 × 3 bits.
In each frame, only part of watermark information is hidden. In selected fast motion frames, from each 8 × 8 blocks, two ac coefficients are taken for watermark insertion.

if $m == -1$ **then**
AC_n==AC_m+Th
else
AC_n==AC_m-Th
end if
where m is watermark bit. The concealing procedure is repeated for all spread spectrum coded watermark bits in all four bands of UDWT.
8. Apply inverse 2D UDWT to obtain watermarked luminance.
9. Concatenate the chrominance components U & V with watermarked luminance Y_{mod} to get signed YUV frame.
10. Transform color space from YUV to RGB.
11. Merge all the frames to obtain the signed video.
endfor

tion of LL_C, LH_C, HL_C, and HH_C is 144 × 176. Hence, 18 × 22, 8 × 8 blocks are available for watermark insertion. If the watermark size is 144 × 176 × 3, then the number of frames require for watermark insertion are 64. Thus, four different logos can be inserted in the cover video of 400 frames using the watermark embedding algorithm.

3.2 *Watermark Extraction*

The flow diagram of watermark extraction is shown in Fig. 3. The proposed algorithm is blind because it requires only the frames of signed video for watermark detection.

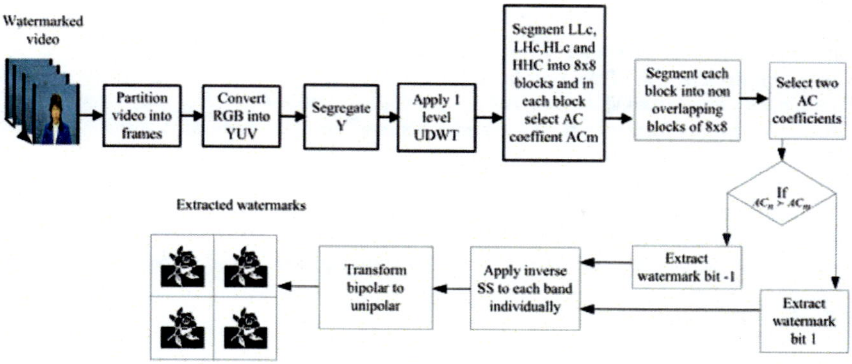

Fig. 3 Watermark extraction from video using UDWT

The watermarked video is partitioned into frames, and later, only the frames with motion which are used in concealing are used for watermark extraction. As shown in Algorithm 2, the watermarks are extracted from LL_w, LH_w, HL_w, and HH_w. The resolution of the watermarks extracted from four bands is 144 × 176.

4 Simulation Results

In this paper, 200 different video sequences are used for experimentation. These video sequences are quarter common intermediate format (QCIF), common intermediate format (CIF), and high definition (HD) videos which are downloaded from video library www.xiph.org. The resolution of QCIF, CIF, and HD videos is 144 × 176, 288 × 352, and 1080 × 1920. The binary logo employed in this watermarking scheme is rose. Claire is the frequently used test video sequence. Now, due to lack of space, the simulations are confined to six video sequences of ducks take off, life, controlled burn, football, Miss America, and flower garden in addition to Claire video. The proposed watermarking scheme is implemented on MATLAB 2013B. The experiment is conducted on Pentium processor I5 on Windows 11 operating system. The Th used for LL_C band is 15, and for the remaining three bands, a Th of 30 is used. The Th controls the imperceptibility and robustness. The higher the Th, the more robust the watermarking scheme and less the transparency. Opposite is true if Th is curtailed. The proposed watermarking scheme imperceptibility, robustness, capacity, and security are discussed in Sects. 4.1, 4.2, 4.3 and 4.4.

Algorithm 2 Pseudo-code for the watermark extraction in video using UDWT

Input : Frames of signed video
Output : Extracted logos from four bands of UDWT
Partition the watermarked video into frames.
Extract watermark information from the frames where watermark is embedded.
for each frame **perform the following steps**
1. Convert the color space of signed video from RGB to YUV. Extract the watermarked luminance.
2. Apply forward UDWT to partition the frame into LL_w, LH_w, HL_w, and HH_w
3. Partition each frame into non-overlapping blocks of 8 × 8.
4. By using the following rule, extract spread spectrum-based watermark information.

if $AC_n \geq AC_m$ **then**
$m == -1$
else
$m == +1$
end if
where m is watermark bit.
5. By using inverse SS, extract the bipolar watermark information.
6. Convert bipolar data to unipolar.
7. Extract watermark from the four bands.
endfor

4.1 Imperceptibility

This experiment uses PSNR to evaluate the imperceptibility. The PSNR and MSE are given in Eqs. 14 and 15 in [6]. The host and watermarked video sequences of HD videos of ducks take off, life, controlled burn, and CIF videos of football, Miss America, and flower garden are shown in Figs. 4 and 5.

The average PSNR of proposed watermarking scheme is 34.44 dB as compared to 39.89 dB of [6]. The low PSNR is due to the increased capacity of the proposed algorithm as imperceptibility and capacity are mutually conflicting (Fig. 6).

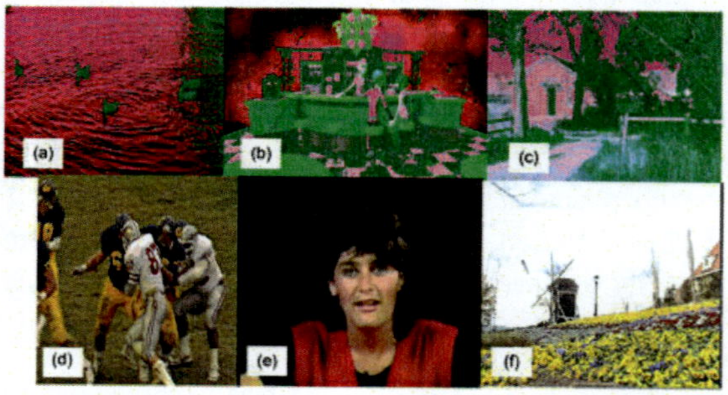

Fig. 4 Frames of host video **a** life, **b** controlled burn, **c** ducks take off, **d** football, **e** Miss America, **f** flower garden

Fig. 5 Frames of watermarked video **a** ducks take off, **b** life, **c** controlled burn, **d** football, **e** Miss America, **f** flower garden

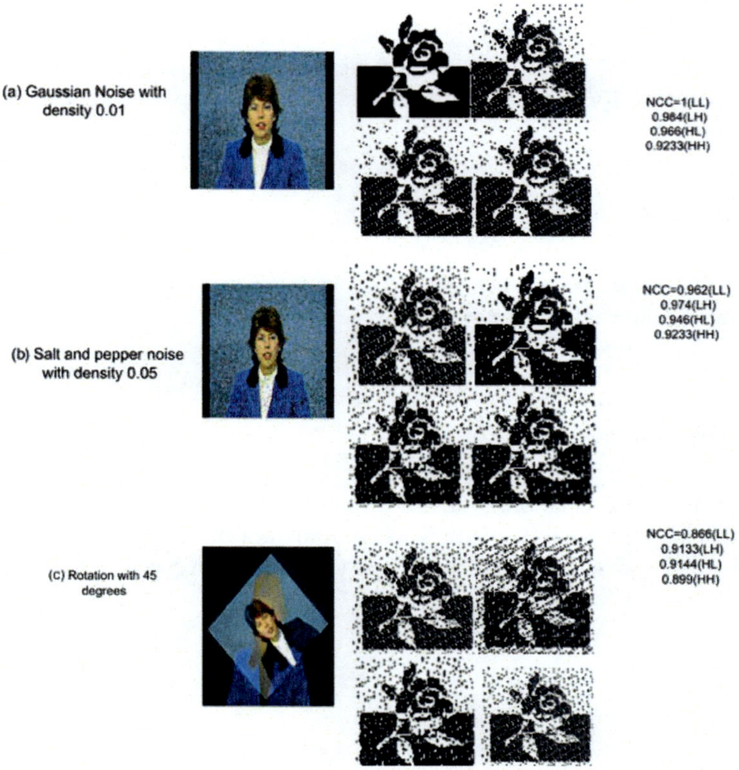

Fig. 6 Results of applying attacks on Claire video **a** Gaussian noise with density 0.01, **b** Salt and pepper noise with density 0.05, **c** Rotation with 45 degrees

4.2 Robustness

To assess the measure of resistance against noise and filtering operations performed on the signed video. The attacks applied are Gaussian noise with 0.01 density, salt and pepper noise with 0.05, and rotation along 45°.

The high NCC obtained for the above attacks show that the method is robust against Gaussian noise with 0.01 density, salt and pepper noise with 0.05, and rotation along 45°. Compared with [6], the average NCC of the three attacks is 0.76, whereas ours is 0.933.

4.3 Capacity

The capacity of [6] is 8×15, whereas the capacity of the proposed watermarking scheme is $144 \times 176 \times 3 \times 4$, and there is significant improvement in capacity.

4.4 Security

The SS ensures security. When attacker hacked, encrypted watermark is available which provides security to the system.

5 Conclusion

A robust video watermarking with UDWT is developed and simulated. The results prove that the proposed watermarking scheme provides all the requirements of watermarking that is imperceptibility, robustness to attacks, high capacity, and security.

References

1. Meenakshi, K., Swaraja, K., & Kora, P. (2019). A robust DCT-SVD based video watermarking using zigzag scanning. In *Soft computing and signal processing* (pp. 477–485). Berlin: Springer.
2. Swaraja, K., Latha, Y. M., & Reddy, V. (2010). The imperceptible video watermarking based on region of motion vectors in p-frames. *Advances in Computational Sciences and Technology, 3*, 335–348.
3. Meenakshi, K., Srinivasa Rao, C., & Satya Prasad, K. (2014). A scene based video watermarking using slant transform. *IETE Journal of Research, 60*, 276–287.
4. Meenakshi, K., Prasad, K. S., & Rao, C. S. (2017). Development of low-complexity video watermarking with conjugate symmetric sequency-complex hadamard transform. *IEEE Communications Letters, 21*, 1779–1782.

5. Meenakshi, K., Rao, C. S., & Prasad, K. S. (2014). A robust watermarking scheme based Walsh-Hadamard transform and SVD using zig zag scanning. In *2014 International Conference on Information Technology (ICIT)* (pp. 167–172). New York: IEEE.
6. Karmakar, A., Phadikar, A., Phadikar, B. S., & Maity, G. K. (2016). A blind video watermarking scheme resistant to rotation and collusion attacks. *Journal of King Saud University-Computer and Information Sciences, 28*, 199–210.
7. Makbol, N. M., Khoo, B. E., & Rassem, T. H. (2018). Security analyses of false positive problem for the SVD-based hybrid digital image watermarking techniques in the wavelet transform domain. Multimedia Tools and Applications, pp. 1–35.
8. Ernawan, F., & Kabir, M. N. (2018). A block-based RDWT-SVD image watermarking method using human visual system characteristics. In *The visual computer* (pp. 1–19).
9. Kaur, K. N., Gupta, I., Singh, A. K., et al. (2019). Digital image watermarking using (2, 2) visual cryptography with DWT-SVD based watermarking. In *Computational intelligence in data mining* (pp. 77–86). Berlin: Springer.
10. Lagzian, S., Soryani, M., & Fathy, M. (2011). A new robust watermarking scheme based on RDWT-SVD. *International Journal of Intelligent Information Processing, 2*, 22–29.

Recognition of Botnet by Examining Link Failures in Cloud Network by Exhausting CANFES Classifier Approach

S. Nagendra Prabhu, D. Shanthi Saravanan, V. Chandrasekar, and S. Shanthi

Abstract Consequently, verifying the cloud from the botnet is required for keeping the administrations from different assaults, for example, distributed denial of service (DDoS), spreading malware and hacking of private data. To identifiy botnet and furthermore to identify bot master from the cloud condition, by examining link failures in cloud network by exhausting CANFES classifier. The link failures between cloud server and client because of bots in cloud network. The probabilistic highlights of the bots and link gain are assessed on each port of the cloud framework. In view of this estimation, the connections between cloud server and client are broken down for the likelihood of disappointment and to recognize the bot master utilizing CANFES. The exhibition of the proposed framework is broke down as far as throughput, path loss and precision rate. The proposed system achieves 7623 bits/second as throughput, 17.04 dB as path loss and 94.6% of precision rate.

Keywords DDoS · CANFES classifier · Botnet · Throughput · Path loss

1 Introduction

These days the botnet is turning into the base of all cybercrime which is performed through the web. bot master utilize various techniques to contaminate a client gadget to make it bot (zombie) like drive by download, email and pilfered programming's are the most well-known method for assaults. As indicated by the past research, loads of the recognition methodologies have been proposed. Be that as it may, the greater part of them are centered around the disconnected identification of botnet; still we have to concentrate on constant location. As society turns out to be progressively

S. Nagendra Prabhu (✉) · V. Chandrasekar · S. Shanthi
Department of CSE, Malla Reddy College of Engineering and Technology, Dhulapally, Secunderabad 500100, India
e-mail: snagendraprabhu@gmail.com

D. Shanthi Saravanan
Department of CSE, PSNA College of Engineering and Technology, Dindigul, Tamil Nadu 624622, India
e-mail: dshan71@gmail.com

S. C. Satapathy et al. (eds.), *Intelligent System Design*, Advances in Intelligent Systems and Computing 1171, https://doi.org/10.1007/978-981-15-5400-1_18

interconnected and more gadgets are getting to be Internet-empowered, digital security is a developing concern. With this expanded level of network comes expanded danger of cybercrime. The commonness of cybercrime is additionally exacerbated by simple access to hacking devices and instructional exercises inside programmer networks and illegal businesses. One especially perilous part of cybercrime is the danger forced by botnets. Botnets are accumulations of contaminated PCs, frequently alluded to as bots, automatons or zombies, which are given directions to complete noxious exercises. Model botnet exercises incorporate distributed denial of service (DDoS) assaults, spam circulation, and the spreading of malware.

The top wide-ranging cases achieved by botnets are DDoS, click fraud, phishing extortion, key logging, bitcoins misrepresentation, spamming, sniffing traffic, spreading new malware, Google AdSense misuse, secret word stealer and mass extensive fraud with bots. Like worms proliferation the botnet likewise engender itself, correspondingly like infection, botnet additionally keep it escaped recognition. Botnet has an incorporated control and order framework that is the reason it assault comparable various institutionalization unaccompanied apparatuses. It spikes with an exceptionally high contaminated by botnet; bots are otherwise called a zombie that is the reason a botnet is likewise called zombie organize. Eggdrop was the first botnet made in 1993 [1]. Gtbot and Spybot were made in 2000 [2]. As indicated by Malaysian Computer Emergency Response Team (MYCERT) insights report of most recent five years shows that the botnet automatons assaults are expanded with a great proportion.

Objectives

- The proposed paper deals the link failure between cloud server and client due to bots in cloud network.
- The probabilistic features of the bots and link gain are estimated on each port of the cloud system.
- Based on this estimation, the links between cloud server and client are analyzed for the possibility of failure and to detect the bot master using CANFES.

The rest of the paper is sorted out as pursues, Sect. 2 represents the current botnet recognition methods. Section 3 portrays the CANFES grouping system. Section 4 depicts the Proposed Methodology. Section 5 portrays results and discourses of the recommended technique and the paper is finished up in segment 6.

2 Botnet Detection Techniques

Botnet location is the utmost imperative errand to progress the digital security beside different digital assaults happens in web these days. As indicated by the past research, botnet identification systems can be ordered into two classifications, honeynets discovery strategies and interruption recognition methods [3, 4]. Intrusion detection system is further separated into subcategories.

2.1 Counter-Based System

Counter-based location is a clear method that counts the full scale traffic volume or the amount of site page requests. Since the DDoS attack with the low volume of traffic, for instance, the HTTP GET divided ambush is regular in these days, the repeat of page requests from clients will be a logically convincing segment.

2.2 Access Pattern-Based Method

The passage configuration-based area strategy acknowledged that clients polluted by a comparative bot lead a comparable direct and that aggressors could be isolated from customary clients. This strategy required more than two MapReduce, [2] employments: the main occupation gets get to plan to the website page between a customer and a web server and appraisals the investing energy and the bytes mean each solicitation of the URL; the subsequent pursuits of employment out tainted has by looking at the entrance arrangement and the investing time among customers attempting to get to a similar server [5].

2.3 Entropy Based Anomaly Detection System

In this, the whole cloud system is divided into various legitimate space, which is controlled autonomously by its own special authentication and certification authority. In this designing, a tree was kept up at each switch, by meaning each group with way modification system, so the harmed individual can pursue the sender of the parcel [3, 4].

2.4 Min-Vertex Cover Method

In Xu et al. [6], the maker has developed a novel P2P botnet area show that joins session-based examination and least vertex spread speculation; this model just looks at sort out header information, paying little personality to extra framework data weight, and takes usage of least vertex spread to test the middle centers of botnet.

2.5 Page Rank Algorithm

Page Rank calculation [7] is an association assessment computation used by the Google web record to weight the general criticalness of website pages on the Internet. It positions each webpage page as showed by the hyperlink structure among site pages. Page Rank is especially fitted for MapReduce and this fragment reviews the stray pieces of executing Page Rank in the MapReduce setting without thinking about the dangling centers or the damping factor for clearness reason.

2.6 Honeynets and Honeypots Based Detection System

Honeynets and honeypots both are meaning the end customer contraptions. These end customers PC's are the best way to deal with accumulate fundamental information about the advanced attacks. This end customer PC is straightforward for bot master to ambush and deal, since it is really feeble against toxic attacks. The computerized security social affair will more likely than not make extraordinary acknowledgment techniques under the assembled information about the botnet attacks through these honeynets. As indicated by the past research, the botnet change their engraving time to time due to the security reason and honeynets are basic for perception these botnet properties [5, 8].

3 CANFES Classifier System

The inspiration for approximating fuzzy frameworks by neural systems depends on the characteristic capacity of neural systems to perform huge parallel handling of data. This is significant in the fluffy controllers or all the more by and large fuzzy master frameworks that are required to process huge quantities of fuzzy surmising rules continuously. At the point when the neural system speaking to a given fuzzy master framework is executed all pertinent fuzzy induction rules are prepared in parallel. These outcomes have high computational effectiveness, which is urgent in numerous applications, for example, analysis, expectation, etc. The unaided methods which develop inner models that catch regularities in their information vectors without accepting any extra data are appropriate to discover groups of information showing the nearness of fuzzy standards [9]. The managed strategies which require an instructor to indicate the ideal yield vector, and the fortification methodology which just require a solitary scalar assessment of the yield are a great idea to adjust the decision of enrollment capacities for the ideal yield in fuzzy rationale frameworks. Consequently, a half breed learning calculation is being utilized for the CANFES model. The learning calculation for this model consolidates solo learning and managed learning methodology to manufacture the standard hubs and train the enrollment capacities. The CANFES model additionally keeps up the idea of human reasoning and thinking as in fuzzy surmising frameworks. Along these lines, specialists' learning can undoubtedly be fused into the structure. The CANFES structure likewise spares the standard coordinating time of the deduction motor in the conventional fluffy surmising framework. This CANFES model system is being created for the exemplification of learning and to improve and energize explore in the field of coactive neurofuzzy demonstrating. The learning architect and human master (area master) have long sessions of exchange so as to inspire information for the master framework. The aftereffect of the correspondence and discourse between the information architect and area master builds up the learning securing module.

4 Proposed Methodology

The architecture representation of Fig. 1 shows the proposed bot master detection and classification system using CANFES classification method. Link gain between all nodes in cloud environment is determined and all these determined link gains are ordered as ascending order.

Then, low link gains are determined and these low link gains are trained and classified using CANFES classification approach, as depicted. Link gain detection

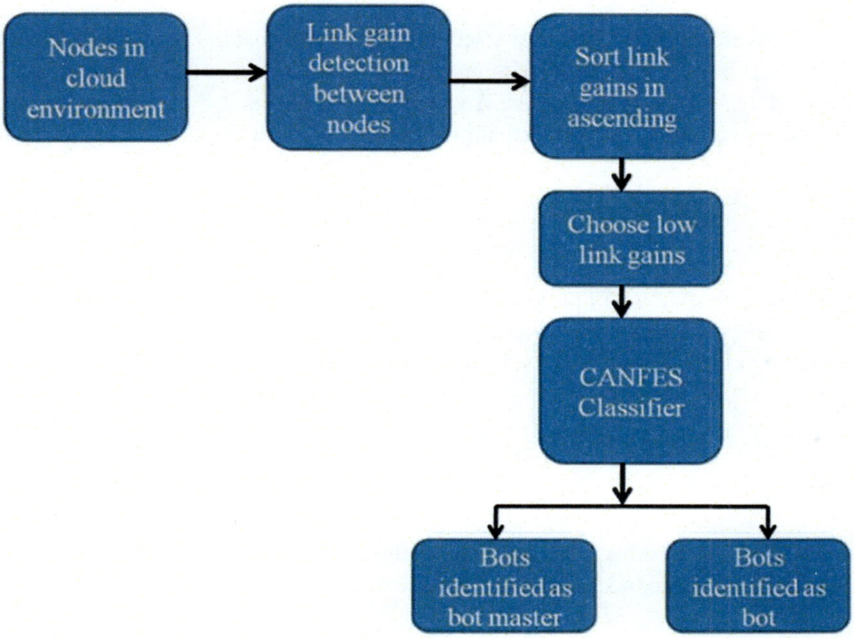

Fig. 1 Proposed bot master detection system

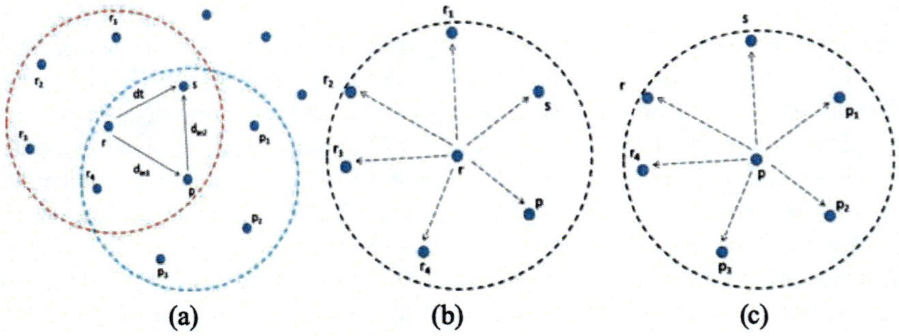

Fig. 2 **a** Link gain estimation of node '*r*' by node '*s*'. **b** Link gain estimation of node '*p*' by node '*r*.' **c** Link gain estimation of node '*r*' by node '*s*' through node '*p*'

between nodes in cloud environment is as shown in the Fig. 2. Figure 2a Link gain estimation of node '*r*' by node '*s*' (b) Link gain estimation of node '*p*' by node '*r*' (c) Link gain estimation of node '*r*' by node '*s*' through node '*p*'.

The direct link gain (d_t) is computed by,

$$d_t = \sum_{i=1}^{N_1} (i - \mu)^2 \times P_i \tag{1}$$

where the probability metric is denoted by P_i, the usual quantity of packets received by r over the time period 't'. The quantity of neighboring nodes over node 'r' is denoted by N_1. The average quantity of packets received is μ.

The probability metric of each individual node is given by,

$$P_i = \frac{\alpha_i - \beta_i}{\alpha_i} \tag{2}$$

where α_i is the number of packets received over time period 't' and β_i is the number of packets transferred over the time period 't'.

The link gain between nodes 'r' and 'p' ($d_{\text{in}1}$) is,

$$d_{\text{in}1} = \sum_{i=1}^{N_1} (i - \mu)^2 \times P_i \times W_i \tag{3}$$

where N_1 is the number of surrounding nodes over the node p.

The weight of individual node with respect to node p can be computed as,

$$w_i = \frac{\sum_{i=1}^{N} P_i \times X_i}{k} \tag{4}$$

Table 1 Link gains for nodes

Nodes	Node p1	Node p2	Node p3	Node r4	Node r	Node s
Link gain	0.2	0.6	0.4	0.3	0.2	0.1

The packets are denoted by X_i and 'k' is kappa factor which is given as,

$$k = \sum P_i \tag{5}$$

The link between nodes 'p' and 's' is,

$$d_{\text{in}2} = \sum_{i=1}^{N_2} (i - \mu)^2 \times P_i \times W_i \tag{6}$$

where N_2 is the number of surrounding nodes over the node s.

The total link gain is given as,

$$d_{\text{in}} = d_{\text{in}1} + d_{\text{in}2} \tag{7}$$

$$\text{Total Trust} = d_t + d_{\text{in}} \tag{8}$$

In this method, the link gains of all other nodes are determined.

Hence, total link gain of the individual node 'r' is given as,

Then, all these computed link gains are ordered as ascending. From these computed link gains, the link gains whose gain value has more than 0.5 are chosen as low link gains. These low link gains are given to the CANFES classifier to train and classify the system in order to detect the bot master by finding the optimum link failure gain.

The link gains for all surrounding nodes of node 's' (Fig. 2a) are given in Table 1.

5 Result and Discussion

The performance of the projected bot master recognition system through the identification of link failures in cloud environment is analyzed using cloudsim simulator in terms of latency and bot master detection rate with respect to quantity of nodes or computers in network. The initial setup of the simulation tool is described in Table 2. The maximum number of packets for each client-end-user used in this research work is 2500 and each node or computer transfers the packets at the rate of 125 kb/s with the energy consumption of 120 mJ per cycle. In this research work, number of bot masters is set to 6 and total number of bots in cloud system is 30. This proposed system is initiated with 100 numbers of nodes as initial parameter setting.

Table 2 Initial network parameters structure

Parameters	Initial value
Maximum packets	2500
Throughput	125 kb/s
Energy consumption	120 mJ per cycle
No. of bot masters	6
No. of bots	30
No. of nodes	100

Table 3 Analysis of throughput for bot master detection

No. of bots	10	20	30	40	50	60	70	80	90	100
Throughput	10,537	9675	9102	8964	8012	7648	6397	5903	5102	4897

In this paper, the following parameters are used to investigate the performance of the proposed system.

- Throughput, Path loss and Precision rate.

5.1 *Throughput*

It is defined as that the rate at which the data can be transferred or received through a port over a particular time period. It is measured in bits per second. The throughput must be high for high performance of the system. Table 3 describes the throughput of the proposed bot master detection architecture for different number of client-end-users. The throughput of the bot master detection system will be decreased when the number of bots in the cloud system increases. The proposed bot master detection system consumes 7623 bits/seconds as average through, when the number of bots in the system varies from 10 to 100.

5.2 *Path Loss*

It is defined as that the difference between transmitted and received powers of transmitting nodes and receiving nodes in cloud, respectively. It is measured in decibels (dB). The path loss should be low for the better cloud environment. Table 4 describes the path loss of the proposed system architecture for different number of bots. The path loss will be increased when the number of bots increases.

The proposed bot master detection system consumes 17.04 dB as average path loss, when the number of bots in the system varies from 10 to 100.

Table 4 Analysis of path loss

No. of. bots	10	20	30	40	50	60	70	80	90	100
Path loss	12.9	13.2	14.7	15.9	16.8	17.1	18.9	19.3	19.9	21.7

Table 5 Analysis of precision

No. of bots	10	20	30	40	50	60	70	80	90	100
Precision(%)	99.7	98.5	97.1	96.7	95.9	94.3	92.1	91.9	90.2	89.6

5.3 *Precision*

It describes the number of data correctly received at the user end side. It is measured in percentage and its value ranges from 0 and end with 100. The available of bots in cloud network will affect the precision rate. The precision rate is indirectly proportional to the number of bots. The precision rate will be increased if the number of bots in cloud network reduced. The precision rate will be decreased if the number of bots in cloud network increased.

Table 5 shows the precision rate of the proposed system with respect to number of bots in cloud network environment. The proposed methodology stated in this thesis achieves 94.6% of average precision rate.

5.4 *Evaluation Among Conventional Methodologies*

Table 6 compares the performance of the proposed bot master detection system with conventional methodologies as [10–12]. The proposed bot master detection system achieves 7623 bits/second of average throughput and 17.04 dB of average path loss, while the conventional methodologies as Lu et al. [10] provided 6382 bits/second of

Table 6 Comparisons between proposed and conventional bot detection system

Methodologies	Throughput (bits/sec)	Path loss (dB)	Precision (%)
Proposed method	**7623**	**17.04**	**94.6**
Lu et al. [10]	6382	21.29	90.1
Jiang et al. [11]	5382	26.86	89.5
Jhawar et al. [12]	5937	25.97	91.6

Bold values indicates the proposed bot master detection system achieves 7623 bits/second of average throughput and 17.04 dB of average path loss and 94.6% of average precision rate, which have shown better efficiency when compare to conventional method

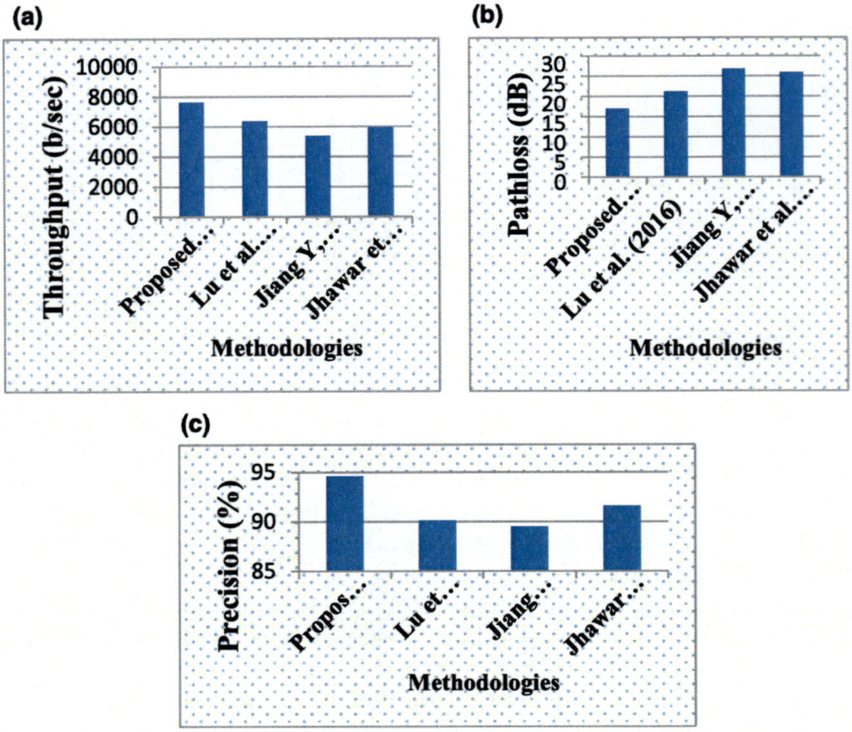

Fig. 3 Comparisons of proposed method with conventional methodologies with respect to **a** Throughput. **b** Path loss. **c** Precision

average throughput, 26.86 dB of average path loss and 90.1% of average precision rate, Jiang et al. [11] achieved 5382 bits/second of average throughput and 26.86 dB of average path loss and 89.5% of average precision rate, Jhawar et al. [12] achieved 5937 bits/second of average throughput and 25.97 dB of average path loss and 91.6% of average precision rate (Fig. 3).

6 Conclusion

In this paper, CANFES classification method is used to detect the bot master using link failure computation method. Initially, the link gains of each node in cloud environment are computed and then they are sorted as ascending order. The lower link gains are computed and these link gains are trained and classified using CANFES classifier in order to detect the bot master in cloud environment. The performance of the projected system is analyzed in terms of throughput and path loss. The proposed system achieves 7623 bits/second as throughput, 17.04 dB as path loss and 94.6% of precision rate.

References

1. Batt, C. (1999). Eggheads. *Food Microbiology, 16*(3), 211.
2. Francois, J., Wang, S., Bronzi, W., & State, R. (2011). BotCloud: Detecting botnets using mapreduce. In *IEEE International Workshop on Information Forensics and Security (WIFS)*, pp. 1–6.
3. Nair, H. S., & Ewards, V. S. E. (2012). A study on botnet detection techniques. *2*(4), 2–4.
4. Sgbau, A. (2013). A review-botnet detection and suppression in clouds. *3*(12), 1–7.
5. Meena, B., Challa, K. A. (2012). Cloud computing security issues with possible solutions. *IJCST, 3*(1).
6. Xu, L., Xu, X. L., & Zhuo, Y. (2012). P2P botnet detection using min-vertex cover. *Journal of Networks, 7*(8).
7. Page, L., Brin, S., Motwani, R., & Winograd, T. (1998). The pagerank citation ranking: Bringing order to the web.
8. Rajab, M. A., Zarfoss, J., Monrose, F., & Terzis, A. (2006). A multifaceted approach to understanding the botnet phenomenon. In *Proceedings of Internet Measurement Conference, October(IMC'06)*, Rio de Janeiro, Brazil.
9. Zang, X., Tangpong, A., Kesidis, G., Miller, D. J. (2011). *Botnet detection through fine flow classification.* Technical Report, Vol. No. 0915552, pp. 1–17.
10. Lu, K., Yahyapour, R., Wieder, P., Yaqub, E., Abdullah, M., Schloer, B., et al. (2016). Fault-tolerant service level agreement lifecycle management in clouds using actor system. *Future Generation Computing Systems, 54*, 247–259.
11. Jiang, Y., Huang, J., Ding, J., & Liu, Y. (2014). Method of fault detection in cloud computing systems. *International Journal of Grid Distribution Computing, 7*(3),205–212.
12. Jhawar, R., Piuri, V., & Santambrogio, M. (2013). Fault tolerance management in cloud computing: A system-level perspective. *IEEE System Journal, 7*(2), 288–297.

Low Power, Less Leakage Operational Transconductance Amplifier (OTA) Circuit Using FinFET

Maram Anantha Guptha, V. M. Senthil Kumar, T. Hari Prasad, and Ravindrakumar Selvaraj

Abstract In various signal processing applications, the performance of the data acquisition circuits is depended upon the performance of the amplifier. On the other hand, the noise signal may saturate the system. The operational amplifier is found to be the most efficient amplifier for signal processing circuits. In literature, several operational amplifiers were designed using Bipolar junction transistors, Field effect transistors and CMOS devices. But the power consumption has limitations on design with BJT and CMOS devices. To overcome this, new alternative devices are required. The minimum parasitic interconnection elements and low supply voltage will enhance the performance. To enhance these requirements, this paper presents a FinFET-based OP-AMP and OTA design with low power, very high unity gain bandwidth (UGB) and high open-loop gain (DC gain). The FinFET-based circuit is designed using 32 nm technology. The proposed circuit has higher driving capacity and is highly stable.

Keywords Low power · Miller capacitance · Cascade technique · FinFET · Leakage current · Operational amplifier · CMOS

1 Introduction

The intention of the work is to design a low power, less leakage amplifier circuit using Multigate Device. All electronics integrated circuits hold OP-AMPs which consume part of the supply power. To reduce power consumption, various aspects are taken in OP-AMP design. The OP-AMPs are used in most of the circuits used for

M. A. Guptha · V. M. Senthil Kumar · T. Hari Prasad
Malla Reddy College of Engineering and Technology, Hyderabad, Telangana, India
e-mail: ananthaguptha007@gmail.com

V. M. Senthil Kumar
e-mail: vmspraneeth@gmail.com

R. Selvaraj (✉)
IRRD Automatons, Karur, India
e-mail: gsravindrakumar7@gmail.com

S. C. Satapathy et al. (eds.), *Intelligent System Design*, Advances in Intelligent Systems and Computing 1171, https://doi.org/10.1007/978-981-15-5400-1_19

computation purposes. High the computation higher will be the consumed power. To reduce the power several methods have to be inculcated [1]. The methods available are improving unity gain of the amplifier circuits. Also, a miller capacitance is used with the gain amplifier to reduce the power consumption; it results in an improvement in unity gain bandwidth. Normally most of the OP-AMPs used in the market use FET or MOSFET. But the MOSFET devices suffer from leakage current and low-speed operation due to lack of current control when the size of the device reduces. So the FinFET-based device is suitable when the size and lower are to be reduced. The system performance improves by using FinFET by reducing the second-order effects.

FinFET Technology:
Even though the improvement in technology scaling has increased, the limitations on maximum operating voltage in CMOS transistors arises. Designing OP-AMP for low-power applications using small channel length devices is a challenge. By reducing the channel length reduces the thermal dissipation and power consumption [2]. The supply voltage used will be low. But working with low supply voltage reduces the performance in CMOS devices. To cope with the same FinFET is the alternative. FinFET design reduces the short channel effects.

2 Literature Survey

Several OP-AMP circuits are designed in literature (Razavi) [3]. As integrated circuits dominated the discrete components, (Gray and Meyer) [4] CMOS-based circuits become famous (Allen Holberg). The CMOS-based circuits have replaced most of BJT and FET based designs. The low-power consumption of the CMOS circuits have helped the engineers to design efficient products. The analog integrated circuits are being the part of several applications especially the signal processing systems. The power is less consumed to amplify a low amplitude signal. Since most of the applications are working in battery power consumption should be low as possible [5]. Sizing the transistors have reduced the power to an acceptable level but the output swing may reduce (Binkley et al. 2003). When operational amplifier was used in these systems, they dominated the electronics industry (Amin et al. 2006). The OP-AMP circuits were enhanced with different features like compact (Carrillo et al. 2009) and high speed (Aslanzadeh et al. 2003). The improvements on noise immunity has come when differential stage and multistage were used (Mahattanakul 2005). Low-power programmable OP-AMP were used in signal processing applications were the swing should be modifiable (Dai et al. 2013). Taherzadeh-Sani M., Hamoui A. A. (2011) presented a OP-AMP design [2] which provides enhanced DC gain and settling behavior. The design was done using 65 nm CMOS technology. Improvising the transconductance will increase the performance of the operational amplifier (Zuo and Islam 2013). For sub 1 V single-supply operation transconductance amplifier are suitable. But the gain stages have to be increased [6] and standby current can be reduced. Adaptive biasing techniques can improve linearity, gain bandwidth product and slew rate. But maintaining the DC bias currents and supply voltage is complex (Saso et al. 2017). Valero et al. (2012) presented an ultralow-power operational amplifier using 180 nm CMOS technology. But the performance can be improved by using FinFET. The feedback compensation improves the frequency response and can drive large capacitance loads (Joao et al. 2004). Several works are also carried out in designing a FinFET-based circuits for various applications [7].

3 Background Methodology

Several OP-AMP configurations are available in literature but in this work, a two-stage OP-AMP and a self-bias OP-AMP are investigated by implementing in CMOS and FinFET.

3.1 Two-Stage CMOS OP-AMP

The block diagram of the two-stage OP-AMP is shown in Fig. 1a which consists of input differential amplifier, second gain stage, compensation circuit and bias circuit. The input differential amplifier removes the additive noise in the system and provides the gain to improve the performance [8]. For suitable operating point, the bias circuit is used and for stability the compensation circuit is used [9]. The second gain circuit provides additional gain to obtain the required output swing. The two-stage OP-AMP is demonstrated in Fig. 1b.

The circuit in Fig. 1b provides low output impedance, high open-loop gain, unity gain bandwidth, slew rate and decreases power consumption. But the circuit suffers

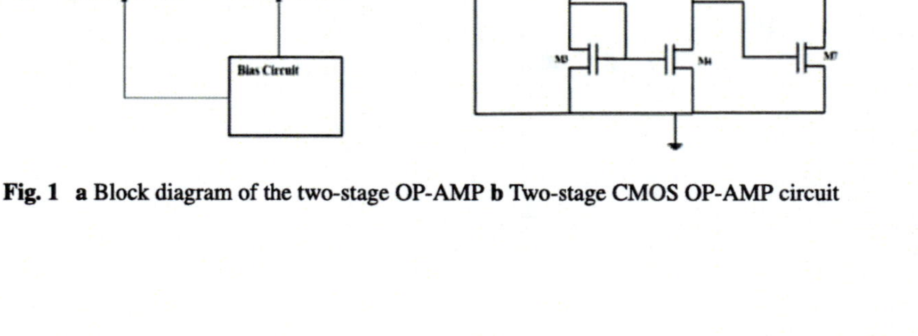

Fig. 1 **a** Block diagram of the two-stage OP-AMP **b** Two-stage CMOS OP-AMP circuit

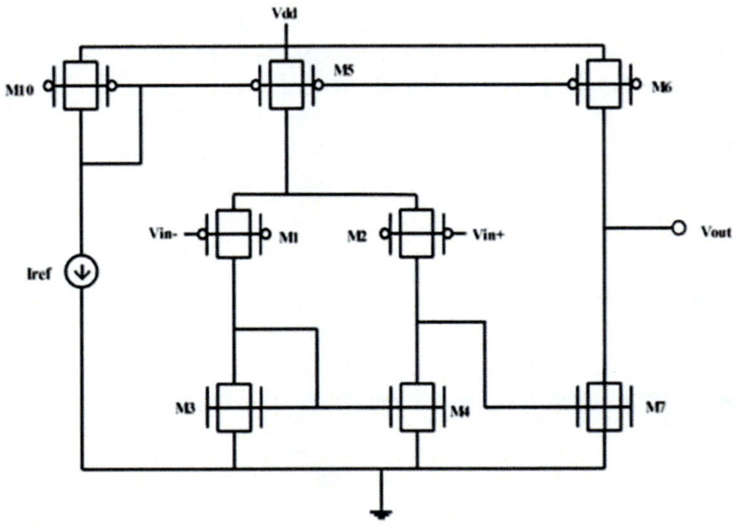

Fig. 2 Two-stage FinFET OP-AMP circuit

from leakage current [6]. To reduce the leakage and power consumption the two-stage OP-AMP circuit is implemented using FinFET device as shown in Fig. 2. This circuit shows better performance when compared to CMOS.

4 Proposed Methodology

The CMOS and proposed FinFET-based miller compensated OTA is shown in Fig. 3a, b. The first stage converts the voltage input into a current. The second stage is the gain enhancement stage for the OTA [10]. This is obtained through a high output

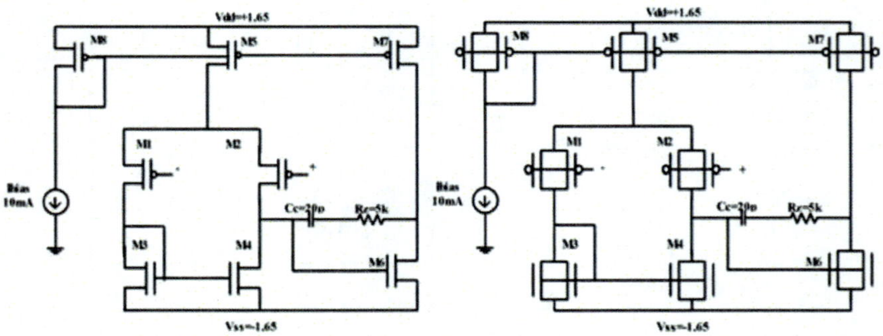

Fig. 3 Two-stage Miller compensated OTA using **a** CMOS **b** FinFET

resistance provided by r06 in parallel with r07, where r0i is the channel resistor of the transistor. The compensation capacitor C_c splits the poles into two sides. It is placed between the input and the output of the second stage [3]. The splitting is done to have one pole at lower frequencies and the other at high frequencies [4]. The compensation capacitance used for this OTA is 20 pF.

To enhance the performance a two-stage operational transconductance amplifier is designed using FinFET 32 nm technology [11] as shown in Fig. 3. In the proposed, design; the cascade technique is employed which results in high open-loop gain owing to excellent frequency response and increased output impedance. In Fig. 4. M1, M2, M3 and M4 form the first stage of OP-AMP. The M7 and M8 transistors current mirrors M1 and gets deducted from the current flowing through M2. The differential current from M1 and M2 FinFETs get multiplied by output impedance and offers the single-ended output which acts as the input for the second stage [12]. The current sink second-stage load inverter obtains large output impedance. Miller compensation technique enhances the close loop stability but creates phase margin which is controlled by the compensation capacitor connected in series with the nulling resistor or with gain stage (common gate).

To further enhance the performance a topology of two-stage Capacitor Multiplier Compensation OTA is proposed using FinFET as shown in Fig. 5. The first stage of this OTA is a capacitor multiplier stage which blocks the feed-forward capacitive path from the output of the first stage to the output of the amplifier. The circuitry boosts the

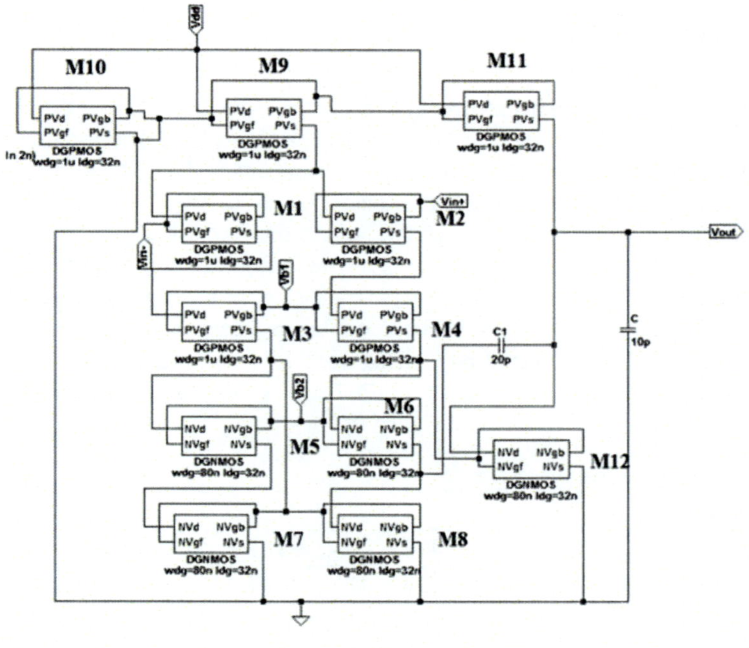

Fig. 4 Two-stage FinFET operational transconductance amplifier

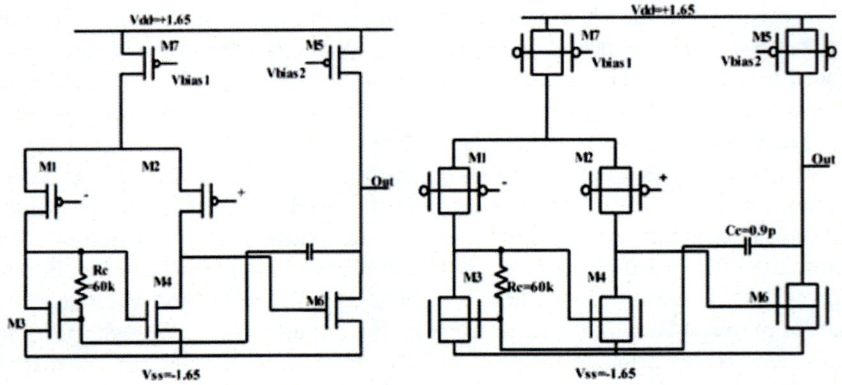

Fig. 5 Two-stage capacitor multiplier compensated OTA

phase margin and improves the stability of the OTA. The amplifying stage is realized by a transconductance stage [13]. Due to the less leakage problems, the FinFET-based Capacitor Multiplier permits the amplifier to drive a very large capacitive load using a small compensation capacitor [14]. The compensation capacitance C_c used is 0.9 pF and R_c is equal to 60 kΩ.

5 Results and Discussion

The SCEs are suppressed by having stronger control in the FinFET over the channel. The dielectric leakage current is suppressed and high switching activity is obtained. The results on power testify the low leakage current and high ON-State current. For further investigation, the electrical characteristics of 32-nm FinFET's and 32-nm Bulk CMOS devices are taken from PTM indicated in Table 1.

The primary performance parameters obtained from the PTM are tabulated in Table 2. FinFET provides flexibility and the front-gate and the back-gate tied together for Short-Gate mode. The proposed circuit can be implemented in Independent-Gate mode and Low-Power mode also [5]. That's left for future work. In this paper the Short-Gate (SG) mode, in which the FinFET acts a three-terminal device is used

Table 1 Primary parameters in PTM

Primary parameters in PTM					
n-type FinFET	$L_{gate} = 32$ nm	$H_{fin} = 40$ nm	$W_{fin} = 8.6$ nm	$T_{ox} = 1.4$ nm	VDD = 1 V
p-type FinFET	$L_{gate} = 32$ nm	$H_{fin} = 50$ nm	$W_{fin} = 8.6$ nm	$T_{ox} = 1.4$ nm	VDD = 1 V
NMOS	$L_{eff} = 32$ nm	$T_{oxe} = 1.4$ nm	$V_{th0} = 0.42$ V		VDD = 1 V
PMOS	$L_{eff} = 32$ nm	$T_{oxe} = 1.5$ nm	$V_{th0} = -0.41$ V		VDD = 1 V

Table 2 Performance of Two-stage Op-amp design

Input (V)	Average current (A)	Average power (W)	Average energy (J)	Delay (S)
0.4	10.607×10^{-6}	1.0516×10^{-6}	530.33×10^{-15}	449.26×10^{-12}
0.7	86.696×10^{-6}	14.069×10^{-6}	7.3006×10^{-12}	259.55×10^{-12}
0.9	158.7×10^{-6}	35.372×10^{-6}	18.077×10^{-12}	133.23×10^{-12}
1	195.88×10^{-6}	50.178×10^{-6}	25.341×10^{-12}	84.634×10^{-12}
1.2	271.64×10^{-6}	91.696×10^{-6}	47.333×10^{-12}	54.601×10^{-12}
1.3	310.12×10^{-6}	124.24×10^{-6}	63.575×10^{-12}	33.85×10^{-12}

which has higher ON-State current, faster-switching speed and strong gate control. The suppression of second-order effects is higher [9]. The circuits in FinFET is a novel replacement for Bulk CMOS devices [15]. If the proposed circuit is implemented in Independent-Gate (IG) mode which is a four-terminal device with each gate of different inputs reduces the transistors count [16]. On the other hand, the OTA in Low-Power (LP) mode reduces the threshold leakage [17]. Here a reverse bias potential is applied to the back-gate. The implementation was done using SPICE predictive technology Models in CMOS and FinFET 32 nm technology. The Average current, Average power, Average energy and delay are measured through experiments for varying input voltage for two stages and transconductance amplifier.

The performance of Two-Stage Miller Compensated OTA Topology (Circuit 1) and Capacitor Multiplier Compensated OTA (Circuit 2) was presented in Table 3. The average power in CMOS capacitor multiplier is lower than the miller compensated OTA. Similarly, when considering the FinFET-based design the driving capability is higher than the CMOS. The power is less in FinFET circuit 1 and 2 when compared to CMOS counterpart. The performance of the modified two stages transconductance OP-AMP design is shown in Table 4 the power consumption in this circuit is nominal when compared to previous circuits presented in Table 3.

Table 3 Performance of two-stage miller compensated OTA topology (Circuit 1) and capacitor multiplier compensated OTA (Circuit 2)

	Designed using CMOS		Designed using FinFET	
	Circuit 1	Circuit 2	Circuit 1	Circuit 2
Avg power (W)	1.4824E−04	1.0585E−04	7.5373E−05	6.0901E−05
Peak power (W)	1.4824E−04	1.0585E−04	7.2210E−05	6.0901E−05
Avg current (A)	5.6979E−05	3.2070E−05	3.5968E−05	4.6847E−05
Peak current (A)	5.6979E−05	3.2070E−05	2.4412E−05	4.6847E−05

Table 4 Performance of modified two-stage transconductance OP-AMP design

Input (V)	Average current (A)	Average power (W)	Average energy (J)	Delay (S)
0.4	297.91×10^{-6}	111.7×10^{-6}	55.87×10^{-12}	770.48×10^{-12}
0.7	732.92×10^{-6}	485.38×10^{-6}	242.76×10^{-12}	903.27×10^{-12}
0.9	995.97×10^{-6}	849.6×10^{-6}	424.93×10^{-12}	671.58×10^{-12}
1	0.0011238	0.0010657	533.01×10^{-12}	661.21×10^{-12}
1.2	0.001384	0.0015683	784.42×10^{-12}	902.02×10^{-12}
1.3	0.0015187	0.0018638	932.13×10^{-12}	377.35×10^{-12}

6 Conclusion

In analog circuits design, the circuit performance depends on the integrated circuits present in the system. OP-AMPs are the power-consuming circuits in the system. To attain better performance a low power-consuming circuits technique based on FinFET is used in the presented design of the OP-AMP circuit. The OTA circuit is better in several factors and if implemented in FinFET the leakage effects will be eliminated. Investigation on the circuit performance was carried out for three different configurations and the results are presented.

References

1. Dai, S., Cao, X., Yi, T., Hubbard, A. E., & Hong, Z. (2013). 1-V Low-power programmable rail-to-rail operational amplifier with improved transconductance feedback technique. *IEEE Transactions on Very Large Scale Integration (VLSI) Systems, 21*, 1928–1935.
2. Taherzadeh-Sani, M., & Hamoui, A. A. (2011). A 1-V process-insensitive current scalable two-stage opamp with enhanced dc gain and settling behavior in 65-nm digital CMOS. *IEEE Journal of Solid-State Circuits, 46*, 660–668.
3. Razavi, B. (2001). *Design of analog CMOS integrated circuits*. New York: McGraw-Hill.
4. Gray, P. R., & Mayer, R. G. (2001). *Analysis and design of analog integrated circuits*. New York: Wiley.
5. Binkley, D., Hoper, C., Trucker, S., Moss, B., Rochelle, J., & Foty, D. (2003). A CAD methodology for optimizing transistor current and sizing in analog CMOS design. *IEEE Transactions on Computer-Aided Design of Integrated Circuits and Systems, 22*(2), 225–237.
6. Cabrera-Bernal, E., Pennisi, S., &Grasso, A. D., et al. (2016). 0.7 V three-stage class-AB CMOS operational transconductance amplifier. *Transactions* on *Circuits and Systems I*, *63*(11), 1807–1815. https://doi.org/10.1109/tcsi.2016.2597440.
7. Senthilkumar, V. M., Ravindrakumar, S. (2018). A low power and area efficient FinFET based approximate multiplier in 32 nm technology. In *Springer-International Conference on Soft Computing and Signal Processing*.
8. Carrillo, J. M., Perez-Aloe, R., Valverde, J. M., Duque-Carrillo, J. F., & Torelli, G. (2009). Compact low-voltage rail-to-rail bulk-driven CMOS opamp for scaled technologies. In: *Proceedings of European Conference on Circuit Theory & Design (ECCTD)* (pp. 263–266).
9. Zuo, L., & Islam, S. K. (2013). Low-voltage bulk-driven operational amplifier with improved transconductance. *Transactions* on *Circuits and Systems I, 60*(8), 2084–2091. https://doi.org/10.1109/tcsi.2013.2239161.

10. Saso, J. M., Lopez-Martin, A., & Garde, M. P., et al. (2017). Power-efficient class AB fully differential amplifier. *Electrons Letter, 53*(19), 1298–1300. https://doi.org/10.1049/el.2017.2070.
11. Allen, P, E., & Holberg, D, R., *CMOS analog circuit design* (2nd ed). Oxford University Press.
12. Razavi, B. (2002). *Design of analog CMOS integrated circuits*. Tata McGraw-Hill.
13. Aslanzadeh, H. A., Mehrmanesh, S., Vahidfar, M. B., Safarian, R., & Lotfi, A. Q. (2003). A 1-V 1-mW high-speed class AB operational amplifier for high-speed low power pipelined A/D converters using slew boost technique. In *ISLPED'03*, August 25–27, Seoul, Korea.
14. Guigues, F., Kussener, E., Duval, B., & Barthelemy, H. (2007). Moderate inversion: Highlights for low voltage design. In PATMOS 2007, *LNCS* (Vol. 4644, pp. 413–422).
15. Valero Bernal, M. R., Celma, S., & Medrano, N., et al. (2012). An ultralow-power low-voltage class-AB fully differential opamp for long-life autonomous portable equipment. *Transactions on Circuits and Systems II, 59*(10), 643–647. https://doi.org/10.1109/tcsii.2012.2213361.
16. Mahattanakul, J. (2005). Design procedure for two-stage CMOS operational amplifiers employing current buffer. *IEEE Transactions on Circuits and Systems, 52*, 766–770.
17. Shameli, A., & Heydari, P. (2006). A novel power optimization technique for ultra-low power RFICs. In *ISLPED'06*, October 4–6, Tegernsee, Germany.

Fast Converging Magnified Weighted Sum Backpropagation

Ashwini Sapkal and U. V. Kulkarni

Abstract In backpropagation (BP), Neuron's output drops either into weakly committed or strongly committed zone. Neuron's weighted sum, referred as a net, if close to zero, neuron is weakly committed, otherwise it is strongly committed. To push the weakly committed neurons in strongly committed zone, additional iterations are required, which causes the poor convergence rate. In this manuscript, the weighted sum entity of the backpropagation is magnified. This variant of the backpropagation is referred as a magnified weighted sum backpropagation (MNBP) algorithm. This net enlarging process of the MNBP makes sure that the neuron produces output in strongly committed space. As the net is magnified, it is gradient and is also magnified. It is noted here that MNBP needs lesser number of epochs for convergence unlike standard backpropagation. But, it may arise the flat spot issue. Hence, the flat spot problem is also studied and the appropriate majors are taken in the proposed algorithm to solve this problem. The implementations are carried out on parity (two bit, three bit and five bit) problem, encoder problem and standard benchmark problems. The outcomes are matched with the standard BP and its two variants named Fahlman approach and MGFPROP. Based on experimentation carried out here, it is concluded that the MNBP needs small amount of epochs for its convergence dissimilar to the standard BP and its two variants.

Keywords Backpropagation neural network · Convergence rate · Flatspot problem

1 Introduction

The backpropagation [1] algorithm is used effectively in various fields like remote sensing, agriculture, mechanical, robotics, medical and many more. Inspite of the huge success of the BP, its convergence speed is very low. Tremendous efforts are

A. Sapkal (✉)
Army Institute of Technology, Dighi Hills, Pune 411015, India
e-mail: asapkal@aitpune.edu.in

A. Sapkal · U. V. Kulkarni
SGGSI & ET, Vishnupuri, Nanded 431605, India

S. C. Satapathy et al. (eds.), *Intelligent System Design*, Advances in Intelligent Systems and Computing 1171, https://doi.org/10.1007/978-981-15-5400-1_20

taken by scientists to tackle this problem of the BP. In [2–5], the better versions of the BP are proposed which converge faster when checked with the standard BP. The basic feedforward architecture is shown by Fig. 1. The backpropagation algorithm is divided into two phases: (1) feedforward phase and (2) backpropagation phase.

1. Feedforward phase: This is the first phase of the backpropagation algorithm. As BP is supervised learning algorithm, the desired output of the given input is already known. In this phase, the output of the neuron is measured by applying input feature vector to the input layer. This measured output is tested with the expected output and error is calculated, which is subtracted value of an actual output from its desired output.
2. Backpropagation phase: In this phase, the error calculated in previous phase is backpropagated so that weights can be adjusted in such a manner that next time for the given input the error must be reduced.

In the proposed MNBP, due to the magnification of net, the chances of flat spot issue are increased. This problem is handled properly by monitoring the whole value of error. Hence, it is essential here to study the flat spot problem.

1.1 Flatspot Problem

As shown in Fig. 1, net_j and net_k are the weighted sum which are feed as a input to the middle and last layer neuron, respectively. The unipolar sigmoid function is utilized here, which is shown in 1. The flatspot issue is created due to the slope of logistic function. The slop of the unipolar logistic function are measured using equations shown in 2.

$$y_j = f(net_j) = \frac{1}{1 + \exp^{-\lambda * net_j}} + \mathbf{v_b}, \quad o_k = f(net_k) = \frac{1}{1 + \exp^{-\lambda * net_k}} + \mathbf{w_b} \tag{1}$$

$$f'(net_k) = f(net_k) * (1 - f(net_k)), \tag{2}$$

$$f'(net_j) = y_j * (1 - y_j) \tag{3}$$

$$\text{here} \quad k = 1 \quad to \quad K \quad \text{and} \quad j = 1 \quad \text{to} \quad J.$$

A slope of sigmoid function is used into weight updating rule of the backpropagation, which is given in Eq. 5 and 4, which updates hidden and output layer weights, respectively.

$$\Delta w_{k,j} = \eta * e_k * f(net_k) * (1 - f(net_k)) * f(net_j), \tag{4}$$

$$\text{here} \quad k = 1 \quad \text{to} \quad K \quad \text{and} \quad j = 1 \quad \text{to} \quad J.$$

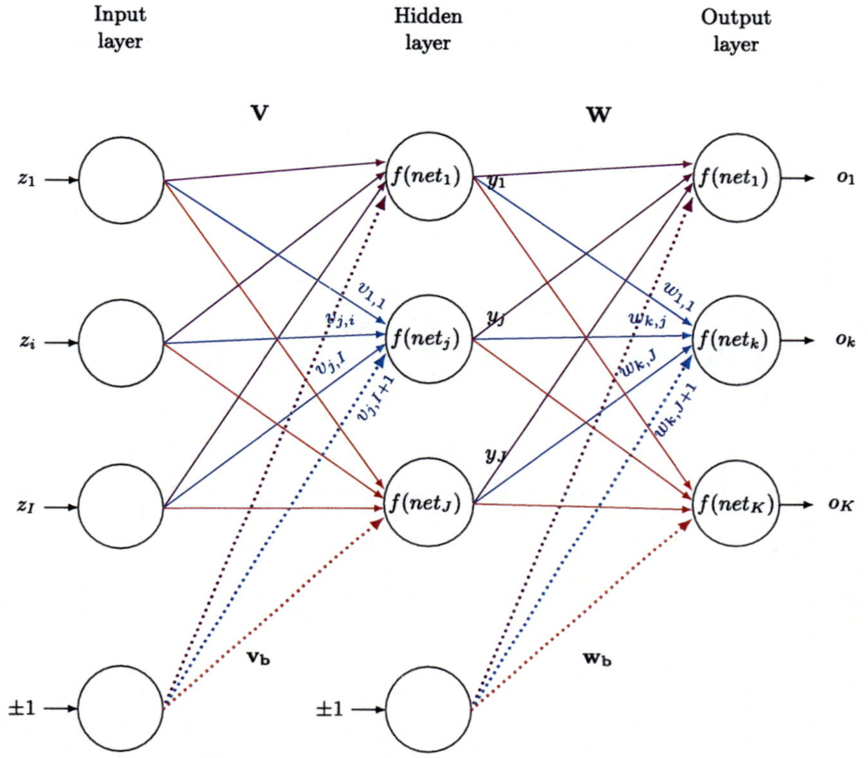

Fig. 1 Multilayer perceptron architecture

$$\Delta v_{j,i} = \eta * f(net_j) * (1 - f(net_j) + `) * (\sum_{k=1}^{K}(e_k * f'(net_k)) * w_{k,j}) * z_i, \quad (5)$$

$$\text{here } j = 1 \text{ to } J \text{ and } i = 1 \text{ to } I.$$

The flat spot situation occurs when the output of the neuron lies in the undesirable strongly committed region. If the value of the either y_j or o_k becomes very close to 0 or 1 or exactly matches with either 0 or 1, then the weights are changed with very small value or remain unaffected. This situation is acceptable if the output of the neurons is equal to its desired output. But in case of the wrong output, i.e., the desired output is either 1 or 0 and actual output is exactly opposite to it, then the weights must be restructured with great values. In such situations also with the weight updating rule shown in Eqs. 5 and 4, Many researchers have investigated and studied this problem and proposed different solution to solve this problem [6–11]. The weights are not modified properly. This scenario is known as a flat spot issue. Here, the logistic function given in Eq. 1 is used for activation of the neuron. The plot of logistic activation is shown in 2.

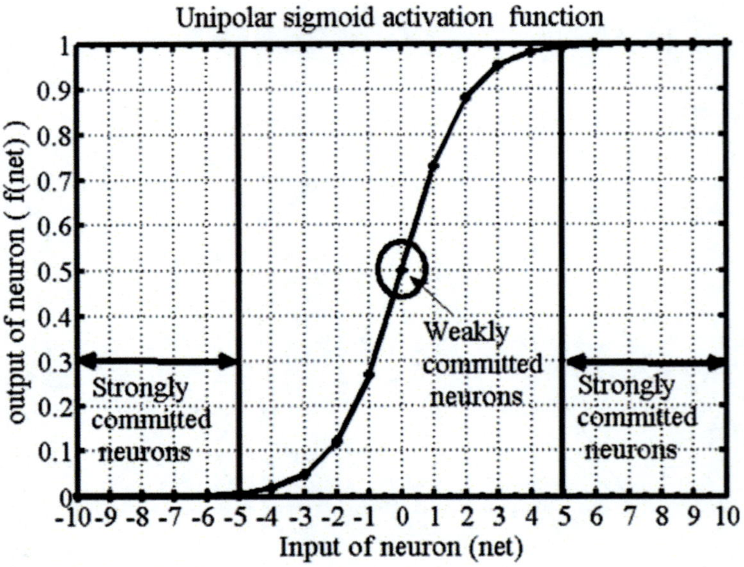

Fig. 2 Different regions of sigmoid function graph

As shown in Fig. 2, there are two main regions where the neuron's output lies, i.e., weakly committed and strongly committed region. If the net value of the neuron is either 0 or close to 0, the result of the neuron becomes 0.5 or nearby to 0.5, which does not show the clear decision about the membership of the class. Thus, it lies in weakly committed region. If neuron's output is either 0 or 1 or close to 0 or 1, it clearly shows the class membership decision of the neuron. Then, it is in strongly committed space. If the neuron's output lies in strongly committed space but exactly opposite to its desired output, then such neuron is termed as a "undesirable strongly committed neurons" which generated the wrong output.

In this manuscript, a revised BP, MNBP is presented that upturns the convergence speed of the BP. The net is enlarged using the two constant min and max. Due to enlarged input, the neurons are forced to generate output which always lies in strongly committed zone. But these changes result in flatspot problem, which can be solved by checking the complete error value. In case of incorrect output of the neuron, the standard weight updating rule is altered in which the slope of the logistic function is skipped besides it is replaced with value one (Fig. 3).

In next section, thorough implementation of the MNBP and its simulation results is discussed.

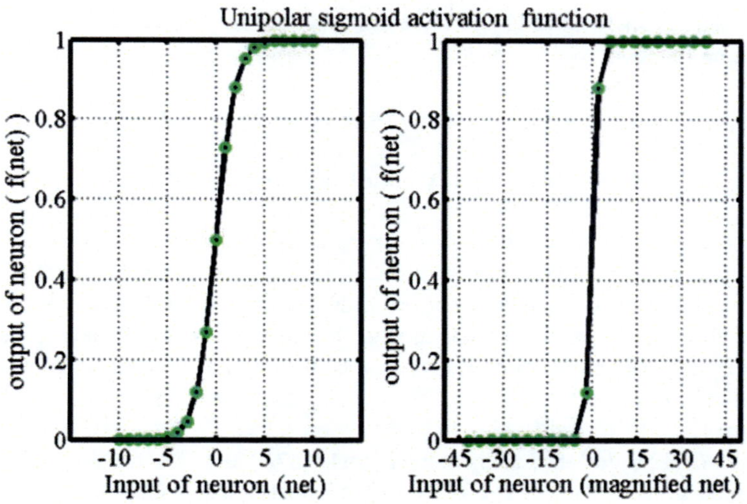

Fig. 3 Effect of magnification on weighted sum entity

2 Magnified Weighted Sum BP

To enlarge the net, in the modified BP, two constant parameters are introduced, i.e., min and max. Different values are experimented to finalize the value of min and max. It is observed based on the implementations carried out, that MNBP outperforms superior, when applied with min and max values as either -2 and $+2$ or -3 and $+3$. In the MNBP, the original calculated net is enlarged using Eq. 7.

$$net_k = \Sigma W_{kj} * f(net_j) \tag{6}$$

$$newnet_k = (max - min) * (\Sigma W_{kj} * f(net_j)) + min \tag{7}$$

As the net is modified, the slope of the logistic function is also modified, and it is also enlarged. The new slope is shown in Eq. 8

$$f'(newnet_k) = (max - min) * f(net_k) * (1 - f(net_k)) \tag{8}$$

As the neurons are emphasized to lie in strongly committed space, there are the chances of the flat spot issue. In MNBP, the whole value of the error is tested, and if it exceeds than 0.5, then the weight updating rule applies Eq. 9 or it implies Eq. 10.

$$\delta_{ok} = \eta * E_k \tag{9}$$

$$\delta_{ok} = \eta * E_k * f'(net_k) \tag{10}$$

Equation 10 is rewritten as 11 and 12, respectively, for unipolar and biopolar sigmoid activation function.

$$\delta_{ok} = \eta * E_k * (max - min) * f(net_k) * (1 - f(net_k)) \quad (11)$$

$$\delta_{ok} = \eta * E_k * (max - min) * 0.5 * (1 - f(net_k)^2) \quad (12)$$

The mean value of epochs necessary for the convergence of the algorithm in 100 runs is termed as a convergence rate. Here, the algorithm terminates if either MSE becomes less than 0.0001 or *epoch* exceeds the value 0.0001. If the MSE does not reach to 0.0001 within 30,000 epochs, then it is stated that the algorithm does not converge. The global convergence rate is measured as the number of turns the algorithm converges in 100 runs. The investigations are carried out on different problems.

The MNBP is designed using momentum term (α) also. The modified BP is also compared with other version of the MNBP. It is observed here that for each problem, the mean epochs required for the convergence of the MNBP is smaller than noted to the standard BP. It also converges within 30,000 epochs. Hence, it can be stated based on the simulation results that the MNBP convergence speed is higher unlike the standard BP and also it always converges to 0.0001 MSE within 30,000 epochs.

2.1 Algorithm's Results

The parity problem with two bit, three bit and five bit is investigated in table 1. Here, each algorithm converges successfully for two bit parity problem, but MNBP converges with less number epochs unlike the other algorithms. In the other two cases, only MNBP converges in all 100 runs with lesser number of epochs. The min and max are assigned value −2 and +2, respectively, (Tables 2, 3, 4, 5 and 6).

Table 1 Parity problem

Method	Two bit parity		Three bit parity		Five bit parity	
	Convergence rate	% of global convergence	Convergence rate	% of global convergence	Convergence rate	% of global convergence
BP	18,199	100	19,254	100	FAIL	0
Fahlman	5276	100	11,502	25	FAIL	0
MGF	1968	100	7160	60	FAIL	0
MNBP	1467	100	2809	100	5535	100

Table 2 Encoder problem

Method	Encoder		Encoder auto association	
	Convergence rate	% of global convergence	Convergence rate	% of global convergence
BP	Fail	0	Fail	0
Fahlman	6853	100	5997	100
MGF	2991	100	2810	100
MNBP	4485	100	3522	100

Table 3 Standard benchmarks problems

Method	Five bit counting		Regression		Char. recognition	
	Convergence rate	% of global convergence	convergence rate	% of global convergence	convergence rate	% of global convergence
BP	Fail	0	4974	100	11,752	100
Fahlman	8491	100	2241	100	1455	100
MGF	2688	100	1009	100	652	100
MNBP	8714	100	761	100	1126	100

Table 4 Parity problem with α

Method	Two bit parity		Three bit parity		Five bit parity	
	Convergence rate	% of global convergence	convergence rate	% of global convergence	convergence rate	% of global convergence
BP	10,536	100	11,055	100	25464	3
Fahlman	3151	100	10,029	23	FAIL	0
MGF	Fail	0	Fail	0	Fail	0
MNBP	888	100	965	100	6597	100

Table 5 Encoder problem with α

Method	Encoder		Encoder auto association	
	Convergence rate	% of global convergence	convergence rate	% of global convergence
BP	24,134	43	26,933	53
Fahlman	4049	100	3513	100
MGF	Fail	0	Fail	0
MNBP	2650	100	2150	100

Table 6 Standard benchmarks with α

Method	Five bit counting		Regression		Character recognition	
	Convergence rate	% of global convergence	convergence rate	% of global convergence	convergence rate	% of global convergence
BP	Fail	0	4867	100	6948	100
Fahlman	5035	100	2054	100	843	100
MGF	Fail	0	Fail	0	24,530	100
MNBP	5729	100	1200	100	1250	100

2.2 *Algorithm's results with momentum*

Like parity problem, the outcomes of the MNBP are improved in case of other applications used here in experimentation. The MGFPROP works better when applied without α. The outcomes of the online MNBP are checked with standard BP, and it is observed that MNBP performs better as compared to standard BP. In all 100 runs of the algorithm, MNBP always converges.

3 Conclusions

In this manuscript, the revised backpropagation, MNBP is presented, that modifies the net in such a way that neuron's will always lie in strongly committed part. These modifications also magnifies the gradient, and hence, it requires a reduced number of epochs. The flat spot issue may occur due to these changes, and it is carefully handled in the revised BP by inspecting the complete value of the error. It is observed here that the MNBP needs small number of epochs, and it successfully converges for all 100 runs. Further, these modifications can be applied to the other variations of the backpropagation such as Rprop [12] and Quickprop [2] to monitor its enactment.

References

1. Rumelhart, D. E., Hinton, G. E., & Williams, R. J. (1986). Learning internal representations by error propagation. In *Parallel distributed processing: Explorations in the microstructure of cognition* (Chap. 8). Cambridge: MIT Press.
2. Fahlman, S. E. (1988). Faster-learning variations on backpropagation: An empirical study. In *Proceedings of the 1988 Connectionist Models Summer School* (pp. 38–51).
3. Chen, J. R., & Mars, Stepsize variation methods for accelerating the back-propagation algorithm. In *Proceedings of the International Joint Conference on Neural Networks* (Vol. 1, pp. 601–604).

4. Ng, S. C., Cheung, C.-C., & Leung, S. H. (2004). Magnified gradient function with deterministic weight evolution in adaptive learning. *IEEE Transactions in Neural Networks, 15*(6), 1411–1423.
5. Gori, M., & Tesi, A. (1992). On the problem of local minima in back-propagation. *IEEE Transactions on Pattern Analysis and Machine Intelligence, 14*(1), 76–86.
6. Vitela, J. E., & Reifman, J. (1997). Premature saturation in backpropagation networks: Mechanism and necessary conditions. *Neural Networks, 10*(4), 721–735.
7. Cheung, C.-C., Ng, S.-C., Lui, A. K., & Xu, S. S. (2013). Solving the local minimum and flat-spot problem by modifying wrong outputs for feed-forward neural networks. In *Proceedings of IJCNN 2013*, pp. 1463–1469, Dallas, Texas, USA, August 2013.
8. Cheung, C.-C., Ng, S.-C., Lui, A. K., & Xu, S. S. (2014). Further enhancements in WOM algorithm to solve the local minimum and flat-spot problem in feed-forward neural networks. In *Proceedings of IJCNN 2014*, pp. 1225–1230, Beijing, China, July 2014.
9. Parekh, R., Balakrishnan, K., & Honavar, V. (1993). An empirical comparison of flatspot elimination techniques in backpropagation network. In *Proceedings of the Third Workshop on Neural Networks: Academic/Industrial/NASA/Defense* (pp. 55–60). San Diego, CA: Society for Computer Simulation.
10. Moallem, P., & Ayoughi, S. A. (2011). Removing potential flat spots on error surface of multilayer perceptron (MLP) neural networks. *International Journal of Computer Mathematics, 88*(1), 21–36.
11. Yang, B., Wang, Y.-D., Sui, X.-H., & Wang, L. (2004). Solving flat-spot problem in backpropagation learning algorithm based on magnified error. In *Proceedings of the Third International Conference on Machine Learning and Cybernetics* (pp. 1784–1788), Shanghai, August 2004.
12. Riedmiller, M., & Braun, H. (February 1993). A direct adaptive method for faster backpropagation learning: The RPROP Algorithm. In *Proceedings of International Conference on Neural Networks* (Vol. 1, pp. 586–591).

Quantum Cryptography Protocols for Internet of Everything: General View

Ch. Nikhil Pradeep, M. Kameswara Rao, and B. Sai Vikas

Abstract Data Security is a significant issue these days in light of the fact that the data contains individual information, organization's exchanges and so forth, and with that data anybody can hurt anybody's life. Each system on the planet attempts to look into the system and need to control or need to get to the information. Notwithstanding this, these days IoT, IoE came into the picture where individuals' close to home information can be assembled. Mostly in IoE, individuals' information is given high significance. Thus, there is extraordinary need of giving security to the information. In this sort of circumstance, Cryptography assumes a key job to ensure the information. Exemplary Cryptography calculations like RSA, ECC are not progressively effective in securing the information as they depend on scientific estimations. These numerically based calculations might be unscrambled in one or other manner. Along these lines, to upgrade the security for information, Quantum came into the image. This Quantum Cryptography is progressively proficient when contrasted with different calculations as it depends on the idea Quantum Entanglement and Heisenberg vulnerability guideline. This paper manages all parts of quantum key dispersion which are significant crude for Quantum Cryptography alongside Quantum Key Distribution Protocols utilized for creating IoE security.

Keywords Data security · Quantum cryptography · IoT · IoE

1 Introduction

The Internet of Everything (IoE) is an idea that expands the Internet of Things (IoT) accentuation on machine-to-machine (M2M) interchanges to portray a progressively

Ch. Nikhil Pradeep (✉) · M. Kameswara Rao · B. Sai Vikas
Department of Electronics and Computer Engineering, KLEF, Vaddeswaram, Guntur Dist., India
e-mail: nikhilpradeepchittoor@gmail.com

M. Kameswara Rao
e-mail: kamesh.manchiraju@kluniversity.in

B. Sai Vikas
e-mail: saivikas7209@gmail.com

S. C. Satapathy et al. (eds.), *Intelligent System Design*, Advances in Intelligent Systems and Computing 1171, https://doi.org/10.1007/978-981-15-5400-1_21

perplexing framework that likewise incorporates individuals and procedures. To state basically, it is another type of data trade in broadcast communications [1].4 This idea was started at Cisco, who characterizes IoE as "the wise association of individuals, procedure, information and things." The more far-reaching IoE idea incorporates machine-to-machine interchanges (M2M), machine-to-individuals (M2P) and creativity have enabled individuals-to-individual (P2P) partnerships. The IoE, then again, likewise incorporates client created interchanges and cooperation related to the worldwide aggregate of organized gadgets (Fig. 1).

As in IoE, individuals additionally are being included it will manage a lot of delicate information. These information need fitting security and respectability. In the present situations, elliptic bend cryptosystems (ECCs) are the well-known decisions for IoE security [1]. Nonetheless, as indicated by cisco, IoE's utilization will be expanded more in the future. As indicated by Gartner, in the coming decades, the quantum PCs are relied upon to show up and that will positively unravel every one of the ECCs effectively. In this way, IoE doesn't have an appropriate security structure over the long haul. That is the primary inspiration for us to ace posture Quantum Cryptography (QC) for IoT as strong security that can continue the dangers from the quantum PCs.

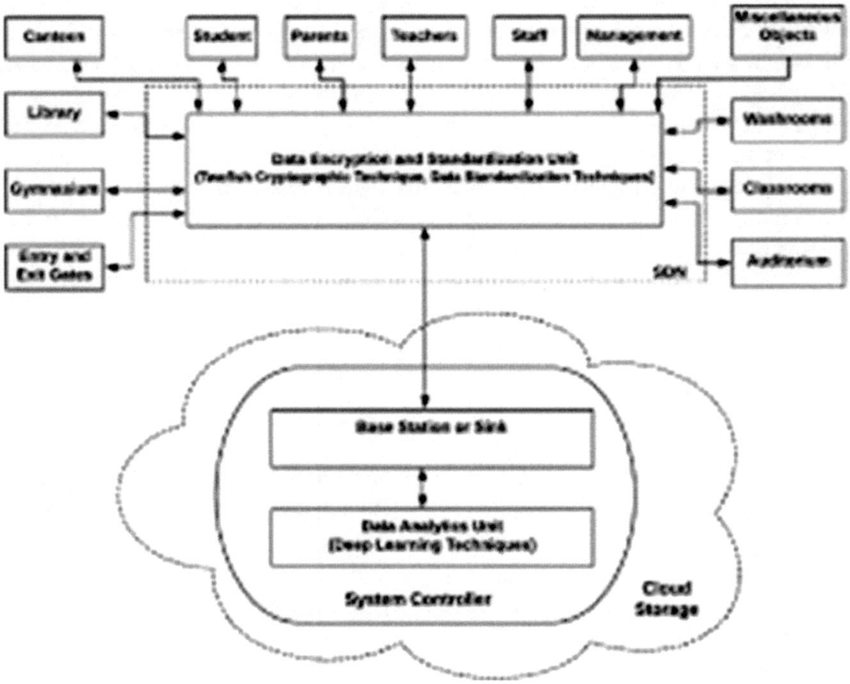

Fig. 1 Architecture of IoE

2 Quantum Cryptography

Quantum Cryptography is the Emerging Technology [2] in the Information Security Field. It depends on the idea of Quantum Mechanics which are making this cryptography strategy remarkable when contrasted with old-style cryptography calculations. This Quantum Cryptography is created by refreshing the current Cryptography Techniques. It depends on quantum mechanics and on the Heisenberg Uncertainty guideline and the standard of photon polarization [3]. At present these are simply speculations and there are no right executions as the quantum processing is still in the newborn child organize. Quantum Cryptography can be applied in both optical and remote correspondence which is basically a fundamental prerequisite for IoE Security [3].

There are more than one Quantum Primitives, however, the most significant among those ideas is Quantum Key Distribution (QKD). By utilizing the quantum properties of light, lasers, and so on, can be utilized for Quantum Key Distribution, with the goal that security can be based on the laws of quantum as it were. This QKD is as of now actualized in arrange structures like SECOQC system and DARPA organize. To increase new security properties QKD is an instrument which is can be utilized in frameworks (Fig. 2).

So as to become familiar with Quantum Cryptography we have to think about their properties which are given underneath.

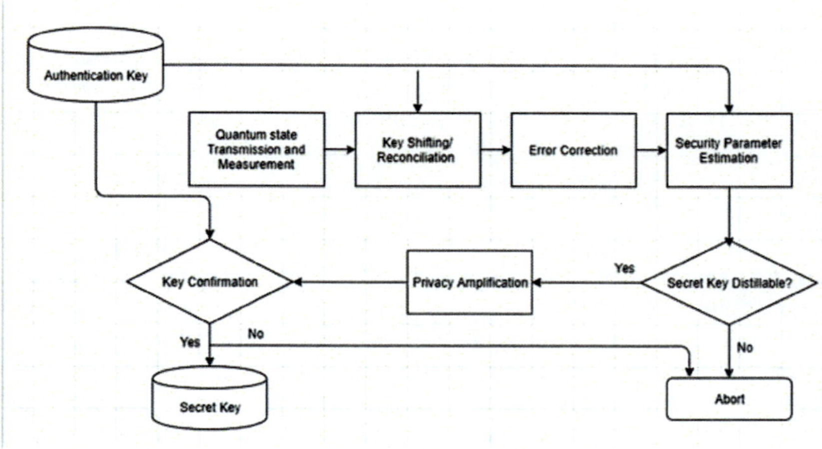

Fig. 2 Flowchart for quantum key distribution protocol

3 Quantum Properties

A Quantum superposition is portrayed [4] by a probabilistic wave work, which finds the probability of quantum in a specific position, yet not its real position which is only like the Schrodinger wave condition. Quantum is having numerous states and it exists in every one of the states at the same time without an eyewitness which is known as quantum superposition.

The Heisenberg Uncertainty Principle states that on the off chance that we need to figure out the [5] quantum position that can be photon, electron, or anything else, we can't know the exact speed of the quantum molecule and the other way around. This vulnerability exists to secure the quantum, its precise position, or its speed.

One of the Quantum properties which are much the same as QKD [3] is Quantum Entanglement. It expresses that quanta sets can be delivered which act as EPR sets. For instance, quanta have a property called 'turn' that is on the off chance that one quantum could have turn up, the other one twists inverse way, so the all out turn is said to be in a killed state. Be that as it may until an estimation is done it isn't clear which has a place with which pair. In the event that the pair is segregated, it causes the other's wave capacity to fall into an opposite state which is known as the "EPR mystery".

Quantum Channel [5] can transmit information between two gatherings. From the Physical viewpoint, it is a light quantum since quantum conditions of photons can be transmitted crosswise over bigger separations without decoherence by other quantum particles. There might be misfortunes because of dissipating, this don't influence the general security of QKD.

The bearing of the vibration of electromagnetic waves is called polarization [4]. Polarization of the photons is made by going typical light through a channel for a particular point of polarization. The likelihood of each channel relies upon the contrast between the approaching photon and the polarization edge of the channel. On the off chance that the estimation of one of the polarizations is randomizing the other polarization then the two bases are said to be conjugate (Fig. 3).

Quantum No-Cloning [6] avoids the formation of duplicates of obscure quantum states. It is another method for securing the Quantum Theory that is duplicating obscure quantum states will give client to gauge the quantum precisely. Due to this method meddler can't make a copy duplicate of quantum data sent through quantum channel. This likewise expresses quantum signal can't be intensified in a channel.

The measurements of quantum states, which is known as Quantum Security, lead to adjusting the quantum framework. In this manner, Eve can't get the data with been distinguished in the system. Regardless of whether Eve utilizes quantum PC, he will be distinguished by the laws of the material science. Lamentably [4], unlimited security can be accomplished just on the off chance that it obeys for these conditions.

- Eve can't get to Alice and Bob gadgets to watch or control the creation or recognition of photons.
- Random number generator which is utilized by Alice and Bob must be really arbitrary.

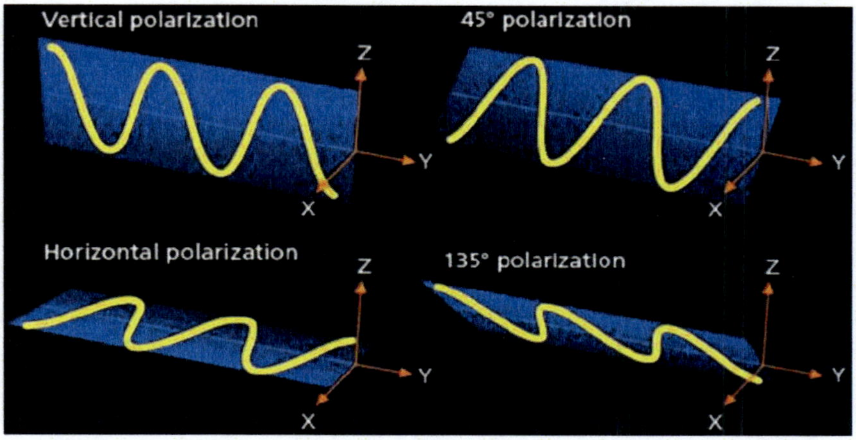

Fig. 3 Photon polarization direction [8]

- Traditional validation should be finished with genuinely secure conventions.
- Eve must obey laws of material science. With these particulars, QKD conventions will be watched and significant security verifications will be recognized.

4 A Comparison of Quantum Cryptography Protocols and Implementation Security

4.1 Comparative Analysis

See Table 1.

Table 1 Comparative analysis of different protocols [8]

Protocols	PNS attacks	Quantum states	Key generation	Polarization
BB84	Yes	4	–	For single particle
B92	Yes	2	Orthogonal	One particle
SARG04	No	Doesn't know the degree of certainty	4 Non-orthogonal quantum states	One particle
E91	No	Based on bells inequality	Estimation of the polarization of photographs are separated into two gatherings	Entangled particles
BBM92	Unconditional security	Contains EPR pairs	Contains EPR pairs	One particle

4.2 *Implementation Security*

Given that the hypothetical security of a cryptographic convention is known, we think about how it tends to be moved into a genuine cryptosystem. This inquiry characterizes the "execution security" of a convention, which especially relies upon the suppositions made about the QKD gadgets. Such gadgets are regularly thought to be "impeccable" (i.e., to carry on precisely as portrayed in the model used to demonstrate QKD security), yet by and by they show flaws, for example, deviations from the admired model, which may debilitate security. Evaluating the size of the deviations between the framework and the perfect and decreasing them adequately is the primary objective of usage security.

5 Significance of Security in IoE

Nearly everything that can be utilized for good can likewise be utilized for terrible. In like manner, while IoE innovation doing great to the individuals it can likewise do awful to the individuals. IoE makes more assault vectors in light of the fact that expanded network makes more attack vectors for awful entertainers to abuse. Over half of the organizations said that they couldn't stop the break since it dodged their current precaution measures. One of the break or assault is Distributed Denial of Service (DDoS) assault against Dyn in October 2016. This episode utilized a botnet named "Mirai", which comprises more than 100,000 IoE has, which likewise incorporates computerized cameras and switches. This botnet propelled DDoS assaults against Dyn and cut down its DNS which has brought about blackout of significant business sites [7]. The majority of the organizations which got assaulted, can't discover how the break has happened in their system. 32% of the organizations have taken over 2 years to discover the purpose of the break. Because of these developing dangers, it is to some degree hard to bring issues to light on potential IoE security dangers among end clients through individual hazard appraisals and representations. Particularly, home clients are helpless in light of the fact that they are constantly encompassed by IoE apparatuses, however, they were inadequate with regards to the assets or aptitudes to recognize their very own dangers.

6 Quantum Cryptography in IoE

So as to give security to IoE, we have to discover the benefits that can be focused for digital assaults. Next, we have to examine the passage focuses that can be utilized by the aggressors. Later on, building risk situations and organizing them. CISCO has just structured and security administration system for IoE. Quantum Cryptography has numerous focal points. Cryptography is the innovation which supports in the quantum world [3]. We are expecting that in next hardly any years' quantum PCs will appear. They will be accessible for registering applications. On the off chance that occurs, all the as of now existing cryptographic advancements like ECC and so on, will come up short aside from Quantum Cryptography. As there is increment in the use of IoE resources, the security provisioning for those advantages is significant. With the approach of quantum PCs, new security dangers can be presented on the grounds that all the current cryptographic calculations can be unscrambled effectively. Along these lines, Quantum Cryptography as it utilizes quantum mechanics can deal with the complexities made by quantum PCs. Indeed, even this Quantum Cryptography can be actualized in optical and remote interchanges which assumes a significant job in IoE engineering.

As IoE has various parts (People, Process, Data, Things) included, Quantum Cryptography can be actualized to every one of the parts independently. Indeed, even in each part, this cryptography can be executed to the accessible layers independently. For instance, it very well may be utilized in physical layer so as to check or distinguish any interruption into the frameworks. A portion of the potential assaults on IoE is Sniffer, DOS, bargained key, Password-based and Man in the center assaults. In light of these assaults likewise, we can settle the layers so we can maintain a strategic distance from these potential assaults. Therefore, in view of the system setups and resources, Quantum Cryptography must be picked cautiously [8].

7 Conclusion

In the present day, information security is turning out to be significant criteria and if quantum PCs appear the circumstance will be much more dreadful. In this way, Quantum Cryptography is the most progressive degree of securing the information to determine the information honesty issues. This is a vigorous Technology, it can deal with the security dangers which are sup-presented to rise up out of the quantum PCs.

Every one of these conventions and ideas are simply hypothetical methodologies for the Quantum Cryptography in the Information Security Field and suits well for IoE related applications in light of the fact that the utilization IoE was expanding more and furthermore going into all the accessible basic parts of associated living and keen condition. Till now there is no definite Implementation of Quantum Cryptography as quantum PCs are in the baby organize, however, when quantum PCs supplant the present age PCs every one of these ideas and conventions assumes a significant job in giving security to the IoE applications.

References

1. Miraz, M. H., Ali, M., Excell, P. S., & Picking, R. (2015). *A review on Internet of Things (IoT), Internet of Everything (IoE) and Internet of Nano Things (IoNT), Internet Technologies and Applications (ITA).*
2. Hughes, R. J., Alde, D. M., Dyer, P., Luther, G. G., Morgan, G. L., & Schauer, M., *Quantum cryptography.* University of California, Physics Division, Los Alamos, National Laboratory Los Alamos, NM 87545. LA-UR-95-806.
3. Routray, S. K., Jha, M. K., Sharma, L., Nyamangoudar, R., & Javali, A. (2017). Quantum cryptography for IOT: A perspective. In *2017 International Conference on IoT and Application (ICIOT).* Bangalore, India: Department of Telecommunication Engineering, CMR Institute of Technology.
4. Cobourne, S.: *Quantum key distribution protocols and applications.* Technical Report RHUL–MA–2011–05.
5. Sharbaf, M. S., *Quantum cryptography: An emerging technology in network security.* Senior IEEE Member. msharbaf@ieee.org. Loyola Marymount University California State University, Northridge Sharbaf & Associates. 978-1-4577-1376-7/11/.
6. Quantum No-Cloning. https://en.wikipedia.org/wiki/No-cloning_theorem. 12-1-2019.
7. Ryoo, J., Kim, S., Cho, J., Kim, H., Tjoa, S., & DeRobertis, C. V. (2017). IoE security threats and you. In *2017 International Conference on Software Security and Assurance (ICSSA).* Penn State Altoona, 3000 Ivyside Park, Altoona, PA, 16601.
8. Nikhil Pradeep, Ch., Kameswara Rao, M., & Sai Vikas, B. (2019). *Quantum cryptography protocols for IOE security: A perspective.* Department of Electronics & Computer Engineering, KLEF, Vaddeswaram, Guntur Dist.
9. Cangea, O., & Oprina, C. S. (2016). Implementing quantum cryptography algorithms for data security. In *ECAI 2016—International Conference, Electronics, Computers and Artificial Intelligence* (8th ed.), 30 June–02 July 2016. Ploiesti, ROMÂNIA: Department of Control Engineering, Computers and Electronics, Petroleum-Gas University.
10. Chen, C.-Y., Zeng, G.-J., Lin, F.-J., Chou, Y.-H., & Chao, H.-C., *Quantum cryptography and its applications over the internet.* IEEE Network.
11. SARG04. https://en.wikipedia.org/wiki/SARG04. 15-1-2019.
12. Haitjema, M., A *survey of prominent quantum key distribution protocols.* https://www.cse.wustl.edu/~jain/cse571-07/ftp/quantum/.

Secure Mobile-Server Communications Using Vector Decomposition Problem

P. Arjun Gopinath and I. Praveen

Abstract Rapid advancement in the communication sector through mobile networks has made handheld devices such as smartphones and tabs as a medium for most of their private transactions and communications. This necessitates the speed and efficiency of communication at a lower cost, without compromising the security. The proposed scheme is an authenticated key agreement protocol for the communication between the user and the service provider. This scheme supports batch verification and reduces the number of verification steps required and is based on the hardness of the vector decomposition problem (VDP).

Keywords Wieless Communication · Batch Verification · Vector Decomposition Pairings

1 Introduction

The fast-growing technology of communication enables users to access network service anywhere and anytime. The number of mobile clients that require temporary as well as permanent connection with a central server has increased in the recent years, and hence, users access these servers through insecure channels, so it faces various security threats and attacks. There has been intense study about various authentication protocols in mobile-server architecture. Traditional authentication protocols based on public key cryptosystem consumes large amount of time and power, so some of these studies were considered as ineffective. Hence, necessity for a much efficient authenticated key agreement scheme between these entities is significant. However, cryptographic protocols based on elliptic curves provide RSA equivalent security with smaller keys which are desirable for mobile-server environments as in

P. Arjun Gopinath · I. Praveen (✉)
Department of Mathematics, Amrita School of Engineering, Coimbatore,
Amrita Vishwa Vidyapeetham, Coimbatore, India
e-mail: i_praveen@cb.amrita.edu

P. Arjun Gopinath
e-mail: arjungopinath.96@gmail.com

S. C. Satapathy et al. (eds.), *Intelligent System Design*, Advances in Intelligent Systems and Computing 1171, https://doi.org/10.1007/978-981-15-5400-1_22

[7]. The proposed scheme is similar to [11] which uses a novel security assumption called VDP assumption. We use VDP assumptions for security, initially suggested by Yoshida [14].

Initially suggested by Yoshida [14], VDP on a higher dimensional provides equivalent security as that of CDHP on a one-dimensional space. Duursma and Kiyavash [2] introduced genus 2 non-supersingular curves in which VDP is hard. Galbraith and Verheul [4] and Duursma and Park [3] further analyzed VDP. Okamoto and Takashima [9] extended VDP into a higher dimensional space, and vector space suitable for this problem was given by [13]. Al-Riyami and Paterson [1] proposed certificateless public key cryptography using a key generating center (KGC). They also proved that trust needed on such KGC is less than that of a public key generator (PKG) used in ID based cryptography and certificate-based schemes. The applications of VDP in different cryptographic primitives can be seen in [5, 8, 10, 12].

KGC acts as a public key generator (PKG) in this scheme. PKG is the master authority, and in each session, entities choose a random number in order to create the session key. Due to the presence of random number, the private key will be a partial private key, even though PKG knows it. Initially, entities will register with PKG. PKG creates its own secret key. This key is called the master secret key. This key is used to provide secret key to each entity who register in the setup. Once the registration is complete, entities can communicate with themselves without the presence of PKG.

2 Preliminaries

2.1 Bilinear Pairing

Let G_1,G_2, and G_3 be three cyclic groups of prime order 'q'. Let $e : G_1\ X\ G_2 \mapsto G_3$ be a map which follows the following properties:

Bilinearity: $for\ x_1, x_2 \in G_1\ and\ y_1, y_2 \in G_2$

- $e(x_1 + x_2, y_1) = e(x_1, y_1)e(x_2, y_1)$ and $e(x_1, y_1 + y_2) = e(x_1, y_1)e(x_1, y_2)$

Non-degeneracy: $e(x, y) \neq 1,\ \forall x \neq y$.
Computability: There exists an efficient algorithm to calculate pairing.
Then, e is a bilinear pairing.

2.2 Vector Decomposition Problem

Generalized Computational Vector Decomposition Problem: Let $\mathbb{V}$ be a vector space over a field $\mathbb{F}_p$. $\{v_1, v_2, \ldots v_n\}$ be a basis for $\mathbb{V}$. Given $A \in \mathbb{V}$, VDP is to compute the element $B \in \mathbb{V}$ such that $B \in < v_1, v_2, \ldots v_m >$ and $A - B \in < v_{m+1}, \ldots, v_n >$, $m < n$ with respect to the basis $\{v_1, v_2 \ldots, v_n\}$.

Generalized Computational Diffie–Helman Problem: Suppose $\mathbb{V}$ be a n-dimensional vector space over a field F_p. Let $(B_m, B_{m+1}, \ldots, B_n)$ and $(B'_{m+1}, B'_{m+2}, \ldots, B'_n) \in \mathbb{V}$, where $m < n$. Consider the vector $a = \sum_{i=m+1}^{n} x_i B_i$ and $b = \sum_{i=m+1}^{n} x_i B'_i$ where $x_1, x_2, \ldots, x_n \in F_p$. Given a, B_i, B'_i, $i = m+1, \ldots, n$ the problem is to find b.

2.3 *Trapdoor for VDP*

Trapdoor function is used for authentication and encryption in this scheme. Okamoto and Takashima [9] proposed the concept of distortion eigenvector space as a trapdoor for VDP.

Distortion Eigenvector Space: A n-dimensional vector space $\mathbb{V}$ is defined over a field F which satisfies the following properties:

- There exists B, F, and $\{\phi_{i,j}\}_{1\le i,j\le n}$ where $B = (B_1, B_2, \ldots, B_n)$ be a distortion eigenvector basis of F_r-vector space $\mathbb{V}$. F is a polynomial-time computable automorphism on $\mathbb{V}$. Here, B_i's are eigenvectors of F with different eigenvalues. $\{\phi_{i,j}\}$ is called a distortion map and is a polynomial-time computable endomorphism of $\mathbb{V}$ such that $\{\phi_{i,j}\}(B_1) = B_1$.
- A skew-symmetric non-degenerate bilinear pairing $e : \mathbb{V}\ X\ \mathbb{V} \mapsto \nu_p$ exists where ν_p is a multiplicative group of order p.
- A polynomial-time computable automorphism ρ on $\mathbb{V}$ exists and $e(v, \rho(v) \neq 1$ for all v except that in a quadratic hypersurface of $\mathbb{V} \cong (F_r)^n$.

Lemma 1 *Suppose $B = (B_1, B_2, \ldots, B_n)$ be a distortion eigenvector basis of $\mathbb{V}$ and let B_i has eigenvalue λ_i of A. Then, the projection operator Po is a polynomial of A such that: $Po_j(v) = (\Pi_{i+j}(\lambda_j - \lambda_i))^{-1}\Pi_{i+j}(A - \lambda_i)(v)$.*
Then,

$$Po_j(B_i) = 0 for i \neq j \text{ and } Po_j(B_j) = B_j.$$

Consider the matrix $M = (m_{ij})$ such that $U_i = \sum_{j=1}^{n} m_{ij} B_j$. Then, $U = (U_1, U_2, \ldots, U_n)$ will also forms a basis for $\mathbb{V}$. Let $v = \sum_{i=1}^{n} c_i U_i$ be a vector in $\mathbb{V}$. Then, we have to decompose v into sum of two vectors where one exists in the subspace generated by $(U_1, U_2, \ldots, U_m) m < n$ and the other exists in the subspace generated by $(U_1, U_2, \ldots, U_n)$. Suppose $M^-1 = (t_{ij})$, then as per lemma 3 of [9], the function

$$Vdeco(v, < U_j >, M, < U_1, U_2, \ldots, U_n >) = \sum_{i=1}^{n}\sum_{k=1}^{n} t_{ij} m_{jk} \phi_{ki}(Po_i(v))$$

accomplishes our aim.

3 Proposed Scheme

The proposed scheme is an efficient mobile-server authenticated key agreement protocol that has three phases: (I) system setup phase (II) user registration phase, and (III) authentication and key agreement Phase.

System Setup Phase: Let $\mathbb{V}$ be a hyper elliptic curve. $F : \mathbb{V} \times \mathbb{V}$ be an endomorphism. $E[n]$ is the set of n torsion points for a prime n. $S = (S_1, S_2, S_3)$ be a distortion eigen basis for $E[n]$. Then, there exist $\lambda_1, \lambda_2, \lambda_3$ such that $F(S_1) = \lambda_1 S_1$,$F(S_2) = \lambda_2 S_2$,$F(S_3)\lambda_3 S_3$. Let $\phi_{i,j}$ be a distortion map such that $\phi_{i,j}(S_j) = S_i$ and $M = (m_{ij})$ be a matrix such that $m_{ij} \in F_r$ and $\det M \neq 0$. Let $P = (P_1, P_2, P_3)$ where P_1, P_2, P_3 are points such that: $P_1 = m_{11}S_1 + m_{12}S_2 + m_{13}S_3$, $P_2 = m_{21}S_1 + m_{22}S_2 + m_{23}S_3$, $P_3 = m_{31}S_1 + m_{32}S_2 + m_{33}S_3$. Then, P forms a basis but will not be a distortion eigenvector basis. $\mathbb{V}$, B, P are public parameters. The transformation matrix M is kept as a secret. PKG selects the distortion eigenvector basis, the matrix $M_p = (m_{ij})$, and creates $P = (P_1, P_2, P_3)$ as above. A random number x is chosen by the PKG which is the master secret key. M chooses the distortion eigenvector basis, its matrix $M_m = (n_{ij})$, and generates $N = (N_1, N_2, N_3)$.

User Registration Phase: A mobile terminal M wants to register itself with PKG. There are two registration keys namely $A_m R$ and $B_m R$ embedded in the device.

- Let y_m be a random number chosen by M, and M calculates $K_g = A_m R P_1 + B_m R P_2 + y_m P_3$. M then sends a request to PKG along with K_g.
- On receiving K_g, PKG calculates

$$Vdeco(K_g, < P_1 >, M_p, < P_1, P_2, N_3 >) \text{ to get } A_m R P_1$$

$$Vdeco(K_g, < P_2 >, M_p, < P_1, P_2, P_3 >) \text{ to get } B_m R P_2$$

- PKG verifies $A_m R P_1$ and $B_m R P_2$, and if it is same as that of M, PKG computes $Sk_m = xN_1$ and fixes it as M's secret key. PKG then encrypts xN_1 in K as $K = xN_1 + A_m R N_2 + B_m R N_3$ and sends to M.
- After receiving K, M uses:

$$Vdeco(K, < N_1 >, M_m, (N_1, N_2, N_3)) \text{ to get } xN_1$$

- M confirms the originality by verifying whether: $A_m R N_2 = Vdeco(K, < N_2 >, M_m, (N_1, N_2, N_3))$ and $B_m R P_3 = Vdeco(K, < N_3 >, M_m, (N_1, N_2, N_3))$ Any mismatch will result in the abortion of process.
- The server S selects its distortion eigenvector basis and its transformation matrix $M_s = s_{ij}$ and generates $S = (S_1, S_2, S_3)$. Then, PKG finds $Q = x_1 S_1 + x_2 S_2 + x_3 S_3$. PKG makes $Q, x_1 Q, hQ$ public where h is a random number chosen by PKG.
- S calculates: $Vdeco(Q, < S_1 >, M_s, (S_1, S_2, S_3))$ to get the secret key $Sk_s = xS_1$ After this, any device can communicate with the server using the session key.

Authentication and Key Agreement Phase: M performs the below procedure for key establishment and authentication with the server.

- M chooses the public key of the server (S_1, S_2, S_3) and selects a random number k_1 to find $k_1 S_1$.
- M finds:
  - $S_c = k_1 S_1 + k_2 S_2 + k_3 S_3$ and fixes the first coordinate as the session key
  - $S_c^{'} = k_1 hQ + Sk_m$
  - $S_c^{"} = S_c + hQ + rS_1 + r_S^{'}2$ for some random elements r and $r^{'}$.
- M sends $(S_c^{'}, S_c^{"})$ to server S.

On receiving the message, S performs the decryption and authentication by the following procedure.

- Finds $S_c^{"} - hQ = Z$ and finds $k_1 S_1$ using $Vdeco(Z, < S_1 > M_S, (S_1, S_2, S_3)$
- Finds u, the X coordinate of $k_1 S_1$.
- Verification is done in the following steps

$$e(S_c^{'}, S_1) = e(hQ, k_1 S_1) e(N_1, Sk_s) \tag{1}$$

$$e(S_c, Q) = e(N_1, x_1 Q) \tag{2}$$

- If the verification fails, restart the procedure.

3.1 Correctness of the Scheme

To find $k_1 S_1$ from $S_c^{"}$, one needs to solve VDP. As Sk_m is the secret key of mobile terminal, no adversary can find k_1 from $S_c^{'}$
During the verification process, LHS of Eq. (1) changes to:
$e(S_c^{'}, S_1) = e(k_1 hQ + Sk_m, S_1) = e(k_1 hQ + x_1 N_1, S_1) = e(hQ, k_1 S_1) e(x_1 N_1, S_1)$ $= e(hQ, k_1 S_1) e(N_1, x_1 S_1) = e(hQ, k_1 S_1) e(N_1, Sk_s)$ which is equal to RHS.
As $e(P, P) = 1$, LHS of Eq. (2) changes to:
$e(S_c^{'}, Q) = e(k_1 hQ + Sk_m, Q) = e(Sk_m, Q) = e(x_1 N_1, Q) = e(N_1, x_1 Q) =$ RHS.

3.2 Batch Verification

The scheme proposed in Sect. 3 can be used for the process of batch verification in the following way. This reduces the time required to verify the message even when there is lots of traffic in the network.

Let there be n mobile terminals in the network and if these n mobiles send $(S_{c1}^{'}), S_{c1}^{"}), (S_{c2}^{'}), S_{c2}^{"}), \ldots, (S_{cn}^{'}, S_{cn}^{"})$ to server S at a time.
Then, Eq. (2) becomes,

$$e\left(\sum_{i=1}^{n}, S'_{ci}, S_1\right) = \prod_{i=1}^{n} e\left(hQ, x_1^i S_1\right) e\left(M^i, Sk_s\right) = e\left(hQ, \sum_{i=1}^{n}, x_1^i S_1\right) e\left(\sum_{i=1}^{n} M^i, Sk_s\right)$$

where M^i and $x_1^i S_1$ are the public key and session key of the i-th device, respectively. Similarly, Eq. (2) becomes

$$e\left(\sum_{i=1}^{n} S'_{ci}, Q\right) = \prod_{i=1}^{n} e\left(S'_{ci}, Q\right) = \prod_{i=1}^{n} e\left(M^i, x_1 Q\right) = e\left(\sum_{i=1}^{n} M^i, x_1 Q\right)$$

Hence, Eqs. (1) and (2) can be used for batch verification which reduces the time.

3.3 Batch Identification

As shown above, batch verification is possible using this scheme. If that fails, the server has to find the device which sent the invalid data packet. Condensed binary identification can be used for this purpose. Let n devices send the data packet $((S'_{c1}, S''_{c2}), (S'_{c2}, S''_{c2}), \ldots, (S'_{cn}, S''_{cn}))$, which contains invalid pairs. Then, atleast one among the equations for verifications fails. Divide the n packets into a set having equal number of elements and do the batch verification separately. Repeat the process until invalid data packets are found. Let n devices send the data packet $((S'_{c1}, S''_{c2}), (S'_{c2}, S''_{c2}), \ldots, (S'_{cn}, S''_{cn}))$, which contains invalid pairs. Then, atleast one among the equations for verifications fails. Divide the n packets into a set having equal number of elements and do the batch verification separately. Repeat the process until invalid data packets are found. Let n devices send the data packet $((S'_{c1}, S''_{c2}), (S'_{c2}, S''_{c2}), \ldots, (S'_{cn}, S''_{cn}))$, which contains invalid pairs. Then, atleast one among the equations for verifications fails. Divide the n packets into a set having equal number of elements and do the batch verification separately. Repeat the process until invalid data packets are found.

4 Security Analysis

The security of the scheme depends upon the hardness of the vector decomposition problem (VDP) and elliptic curve discrete logarithmic problem (ECDLP). As the curve used satisfies conditions proposed by Yoshida (1), VDP is atleast as hard as computational Diffie–Helman problem (CDHP) defined on one-dimensional space.

4.1 Man-In-The-Middle Attack

- Suppose an adversary A wants to disturb the scheme, one method is to change the data packet sent to server (S). That is to change $(S_c^{'}, S_c^{''})$ to $(aS_c^{'}, bS_c^{''})$ where a and b are some random elements. When that happens, S calculates $Z = bS_c^{''} - hQ$ and then uses:
  $Vdeco(Z, < S_1 >, M_S, (S_1, S_2, S_3))$ which yields $bk_1S_1(b-1) + hx_1S_1 = (bK_1 + (b-1)hx_1)S_1$.
  Then, in Eq. (1), LHS will become:
  $e(aS_c^{'}, S_1) = e(ahk_1Q + ax_1)N_1, S_1) = e(a(hk_1Q + x_1N_1), S_1) = e(hk_1Q + x_1 N_1, aS_1) = e(hQ, ak_1S_1)e(N_1, ax_1S_1) = e(hQ, k_1aS_1)e(N_1, aSk_s) = e(hQ, k_1 S_1)^a e(N_1, Sk_s)^a$ RHS $= e(hQ, (bk_1 + (b-1)hx_1)S_1)e(N_1, Sk_s) = e(hQ, S_1)^{bk_1+(b-1)hx_1} e(N_1, Sk_s)$ To find a and b satisfying, LHS = RHS is very difficult since the adversary do not know about k_1 and x_1
  Also, if we take Eq. (2), we get
  LHS = $e(aS_c^{'}, Q) = e(N_1, ax_1Q) = e(N_1, x_1Q)^a$
  RHS = $e(N_1, x_1Q)$
  LHS $\neq$ RHS
- If an adversary adds aQ to $S_c^{'}$ and bS_1 to $S_c^{''}$, for some random a and b. Then, server S gets:
  $$((S_c^{'} + aQ), (S_c^{''} + bS_1))$$
  Then, after $Vdeco(Z, < S_1 > M_S, (S_1, S_2, S_3)$, server S gets $(k_1 + b)S_1$ as the output.
  In the verification procedure, LHS of Eq. (1) becomes:
  $e((S_c^{'} + aQ), S_1) = e((k_1h + a)Q, S_1) = e(x_1N_1, S_1) = e(Q, S_1)^{k_1h+a} e(N_1, Sk_s)$.
  RHS becomes $e(hQ, (k_1 + b)S_1)e(N_1, Sk_s) = e(hQ, S_1)^{k_1+b} e(N_1, Sk_s)$
  So, LHS = RHS if $a = hb$, but h is unknown.
- If an adversary changes to $S_c^{'}$ to $S_c^{'} + dS_1$ for some random d, then server receives $(S_c^{'} + dS_1, S_c^{''}$. Then, Eq. (1) will be satisfied, but Eq. (2) will become:
  $e((S_c^{'} + dS_1), Q) = e(S_c^{'}, Q)e(dS_1, Q) \neq e(N_1, x_1Q)$ which will result in the abortion of the process.

4.2 Key Substitution Attack

This attack was developed by Lim et al. on schemes by Okamoto and Takashima [6].

Proposition *Let* $S_1^{'}, S_2^{'}, S_3^{'}$ *be the public key such that for some arbitrary* w_{ij} *for* $i, j = 1, 2, 3$. *Then, let* $S_1^{'} = w_{11}S_1 + w_{12}S_2 + w_{13}S_3$, $S_2^{'} = w_{21}S_1 + w_{22}S_2 + w_{23}S_3$, $S_3^{'} = w_{31}S_1 + w_{32}S_2 + w_{33}S_3$. *Then* $S_1^{'}, S_2^{'}, S_3^{'}$ *cannot be used to decrypt* S_1, S_2, S_3.

Proof The encryption scheme is as follows:
$S_c = k_1S_1 + k_2S_2 + k_3S_3$ where S_1, S_2, S_3 are the public keys of the server. Then, the ciphertext $B_c = k_1S_1' + k_2S_2' + k_3S_3'$ obtained after encryption using S_1', S_2', S_3' as the public keys.
$B_c = k_1S_1' + k_2S_2' + k_3S_3' = k_1(w_{11}S_1 + w_{12}S_2 + w_{13}S_3) + k_2(w_{21}S_1 + w_{22}S_2 + w_{23}$ $S_3) + k_3(w_{31}S_1 + w_{32}S_2 + w_{33}S_3)$.
Let $M_{'}$ be the trapdoor associated with S_1', S_2', S_3'.
Then, $Vdeco(S_c, < S_1' >, M_{'}, (S_1', S_2', S_3')) = k_1S_1'$ and $Vdeco(S_c, < S_1 >, M_s,$ $(S_1, S_2, S_3)) = K_1S_1$.
Suppose X coordinate of k_1S_1 and k_1S_1' be x and x', respectively. Then, $x \neq x'$.

4.3 *Adaptive Chosen-Ciphertext Attack*

Indistinguishability under adaptive chosen-ciphertext attack or IND-CCA2 is an attack scenario in which an adversary sends ciphertexts to be decrypted and then uses the results to decrypt a chosen ciphertext. Also, the attacker is allowed to ask adaptive queries. To prevent this attack, one has to restrict ciphertext malleability. Ciphertext malleability is a property by which we can transform one ciphertext into another which can be decrypted. In this scheme 3, suppose the adversary A gets the corresponding secret key $(x_{11}S_1, x_{12}S_1, x_{13}S_1, \ldots, x_{1n}S_1)$ for the data packets $((S_{c1}', S_{c1}''), (S_{c2}', S_{c2}''), \ldots, (S_{cn}', S_{cn}''))$, the attack is possible when A can uses this information to decrypt $x_{1m}S_1$ from (S_{cm}', S_{cm}''). This is not possible since the mobile terminal M chooses random k_m in each session. Ciphertexts S_c' and S_c'' are obtained by the following procedure:
$S_c' = k_1hQ + Sk_m$ $S_c'' = S_c + hQ + rS_1 + r_S'2$ for some random elements r and r' where $S_c = k_1S_1 + k_2S_2 + k_3S_3$. Since r and r' are random, ciphertext malleability can be prevented. Thus, the scheme proposed is IND CCA2 secure.

5 Conclusion

We propose an authentication and key agreement protocol which can be used in mobile-server environments using vector decomposition problem. A batch verification process which helps the server for authentication as a batch has been introduced. We also suggest a method for identification if batch verification fails. The short keys of this scheme make it suitable for mobile-server environments.

References

1. Al-Riyami, S. S., & Paterson, K. G. (2003). Certificateless public key cryptography. In *Advances in Cryptology-ASIACRYPT 2003* (pp. 452–473). Berlin: Springer.
2. Duursma, I. M., & Kiyavash, N. (2005). The vector decomposition problem for elliptic and hyperelliptic curves. *IACR Cryptology ePrint Archive, 2005*, 31.
3. Duursma, I. M., & Park, S. K. (2006). Elgamal type signature schemes for n-dimensional vector spaces. *IACR Cryptology ePrint Archive, 2006*, 312.
4. Galbraith, S. D., & Verheul, E. R. (2008). An analysis of the vector decomposition problem. In *Public Key Cryptography–PKC 2008* (pp. 308–327). Berlin: Springer.
5. Kumar, M., & Praveen, I. (2015). A fully simultable oblivious transfer scheme using vector decomposition. In *Intelligent computing, communication and devices* (pp. 131–137). Berlin: Springer.
6. Lim, S., Lee, E., & Park, C.-M. (2014). Equivalent public keys and a key substitution attack on the schemes from vector decomposition. *Security and Communication Networks*, *7*(8), 1274–1282.
7. Mo, J., Zhongwang, H., & Lin, Y. (2018). Remote user authentication and key agreement for mobile client-server environments on elliptic curve cryptography. *The Journal of Supercomputing*, *74*(11), 5927–5943.
8. Nidhin, D., Praveen, I., & Praveen, K. (2016). Role-based access control for encrypted data using vector decomposition. In *Proceedings of the International Conference on Soft Computing Systems* (pp. 123–131). Berlin: Springer.
9. Okamoto, T., & Takashima, K. (2008). Homomorphic encryption and signatures from vector decomposition. In *Pairing-Based Cryptography–Pairing 2008* (pp. 57–74). Berlin: Springer.
10. Okamoto, T., & Takashima, K. (2009). Hierarchical predicate encryption for inner-products. In *Advances in Cryptology–ASIACRYPT 2009* (pp. 214–231). Berlin: Springer.
11. Praveen, I., Rajeev, K., & Sethumadhavan, M. (2016). An authenticated key agreement scheme using vector decomposition. *Defence Science Journal*, *66*(6), 594.
12. Praveen, I., & Sethumadhavan, M. (2012). A more efficient and faster pairing computation with cryptographic security. In *Proceedings of the First International Conference on Security of Internet of Things* (pp. 145–149). New York: ACM.
13. Takashima, K. (2008). Efficiently computable distortion maps for supersingular curves. In *Algorithmic Number Theory* (pp. 88–101). Berlin: Springer.
14. Yoshida, M. (2003). Vector decomposition problem and the trapdoor inseparable multiplex transmission scheme based the problem. In *The 2003 Symposium on Cryptography and Information Security SCIS'2003*.

A Hybrid Full Adder Design Using Multigate Devices Based on XOR/XNOR Logic

V. M. Senthil Kumar, Ciddula Rathnakar, Maram Anandha Guptha, and Ravindrakumar Selvaraj

Abstract Adders form the basic building blocks of several signal processing applications. Power optimization is an important requirement of the design today. Several hybrid circuits using XOR–XNOR or XOR/XNOR are implemented using CMOS devices. In this paper, FinFET device based XOR/XNOR and simultaneous XOR–XNOR functions are proposed and implemented. The proposed circuits reduce the power consumption and delay. The FinFET full swing XOR–XNOR or XOR/XNOR gates are used to implement the full adder (FA) circuits. The circuits showcase better performance in power consumption. The experimental simulation was carried out in 32-nm CMOS and FinFET process technology. The proposed FinFET hybrid adder showed superior performance when compared to CMOS. Out of the six types of adders Hybrid Full Adder of 22 transistors FinFET circuit is 90% efficient than CMOS circuit.

Keywords CMOS · FinFET · XOR · XNOR · Full adder · Low power · 32 nm

1 Introduction

Today, digital circuits like microprocessors, communication devices, and signal processors are designed mainly using CMOS devices. Increase in integration level scales down the size of the CMOS transistors. The usability of circuits is controlled by its power consumption. To reduce the power consumption and area, the device

V. M. Senthil Kumar · C. Rathnakar (✉) · M. A. Guptha
Malla Reddy College of Engineering and Technology, Hyderabad, Telangana, India
e-mail: rciddula@gmail.com

V. M. Senthil Kumar
e-mail: vmspraneeth@gmail.com

M. A. Guptha
e-mail: gsravindrakumar7@gmail.com

R. Selvaraj
Sri Shakthi Institute of Engineering and Technology, Coimbatore, Tamil Nadu, India
e-mail: ananthaguptha007@gmail.com

S. C. Satapathy et al. (eds.), *Intelligent System Design*, Advances in Intelligent Systems and Computing 1171, https://doi.org/10.1007/978-981-15-5400-1_23

size or new data analyzing method are to be adopted. Optimizing the W/L ratio can reduce the power-delay product (PDP) but the driving capacity reduces. The supply voltage reduction also reduces the driving capacity [1, 2]. The performance of many digital applications is dependent on the design of logic gates adders, etc. Due to the basic role of addition, many researches are conducted to find an efficient adder [3–5].

2 Literature Survey

In literature, adders were implemented using different configurations and circuits. The survey shows that the number of transistors used is reduced for applications were the output swing can be low. Hybrid adders are designed to meet the requirements of higher output swing and minimum power. Hybrid FAs are divided into 2-to-1 multiplexer and two input XOR/XNOR gate which are coupled [3]. In full adder cell, maximum power is consumed by the XOR/XNOR gate. Therefore, the design of XOR/XNOR gate needs additional attention. Pass transistor logic-based CMOS circuits and its operation are discussed in [6]. The applications of XOR/XNOR gates in many digital circuits are presented in literature especially the adder circuits which form the main part of a signal processing architecture. To analyze 1-bit full adder circuits it is implemented and discussed in [7, 8]. Figure 1 shows the different circuits implemented for XOR/XNOR gate [9–12].

The pass transistor logic-based XOR/XNOR gate [12] is shown in Fig. 1a. The circuit provides maximum swing at output. But the problem is the addition power consumption by the NOT gate present in the critical path. So to reduce the delay in critical path the scaling down of transistors is done. The other issue is the short circuit power. The circuit in Fig. 1b has improved power consumption and delay [9] when compared to Fig. 1a, b. The PTL logic circuit in Fig. 1b has the problem of additional power consumption due to the NOT gates present in the critical path.

2.1 Background Methodology

For the full adder implementation MOS based XOR–XNOR circuits were used [3, 7, 11, 13]. For hybrid full adders designed using XOR–XNOR circuit, the select lines of MUX are corrected with XOR–XNOR signals. But the main challenges faced are to keep the delay of concurrent signals equal to avoid glitches. In [14] and [15], full adders designed for arithmetic operations with low-power consumption are presented. A CPL logic-based simultaneous XOR–XNOR circuits [16] implemented using ten transistors is shown in Fig. 1c. The output-level voltages are improved by the NMOS transistors driving the output and two PMOS transistors linked to outputs. The PMOS transistors are cross-coupled for improving the driving capacity. The delay and power due to short circuit increases because of the feedback.

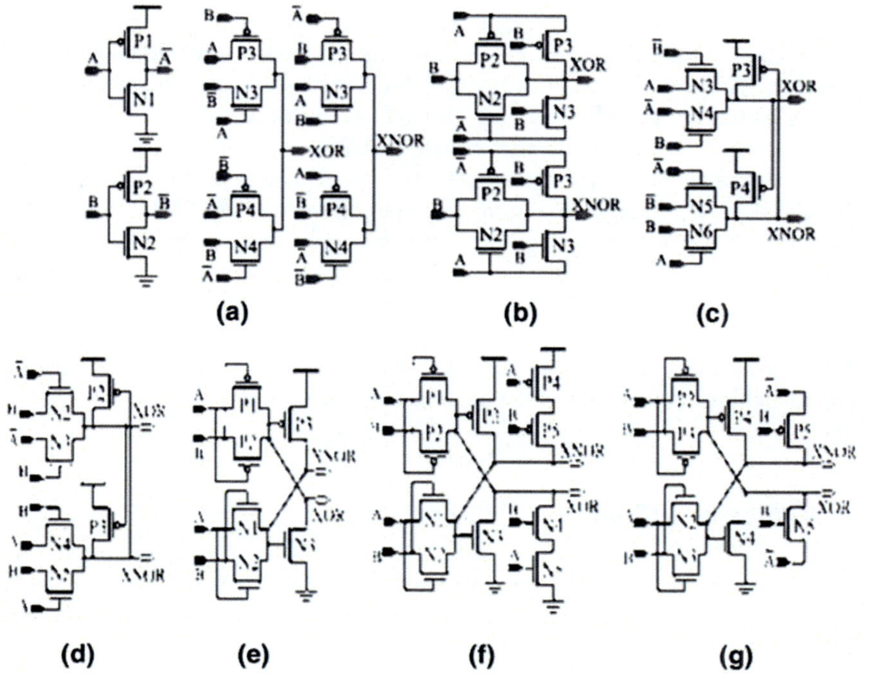

Fig. 1 **a**, **b** Full swing XOR/XNOR and **c–g** XOR–XNOR circuits. **a** [16], **b** [11], **c** [16], **d** [3], **e** [7, 13], **f** [18], **g** [23]

The XOR–XNOR circuit Fig. 1c reduces the power dissipation when implemented in a NOT gate [1]. Compared to the circuit in Fig. 1c, d has higher critical path delay. This happens due to the fact that the logic "0" conceded through N2 to XOR for AB = 0. The XNOR output is charged to VDD through XOR for "0" logic. For transistors on AB from 01 to 00, the short circuit current passes in circuit. For AB = 01, the passing of "0" logic to XOR output via N4 and logic "1" to XOR output through N2, N3, and P2 happens. All transistors get OFF for AB = 00 (except N2 and P2). So through P2 and N2, the short circuit current passes. The switching of XOR and XNOR output avoided due to this current which is drawn from VDD. Functioning of the circuit is maintained by increasing the resistance of PMOS transistors than the PMOS transistors, i.e. RP2 > RN2 and RP3 > RN5. The circuit functionality is affected more and often fails for irregular scaling of transistors. The full swing XOR/XNOR gate is shown in Fig. 1e [4, 13]. The output nodes of XOR and XNOR are reinstated by the feedback loop for weak logic made by the NMOS N3 and PMOS P3 complementary pair. This state happens for input combinations AB = 00 and AB = 11. For the transition of AB from 01 to 11 and 10 to 00, the output takes two steps to reach final output with increased delay. The circuit in Fig. 1e is modified using instantaneous XOR–XNOR gate using six transistors (Fig. 1f).

As the circuit in Fig. 1e suffers from slow response, the circuit in Fig. 1f was presented by Chang et al. [16]. The SoC transistor circuit was modified by adding two gates (XOR and XNOR) and NMOS/PMOS transistor pairs. This improves the driving capability, output swing and immune towards improper scaling of transistors. The supply voltage scaling reduces the robustness of circuit [17]. But adding compensation elements can be improved, but the delay and power increase due to the additional capacitance. To improve the speed instead of this PMOS and NMOS pairs a NOT gate is used in Fig. 1g. This circuit provides path to VDD and GND through N5 and P5 for few logic states. Similar to previous circuit this circuit also suffers from additional parasitic capacitance which is more than the capacitance of the circuit in Fig. 1e [18–20]. So from the analysis, it has been found that the circuits in Fig (a–g) have its own advantages and disadvantages. The adders are the most important block of several application circuits [21–25].

3 Proposed Design

The six FA circuits for different applications are shown in Fig. 2. These FAs are designed using MOS devices and are having a hybrid logic style. The designed logic circuit is designed using CMOS (Fig. 2) and proposed FinFET (Fig. 3).

The hybrid FA (HFA-20T) shown in Fig. 2a designed by two 2-to-1 MUX gates and the XOR–XNOR gate using CMOS. The FinFET design is shown in Fig. 3. The proposed FinFET-based 20T circuit has low power, low leakage and full voltage swing. The advantages of the design are its robustness towards scaling, high-speed operation and sizing of transistors. But the HFA-20T suffers from low output driving capability. So not suitable to be used in circuits were chain structure is available. The FA structures power consumption can be reduced as discussed in the review by the use of NOT and XOR/XNOR logic gates. The XOR gate in Fig. 1b is used to design the existing CMOS and FinFET adder cell (HFA-17T) shown in Figs. 2b and 3b. This structure is made by 17 transistors but the delay increases when compared to HFA-20T due to NOT gates on the critical path. The power consumption is reduced than that of HFA-20T due to the transistor count reduction. The short circuit power is more due to the critical path. So the power dissipation of the HFA-17T is moderate. Figure 2c presents the 26 transistors based FA with buffers. The critical path has one 2-1-MUX, XOR–XNOR gate and NOT gates. The driving problems are reduced by the NOT gate. The power consumption and delay of this circuit are more than that of Fig. 2a, b. So the FinFET design is not considered for analysis in this paper. Some of the other conventional CMOS circuits are shown in Fig. 2d which has 26 transistors. The final value in the data input nodes of 2-1-MUXs should reach VDD or GND before the results of XOR and XNOR stays valid. The critical path of this circuit constitutes the 2-1 MUX and XOR–XNOR gates. The delay is nominal when compared to 26 transistor circuit HFA-B.

The HFA-NB circuit suffers from lower driving capability when compared to the HFA-B-26T circuit. The MUX gate is the main reason for this drawback. This design

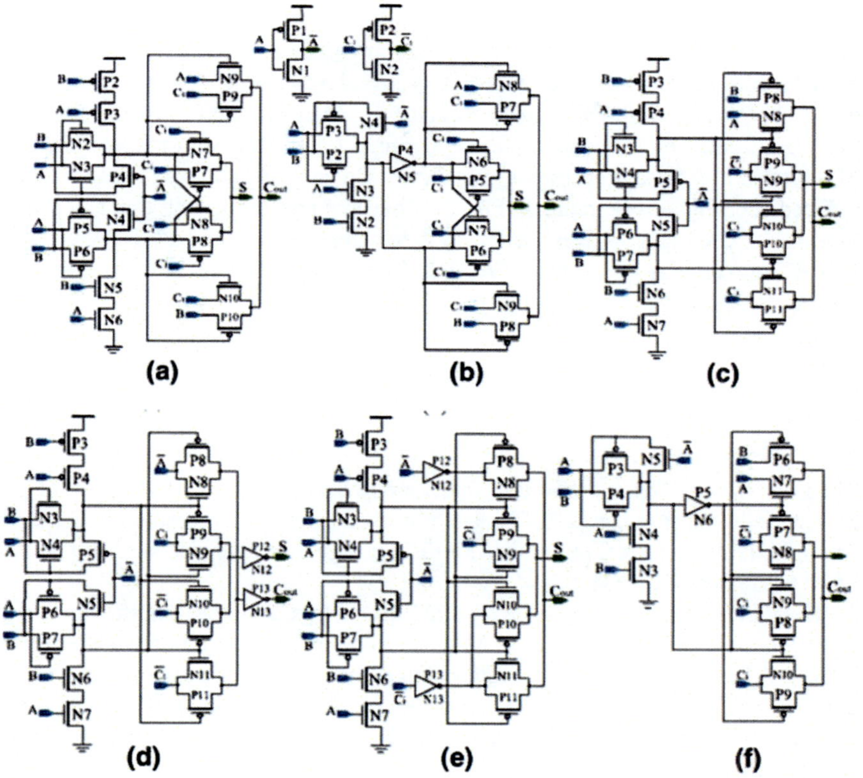

Fig. 2 The six hybrid FA circuits **a** HFA-20T, **b** HFA-17T, **c** HFA-B-26T, **d** HFA-NB-26T, **e** HFA-22T, **f** HFA-19T using CMOS transistors 32 nm technology

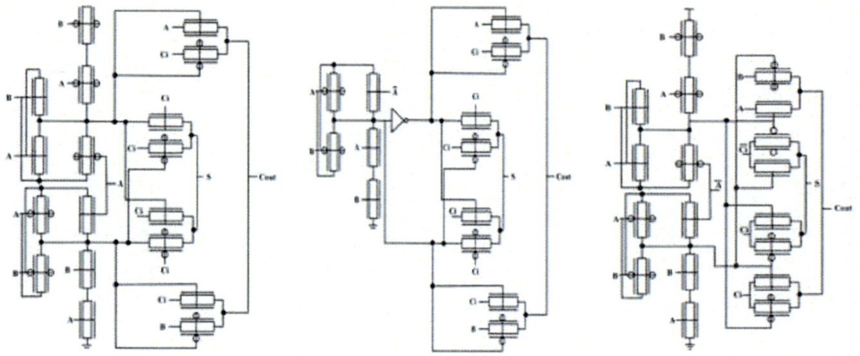

Fig. 3 Proposed new hybrid FA circuits using FinFET **a** HFA_20T **b** HFA_17T and **c** HFA_22T

also not considered for implementation in FinFET. So the circuits of HFA-20T and HFA-17T are chosen for their efficiency towards area and power. The advantage of NOT gate reduction in the output carry signal generation and sum generation using XOR, XNOR and C signals reduce the power and delay. To improve the delay and reduction in capacitance the C signal is used for sum output. The TG Multiplexer is not used for driving sum signal. On analysis, it can be found that the reduced capacitance of circuits in Fig. 2e, f has less power consumption and delay. In addition, C signal improves the driving capability of HFA-22T and HFA-19T. The FinFET implementation of the HFA_20T, HFA_17T and HFA_22T are shown in Fig. 3a–c, respectively.

4 Result and Discussion

The Full Swing XOR and XNOR circuits are implemented in CMOS and FinFET 32 nm Technology using HSPICE. The observed results are presented in Table 1. From the table, it's been found that FinFET at higher frequencies performs better when compared to CMOS.

The FinFET- and CMOS-based Hybrid Full Adder Circuits with Different Transistor Count is implemented using 32 nm Technology and results are tabulated in Table 2. The investigation of the adder circuit shows a mixed response for CMOS and FinFET. The performance varies between the circuits. When these circuits were used in biomedical applications like prosthetic devices the battery power reduction will be helpful.

Table 1 FinFET and CMOS implementation of Full Swing XOR and XNOR circuits with 32 nm technology

Hybrid full adder circuits with different transistor count	Frequency (GHz)	Average power (W) CMOS	Average power (W) FinFET
Full Swing XOR and XNOR1 (a)	1	2.7949E−07	1.0520E−06
	2	1.4609E−06	1.2565E−06
Full Swing XOR and XNOR2 (b)	1	1.1918E−06	1.9364E−06
	2	2.0549E−06	7.8820E−07
XOR–XNOR circuits (c)	1	2.2930E−07	2.9847E−08
	2	2.7534E−06	9.8980E−07
XOR–XNOR circuits (d)	1	2.8142E−07	5.1199E−08
	2	4.3298E−06	1.8936E−06
XOR–XNOR circuits (e)	1	1.7617E−07	2.8950E−08
	2	4.7891E−05	4.0011E−05
XOR–XNOR circuits (f)	1	1.7617E−07	2.8950E−08
	2	4.7891E−05	4.0011E−05
XOR–XNOR circuits (g)	1	2.4579E−07	3.4667E−08
	2	4.7891E−05	4.0123E−05

Table 2 FinFET and CMOS Hybrid Full Adder Circuits with different transistor count using 32 nm technology

Hybrid Full Adder Circuits with different transistor count	Frequency (GHz)	Average power (W) CMOS	Average power (W) FinFET
HFA_17T	1	1.1463E−07	3.7787E−08
	2	3.5904E−05	4.0971E−05
HFA_19T	1	5.7330E−05	5.0326E−05
	2	1.2089E−04	1.0198E−04
HFA_20T	1	1.4036E−07	1.3434E−−08
	2	4.0802E−06	1.9493E−06
HFA_22T	1	6.5244E−07	1.2459E−07
	2	6.0681E−06	3.3449E−06
HFA_B26T	1	1.0051E−06	5.9984E−08
	2	7.0825E−06	4.5096E−06
HFA_NB26T	1	6.5206E−07	1.2741E−07
	2	6.1386E−06	3.6091E−06

5 Conclusion

In this paper, the XOR/XNOR and XOR–XNOR circuits using CMOS and FinFET devices are implemented and the results are tabulated. From the results, it's been found that the FinFET-based design performance is better. By using feedback and NOT gates, delay, output capacitance, and power consumption increases. The multigate XOR/XNOR and XOR–XNOR gates are useful for future applications if implemented in multipliers, processing elements, etc.

References

1. Goel, S., Kumar, A., & Bayoumi, M. (2006). Design of robust, energy-efficient full adders for deep-submicrometer design using hybrid-CMOS logic style. *IEEE Transaction on Very Large Scale Integration (VLSI) System, 14*(12), 1309–1321.
2. Bui, H. T., Wang, Y., & Jiang, Y. (2002). Design and analysis of low-power 10-transistor full adders using novel XOR-XNOR gates. *IEEE Transactions on Circuits and System II, Analog and Digital Signal Processing, 49*(1), 25–30.
3. Timarchi, S., & Navi, K. (2009). Arithmetic circuits of redundant SUT-RNS. *IEEE Transactions on Instrumentation and Measurement, 58*(9), 2959–2968.
4. Rabaey, M. J., Chandrakasan, A. P., & Nikolic, B. (2002). *Digital integrated circuits* (Vol. 2). NJ, USA: Prentice-Hall, Englewood Cliffs.
5. Radhakrishnan, D. (2001, February). Low-voltage low-power CMOS full adder. *IEEE Proceedings—Circuits, Devices and Systems, 148*(1), 19–24.
6. Yano, K., Shimizu, A., Nishida, T., Saito, M., & Shimohigashi, K. (1990). A 3.8-ns CMOS 16 × 16-b multiplier using complementary pass-transistor logic. *IEEE Journal of Solid-State Circuits, 25*(2), 388–395.

7. Shams, A. M., Darwish, T. K., & Bayoumi, M. A. (2002). Performance analysis of low-power 1-bit CMOS full adder cells. *IEEE Transactions on Very Large Scale Integration (VLSI) Systems, 10*(1), 20–29.
8. Zhuang, N., & Wu, H. (1992). A new design of the CMOS full adder. *IEEE Journal of Solid-State Circuits, 27*(5), 840–844.
9. Weste, N., & Eshraghian, K. (1985). *Principles of CMOS VLSI design.* New York, NY, USA: Addison-Wesley.
10. Bhattacharyya, P., Kundu, B., Ghosh, S., Kumar, V., & Dandapat, A. (2015). Performance analysis of a low-power high-speed hybrid 1-bit full adder circuit. *IEEE Transaction on Very Large Scale Integration (VLSI) Systems, 23*(10), 2001–2008.
11. Vesterbacka, M. (1999, October). A 14-transistor CMOS full adder with full voltage swing nodes. In *Proceedings of IEEE workshop on signal processing system (SiPS)* (pp. 713–722).
12. Alioto, M., Di Cataldo, G., & Palumbo, G. (2007). Mixed full adder topologies for high-performance low-power arithmetic circuits. *Microelectronics Journal, 38*(1), 130–139.
13. Shams, A. M., & Bayoumi, M. A. (2000). A novel high-performance CMOS 1-bit full-adder cell. *IEEE Transaction on Circuits Systems II, Analog and Digital Signal Processing, 47*(5), 478–481.
14. Aguirre-Hernandez, M., & Linares-Aranda, M. (2011). CMOS full-adders for energy-efficient arithmetic applications. *IEEE Transaction on Very Large Scale Integration (VLSI) Systems, 19*(4), 718–721.
15. Hassoune, I., Flandre, D., Connor, I. O., & Legat, J. D. (2010). ULPFA: A new efficient design of a power-aware full adder. *IEEE Transactions on Circuits Systems I, Reg. Papers, 57*(8), 2066–2074.
16. Chang, C. H., Gu, J., & Zhang, M. (2005). A review of 0.18-μm full adder performances for tree structured arithmetic circuits. *IEEE Transaction on Very Large Scale Integration (VLSI) Systems, 13*(6), 686–695.
17. Kumar, P., & Sharma, R. K. (2016). Low voltage high performance hybrid full adder. *Engineering Science and Technology, an International Journal, 19*(1), 559–565.
18. Ghadiry, M., Nadi, M., & Ain, A. K. A. (2013). DLPA: Discrepant low PDP 8-bit adder. *Circuits, Systems, and Signal Processing, 32*(1), 1–14.
19. Wairya, S., Nagaria, R. K., & Tiwari, S. (2011). New design methodologies for high-speed low-voltage 1 bit CMOS Full Adder circuits. *International Journal of Computer Technology and Applications, 2*(2), 190–198.
20. Chowdhury, S. R., Banerjee, A., Roy, A., & Saha, H. (2008). A high speed 8 transistor full adder design using novel 3 transistor XOR gates. *International Journal of Electronics, Circuits and Systems, 2*(4), 217–223.
21. Senthilkumar, V. M., & Ravindrakumar, S. (2018). A low power and area efficient FinFET based approximate multiplier in 32 nm technology. In *Springer-International Conference on Soft Computing and Signal Processing.*
22. Sujatha, V., Senthilkumar, V. M., & Ravindrakumar, S. (2018). Design of adiabatic array logic adder using multigate device in 32 nm FinFET process technology. *Journal of Advanced Research in Dynamical and Control Systems, 13,* 464–472.
23. Senthilkumar, V. M., Muruganandham, A., Ravindrakumar, S., & Gowri, G. N. S. (2019). FINFET operational amplifier with low offset noise and high immunity to electromagnetic interference. *Microprocessors and Microsystems Journal, 71,* 102887.
24. Senthilkumar, V. M., Ravindrakumar, S., Nithya, D., & Kousik, N. V. (2019). A vedic mathematics based processor core for discrete wavelet transform using FinFET and CNTFET technology for biomedical signal processing. *Microprocessors and Microsystems Journal, 71,* 102875.
25. Prasanth, K., Ramireddy, M., Keerthi Priya, T., & Ravindrakumar, S. (2020). High speed, low matchline voltage swing and search line activity TCAM cell array design in 14 nm FinFET technology. In T. Hitendra Sarma, V. Sankar, & R. Shaik (Eds.), Emerging Trends in Electrical, Communications, and Information Technologies. *Lecture Notes in Electrical Engineering* (Vol. 569). Singapore: Springer.

Examining Streaming Data on Twitter Hash Tags with Relevance to Social Problems

S. Shanthi, D. Sujatha, V. Chandrasekar, and S. Nagendra Prabhu

Abstract During the most recent years, usage of a variety of social networking sites, such as Twitter, Facebook, has been dramatically used by many. Due to these networking sites, an enormous volume of information is generated and there is an urge to investigate and process these data. At the present moment, most of the people accurately articulate their own opinions and views on a diverse of topics such as politics, movies, industries, etc. all the way through personal blog websites. Our paper will analyze the diverse posts of users who had expressed their opinions and views in Twitter and we try to extract the sentiment polarity as positive, negative and neutral. We are using R programming to bring out the sentiment analysis and analyze the data usually in the form of criticisms and reviews. We characterize the outcomes of sentiment analysis of Twitter statistics as positive, neutral and negative sentiments.

Keywords Sentiment analysis · Twitter · Maximum entropy · R programming

1 Introduction

Nowadays, everybody is interested in using social networking websites. The population needs to get satisfied with the various products and services which have been marketed by various companies. Remarks, comments, analysis and personal opinion of the diversified people play a very significant role in concluding whether the known population has been satisfied with the product and services provided by various

S. Shanthi (✉) · D. Sujatha · V. Chandrasekar · S. Nagendra Prabhu
Malla Reddy College of Engineering and Technology, Hyderabad, Telangana 500050, India
e-mail: shanu_shivak@yahoo.com

D. Sujatha
e-mail: sujathadandu@gmail.com

V. Chandrasekar
e-mail: drchandru86@gmail.com

S. Nagendra Prabhu
e-mail: snagendraprabhu@gmail.com

S. C. Satapathy et al. (eds.), *Intelligent System Design*, Advances in Intelligent Systems and Computing 1171, https://doi.org/10.1007/978-981-15-5400-1_24

companies. It facilitates in forecasting the sentiment and opinion of an ample range of people on a meticulous event of their interest such as review of a particular movie, their opinion on various topics such as citation analysis, similarity among documents, medical documents, etc. in and around the earth. These data play a very essential role in sentiment analysis. We need to find out the overall sentiment of people throughout the world, and in order to perform these operations, we are in need of the required data from various sources such as Facebook, Twitter, Blogs to perform analysis. For performing sentiment analysis, our attention has been toward Twitter, a micro-blogging social networking website. A huge amount of data is being generated by Twitter, and it could not be handled manually to dig out a little constructive information, and hence, we need few automatic classification techniques to handle the huge volume of data. A maximum of 140 characters have been allowed for tweets and they are short text messages. Throughout the world by making use of Twitter, many people such as their friends, colleagues and relatives came into close communication via their mobiles and computers. Twitter data grows day by day and any user post short text messages and any user can read it. Due to the popularity and enormous users, Twitter has been chosen as the source for performing sentiment analysis.

The traditional approach is not able to handle and process a huge amount of data. It has many drawbacks such as more processing time and limited data only could be processed. If a huge volume of data is there, it could not process and give the output in time. So we need to use some of the current technologies such as Hadoop, R programming, etc. so that we can able to solve the drawbacks of the existing system.

Due to the development of Web 2.0, web-based business; the measure of surveys carried out through online has achieved a remarkable volume development. Individuals might want to compose surveys for a few parts of the item, on which some significant sentiments are certain, and some are negative. These remarks for dealers and purchasers have incredible business esteem. The extensive count of audits is in the form of an unstructured content arrangement which is rigid to record client suppositions naturally. Airtel Tweet investigation utilizes every one of the tweets from the Airtel hash labels, utilizing the library Twitter R. This investigation gives us the client audits on various results of Airtel. We produce a world cloud in light of the tweets. We have utilized content mining library to direct the content and create content free tweets. We build up this code by utilizing R dialect.

Sentiment analysis otherwise called as opinion mining. It is the process of analyzing an e-text to be positive, negative or neutral. E-text represents reviews, messages or comments in the electronic form. Sentiment analysis finds its application in many fields such as customer service, reviews, marketing, etc.

The remaining paper is arranged as given below. Section 2 explains about the diverse existing sentiment analysis used. Section 3 presents the problem statement and the algorithm proposed and Sect. 3.2 discusses the simulation results. Lastly, Sect. 4 provides the conclusion of the paper.

2 Related Works

Sentiment analysis plays a major role in recent years. Several researchers are in progress working on sentiment analysis on Twitter because of the giant volume of data and there is an urge to examine and process these huge data. In the initial stage, barely binary classification is being used to analyze the opinions, reviews and the results are positive or negative.

Pak and Paroubek [1] worked on classification of tweets and they classified as positive, objective and negative. Using Twitter API, they collected tweets from Twitter and created a Twitter corpus. These tweets are automatically annotated with emoticons. A multinomial Naive Bayes method is used to classify the sentiments and it uses the features of N gram and POS-tags. Tweets included only emoticons and the training set considered was not efficient in this model.

Parikh and Movassate [2] initiated two such models, first one being called as Naive Bayes bigram model and next one maximum entropy model used to categorize a variety of tweets. It is implicit with the intention that Naive Bayes classifiers are widely used than maximum entropy model.

Huang [3] used distant supervision for providing a solution to sentiment analysis. Here, the emoticons were only used in the training data which have been considered for analyzing the tweets. Models like SVM, maximum entropy and Naïve model have been built. The features such as unigrams, POS and bigrams are being considered. SVM was much better when compared to other models and unigram feature was better than other features considered.

Barbosa et al. [4] suggested a sentiment analysis method which consists of two phases for analyzing the tweets in Twitter. During the initial phase, mainly the tweets have classified into objective and subjective. During the second phase, the tweets under the subjective phase are in turn further categorized as positive and negative. Some of the features used were hash tags, re-tweets, links, exclamation marks punctuation, etc.

Bifet and Frank [5] utilized a stochastic gradient descent model and found to be good at times with the proper learning rate. Agarwal et al. [6] proposed a three-way model that performs the sentiment classification as negative, positive or neutral classes. Some of the features used include a tree kernel-based model, a feature-based and unigram-based model. It is found that tree based model is better and efficient than other two models used. Davidov et al. [7] used K-Nearest Neighbor strategy to assign labels and used user-defined hash tags in tweets for the classification of sentiments. Liang et al. [8] used three different types of training data such as the camera, mobile and movie. They have used unigram Naive Bayes model.

Pablo et al. [9, 10] make use of two different Naive Bayes classifiers for analyzing the tweet's polarity. Baseline classifies the tweets into four types such as positive, negative and neutral and Binary. Turney et al. [11] used a collection of words and there are no relationships between words. A document is simply characterized as a group of words. Kamps et al. [12] suggested WordNet, a lexical database to administer the emotional contents present in a word. Xia et al. [13] used feature sets like Part-of-speech, word relations and base classifiers such as Naive Bayes, maximum entropy and support vector machines. Luo et al. [14] emphasized on the challenges and it provided a well-organized technique to retrieve opinions of various tweets in Twitter.

3 Proposed Work

This section mainly discusses about the analyzing of data streams on Twitter hash tags. Usually, the search begins with the hash tags that are generally preceded by # and it explores for a meticulous keyword and then further evaluates the polarity of tweets as negative and positive. In this paper, chi-square test is used to select the best features and for training and testing the feature's maximum entropy classifier is used. Sentimental polarity is also evaluated using maximum entropy classifier as positive or negative. The proposed system is implemented using R programming. The detailed algorithm is discussed in Sect. 3.2.

3.1 Problem Statement

The proposed system tries to collect all pertinent data, analyze and sum up the whole sentiment on a topic. It further classifies the polarity of tweets or it can be a word or a sentence or a document or a feature as positive, negative or neutral. Some of the problems faced during sentiment analysis include the searching problem, tokenization and identification and reliable content identification.

Two major techniques have been used for analyzing the sentiment in Twitter data:

A Machine Learning Approach:
Mainly supervised learning and unsupervised learning features considered play a vital role in identifying the sentiments. The machine learning approach classified under supervised classification for detecting sentiment analysis. Basically two sets of data are being used:

- Training Set
- Test Set.

There are a variety of machine learning techniques that have been framed to analyze and detect and classify the tweets. The commonly used machine learning techniques include NB—Naive Bayes, ME—maximum entropy, and SVM—support vector machines. These techniques played a major role in carrying out sentiment analysis efficiently. Firstly, the machine learning starts collecting the training dataset followed by a classifier employed to train the training data.

A Lexicon-Based Approach:
Lexicon-based approach [20] makes use of a dictionary for each and every word and tries to match it with the data to determine the polarity. For each and every opinion words, a sentiment score is attached and has classified into objective, positive and negative. It mainly depends on factors such as a sentiment lexicon, a collection of idioms, phrases, etc. Sentiment terms have been used to perform the sentiment analysis. This approach has been further classified into two types:

Dictionary-Based: The words are collected manually and it tries to search the antonyms and synonyms and tries to expand the number of words. Example is Wordnet. The main drawback is it copes with context and domain specific orientation.

Corpus-Based: It works on dictionaries based on a specific domain. By using statistical or semantic techniques, the related words are searched and the set of words is increased.

- Statistics Methods: Latent Semantic Analysis (LSA).
- Semantic methods: makes use of synonyms, antonyms, relationships from thesaurus.

3.2 Proposed Algorithm

In this section, we discuss the analysis of Twitter hash tags based on maximum entropy classification in the proposed algorithm. The proposed model is depicted in Fig. 1.

Proposed Algorithm

1. Receive the input tweets (hashtags) provided by the user.
2. Assume N[i] the number of features varying from 10,100 up to 1000
3. Preprocessing tweet:

   (a) Normalize the character set(converting alpha characters to lower case letters)

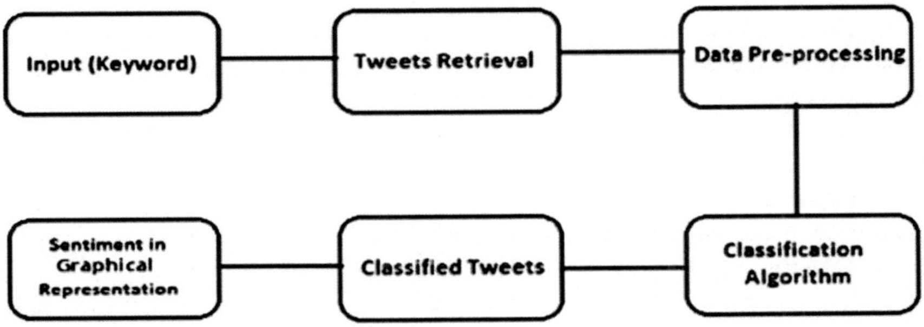

Fig. 1 Proposed system

   (b) Removing punctuation
   (c) Performing normalization—Grouping the words with the same meaning
   (d) Classification of words—Removing stop words, removing less than 3 characters, removing words with numbers
   (e) Remove the suffix and prefix of tweets.

4. Using the chi-square test build the dictionary of the word score
   (a) Examine both positive and negative word
   (b) Analyze the frequency Distribution of words
   (c) Count the total number of words, positive and negative words.
5. Sort the word score, best words are received.
6. Evaluate the features of selected best words.
7. Train the features using Maximum Entropy method.
8. Tweets are categorized as negative and positive, depending on score words.

There are various stages involved in the sentiment analysis of Twitter and it includes the subsequent steps as follows: Data streaming is used to collect the raw tweets and have been provided as inputs to generate the sentiment polarity. Two kinds of API's are usually used in Twitter: Search and Streaming API. Using Search API, the necessary training data set is created and displayed using the streaming API.

The next step involves preprocessing of extracted data. The unstructured textual data is converted into structural textual data. We convert alpha characters to lowercase letters and remove punctuation. It may lead to inefficiencies so to avoid and have better results few more steps are involved in preprocessing.

- Filtering: Involves removing user names and special words.
- Tokenization: Words are formed after removing spaces, punctuations.
- Removal of Stop words: Articles and other stop words are removed.

Using the chi-square method we compute the scores of the word of the various features considered. A list of all negative and positive words is created. Then compute the distribution of words as well as words with negative and positive labels. Lastly, based on the chi-square test, the number of negative and positive words and total words and the word score is found. This test method gives good result for both positive and negative classes and it is also used to select a feature from high dimensional data. So that word scores are found and the best number of words based on word scores are also extracted.

Maximum entropy is used to classify the sentiment. There are other methods which have been proposed such as Naives Bayes and SVM. Maximum entropy has been used in this paper to classify the sentiment. They do not consider the random variables (statistical independence) which serve as predictors. In this, using maximum aposteriori (MAP) estimation, the weights typically maximized and it should be made to learn via an iterative procedure. Accuracy achieved is more when compared to other methods (Figs. 2, 3, and 4).

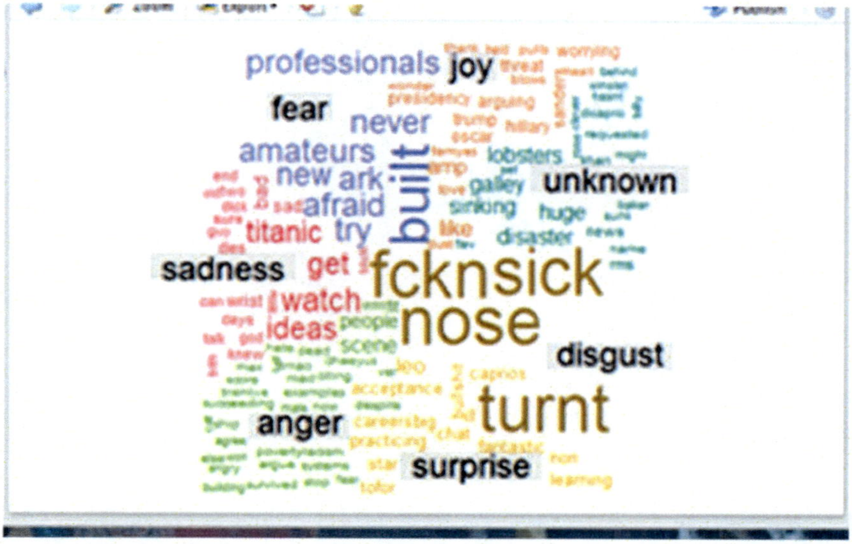

Fig. 2 Analysis of words

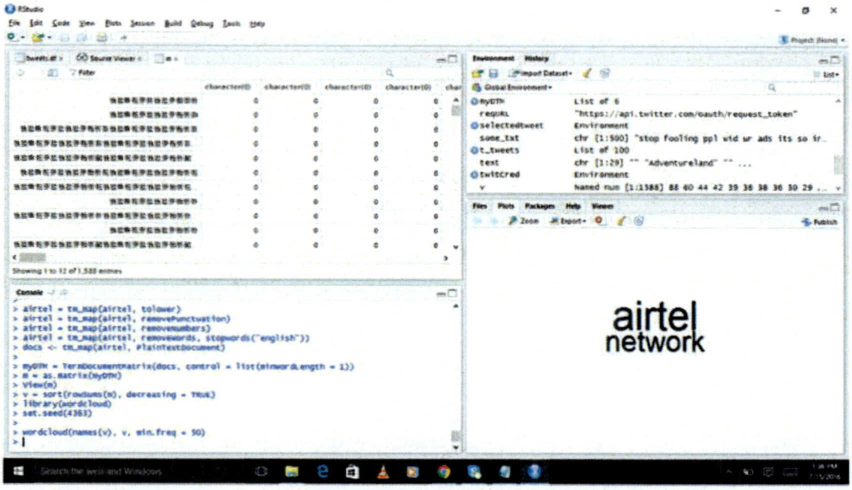

Fig. 3 The final result about cloud services from people's opinion

4 Conclusion

We introduced comes about for Airtel Tweet Analysis on Twitter. We utilize already proposed progressive unigram display as our pattern. We displayed a widespread understanding of analysis for both the two-way and three-way undertakings that is a

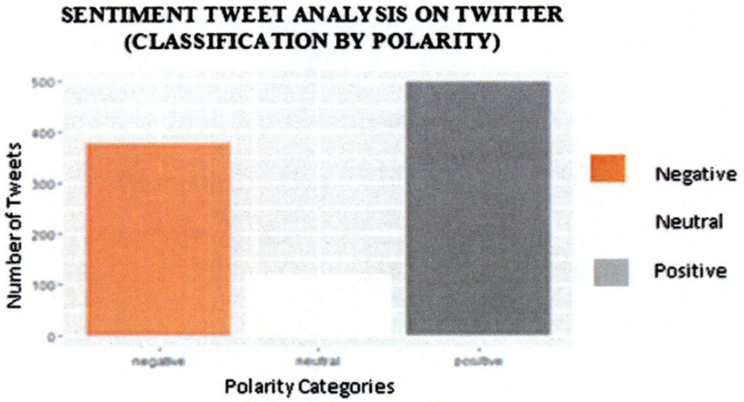

Fig. 4 Sentiment analysis

random specimen of a stream of tweets. We probably conclude that Airtel Tweet Analysis for Twitter is not a unique one in relation to estimation investigation for different classifications. We have analyzed the Airtel tweets and performed the sentiment analysis and classified them as positive, negative and neutral.

References

1. Ouyang, C., Zhou, W., & Yu, Y., et al. (2014). Point sentiment analysis in Chinese News. *Global Journal of Multimedia and Ubiquitous Engineering, 9*(11), 385–396.
2. Pang, B., Lee, L., & Vaithyanathan, S. (2002). Thumbs up? Supposition classification utilizing machine learning techniques. In *Proceedings of the 2002 Conference on Empirical Methods in Natural Language Processing* (pp. 79–86).
3. Turney, P. (2002) Thumbs up or thumbs down? Semantic introduction connected to unsupervised arrangement of reviews. In *Proceedings of the 40th Annual Meeting of the Association for Computational Linguistics* (pp. 417–424). Morristown, NJ, USA: Association for Computational Linguistics.
4. Dong, R., & O'Mahony, M., et al. (2015) Consolidating similitude and notion in sentiment digging for item proposal. *Diary of Intelligent Information Systems, 17*(1), 28.
5. Quan, C., & Ren, F., et al. (2014). Target based survey order for fine-grained assumption analysis. *Global Journal of Innovative Computing, Information and Control, 10*(1), 257–268.
6. Kamps, J., Marx, M., Mokken, R. J., & Rijke, M. D. (2004). Utilizing WordNet to gauge semantic introduction of adjectives. In *Proceedings of LREC04, 4th International Conference on Language Resources and Evaluation*, Lisbon (pp. 1115–1118).
7. Miao, Q., et al., Fine-grained conclusion mining by incorporating various audit sources. *JASIST*, 2010(61), 2288–2299.
8. Zhang, L., Qian, G. Q., Fan, W. G., Hua, K., & Zhang, L. (2014). Conclusion examination in light of light audits. Ruan Jian XueBao/*Journal of Software, 25*(12), 2790–2807 (in Chinese).
9. Balazs, J. A., & Velásquez, J. D. (2015). Conclusion mining and information fusion: An overview. *Data Fusion, 27*, 95–110.
10. Luof. (2011). *Inquires about on key issues in opinion mining*. Wuhan University of Technology.
11. Introduction to TF-IDF [EB/OL]. [2015–08-06]. http://baike.baidu.com/see/1228847.htm.

12. Introduction to TF-IDF [EB/OL]. [2015-08-06]. https://en.wikipedia.org/wiki/Tf-idf.
13. Liu, Z., & Yu, W., et al. (2010). A feature selection method for document clustering based on part-of-speech and word co-occurrence. In *Procedures—2010 Seventh International Conference on Fuzzy Systems and Knowledge Discovery* (pp. 2331–2334).
14. Peñalver-Martinez, I., et al. (2014). Highlight based sentiment mining through ontologies. *Expert Systems with Applications, 41*(13), 5995–6008.

A Semantic-Aware Strategy for Automatic Speech Recognition Incorporating Deep Learning Models

A. Santhanavijayan, D. Naresh Kumar, and Gerard Deepak

Abstract Automatic Speech Recognition (ASR) is trending in the age of the Internet of Things and Machine Intelligence. It plays a pivotal role in several applications. Conventional models for automatic speech recognition do not yield a high accuracy rate especially in the context of native Indian Languages. This paper proposes a novel strategy to model an ASR and Speaker Recognition system for the Hindi language. A semantic-aware strategy incorporating acoustic modeling is encompassed along with Deep Learning Techniques such as Long Short-Term Memory (LSTM) and Recurrent Neural Networks (RNNs). MFCC is further imbibed into the model strategically for feature extraction and the RNN for pattern matching and decoding. The proposed strategy has shown a promising performance with the speaker recognition system yielding 97% accuracy and speech recognition system with a very low word error rate of 2.9 when compared to other existing solutions.

Keywords Automatic speech recognition · Deep learning · Feature extraction · LSTM

1 Introduction

Automatic speech recognition is a method where a given audio signal is converted into text. An Automatic Speech Recognizer behaves like an interface for making the use of the machines much easier. For the communications in Indian language such as Hindi, machine models as a huge help to the public in the native speaking countries. The Majority of the population of India is unaware of other languages like English. So, Automatic Hindi speech recognition is the necessity of the citizens of our country. There are several applications in which automatic Hindi speech recognition

A. Santhanavijayan · G. Deepak
Department of Computer Science and Engineering, National Institute of Technology Tiruchirappalli, Tiruchirappalli, India

D. Naresh Kumar (✉)
Department of Mechanical Engineering, National Institute of Technology Tiruchirappalli, Tiruchirappalli, India
e-mail: d.nareshkumar1040@gmail.com

S. C. Satapathy et al. (eds.), *Intelligent System Design*, Advances in Intelligent Systems and Computing 1171, https://doi.org/10.1007/978-981-15-5400-1_25

can be put to great use, such as government offices, railway stations, information retrieval systems at railway stations, bus stations, aviation stations, etc. by providing the people with the answer to their problems faced by them. Healthcare sectors also find a good variety of use for ASR like documentation processes of patients' records. It has shown significant importance for helping people with short-term memory loss by treating them with prolonged speech to help them remember the required details thus reinforcing their brain to retain memory. It also finds use in military technologies like training air traffic controllers and helicopters. Education and daily life is another domain which can make use of the ASR systems. It helps in the education of students who are physically disabled or blind.

Organization: The remaining paper organization is as follows. Section 2 provides a brief overview of the related work of research that has been conducted. Section 3 illustrates the proposed architecture. Section 4 describes the Results and Performance Evaluation and finally, the paper is concluded in Sect. 5.

2 Related Work

Kuldeep et al. [1] have formulated a system for speech recognition employing the strategy of Hidden Markov Model. The methodology was adopted for the processing of continuous speech signals. The feature extraction method used was relative cepstral transform and the word level segmentation was used for modeling the system. The model was for speech in Hindi language only and performance was found to be good for audio files which were in its database used for training. Mohit et al. [2] have implemented a model for recognition of speech modeled by the usage of Hidden Markov Model. The model uses a method called discriminative training and feature extraction is done using perpetual linear prediction along with the Mel frequency cepstral coefficients. It mainly is using heterogeneous feature vectors. N. Rajput Kumar et al. have proposed a continuous system for recognition of speech using the Gaussian mixture models. Trigram model was used for the language model. The speech segmentation was done phoneme based. The feature extraction technique used was the Mel frequency cepstrum coefficient. Wu et al. [3] have proposed a scheme for ASR incorporating joint learning in front end speech processing. Tang et al. [4] have proposed a strategic model ASR by employing a multitask model comprising of neural networks. Tang et al. [5] have proposed a joint collaborative model for speech and speaker recognition that incorporates multitask training in it where the outputs of each task are backpropagated. Tarun et al. [6] have proposed a scheme that amalgamates Vector Quantization and HMM model for isolated word recognition for the Hindi language. This method also imbibes peak picking and is suitable for the Hindi language. Mishra et al. [7] have proposed a speaker-independent strategy hybridizing revised perceptual linear prediction (RPLP), Bark frequency cepstral coefficients (BFCC), and Mel frequency perceptual linear prediction (MF-PLP) for recognizing Hindi digits. Sinha et al. [8] have proposed a strategy for context-dependent speech recognition for Hindi language using Hidden Markov Model with continuous density.

Also, HLDA is incorporated for feature reduction. Certain Semantic approaches as depicted in [9–21] can be imbibed along with Deep Learning Models especially for the rearrangement of phonemes in order to achieve a much efficient technique, as semantic-aware systems are more intelligent and efficient.

3 Proposed System Architecture

The proposed system architecture is depicted in Fig. 1. The process of feature extraction is a method of extracting a set of properties of an utterance having certain acoustic relations and matching with the speech signal. A feature extractor removes irrelevant properties from the input and retains the required properties. For doing this a part of the speech signal is taken into consideration for the purpose of processing. This portion is called window size. A frame is the required data received from a window. The range of frame is usually between 10 and 26 ms. This range has an overlap of nearly 50–70% between two sequential frames. The data obtained from this analysis interval is then required to be multiplied with a windowing function. The proposed model is using MFCC for extracting features. For MFCC, an audio signal is constantly changing, and the short time scales the audio signals do not change much. This is the reason the frame of the signal is presumed into 20–40 ms frames. Power spectrum calculation is the next step of every frame. The periodogram

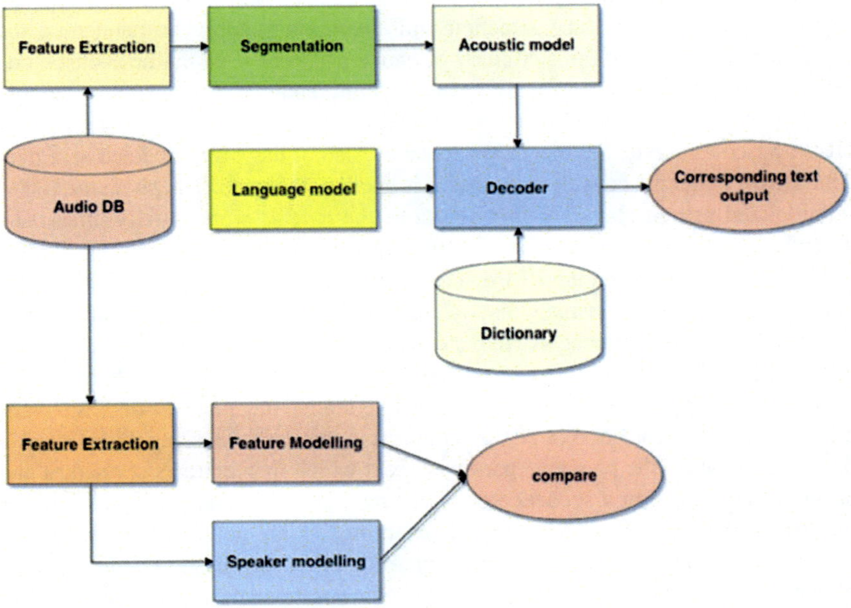

Fig. 1 Proposed system architecture

in the proposed approach performs a similar job by identifying which frequencies are present in the frame.

There is a lot of information in periodogram spectral that is actually not required for the computation of automatic speech recognition (ASR). Due to this, periodogram bin clumps are taken and are summed up in order to obtain an awareness of the existing energy in various regions of frequency. Mel filterbank performs this process: the initial filter gives an indication of the energy existing near 0 Hz and is very narrow. These filters get wider following with the trend of frequencies getting higher. The width that is present is directly proportional to the frequency available. At these high frequencies, these filters become less concerned about the variations. The main concern is the spot at which the maximum energy occurs. The method of spacing of the filterbanks is clearly depicted by the Mel scale. As soon as the filterbank energies are obtained, logarithm is applied and this application facilitates the use of cepstral mean subtraction.

This is a technique that involves channel normalization. Computation of DCT of log filterbank is the final process. The main motives behind these processes can be stated as two. One being the fact that the filterbanks are overlapping and the other reason being the energies of these filterbanks are correlated with one another. The decorrelation of these energies is done by DCT. It should be noticed that only 12 of the present 26 DCT coefficients. The reason for this is that fast changes are represented in higher values of DCT coefficients and these higher values are responsible for the degradation of ASR performance. Hence improvement in performance can be achieved by dropping these values.

The process of identifying the boundaries between spoken words is called speech segmentation. The boundary detection will enable us to form the phonemes and which can be used form the dictionary of phonemes to be used by the acoustic and language models. It is necessary to take into account semantics as well as the grammar while processing the natural languages. The speech segmentation is achieved using HMM-based Phonetic. Segmentation scheme. Also, it has been modeled as a five-state HMM under the consideration that it is another phoneme. The phoneme HMM is initialized using a flat start training where the models were equally initialized. Therefore, there was no necessity of manually segmented database for training for the purpose of initializing the HMM. The segmentation result is probabilistic and is still a major challenge for many natural languages. Text Transcriptions are carried out using Unicode Devanagari characters is to be formed by converting the raw input given in the form of text which is further converted into Roman characters by further processing since the keywords available are having a corresponding ASCII code generated. Translation of all the input text in our database using this method is done. The input to the phoneme parser is a text which then extracts phonemes and arranges them to form a sequence of words from the given sentence and produce the list of phonemes as output. The acoustic models are put to use to match the features observed of the given speech signal with the hypothesis for the expected phonemes. The acoustic model is one of the main components of ASR and accounts for the computational load and system. One of the common implementations is using a probabilistic method that uses the hidden market models. A training process is used for the purpose of generating a mapping of the speech units like phonemes and observations found from the acoustics. For the purpose of training, a pattern representing the features is created for each class that should correspond to the speech utterances of the corresponding class. A database rich in phonetics and database which is balanced is required for the purpose of training the acoustic models. For the purpose of transcriptions into linguistic units from acoustic features, a large variety of representations are used as whole words and syllables.

The estimation of convergence of class posteriors due to the application of cross-entropy is done by using recurrent neural networks which have SoftMax output layer. The training method generally followed is by using targets along with some alignments. Bootstrapping or flat start can be used for the purpose of aligning the acoustic feature sequences with the help of an existing model. The detection of phonetic units can be predicted using the phonemes that occur preceding it. Different models having different contexts can be modeled to enhance the modeling power as well. A three-layer long short-term memory cell recurrent neural networks have been used in the proposed system. There are eight hundred memory cells in each of long short-term memory layers. The recurrent projection layer is five hundred and twelve units. The SoftMax activation function is used for the output layer. A tangent activation function is also present in each layer along with a sigmoid function for calculation. The long short-term memory is given an at an interval of twenty-five-millisecond frame of 40-dimensional Log Mel filterbank features. Better decisions were formed from the information got from previous frames and future frames and hence were able to make better decisions by the neural network. Computation of DCT for the energies of log filterbanks is the final step.

4 Results and Performance Evaluation

The experimentation was realized in a Python 3.6 environment. The system imbibed a RAM specification of 4 GB under the Ubuntu operating system. The dataset used is from Hindi corpus-Technology for Development of Indian Languages (TDIL). The dataset had 150 speakers and each speaker has roughly around 95 audio files. The performance is evaluated using the word error rate (WER) and accuracy as potential metrics. Figure 2 shows a graph depicting the average word error rate of the proposed model during training through different epochs. The system was trained for 350 epochs. The graph shows there is a decrease in the WER with each epoch passed. The first epoch of the system showed very high word error rate as the neural network was not having enough knowledge and loss from which it can predict the speech. As the model passes through the iteration in each epoch the knowledge of the neural network is becoming more and better. The neural network slowly will learn to use the parameter inputs got from the language and acoustic model and predict better. Thus, as the epochs increases, there is a decrease in the word error rate. The proposed model is outperforming because of the incorporation of attention mechanisms in the sequence to sequence model of the recurrent neural network.

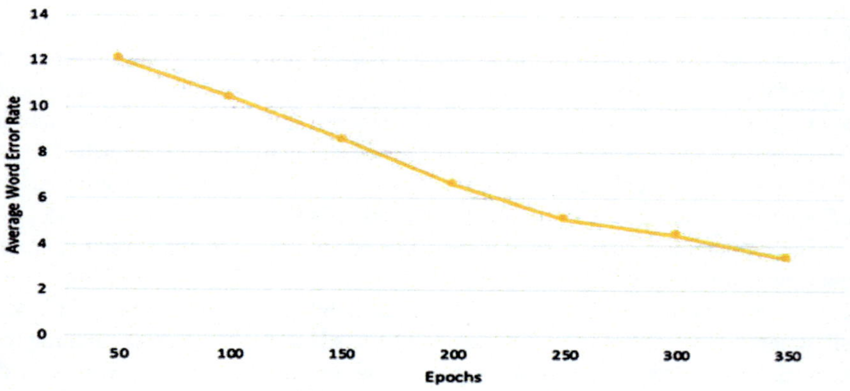

Fig. 2 Graph of average WER of the proposed model

Figure 3 shows the graph plotted during the training period for the accuracy of the system. Speaker recognition system was trained for 350 epochs using recurrent neural networks. A total of 150 speakers and each speaker has 95 audio files spoken. The accuracy of the system was observed to increase with each subsequent epoch. The final accuracy was found to be 97%. Figure 4 depicts a comparison of accuracies of different models trained in the same environment as the proposed system. The proposed system is making use of recurrent neural networks. The proposed model is better because of the incorporation of attention mechanism which helps in training the neural network in such a way that it learns from previous iterations and thus learning from the errors made in prediction in the previous iteration.

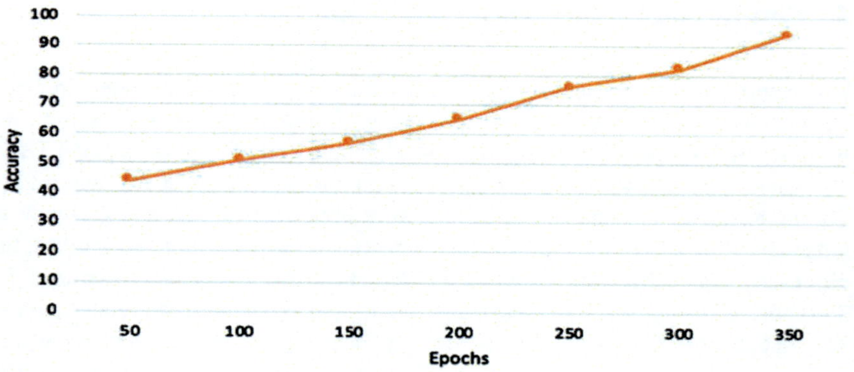

Fig. 3 Graph of the accuracy of the proposed speaker recognition system

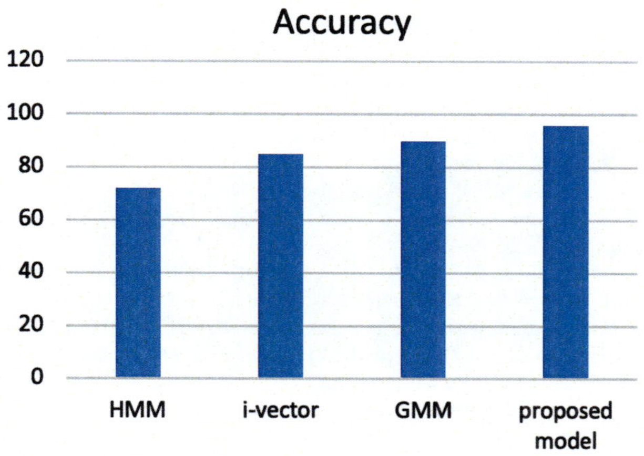

Fig. 4 Comparison of existing speaker recognition systems

5 Conclusions

A novel strategic approach for automatic speech recognition as well as speaker recognition is encompassing a semantic-aware deep learning model. This paper devises a mechanism for continuous speech and speaker recognition of Hindi language. A triphone model is incorporated as acoustic model and bigram model is encompassed as the language model. The formulated strategy yielded promising performance with an accuracy of 97%, which is much more efficient when compared to the existing models. The model can be implemented on cross-platform operating systems and the results can be successfully obtained. In addition, the proposed model also yielded a very low word error rate of 2.9 which is vastly better than the other existing models. LSTM techniques incorporating Deep Learning has proven to be much more robust in speech recognition and have played a major role in obtaining the best in class accuracy in the context of Indian Languages.

References

1. Kumar, K., Aggarwal, R. K., & Jain, A. (2012). A Hindi speech recognition system for connected words using HTK. *International Journal of Computational Systems Engineering, 1*(1), 25–32.
2. Mohit, K., Rajput, N., & Verma, A. A. (2016). A large vocabulary continuous speech recognition system for Hindi. *IBM Journal of Research and Development, 48*(5.6), 723–755.
3. Wu, B., Li, K., Ge, F., Huang, Z., Yang, M., Siniscalchi, S. M., & Lee, C.-H. (2017). An end-to-end deep learning approach to simultaneous speech dereverberation and acoustic modeling for robust speech recognition. *IEEE Journal of Selected Topics in Signal Processing, 11*(8), 1289–1300.
4. Tang, Z., Li, L., & Wang, D. (2016). Multitask recurrent model for speech and speaker recognition. *IEEE/ACM Transactions on Audio, Speech, and Language Processing, 20*(2), 493–504.

5. Tang, Z., Li, L., Wang, D., Vipperla, R., Tang, Z., et al. (2017). Collaborative joint training with multitask recurrent model for speech and speaker recognition. *IEEE/ACM Transactions on Audio, Speech and Language Processing (TASLP), 25*(3), 493–504.
6. Pruthi, T., Sakssena, S., & Das, P. K. (2016, December). Swaranjali: Isolated word recognition for Hindi language using VQ and HMM. In *International Conference on Multimedia Processing and Systems (ICMPS)*, IIT Madras.
7. Mishra, A. N., Chandra, M., Biswas, A., & Sharan, S. N. (2011). Robust features for connected Hindi digits recognition. *International Journal of Signal Processing, Image Processing and Pattern Recognition, 4*(2), 79–90.
8. Sinha, S., Agrawal, S. S., & Jain, A. (2013, August). Continuous density Hidden Markov Model for context-dependent Hindi speech recognition. In *2013 International Conference on Advances in Computing, Communications and Informatics (ICACCI)* (pp. 1953–1958). IEEE.
9. Gerard, D. & Gulzar, Z. (2017). Ontoepds: Enhanced and personalized differential semantic algorithm incorporating ontology-driven query enrichment. *Journal of Advanced Research in Dynamical and Control Systems, 9*(Special), 567–582.
10. Giri, G. L., Deepak, G., Manjula, S. H., & Venugopal, K. R. (2018). OntoYield: A semantic approach for context-based ontology recommendation based on structure preservation. In *Proceedings of International Conference on Computational Intelligence and Data Engineering* (pp. 265–275). Singapore: Springer.
11. Deepak, G., & Priyadarshini, J. S. (2018). Personalized and Enhanced Hybridized Semantic Algorithm for web image retrieval incorporating ontology classification, strategic query expansion, and content-based analysis. *Computers & Electrical Engineering, 72*, 14–25.
12. Deepak, G., Priyadarshini, J. S., & Babu, M. H. (2016, October). A differential semantic algorithm for query relevant web page recommendation. In *2016 IEEE International Conference on Advances in Computer Applications (ICACA)* (pp. 44–49). IEEE.
13. Pushpa, C. N., Deepak, G., Thriveni, J., & Venugopal, K. R. (2015, December). Onto Collab: Strategic review oriented collaborative knowledge modeling using ontologies. In *2015 Seventh International Conference on Advanced Computing (ICoAC)* (pp. 1–7). IEEE.
14. Deepak, G., Ahmed, A., & Skanda, B. (2019). An intelligent inventive system for personalized webpage recommendation based on ontology semantics. *International Journal of Intelligent Systems Technologies and Applications, 18*(1/2), 115–132.
15. Deepak, G., & Priyadarshini, S. (2016). A hybrid framework for social tag recommendation using context driven social information. *International Journal of Social Computing and Cyber-Physical Systems, 1*(4), 312–325.
16. Deepak, G., Shwetha, B. N., Pushpa, C. N., Thriveni, J., & Venugopal, K. R. (2018). A hybridized semantic trust-based framework for personalized web page recommendation. *International Journal of Computers and Applications*, 1–11.
17. Pushpa, C. N., Deepak, G., Thriveni, J., & Venugopal, K. R. (2016). A Hybridized Framework for Ontology Modeling incorporating Latent Semantic Analysis and Content based Filtering. *International Journal of Computer Applications, 150*(11).
18. Deepak, G., & Priyadarshini, J. S. (2018). A hybrid semantic algorithm for web image retrieval incorporating ontology classification and user-driven query expansion. In *Advances in Big Data and Cloud Computing*, p. 41.
19. Gulzar, Z., Leema, A. A., & Deepak, G. (2018). PCRS: Personalized course recommender system based on hybrid approach. *Procedia Computer Science, 125*, 518–524.
20. Deepak, G., & Priyadarshini, S. J. (2016). Onto tagger: Ontology focused image tagging system incorporating semantic deviation computing and strategic set expansion. *International Journal of Computer Science and Business Informatics, 16*(1).
21. Pushpa, C. N., Deepak, G., Zakir, M., & Venugopal, K. R. (2016). Enhanced neighborhood normalized pointwise mutual information algorithm for constraint aware data clustering. *ICTACT Journal on Soft Computing, 6*(4).

A Novel Encryption Design for Wireless Body Area Network in Remote Healthcare System Using Enhanced RSA Algorithm

R. Nidhya, S. Shanthi, and Manish Kumar

Abstract Most emerging field of WSN is Wireless Body Area Network (WBAN). WBAN is a group of sensors in body carried by the human to gather frequent health information transmit those health data to medical server via wireless medium of communication. The cloud-based WBAN will be helpful to save patient life in case of emergency because it providing anywhere/anytime access to the patient data. Since the patient's information is private and highly sensitive, it is vital to afford high-level security and guard to the patient's medical information over an insecure public communication medium. In this paper, a secure architecture for accessing and observing health data gathered by WBAN is proposed. For providing a strong security Enhanced RSA (E-RSA) authentication mechanism is designed. While generating secret key for the user in addition to the master key, some of the user attributes are also considered to improve the security of the health data transmission process, and also random value for secret key generation is generated based on the bilinear mapping concept. The simulation model demonstrates how efficiently the proposed system preserving the confidentiality and privacy to the patient health data in remote health monitoring system.

Keywords Wireless body area network · Remote medical care system · E-RSA algorithm · Attribute-based encryption · Body sensor

1 Introduction

The rapid advancement in wireless sensor networks in medical field is wireless body area networks (WBANs) [1] still need a lot of improvement in security and energy consumption during transmission. A WBAN is a collection of wearable or implantable [2], sensors, and a controller [3]. The sensors are used to observe a

R. Nidhya (✉) · M. Kumar
Madanapalle Institute of Technology & Science, Madanapalle, Andhra Pradesh, India
e-mail: nidhuraji88@gmail.com

S. Shanthi
Malla Reddy College of Engineering and Technology, Hyderabad, Andhra Pradesh 500050 , India

S. C. Satapathy et al. (eds.), *Intelligent System Design*, Advances in Intelligent Systems and Computing 1171, https://doi.org/10.1007/978-981-15-5400-1_27

patient's health information such as heart rate, breath rate, electrocardiogram and pressure and environmental parameters such as light, humidity and temperature. The sensor nodes transfers collected information to the controller and the controller plays a role of gateway that forwards the collected health information to the care giver through medical healthcare servers. The emergency medical care for the patient is also developed by the WBANs. So, the WBAN plays a major part in generating a highly consistent pervasive healthcare model. A detailed survey about the recent trends of WBANs is described by author [4].

2 Related Work

Before start the real development in WBAN, security issues must be solved [5]. In paper [6], the author described the communication protection between WBAN and external users. The major solution for security in WBAN is attribute-based encryption (ABE) [7]. Still, ABE cannot be considered as a better choice because it needed expensive cryptographic functions. These expensive operations are serious trouble for limited resource sensors. In paper [8], the author designed a privacy-protecting method for the WBANs. By disclosing minimal privacy the author obtained reliable data process and transmission in this method. In the paper, [9] discussed the problem of key management in WBANs. To decrease the energy consumption in data transmission author adopted multihop energy-based route method along with biometrics synchronization system. In paper [10], the author described an establishment of secure transmission channel in the WBANs in modern healthcare system. Channel is secured using the anonymous lightweight authentication protocol.

In paper [11], the author described IBE (identity-based encryption)-Lite scheme for the WBANs. In this proposed method IBC (identity-based cryptography) [12], unlike traditional PKI (public key infrastructure) does not need digital certificates to do encryption. Based on the user's personal information like e-mail id, phone number or IP address public key of user will be generated. Trusted third-party PKG (Private Key Generator) generates the user's private key. Without the attached digital certificate, Public Key Authenticity is openly achieved. So unlike traditional PKI, the proposed IBC method eradicates trouble of certificate management such as generation, allocation, storage, authentication and revocation. Even though the lightweight identity-based cryptography is appropriately suitable for source controlled WBANs, it has a problem in key generation because the PKG find out users' private keys. PKG can of decrypt a cipher text in an IBE method and create an identification for source information in an IBS (identity-based signature) method. Hence IBC suits for tiny networks like WBAN and not suitable for large networks like the internet. Still, WBANs access control goal has limited the access of WBAN by the Internet users. So IBC couldn't assure the goal.

In paper [13], the author designed two anonymous authentication models using certificateless signature for the WBANs. Before accessing the store health data in server user should be authenticated. The advantage of this proposed scheme is to use the CLC (certificateless cryptography) so that it doesn't have key escrow problem or public key certificates [14]. The CLC needs a key generating center which is a trusted third-party who can produce a part private key using user's identity and master key. By merging the part private key with secret key, complete private key is generated by the user. Since the key generating center does not have the secret value, it cannot find full private key. So the key escrow problem is neglected. However, this proposed scheme is designed to limit the access of a network server by the user not the WBANs. There are some significant works related to WBANs access control. In paper [15], the author used identity-based signcryption (IBSC) to develop an enhanced access control scheme for the WBANs. The novelty of the proposed scheme is the signcryption use in which able to authenticate the users simultaneously and used to protect the query messages. Therefore, signcryption scheme [16] is capable of concurrently accomplishing integrity, non-repudiation, confidentiality and authentication with a lower cost. However, this method also has the key escrow problem because it is also dependent on the IBC.

3 Proposed Work

3.1 System Architecture

We proposed a new architecture which includes a remote health care center such as clinic or hospital which is used to manage the patient health data, a patient with set of body area sensor (BS) to record the vital signs from body and health care professional. The proposed architecture can store the great volume of health care data produced by body area sensors and its scalable architecture. Since medical information needs high security because of sensitiveness, we propose a new security system to assure the confidentiality, integrity of data generated by BS.

Figure 1 describes the proposed architecture. The architecture is composed of the following elements: 1. Patient with body area sensors. It is used to gather information

Fig. 1 Proposed architecture

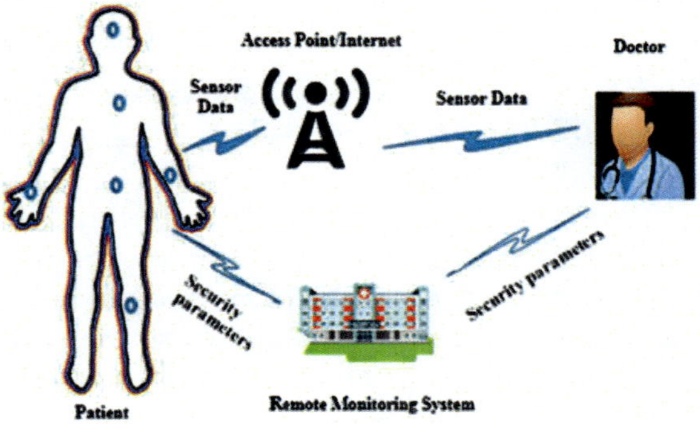

from the patient body. 2. Remote Monitoring system which will generate the security parameters for both patient and doctor. It ensures data security for sensor data. 3. Doctor or healthcare professional to provide proper treatment to the illness of patient. In addition, the architecture is provided with cloud storage. The data generated by sensor devices are heterogeneous in nature and also huge in volume. This can't be stored by an ordinary server hence the hospital server is linked with cloud storage based on the concept of on-demand provision.

To achieve data security the concept of E-RSA cryptosystem has been applied in our architecture. E-RSA scheme contains four major phases. They are Setup phase, Encryption phase, Key generation Phase and Decryption phase.

Setup: First phase describes the attributes set (U) and computation of public key (PuK) and master key (MK). The Key generation algorithms are generating secret keys using master key (MK). The random value for the key generation is produced by the bilinear mapping method. The following two properties of bilinear mapping play a major role to generate the input for key generation process. They are as follows:

- Property 1:

$$e(i + j, k) = e(i, k) * e(j, k)$$

where

$$e(k, i + j) == e(i + j, k)$$

Here *i*, *j*, *k* are positive random values.

- Property 2:

Select any two points as *x*, *y*

$$e(x, y) = (x - y) * (x + y)$$

In the above-mentioned property, randomly generated value is multiplied

$$\begin{aligned} e(x * i, y * i) &== e(i, y * i)^x \\ e(i, y * i)^x &== e(x * i, i)^y \\ e(x * i, i)^y &== e(i * i)^{x*y} \end{aligned}$$

So random value generated for the key generation provides efficient security when compared with the normal key generation process.

Encryption (PuK, M, A): This phase takes public key PuK which is generated in the previous phase, message *M* and access structure A created from the attributes set (*U*). Encryption algorithm converts the *M* into cipher text CT based on the A value.

The decryption process is performed to the users who have an attribute set of access structure *A*.

Key generation (MK, S): It considers MK and the set of user attribute *S* as inputs and produces the secret key of user SeK.

Decryption (PuK, CT, SeK): This algorithm takes input as the public key (PuK), the ciphertext (CT) and a secret key (SeK). It converts the CT into message *M* which is plaintext of CT. It can be generated only if the group of attributes of user belongs to SeK satisfies the access structure of cipher text.

(a) **Security services**

Our proposed architecture ensures the below security services.

Access control: System guarantees patient data confidentiality and scalable access control to stored data.

Authenticity and Integrity: System confirms message reliability while exchanging between two ends.

Scalability and Availability: System certifies availability of service for justifiable users in needed time.

(b) **Security Implementation**

3.2 System Initialization

During the initialization phase, hospital server (HS) generates common set of attributes and E-RSA algorithm to produce master key (MK) and the public key (PuK). The MK should be kept secret and PuK can be informed to all users to do the encryption and decryption process. The PuK is shared from HS to user in secure way. The PuK is encrypted using HS's private key and forwards the same with signature to the cloud servers. After PuK is forwarded to cloud for storage, users can use the data from anywhere by signature-based authenticity. Public Key Infrastructure (PKI) is used to generate both private and public key for key generation.

Adding a New User

When a new user is added with the system the HS will provide an access structure and secret key to the user. *A* is used to do the encryption process of user data prior to storing in on the cloud. There are two different users in our architecture. 1. Patient 2. Doctor. The security parameters of a patient and doctor are different based on the access structure. A patient always encrypts the data which has been generated from sensor devices, it's only in a read mode. But doctor requires encrypting medical data which should be in read-write mode.

The algorithm for adding new user in the HS system:

1. The PKI produces public and private (PuP, PriP) for the *P* patient.
2. The HS calls the key generation algorithm E-RSA to produce the secret key of patient SeKP. It also decides the access structure A for patient.

3. The HS also updates the cloud by adding patient *P* to the users list.
4. After performing the patient adding process, Patient details will be updated in users list.
5. During the first time connection between patient's gateway and HS, it receives the respective secret key SeKP, private key PriP and access structure.

The following procedure is used when a new doctor joins in HS the system:

1. The PKI produces a public and private (PubD, PriD) for the Doctor.
2. The HS calls the key generation algorithm of E-RSA to produce the secret key SeKD. It also built an access structure to encrypt the medical data.
3. The HS asks to add the Doctor to the user's list in cloud.
4. After updating the doctor addition request the doctor's public key PubD is added to the user's list in cloud.
5. For the first time when doctor establishes a connection to HS, it obtains the corresponding secret key SeKD, private key PriD and access structures.

3.3 Health Data Management

The information collected from body sensor network is called health data file. It should be accessed in reading mode. The sensor device continuously sends data to gateway. When data is ready to store in cloud the gateway executes the following algorithm:

1. Assign a unique value UV to the health information file *F*. It acts as a key to find the file.
2. Generate a random secret key RSeK for a symmetric cryptography algorithm.
3. Calculate the hash value Hs of file *F*.
4. To encrypt the file combine with hash value Hs, use RseK.
5. Encrypt RSeK with E-RSA encryption algorithm based on the access structure *A*.
6. Send the data to cloud in the following format is mentioned in Fig. 2.

Once the data stored on the remote server, data can be accessed by doctors remotely to monitor the patient health information. This data can also be accessed by patient itself.

Fig. 2 Encrypted health data

UV	$\{RSeK\}_A$	$\{DATA+Hs\}_{RSeK}$

4 Performance Analysis

To evaluate the proposed model's performance, several scenarios were simulated to analyze their impact by varying multiple parameters. For simulation process, we have chosen NS-2 simulator environment. For performance evaluation, the following parameters are considered. They are 1. Throughput, 2. Delay time, 3. No. of Attributes, 4. Jitter, 5. Dropping ratio, 6. Average Waiting Time.

ABE is used to encrypt data based on access structure which is the access policy's logical expression. In ABE, the rely is not based on storage server so that unauthorized data access is not possible. This is the advantage of ABE. Still ABE based encryption systems are always patient-centric approaches that will not be suitable for our application. Our proposed system combines the attribute-based encryption with E-RSA algorithm to provide better performance. Our system provided better results than ABE based system.

Figure 3 describes the performance evaluation based on throughput parameter between the E-RSA algorithm and ABE encryption process. Throughput is a number of data transmitted during a particular amount of time. In our proposed system number of data transmitted in a unit amount of time is significantly high when compared to the attribute-based encryption process.

Figure 4 describes the performance evaluation based on the dropping ratio parameter between the E-RSA algorithm and ABE encryption process. Dropping ratio is the number of data packets dropped based on the number of data packets transmitted. In our proposed system number of data packets dropped is considerably less when compared to the attribute-based encryption process because of the high security provided in key generation. During the data transmission, the proposed system ensures the integrity of data.

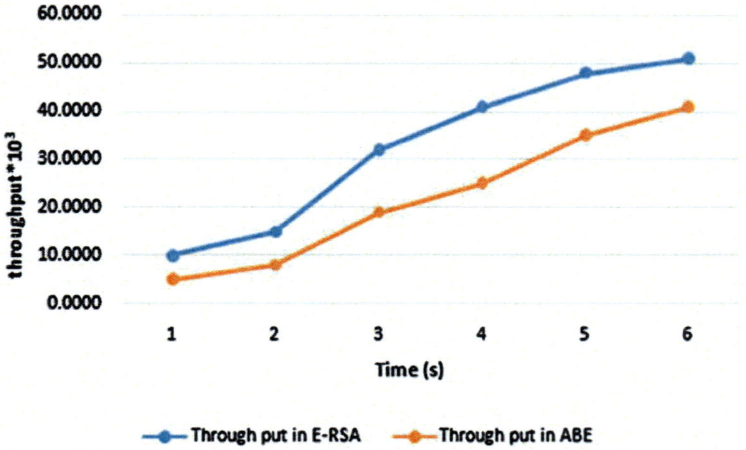

Fig. 3 Performance evaluation based on throughput

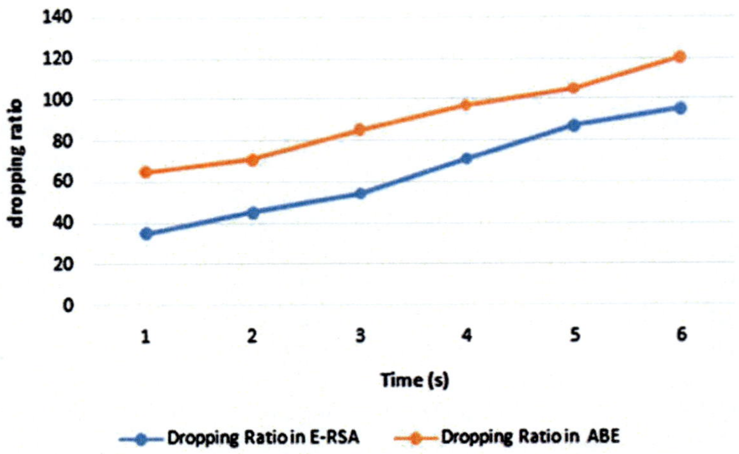

Fig. 4 Performance evaluation based on dropping ratio

Jitter is the consecutive packet delay in the wireless data transmission. It has been comparatively reduced in our proposed system. Average waiting time is the average amount of time the packets waited to transmit is also reduced much in our proposed solution. Delay time is the time difference between starting and ending time of packet transmission. Delay time parameter is much reduced in our proposed system when compared with existing methodologies. The number of attributes taken for the encryption and decryption process is also much improved with our solution.

5 Conclusion

In this paper, we discussed the challenge of security and data handling in medical WSN for patient monitoring process. When the data is transmitted in wireless medium, security is the main issue. To overcome the aforementioned issue we proposed a novel security model that removes possible security extortions of medical data transmission and it also guarantees integrity and confidentiality without intervention of patients or doctors. To provide energetic security policies for healthcare applications, we provided an efficient access control system that enhances the RSA cryptography method. This combination reduced time delay and dropping ratio in encryption/decryption process have been shown in the performance evaluation. At the end, we have performed wide-range of simulations that proved that our model provides competent and scalable access control to medical applications. In the future, ECC along with attribute-based encryption in a distributed architecture to have multiple accessing authorities can be considered.

References

1. Wu, T. Y., & Lin, C. H. (2015). Low-SAR path discovery by particle swarm optimization algorithm in wireless body area networks. *Sensors Journal, 15*(2), 928–936.
2. He, J., Geng, Y., Wan, Y., Li, S., & Pahlavan, K. (2013). A cyber physical test-bed for virtualization of RF access environment for body sensor network. *Sensors Journal, 13*(10), 3826–3836.
3. He, D., Chan, S., & Tang, S. (2014). A novel and lightweight system to secure wireless medical sensor networks. *IEEE Journal of Biomedical Health Information, 18*(1), 316–326.
4. Nidhya, R., & Karthik, S. (2019). Security and privacy issues in remote healthcare systems using wireless body area networks. In R. Maheswar, G. Kanagachidambaresan, R. Jayaparvathy, & S. Thampi (Eds.), Body area network challenges & solutions. *EAI/Springer Innovations in Communication and Computing*. Springer.
5. Li, M., Lou, W., & Ren, K. (2010). Data security and privacy in wireless body area networks. *IEEE Wireless Communication, 17*(1), 51–58.
6. Hu, C., Zhang, F., Cheng, X., Liao, X., & Chen, D. (2013). Securing communications between external users and wireless body area networks. In *Proceeding of 2nd ACM Workshop on Wireless Network Security Privacy (HotWiSec)*, Hungary (pp. 31–35).
7. Lai, J., Deng, R. H., Guan, C., & Weng, J. (2013). Attribute-based encryption with verifiable outsourced decryption. *IEEE Transaction Information Forensics Security, 8*(8), 1343–1354.
8. Lu, R., Lin, X., & Shen, X. (2013). SPOC: A secure and privacy-preserving opportunistic computing framework for mobile-healthcare emergency. *IEEE Transaction on Parallel Distribution System, 24*(3), 614–624.
9. Zhao, H., Qin, J., & Hu, J. (2013). An energy efficient key management scheme for body sensor networks. *IEEE Transaction Parallel Distributed System, 24*(11), 2202–2210.
10. Nidhya, R., Karthik, S., & Smilarubavathy, G. (2019). An end-to-end secure and energy-aware routing mechanism for IoT-based modern health care system. In Wang, J., Reddy, G., Prasad, V., & Reddy, V. (Eds.), Soft computing and signal processing. *Advances in Intelligent Systems and Computing* (Vol. 900). Singapore: Springer.
11. Tan, C. C., Wang, H., Zhong, S., & Li, Q. (2009). IBE-Lite: A lightweight identity-based cryptography for body sensor networks. *IEEE Transaction on Information Technology Biomedical, 13*(6), 926–932.
12. Boneh, D., & Franklin, M. (2003). Identity-based encryption from the Weil pairing. *SIAM Journal of Computer, 32*(3), 586–615.
13. Liu, J., Zhang, Z., Chen, & Kwak, K. S. (2014). Certificateless remote anonymous authentication schemes for wireless body area networks. *IEEE Transaction on Parallel Distributed System, 25*(2), 332–342.
14. Al-Riyami, S. S., & Paterson, K. G. (2003). Certificateless public key cryptography. In (Lecture Notes in CS), *Advances in Cryptology* (Vol. 2894, pp. 452–47). Springer-Verlag.
15. Cagalaban, G., & Kim, S. (2011). Towards a secure patient information access control in ubiquitous healthcare systems using identity-based signcryption. In *Proceedings of 13th International Conference on Advanced Communication Technology (ICACT)*, Korea (pp. 863–867).
16. Zheng, Y., Digital signcryption or how to achieve cost (signature & encryption) cost (signature) + cost (encryption). In (Lecture Notes in CS) *Advances in Cryptology* (Vol. 1294, pp. 165–179).

Sentimental Analysis on Twitter Data Using Hadoop with Spring Web MVC

RaviKiran Ramaraju, G. Ravi, and Kondapally Madhavi

Abstract This paper addresses the sentiment analysis on Twitter data, i.e., obtaining user sentiments by classifying tweets according to their opinions positive, negative or neutral. This paper primarily consists of live stream data generation, process and visualization through GUI based application for the results. Output will be visualized through Google Charts. Opinion tweet results will be displayed in the form of pie charts. Twitter is an online microblogging on social-networking platform which allows users to express their ideas. It is a rapidly expanding service with over 555 million users, generating millions of tweets per day. We can obtain public sentiment through this data, it is crucial for many applications such as predicting political elections suppose public opinion on the prime minister and for obtaining product reviews of firms. The aim of this project is to find sentiments from the Twitter live streaming data using Apache Flume, Hadoop, and visualize the results in the web application.

Keywords Apache Hadoop · Flume · HDFS · Spring · Bootstrap · JQuery · Google chart · Maven · Twitter

1 Introduction

In this era with the invention of smart mobiles, there is a revolution in the usage of the internet and social media. Social media sites like Facebook, Twitter, or Whatsapp become very popular [1]. Irrespective of educational background now people are expressing their views on these social sites about the products they are purchasing,

R. Ramaraju (✉) · G. Ravi
Department of IT, B.V Raju Institute of Technology, Narsapur, Medak, Telangana, India
e-mail: rrk.wipro@gmail.com

G. Ravi
e-mail: g.raviraja@gmail.com

K. Madhavi
Department of CSE, MallaReddy College of Engineering, Hyderabad, India
e-mail: madhavi.ramaraju0519@gmail.com

S. C. Satapathy et al. (eds.), *Intelligent System Design*, Advances in Intelligent Systems and Computing 1171, https://doi.org/10.1007/978-981-15-5400-1_28

people they are voting and the incidents happening in and around, the data, i.e., produced by these sites are large called as Big Data. If the data is small, then it is easy to analyze otherwise if it is Big and unstructured, very difficult to analyze. With the advancement of technology, with the invention of Hadoop and Hadoop echo system, Data generation, Data processing becomes easy. Now people are openly sharing their views and opinions. Marketing agencies, firms, and political parties depend on these reviews and opinions made by the user [2].

Twitter, the popular microblogging forum where millions of tweets are produced in every minute, analyzing these tweets and identifying what they express is a difficult job [3]. Jallikattu is a traditional sport played in Tamil Nadu, animal rights organization called for a ban to this sport as there are events of injury and death associated with it and animals has been forcing into it. However, there is a protest from a few people against the ban; in this scenario, it is difficult to track people's opinions and views with the traditional approach [4]. Sentiment analysis plays a very crucial role to know people sentiments about Jallikattu. Another scenario an organization wants to identify customer's opinions of their products; this can be solved by sentiment analysis. The data generated from Twitter is unstructured data. It is the biggest challenge to analyze and visualize this data. To store and analyze the data we use Hadoop Framework to visualize this data we use Google Charts, JQuery, and Maven. We will further discuss this topic in methodology and Implementation.

This paper is organized in the following way. Section 2 Literature Survey, Sect. 3 explains methodology, Sect. 4 Results, Sect. 5 Conclusion and Future Work.

2 Literature Survey

1. Anisha [5] discussed the importance of Hadoop Technologies for Big Data, used Apache Flume for streaming on-going Twitter data into HDFS, PIG scripts are used to extract opinions from raw Twitter data, The dictionary-based approach used for sentiment analysis.
2. Divyasehgal [6] also used Hadoop Framework to find out opinions of people towards blogs and movies. Total accuracy was 72.22%.
3. Monu Kumar [7] proposed a model that uses Hadoop for storing Twitter data and for intelligent analysis. The subject they have used is #Airtel to analyze the customer feedback on #Airtel, HDFS and MapReduce engine used for storage and analysis. Apache Mahout is a machine learning tool used for clustering and classification. They proposed an algorithm Hadoop with Mahout for analysis.
4. Umit [8] proposed an approach to automatically classify Twitter data obtained from BGS (British Geological survey) collected using specific keywords using MapReduce algorithm and a model to distinguish tweets using Naive Bayes machine learning algorithm with N-gram language model on Mahout.

3 Methodology

3.1 Data Storage

In Big Data technologies, storage is a critical aspect to support large volumes of data. Hadoop is supporting by commodity hardware. Due to its distributed architecture approach, the data stores in HDFS across multiple nodes.

3.2 Data Generator

Initially, data generator is a shell script. It starts the Apache Flume server. Finally, it triggers the MapReduce program.

3.3 MapReduce

MapReduce programs written in Java language which is used to process high volumes of data from HDFS. In a high level, a MapReduce program will be stated as a processing unit within the Hadoop environment. MapReduce is flexible to integrate with any available API, i.e., Afinn dictionary.

3.4 Analysis Process

The streaming tweets are getting processed through MapReduce programs, and it performs text mining operations with the help of latest dictionary called “AFFIN”. AFFIN is the most popular dictionary when compare with other available dictionaries.

3.5 Persisting Output Result

MySql database is being used to persist in the outcome of the analysis. MySql is an open-source database server.

3.6 Web User Interface or GUI

The web application is used as the user interface to monitor results for sentiment analysis. Apache Spring is a light-weight web MVC framework. Spring can be integrated with any other Java APIs. Spring supports all dependency injections and having an inversion of control mechanism.

3.7 Open-Source Tool Kit for Developing HTML

Bootstrap is an open-source library for developing web user interfaces with HTML, CSS, and Javascript. Applications built through bootstrap are very much responsive, and they are mobile-first websites.

3.8 Visualization/Charts

The outcome of analysis results will be visualized in rich graphical notations. Google chart is one of open-source API. It is flexible and acts as a graphical injection in spring-based applications.

3.9 Maven

Maven is used to building the project. It is being used in any continuous integration environment. All the project artifacts referred to the concept of a central repository system.

3.10 Web/Application Container

Apache Tomcat is an open-source web application server. Web applications are hosted on Tomcat web server. All Java EE applications are supported by the Tomcat container. Catalina is also known as Servlet container in Tomcat. This web server uses Coyote as a connector to support HTTP protocol. Jasper is popular as JSP engine in Tomcat.

High-Level Architecture:
Users tweet their opinion on various subjects through Twitter. A data generator script will be triggered by an end user. The given subject is passed as an argument to the script. The shell script in terms invokes Apache Flume agent. The Flume agent

fetches the tweets into HDFS in the form of JSON format. Once the data is available on HDFS, Script triggers MapReduce. Afinn dictionary is loaded into MapReduce in the form of input stream. The Latest Afinn dictionary contains about 2,400 words. Each word in the dictionary is associated with rating value. The rating values are range between −5 and +5. The rating value with negative represents a bad opinion. The rating value with positive represents a good opinion. The rating value with zero represents a neutral opinion [9]. MapReduce performs analysis process on tweets and calculates the rating value associated with each word. The average rating value of a given tweet text will be inserted into database. Once the MapReduce process is completed then the program will execute a JDBC method. The JDBC method reads all the inserted records from database. The JDBC method calculates the total number of positive, negative, and neutral opinions and then inserts the outcome into a history table (Fig. 1).

A Spring MVC application is pointed to the same database schema as that of MapReduce program. An end user accesses the web URL in the browser. The URL request goes to the web container [10]. Web container invokes the dispatcher Servlet. Spring container resolves the given request through JSP resolver. Each Spring Controller is represented with a unique request mapping value in the form of an annotation. The controller then invokes the model objects. Once the model objects are available then container invokes service layer. Service layer invokes processor and DAO methods. The DAO layer is integrated with the database. Google chart API is

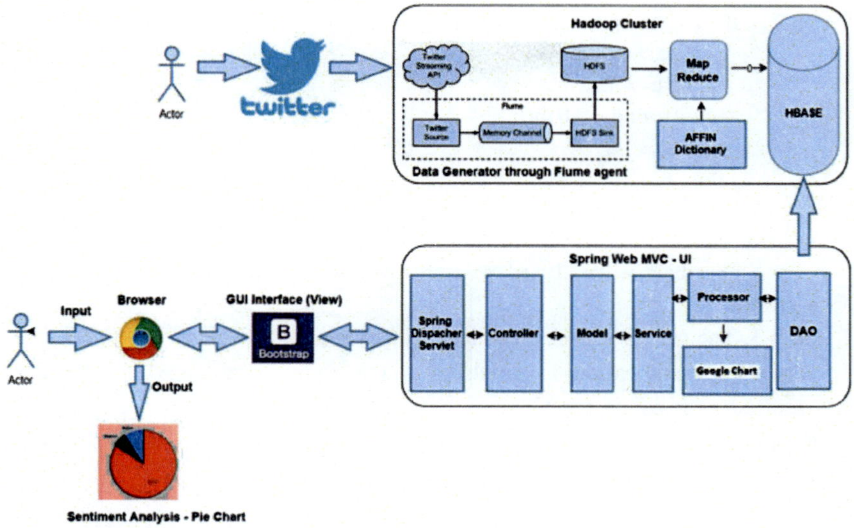

Fig. 1 Hadoop high-level architecture

used to render the pie chart. The end user is able to see the opinion result in the form of a pie chart.

Algorithm OR Pseudo Code

Step 1. Begin
Run Data Generator script.
Step 2. Evaluating sentiment ratings and opinion by MapReduce

Data Sets: Collection of Twitter data on HDFS
Result classification:= positive, negative and neutral

Load Afinn dictionary as an input stream in MapReduce program.

Configure and initialize DB Input and Output format classes in MapReduce program.
while tweets in Data Sets
Read Twitter id attribute. [The attribute "id" represents Twitter id on HDFS data sets.]
Read the text attribute. [The attribute "text" represents Twitter text posted by the user.]
Rate each word of the tweet as per the defined rating value in Afinn dictionary.
rating:= Round value of (Total rating of all words in tweet text/Total number of words in tweet)
if rating > 0.0 then
opinion:= Positive
else if rating < 0.0 then
opinion:= Negative
else
opinion:= Neutral
endif
Set the "Twitter id" and "opinion" values to DB output format classes.
end while
Step 3. The results will be visualized in web application.
[The Web application is pointing to the same database in which the MapReduce is connected.]
Step 4: End

4 Results

Table 1 gives you a few tweets and ratings of tweets either positive or negative based on Afinn dictionary.

Table 1 Analysis process on Twitter Data—the Twitter keyword used—"Brexit" data collected through Apache Flume

Twitter ID	Text	Rating
1058635743629844480	RT @Peston: Government has conceded that the city of London will be EU rule-taker—more salt rubbed by @theresa_may into the wounds of @Jacob…	0
1058635745479536640	RT @campbellclaret: More humiliation for U.K. more evidence of broken promises by all re "same benefits" RIP https://t.co/L7gX5BV7jv	– 3
1058635781873504256	RT @hayretkamil: Suck on that @DailyMailUK https://t.co/zVYpatH1Eu	– 1
1058636032395169792	RT @FaranakAzad1: World should take urgent action! #Iranian environmentalists are in danger. Two of them vanished now. They may	2
1058636032877518850	RT @MickeyK69: Congratulations @touchlinefracas for bringing the pod up north. Now Londoners can now relate to Mancs supporting & discussion	3

Twitter ID represents the account of a given user. Text represents the given tweet posted by the user. The rating column represents the opinion rating value for a given tweet.

Consider the text "RT @Peston: Government has conceded that the city of London will be EU rule-taker—more salt rubbed by @theresa_may into the wounds of @Jacob…". In the above text, none of the words are associated with any rating value from Afinn dictionary. This reason the MapReduce program marked the rating value as 0 for the above Tweet ID 1058635743629844480.

Consider the text "RT @campbellclaret: More humiliation for the U.K. more evidence of broken promises by all re "same benefits" RIP https://t.co/L7gX5BV7jv".

In the above text, the below words are associated with rating values from Afinn dictionary. The word "humiliation" is representing with a rating value as −3. The word "broken" is representing with a rating value as −1. The word "promise" is representing a rating value 1. The word "benefit" is representing a rating value 2. The word "RIP" is representing a rating value −2. The total rating of all these words becomes −3 for the above Tweet ID 1058635745479536640.

Twitter data stored to HDFS using Apache Flume as shown in Fig. 2. The block size of HDFS is 64 MB. The data collected is real-time Twitter data; flume generated multiple files. After a few seconds, we stopped the flume otherwise number files will be increased [11], which will waste the storage space.

After running MapReduce program Twitter sentiments are obtained. The output is stored in MYSQL using Maven, now it is directly connecting to web application developed using Spring MVC. Figure 3 shows the positive, negative, and neutral opinions were reflected under the web application.

The final output of sentiment analysis is visualized using Google Chart. The keyword we used is Brexit. We got 16.7% positive opinion, 81.7% negative opinion, and 1.6% as neutral opinion (Fig. 4).

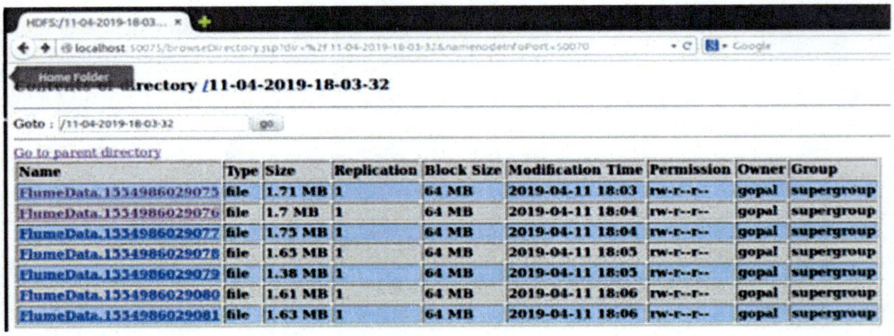

Fig. 2 Twitter data loaded on HDFS by Apache Flume

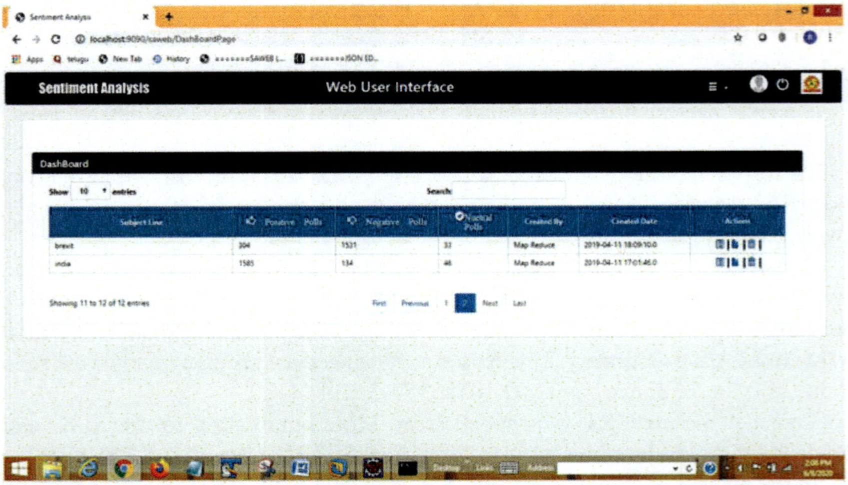

Fig. 3 The positive, negative, and neutral opinions are being reflected under web application

5 Conclusion and Future Work

The given web application will be used as a replacement for existing survey websites, i.e., the current solution is more superior though it involves social media data when compared with existing survey sites. Currently, data generator, process, and web applications are three separate components in the entire analytics. As future work, it is recommended to trigger data generator through web application itself. In this way, the Big Data system can be simplified and operates through light-weight web MVC frameworks, i.e., make use of restful web service calls to invoke data generator script from web application.

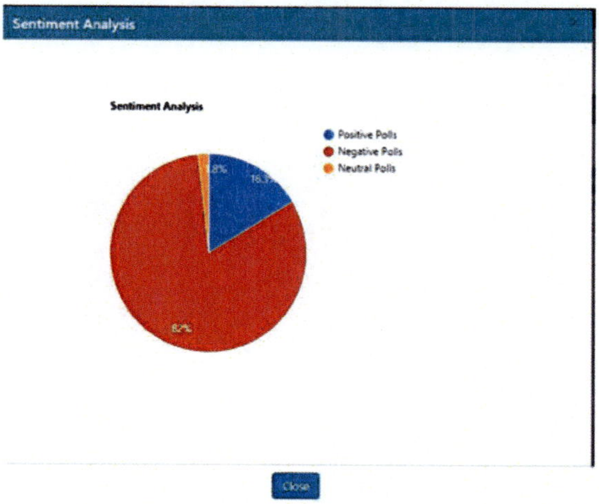

Fig. 4 The final output of sentiment analysis in Google Pie chart

References

1. Li, Z., Fan, Y., & Jiang, B., A survey on sentiment analysis and opinion mining for social multimedia. *An International Journal of Multimedia Tools and Applications* (Springer). ISSN: 1380-7501 (Print) 1573-7721 (Online).
2. Choudhury, J., Pandey, C., & Saxena, A., Sentimental analysis of Twitter data on Hadoop. In *Advances in Intelligent Systems and Computing* (Vol. 810). Singapore: Springer.
3. Yousif, A., Niu, Z., Tarus, J. K., & Ahmad, A. (2017). A survey on sentiment analysis of scientific citations: Artificial intelligence review. *An International Science and Engineering Journal.* ISSN: 1573-7462 (Online).
4. Powar, S., & Shinde, S. (2017). Named entity recognition and Tweet sentiment derived from Tweet segmentation using Hadoop (pp. 194–198). 978-1-5090-4264-7/17/$31.00©2017. IEEE.
5. Rodrigues, A. P., & Rao, A. (2017). Sentiment analysis of real-time Twitter data using big data approach. In *2nd IEEE International Conference on Computational Systems and Information Technology for Sustainable Solutions* (pp. 175–180).
6. Sehgall, D., & Agarwal, A. K. (2016). Sentiment analysis of big data applications using Twitter data with the help of HADOOP framework. In *2016 International Conference System Modeling & Advancement in Research Trends (SMART)*-2016 (pp 251–55). IEEE Conference ID: 39669. Moradabad, India: Teerthanker Mahaveer University.
7. Kumar, M., & Bala, A. (2016). Analysing Twitter sentiments through big data. In *2016 International Conference on Computing for Sustainable Global Development (INDIACom)* (pp. 2628–2631).
8. Demirbaga, U., & Jha, D. N. (2018). Social media data analysis using MapReduce programming model and training a Tweet Classifier using Apache Mahout. In *2018 IEEE 8th International Symposium on Cloud and Service Computing (SC2).*
9. RamakrishnaMurty, M., Murthy, J. V. R., Prasad Reddy, P. V. G. D., Sapathy, S. C. (2012). A survey of cross-domain text categorization techniques. In *International conference on Recent Advances in Information Technology RAIT-2012*, ISM-Dhanabad. 978-1-4577-0697-4/12. IEEE Xplorer Proceedings.
10. EI Alaoui, I., & Gahi, I. (2018). A novel adaptable approach for sentiment analysis on big social data. *Journal of Big Data.* https://doi.org/10.1186/s40537-018-0120-0.
11. Fang, X., & Zhan, J. (2015). Sentiment analysis using product review data. *Journal of Big Data* (Springer).

Slicing Based on Web Scrapped Concurrent Component

Niharika Pujari, Abhishek Ray, and Jagannath Singh

Abstract In this paper, we have proposed an algorithm for dynamic slicing of concurrent COPs that consist of multiple threads. In order to portray the concurrent COP effectively, an intermediate representation graph called concurrent component-oriented dependency graph (CCmDG) is developed based on the dependencies among all the edges. The intermediate representation graph developed for component-oriented programs consists of system dependence graph (SDG) for individual components, along with that some new dependence edges are introduced that are used to connect each system dependency graph to the interface. Based on the intermediate graph created in first step, a dynamic slicing algorithm is proposed for undertaken case study for the concurrent component-oriented program. The resultant is a graph marking and unmarking the executed nodes in concurrent components dynamic slicing (CCmDS) appropriately during run-time.

Index Terms Dynamic slicing · Concurrent component-oriented programming · Web scrapping · Thread connect edge (TCE)

1 Introduction

Program slicing [1–2] is a practical fragmentation methodology which overlooks the modules of the program which are immaterial to the specific computation procedure grounded on a condition called slicing criterion [3]. Using this slicing process, we can determine a module's significance for a particular computation. So, program slicing is the best suitable for bridging the gap between complex programs and program comprehensibility, debugging and testing. A static slice of a given program contains

N. Pujari (✉) · A. Ray · J. Singh
School of Computer Engineering KIIT, Deemed to Be University, Bhubaneswar, Odisha, India
e-mail: niharika.pujari06@gmail.com

A. Ray
e-mail: arayfcs@kiit.ac.in

J. Singh
e-mail: jagannath.singhfcs@kiit.ac.in

S. C. Satapathy et al. (eds.), *Intelligent System Design*, Advances in Intelligent Systems and Computing 1171, https://doi.org/10.1007/978-981-15-5400-1_29

the statements that may affect the value of a variable at a point of interest for every possible inputs. Dynamic slicing was introduced by Korel et al. [4, 5]. A dynamic slice marks out with those line of codes that are affected by a given input value [6], based on the slicing criterion < s, V > for a particular execution.

Some of the issues and challenges that may appear in implementation of concurrency programs in COP are testing, debugging and synchronization among components in real-time programming and parallel programming. The real and flawless digital world is made possible by a programming paradigm called component-oriented programming (COP) in which the integration of different components is carried out through component's interface, such that internal structure is hidden from outside world [7, 8]. In order to achieve multi-core architecture and faster execution speed of the program, the sequential COP is needed to be extended to concurrent COP (CCOP). Program slicing overcomes the obstacles in testing the CCOP along with its complexity. Concurrency can be achieved by multithreading which is embedded to different types of programs like OOP, AOP to have concurrent OOP, AOP [9–10] and COP. Based on our literature survey, a clear outline can be drawn that no concurrency principle has been embedded to COP and found a gap in slicing of concurrent component-oriented programs.

We propose a dynamic slicing algorithm for CCOP. We have extended our existing CmDG into a dependence-based intermediate representation graph called concurrent component dependency graph (CCmDG) for CCOP. Then, we have developed a slicing algorithm called concurrent components dynamic slicing (CCmDS) to handle the concurrency mechanism in COP. The rest of the paper is organized as such: In Sect. 2, the introductory concept of Web scrapping and concurrent COP is dealt. In Sect. 3, the literature survey of correlated work is defined. Section 4 defines the concurrency model of COP, through the case study, that is considered in this paper. Sect. 5 explains our proposed intermediate program representation (CCmDG) of the case study. Sect. 6 gives the overview working of the proposed algorithm concurrent components dynamic slicing (CCmDS), and also the resultant slice of the proposed algorithm is shown. Section 7 concludes the paper.

2 Basic Concepts

This section provides the preliminary concepts of Web scrapping, concurrent component-oriented programming that is needed for getting the clear idea about our proposed approach.

Web scrapping is also referred as data scrapping used for extracting data from websites. Web service is a new paradigm to deliver application services on Web and enabling it as programmable Web rather than just an interactive Web [8]. Though features of COP can be achieved by many component technologies/frameworks such as CORBA, NET, EJB, Web service is the only technology that provides cross-platform, cross-programming language and Internet-firewall solutions to the interoperable distributed computing. The viewpoint behind Web services is to develop distributed software by composition replacing programming that is building new application from existing components either by purchasing or by reusing the existing components. This is the exact motive behind COP and component-based software development (CBSD). One of the key features of COP is self-deployability of components, which is very difficult to acquire in case of other component technologies but easy to deploy, publish, discover and invoke a Web service on Internet. Many Web services' development tools are there such as Microsoft. NET Web service based on IIS server. Java Web service development pack (JWSDP), but the tools we have considered in this paper is XAMPP. XAMPP is a free and open-source cross-platform Web server solution stack package developed by Apache friends. XAMPP stands for Cross-Platform (X), Apache (A), MariaDB (M), PHP (P). It is a simple, lightweight Apache distribution that makes it easy for developers to create a local W server for testing and deployment purposes.

2.1 Concurrent COP

COP can be referred as an interface-based programming as other programming paradigms like OOPs emphasizes on classes and objects, COP emphasizes on composition and interfaces [8]. Major advantages of using COP:

(1) Abstraction: reduces abstraction.
(2) Complexity: decreases complexity
(3) Reusability: increases reusability

Concurrency is the critical aspect which has to be taken into concern. Concurrency can be put into picture by using threads in the component-oriented programs. Multithreaded programs run faster than single threaded programs giving a better throughput [11]. The challenges that are needed to be handled while implementation of multithreaded program and putting existing code to concurrent code are debugging, testing, synchronizations. We have used Web application slicing technique for overcoming the challenges. The component technology tools used here is XAMPP, and the scripting language used is PHP. Figure 1 is a simple example showing how concurrency is handled in Web service applications (here PHP scripts) programming.

3 Reviews of Related Works

In this section, due to the absence of any work related to concurrent COP, we present here reviews of only the researches related to concurrent OOP, AOP.

Program slicing was coined by Mark Weiser in 1979 [3]. According to Weiser, program slicing is the process of extracting set of statements from a given program that is affected by a point of interest. The point of interest is denoted as Slicing criterion that is a set of < s,V> , where 's' represents the statement number of the

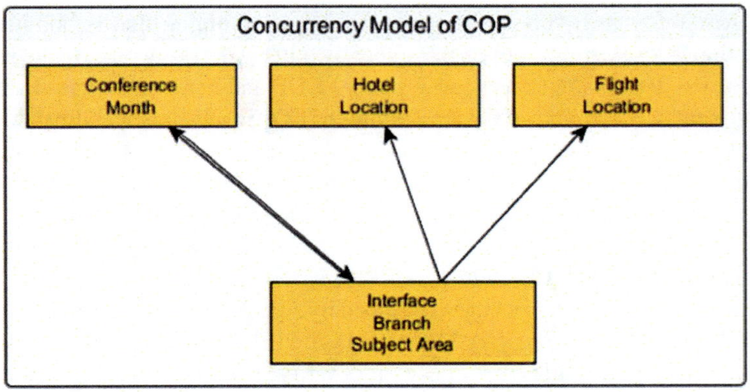

Fig. 1 Concurrency model of sample COP

point of interest and 'v' represents the set of variables present in the point of interest. Similar work has been carried out by Ottenstein et al. [12] that compute program slicing by traversing through the graph which represents given program. They have introduced another graph reachability representation called program dependence graph (PDG).

In 1990, Horwitz et al. [1] introduced another graph called system dependence graph (SDG), to represent the control and data dependencies among subprograms. He developed inter-procedural slicing algorithm called two-phase slicing algorithm to compute desired slices. They have also proposed an algorithm that traverses in backward manner phase-wise on the SDG to find inter-procedural program slices.

Krinke [13] gave a new approach for slicing concurrent programs by taking time-sensitive information for computing slice. Krinke has developed a new intermediate graph called threaded inter-procedural dependency graph (tIPDG) by introducing a new edge called interference edge. The problem of shared variable in different threads is solved in this approach.

Mohapatra et al. [9] developed a new intermediate representation graph called concurrent system dependence graph (CSDG) and proposed a dynamic program slicing algorithm for concurrent OOPs called as marking-based dynamic slicing (MBDS) depending on the edges traversed. The graph works by marking and unmarking the edges in the graph as per the associated dependencies execute or cease in the run-time. Zhao et al. [14, 15], introduced a intermediate representation graph called aspect-oriented system dependence graph (ASDG), which is the extended version of their existing dependence graph to represent aspect-oriented software. They also proposed a slicing algorithm for aspect-oriented program slicing. Similar work is done by Singh et al. [16] presented a dynamic parallel context-sensitive slicing algorithm for distributed AOPs. The preciseness and accuracy of the computed slice depends on its context sensitivity. In order to make the computation of slice faster, the concept of parallelism has been introduced in their algorithm. A novel technique is used to invent a tool called D-AspectJ slicer to compute dynamic slices for distributed AOPs. Similarly, Ray et al. [10] presented a novel approach for slicing concurrent AOP. They have introduced dynamic slicing algorithm for computing slice of concurrent AOP and also proposed a dependence-based intermediate program representation called concurrent aspect-oriented system dependence graph (CASDG).

4 Concurrency Model of Cop

Concurrency creates a vital impact factor for modern codes. Concurrency is achievable in OOP by representing methods as threads for executing it concurrently. These types of programs are called concurrent OOP. In order to handle "software crisis," software engineers make wide use of COP as the significant technology. COP provides software that are built by assembling existing components. For achieving better performance throughput from Web services, concurrency plays a lead role in COP. But putting concurrent code to Web codes (script code) is a challenging work do be done. As per the best of our literature survey, no existing work or model is found in the field of slicing to address the concurrency aspect of COP. We have proposed an algorithm for computing dynamic slicing of concurrent COP.

The concurrency model is explained by taking a simple real-time case study shown in Fig. 1. The component used here is Web service component, where Web scrapper involves fetching and extracting the contents from the website. Web scraping can be done manually but the automated version of it is Web crawler. Web crawling is a main component of Web scrapping, to fetch pages for later processing. The content of the page may be parsed, searched, reformatted. The Web crawler used here is CURL.

In Fig. 1, we have created three APIs named as conference, flight and hotel. These are three independent API components, which perform to extract valuable information, such as: conference component, scrap the detailed data about conference and shows the result w.r.t. Scopus-related conferences. In the same way, two threads named as Hotel and Flight are fired concurrently to scrap out only five starrer hotels and direct communication to the conference location is available or not. Here we are creating three threads; two threads hotel and flight are independent to each other, and hence, they are fired concurrently from the interface, making two processes to work simultaneously. Interface collects the conference details based on branch and month by executing conference API. Then, based on the location of conference details about hotel and direct flight information is collected by executing Hotel and Flight API.

Component1-Conference

3. *Function.php*

```
$cl = curl_init(); $timeout = 3;
curl_setopt($cl,CURLOPT_URL, $url);
curl_setopt($cl,CURLOPT_RETURNTRANSFER, 1);
curl_setopt($cl,CURLOPT_CONNECTTIMEOUT,$timeout);
```

```
$res = curl_exec($cl);
curl_close($cl);.
return $res;
.
.
.
```

```
4. $returned_content = trim(get_data("https://conferencealerts.com/advancedSearch?startDate=2018-07-1&endDate=&searchCountry=100_India&advancedSearchTerm= SCOPUS&x = 6&y = 3"));
5. $ret_arr = explode (' < td colspan = "2" width = "80%" id = "eventMonth-Heading" class = "textLeft">',$returned_content);
6. for($i = 1,$j = 0;$i <=count($ret_arr);$i ++,$j ++) {
7. $ret_arr2 = explode(' </td > </tr>',$ret_arr[$i]);
8. $con_month_year = $ret_arr2[0];...
.
.
.
23. mysqli_query ($conn,"insert into conference (conference_name, conference_date,conference_place) values ('".$conference_arr[$j] ['confer name']."','".$conference_arr [ $j] ['conference_date']."','".$conference_arr[$j] ['conference_place']."')");}}}
24. $month = $_REQUEST['month'];$year = date("Y");
25. $sql1 = mysqli_query($conn,"SELECT * FROM conference WHERE MONTH (conference_date) ='".$month."' AND YEAR(conference_d='".$year."'");
26. $num1 = mysqli_num_rows($sql1); $i = 0;$resArray = array();
27. while($a1 = mysqli_fetch_assoc ($sql1)) {
28. $resArray[$i] = $a1;
29. $i ++;}}
30. echo $json_response = json_encode($resArray);? >}.
```

Code1: First component (Conference) of case study

In the first component, named as Conference, statement from 4-23 extracts the data from the website Conferenceal- erts.com based on the month specified and uses CURL as a Web scrapper. Statement 24-30 results with all the conference details filtered based on specified month and Scopus indexed. This API gives filtered out data from website as per client requirement which is further reused by other components (here used by the interface) to build up new software.

Component2-Hotels Function.php

```
$cl = curl_init();$timeout = 3;
curl_setopt($cl,CURLOPT_URL,$url);
curl_setopt($cl,CURLOPT_RETURNTRANSFER,1);
curl_setopt($cl,CURLOPT_CONNECTTIMEOUT,$timeout);
$res = curl_exec($cl);
```

```
curl_close($cl);.
return $res;
.
.
class Hotels extends thread{
public function_construct(Hotel $place)
{
$this- > place = $place;
}
public function run()
{
1.  require_once('function.php');
2.  $place = $_REQUEST['place'];
3.  $sql1 = mysqli_query($conn,"SELECT h.* FROM hotels h WHERE
    h.hotel_place ='". $this- > place- > fetch()."'" and h.hotel_star- >=5);
4.  $num1 = mysqli_num_rows($sql1);$i = 0;
    $resArray = array();
5.  while($a1 = mysqli_fetch_assoc($sql1)) {
6.  $j = 0;$resArray[$i] = $a1;
7.  $sql2 = mysqli_query($conn,"SELECT * FROM hotel_details WHERE
    hotel_id ='".$a1['hotel_id']."'");
8.  while($a2 = mysqli_fetch_assoc($sql2)) {
9.  resArray[$i]['hotel_details'][$j] = $a2;
10. . $j ++;}
11. $i ++;}
12. echo $json_response = json_encode($resArray);?>
```

Code 2: Second component (Hotel) of case study

The second component, namely Hotels, scraps the data from hotel website and comes out with only five or above starrer hotels in the location specified. This API is an individual component that further be accessed by any other component for reusability.

Component 3 - Flight

```
Function.php
$cl = curl_init();$timeout = 3;
curl_setopt($cl,CURLOPT_URL,$url);
curl_setopt($cl,CURLOPT_RETURNTRANSFER,1);
curl_setopt($cl,CURLOPT_CONNECTTIMEOUT,$timeout);
$res = curl_exec($cl);
curl_close($cl);.
return $res;
.
.
class Flights extends thread {
```

```
public function_construct(hotel $place)
{
$this- > place = $place;}
public function run()
{
```

```
1. require_once('function.php');
2. $place = $_REQUEST['place'];
3. $sql1 = mysqli_query($conn,"SELECT * FROM flight WHERE flight_place =
   '". $this- > place- > fetch()."'");
4. $num1 = mysqli_num_rows($sql1);
5. $resArray = array();
6. if($num1 > 0) {
7. $resArray[0] = 1;} else {
8. $resArray[0] = 0;}
9. echo $json_response = json_encode($resArray); ? >}
```

Code 3: Third component (Flight) of case study

In this, API name as Flight, and the availability of direct communication to a given location is analyzed by scrapping the data from real site. This API is reusable by integrating it with components for getting desired client requirement.

Interface

```
7.  $month = $_POST['month'];
8.  $branch = $_POST['branch'];
9.  $post_fields = array("month" => $month,"branch" => $branch);
10. $r_file = "http://localhost/conference1/find_conference.php";
11. $curl = curl_init($r_file);
12. curl_setopt($curl, CURLOPT_VERBOSE- > true, CURLOPT_RETURN
    TRANSFER- > true, CURL OPT_SSL_VERIFY PEER- >
    false,CURLOPT_POST- > true,CURLOPT_POSTFIELDS- > $post_fields);
13. $result = curl_exec($curl);
14. $resultSet = json_decode($result);} else {
16. for($i = 0;$i < count($resultSet);$i ++) {
17. if($resultSet[$i]- > conference_branch == $_POST['branch']){
18. $post_fields2 = array("place" => $resultSet[$i]- > conference_place);
19. $hotel = new Hotel($post_fields2)
20. $hotels = new Hotels($hotel);
21. $flight = new Flight($post_fields2)
22. $flights = new Flights($flight);
23. $hotels- > start();
24. $flights- > start();
25. $r_file1 = $hotels('http://localhost/conference/find_hotel.php');
26. $curl1 = curl_init($r_file1);
```

27. *curl_setopt($curl1, CURLOPT_VERBOSE->true, CURLOPT_RETURN TRANSFER- > true,CURLOPT_SSL_VERIFYPEER- > false,CURL > true, CURLOPT_POSTFIELDS, $post_fields2);*
28. *$result1 = curl_exec($curl1);*
29. *$resultSet1 = json_decode($result1);*
30. *$r_file2 = $flight = "http://localhost/conference/find_flight.php";*
31. *$curl2 = curl_init($r_file2);*
32. *curl_setopt($curl2, CURLOPT_VERBOSE- > true,CURLOPT_RETURN TRANSFER- > true,CURLOPT_SSL_VERIFYPEER- > false,CURL > true, CURLOPT_POSTFIELDS, $post_fields2); 33. $result2 = curl_exec($curl2);*
34. *$resultSet2 = json_decode($result2);*

Code 4: Case study (Main program)

Here, we come with the main part of our example. In this interface, we integrate the scrapped data from three components: Conference, Hotels and Flight. Interface accesses the result from conference based on criteria of branch and month. Then, based on the location of conference, both the Hotel and Flight components are fired concurrently. This concurrency is achieved by making both the component as pthread. Since pthread can enable two process to execute independently improvising decrease in response time.

The resultant is a new component, where we can acquire the details of the conferences held on a particular month and belongs to a particular branch, along with its correlated parameters like showing the direct communication availability to the location where conference is held and all the hotels available in that location.

5 Proposed Work

Intermediate Program Representation

In this section, we have discussed the method to handle concurrency in component-oriented program by defining concurrent component dependency graph (CCmDG).

5.1 Concurrent Component-Oriented Dependency Graph (CCmDG)

Concurrent component-oriented dependency graph (CCmDG) is the extension of our intermediate graph component- oriented dependency graph (CmDG), to handle the concurrency aspects of component-oriented programs. In CCmDG, the edges introduced are component connect edge (CCE) that interprets the dependencies between components and interfaces, and another new edge called component thread edge, when we need to connect more than one component simultaneously, for more efficient result. Concurrent component-oriented slicing is explained by considering a simple case study, where an interface communicates with three components concurrently to access data. In order to achieve this, two components are defined as thread, for concurrent execution of both. The CCmDG is a directed graph G = (N, E). Each node n belongs N (collection of nodes) represents the statement number in the program, and e exists in E (collection of edges, represents the edges along with their dependencies, present in CCOP. Figure 2 shows the CCmDG contains the following types of edges

- **Data Dependency Edge**: x1 $\overrightarrow{dd}$ x2 ∈ E is known as data dependency edge, and if there exist two vertices x1 and x2, where x1 and x2 ∈ N, transmission of data takes place from x1 to x2.

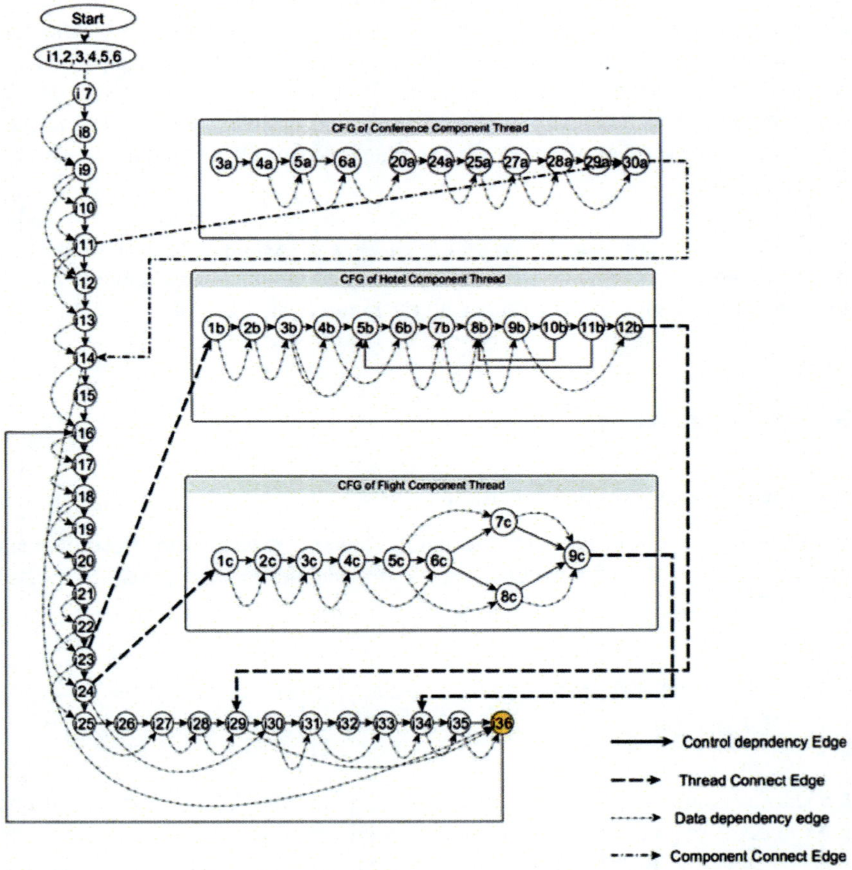

Fig. 2 CCmDG of the concurrent COP

- **Control Dependency Edge**: $x1 \xrightarrow{cd} x2 \in E$ is known as control dependency edge, and if there exist two vertices x1 and x2, where x1 and x2 $\in$ N, transmission of control takes place from x1 to x2.
- **Component Connect Edge**: $x1 \xrightarrow{cce} x2 \in E$ is known as component connect edge, and if there exist two vertices x1 and x2, where x1 and x2 $\in$ N such that x1 $\in$ interface, x2 $\in$ component and transmission of control takes place from x1 to x2, based on a criteria.
- **Thread Connect Edge**: $x1 \xrightarrow{tce} x2 \in E$ is known as thread connect edge. If there exist two nodes x1, x2 such that x1 represents a thread calling node and x2 represents a entry node of the component to execute its run() method, then the edge is said to be thread connect edge. This dependency exists between the last statement of a thread and the statement of the main thread after which thread is called.

6 Concurrent Component-Oriented Dynamic Slicing (CCMDS)

Concurrent component-oriented dynamic slicing is the extended version of our existing algorithm, where components connectivity is handled by traversing from the point of interest node in backward manner. The proposed algorithm is unable to handle concurrency within components, though it could efficiently handle self-deployable components. Our proposed algorithm improvises by adding a new phase to enable slicing of concurrent components, using threads.

6.1 Proposed Algorithm of CCOP Slicing

Algorithm 1 INPUT: CmDG $G < N,E >$, s-slice criterion
OUTPUT:The Slice S for slicing criterias

1. Consider three Traversal List L1,L2 and L3 and Initialise as $L1 = \{s\}, L2 = \{\}$, $L3 = \{\}, S = \{s\}$
a. n represents current node.
b. Remove the current node *n* from *L1*
2. Repeat the below steps until $L1! = \varphi$.//STEP1
a. For each a, n//a is the source node to current node n
i. If source node *a* is not in the slice *S*, then add source node *a* to the slice list *S*.
ii. If the current edge *e* does not belong to component connect edge (*cce*) or thread connect edge (*tce*), then add the current node a to the traversal list *L1*.
iii. Else if the current edge *e* belongs to then
a. Add the component connect node *P* to the traversal List *L2*.
b. Add the current node *a* in traversal list *L1*.

iv. Else if the current edge *e* belongs to thread connect edge (*cce*) then
a. Add the Thread connect node *Q* to the traversal list *L3*.
b. Add the current node *a* in traversal list *L1*.
3. Repeat the below steps until $L2! = \varphi$.//STEP2
b. a.Remove the current node *n* from *L2*
c. For each b, n//b is the source node for incoming edge to current node n
i. If node *b* is not in the slice *S*, then add node *b* to the slice list *S*.
4. Repeat the below steps until $L3! = \varphi$.//STEP3
a. Remove the current node *n* from *L3*
b. For each c, n//c is the source node for incoming edge to current node n
i. If node *c* is not in the slice *S*, then add node *b* to the slice list *S*.
5. Finally, the resultant slice *S* gives all relevant nodes as per slicing criteria *s*.

6.2 Slice Computation

In our proposed intermediate graph CCmDG, the components and interface are interlinked by the use of a new edge called thread edge. Working of the slicing algorithm is explained by considering a case study as shown in Codes 1, 2, 3, 4. Let us consider node i36 as the point of interest. In order to identify the nodes of different components, the alphabets a, b, c are followed with every node number (alphabets a, b, c are for statement numbers in component 1, 2 and 3, respectively. Thus, the list of nodes traversed in initial step is:

S = i36

L1 = i36

$L2 = \varphi$

$L3 = \varphi$

In the first step, the algorithm takes slicing criteria or point of interest into consideration and traverses backward along control and data dependence edges except thread connect edge and component connect edges. During the first step, all the vertices encountered during traversal are added to the slice. Then in Step 2, it traverses CCmDG in backward direction considering only component connect edges and pushes all vertices encountered during this traversal. Then in Step3, the algorithm traverses backward from all vertices along the component and thread connect edges. The final slice found is unification of all the vertices marked during the three steps.

In Step 1, the vertex that is reached is removed from L1, and if the vertex is not present in S, then it is pushed to S. The edges coming to the current vertex/node is verified. If the current edge does not belong to component connect edge (CCE) or thread connect edge, then the corresponding node is pushed to L1 for next operation in Step 1, else the node is pushed to L2 and L3, respectively, for next operation. After completion of Step 1, the vertices/nodes in the lists are:

L1 $= \varphi$

L2 = i11

L3 = i23, i24

S = i36, i35, i29, i14, i34, i28, i13, i33, i27, i11, i31,

i25, i10, i30, i23, i9, i24, i20, i8, i7, i22, i19, i6, i21, i18, i5, i17, i4, i16, i3, i2, i1

In Step 2, the current node is popped up from L2, pushed into S and all the edges entering the current node are verified. The current node is pushed to L2, if the corresponding edge is a CCE edge. Similar process is repeated while L2 is empty. After Step 2, the nodes in the list are:

L1 $= \varphi$

L2 $= \varphi$

L2 = i23,i24

S = i36, i35, i29, i14, i34, i28, i13, i33, i27, i11, i31, i25, i10, i30, i23, i9,

i24, i20, i8, i7, i22, i19, i6, i21, i18, i5, i17, i4, i16, i3, i2, i1, 30a, 28a, 27a, 25a,

24a, 20a, 6a, 5a, 4a

In Step 3, the current node is popped up from L3, pushed into S and all the edges entering the current node are verified. The current node is pushed to L3, if the corresponding edge is a TCE edge. Similar process is repeated while L3 is empty. After Step 3, the nodes in the list are:

L1 $= \varphi$

L2 $= \varphi$

L3 $= \varphi$

S = i36, i35, i29, i14, i34, i28, i13, i33, i27, i11, i31, i25, i10, i30, i23, i9,

i24, i20, i8, i7, i22, i19, i6, i21, i18, i5, i17, i4, i16, i3, i2, i1, 30a, 28a, 27a,

25a, 24a, 20a, 6a, 5a, 4a, 1b, 2b, 3b, 4b, 5b, 6b, 7b, 8b, 9b, 12b, 1c, 2c, 3c,

4c, 6c, 7c, 9c

Hence for the given slicing criterion NDi36, the slice found is

S = i36, i35, i29, i14, i34, i28, i13, i33, i27, i11, i31, i25, i10,

i30, i23, i9, i24, i20, i8, i7, i22, i19, i6, i21, i18, i5, i17, i4, i16,

i3, i2, i1, 30a, 28a, 27a, 25a, 24a, 20a, 6a, 5a, 4a, 1b, 2b, 3b,

4b, 5b, 6b, 7b, 8b, 9b, 12b, 1c, 2c, 3c, 4c, 6c, 7c, 9c

6.3 Resultant Slice

The resultant slice of the intermediate representation graph CCmDG of the concurrent COP is shown in Fig. 3. The marked nodes represent the resultant slice of the case study undertaken.

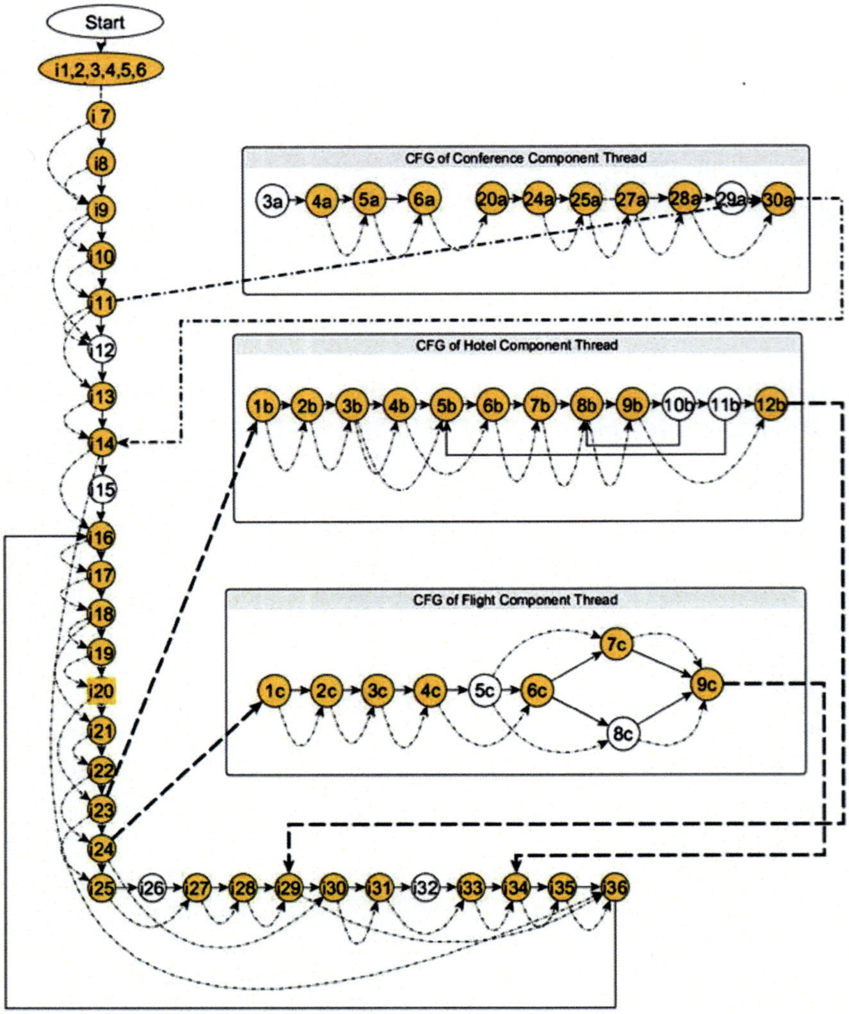

Fig. 3 Resultant slice

7 Conclusion

In this paper, we proposed a dynamic slicing algorithm for concurrent component-oriented program (CCmDS). In order to achieve the same, first of all we have developed an intermediate graph CCmDG for concurrent component-oriented programs over a case study shown in Codes 1,2,3,4. Taking the graph as input, the resultant slices are computed by executing the slicing algorithm. CCmDG has introduced a new data dependency edge named as thread connect edge (TCE). TCE connects components that act as threads and interface by executing it simultaneously. We have extended our existing work of slicing dynamic slicing algorithm (CmDS) to concurrent dynamic slicing algorithm slicing (CCmDS).

References

1. Horwitz, S., Reps, T., & Binkley, D. (1990). Interprocedural slicing using dependence graphs. *ACM Transactions on Programming Languages and Systems (TOPLAS), 12*(1), 26–60.
2. Xu, B., Qian, J., Zhang, X., Wu, Z., & Chen, L. (2005). A brief survey of program slicing. *ACM SIGSOFT Software Engineering Notes, 30*(2), 1–36.
3. Weiser, M. (1984). Program slicing. *IEEE Transactions on Software Engineering, 10*(4), 352–357.
4. Korel, B., & Laski, J. (1988). Dynamic program slicing. *Information processing letters, 29*(3), 155–163.
5. Larsen, L., & Harrold, M. J. (March 1996). Slicing object-oriented software. In Software Engineering, Proceedings of the 18th International Conference on IEEE, (pp. 495–505).
6. Binkley, D. W., & Gallagher, K. B. (1996). Program slicing. In *Advances in computers*, (vol. 43, pp. 1–50.) San Diego, CA: Elsevier, Academic Press.
7. Component-Oriented Programming, www.wikipedia.org.
8. Wang, A. J. A., & Qian, K. (2005). Component-Oriented Programming, Inc., publication, Wiley-Interscience, Wiley.
9. Mohapatra, D. P., Mall, R., & Kumar, R. (2005). Computing dynamic slices of concurrent object-oriented programs. *Information and Software Technology, 47*(12), 805–817.
10. Ray, A., Mishra, S., & Mohapatra, D. P. (2014). An approach for computing dynamic slice of concurrent aspect-oriented programs. arXiv preprint: 1404. 3382.
11. Singh, J., Munjal, D. & Mohapatra, D.P. (2014). Context-sensitive dynamic slicing of concurrent aspect-oriented programs. In *Software Engineering Conference (APSEC), 21st Asia-Pacific 2014 Dec 1* (vol. 1, pp. 167–174) IEEE.
12. Ottenstein, K. J., & Ottenstein, L. M. (1984). The program dependence graph in a software development environment. In *ACM SIGSOFT/SIGPLAN Software Engineering Symposium on Practical Software Development Environments.* (Vol. 19. No. 5. pp. 177–184) ACM.
13. Krinke, J. (2003). Context-sensitive slicing of concurrent programs. *ACM SIGSOFT Software Engineering Notes, 28*(5), 178–187.
14. Zhao, J. (May 1998). Dynamic slicing of object-oriented programs. Technical Report: SE-98-119 (pp. 17–23). Information Processing Society of Japan (IPSJ).
15. Zhao, J. (2002). Slicing aspect-oriented software. In *Program Comprehension, IEEE.Proceedings 10th International Workshop on Program comprehension* (pp. 251–260).
16. Singh, J., Khilar, P. M., & Mohapatra, D. P. (2017). Dynamic slicing of distributed aspect-oriented programs: A context-sensitive approach. *Computer Standards & Interfaces, 52,* 71–84.
17. Tip, Frank. (1995). A survey of program slicing techniques. *Journal of Programming Languages, 3*(3), 121–189.

Service Layer Security Architecture for IOT Using Biometric Authentication and Cryptography Technique

Santosh Kumar Sharma and Bonomali Khuntia

Abstract Data security and authentication mechanism is a very challenging job for smart devices. And more ever IOT is suffering with login and verification process. Here in our paper, we have focused on human characteristics base security system which cannot be pinched easily such as iris, thumb, palm, DNA and voice base authentication system. Using biometrics authentication theory, we have presented that how biometric systems are the boundless computational resources and prospective of flexibility, reliability and cost reduction along with high security performances resources. To maintain the security of biometric traits over the Internet channel, end user can apply cryptography algorithm such as Elgamal, MAC Omura, Cramer Shoup, RSA. As a final point, this paper is contributed for evidencing the strength of integrating the biometrics authentication system with cryptography techniques and its application on Internet base applications. In order to develop strong security, we have proposed integrated approach of three mechanism using biometrics, OTP and cryptography. The work is validated for biometrics through AVISPA (SPAN) security tool which is worldwide acceptable for approving the security architecture.

Keywords Service layer · Biometrics · OTP · Cryptography · AVISPA

1 Introduction

This paper has presented a dynamic verification process using finger scan-based authentication scheme, which provides run-time authentication among end users, and both side verification will be done through dynamic verification with additional key security exchange between data server and user as a final security verification process. Biometric system is having very high efficiency during identity of any human being

S. K. Sharma
Department of MCA, Vignan's Institute of Information Technology, Visakhapatnam, AP, India
e-mail: sharma.santosh83@gmail.com

B. Khuntia (✉)
Department of Computer Science, Berhampur University, Berhampur, Odisha, India
e-mail: Bonomalikhuntia@gmail.com

S. C. Satapathy et al. (eds.), *Intelligent System Design*, Advances in Intelligent Systems and Computing 1171, https://doi.org/10.1007/978-981-15-5400-1_30

and due to this reason many government and private sector organizations are using the thumb biometric system to maintain proper attendance management system without any bias. Working principle of biometrics is to take the input of small portion of finger surface showing to the sensor for feature mining and comparison, thus leading to relatively high matching speed and accuracy with moderate cost. Even though biometric is reliable authentication and identity system, but it cannot be assured as secure and concerns to worry about. The major problem of biometrics is pirate the biometrics key which may happen only through stolen biometrics, replacing compromised biometrics, frauds done by administrators, denial of service and intrusion, etc. Biometric is suffering with data leakage due to vulnerable insecure shipper. In this regard, we have integrated the biometric technique with OTP and powered by secret key exchange as a final phase for security verification. Here, security verification work is divided into three steps, in which first step is user authentication using finger base biometric authentication; in second step, we are using one-time password for registered MAC address verification; and in third phase verification using cryptography. In service layer, security service layer receives the data from element layer and secures data from insider attack which restricts unauthorized access along with protecting the data from malicious, unauthorized access attacks and denial of service attacks.

2 Related Works

Author [1, 2] has investigated the different types of attacks and used BAN logic to solve the synchronization problem where security analysis has been done formally and informally to check the efficiency of proposed protocol. Here the author has use pseudrandom number generator with hash function for securing biometric data for which they have validated the entire work using AVISPA tool to verify the authentication architecture. Author discussed smart card and its application [3, 4] with smart security technology against the Ann's scheme which is scheme during mutual authentication; further new scheme is validated through SPAN tool to verify this scheme for evaluating the passive attacks followed by active out breaks [5, 6]. In this contributed paper, author has monitored many devices which are very sensitive and prone to risk; in this regard, the author proposed the scenario for how to handle the future IOT application and connected devices. Here author has proposed framework on contract basis which is dividing the entire framework into access control contract—ACC, judge contract—JC, register contract—RC to monitor IOT devices for different purposes [7, 8]. Here author has keep focus on the VoIP on application layer for managing and controlling the participants by session initiation protocol; SIP can be implemented in TCP/UDP networks [9, 10]. At presently, IOT technology is promising to human society for making smart world to convert each physical object in smart gadget that can make control of daily needs activity on finger tip by using Internet from remote location. Since the introduction of the Internet base devices, it is having vital impact on the daily lives of human being, but apart from all the feature such as easy to access invited the multiple types of threats to the devices which raise the serious damages to the system and inspire the researchers to create optimal security architecture with reliable security protocol that can face and stop maximum security challenges related to data integrity and privacy of IOT. The author [11–13] projected the future of IOT by estimating the subjugated heavy content-oriented traffic and cohesive conversations. The author [14–16] has gone through the ON-OFF attacks and observed the impact of Kalman filter technique for analyzing the different behaviors of attacks. Lionel, Abderrahim, Hamdi extended the work with new identical homogeneous encryption proposal and adoptable introduction analysis for IOT security services. Threat assessment [18–20] for IOT-oriented secure information network, and end-to-end security for federated Massimo-IOT network. Sameera and Yutaka [21, 23, 25] touched the security architecture for security services and discussed the issue for providing secure framework of cloud and how to maintain defense system with UES algorithm. Heydam [27–30] contributed their work verification and identity technique using reverse engineering. Jadhav has handled variety of OS threats related to application layer for things and kept focus on other standards which is creating threat to the smart devices, professional and social projects such as authentication identity, data storage and recovery management for run-time data along with other vulnerable flaws. The author [31–34] kept focus on physical layer security to protect the devices access itself using jamming technique and reduced the possibility of any type of machines use of services and [36–38] proposed the channel-based mechanism to provide the corporal security from any eavesdropping by analyzing the noise ratio [39, 40].

2.1 System Architecture

(See Fig. 1)

2.1.1 Biometric Authentication Feature

Managing identity and accessibility control is very important issue in several applications which should be handle very carefully regarding this physical and virtual access control for every end users, and participated device must be maintained regularly. This type of protocol checks two things: First is for the object "who is the user and what is his/her role, and second is that particular user is eligible to proceed. In order construct resilient security framework, we have proposed integrated approach of three mechanisms using biometrics, OTP and cryptography.

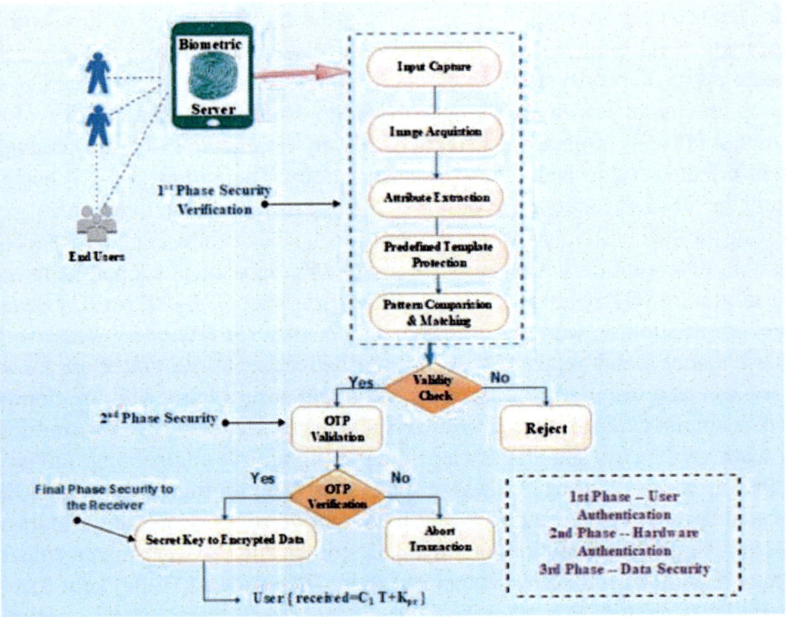

Fig. 1 Three-phase architecture for security management

2.1.2 One-Time Password

OTP is hardware authentication password that is applied for single session transaction on different devices and OTP is generally accompanied by additional layer of authentication. End user makes request to authorize him using user name and password before proceeding verification with the OTP. The main purpose of the OTP is the physical verification process of the device which is registered to OTP server with the unique ID, and objective of the OTP token is user need to physically carry it, so it is more secure for remote user and other stakeholder's authentication (Fig. 2).

2.1.3 Cryptography

2.2 Algorithm (Honey Encryption)

This encryption technique is used for shielding the password of customer who participates in communication against security breaches. The reason behind the development of honey encryption is to generate fake, probable and duplicates passwords which can be attempt by malicious user and misguide them for a given session, so honey encryption can stop the BF attack completely, so honey encryption helps to reduce the vulnerability. Honey encryption makes secure those groups of messages

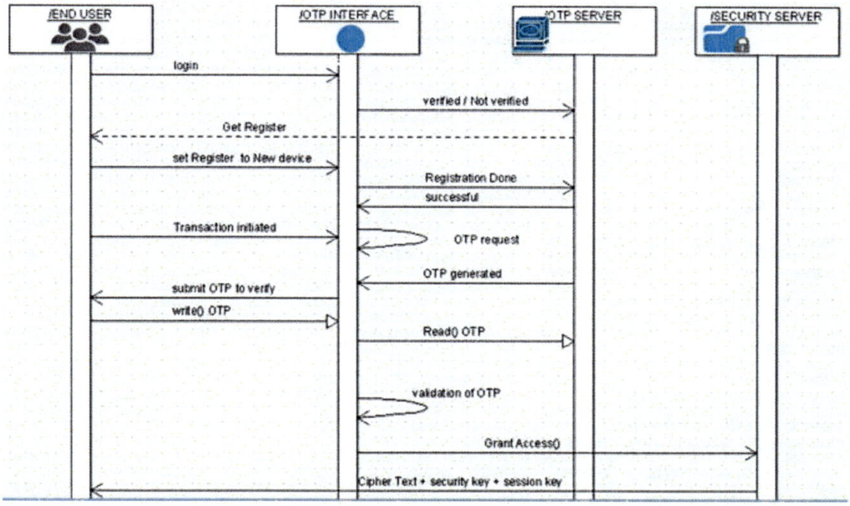

Fig. 2 Interaction diagram for the objects

Table 1 Honey encryption and decryption algorithm description

Algorithm keywords description:	Step-by-step process for encryption	Decryption process
1. HF-Hash Function 2. K-key 3. M-Message 4. RN-Random number 5. S-Seed value 6. CT-Cipher Text	HF ← Enc(K,M) Se ← $encode(M) RN ← ${0,1}n Se ← HF(RN,K CT ← Se' ⊕ Se	HF ← Dec(K,(RN,CT)) Se" ← HF(RN,K) Se ← CT ⊕ Se" M ← decode(Se) Return (M)

those have common property which is defined as message space, and these message spaces must be defined before applying encryption on messages (Table 1).

3 AVISPA Code

role user(Ui,Rs,AS:agent, SKuirs:symmetric_key,

```
            H:hash_func,
           SEND,RECV:channel(dy))
   Played_by Ui
   def=
          local State=nat,

            F:hash_func
   const
   ui_as_xu,B,Au,PIDu,Pwu:
   text,
   USK,PSK,P,SIDs.Xu,RU,
   Ruu:text,
   AIDu,M1,M4,Xs,N1:text,
        s1,s2,s3:protocol_id
init State:=0
transition
%user registration phase
1.State=0/\RECV(start)=|>
Sate':=1/\B':=new()/\Au':=H(PWu.B')
        /\PIDu':=H(IDu,B')
        /\secret({IDu,B'PWu},s1,{Ui})
        /\SEND({PIDu'.Au'}_SKuirs}
        2.State=1/\RECV({H(H(IDu.B').H(PWu,B').USK).SIDs.xor(H(PWu.B'),H(H(IDu.B').H(SIDs.P

SK))).H(SIDs,PSK))).H((SIDs,PSK).P.H)_SKuirs)
=|>
            State':=2/\secret({USK,PSK},s2,RS)
                  /\Xu':=new()/\Ru';=F(Xu'.P)
        /\Ruu':F(Xu',F(Xs'.P))
        /\AIDu':=xor(H(IDu.B'),Ruu')
        /\M1':=H(H(IDu.B').H(H(IDu.B').H(SIDs.PSK)).Ru'.Ruu')
        /\secret(Xs',s3,{RS,AS})
        /\SEND(AIDu',M1'.Ru')
3.State=2/\RECV(xor(H(IDu.B'),N1').H(H(F((Xu'',P)).H(H(IDu,B').H(SIDs.PSK)).SID.N1').H(IDu.B')
.N1'.H(H(IDu.B').H(SIDs,PSK)).F(Xu'.P)))=|>
State':=3/\M4:=H(H(F(Xu'.F(Xs'.P)).H(IDu.B').H(SIDs,N1'),N1)/\SEND(M4')/\request(AS,Ui,as_ui_n1 ,N1')

        end role
        role applicationserver(Ui,RS,AS:agent,
                              H:hash func
                              SEND,RECV:channel(dy))
        played_by AS
        def=
              local State:nat,
              B,Bu,SIDs,IDu,PWu,USK,PSK,P,P,Cus:text,
              Dus,Xu,Xs,Rs,SK,M2,N1,M3:text,
        F:hash_fun
        const as_rs_xs,ui_as_xu,as_ui_n1,
              s1,s2,s3:protocol_id
        init Sate:=0

        transition
        %application registration phase
        1.Sate=0/\RECV(start)+|>
        State':=3
```

```
        /\Xs':new()
        /\Rs':F(Xs.P)
                /\SEND(SIDs,Rs')\
        /\witness(AS,RS,as_rs_xs,Xs')
2.State=3/\RECV(xor(H(IDu,B');F(Xu'.F(Xs'.P))).H(H(IDu.B').H(H(IDu.B').H(SIDs.PSK)).F(Xu'F(Xs' .P))).F(Xu'.P))=|>

State':=6/\secret({IDu,B'PWu},s1.{Ui})
        secret({USK,PSK},s2,RS)
        secret(Xs's3,{RS,AS})
        /\N1:=new()
        /\Sk':H(F(Xu'.F(Xs'P)).H(H(IDu.B').H(SIDs.PSK)),SIDs.N1')
        /\M2':xor(H(IDu.B').N1')
        /\M3':H(SK'.H(IDu.B').N1'.H(H(IDu.B').H(SIDs.PSK)).F(Xu',P))
        /\witness(AS,Ui,as_ui_n1,N1)
3.State=6/\RECV(H(H(F(Xu'.F(Xu'.F(Xs'.P)).H(H(IDu.B').H(SIDs,PSK)).SIDs.N!'))=|>
State':=8/\request(Ui,AS,ui_as_xu,Xu') end
role
role registrationserver(Ui,RS,AS:agent,
        SKuirs:symmetric_key,
        H:hash_func,
        SEND,RECV:channel(dy))
played_by RS
def=
        local State=:nat,
                        B,Bu,SIDs,IDu;PWu,USK,PSK,P,Cus,Dus,Xs:text,
                F:hash_func
const as_rs_xs,s1,s2,s3:protocol_id
init State:=0
transition

%user registration phase
%Receive request message <PIDu,Au> from Ui securely

1.State=0/\RECV({H(IDu.B').H(PWu.B')}_SKuirs=|>
  State':1/\secret({USK,PSK},s2,RS)
        /\Bu'=H(H(IDu.B').H(PWu.B').USK)\
        /\Cus':=H(H(IDu.B').H(SIDs.PSK))
        /\Dus':=xor(H(PWu.B'),Cus')
        /\SEND({Bu'.SIDs.Dus'.H(SIDs.PSK).P.H}_SKuirs)
1.State=0/\RECV(SIDs.F(Xs'.P))=|>
  Stat':4/\secret(Xs',s3,{Rs,AS})
        /\request(AS,RS,as_rs_xs,Xs')
end role
role session(Ui,RS,AS:agent,
          SKuirs:symmetric_key,
          H:hash_func)
def=
        local TX1,TX2,TX3,RX1,RX2,RX3:channel(dy)

composition
        user(Ui,RS,AS,SKuirs,H,TX1,RX1)
                /\registrationserver(Ui,RS,AS,SKuirs.H,TX2,RX2)
                /\applicationserver(Ui,RS,AS,H,TX3,RX3)
```

```
end role
role environment()
def=
        const  ui,rs,as:agent,
                skuirs:symmetric_key,
                h,f:hash_func,
                sids,p,ru,aidu,m1,m2,m3,m4:text,
                as_rs_xs,ui_as_xu,as_ui_n1,
                s1,s2,s3:protocol_id
        intruder_knowledge={ui,rs,as,h,f,sids,p,ru,aidu,m1,m2,m3,m4}
        composition session(ui,rs,as,skuirs,h)
                /\session(I,rs,as,skuirs,h)
         /\session(ui,i,as,skuirs,h)
        /\session(ui,,rs,i,skuirs,h)
end role
goal
        secrecy_of   s1
        secrecy_of   s2
        secrecy_of s3
        authentication_on     as_rs_xs
        authentication_on    ui_as_xu
        authentication_on as_ui_n1
end          goal
environment()
```

4 Result Analysis

Result. 1 Protocol simulation for identifying the different types of attacks.

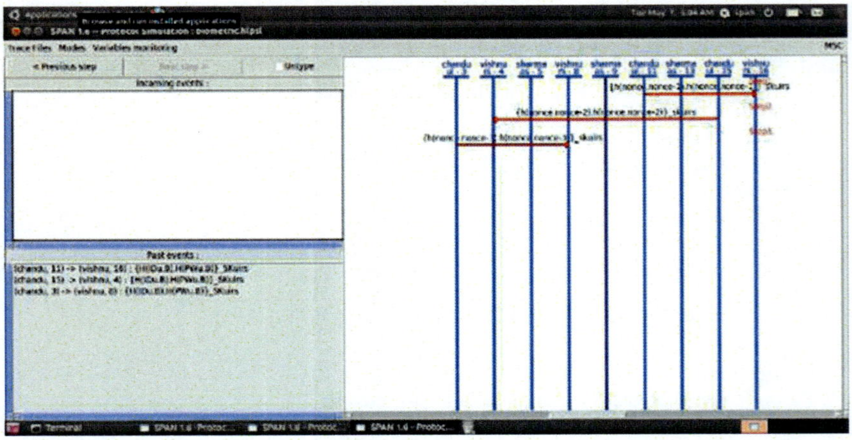

Result. 2 Intruder attacks analysis attempts at multiple sources

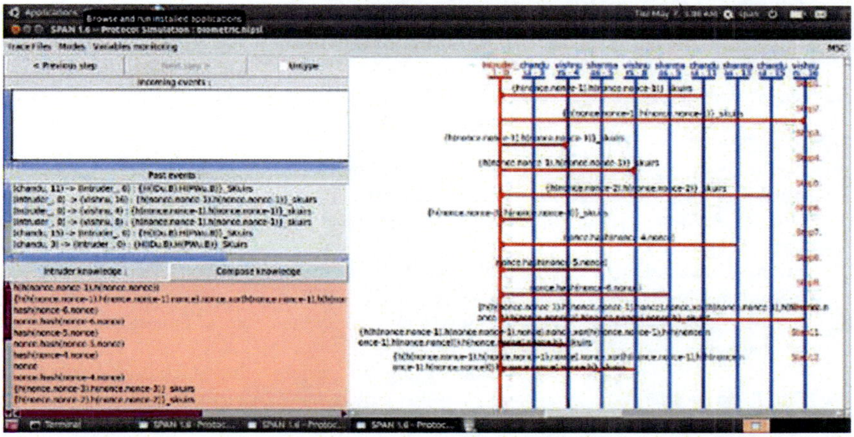

5 Conclusion and Future Work

In this paper, we have discussed strong authentication and cryptography mechanism for handling futuristic security challenges in IOT domain using integrated framework of different techniques. We have developed resilient security using integrated approach of three mechanisms using biometrics, OTP and cryptography. From the result, it is concluded that biometrics is also under the threat of different types of attacks, which has motivated the researcher to enhance the exiting mechanism of biometrics for better security. So our research work has enhanced the exiting biometrics mechanism with the one-time password for authenticating the individual physical devices associated with cryptography to provide the security for our confidential data. At last, our work is providing the security at three different phases to provide strong security system. The biometric strength is validated through AVISPA (SPAN) security tool which is worldwide acceptable for approving the security architecture. Our future work is to develop multilayer security framework protocol to handle diverse security challenges for IOT environment.

References

1. Chikouche, Cherif, F., & Mohamed. An authentication protocol based on combined RFID-biometric system IJACS.
2. Das, & Goswami, A. (2014). A robust anonymous biometric-based remote user authentication scheme using smart cards (pp. 3–19). Elsevier.
3. Reddy, A., Ashok, Odelu, V., & Yoo, K.-Y. (2016). An enhanced biometric based authentication with key-agreement protocol for multi-server architecture based on elliptical curve cryptography, *PLOS ONE*, 3–10.

4. Chaudhry, S. A., Mahmood, & Naqvi1, H. An improved and secure biometric authentication scheme for telecare medicine information systems based on elliptic curve cryptography.
5. Tyagi, N., Wang, J., Wen, K., & Zuo, D. (2015). Honey encryption applications implementation of an encryption scheme resilient to brute-force attacks. Spring.
6. Murty, S., & Mulchandani, M. (2017). Improving security of honey encryption in database: Implementation (ICSESD).
7. Yin, W., Jadwiga, & Zhou, H. (2017). Protecting private data by honey encryption. Hindawi.
8. Jaeger, J., & Thomas. (2016). Honey encryption beyond message recovery security, In *EUROCRYPT*.
9. Lourdu Gracy, P., & Venkatesan, D. (2018). An honey encryption based efficient security mechanism for wireless sensor networks, In *IJPAM*.
10. Sahu, R., & Ansari, S. (2017). A secure framework for messaging on android devices with honey encryption, *IJECS*.
11. Sravani, Wazid, Ashok, & Neeraj. (2016). Secure signature-based authenticated key establishment scheme for future IoT applications (pp. 1–16). IEEE.
12. Sain, M., Kang, Y., & Lee, H. (2017). Survey on security in internet of things: State of the art and challenges, In *ICACT* (pp. 1–6).
13. Huh, S., Cho, S., & Kim, S. (2017). Managing IoT devices using blockchain platform, In *ICACT* (pp. 1–4).
14. Berrehili, F., & Belmekki, A. (2017). Risk analysis in internet of things using EBIOS (pp. 1–7). IEEE.
15. Majeed, A. (2017). Internet of things (IoT): A verification framework (pp. 1–3). IEEE.
16. Abels, T., Khanna, R., & Midkiff, K. (2017). Future proof IoT: Composable semantics, security, QoS and reliability (pp. 1–4). IEEE.
17. Tewfiq, Seigneur, J.-M. (2016). Efficient security adaptation framework for internet of things (pp. 1–6). IEEE.
18. Mohsin, M., & Anwar, Z. (2016). IoTSAT: A formal framework for security analysis of the internet of things (IoT) (pp. 1–9). IEEE.
19. Zada, W., Zangoti, H., & Aalsalem, Y. (2016). Mobile RFID in internet of things: Security attacks, privacy risks, and countermeasures (pp. 1–6). IEEE.
20. Baldini, G., & Le, F. (2015). Security certification and labelling in internet of things, ICT (pp. 1–6).
21. Mukrimah, Amiza, & Naimah. (2016). Internet of things (IoT): Taxonomy of security attacks, ICED (pp. 1–6). Thailand.
22. Sklavos, N., & Zaharakis, D. (2016). Cryptography and security in internet of things (IoTs): Models, schemes, and implementations (pp. 1–2). IEEE.
23. Abderrahim, O., & Elhdhili, M. H. (2017). TMCoI-SIOT: A trust management system based on communities of interest for the social internet of things (pp. 1–6). IEEE.
24. Metongnon, L., Ezin, E., & Sadre, R. (2017). Efficient probing of heterogeneous IoT networks (pp. 1–7). IFIP.
25. Abderrahim, O., & Housine M. (2017). CTMS-SIOT: A context-based trust management system for the social internet of things (pp. 1–6). IEEE.
26. Zouari, J., Hamdi, M., & Tai-Hoon. (2017). A privacy-preserving homomorphic encryption scheme for the internet of things (pp. 1–6). IEEE.
27. Midi, D., Mudgerikar, A., & Bertino. (2017). Kalis—A system for knowledge-driven adaptable intrusion detection for the internet of things. In *ICDCS* (pp. 1–11).
28. Dorsemaine, B., Gaulier, J. –P., & Kheir, N. (2017). A new threat assessment method for integrating an IoT infrastructure in an information system. In *ICDCSW* (pp. 1–8); Sabrina, Grieco, & Coen-Porisini. (2017). A secure ICN-IoT architecture (pp. 1–6). IEEE; Massonet, P., Deru, L., Achour, & A., Livin, A. (2017). End-to-end security architecture for federated cloud and IoT networks (pp. 1–6). IEEE.
29. Batool, S., Nazar, A., Saqib, & Khan, M. A. (2017). Internet of things data analytics for user authentication and activity recognition. In *FMEC* (pp. 1–5)8; Jerald, V., & Rabara, A. (2016). Algorithmic approach to security architecture for integrated IoT smart services environment. In *WCCCT* (pp. 1–6).

30. Stergiou, C., & Psanni, K. E. (2017). Architecture for security monitoring in IOT environments (pp. 1–4). IEEE.
31. Nakagawa, I., & Shimojo, S. (2017). IoT agent platform mechanism with transparent cloud computing framework for improving IoT security, In *COMPSAC* (pp. 1–6).
32. Praveena, A. (2017). Achaqieving data security in wireless sensor networks using ultra encryption standard version—IV algorithm, In *ICIGEHT* (pp. 1–5).
33. Tellez, M., El-Tawab, & S., Heydari, H. (2017). IoT security attacks using reverse engineering methods on WSN applications (pp. 1–6). IEEE.
34. Ahmed, I., Beheshti, B. Khan, Z. A., & Ahmad, I. (2017). Security in the internet of things (IoT) (pp. 1–7). Dubai: ITT.
35. Sowmya, & Kulkarni. (2017). Security threats in the application layer in IOT applications, I-SMAC (pp. 1–4).
36. Hu, L., & Xiumin. (2018). Cooperative jamming for physical layer security enhancement in internet of things (pp. 1–10). IEEE.
37. Sasi, Jindal, & Ranjan. (2017). Channel-based mapping diversity for enhancing the physical layer security in the internet of things (pp. 1–6). IEEE.
38. Burg, A., & Lam, K.-y. (2018). Wireless communication and security issues for cyber–physical systems and the internet-of-things (pp. 1–23). IEEE.
39. Atat, R., Ashdown, & Yi, Y. (2017). A physical layer security scheme for mobile health cyber-physical systems (pp. 1–15). IEEE.
40. Sasi, Jindal, & Ranjan. (2017). Channel based mapping diversity for enhancing the physical layer security in the internet of things (pp. 1–15). IEEE.

An Effective Modeling and Design of Closed-Loop High Step-up DC–DC Boost Converter

Baggam Swathi and Kamala Murthy

Abstract A coupled inductor whose principle is based on a boost inverter connected to a 3-ϕ grid. This maintains a zero voltage switching with a coupled inductor which is based on dc–dc boost converter which acts as a fuel cell. As it acts as a battery, it is known as the battery-based circuit system. The above circuit module unit has a coupled inductor connected to a zero voltage source inverter to generate a large AC output voltage when compared to DC input voltage. The major usage of a coupled inductor-type boost convertor is that it operates at a duty cycle closer to 0.4 with a PWM modulator. This PWM modulation achieves very high voltage variation gains. So, the above proposed power conversion unit replaces the usage of a transformer from its configuration, which reduces the cost of the operating unit and runs the system efficiency and makes the cost economical. The above-designed module is stimulated by using the technical software package of MATLAB/SIMULINK and SimPowerSystems blockset Simulink tool.

Index Terms DC–DC power conversion · Capacitor modules · Interleaved methodology · PI controller

1 Introduction

Electrical utilities and end-users are gradually more concerned about meeting rising energy require. Fossil fuel burning meets 75% of total global demand for energy. Nevertheless, growing air pollution, concerns about global warming, decreasing fossil fuels and rising costs have made it essential to look at renewable sources as a potential energy substitute. In many countries, since the last decade, there has been a huge interest for renewable energy for electricity production. Market liberalization and government incentives have further boosted the renewable energy sector's

B. Swathi (✉) · K. Murthy
Malla Reddy College of Engineering and Technology, Hyderabad, India
e-mail: baggamswathi27@gmail.com

K. Murthy
e-mail: erkamaleee@gmail.com

S. C. Satapathy et al. (eds.), *Intelligent System Design*, Advances in Intelligent Systems and Computing 1171, https://doi.org/10.1007/978-981-15-5400-1_31

growth [1–3]. Renewable energy supply (RES) implemented at the transmission stage is referred to as distributed generation (DG). However, an AC module-based system has a few problems also, including potential reduction in system efficiency and possible cost increase in the overall system cost due to the use of several low power inverters. Greater maintenance may also be required, though the overall reliability may actually improve due to the use of several parallel inverters. Among the variety of circuits and control methods that have been developed for the AC module application [4], the flyback discontinuous conduction mode (DCM) inverter is one of the favored topologies because of its simplicity and potential low cost. The component count is low, and the inverter requires only a simple control scheme. This work is based on low voltage direct current shared in between two step-up DC–DC converters, and those two converters are connected in parallel. The main objectives of this project are the creation of Simulink model for DC–DC boost converter with the help of Simulink blocks and analyzation of designed model. Analyzing the output voltages of DC–DC boost converters in both open- and closed-loop [5]. Observation of the output voltages the DC–DC converter by using switching power electronic devices. The output voltages have the waveforms which are verified at different voltages. In power electronics, the DC–DC converter is a simple circuit which modulates one level of input voltage into another level of output voltage at low power ratings. These are applicable in mobile charger power supply systems in computers, batteries in offices, telecommunications, etc. Many industries are in demand to use this DC–DC converter at high power ratings. To participate in those large industries, the converters use different controlling techniques to improve the output. We used PWM technique to improve the output at higher values. The PWM controller is used to control the switching mode of operation at different high power ratings to maintain good efficiency. We are using a DC–DC converter for its simple structure and operation endless losses. But, it gets operated at low powers, which confined to low rating applications only. This problem is overcome by using PWM technique with the switching mode of operation. Sect. 2 covers the proposed converter topology; Sect. 3 presents stages of ZVS operation [6]. Section 4 explains the modeling and simulation results, and lastly, Sect. 5 finalizes the conclusion of the paper with the future scope.

2 Proposed Converter Topology

2.1 ZVS Inverter

An inverter takes in direct current from batteries or solar systems and generates alternating current. An inverter is a reliable source which is used for home and business applications. The inverter can generate high AC currents or low AC currents for required voltage and frequencies by using appropriate rated transformers [7]. Static inverters have no affecting elements and are applied in various appliances, from

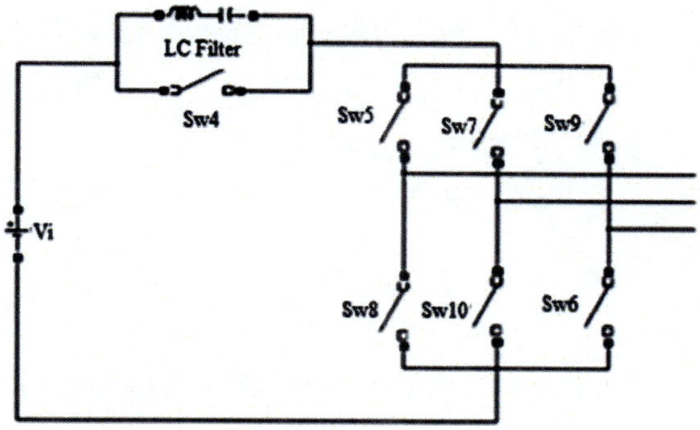

Fig. 1 ZVS inverter model used in the proposed circuit

little switching power provisions in systems, to huge electro-efficacy high-voltage DC current appliance that transfers immensity power. The solar system stores the DC by observing sun light; the stored energy can be converted to AC with the help of an Inverter for low or high voltages. Inverters are majorly distinguished into two types. One is square wave-type response that implies the output varies between positive and negative currents in this for a particular period of time the current becomes zero [8]. The second type of inverter is sine wave; in this, the current fluctuates between positive and negative. The output is identical to the input sine wave so it has very low harmonic distortions. So, it is flexible for any type of AC electronic devices [9]. Its plan is added compound and expenses 6 or 12 times extra per every unit of power. The inverter oscillates at very high power. It is so called because at some instances it needs to convert AC to DC; to act as an inverter it requires DC source so at that instance it acts as a rectifier making the source reverse.

Figure 1 shows a structural combination of clamping branch with a PWM inverter the clamping branch has a series inductor L_{re} and a capacitor C_{ca} in parallel to a switch S_{w4}, the total combination is a LC filter. When Vi is in action, the switch S_{w4} is in action, then the conduction starts and the energy gets circulated in the LC filter branch. When the switch Sw4 is turned OFF, the current stored in Lre gets active and releases the parallel capacitor and the PWM equal bridge gets activated and starts conduction below zero voltage. Here there are three legs in the major equal bridge [10]. Generally, the switch Sw4 nothing but a supporting switch must stimulate all three bridges 3 times for cycle may be at different frequencies to overcome this complexity we use a highly supporting switch in such a way that it works at synchronous frequency for all three bridges by a extremely high SVM system to control the inverter.

2.2 *Operation of ZVS Inverter*

The topological construction in Fig. 2 is to add a SVPWM inverter behind the clamping branch nothing but a series significant inverter and a clamping capacitor and a supporting switch Sw4 in parallel when the supporting switch Sw4 is in turn ON

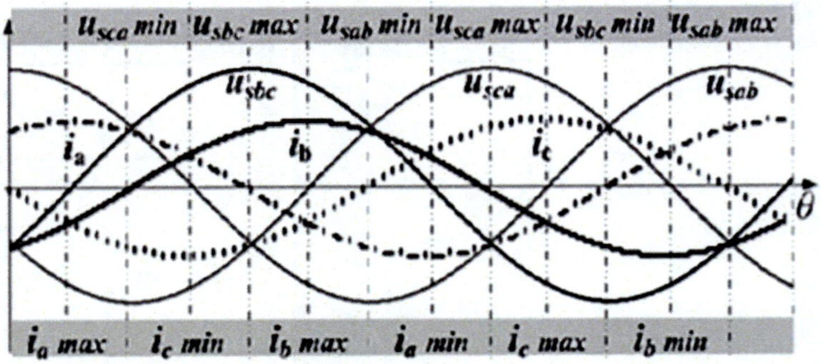

Fig. 2 Grid line voltage and inverter output current waveform

position the energy circulates in the circuit. But when the significant switch Sw4 is turned OFF, then the Sw4 switch gets opened and the Lre inductor releases the stored energy and frees the capacitor which conducts the equal bridge; since the bridge has three legs, PWM control must be done for three times and the inductor opposes the reverse current from the anti-parallel diode and conducts single bridge at a time. So, the PWM cycle is alone three times every bridge. Since lack of synchronous switch working at on simulation frequency. To overcome this problem, we should allow the supporting switch in such a way that it also has the simulating frequency as the major switch. To attain this technique, we use a SVM system to control the inverter [11]. The topological construction in Fig. 2 is to add a SVPWM inverter behind the clamping branch nothing but a series significant inverter and a clamping capacitor and a supporting switch Sw4 in parallel; when the supporting switch Sw4 is in turn ON position, the energy circulates in the circuit. But when the significant switch Sw4 is turned OFF, then the Sw4 switch gets opened and the Lre inductor releases the stored energy and frees the capacitor which conducts the equal bridge since the bridge has three legs; PWM control must be done for three times, and the inductor opposes the reverse current from the anti-parallel diode and conducts single bridge at a time. So, the PWM cycle is alone three times every bridge [12]. Since lack of synchronous switch working at on simulation frequency. To overcome this problem, we should allow the supporting switch in such a way that it also has the simulating frequency as the major switch. To attain this technique, we use a SVM system to control the inverter.

Let us imagine that the connected bridge circuit gets controlled at power factor equal to 1; the single leg voltage and inverter output current waveforms are in Fig. 2. In SVM system, the total voltage cycle is divided into six sectors and each sector is still divided into two sectors according with the phase currents in the inverter.

As already discussed, Sector I is divided into two sectors, Sect1 and Sect 2. Sect. 1 has a highest current value in Phase A. Sect. 1-2 has highest current values in phase C. The currents of Phase A and Phase C has similar value because of the inverter showing similar values at every 30. Suppose the inverter is controlling the Sect. 1-1. If the grid bridge-linked inverter maintains the power factor as 1, ia > 0 and ic <

ib < 0 Sect. 1-1. So, by this operation, the Phase A voltages and currents get top most values and creates four switching states as attained in Fig. 3 110, 100, 110 and 000. If the switching states circulate from 111 to 100, the switches Sw10 and Sw6 are closed and allow the energy circulation simultaneously. In this process, the supporting switch comes into action and opposes the reverse currents anti-parallel diode of switches Sw7 and Sw9 and propagates the ZVS condition for Sw10 and Sw6. In the next state from 100 to 110, the current in Sw10 at first will flow through this anti-parallel diode of Sw7, and in the state from 110 to 111, the switch Sw8 flows through anti-parallel diode Sw9. So, by this, it is clear that the supporting branch is required to act once at every switching cycle to DC bus to zero by owning ZVS condition for switches and diode reverse in both phases. By this technique, the supporting switch works at simultaneous frequency just like major switch. By the SVM control technique, the major switch gets operated as turn ON or OFF. The simple expectations are made to study the ZVS inverter: Since anti-parallel diodes are connected to switches Sw5–Sw10, they are called as model switches [13]. Cre5–Cre10 capacitors are parallel with switches Sw5–Sw10 and correspondingly consist of free loading capacitors and exterior capacitors. The number of current ripples in an inductor is negligible in one switching cycle so it is considered as stable current source [14]. As the clamping capacitors Cca has high capacitance so its voltage ripples is minute. Hence, it is considered as voltage source. The action frequency is always higher than the significant frequency of Cca and Lre.

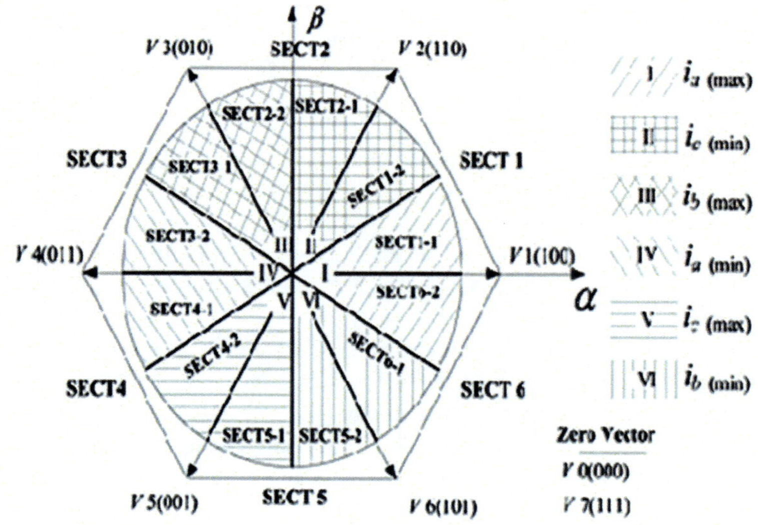

Fig. 3 Grid voltages and inverter current space vector diagram

3 Stages of ZVS Operation

Level 1(t_0–t_1): Major switches Sw5, Sw7 and Sw9 and supporting switches Sw4 get turned ON and start conduction at states 111. The voltage of Lre is got clamped with the capacitor Cca, and current Lre is enhanced.

Level 2(t_1–t_2): In t1, Sw4 stops conduction since it is turned OFF, and the significant inductor Lre drops the capacitors Cre8, Cre10 and Cre6 and accusation parallel capacitors Cre4 of the supporting switch Sw4. Sw4 is twisted OFF below the ZVS states because of the snubber capacitor Cre4.

Level 3(t_2–t_3): The voltage across Sw4 reaches Vdc, and S4 is OFF, and the voltages cross ways Cre8, Cre10 and Cre6 drains to zero also opposed to analogous diode of those major switches begin to perform. Due to these criteria, the resonant between the capacitors get discontinued. In the SVS state, the significant switches Sw6 and Sw10 are turned ON.

Level 4(t_3–t_4): The currents in Phase B and Phase C are shifted to parallel diodes of Sw7 and Sw9 to Sw6 and Sw10, respectively. The major switch change procedure components, the anti-similar diodes of Sw7 and Sw9 experience diode overturn revival. This overturned revival voltages are suppressed by the significant inductor, and the current passed through Lre changes iLre [15].

Level 5(t_4–t_5): The OFF mode switches in this level 100 are the major switches Sw7, Sw8 at time t4. Lre, Cre7, Cre6, Cre9 and Cre7 begin to be in importance. The voltages across Sw7, Sw8 and Sw9 begin to amplify, and the voltage across Sw4 begins to amplify, and the voltages across Sw4 begins to reduce. At t5, the voltage across Sw6, Sw8 and Sw10 attains to Vdc, the voltages across Sw4 reduce to zero and the Sw4 starts conduction under ZVS states. Lre, Cre8, Cre9 and Cre4 discontinue being in importance. The time between t2 and t5 is the duty cycle defeat of Sw6 and Sw10.

Level 6(t_5–t_6): At t5, it is confirmed that the circuit is in 100 state. The major switches Sw5, Sw6 and Sw10 and the switch Sw4 start conducting the clamping capacitor in the LC filter gets charged by the resonant induction.

Level 7 (t_6–t_7): In this level, the switch Sw10 gets turned OFF. C6 gets charged, and C7 gets discharged by the inductor. ZVS state controls the switch S10 to get turned OFF.

Level 8 (t_7–t_8): At the time t7, the circuit stays in state 110. The supporting switch Sw4 and the major switch Sw5, Sw7 and Sw6 get turned ON. The clamped capacitor Cca gets charged by the inductor [16].

Level 9 (t8–t9): At t7, Sw6 gets OFF. C6 gets charged by the induction Lc and release Ca due to substance of Ca and C6, S6 gets OFF under ZVS state.

4 Modeling and Simulation Results

Here the simulation is carried out by two cases, in that 1. Proposed DC/DC Converter Operating Under Open-Loop Condition, 2. Proposed DC/DC Converter Operating Under Closed-Loop Condition, Case 1: Proposed DC/DC Converter Operating Under Open-Loop Condition.

Figure 4 shows the MATLAB Simulink model of proposed DC/DC converter working under open-loop condition using Simulink tool in MATLAB. Fig. 5 shows the output voltage of proposed DC/DC converter operating under open-loop condition, due to non-presence of feedback system attains low stable operation, attains 0.02 s for fast response.

Case 2: Proposed DC/DC Converter Operating Under Closed-Loop Condition. Fig. 6 shows the MATLAB/Simulink model of proposed DC/DC converter operating under closedl-Loop condition using MATLAB/Simulink tool.

Figure 7 shows the output voltage of proposed DC/DC converter operating under closed-loop condition, due to presence of feedback system attains high stable operation, attains 0.01 s for fast response.

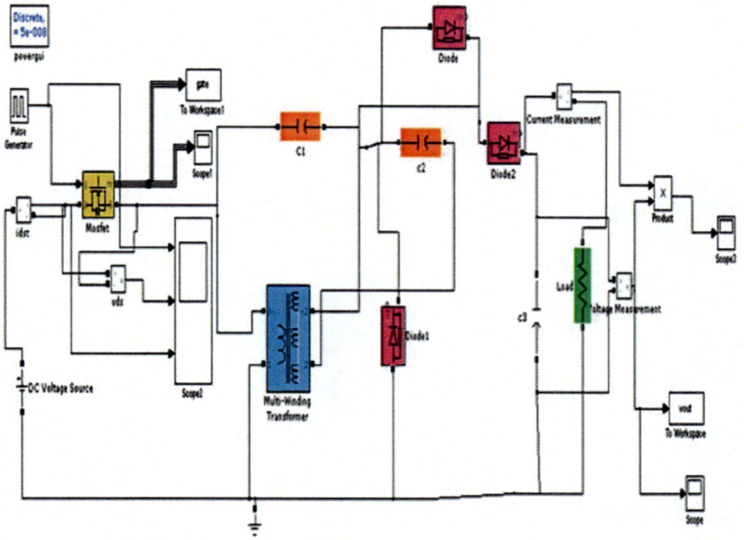

Fig. 4 MATLAB/ Simulink model of proposed DC/DC converter working under open-loop condition

Fig. 5 Output voltage

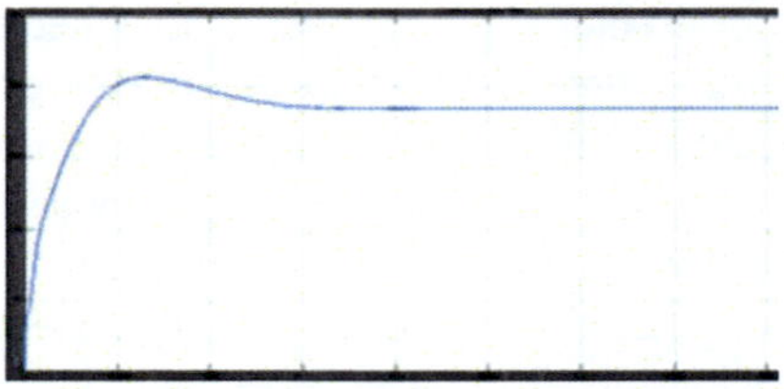

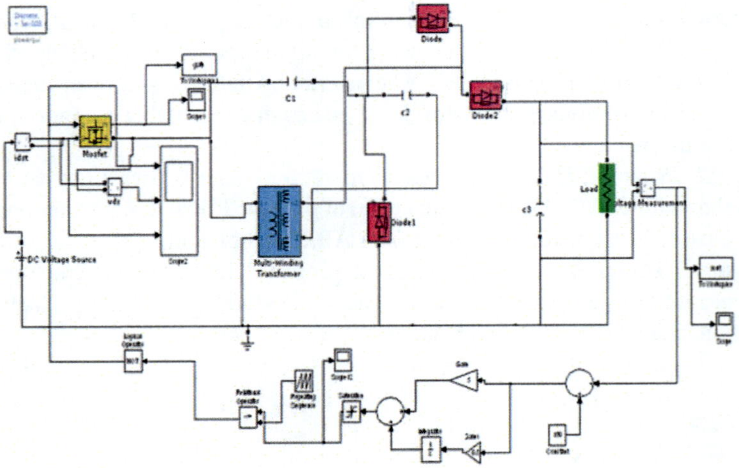

Fig. 6 MATLAB/Simulink model of proposed DC/DC converter operating under closed-loop condition.

Fig. 7 Output voltage

5 Conclusion

The simulation investigation of the joined inductor is completed, and the results are presented based on the zero voltage switching 3-π inverter linked to the grid. To achieve high alteration ratio, the DC–DC controller used in the converter uses a joined inductor. The inverter is connected to an additional zero voltage switch branch and a three-phase switching branch controlled by SVM in series with the inverter and SVPWM technique controlling the inverter. By holding the power factor as 1 or less than 1, the ZVS technique is obtained. This experiment's main motto is to cross-check that in all switching instances, the switching grid connected to the inverter is opposed to the reverse recovery current in anti-parallel diodes. The SVM technique allows fixed frequencies to work on the inverter grid. The future scope is the study of further extension of this project. This technique is also used to DC–DC buck converter by using PI blocks in order to reduce the nonzero study error. Due to time component, the tap out got missed in the project which results the real measurement is not done in DC–DC converter.

References

1. Yaang, L.-S., Lianng, T.-J., & Chen, J.-F. (2009). Transformerless DC–DC converters with high step-up voltage gain. *IEEE Trans. Industrial Electronics, 56*(8), 3144–3152.
2. Li, Ruui, Ma, Zhhiyuan, & Dehhong, X. (2012). A ZVS grid-connected three-phase inverter. *IEEE Transactions Power Electronics, 27*(8), 3595–3604.
3. Mohain, N., Undeeland, T., & Robbins, W. (2003). *Power electronics: Converters, applications and design* (pp. 524–545). New York: Wiley.
4. Bellar, M. D., Wul, T. S., Tchamdjou, A., Mahdavi, J., & Ehsani, M. (1998). Preview of soft switched DC–AC converters. *IEEE Transactions on Industry Applications, 34*(4), 847–860.
5. Divvan D. M. (1989). Static power conversion method and apparatus having essentially zero switching losses and clamped voltage levels, U.S. Patent 48 64 483.
6. Nakaok M., Yonemori H., Yurugi K. (1993). Zero-voltage soft-switched PDM three phase AC–DC active power converter operating at unity power factor and sinewave line current. In *Proceedings IEEE Power Electronics Specialist Conference* (pp. 787–794).
7. Yonemori H., Fukuda H., Nakaoka M. (1994). Advanced three-phase ZVS-PWM active power rectifier with new resonant DC link and its digital control scheme, In *Proceedings IEE Power Electronics Variable Speed Drives*, pp. 608–613.
8. Venkataramanan, G., Divan, D. M., & Jahns, T. (1993). Discrete pulse modulation strategies for high frequency inverter system. *IEEE Transactions on Power Electronics, 8*(3), 279–287.
9. Venkatararmanan G., Divan D. M. (1990). Pulse width modulation with resonant dc link converters, In Proceedings Conference Record IEEE Industry Applications Society. Annual Meeting (pp. 984–990).
10. Cheen, Y. (1998). A new quasi-parallel resonant dc link for soft-switching PWMinverters. *IEEE Transactions on Power Electronics, 13*(3), 427–435.
11. Jafarr, J. J., & Fermandes, B. G. (2002). A new quasi resonant DC-link PWM inverter using single switch for soft switching. *IEEE Transactions on Power Electronics, 17*(6), 1010–1016.
12. Mezaaroba, M., Maartins, D. C., & Barbi, I. (2007). A ZVS PWM half-bridge voltage source inverter with active clamping. *IEEE Transactions on Industry Applications, 54*(5), 2665–2762.
13. Shimizu, T., Wada, K., & Nakamura, N. (2006). Flyback-type single-phase utility interactive inverter with power pulsation decoupling on the dc input for an ac photovoltaic module system. *IEEE Transactions on Power Electronics, 21*(5), 1264–1272.
14. Rodriguez, C., & Amaratunga, G. A. J. (2008). Long-lifetime power inverter for photovoltaic ac modules. *IEEE Transaction Industry Applications, 55*(7), 2593–2601.
15. Li, W., & He, X. (2011). Review of non-isolated high step-up dc/dc converters in photovoltaic grid-connected applications. *IEEE Transaction Industry Applications, 58*(4), 1239–1250.
16. Kjaer, S. B., Pedersen, J. K., & Blaabjerg, F. (2005). A review of singlephase grid-connected inverters for photovoltaic modules. *IEEE Transaction Industry Applications, 41*(5), 1292–1306.

Sales Analysis on Back Friday Using Machine Learning Techniques

Somula Ramasubbareddy, T. A. S. Srinivas, K. Govinda, and E. Swetha

Abstract Amid the deals on Black Friday, all the retailers are packed. Many items are discounted with limits and clients surge in purchase of the items. It is hard for clients to purchase the items even with a strong arrangement. In any case, the shop proprietors face significantly more trouble on controlling the group with constrained staff and in focusing on imminent clients. A few methods have been utilized to handle this issue, yet they are not unreasonably effective. A model for predicting is a system that demonstrated promise in taking care of this issue. I focus on predicting models to build an exact and proficient algorithm to analyze the clients spending before and yield the future spending of the client with similar features. Distinctive data analyzing techniques such as regressors, NN and classifiers to develop a model for prediction are actualized and an examination is implemented dependent on their exhibition and exactness of predicting. These machine learning techniques are executed using various algorithms to find the optimal prediction. I applied six distinct ML algorithms. Furthermore, I apply the visuals and data preprocess to obtain the optimum results.

Keywords Black Friday · Model for predicating · Regressors · NN · ML

1 Introduction

In earlier days, the customers used to depend on small-scale business stores for their needs. The small business owners aware about customers' needs and their spendings pattern and interests. After few decades, the small business became big franchise and

S. Ramasubbareddy (✉)
IT, VNRVJIET, Hyderabad, Telangana, India
e-mail: svramasubbareddy1219@gmail.com

T. A. S. Srinivas
CSE, RGMCET, Nandyal, Andhra Pradesh, India

K. Govinda
SCOPE, VIT University, Vellore, Tamilnadu, India

E. Swetha
CSE, SVCE, Tirupathi, Andhra Pradesh, India

S. C. Satapathy et al. (eds.), *Intelligent System Design*, Advances in Intelligent Systems and Computing 1171, https://doi.org/10.1007/978-981-15-5400-1_32

spread across country. It became difficult for business owners to predict customer choices [1]. Some of the popular stores are Costco, Walmart and Wholefoods. Some examples of such franchises are Costco, Walmart and Wholefoods. These stores with no legitimate information of their client base are attempting to fulfill the client needs. Along these lines, predicating models are expected to all the more likely comprehend client inclinations.

Building a predicating model relies upon different features, for example, the location and the time [2]. The day, Black Friday, after thanks giving is the biggest day for shopping cheaply in USA [3]. The customers pay more attention on discounts during Black Friday than normal day because high-price detection on products. For that, a prediction model required to predict user interest during Black Friday [4, 5]. Finally, visualization methods can provide information about customer preferences to business owners.

2 Literature Survey

Regression analysis is a type of prescient displaying strategy which explores the connection between a dependent (target) and independent variable (predictor). This system is utilized for forecasting, time series modeling and finding the casual impact connection between the variables. For example the shape of regression line, type of dependent variable and number of independent variable [6].

2.1 Linear Regression

Linear regression is machine learning algorithm which established linear relationship between variable (Y) and independent variable (X). Therefore, the equation $Y = a+b*X + e$, defines the linear regression, where a is intercept, b is slope of regression line and e represents error learn. This condition can be utilized to anticipate the estimation of target variable dependent on given indicator variable (*s*) [7].

2.2 Ridge Regression

Ridge regressor is a method utilized when the information experiences multicollinearity (autonomous factors are exceptionally related). In multicollinearity, despite the fact that the least squares estimates are unbiased, their variances are large which deviates the observed value far from the true value. By adding a degree of bias to the regression estimates, ridge regression reduces the standard errors.

2.3 Decision Tree

Decision tree is machine learning algorithm which is used to develop an effective model. The future patterns can be predicted by using time series analysis and future spending of customer. Generally, this technique is classified into two types, mathematical and statical models.

2.4 XGBoost

The XGB model executes the step-by-step, ridge regression internally which progressively chooses the features and removes the multicollinearity with the features.

The serious issue with the current predicting model is that the information utilized for advancement contains several anomalies such as unaccounted for values or wrong data. Additionally, choice of optimal algorithm plays a major role in building a precise model.

3 Proposed Method

Our system involves the application of machine learning techniques to predict the testing values in four steps.

3.1 Data Analyzing

In this stage, we will just analysis various components of our data like mean, median, standard deviation, frequency, etc. Also, I will find skew, kurtosis followed by correlation matrix with respect to purchase values.

3.2 Data Preprocessing

This stage will involve calculating all the NA values and replacing them with some integer so that processing can be carried out. Later, I can also display the unique value frequencies of all the columns and finally send this data into a modified train and test CSV files.

3.3 Machine Learning Techniques

After preprocessing the data, we will use the modified train and test files to apply the above given algorithms to find which is the best algorithm to calculate the purchase values by calculating the minimum RMSE values.

3.4 Rule-Based Learning

At the end, I will apply rule-based learning to get the best possible rules and apply them to the dataset and finally apply it to the best obtained algorithm to get the perfect RMSE scores.

Dataset Used
Our dataset is the Black Friday Sales Dataset in Kaggle. In this dataset, I have the information about the age, occupation, city, duration stayed, marital status, the quantity of products bought of various types and the total amount spent. I am using these inputs to find the most necessary attributes, potentially excluding some attributes. Finally, we arrive at the conclusion from applying these models to find which model is best suited to predict purchase trend of customers [8, 9].

Data Analyzing
In this step, we just use common descriptive statistics techniques and apply them on our existing data like mean, median, standard deviation, frequency, etc. Also, we find skew, kurtosis followed by correlation matrix with respect to purchase values.

The output obtained from the given codes in the appendix is

	User_ID	Product_ID	Gender	Age	Occupation	City_Category
0	1000001	P00069042	F	0-17	10	A
1	1000001	P00248942	F	0-17	10	A
2	1000001	P00087842	F	0-17	10	A
3	1000001	P00085442	F	0-17	10	A
4	1000002	P00285442	M	55+	16	C

Figure 1 shows the graphical representation of occupation versus count; Figure 2 shows the relationship between marital status of the person with count. Figures 3 and 4 represent product category versus count and gender versus count relationships in graphical representation, respectively.

Fig. 1 Occupation versus count

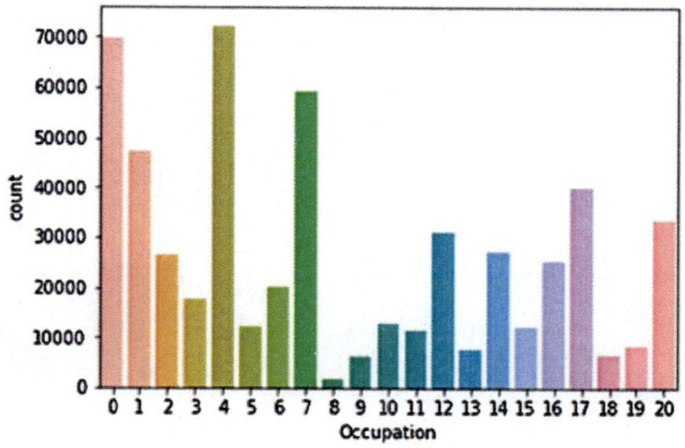

Fig. 2 Marital status versus count

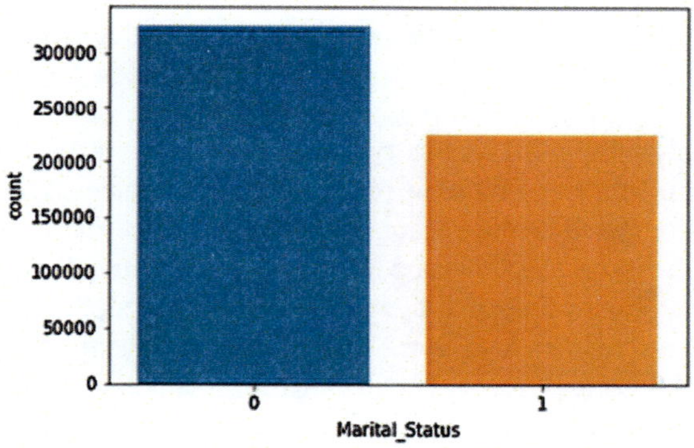

Fig. 3 PC3 versus count

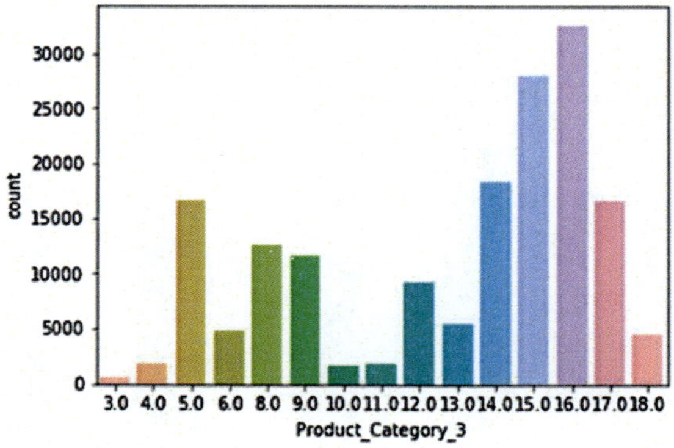

4 Results and Discussion

Before selecting the models to use for the training process, I need to decide the columns/features that can be used as predictors and drop the others. This decision needs to be made on the basis on the data analysis done at the first stage of this

Fig. 4 Gender versus count

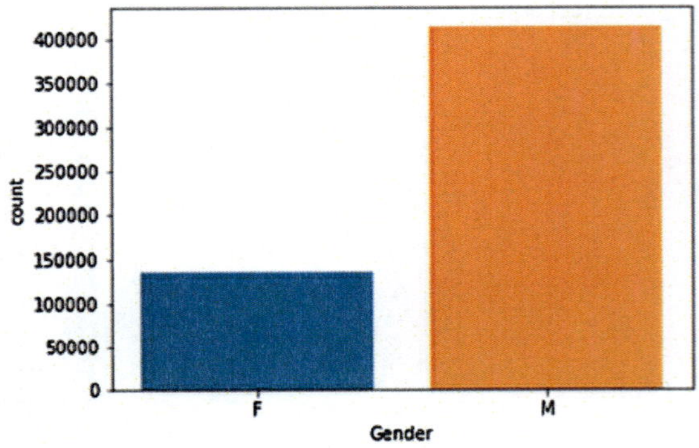

process. For e.g.: when I made the correlation matrix of the dataset, I found that the column "purchase" was highly correlating with the column "occupation." This implies the column occupation should be included in the predictors.

The next step is to use the modified train and test data and apply machine learning algorithms of various types as read about in the survey. To predict the purchase amount using multiple regressions, I implemented machine learning algorithms and compared them on accuracy and performance metric. Since it is a regression problem, the loose function used is the root mean squared error (RMSE).

Figure 5 depicts the plot of RMSE for all the above implementation for visual comparison. Lower the value of RMSE better the prediction by the algorithm. According to the obtained RMSE value, rule-based decision tree is the most optimized algorithm to analyze the Black Friday sales.

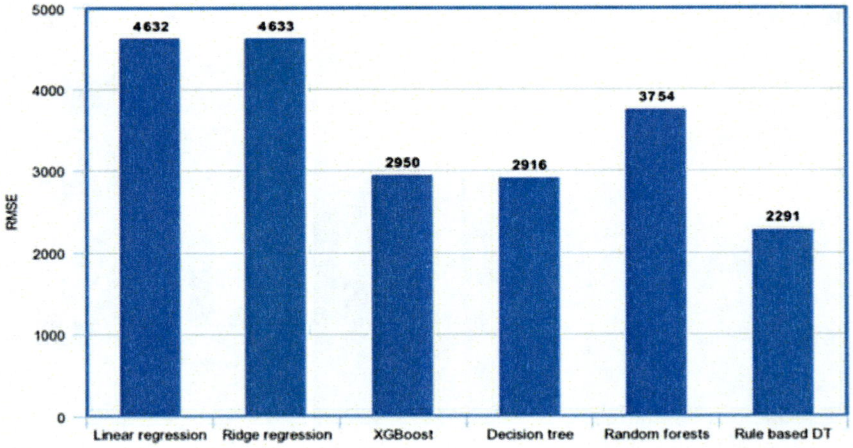

Fig. 5 RMSE values of different classifiers AND regressors

5 Conclusion

The developers adapted neural network even for small problems like regression and data preprocessing. In recent days, the customers are becoming more and more on Black Friday sales. For that, machine learning algorithms can produce better models to predict customers' interest on products, which is helpful for business owners to increase their business. For better accuracy, the data need to be processed by using effective techniques. The dataset with sufficient features can produce better result. Further, the existing machine learning algorithms can be improved by performing on real-time data.

References

1. Guo, Z. X., Wong, W. K., & Li, M. (2013). A multivariate intelligent decision-making model for retail sales forecasting. *Decision Support System, 55*, 247–255.
2. Guyon, I., & Elisseeff, A. (2003). An introduction to variable & feature selection. *Journal of Machine Learning Research, 3*, 1157–1182.
3. Petrescu, M., & Murphy, M. (2013). Black Friday & cyber Monday: A case study. *International Journal of Electronic Marketing & Retailing (IJEMR), 5*(3).
4. Soroush, A., Bahreininejad, A., & van den Berg, L. (2012). A hybrid customer prediction system based on multiple forward stepwise logistic regression mode. *Intelligent Data Analysis, vol. 16*, 265–278.
5. Bing, L., & Yuliang, S. (2016). Prediction of user's purchase intention based on machine learning. In *3rd International Conference on Soft Computing Machine Intelligence (ISCMI)*, pp. 99–103.
6. http://scikit-Ieam.org/stable/auto_examplesilinear_modellplot_ols.html.
7. Barroso, L. P., Bussab, W. O., & Knott, M. (1998). Best linear unbiased predictor in the mixed model with incomplete data. *Communications Statistics Theory Methods, 27*(1), 121–129. Doi: 10.108010361092980883265 4 J.
8. Qin, Y., & Li, H. (2011). Sales forecast based on BP neural network. In *2011 IEEE 3rd International Conference on Communication Software & Network*, pp. 186–189.
9. Singh, K., & Wajgi, R. (2016). Data analysis & visualization of sales data. In *2016 World Conference on Futuristic Trends in Research & Innovation for Social Welfare (Startup Conclave)*, Coimbatore, pp. 1–6.

GMusic Player Using Self-adjusting Graph Data Structures

S. Aravindharamanan, Somula Ramasubbareddy, K. Govinda, and E. Swetha

Abstract Music player is not just founded on craftsmen or collections, yet in addition dependent on the envelope structure. Music player will direct you discover all the music documents in a moment or two, and supports you to speedily hunt relative music recordings. G music is a music application which relies upon various sorts of graph data structures to work and fill social networking purposes. It use ORM graph databases and splay trees with it in a musical social network suite, for accessing the particular user-defined query from the database. For utilizing abstract database access with a picked object-oriented language, object relational mappers (ORM) are also utilized. In this paper, we analyze and implement *splay* tree, a self-conform plan of binary search tree for accessing the database in an efficient and effective manner.

Keywords Self-adjusting · Memorization · Dynamic algorithms

1 Introduction

In numerous applications, it very well may be significant to effectively revise the calculation if the information switches, outwardly retaining to enumerate it. Data structures that productively keep up the yield for a particular issue as the input are refreshed. The above are alluded as powerful algorithms or data structures [1, 3].

S. Aravindharamanan
SCSE, VIT University, Chennai, Tamil Nadu, India

S. Ramasubbareddy
IT, VNRVJIET, Hyderabad, Telangana, India

K. Govinda
SCOPE, VIT University, Vellore, Tamil Nadu, India

E. Swetha (✉)
CSE, SVCE, Tirupathi, Andhra Pradesh, India
e-mail: swethaevakattu@gmail.com

S. C. Satapathy et al. (eds.), *Intelligent System Design*, Advances in Intelligent Systems and Computing 1171, https://doi.org/10.1007/978-981-15-5400-1_33

Data structures that can rearrange themselves after operations are committed to it. This is typically done to enhance expertise on forthcoming operations. The rearrangement can be heuristic in its nature and typically happens in every operation (even if it was only accessing an element).

This data structures are expected to reduce the most worst-scenario time per an errand. Regardless, in common utilizations of search trees, rather a succession of tasks is performed. Previous utilizations are aimed to deduce the time consumption by performing the operations infrequently [2]. An approach to get amortized productivity is to utilize a "self-adjusting" data structure. We affix a smooth restructuring decree expected to enhance the efficiency of forthcoming procedures.

The above data structures ensure O(log n) time (perhaps odd) to get every hub of the tree, as they are proficient in the most pessimistic scenario [2]. It is important to know whether the data structure can accomplish considerably higher degrees of productivity by misusing the way that a few hubs are significantly more liable to be gotten to than others.

2 Literature Survey

Raw SQL statements are an example of direct way in communicating with the database. These statements can be present in two ways like embedding into database access layer or storing in as a stored procedure in the database. There are certain advantages and disadvantages in raw SQL statements.

Advantages like

- It is able to switch and upgrade database in an easy manner.
- CRUD operations can be effectively applied.
- Generation of objects from database table is possible.
- The database is resistant to SQL injection attacks.
- Storage is easy in source control.

But, the main disadvantages of this method are being slower than stored procedure method because of lack of compilation and sending raw SQL over network.

A chart database utilizes diagram texture for phonological request amidst centers, edges, and equity to address and stock data. A key thought of the structure is the chart (or edge or relationship), which explicitly relates data things in the store. The associations empower data in the store to be associated together explicitly, and a great part of the time recuperated with one errand. This shows up diversely in connection to social databases that, with the guide of social database the board structures, permit managing the data without compelling execution perspectives like physical record chains.

Chart databases resemble 1970s framework, exhibit databases in that both address general diagrams, anyway organize and show databases work at a lower measurement of abstraction, and need straightforward traversal over a chain of edges. Some depend upon a social engine and "store" the chart data in a table (regardless of the way that a table is a reasonable part, and consequently this strategy powers another component of reflection between the diagram database, the chart database, the officials system, and the physical devices where the data is truly secured). Others use a key-regard store or document-organized database for limit, making them normally NoSQL structures. Most chart databases are reliant on non-social storing engines and also incorporate marks or properties, which are fundamentally associations that having a pointer to another file.

3 Proposed Method

The music player does the essential control of stacking and playing music; anyway, there are a couple of features particular to this, the tunes are secured as Music Item Objects and are secured in a spread tree, which allows us to execute hold. We can get manner benefit by spread tree with our chase helpfulness standard one may declare that with the interest center point, the songs like it or on a near measurement [4].

In our proposed model, the related concept of increased predictability is used along with binary search trees which help in accessing the data repository in GMusic application. The succession of orbit across the way from hub to root is due to the restructuring formula used in splay trees. The splaying (spreading) nature of the splay is the reason for successional moves of orbit from one level to another in the tree. Spread trees efficiency emulate both the powerfully adjusted trees and static ideal trees along with high optimality properties. Spread trees also stand equally with any type of progressively paired hunt tree.

Self-conform data structures have successive Pro's:

(i) As there is an impression on ignorance of constant factors which are not much worse than the awkward framework, their utilization design is changed to make it more efficient.
(ii) Since no balance or other requirement data is skewed, they require less space.
(iii) It is simple to use as their updated calculations that are skillfully straightforward.

Self-adjusting structures have two potential cons:

(i) They need increasingly nearby changes, particularly amid gets too.
(ii) Specific procedure inside the arrangement can be costly.

An item is the input to every operation; the yield of the operation means whether the thing is in the set, alongside the relevant information is the event that it is. A binary search tree is an approach to resolve this issue to set by. This represents binary search tree which consist of set of things or elements. Each thing is assigned to each hub. Let us assume that a hub h contains the element e. As per the binary search tree, the value assigned to the left subtree should be less than h, and the value in the right subtree must be greater the element in the hub. The search operation in binary search tree follows the top-down approach. In this approach, the search operation is done from root node to the leaf nodes or hub. In this process, each and every node that belongs to the tree will be visited at least once until the seeking element is identified. To complete this whole process, it consumes 8 (d) time where d represents profundity of the hub that contains the required element [5–7].

On the off chance, getting to things is the main activity required, and at that point, there are preferable arrangements over binary search trees, e.g., hashing. Besides, we can stretch out double pursuit trees to help get to symmetric-request position. It can be achieved through storing the quantity of relative hub in each of the hubs and placing the quantity of the hubs in the left subtree. Finally, all the quantity of hub that belong to the whole tree are maintained in a tree header. In case of paired hunt tree, the most signified part of the tree is pointed by a pointer to its root. And a vacant tree is indicated by a pointer to the invalid hub.

Operations on binary tree are examined through using variable n for indicating the quantity of hubs an m for representing the total number of tasks where in case arrangements in paired pursuit tree is required for accessing the tasks. For all out access time to be little, regularly gotten to things ought to be close to the base of the tree frequently. We will likely devise a basic method for rebuilding the tree after each entrance that draws the got to thing nearer to the root, and on the conceivable supposition, this thing is probably going to be gotten to again soon.

Single rotation. In the wake of getting to a thing i in a node x, swing the fringe and adhere x to its parent.

Move to root. In the wake of getting to a thing i in a node x, rotate the edge joining x to its parent, and do this progression until x becomes the root (Figs. 1 and 2).

Our rebuilding heuristic, called spreading, is like move-to-establish in that it does revolutions base up along the entrance way and moves the got to thing right to the root. In any case, it contrasts in that which does the revolutions two by two in a request that relies upon the structure of the entrance way. To spread a tree at a hub x, we rehash the accompanying splaying step until x is the foundation of the tree.

Case 1 (*zig*) Whirl the fringe x amidst p(x), in case of p(x) is the parent (Fig. 3).
Case 2 (*zig-zig*) Whirl the fringe p(x) with its grand parent g(x) and then whirl the fringe with merging x with p(x) (Fig. 4).
Case 3 (*zig-zag*) If p(x) is not the root and x is a left child and p(x) a right child, or vice versa, whirl the fringe joining x with A(x) and then whirl the fringe merging x with the p(x) (Fig. 5).

Following are the details of data structures used in the project [5] (Table 1).

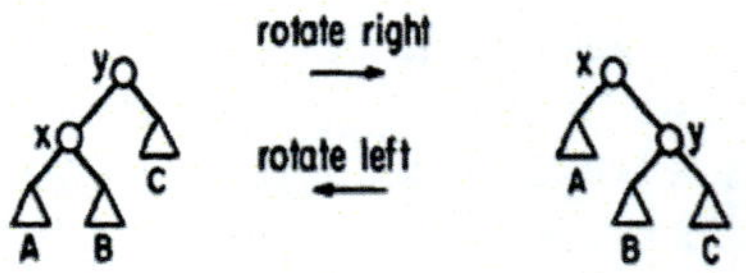

Fig. 1 Revolution of the edge joining hubs x and y. Triangles indicate subtrees. The tree indicated can be a subtree of a bigger tree

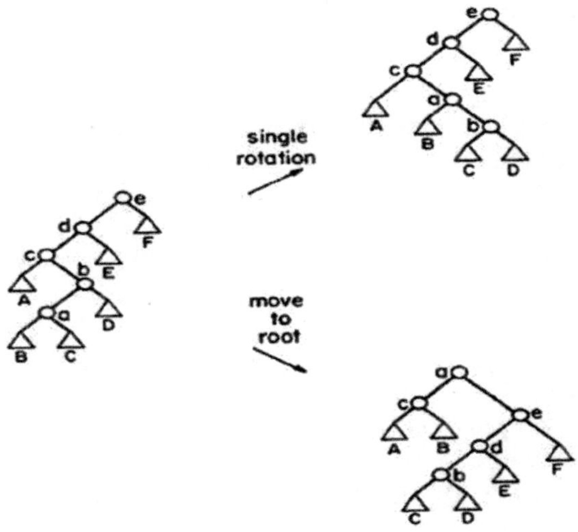

Fig. 2 Reestablishment heuristics

Fig. 3 Implementation of splay tree in zig way

Fig. 4 Implementation of splay tree in zig-zig way

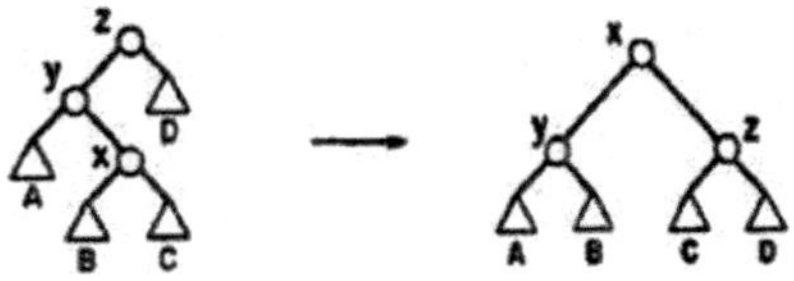

Fig. 5 Implementation of splay tree in zig-zag way

Table 1 Utilization of different data structures in GMusic

Data structure	Use case
Splay trees	Implementing song cache
Hash maps	Accounts validation/storage
Linked lists	Populating list views/adjacency list
Queues	Breadth first search for suggestions

4 Results and Discussion

In this paper, we are going to test the various databases like ArangoDB, MongoDB, Neo4j, and OrientDB with various parameters like shortest path, neighbors, single read, single write, aggregation, and memory, and it is represented in Fig. 6.

With this Fig. 6, we are going to understand that in the parameter of the shortest path, the performance of OrientDB is splendid, in the aspect of neighbors both Neo4j and Neighbors is preferable, but in the case of single write, Neo4J grabs a good position and in the aspects of aggregation and memory usage the OrientDB is the best choice.

Neo4j is a graph database management system developed by Neo4j, Inc. In Neo4j, everything is kept as an edge, node, or attribute. Nodes and edges can be labeled. Labels can be utilized to limit the searches. As of variant 2.0, with the introduction of schemas, indexing was adjoining to Cypher. Beforehand, indexes were bolstered separately from Cypher. The calculation of the shortest path with time of Neo4j is represented in Fig. 7.

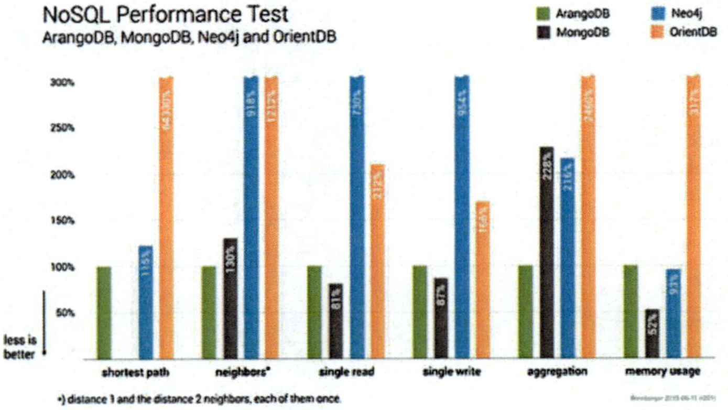

Fig. 6 Comparision of databases with various parameters

Fig. 7 Mean runtime for shortest path on Neo4j

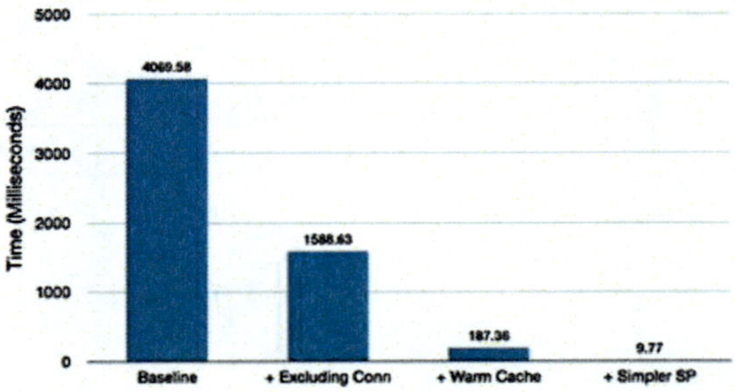

5 Conclusion

The Internet has made a great and huge affect on the music industry on a very massive scale over the recent decades. With the help of Internet, music applications made music more accessible to the public. By using the G music application on social media, it can provide the services to the users by accessing the database in an effective manner. For accessing the user-defined query from the database, splay trees have splendid territory properties. When complete running time is the proportion of intrigue, splay trees are as effective as balanced trees. Also, for adequately long access groupings, spread trees emulate static ideal hunt trees in efficiency from inside the steady factor. The productivity of spread trees comes not from an unequivocal auxiliary limitation, likewise with adjusted trees, yet from applying a straightforward rebuilding heuristic, called spreading, at whatever point the tree is gotten to.

References

1. Acar, U. A., Blelloch, G. E., Blume, M., Harper, R., & Tangwongsan, K. (2009). An experimental analysis of self-adjusting computation. *ACM Transactions on Programming Languages and Systems (TOPLAS), 32*(1), 3.
2. Sleator, D. D., & Tarjan, R. E. (1985). Self-adjusting binary search trees. *Journal of the ACM (JACM), 32*(3), 652–686.
3. Park, E., & Mount, D. M. (2012). A self-adjusting data structure for multidimensional point sets. In *European Symposium on Algorithms* (pp. 778–789). Berlin, Heidelberg: Springer.
4. Tarjan, R. E. (1985). Sequential access in splay trees takes linear time. *Combinatorica, 5*(4), 367–378.
5. Pramod Reddy, A., Ramasubbareddy, S., & Kannayaram, G. (2019). Prediction models for ozone gas estimation. *Journal of Computational and Theoretical Nanoscience, 16*(5–6), 2001–2005.
6. Nalluri, S., Ramasubbareddy, S., & Kannayaram, G. (2019). Weather prediction using clustering strategies in machine learning. *Journal of Computational and Theoretical Nanoscience, 16*(5–6), 1977–1981.
7. Bhukya, K. A., Ramasubbareddy, S., Govinda, K., & Srinivas, T. A. S. (2020). Adaptive mechanism for smart street lighting system. In *Smart Intelligent Computing and Applications* (pp. 69–76). Singapore: Springer.

Plants Nutrient Deficiency Identification Using Classification

Somula Ramasubbareddy, M. Manik, T. A. S. Srinivas, and K. Govinda

Abstract Agriculture is the major for the people in India, where the loss of irrigation is due to deficiency in plants. This project is to identify the nutrient deficiency in plants like calcium (Ca), magnesium (Mg), nitrogen (N), phosphorus (P), potassium (K), and sulfur (S). A demanding component of plant health is decided by plant stress responses, which are characterized by a suite of molecular and cellular processes that are triggered by the plant in some form of extraction. Extraction can be abiotic, for example, dry season or abundance light, or biotic, for example, herbivores. Studies have appeared, while plants do demonstrate their pressure by means of variety of leaf in synthetic substance and atomic changes, the shades of leaves are additionally clear markers of plant pressure. In general isolation, deficiency of nutrients such as nitrogen, phosphorous, potassium, calcium, and magnesium in soils results in changes in coloring patterns of leaves. Healthier/stress-free leaves are typically greener in color, while increased yellowing indicates progressively unhealthier leaves. Unfortunately, existing techniques to detect plant health incur for excess exorbitant infrastructure. In this paper, the features of different leaf images are captured, then train and test those images to identify mineral deficiency through machine learning.

Keywords Business analysis · K-means · Clustering techniques · Custom relation

S. Ramasubbareddy (✉)
IT, VNRVJIET, Hyderabad, Telangana, India
e-mail: svramasubbareddy1219@gmail.com

T. A. S. Srinivas · K. Govinda
SCOPE, VIT University, Vellore, Tamilnadu, India

M. Manik
CSE, RGMCET, Nandyal, Andhra Pradesh, India

S. C. Satapathy et al. (eds.), *Intelligent System Design*, Advances in Intelligent Systems and Computing 1171, https://doi.org/10.1007/978-981-15-5400-1_34

1 Introduction

Plant ailments cause occasional episode of illnesses which prompts extensive scale passing and starvation. It is evaluated that the scene of helminthosporiose of rice in northeastern India in 1942 caused a significant loss of sustenance grains and going of a million people. Since the effects of plant illnesses were pulverizing, a segment of the gather advancement has been surrendered. It is evaluated that 2006 plant affliction incidents in Georgia (USA) are around $563.01 million. In India, no estimation has been made yet it is more than USA in light of the way that the preventive advances taken to verify our harvests are for no situation one-tenth of that in USA.

Research develops the advancement enrolling condition to perceive the illnesses using debased pictures of various leaves. Pictures of leaves are taken from advanced camera, PDAs and prepared using picture developing, at that point the piece of the leaf sport has been utilized for the ordering reason for the train and trial of ailment.

In this idea, the leave pictures are prepared and tried to identify whether the picture is sick or not alongside the infection name. Ailments in harvests for the most part on the leaves influence on the decrease of both quality and amount of rural items. Impression of human eye is not grounded in order to watch minute variety. In this paper, we are giving programming answer to naturally recognize and order plant leaf sicknesses. In this, we are utilizing picture handling methods to group maladies and rapid analysis can be done according to infection.

Plant infections cause occasional flare-up of maladies which prompts vast scale passing and starvation. It is noted that the episode of helminthosporiose of crop in India in 1953 caused an overwhelming loss of nourishment crops and passing of a trillion people. Because of the impacts of plant ailments were decimating, a portion of the yield development has been surrendered. It is evaluated that 2005 plant illness misfortunes in Georgia (USA) are roughly $653.06 million. In India, no estimation has been made yet it is more than USA on the grounds that the preventive advances taken to ensure our harvests are not in any case one-fifth of that in USA. Research work builds up the development figuring condition to recognize the sicknesses utilizing contaminated pictures of different leaves. Pictures of leaves are taken from computerized camera, advanced cells, and prepared utilizing picture developing, and at that point, the piece of the leaf sport has been utilized for the characterizing motivation behind the train and trial of ailment.

Sickness of the board is a difficult undertaking. For the most part, sicknesses are clearly observed on the leaves of the plant. Exact evaluation of these outwardly watched ailments, bugs, and characteristics has not examined yet as a result of the multifaceted nature of visual examples. Consequently, there has been expanding interest for increasingly explicit and refined picture design understanding. In organic science, here and there a huge number of pictures are created in a solitary examination.

Practically these errands are prepared physically or with particular programming bundles. It is not just colossal measure of work yet in addition experiences two noteworthy issues: over the top handling time and abstraction ascending from various people. Hence to direct high throughput tests, plant researcher needs productive PC programming to consequently separate and break down huge substance. Here, picture handling assumes essential job.

2 Literature Survey

The authors propose and tentatively assess a product answer for programmed recognition and arrangement of plant leaf infections. Investigations of plant quality/illness allude to the investigations of outwardly detectable examples of a specific plant. These days crops face numerous qualities/ailments. Harm of the creepy-crawly is one of the real attributes/illnesses. Bug sprays are not constantly demonstrated effective in light of the fact that bug sprays might be lethal to some sort of winged animals [1]. It likewise harms normal creature natural ways of life. The accompanying two stages are included progressively after the division stage. In the initial step, we distinguish the most parts of green hued pixels. Next, these pixels are covered dependent on explicit edge esteems that are figured utilizing Otsu's technique, at that point those for the most part pixels which are green are concealed. The other extra advance is that the pixels with 0 red, green, and blue (RGB) qualities and the pixels on the limits of the contaminated group were totally evacuated. The exploratory outcomes exhibit that the proposed strategy is a vigorous system for the identification of plant leaves illnesses. The built-up calculation's proficiency can effectively recognize and characterize the analyzed illnesses with an accuracy somewhere in the range of 83% and 94% and can accomplish 20% [2].

Surface features that rely upon the close-by power go obtained by a bank of channels are examined. The features differentiate in the kind of nonlinear post-taking care of which is associated with the adjacent power extend. The going with features is considered: Gabor imperativeness, complex minutes, and granulating cell executive features. The capacity of the relating directors to convey specific component vector bunches for different surfaces is investigated using two procedures: the F premise and the request result examination. The two procedures give unsurprising results. The crushing cell overseer gives the best detachment and division results. The surface area limits of the executives and their healthiness to non-texture features are furthermore examined. The crushing cell overseer is the unique case that explicitly responds just to surface and does not give false response to non-texture features, for instance, object shapes [3, 4].

In agriculture research of customized leaf, disease disclosure is major research point as it may exhibit benefits in checking significant fields of yields, and along these lines normally recognizes appearances of sickness when they appear on plant leaves. There are essential steps for disease ID of image acquisition, image preprocessing, image segmentation, feature extraction, and statistical analysis. This proposed work is in first picture isolated using center channel and converts the image to gray-level scale or RGB, in second step picture is partitioned using the k-medoid methodology, in following stage veiling green pixels and removal of secured green pixels, after in consequent stage figure the texture features statistics, and at last these features go in neural framework. The neural network gathering performs well and could viably recognize and describe the attempted disease [5, 6].

Pictures structure vital information and data in natural sciences. Plant maladies have transformed into an issue as it can cause noteworthy decrease in both quality and amount of farming items. Programmed identification of plant illnesses is a basic research subject as it might give advantages in observing huge fields of crops, and these lines consequently distinguish the manifestations of maladies when they show up on plant leaves [7]. The proposed framework is a product answer for programmed discovery and calculation of surface measurements for plant leaf maladies. The created handling plan comprises of four principle steps: initially, a shading change structure for the info RGB picture is made; the green pixels are covered and expelled utilizing explicit edge esteem; at that point, the picture is divided and the valuable fragments are extricated; at long last, the surface measurements are processed. From the surface measurements, the nearness of illnesses on the plant leaf is assessed. Trial results on a database of around 250 plant leaves of 30 distinct plants affirm the power of the proposed methodology [8].

3 Proposed Method

In this procedure of leaf illness identification right off the bat, we are going to gain typical picture by the methods for any advanced camera. Getting the leaf picture is the initial phase in leaf infection identification. Besides, we are going for pre-preparing of picture. In this, we are going to upgrade the picture quality. In the wake of improving the picture, we separate the picture in various bunches. In that, green pixels are covered and just contaminated segment left. We did this veiling in light of the fact that to get more exactness. This procedure can be called as division. After division, we include extraction is finished. At that point, we do measurable examination, and finally, we utilize a classifier. Characterization is the last advance and it is finished by utilizing neural system.

- GLCM.
- Random forest.
- Based on GLCM, division and characterization can be effectively performed. Subsequently, time utilization will be less.
- The exactness will be high for Random Forest because of its precise arrangements (Fig. 1).

In the proposed methodology, the strategy received for removing the list of capabilities is called GLCM technique. In the event, the highlights extraction is painstakingly picked to expect that the highlights set will extricate the important data from the information so as to play out the ideal assignment.

GLCM the dim dimension co-event grid is created. Also, we get diverse highlights esteem, i.e., esteem for differentiation, vitality, homogeneity, relationship, and so forth.

Fig. 1 Block diagram

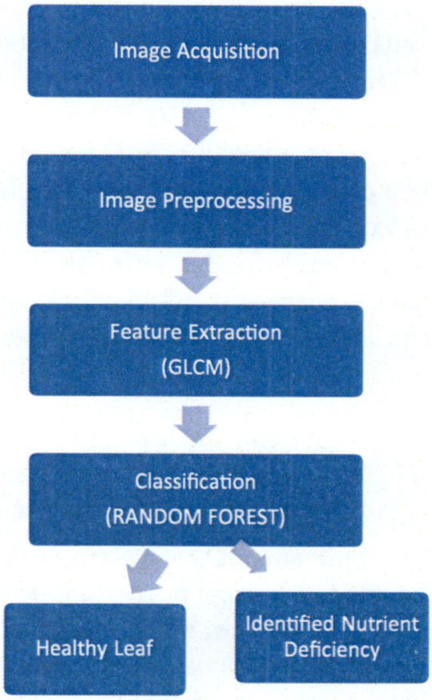

Feature extraction incorporates streamlining the proportion of benefits required to delineate a tremendous course of action of data absolutely. When performing examination of complex data, one of the difficult issues originated from the amount of elements is included. Examination with incalculable overall requires a ton of memory and count control or a portrayal computation which overfits the planning test and wholes up incapably to new precedents. Feature extraction is a general term for procedures for creating blends of the elements to get around these issues while so far depicting the data with satisfactory exactness. Surface investigation points in finding a remarkable method for speaking to the hidden attributes of surfaces and speaks to them in some less complex however one of a kind structure, with the goal that they can be utilized for vigorous, exact grouping and division of articles. In spite of the fact that surface assumes a critical job in picture investigation and example acknowledgment, just a couple of structures actualize locally available textural highlight extraction. In this paper, gray dimension co-event framework is figured to get factual surface highlights.

The gray-level co-event matrix (GLCM) strategy is a method for separating second request measurable surface highlights. The methodology has been utilized in various applications, third and higher request surfaces think about the connections among at least three pixels. These are hypothetically conceivable yet not generally actualized because of count time and understanding trouble. A GLCM is where the quantity of lines and segments is equivalent to the quantity of dim dimensions, *G*, in the picture. The framework component $P(i, j|\Delta x, \Delta y)$ is the ideal solution with which two pixels, isolated by a pixel remove $(\Delta x, \Delta y)$, happen inside a given neighborhood, one with force "*i*" and the other with power "*j*." The network component P (*i*, *j*|*d*, ө) contains the second request likelihood esteems for changes between dim dimensions "*I*" and "*j*" at a specific relocation separate *d* and at a specific point (ө). Utilizing countless dimensions G suggests putting away a great deal of transitory information, for example, a $G \times G$ lattice for every mix of (Δx, Δy) or (*d*, ө). Because of their huge construction, the GLCMs are exceptionally touchy to the measure of the surface examples on which they are assessed. Along these lines, the quantity of dark dimensions is regularly diminished. Here, one pixel balance is utilized. In the event that the window is sufficiently extensive, utilizing a bigger balance is conceivable. The upper left cell will be loaded up with the occasions and the mix (0, 0) happens.

3.1 Classification

First correlation is made between the test and the sound example. On the off chance that the test histogram coordinated with the solid leaf test created the outcome that is plant is not sick. In the event that it does not coordinate with the sound leaf test, the correlation is made between the test and the infected example. On the off chance that the test is coordinated with the sick example, it produces plant is ailing.

In the arrangement part, the preparation and testing part will be independent and the example pictures are finished.

Random Forest

Random boondocks or self-assertive decision forests are an outfit learning technique for course of action, backslide, and diverse errands that works by structuring an immense number of decision trees at planning time and yielding the class that is the strategy for the classes (gathering) or mean desire (backslide) of the individual trees. Arbitrary decision boondocks directly for decision trees' penchant for overfitting to their readiness set.

Points of interest of utilizing random forest: By utilizing this calculation, there are numerous points of interest.

- The random timberland calculation is not one-sided, since there are numerous trees and each tree is prepared on a subset of information. Essentially, the arbitrary timberland calculation depends on the intensity of "the group"; in this way, the general biasedness of the calculation is diminished.
- This calculation is truly steady. Regardless of whether another information point is presented in the dataset, the general calculation is not influenced much, since new information may affect one tree; however, it is hard for it to affect every one of the trees.
- The random woods calculation functions admirably when you have both absolute and numerical highlights.
- The arbitrary backwoods calculation likewise functions admirably when information has missing qualities or it has not been scaled well.

4 Implementation

Our main method includes three main steps:

- Preprocessing,
- Feature extraction,
- Classification.

An overview of our proposed signature verification method. Each method we have used is detailed further.

4.1 Preprocessing

Input image what we caught is not continually fulfilling dependably; there is some clamor included in that picture, so for evacuation of commotion and getting educational picture, we apply preprocessing method. In this procedure, right off the bat the picture is upgraded in the wake of smoothing. While gathering picture, numerous data is gathered which incorporate commotion. To expel commotion, we utilize distinctive kind of sifting procedures: predominantly smoothing and cutting procedure.

It is used to get some important data from divided picture. We are utilizing k-mean grouping. In this, we are going to cover green pixels of picture and stay with the contaminated part of picture. The first thing is to characterize k-centroids, one for each group. These centroids should be set in a guile route in view of various area causing distinctive outcomes. In this way, the good decision is to put them in a random fashion. The ensuing stage is to take each point having a spot with a given enlightening list and accomplice it to the nearest centroid. Exactly when no point is pending, the underlying advance is done and an early groupage is done. Presently, we need to re-discover k new centroids as barycenters of the groups coming about due to the past development. After we have these k new centroids, another coupling must be done between comparable enlightening gathering centers and the nearest new centroid.

5 Results and Discussion

Figure 2 shows the Sulphur deficiency for the plant, Fig. 3 illustrates magnesium in plants and Figs. 4 and 5 show Nitrogen and phosphorous deficiency identified using classification techniques.

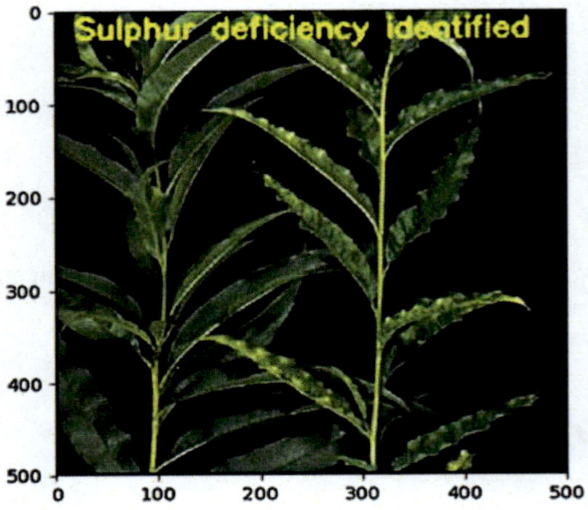

Fig. 2 Sulfur deficiency identified

Fig. 3 Magnesium deficiency identified

6 Conclusion

To satisfy the need of growing population, it is high time to focus maximum yield. This is only possible if plants get sufficient nutrient for growth. Plant nutrient content is neglected many times, though it should be given the importance. This emphasizes the identification of nutrient content using image processing and machine learning techniques. This will help farmers to get the idea about the health of their crops and the preventive measures to be taken. Also, this will be helpful in vertical farms where nutrient supplements are provided to plants.

Fig. 4 Nitrogen deficiency identified

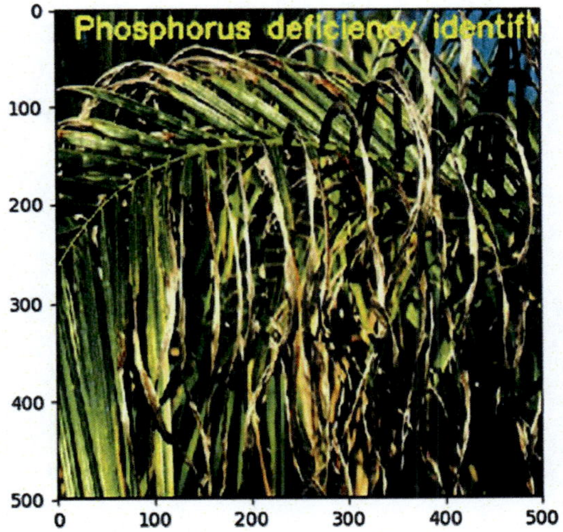

Fig. 5 Phosphorus deficiency identified

References

1. Mainkar, P. M., Ghorpade, S., & Adawadkar, M. (2015). Plant leaf disease detection and classification using image processing techniques. *International Journal of Innovative and Emerging Research in Engineering, 2*(4), 139–144.
2. Patil, J. K., & Kumar, R. (2011). Advances in image processing for detection of plant diseases. *Journal of Advanced Bioinformatics Applications and Research, 2*(2), 135–141.
3. Ganapathy, S., Mallidi, S. H., & Hermansky, H. (2014). Robust feature extraction using modulation filtering of autoregressive models. *ACM Transactions on Audio, Speech, and Language Processing, 22*(8).

4. Telagarapu, P., & Poonguzhali, S. (2019). Filtering and statistical feature based automatic detection of breast lesions in ultrasound. *International Journal of Innovative Technology and Exploring Engineering, 8*(3).
5. Sladojevic, S., Arsenovic, M., Anderla, A., Culibrk, D., & Stefanovic, D. (2016). Deep neural networks based recognition of plant diseases by leaf image classification. *Computational Intelligence and Neuroscience, 2016,* 11. Article ID 3289801.
6. Vishnu, S., & Ranjith Ram, A. (2016). Plant disease detection using leaf pattern: A review. *International Journal of Innovative Science, Engineering and Technology, 2*(6).
7. Ashish, P., & Tanuja, P. (2017). Survey on detection and classification of plant leaf disease in agriculture environment. *International Advanced Research Journal in Science, Engineering and Technology, 4*(4).
8. Singh, M. K., & Chetia, S. (2017). Detection and classification of plant leaf diseases in image processing using MATLAB. *International Journal of Life Sciences Research, 5*(4), 120–124.

Maximize Power Generated Using Solar Tracking Mechanism

G. Naga Chandrika, Somula Ramasubbareddy, K. Govinda, and C. S. Pavan Kumar

Abstract Due to the low and decreasing availability of non-renewable resources, the concentration is shifted towards renewable resources. Solar energy is one of the largest and most abundant resources among renewable resources. Thus, there has been an increase in research in the field of solar energy. The main issue is that the sun rises from the east and sets in the west, that is, the sun's rays move from east to west. The solar panels can harvest a maximum amount of energy only if the rays are exactly perpendicular to the surface of the panels. Thus, the panels also should rotate from east to west in such a way that the rays are always perpendicular to the surface of the panels. This paper proposes a mechanism to extract the maximum energy from renewable resources like solar energy.

Keywords Renewable resource · Non-renewable resource · Solar energy

1 Introduction

The importance given to renewable resources is very high these days due to the well-known harmful effects and also due to the decrease in the reserves or non-renewable resources. Renewable resources are those resources that can regenerate themselves or replenish themselves and cause less pollution to our environment. Non-renewable resources are those resources that cannot replenish themselves. If we think of it in this way, our natures by some biological processes replenish resources like gasoline, petrol and coal. But the rate at which we consume these resources is much faster than the rate at which it is replenished by nature. Thus, there is a shortage of these resources. But in the case of renewable resources, there is no such issue. Hence by harvesting energy from these resources would be fruitful and beneficial.

G. Naga Chandrika · S. Ramasubbareddy (✉)
IT, VNRVJIET, Hyderabad, Telangana, India
e-mail: svramasubbareddy1219@gmail.com

K. Govinda · C. S. Pavan Kumar
SCOPE, VIT University, Vellore, Tamil Nadu, India

S. C. Satapathy et al. (eds.), *Intelligent System Design*, Advances in Intelligent Systems and Computing 1171, https://doi.org/10.1007/978-981-15-5400-1_35

This paper mainly discusses the well-sustained natural resource viz., Solar Energy. Solar energy is the sunlight received by the earth from the sun. Solar energy is harvested by a system called photovoltaic system. A survey conducted by a group researchers found that the total potential energy harvested by us is about 700 KW but the total potential received by the earth from the sun is somewhere about 16,000 KW, thus by harvesting around 80% of the total potential energy, we will be able to eradicate the whole of energy crises all around the world.

Photovoltaic system mainly consists of solar panels, solar cells, battery and solar inverter. Solar panels mainly absorb the sunlight and convert it into electrical energy. Here, a solar inverter is used mainly to transform energy from one form to another form and the harvested solar energy is stored in the batteries [3].

The total energy generated by a solar panel depends mainly on the following two factors:

- Intensity of sunlight and,
- Angle of incidence of the rays of light from the sun.

This means that the panels should be exposed to the sun throughout the day and as the sun rises and sets from east to west, so the panels should also move along with the sun in such a way that the rays from the sun are always perpendicular to the surface of the panels.

The method discussed in the paper is a self-governing and self-controlled system. It does not require extra power to run the mechanism and to carry out the function. It has only one axis of rotation about which the panel rotates from east to west [1, 3, 4].

2 Literature Survey

Many different methods are proposed by many people in order to maximize the total energy harvested by a solar panel and to improve the tracking ability of the system employed.

Few of the projects are

Sun Light Tracking Strategies to Optimize the Energy

A dual-axis tracking mechanism is employed along with a fuzzy logic neural controller. This system can precisely trace the sun's path and rotate the panels along with it, which makes it more efficient.

The disadvantages mainly include the employment of more mechanism and more instruments which makes it more complicated and difficult for normal people to afford it and also the power consumed by the system to carry out the task is also very high which might even result in a loss of the total energy harvested [4, 8].

Robot Tracker

This technique uses a microcontroller called STR which has the ability to turn or rotate the panels, in the required direction along one axis only. An additional motor is used to rotate the panel since; the size of the panel is less and also due to the high torque acting on it. The microcontroller is well-programmed so that it can adjust the robot so that it faces the same direction as the panels and ensure that maximum sunlight is received by the panels.

The drawback of this method is that the increase in the complexity of a simple project which can be completed successfully by using just two light sensors and a simple circuit. Even the cost of this project would be high because along with panels a robotic arm is also to be constructed and fitted to the panels [6, 7, 11].

Microprocessor Tracking System Along With Stepper Motor

Microcontroller allows multiple functions to be carried out by the system. It is used to supervise the tracking system. As we are using a microcontroller we need some external components in order to implement the memory, i.e. RAM and ROM, input and output ports [9].

The external components make the project very expensive and also regular supervision on these components is required which would increase the complexities faced in the project [2, 10].

Miniature Solar Tracker

This controller uses a single axel DC motor to rotate the panel in accordance with the angle of the sun. Initially, the controller starts the panels from the east and ends in the west by the sunset. But the main drawback is that the system cannot track if the panel is out of positioning thus cannot orient in such a way to acquire maximum sunlight [1, 5].

3 Proposed Method

In the model, we use four light-dependent resistors (LDR), two placed on the top of the panel and two placed at the bottom of the panel. These two LDR's are placed as a set on a cardboard plate and separated by a vertical cardboard piece. The intensities received by both the sensors will be equal only if the incident light is perpendicular to the plane of the sensors which in turn will be perpendicular to the panels. Thus, increasing the energy harvested by the panels form the sunlight.

Here the LDR used has a special characteristic in it. The resistance offered by the resistor is always inversely proportional to the intensity of the incident light. It means that the resistance decreases with an increase in the intensity of the light. The resistance offered by the sensors would be about a hundred ohms if the intensity of the light is high and will go up to mega ohms in the absence of light.

$$R \propto \frac{1}{\text{Intensity of Incident Light}}$$

The rise and fall in the resistance cause a potential difference in the circuit which in turn develops a current which is sufficient to supply the required power to the motor in order to rotate the panels. Thus, we use a light sensor made of cadmium sulphide (CdS). This sensor conducts electricity in both the direction mainly because of the zig-zag location of the cadmium sulphide track located in the sensor.

The flow chart given below depicts the overall mechanism of the process (Fig. 1). The main principle and working of the circuit are shown below.

Figure 2 depicts the circuit employed to track the sun's rays. This system mainly consists of a comparator, an H-bridge driver and some other devices and instruments. Mainly four LDRs, LDR1–LDR4, are employed to detect the angle of the sun's light and rotate the panels according to the incident rays. These act as indicators for the motor to rotate the panels. VR1 and VR2 are adjusted to the low output at the pins of the comparators A1 and A2. These comparators are given the task to stop the rotation of the panels, once they reach the position where the incident light is at 90°.

Fig. 1 Flow chart

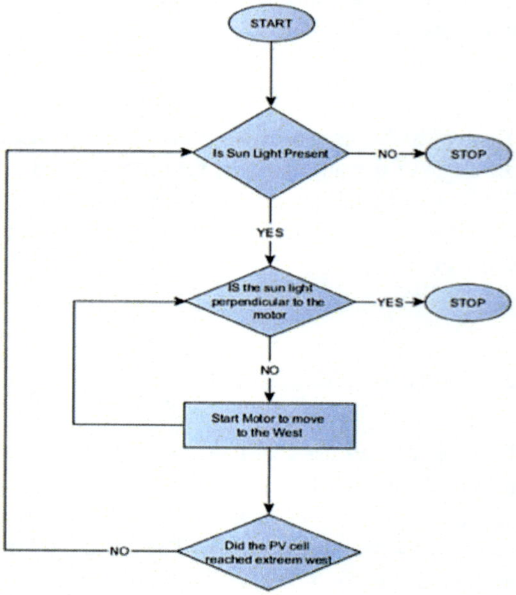

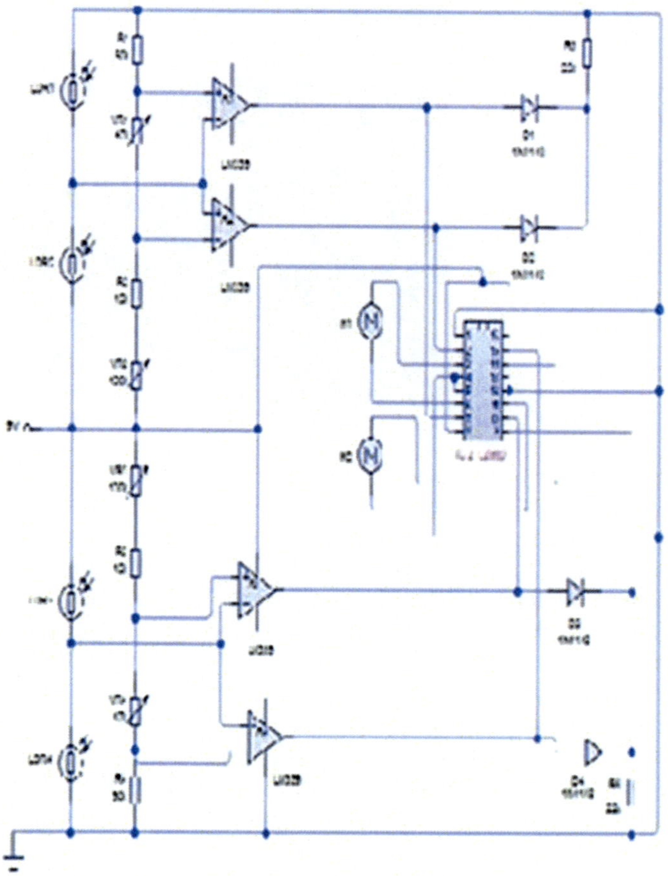

Fig. 2 Circuit of solar tracking system

The sensor LDR1 and the sensor LDR2 are arranged along *x*-axis. If the intensity of light received LDR2 is higher than that received by LDR1, which is allowing a high input to the pins of the comparators A1 and A2 at 4 and 7. This results in rotating the motor in one direction and turns the panel. And the same happens if LDR1 receives higher intensity in comparison with the resistor LDR2, then the resistance offered is less which supplies input to the pins 4 and 7 of comparators A1 and A2. This rotates the panels in the reverse direction (Fig. 3).

In a similar fashion the other two resistors LDR3 and LDR4 and placed along the y-axis to find the exact position of the panels.

If viewed in mathematical terms, the voltage of the motor is 12 V and the R. P. M is 10 along the direction east to west and vice versa.

The current which is used by the circuit is 24 mA.

Thus, the power required is 288 mW/sec (24 mA * 12 V).

Fig. 3 Proposed assembly for the solar tracking system

For the time period of 6 s, the total power required is 108 W (288 * 6 = 1728 mW = 1.8 W).

Suppose for a day the panels rotate from east to west and again back to east. About 10 rotations of motor take a time of 1 min/60 s.

Thus for one rotation, i.e. 360°, the motor takes a time period of 6 s (60/10). In order to turn from east to west time consumed is about 3 s (180° only).

So, for 10° displacement it takes for rotating the panels about 10° time taken is about 167 ms (3000 ms * 10°/180°).

The time taken by the sun moves from east to west is about 12 h, i.e. 720 min. Hence for 10°, the time taken is 40 min.

In 2 h, the sun travels about 30°. In order to turn panels by 30° the time taken by the panel is 501 ms.

4 Results and Discussion

The proposed method is tested along with a normal panel with no tracking system. Both the panels, one with the tracking mechanism and the normal one are placed under the sun and the energy harvested observed and noted. The total energy produced by both the panels are compared and the result is found.

There is nearly a 50% increase in the total energy produced by the panel with the tracking mechanism than the solar panel with no additional mechanism. Graphs are also drawn with respect to the energy readings noted. The graph below depicts the energy harvested by the panels with and without the above-mentioned mechanism.

The golden line is the energy harvested by a normal panel and the red line is the energy harvested by the panel with the tracking mechanism (Fig. 4).

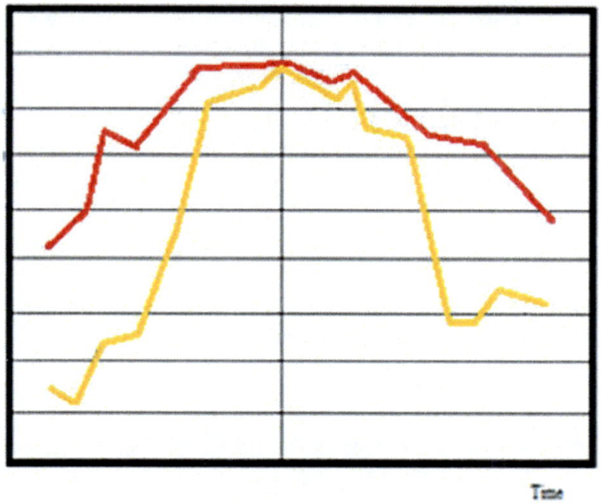

Fig. 4 Power generated versus time

5 Conclusion

From the above results, it shows that by using this method we can easily harvest more amount of energy than a normal solar panel. Moreover, this mechanism is not a costly one at the same time there is no need for constant supervision by a person to look after it. The circuit with the motor does not require an excess of power from the panel because the power obtained from the sensors is sufficient to supply the required power to the motor to rotate the motor.

This would help us overcome power crises not only in India but also all around the world. And also the harm done by mankind towards environment will also be less as we are using renewable resources there would be no emission of harmful gases like carbon monooxide and carbon dioxide hence, in turn, reduce global warming and protect our environment.

Thus, by employing this mechanism with the normal solar panels, we can harvest a major part of the solar light radiated from the sun and overcome all the energy crises.

References

1. Peterson, T., Rice, J., & Valane, J. (2005). Solar tracker. In *ECE 476*, Spring.
2. Saxena, A. K., & Dutta, V. (1990) A versatile microprocessor-based controller for solar tracking. In *IEEE Proceedings*, pp. 1105–1109.
3. Goswami, Y., Kreith, F., & Kreder, J. (1999). Chapter 2: Fundamentals of solar radiation. In *Principles of solar engineering*. Philadelphia: Taylor & Francis.

4. Jaen, C., Pou, J., Capella, G., Arias, A., & Lamich, M. (2009). On the use of sun trackers to improve maximum power point tracking controllers applied to photovoltaic systems. In *The IEEE conference on Compatibility and Power Electronics*, pp. 67–72.
5. Stjepanovic, A., Softic, F., Bundalo, Z., & Stjepanovic, S. (2010). Solar tracking system and modelling of PV module. In *MIPRO 2010*, Opatija, Crotia.
6. Barsoum, N. (2009). Implementation of a prototype for a traditional solar tracking system. In *2009 Third UKSim European Symposium on Computer Modeling and Simulation.*
7. *International Journal of Enhanced Research in Science Technology & Engineering*, 2(8). ISSN: 2319-7463. www.erpublications.com.
8. International Energy Agency, [Online]. Available: http://www.iea.org/extbase/nppdf/free/2009/key-stats-2009.pdf.
9. Gadewadikar, J. (1997). *Microprocessor based solar tracking system using stepper motor*. Indore: S G S Institute of Technology and Science.
10. Piao, Z. G., Park, J. M., Kim, J. H., Cho, G. B., & Baek, H. L. (2005). A study on the tracking photovoltaic system by program type. In *International Conference on Electrical Machines and Systems* (Vol. 2, pp. 971–973).
11. Marsh, A., & Raines, C. (2000). Solar position calculator. In *Square One research and the Welsh School of Architecture at Cardriff University.*

An Automated Glaucoma Detection in Fundus Images—A Survey

V. Priyanka and D. Vaishnavi

Abstract Glaucoma is the second most common reason of blindness in the globe. It is a chronic disease often called "silent thief of sight" in which optic nerve is progressively damaged, as it has no symptoms. It may cause permanent loss of sight, if it is not recognized at an initial stage. The flow leads few structural changes in the eye, which helps the ophthalmologists to identify the glaucoma at an initial phase and stop its further damage. Results produced by ophthalmologist are fairly moderate and accuracy also subjective to the expertise of the professionals. So, this article discusses about the various automated methods exists in state-of-art to detect glaucomatous eye from normal eye using image processing techniques.

Keywords Glaucoma · Intraocular pressure · Disk to cup · NRR

1 Introduction

Glaucoma is a chronic ocular disease which might cause visual impairment, if not recognized at an initial stage. This eye disease is declared as the second largest reason for visual impairment across the world by World Health Organization (WHO), and it includes about fifteen percent of the blindness causing cases around the globe that makes 10 million of the world's population, and therefore, the range is predicted up to eighty million before 2022. This eye disease arises due to an increase in Intra Ocular Pressure (IOP) caused through aqueous humor, which is a liquid continuously flows inside the eye. Figure 1 depicts the changes in human vision of normal eye and glaucomated eye.

In regular, a balance in the eye is maintained as the quantity of fluid made inside the eye equivalent to the quantity of fluid discharged from the human eye. In a normal eye, this pressure range is 14–20 mmHg. If it is 20–24 mmHg, then it shows

V. Priyanka (✉) · D. Vaishnavi
Vardhaman College of Engineering, Hyderabad, Telangana, India
e-mail: priyankavanguru04@gmail.com

D. Vaishnavi
e-mail: vaishume11@gmail.com

S. C. Satapathy et al. (eds.), *Intelligent System Design*, Advances in Intelligent Systems and Computing 1171, https://doi.org/10.1007/978-981-15-5400-1_36

Fig. 1 Normal vision and glaucoma vision

the slightly affected by glaucoma, and above 24 mmHg, it confirms that the eye is fully affected by glaucoma. However, in this case, the liquid does not ooze out of the eye, and stress will be increased in the eye, which leads to the injury of cranial nerves; which connects the cerebrum and the human eye. An increase in pressure within a period of time ends up in severe damage to the cranial nerve and can lead to irreversible visual impairment. This is shown as pictorially in Fig. 2.

An increase in IOP causes no early symptoms or pain, it makes more complex for recognizing glaucoma at early stage itself, which makes the early recognition of glaucoma complex and thus often called "Silent thief of sight". It also increases the cup size and consequently increases the cup-to-disk ratio (CDR). Pictorized illustration of the optic cup and optic disk is given in Fig. 3.

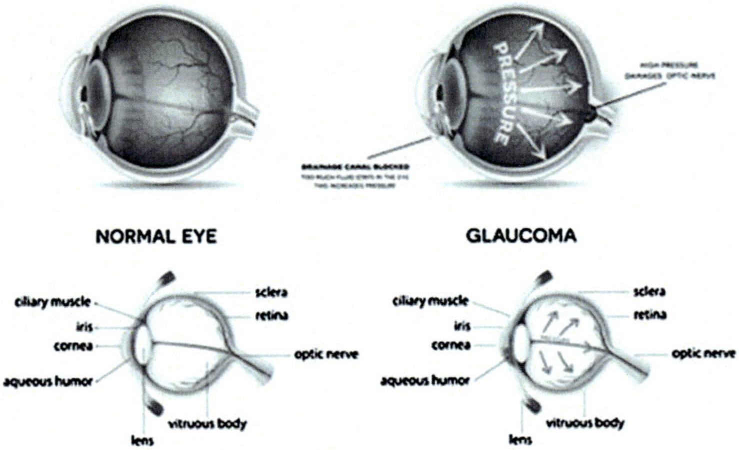

Fig. 2 Representation of pressure in normal and glaucoma eye

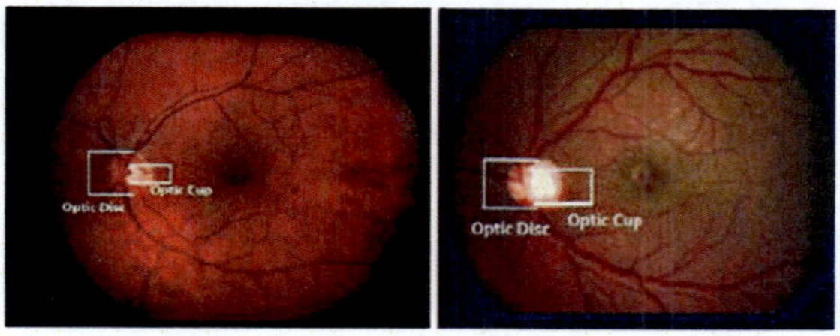

Fig. 3 **a** Normal disc. **b** Glaucomatic disk

Optic cup is bounded inside the optic disk, which is slightly brighter, whereas optic disk is yellowish. The area between optic cup and optic disk is called as neuro-retinal disk (NRR). One of the structural change features occurred in the process is CDR, the ratio between optic cup and optic disk. In glaucomatous eyes, the cup size gradually increases, which results in an increase of CDR ratio and decrease of NRR. If CDR ratio is below 0.45, then it indicates a normal eye, if it is greater than 0.45, then there is a slight chance of glaucoma. Suppose, if it is greater than 0.6, then it is confirmed that the eye is severely affected with glaucoma.

The left over part in this article is structured in such a way: Sect. 2 illustrates the causes of glaucoma and types of glaucoma. Section 3 furnishes a typical glaucoma detection scheme. Section 4 discusses the various feature extraction methods exists in literature, and finally, it concludes an article.

2 Causes of Glaucoma

To keep the eye ball in perfect structure, the eye needs some amount of pressure to work properly. In some cases, the damage occurs due to stress in the eye. Some cases have normal eye pressure, but due to the optic nerve weakness, the damage occurs. In most of the cases, both prototypes are possible, but at different extent. Eye pressure is no way related to the blood pressure.

Aqueous humor is a watery liquid produced behind the iris. The liquid flows from the pupil, i.e., center of the iris and leaves through minute drainage channels in the eye. There is an angle between the cornea, i.e., front of the eye and the iris and returns the fluid back to the blood stream. Generally, the liquid produced inside the eye is balanced to the liquid drained out, but if it is not drained out or if too much liquid produced inside the eye, then there will be a rise in the eye pressure. The aqueous humor is completely different from tears in the eye.

If there is lots of pressure on the optic nerve, then it can cause damage to the eye. The damage depends on pressure in the eye and how long it is from, whether this is due to poor blood flow or any other optic nerve weakness. Optic nerve damages immediately if the pressure is really high. If pressure is low level, then damage will be slow, and gradually, there will be lose in sight if not treated early.

2.1 Types of Glaucoma

Though the glaucoma is classified as various types, angle-closure glaucoma and primary open-angle glaucoma are widely affected one [1]. They are illustrated in the following subsections.

2.2 Primary Open-Angle Glaucoma

This type of glaucoma is very well-known and mostly found in 90% American population. This glaucoma slowly starts to block the drainage canal and increases the IOP, and finally, it affects the optic nerve cells. It leads to a blind spot in optic lens and affects the vision. It cannot be cured once the vision is lost [2]. Figure 4 presents the open-angle glaucoma.

2.3 Angle-Closure Glaucoma

It is less usual than open-angle glaucoma among Americans. In this type, the drainage canals are blocked due to sudden stop of the fluid called aqueous humor flows between the iris and lens. It also results in hasty raise of IOP. This requires immediate medical treatment as early as possible [3]. Figure 4 shows the angle-closure glaucoma.

2.4 Normal Tension Glaucoma

This is also known as low-tension glaucoma. In this, the IOP is normal, but it causes optic nerve damage and visual field loss. This type of glaucoma occurs due to the poor blood flow to the optic nerve, and it leads to the demise of optic cells [4]. Figure 4 shows the normal tension glaucoma.

2.5 Congenital Glaucoma

This type of glaucoma is very more common even in children also. So, it is also called as children glaucoma. It classifies as two types: Primary congenital glaucoma is the result of incomplete or abnormal development of the eye's drainage canal. Secondary congenital glaucoma causes the disorders in the eye or body. Moreover, it may lead as genetic disorder [5]. Figure 4 presents the congenital glaucoma.

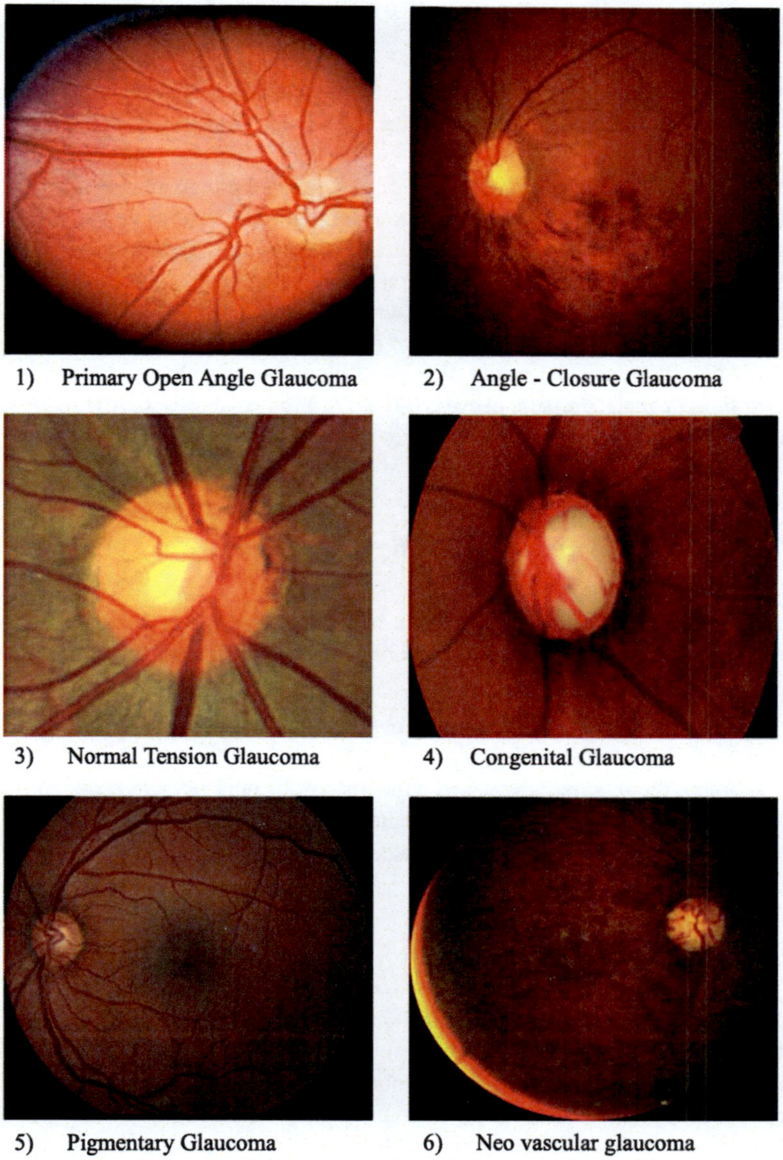

Fig. 4 Types of glaucoma

2.6 Secondary Glaucoma

Glaucoma that results from another disease is called secondary glaucoma. Pigmentary and neovascular glaucoma are secondary glaucoma. These are described in detail below:

Pigmentary Glaucoma: In this type of glaucoma, the pigment granules present in the back of the iris enter into the aqueous humor that flows inside the eye. These flow into the drainage canals and clog them slowly, which leads to increase in eye pressure [6]. Figure 4 represents the pigmentary glaucoma.

Neovascular Glaucoma: Abnormal development of blood vessels over the iris and the drainage canals causes this type of glaucoma. It always associated with other diseases like diabetes. The newly developed blood vessels block the liquid to exit from drainage canals; pressure in eye is raised [7]. Figure 4 represents the neovascular glaucoma.

2.7 A Typical Glaucoma Detection Technique

A conventional framework of glaucoma detection technique is depicted in Fig. 5. The steps involved in detection of glaucoma are characterized as follows:

(a) Read the given image.
(b) Preprocess the given image based on the extraction techniques like region of interest(ROI) and sampling.
(c) Segment the required areas, i.e., optic disk, optic cup, retinal vessel, etc.
(d) Extract the features from the segmented potions.
(e) Recognize or classify the glaucomated eye from normal one.

3 Feature Extraction Methods

Optic nerve head (ONH) is an elliptical-shaped area located at the entrance of the optic nerve at the back of the eye [8]. There are many parts of ONH such as disk, cup, neuro-retinal rim, blood vessels, retinal nerve fiber layer (RNFL) and parapapillary atrophy (PPA) [9] that can be observed to distinguish between normal and glaucoma eyes by extracting the features from it. The following are the few features that can be extracted to detect glaucoma:

3.1 CDR (Cup-to-Disk Ratio)

The CDR can be used as a factor to detect glaucoma disease [10]. Commonly, the CDR value should be less than 0.45 for normal people, but for glaucoma patients, it

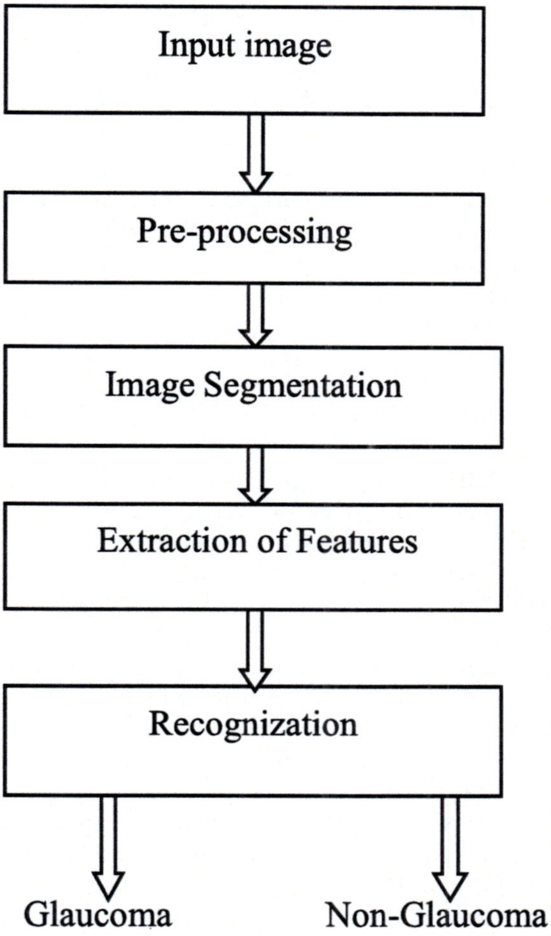

Fig. 5 Block diagram for glaucoma detection

will be greater than 0.45 [11]. If the size of the cup increases, then it will also affect the neuro-retinal rim (NRR).

In this work [12], the authors have found vasculature within the optic disk, and then, ISNT ratio is measured. With the help of ISNT rule, the blood vessels ratio in each quadrant is found. Finally, the results are evaluated like the blood vessels are nasalized (violated) or obeyed ISNT rule.

The authors [13] used active contour method to determine the pathological process of glaucoma from fundus color images to find CDR. The blood vessels are removed, and ONH normalization is done to determine the cup and disk areas so that CDR ratio can be calculated. Although it finds pathologies, it does not work for some fundus images due to the presence of other additional pathologies. The approach in this article can be improved by using some other preprocessing steps.

The work proposed in [14] detects the location of the optic cup properly. The disk segmentation and vessel segmentation are done by thresholding and edge detection, respectively, and for cup segmentation, a method is used for cup quality and vessels. The CDR values obtained from proposed areas, specialist doctors and ratios of ratio are compared and obtained an accuracy of 88.5%.

In this work [15], the author proposed an efficient tool to detect glaucoma using CDR ratio from G-R color channels. The CDR values are computed for four different color channel combinations. The four color channel combinations comparatively represent a large difference in CDR values for glaucoma and non-glaucoma fundus images.

3.2 *Neuro-Retinal Rim (NRR)*

The NRR is the region between the rims of the optic cup to the optic disk. In glaucomated eye, the ratio between the NRR region covered by nasal and temporal is thick than the NRR region covered in the inferior and superior regions [16].

The work proposed in [17] calculated the BM0-MRW from SDOCT using DL prediction. The deep learning (DL) algorithm was compared with actual SDOCT values to quantify the amount of NRR damage. The work obtained the accuracy 0.945 and 0.933 using DL and SDOCT global BMO-MRW measurements (Fig. 6).

The proposed work presented in [18] evaluated the BMO-MRW and RNFLT thickness to identify the glaucoma. The proposed work outperformed well using supero-temporal and infero-temporal RNFLT. It produced a higher sensitivity in detecting glaucoma 98.4%.

In this work [19], the authors have proposed a novel method to perform severity of glaucoma using RDR. In this work, cup–disk area is segmented and computed the RDR and DDLS (disk damage likelihood scale) to identify severity of glaucoma assessment to capture loss of NRR fiber accurately. The work obtained a 8–10% higher performance in terms of RDR.

Fig. 6 NRR

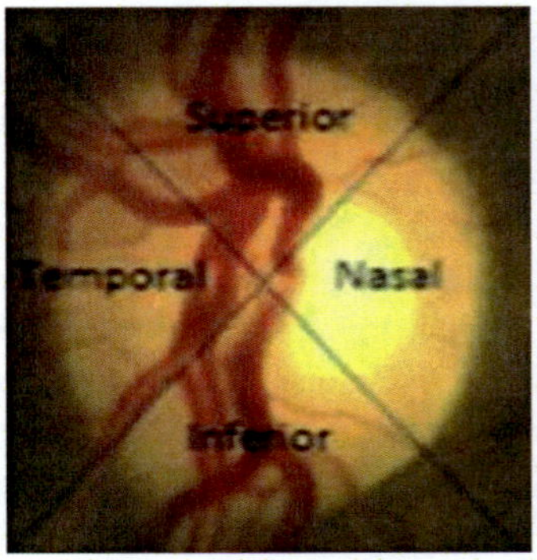

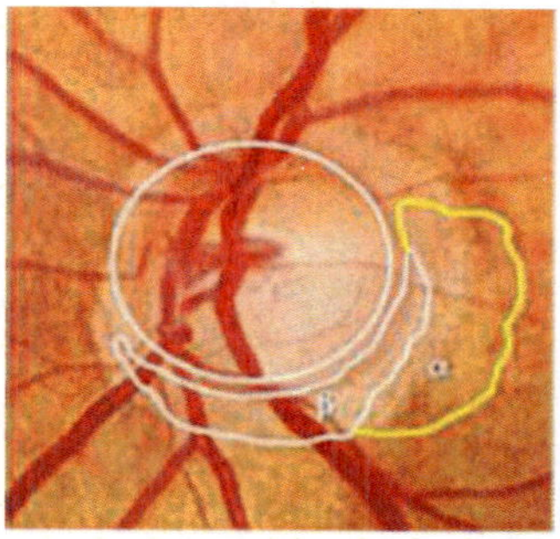

Fig. 7 PPA with α-zone and β-zone

3.3 Parapapillary Atrophy (PPA)

PPA is a part with crescent-shaped, which consists of alpha zone and beta zone. Alpha zone is not affected by glaucoma, whereas beta zone is affected more frequently in patients with glaucoma, which causes a white color layer of the eye. The beta zone directly makes a bordering the disk on the temporal side, and the alpha zone directly occurs adjacent to the beta zone [10] (Fig. 7).

In this work [20], the authors recognized the glaucomated eye using different axial lengths (AL) of beta zone PPA in enhanced-depth imaging OCT ONH B-scans. The region between the BMO and RPE tips inside SDOCT radial B-scans is identified as β-PPA area. The larger β-PPA area indicates that glaucoma diagnosis and has a larger axial length.

The work proposed in [21] has associated OCT-defined PPA zone values in Enface reconstructed SS-OCT fundus images of highly myopic eyes. The capability of each PPA zone is obtained with receiver operating characteristic (ROC) curve and obtained β-zone and the gamma zone areas as 0.686 and 0.560, respectively.

3.4 Retinal Nerve Fiber Layer (RNFL)

This layer appears as a bunch of scratches (like a colored light) is been dispersed evenly on the eye. In a normal eye, RNFL mostly located within the inferior nasal area, followed by the superior nasal, inferior temporal and superior temporal [22].

The work proposed in [23] tree is extracted using unsupervised segmentation algorithm—like grid from retinal blood vessel structure. Then, quantify structural changes among graphs extracted from fundus images. The fractal analysis is also applied to characterize the extracted group. By applying these techniques on three fundus images, there will be significant changes among the normal and glaucoma groups (Fig. 8).

In this work [24], the author proposed a computerized method to extract blood vessels from retinal images. Median filter and CLAHE are used to enhance the image.

Fig. 8 Structure of RNFL

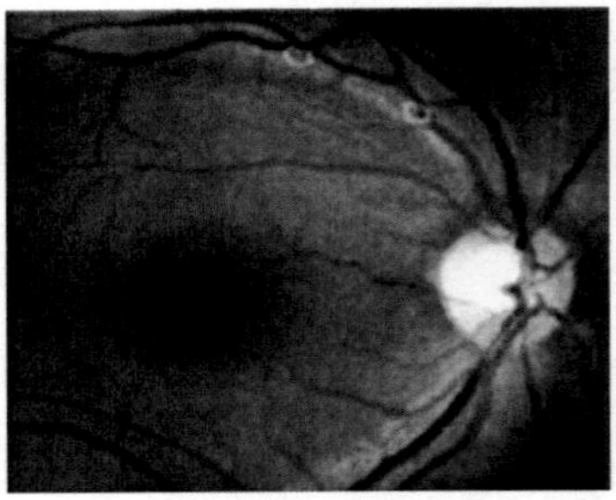

Mean-C thresholding technique is used to segment retinal blood vessels. Lastly, the isolated pixels are removed by morphological cleaning operation. Finally, the DRIVE and CHASE_DB1 databases obtained accuracies of 0.955 and 0.954, respectively.

In the proposed work [25], the authors have presented the variations in thickness of RNFL based on SDOCT in multicultural Asian residents (Indian, Chinese and Malay). The RNFL thickness is calculated using SDOCT, and the people with poor quality are eliminated. The association of ocular and systemic factors used linear regression model to average RNFL thickness. Increase in disk area, eye sight strength and low cholesterol are linked to the larger RNFL thickness. The Indian has thinner average RNFL compared with Chinese, whereas for Malay and Chinese have no difference in average RNFL.

Detection of glaucoma using fundus images is carried out by measuring CDR ratio, ISNT rule, NRR, RNFL, PPA, which are obtained by extracting the cup and disk to calculate their vertical height ratios. In order to calculate the cup and disk ratio, various methods of glaucoma detection schemes available in literature are studied and tabulated its inferences in Table 1.

4 Conclusion

One of the major causes of blindness nowadays is glaucoma. It is observed that the structure of retina gets change, and it results in intensity and texture changes. This article has presented the recognition methods based on the quantifiable features extracted from the structure of the retinal eye. This article has discussed causes of glaucoma and its various types. Also, it discussed the various features which are helping in recognizing the glaucoma affected eye from the normal one. This article has discussed various methods of glaucoma detection using especially cup and disk ratio features and its performances. This study shows that to improve the classification performance, there are still more features. Optic cup segmentation considers more tiresome compared to optic disk because of merge of optic cup with neighboring tissues and blood vessels. Hence, improved segmentation and classification techniques are needed to be proposed to diagnose the glaucoma disease.

Table 1 State of art of various glaucoma detection schemes and its performances

References	Authors and year	Proposed method	Performance
[14]	Carrillo et al. (2019)	Optic cup, disk, blood vessels are segmented, and CDR ratio is measured	It achieved 88.5% of accuracy
[12]	Khan et al. (2019)	Blood vessels are found in optic disk is segmented and ISNT ratio calculated	It was tested on different image dataset to verify whether the retinal images are obeying ISNT rule
[15]	Patel et al. (2018)	The areas of cup, disk are segmented, and the CDR ratio of each channel (G-R combinations) is measured	Combination of both G-R channels produced more difference in CDR values
[17]	Thomsan et al. (2019)	BMO-MRW from SDOCT was predicted in CNN algorithm	High accuracy and eliminates human participation
[13]	Mishra et al. (2011)	Blood vessels are removed. Cup and disk are set by thresholding and active contour methods. CDR ratio is measured	For all pathological images, the CDR ratio is measured correctly and detected glaucoma accurately by 97%
[18]	Zheng et al. (2018)	BMO-MRW and RNFLT with OCT are used to detect glaucoma	Sensitivities and specificities of mild glaucoma (97.9 and 88.4%) and glaucoma (98.4 and 93.6%)
[19]	Kumar et al. (2019)	The areas of cup–disk are segmented, and RDR and DDLS are computed	The results of RDR posses 8–10% more accuracy compared to CDR technique
[23]	Amil et al. (2019)	Tree-like graph is extracted from unsupervised segmentation algorithm, and fractal analysis is applied to group	Segmentation methods and image resolution are sensitive
[24]	Dashi and Bhoi (2017)	CLAHE are identified, and noise and blood vessels are removed with median filter and C mean-thresholding	DRIVE and CHASE_DB1 databases obtained an efficiency (1.66 and 1.98) and accuracy (0.955 and 0.954), respectively
[20]	Manalastas et al. (2018)	Depth imaging OCT ONH radial B-scans are associated with AL	Larger βPPA will lead to risk in glaucoma with range of AL

(continued)

Table 1 (continued)

References	Authors and year	Proposed method	Performance
[21]	Miki et al. (2019)	SS-OCT volumetric scans are done to measure the beta and gamma zone PPA	OCT was associated with beta zone to detect glaucoma and AL, whereas gamma zone only detects AL
[25]	Ho et al. (2019)	Measure RNFL thickness based on SDOCT for three different regional people	RNFL thickness for Indians eyes is comparatively less than Chinese and Malay eyes. Sensitivity and specificity are improved
[26]	Sarkar et al. (2017)	Threshold-based approach	Accuracy 97.58% Improved performance than the existing methods
[27]	Thakur et al. (2017)	Clustering-based approach Extracted the CDR features	Accuracy of optic cup 91.9 Accuracy of optic disk 94.46
[28]	Tan et al. (2015)	Clustering-based approach Improved intersection region, cup location and reduced intersection error	Obtained 95.0% of accuracy Increased computation cost

References

1. Nayak, J., Acharya, U. R., Bhat, P. S., Shetty, N., & Lim, T. C. (2009). Automated diagnosis of glaucoma using digital fundus images.
2. Rhee, D. J. (2017). Primary open angle glaucoma.
3. https://doi.org/onlinelibrary.wiley.com/doi/abs/https://doi.org/10.1111/aos.13879.
4. www.aao.org/eyenet/article/normal-tension-glaucoma-part-one-diagnosis.
5. www.imo.es/en/congenital-glaucoma.
6. Hager, J., & Alward, W. L. M. (2014). Pigmentary glaucoma.
7. Mangan, R. (2015). Neovascular glaucoma stages.
8. Marsh, B. C., Canntor, L. B., WuDunn, D., Hoop, J., Lipyanik, J., Patella, V. M., Budenz, D. L., Greenfield, D. S., Savell, J., Schuman, J. S., & Varma, R. (2012). Optic nerve head (ONH) topographic analysis by stratus OCT in normal subjects: Correlation to disc size, age and ethnicity.
9. Khan, F., Khan, S. A., & Yasin, U. U. (2013). Detection of glaucoma using retinal fundus images.
10. Choplin, N. T., & Lundy, D. C. (Eds.). (2007). *Atlas of glaucoma* (2nd ed.). London: Informa Healthcare.
11. http://www.optic-disc.org.
12. Khan, F., Sharif, S., Khan, F. M. A., & Haq, I. U. (2019). Blood vessels quantification to detect glaucoma using retinal fundus images.
13. Mishra, M., Nath, M. K., & Dandapat, S. (2011). Glaucoma detection from color fundus images.
14. Carrillo, J., Bautista, L., Villamizar, J., Rueda, J., Sanchez, M., & Rueda, D. (2019). Glaucoma detection using fundus images of the eye.
15. Patel, S. C., & Patel, M. I. (2018). Analysis of CDR of fundus images for glaucoma detection.

16. Kavitha, S., & Karthikeyan, S. (2010). Neuroretinal rim quantification in retinal images to detect glaucoma.
17. Thompson, A. C., Jammal, A. A., & Medeiros, F. A. (2019). A deep learning algorithm to quantify neuroretinal rim loss from optic disc photographs.
18. Zheng, F., Yu, M., & Leung, C. K. S. (2018). Diagnostic criteria for detection of retinal nerve fibre layer thickness and neuroretinal rim width abnormalities in glaucoma.
19. Kumar, J. H. R., Seelamantula, C. S., Kamath, Y. S., & Jampala, R. (2019). Rim-to-disc ratio outperforms cup-to-disc ratio for glaucoma prescreening.
20. Manalastas, P. I. C., Belghith, A., Weinreb, R. N., Jonas, J. B., Suh, M. H., Yarmohammadi, A., Medeiros, F. A., Girkin, C. A., Liebmann, J. M., & Zangwill, L. M. (2017). Automated beta zone parapapillary area measurement to differentiate between healthy and glaucoma eyes.
21. Miki, A., Ikuno, Y., Weinreb, R. N., Asai, T., Usui, S., & Nishida, K. (2019). En face coherence tomography imaging of beta and gamma parapapillary atrophy in high myopia.
22. Morrison, C. J., & Pollack, P. I. (2003). *Glaucoma science and practice*. New York: Thieme Medical Publishers.
23. Amil, P., Reyes-Manzano, C. F., Guzman-Vargas, L., Sendina-Nadal, I., & Masoller, C. Network based features for retinal fundus vessel structure analysis.
24. Dashi, J., & Bhoi, N. (2017). A thresholding based technique to extract retinal blood vessels from fundus images.
25. Ho, H., Tham, Y. -C., Chee, M. L., Shi, Y., Tan, N. Y. Q., Wong, K. -H., Majithia, S., Cheung, C. Y., Aung, T., Wong, T. Y., & Cheng, C. Y. (2019). Retinal nerve fiber layer thickness in a multiethnic normal Asian population: The Singapore epidemiology of eye diseases study.
26. Sarkar, D., & Das, S. (2017). Automated glaucoma detection of medical image using biogeography based optimization. In *Proceedings of Advances in Optical Science and Engineering, Springer Singapore*, pp. 381–388.
27. Thakur, N., & Juneja, M. (2017). Clustering based approach for segmentation of optic cup and optic disc for detection of glaucoma. *Current Medical Imaging, 13*(1), 99–105.
28. Tan, N. M., Xu, Y., Goh, W. B., & Liu, J. (2015). Robust multi-scale super pixel classification for optic cup localization. *Computerized Medical Imaging and Graphics, 40,* 182–193.

TEAP-Technically Equipped Agricultural Practice

N. Mangathayaru, M. Lakshmi Lashita, D. Sri Pallavi, and C. Mithila Reddy

Abstract Agriculture is one of the major sectors of the Indian economy. More than 50% of the entire population of India depends on agriculture. The main aim of this study is to improve farm produce by using analytics and real-time data monitoring with the help of Internet of Things (IoT). The Internet of Things has given an opportunity to farmers to cultivate soil with the use of cheap, easy-to-install sensors and an abundance of intuitive data that is offered by the sensors. Expanding on this yielding buildup of the Internet of Things in agriculture, smart farming applications are increasing ground with the promise to deliver 24/7 visibility into soil and crop health.

Keywords Agriculture · Farmers · Internet of things · Analytics

1 Introduction

Agriculture plays a vital role in India's economic prosperity. The conventional methods in agricultural practices have become grossly inadequate to cater to the increasing needs. Thus, it has become imperative to adopt novel technologies to raise agricultural standards. This project proposes a smart farming system in a limited, enclosed area wherein different sensors are strategically positioned to measure parameters such as moisture content, temperature, pressure, light intensity and pH of the soil. Smart Farms is a NodeMCU-ESP8266 based remote sensor data measurement and monitoring system with a cloud interface (MQTT-broker) and user control [1, 2]. The real-time data from the remote sensors is collected by the NodeMCU and is sent to the remote server through wireless connection from which it is sent to the mobile application. With the help of this mobile application, the user will be able to gather necessary information about the soil. Smart Farms helps the users in taking quick and informed decisions regarding plant control by providing all the relevant data in an intuitive format.

N. Mangathayaru (✉) · M. Lakshmi Lashita · D. Sri Pallavi · C. Mithila Reddy
IT, VNRVJIET, Hyderabad, Telangana, India
e-mail: mangathayaru_n@vnrvjiet.in

S. C. Satapathy et al. (eds.), *Intelligent System Design*, Advances in Intelligent Systems and Computing 1171, https://doi.org/10.1007/978-981-15-5400-1_37

2 Literature Survey

Agriculture is a growing activity that largely serves the human population, improving agriculture to make the lives of farmers better and also secure the future of agriculture.

The idea of smart farming began with the introduction of a water motor controlled by phone [3].

This was done through GSMA technology where an SMS would go from the farmer's mobile to the motor setup with a message receiver; this would enable the farmer to switch on and off his motor without actually going to the field.

Now there is more research going on precision agriculture and smart farming, which tries to fulfill the need of reducing the usage of pesticides and fertilizers. Precision agriculture enables the farmer to know how much amount of fertilizer needs to be added to the soil. Smart farming enables the usage of applications over the network of things and lets the farmer do multiple things at a time [5].

This is one such study where we can bring a change in both health of the soil, health of the crop and health of the consumer [6].

3 Proposed Method

A model of the system that is a combination of hardware and software components. The hardware of the system includes the sensors for the data acquisition, processing node and the server for the processing and a smart phone for the control and the display of the output. The software of the system includes the Arduino IDE used for programming the sensors for the data collection, Android Studio for building the mobile application, Jypyter Notebook for data analytics using python and the Cloud MQTT Server which acts as the broker between the clients [4].

The application determines the quality of the soil based on Nitrogen, Phosphorous and Potassium values. The temperature, humidity and moisture content in the soil along with pH values are obtained from the sensors. According to the pH value, the values of NPK are determined and the application shows whether the NPK values are favourable or not for that particular crop. Monitoring the moisture content in the soil will help in the automation of irrigation. By using a moisture sensor one can irrigate the crops with a sufficient amount of water neither with less amount nor with high amount. Also, the application provides an option for switching on/off the motor.

4 Procedure

4.1 STAGE 1: Programming the NodeMCU

The NodeMCU is programmed to retrieve parameter values from the DHT-22 sensor, pH sensor, Moisture Sensor for temperature and humidity, pH and moisture content values. Arduino IDE is used for programming and the MQTT publish and subscribe methods are used for sending the data collected and receiving control instructions from the Cloud MQTT-Broker.

4.2 STAGE 2: Setting up the Cloud MQTT

A Cloud MQTT instance is created. The instance information contains the server name, username, password, port number which will be used in both NodeMCU programming as well as in the development of Android mobile application to connect both the clients with the broker for the purpose of communication. The core of the project is the MQTT protocol. MQTT stands for Message Queuing Telemetry Transport Protocol. It has two components—Client and Broker. In the project, the NodeMCU and the mobile application act as clients, and CloudMQTT is the Broker that manages the communications between the clients with help of publish and subscribe functions. The mobile application subscribes for topics such as temp, humidity, moisture and ph whose values are published by the NodeMCU from the sensors.

4.3 STAGE 3: Analysis

The process of evaluating data using analytical and logical reasoning to examine each component of the data provided. This form of analysis is just one of the many steps that must be completed when conducting a research experiment. Data from various sources is gathered, reviewed, and then analyzed to form some sort of finding or conclusion.

By making use of the concept of correlation and the scatter plot graphs for ph versus crop production, Nitrogen level versus crop production, Phosphorous versus crop production, Potassium versus crop production, the range at which these values are suitable for the high crop production is determined and this is using while testing whether the real-time values obtained are favourable for the crop or not.

4.4 STAGE 4: Mobile Application Development

The mobile application is developed in Android and it makes use of Paho Library which integrated MQTT functions into Android. The publish method is used to send ON/OFF instructions to the relay to switch on/off the motor and the subscribe method is used to get the temperature, humidity, ph and the moisture values and these values are stored with help of SQLite database and are displayed along with the *N*, *P*, *K* values obtained after correlation. Graphs are also generated showing the change in the values over a period of time. Finally, the output determines whether the values of Nitrogen, Potassium, Phosphorous and pH are favourable or not for that particular crop.

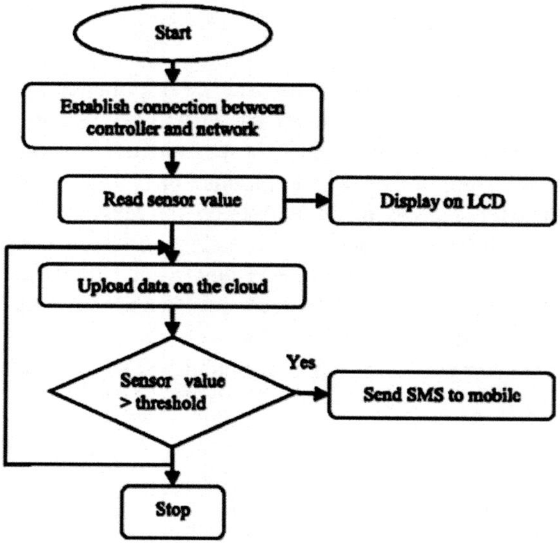

5 Results

See Figs. 1 and 2.

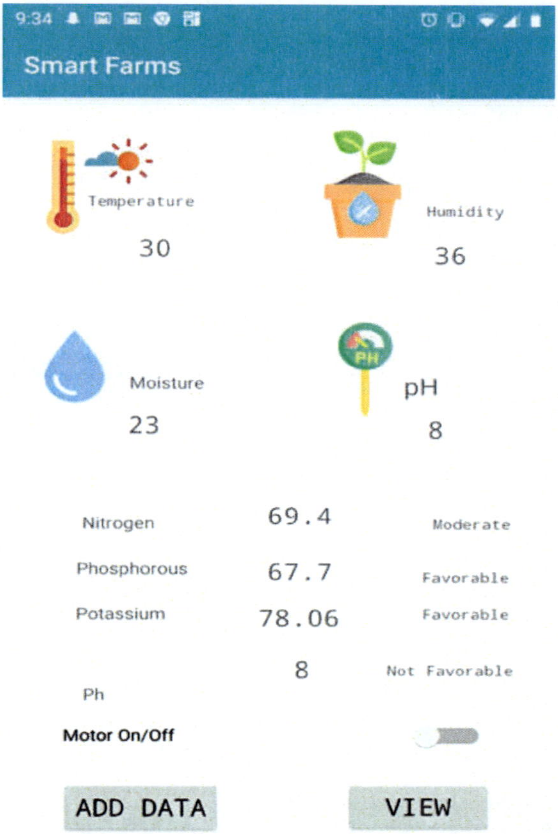

Fig. 1 Display of the data collected by the sensors and prediction of NPK values

6 Conclusion and Future Work

The project focuses on the automation of conventional methods of agriculture practices and it is devised in such a way that it could be setup by any individual at minimum cost. Thus, productivity is enhanced as the favourable environmental conditions required for the plants can be identified. The mobile application simplifies the complete process and provides a way to remotely operate and collect data.

In future, we intend to expand our project in:

- Crop analysis
- Supply chain management.

Crop Analysis:
Further study will attempt to quantify past weather and technological factors, analyze the effects of NPK factors on yields, predict future crop yields based on projected

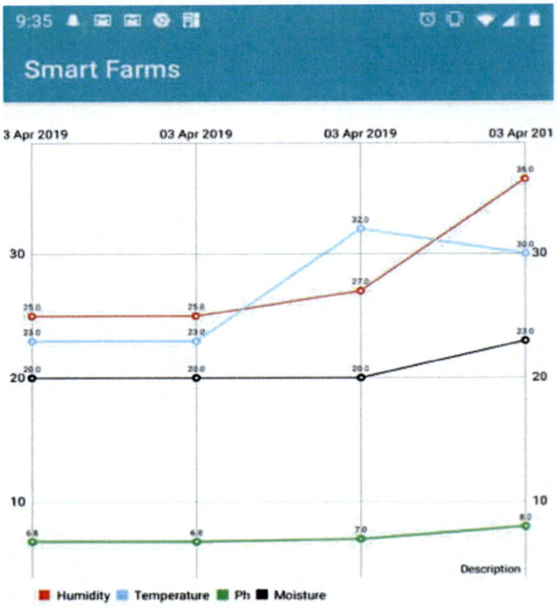

Fig. 2 Display of temperature, humidity, moisture and pH values in a form of graph over the time period

levels of these factors and will also derive estimates of the parameters of the elements associated with these yield predictions.

Supply chain management:
The usage of provision services in several applications like producing, retail and trade government utilities like soil testing is anticipated to rise significantly throughout the forecast amount that, successively, is calculable to strengthen the expansion of the provision market globally.

Bringing such improvements in our project would help a lot of farmers, it would also help to eradicate the middle person between the farmer and the buyer in the supply chain management, which will result in the prosperity of the farmer and also decrease in the pricing of natural products.

Predicting future crops based on the amounts of elements present in the soil will help the farmers produce more organic and healthier crops, which benefits both the farmer and all the people who consume natural products.

Predicting future crop yields has become a necessity as the minerals are getting depleted after every crop growth, hence we need to practice crop rotation, few crop use a specific kind of mineral and others use a different kind, but each crop produces a different mineral when it uses some minerals from the soil, hence when crop rotation is practiced, these factors are taken into consideration and then the analysis is done to produce a flawless report.

References

1. Lin, J. S., & Liu, C. Z. (2008). A monitoring system based on wireless sensor network and an SoC platform in precision agriculture. In *2008 11th IEEE International Conference on Communication Technology* (pp. 101–104). IEEE.
2. Kaewmard, N., & Saiyod, S. (2014). Sensor data collection and irrigation control on vegetable crop using smart phone and wireless sensor networks for smart farm. In *Proceedings of the International Conference on Wireless Sensors (ICWiSE)*, pp. 106–112.
3. Kassim, M. R. M., Mat, I., & Harun, A. N. (2014). Wireless sensor network in precision agriculture application. In *2014 International Conference on Computer, Information and Telecommunication Systems (CITS)* (pp. 1–5). IEEE.
4. Taylor, K., Griffith, C., Lefort, L., Gaire, R., Compton, M., Wark, T., et al. (2013). Farming the web of things. *IEEE Intelligent Systems, 28*(6), 12–19.
5. Gondchawar, N., & Kawitkar, R. S. (2016). IoT based smart agriculture. *International Journal of advanced research in Computer and Communication Engineering, 5*(6), 838–842.
6. Rajalakshmi, P., & Mahalakshmi, S. D. (2016). IOT based crop-field monitoring and irrigation automation. In *2016 10th International Conference on Intelligent Systems and Control (ISCO)* (pp. 1–6). IEEE.

A Comparatives Study of Protein Kinase Domain Regions for MAPK 1-14

Deepak Nedunuri, S. M. B. Chowdary, S. Jaya Prakash, and M. Krishna

Abstract Most of the MAPKs activity is to stimulate a wide variety of signals which include mitogens, growth factors, cytokines, T cell antigens, pheromones, UV and ionizing radiations, osmotic stress, heat shock, oxidative stress and various other factors. MAPKs acts on most non-nuclear oncogenes and is found to be liable for growth factors response towards cellular stimuli. MAPK proteins participating in different pathways are extremely crucial in various dental and periodontal disorders with induced chronic pain. Centred on evolutionary relationships, domains detection, phosphorylated amino acid sites identification and evaluation of amino acids within the active site lining cavity, the role of MAPK 1 and 3 contributing in growth factor pathway, MAPK 8, 9 and 10 participating in stress-related pathway and MAPK 11, 12 and 13 appearing in inflammatory pathway are reported. Most of the research has concentrated on focusing MAPK 14 which is p38 MAPK isoform alpha.

Keywords MAPK's · Oncogenes · Pathways · P38 MAPK isoform

1 Introduction

Most MAPK's activity is transiently (minutes-hours) enthused by a wide variety of signalling event by mitogens specifically, few growth factors, a number of cytokines, T cell mediated antigens, pheromones, a minute amount of phorbol esters, UV as well as ionizing radiation, osmotic external stress and heat mediated shock, oxidative involved stress and other factors. Like the degree of sensitivity, various metrics were used to characterize the MAPKs. These MAPKs were based on serine/threonine kinases which include phosphorylation events on substrates particularly containing proline amino acid in the consensus motif X [ST] P involving $P + 1$ substrate recognition site. Hence, these sort of protein kinases are integral transduction modules which are commonly having the additional elements such as MAPK kinases also called as MKKs in turn called as MEKs, SAPK kinases called as SKKs, in turn,

D. Nedunuri (✉) · S. M. B. Chowdary · S. J. Prakash · M. Krishna
Department of CSE, Sir C R Reddy College of Engineering, Eluru, India
e-mail: nedunurideepak@gmail.com

S. C. Satapathy et al. (eds.), *Intelligent System Design*, Advances in Intelligent Systems and Computing 1171, https://doi.org/10.1007/978-981-15-5400-1_38

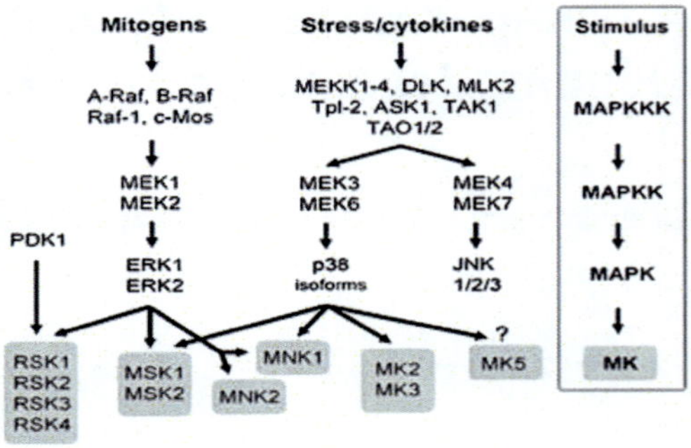

Fig. 1 Cascade of signalling events activating MKs

called as MKKs or SEKs, which were dual specific phosphorylating MAPK's on amino acid threonine and tyrosine residue of a conserved motif, TXY. MEKs and SEKs were collectively called as MKKs in this context. MKKKs were the results of MKK kinases which were phosphorylated by MKKs. For transude signals core units like MKKK-MKK-MAPK generate the cell surface to the core nucleus. For the additional switching elements, they provide an intermediate switch. Most of the inducible transcription factors and products of early genes were the substrates of MAPK. MAPKs were also included in cytoplasmic proteins, e.g. phospholipase A2 and transmembrane proteins, e.g. EGF receptor [1, 2]. Protein phosphorylation mechanism, the well-known post-translational variations, plays a central part in all signalling mechanisms. This further involves the transmission of a phosphate ion to the objective protein from adenosine triphosphate (ATP). The activity of MAPKs was regulated in three tired cascades which were composed of MAPK kinase (MAPKK, MKK or MEK) and MAPKK kinase or MEK kinase (MAPKKK or MEK). These cascades culminate the activity of MAP kinase. MAP kinase can phosphorylate many different proteins after the translocation of nucleus. MAPK activation process in diverse cells might result in a like or dissimilar response. The rationale of the work is to evaluate multiple sequence alignments of all MAPKs to discover the dissemination of specificity of MAP kinases in specified pathways [3] (Fig. 1).

2 Materials and Methods

Different MAP kinase proteins were involved in various factors like growth factor, stress and inflammatory-related pathways which were obtained from biocarta [4]. Of different MAP kinases (MAP4K to MAPK) that take part in three pathways, MAPK with 12 proteins are selected as shown in Fig. 2 [5].

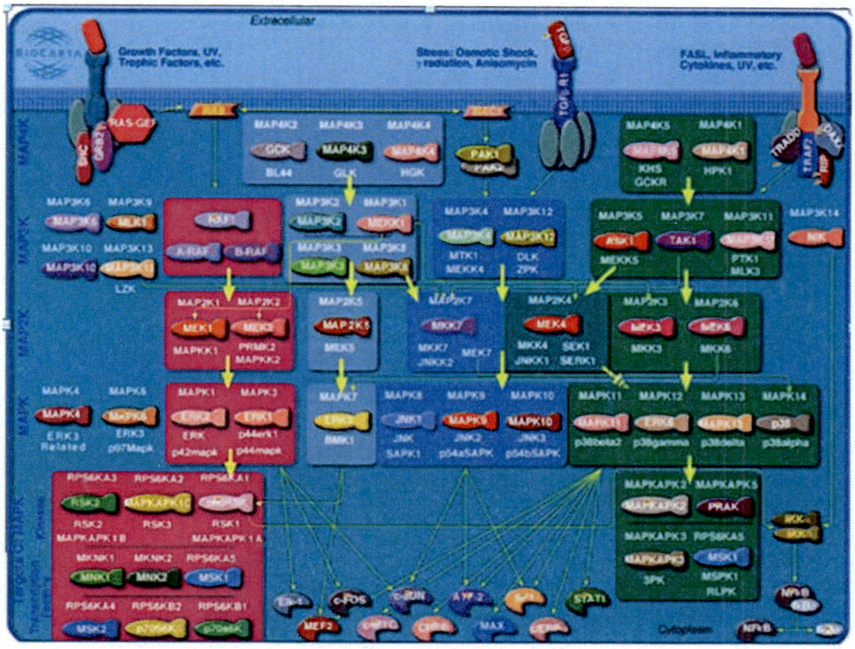

Fig. 2 MAP kinase proteins involved in various factors

2.1 MAPK Sequences in Fasta Format

MAPKs protein sequences were extracted from ExPASy server [6, 7].

2.2 Sequence Alignments

A validation has been written in the program in such a way that a warning message appears if the lengths of the input sequences are unequal. Major emphasis has been made in making the python program to read the scoring matrices, PAM and BLOSSUM. This is because the matrix file is created in a tab-delimiting format. In other words, five spaces between each value of the corresponding amino acids are presented in the form of a matrix. These values are taken from NCBI or EBI websites. The ambiguity codes of bases or residues are taken from IUPAC and IUBMB ambiguity codes. An image of entry P28482 is given in Fig. 3.

Fig. 3 Entry P28482 as seen in Swiss-Prot database

3 Results and Discussion

Major unit of MAPK pathway is a tired cascade which includes three stages, within the stages, they include kinase modules. The MKKs were activated and phosphorylated by the MAPKs. By nature, these MKKs were dual-specificity kinases that can catalyze both tyrosine and the onine residues by phosphorylation. This serine/the onine can be used to activate the MKKs themselves [8]. The gene expressions, metabolism, cytoskeletal functions and other cellular regulatory events controlled by the activation of MAPKs. Most of the MAPKs present the difficult events which include mitogenesis, differentiation, survival and migration [9].

3.1 MAPK 1 and 3

From both MAPK 1 and 3, the pattern of substitution is identical, which defines the role in growth factor related pathway. The participation of these substitutions appeared in MAP kinase domain region.

3.2 MAPK 8, 9 and 10

Stress-related pathway is suggested by MAPK 8, 9 and 10 proteins in the substitution at six functionally active site residues. In the comparison of 1YQJ the properties of residue active site is changed significantly but one substitution at Leu104 is not changed.

3.3 *MAPK 11, 12, 13 and 14*

Since the protein 1YQJ was the presentational member of MAPK 14, as a reason for remaining proteins nearer and available substitutions are observed. Though a single residue substitution appears at His107 in MAPK 14 and threonine in MAPK 11, both of them represent similar function. On the other hand, five residue substitutions in MAPK 12 and 13 at positions Thr106, His107, Leu108, Ala111, Ser154 and Gly110 (only in MAPK 13) hint that they possibly represent a sub-group under inflammatory pathway. Though MAPK 11, 12, 13 and 14 are clustered as a single group, the multiple alignment study refers MAPK 12 and MAPK 13 which are diverged and mostly they have various functionalities.

3.4 *MAPK 4 and 6*

MAPK 4 and 6 are not determined to any pathway even though they perform kinases activity as they also belong to the ATP kinase and Ser/Thr kinase domain regions. Several mutations were identified to describe that they are similar proteins.

3.5 *MAPK 7*

From Fig. 2, it is observed that the MPAK 7 pathway appears to be nearer to growth or stress pathway. Based on the residues MAPK 1 and 3 versus MAPK 7 it can be explained. Such similar observation is observed at MAPK 4 and 6 of 1YQJ residues Leu104 and Gly110. From the observation, it is suggested that MAPK 7 can be a little bit closer group to MAPK 1 and 3 and as well as MAPK 4 and 6.

3.6 *Multiple Sequence Alignments of MAPK 1-14*

```
sp|P28482|MK01_HUMAN      -----------------------MAAAAAAG-----------AGP------EMVRGQVFD 20
sp|P27361|MK03_HUMAN      -----------------------MAAAAAQGGGGGEPRRTEGVGPGVPGEVEMVKGQPFD 37
sp|P53778|MK12_HUMAN      --------------------------------------MSSPPPARSGFYRQEVTKTAWE 22
sp|O15264|MK13_HUMAN      --------------------------------------MS--LIRKKGFYKQDVNKTAWE 20
sp|Q15759|MK11_HUMAN      --------------------------------------MS---GPRAGFYRQELNKTVWE 19
sp|Q16539|MK14_HUMAN      --------------------------------------MS---QERPTFYRQELNKTIWE 19
sp|P45983|MK08_HUMAN      --------------------------------------MSR-SKRDNNFYSVEIGDSTFT 21
sp|P53779|MK10_HUMAN      MSLHFLYYCSEPTLDVKIAFCQGFDKQVDVSYIAKHYNMSK-SKVDNQFYSVEVGDSTFT 59
sp|P45984|MK09_HUMAN      --------------------------------------MSD-SKCDSQFYSVQVADSTFT 21
sp|Q13164|MK07_HUMAN      ----------MAEPLKEEDGEDGSAEPPGPVKAEPAHTAASVAAKNLALLKARSFDVTFD 50
sp|P31152|MK04_HUMAN      ---------------------------------------------MAEKGDCIASVYGYD 15
sp|Q16659|MK06_HUMAN      ---------------------------------------------MAEKFESLMNIHGFD 15
                                                                                :

sp|P28482|MK01_HUMAN      VGPRYTNLSYIGEGAYGMVCSAYDNVNKVRVAIKKIS-PFEHQTYCQRTLREIKILLRFR 79
sp|P27361|MK03_HUMAN      VGPRYTQLQYIGEGAYGMVSSAYDHVRKTRVAIKKIS-PFEHQTYCQRTLREIQILLRFR 96
sp|P53778|MK12_HUMAN      VRAVYRDLQPVGSGAYGAVCSAVDGRTGAKVAIKKLYRPFQSELFAKRAYRELRLLKHMR 82
sp|O15264|MK13_HUMAN      LPKTYVSPTHVGSGAYGSVCSAIDKRSGEKVAIKKLSRPFQSEIFAKRAYRELLLLKHMQ 80
```

```
sp|Q15759|MK11_HUMAN      VPQRLQGLRPVGSGAYGSVCSAYDARLRQKVAVKKLSRPFQSLIHARRTYRELRLLKHLK 79
sp|Q16539|MK14_HUMAN      VPERYQNLSPVGSGAYGSVCAAFDTKTGLRVAVKKLSRPFQSIIHAKRTYRELRLLKHMK 79
sp|P45983|MK08_HUMAN      VLKRYQNLKPIGSGAQGIVCAAYDAILERNVAIKKLSRPFQNQTHAKRAYRELVLMKCVN 81
sp|P53779|MK10_HUMAN      VLKRYQNLKPIGSGAQGIVCAAYDAVLDRNVAIKKLSRPFQNQTHAKRAYRELVLMKCVN 119
sp|P45984|MK09_HUMAN      VLKRYQQLKPIGSGAQGIVCAAFDTVLGINVAVKKLSRPFQNQTHAKRAYRELVLLKCVN 81
sp|Q13164|MK07_HUMAN      VGDEYEIIETIGNGAYGVVSSARRRLTGQQVAIKKIPNAFDVVTNAKRTLRELKILKHFK 110
sp|P31152|MK04_HUMAN      LGGRFVDFQPLGFGVNGLVLSAVDSRACRKVAVKKIA--LSDARSMKHALREIKIIRRLD 73
sp|Q16659|MK06_HUMAN      LGSRYMDLKPLGCGGNGLVFSAVDNDCDKRVAIKKIV--LTDPQSVKHALREIKIIRRLD 73
                          :         :* *  * * :*       .**:**:   :      ::: **: ::  .

sp|P28482|MK01_HUMAN      HENIIGINDIIR-APT--------IEQMKDVYIVQDLMETDLYKLLKT-QHLSNDHICYF 129
sp|P27361|MK03_HUMAN      HENVIGIRDILR-AST--------LEAMRDVYIVQDLMETDLYKLLKS-QQLSNDHICYF 146
sp|P53778|MK12_HUMAN      HENVIGLLDVFTPDET--------LDDFTDFYLVMPFMGTDLGKLMKH-EKLGEDRIQFL 133
sp|O15264|MK13_HUMAN      HENVIGLLDVFTPASS--------LRNFYDFYLVMPFMQTDLQKIMG--MEFSEEKIQYL 130
sp|Q15759|MK11_HUMAN      HENVIGLLDVFTPATS--------IEDFSEVYLVTTLMGADLNNIVKC-QALSDEHVQFL 130
sp|Q16539|MK14_HUMAN      HENVIGLLDVFTPARS--------LEEFNDVYLVTHLMGADLNNIVKC-QKLTDDHVQFL 130
sp|P45983|MK08_HUMAN      HKNIIGLLNVFTPQKS--------LEEFQDVYIVMELMDANLCQVIQ--MELDHERMSYL 131
sp|P53779|MK10_HUMAN      HKNIISLLNVFTPQKT--------LEEFQDVYLVMELMDANLCQVIQ--MELDHERMSYL 169
sp|P45984|MK09_HUMAN      HKNIISLLNVFTPQKT--------LEEFQDVYLVMELMDANLCQVIH--MELDHERMSYL 131
sp|Q13164|MK07_HUMAN      HDNIIAIKDILRPTVP--------YGEFKSVYVVLDLMESDLHQIIHSSQPLTLEHVRYF 162
sp|P31152|MK04_HUMAN      HDNIVKVYEVLGPKG---TDLQGELFKFSVAYIVQEYMETDLARLLEQ-GTLAEEHAKLF 129
sp|Q16659|MK06_HUMAN      HDNIVKVFEILGPSGSQLTDDVGSLTELNSVYIVQEYMETDLANVLEQ-GPLLEEHARLF 132
                          *.*:: : :::                :   *:*   * ::* .::     :  ::   :

sp|P28482|MK01_HUMAN      --------
sp|P27361|MK03_HUMAN      --------
sp|P53778|MK12_HUMAN      --------
sp|O15264|MK13_HUMAN      --------
sp|Q15759|MK11_HUMAN      --------
sp|Q16539|MK14_HUMAN      --------
sp|P45983|MK08_HUMAN      --------
sp|P53779|MK10_HUMAN      --------
sp|P45984|MK09_HUMAN      --------
sp|Q13164|MK07_HUMAN      DLPDLQDP 816
sp|P31152|MK04_HUMAN      --------
sp|Q16659|MK06_HUMAN      --------
```

4 Conclusions

All the above-discussed features are evident in Fig. 4 showing phylogenetic analysis of MAP kinases 1–14. Further evidence is obtained from domain architecture analysis of all proteins using Prosite database. ATP domain, MAP kinase domain and serine–threonine domain, respectively, and the domain architecture is represented in Fig. 4. Striking difference exists between MAPK 11–14 and rest of the group, where, the later contain serine-threonine kinase domain, a feature not observed with MAPK 11–14. Similarly, MAPK 8, 9 and 10 are devoid of ATP kinase domain.

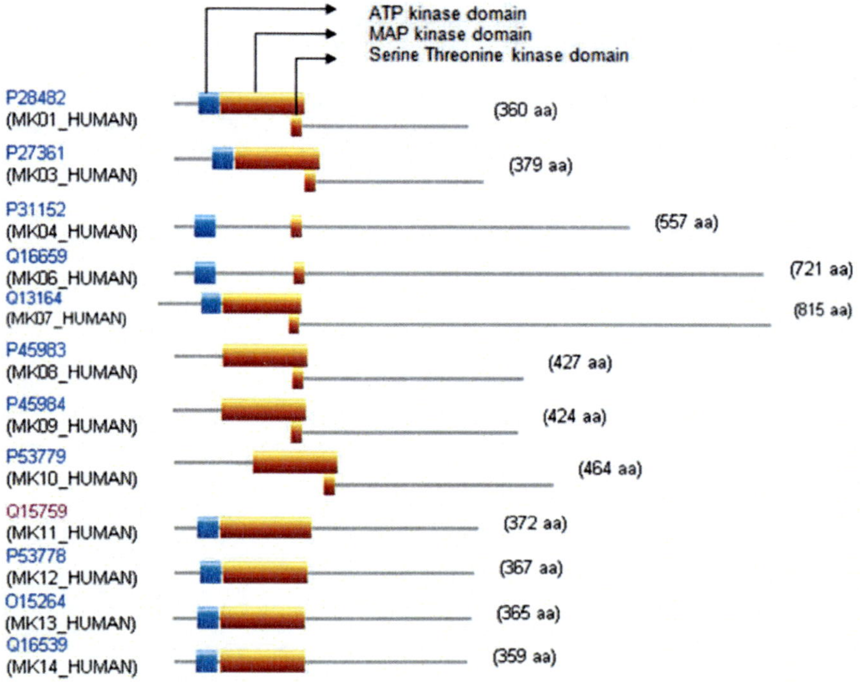

Fig. 4 Comparative view of protein kinase domain regions for MAPK 1–14

References

1. Wood, K., Sarnecki, C., Roberts, T. M., & Blenis, J. (1992). c-Ras mediates nerve growth factor receptor modulation of three signal-transducing protein kinases: MAP kinase, Raf-1 and RSK. *Cell, 68,* 1041–1050.
2. Nedunuri, D., Yesubabu, A., & Reddi, K. K. (2016). In silico pathway analysis of MAPKs and computational investigation of bipyrazole analogues as novel p38 alpha MAPKs inhibitors. *International Journal of Bioinformatics research and applications, 12,* 1–18.
3. Kyriakis, J. M., & Avruch, J. (2001). Mammalian mitogen-activated protein kinase signal transduction pathways activated by stress and inflammation. *Physiological Reviews, 81,* 807–869.
4. Nedunuri, D., Yesubabu, A., & Reddi, K. K. (2016). Selective inhibition of p 38 MAPK isoforms using bipyrazole derivatives: An in silico approach. *International Journal of Computer Applications, 130,* 34–40.
5. Nedunuri, D., Yesubabu, A., & Reddi, K. K. (2016). Pathway analysis of highly conserved Mitogen activated protein kinases (MAPKs). *Springer Briefs in Applied Sciences and Technology Forensic and Medical Bioinformatics,* 71–82.
6. Gasteiger, E., Gattiker, A., Hoogland, C., Ivanyi, I., Appel, R. D., & Bairoch, A. (20013). ExPASy the proteomics server for in-depth protein knowledge and analysis. *Nucleic Acids Research, 31,* 3784–3788.
7. Berman, H. M., Westbrook, J., Feng, Z., Gilliland, G., Bhat, T. N., Weissig, H., Shindyalov, I. N., & Bourne, P. E. (2000). The protein data bank. *Nucleic Acids Research, 28,* 235–242.

8. Xia, W., Longaker, M. T., & Yang, G. P. (2006). P38 MAP kinase mediates transforming growth factor-beta2 transcription in human keloid fibroblasts. *Am J Physiol Regul Integr Comp Physiol, 290*, R501-8.
9. Schiller, M., Böhm, M., Dennler, S., Ehrchen, J. M., & Mauviel, A. (2006) Mitogen- and stress-activated protein kinase 1 is critical for interleukin-1-induced, CREB-mediated, c-fos gene expression in keratinocytes. *Oncogene, 25*, 4449–4457.

Predicting Type of Lung Cancer by Using K-MLR Algorithm

Shameena Begum, T. Satish, Chalumuru Suresh, T. Bhavani, and Somula Ramasubbareddy

Abstract Health care is the maintenance of health via the prevention, diagnosis, and treatment of disease. The disease that persists over a long period of time is known as chronic disease. Chronic diseases may create additional activity restrictions. Common chronic conditions include lung disease, heart stroke, cancer, obesity, and diabetes. Chronic diseases usually show no symptoms and hence not diagnosed in advance. Hence, it is necessary to predict the patient-specific chronic diseases in early stage for effective prevention. Machine learning being the vital component of data analytics that facilitates the medical domain for malignancy predictions. Patients suffering from misdiagnosed and undiagnosed chronic diseases can be easily recognized with the help of these hospital systems. These systems enable the doctors to take precautionary measures and thereby minimizing the chances of a patient being affected. A new hybrid K-MLR framework using K-means and multiple linear regression has been proposed for diagnosing the type of lung cancer among the patients. As most of the real datasets are high-dimensional, this hybrid framework uses K-means clustering algorithm that eliminates the noise from the image-based dataset at the initial stage. Afterward to reduce the dimensionality, it detects the features of nodules in 3D lung CT scans and partitions the data to form the clusters. Finally, it reads the new patient data with malignant nodules to predict the type of associated cancer

---

S. Begum
CSE, Ramachandra Engineering College Eluru, Vatluru, Andra Pradesh, India

T. Satish
Department of CSE, Aditya College of Engineering and Technology, Tadepalligudem, Andra Pradesh 534101, India

C. Suresh
Department of CSE, VNR VJIET, Hyderabad, Telangana, India

T. Bhavani
Department of CSE, Sasi Institute of Technology and Engineering, Tadepalligudem 534101, Andra Pradesh, India

S. Ramasubbareddy (✉)
Department of IT, VNR VJIET, Hyderabad, Telangana, India
e-mail: svramasubbareddy1219@gmail.com

S. C. Satapathy et al. (eds.), *Intelligent System Design*, Advances in Intelligent Systems and Computing 1171, https://doi.org/10.1007/978-981-15-5400-1_39

based on the intensity of the nodule features extracted from each CT scan report using multiple linear regression analysis. Clustering prior to classification makes the hybrid approach beneficial.

Keywords Lung cancer · Pulmonary nodules · CT scan · Prediction · K-means · Regression

1 Introduction

Lung cancer is identified as the primary cause of cancer deaths in the United States. Approximately, 225,000 people are suffering from lung cancer every year, and the health care cost also reached to $12 billion this year in United States. Compared over the deaths caused because of colon, prostate, pancreas, and breast cancer jointly, the death rate of lung cancer is pretty high [1]. Recent researches concluded that the death rate can be considerably reduced with the help of chest CT (computed tomography) scan [2]. The nodules in the lung can be recognized utilizing a chest CT scan. A pulmonary nodule is a single round or oval growth in the lung that is also known as a coin lesion [3]. Nodule causes no symptoms but usually found during a chest X-ray or CT. Based on the imaging characteristics, the probability that a nodule is malignant is assessed. This insisted the researchers to build various image analysis research tools like nodule classifiers and nodule detectors to aid the radiologists in generating accurate assessments of the cancer risk. It is necessary to anticipate efficient algorithm that can detect whether the lesions in the lungs are cancerous and give patients the best chance to survive and recover with a proper diagnosis to prevent the cancer [4].

In this work, a new machine learning algorithm has been proposed to predict the risk with lung cancer by utilizing CT scan reports. The proposed structure detects a nodule in prior considering its features as well as image area around the nodule. The proposed algorithm proved to perform better in terms of accuracy, complexity, and error rate.

2 Literature Survey

Wang et al. deliberated malignant nodules in 185 patients, 171 benign, and texture analysis on NCE CT in 2 patients and observed that there is a differentiation amid benign and malignant nodules with respect to entropy and sum entropy ($P < 0.05$). The restraint of this work is additional characteristics of patients who were not considered in initiating the multilevel binomial logistic model. Future work includes the study of pulmonary nodules in the CT image to aid diagnosis of earlier period lung cancer [5]. Gibbs et al. presented a recent study in which he utilized MRI texture analysis [6] to exhibit high entropy in malignant lesions. The experimental results showed that this technique succeeded in distinguishing a malignant from benign lesions with 0.81 ± 0.07 accuracy. The study has shown the robustness of the logistic regression models using conventional clinical imaging protocols. Textural analysis has been used to discriminate the breast lesions [7].

Cavouras et al. applied texture analysis merged with CT density matrix analysis on 51 nodules confirmed by NCE CT and achieved 90.2% accurate results in differentiating amid malignant from benign lung lumps. The study was based only on features computed from the density values of the SPN CT image. If nodule's contour and size were employed, then the system could possibly achieve higher classification accuracy [8].

McNitt-Gray et al. used NCE CT to analyze 32 lung nodules that includes 14 benign and 14 malignant. This technique employed four textural features along with sum entropy to yield an area under the ROC curve (Az) of 0.992. The study did not observe trends with respect to distance and angle measures [9]. Dujardin et al. did a lot of research on borderline tumors and benign and achieved same kind of outputs. The authors assessed that the distinction may present because of unclear textural differences amid malignant and benign lesions [10].

Chae et al. presented the research done on texture analysis. This paper proposed a technique which excludes CE CT in differentiating amid invasive and preinvasive lung adenocarcinomas. The limitation of the study is that preinvasive lesions can be accurately differentiated from IPAs by using computerized texture analysis [11]. Ganeshan et al. studied texture analysis technique that includes NE CT alone and proposed technique to assess the association of glucose metabolism with nonsmall cell lung cancer. The same authors came up with a latest publication in which they considered the same patients and did texture analysis on both NCE CT as well as CE [12]. This novel technique also successfully assessed the relationship between texture in nonsmall cell lung cancer and histological markers [13, 14].

Krishnaiah et al. proposed an automatic technique for tumor delineation by using the single frame of images of the PET lung. SVM and K-nearest neighbor classifiers are used to discover the performance of the topography. This technique enables the physicians in minimizing the radiotherapy treatment and to forecast lung tumors [15]. Zakaria Suliman Zubi proposed various segmentation and improvement techniques to get accurate results. The proposed method uses MATLAB to generate each measure. The methods such as image preprocessing, processing, feature extraction, and segmentation were discussed in image processing techniques in detail. To improve the quality of the image, the techniques like fast Fourier transform methods, auto improvement, and associate Gabor filter are being used. In the segmentation stage, the watershed and thresholding segmentation is cast off, and the contrast has been made [16]. Balachandran proposed a novel method named FIS (fuzzy inference system) which distinguishes between benign, malicious, and advanced lung nodes. FIS uses artificial neural networks to make these differentiations. The author also suggests to perform linear filtering as a preprocessing step to eradicate the disturbances and to improve the quality of the images [17]. Tariq et al. presented a neuro-fuzzy classifier to identify the lung nodules by assessing the CT scan images.

The technique includes feature classification, extraction, lung division, and enrichment. Threshold segmentation is being used to extract the nodules from the image and to remove contextual smear. This method enables in identifying even the tiny nodules in lungs at an early stage [18]. Ada et al. introduced a technique to identify tumors in lungs based on the images. This tool also enables the physicians to easily differentiate amid normal and abnormal lungs as well as to predict the probability of the survival of a sufferer so that the cancer patients' lifetime can be increased [19].

Chen et al. proposed a novel clustering algorithm EACCD. It works in two steps. At first stage, partitioning among medoids is considered to calculate a different measure. Secondly, the dissimilarity found is analyzed to attain patients' clusters. These patients' cluster arrangement is a basis of a predictive system [20]. The above study has described the various efforts carried out by the researchers with dissimilar computational complexities and shown the significance of considering the relevant features to accurately classify and predict lung cancer patients into high or low risk groups. Hence, a hybrid K-MLR algorithm with low computational complexity has been proposed for cancer risk prediction.

3 Research Methodology

A dataset including thousands of high-quality lung scans has been provided by National Cancer Institute. Lung scans helps in assessing lung nodules which can either be a cancerous or noncancerous. A noncancerous nodule is known as benign, whereas a cancerous lung nodule is considered as a malignant. Lung nodules are usually about 5–30 mm in size. Larger lung nodules above 30 mm have a higher chance of being cancerous [21].

HU, Hounsfield unit, is calculated by considering regions of interest (ROI). ROI is usually drawn around the lesion on every image of the CT scan. Hounsfield measurement of a pulmonary nodule can be used to reliably identify benign or malignant nodules [22]. Based on the Lesion diameter, tumor size is characterized, and intensity of cancer is identified accordingly. Pathological invasiveness for pulmonary nodules is predicted by considering both tumor size as well as CT attenuation into account [23].

The stages of cancer are labeled as adenocarcinoma [24], squamous cell carcinoma, metastatic carcinoma, bronchoalveolar carcinoma, and mucoepidermoid carcinoma.

3.1 Proposed Methodology

1. Capture the cross-sectional images of the lung tissue and detect the abnormalities.
2. Preprocess the dataset to eliminate the noise and transform into consistent unlabeled data.

3. Extract the relevant features like lungareapx, lungareamm2, lungvolumefraction, lungmeanhu, lungpd95hu, and lungpd05hu that help to identify the disease with which the patient is suffering.
4. Using segmentation technique like K-mean clustering, detect the different types of lung diseases. (adenocarcinoma, squamous cell carcinoma, metastatic carcinoma, bronchoalveolar carcinoma, and mucoepidermoid carcinoma).
5. Classify the records of the patient test data to predict the dependent variable, **type of disease** based on the extracted features or say, independent variables using multiple linear regression analysis.

3.2 K-MLR (K-Means and Multiple Linear Regression) Algorithm

Step1: Initial guesses are made for the means $m_1, m_2, \ldots, m_k$
Step2: Until, there are no changes in any mean
Step3: Samples are classified into clusters utilizing the estimated means
Step4: For, i from 1 to k
Replace m_i with the mean of all of the samples for cluster i
End For
Step 5: $y = \beta 0 + \beta 1 \times 1 + \beta 2 \times 2 + \ldots.. + \beta_i x_n$ for $i = 1 \ldots n$.
Step 6: Make predictions
End Until

4 Experimental Results

C# language built on .NET framework, WinForms, Visual Studio 2013—an IDE, and SQL Server, a SQL-based relational database management system and a graphical GUI technology, is used to implement the system.

National Cancer Institute [25] provided a real-time lung cancer dataset. K-means clustering algorithm has been applied on dataset of 267 patient records with six appropriate characteristics of malignant solitary pulmonary nodules to create accurate and distinct clusters of patients with chronic lung disease.

Figure 1 shows that after applying K-means algorithm, it forms five types of cluster. It is determined by distance between attributes and forms the clusters from the given dataset which is represented as a Table 1. After each cluster, data represent

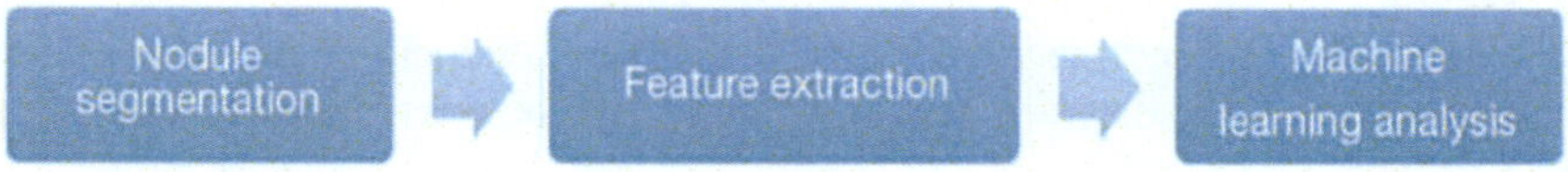

Fig. 1 Architecture of the proposed system

Table 1 Sample lung cancer dataset with six appropriate features

lungareapx	lungareamm2	lungvolumefraction	lungmeanhu	lungpd95hu	lungpd05hu
57,342	40,318.59	0.218742	−644.751	−368	−798
95,955	67,468.36	0.366039	−720.043	−544	−847
60,782	42,737.34	0.231865	−616.723	−294	−777
52,852	37,161.56	0.201614	−664.036	−267.55	−851
81,260	57,135.94	0.309982	−675.348	−295	−842
52,456	36,883.13	0.200104	−596.439	−356	−749
59,331	41,717.11	0.22633	−625.406	−392	−777
75,752	53,263.13	0.288971	−641.878	−356	−813
75,298	52,943.91	0.287239	−675.118	−340	−841

Table 2 Result of K-means clustering with five clusters each of size 41, 55, 95, 29, and 47

lungareapx	lungareamm2	lungvolumefraction	lungmeanhu	lungpd95hu	lungpd05hu	Cluster
57,342	40,318.59	0.218,742	−644.751	−368	−798	4
95,955	67,468.36	0.366039	−720.043	−544	−847	3
60,782	42,737.34	0.231865	−616.723	−294	−777	4
52,852	37,161.56	0.201614	−664.036	−267.55	−851	1
81,260	57,135.94	0.309982	−675.348	−295	−842	3
52,456	36,883.13	0.200104	−596.439	−356	−749	4
59,331	41,717.11	0.22633	−625.406	−392	−777	4
75,752	53,263.13	0.288971	−641.878	−356	−813	5
75,298	52,943.91	0.287239	−675.118	−340	−841	5

the by represent the disease name and make it as a training dataset as given as a Table 2, (Figs. 2, 3 and Table 3).

```
Applying Multiple Linear Regression on Training
Dataset
lm(formula = train$Disease ~ train$lungareapx +
train$lungareamm2 + train$lungvolumefraction +
train$lungmeanhu + train$lungpd95hu + train$lungpd05hu)
```

In Figure 4, Q-Q plot is standardized after applying of training data set to multiple linear regression algorithm, it calculated the confusion matrix, it consists performance metrics like residuals and fitted values, and all the attributes versus standardized residuals for Q-Q graph are represented (Table 4).

Figure 5 shows a bar graph about the comparison between the kNN, MLR, and K-MLR algorithms' test records and their performance metrics of time complexities.

Figure 6 shows a bar graph about the comparison between the kNN, MLR, and K-MLR algorithms' test records and their performance metrics of space complexities (Table 5).

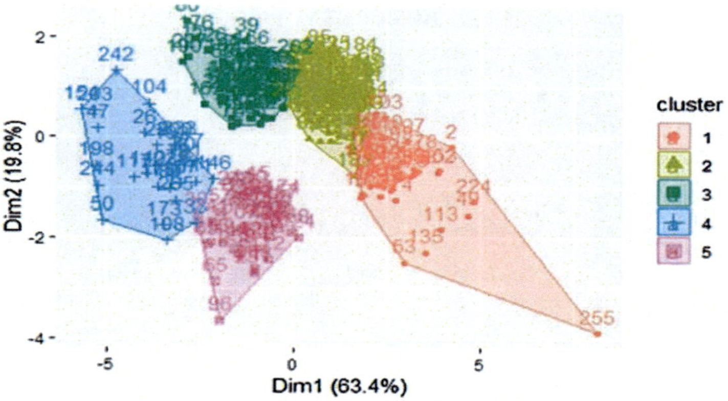

Fig. 2 Plotting of cluster data by K-means algorithm

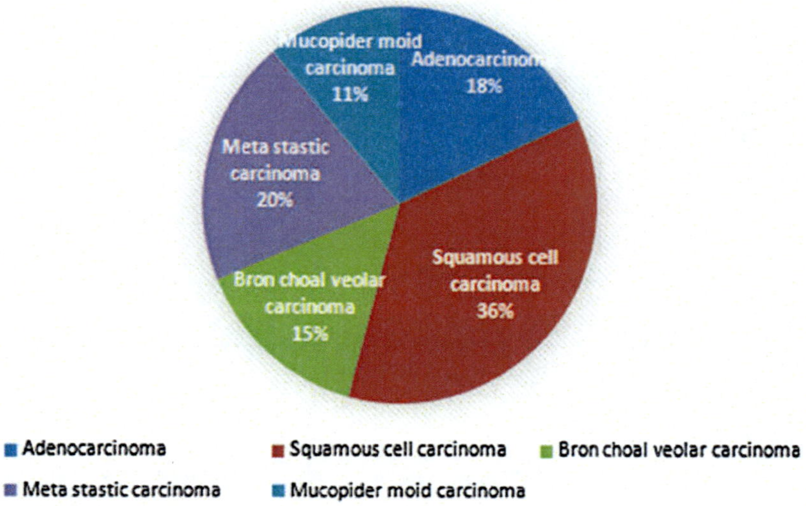

Fig. 3 Patient ratio based on lung disease type

Table 3 Specification of diseases are specified based on the six independent variables

Disease name	Cluster	Record count
Adenocarcinoma	1	47
Squamous cell carcinoma	2	95
Bronchoalveolar carcinoma	3	38
Metastastic carcinoma	4	53
Mucopidermoid carcinoma	5	29

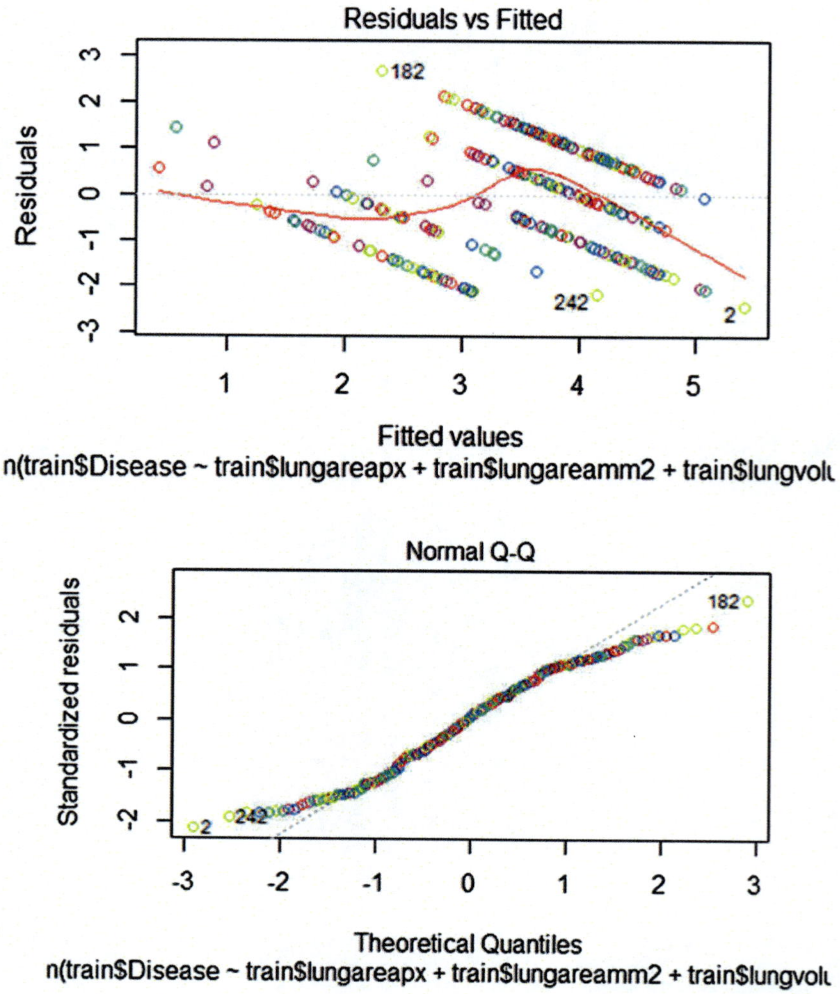

Fig. 4 Q-Q plot of standardized residuals versus theoretical quantities

Table 4 Comparison of algorithms with the proposed one in terms of time and space complexities

Algorithm	No. of records	Time in ms	Space in Kb
kNN	267	4256	8955
MLR	267	3456	6895
K-MLR	267	1218	3458

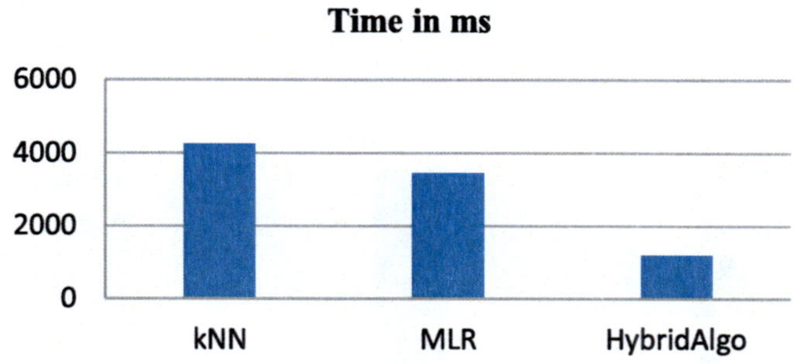

Fig. 5 Comparison graph showing the time complexities of existing and proposed algorithm

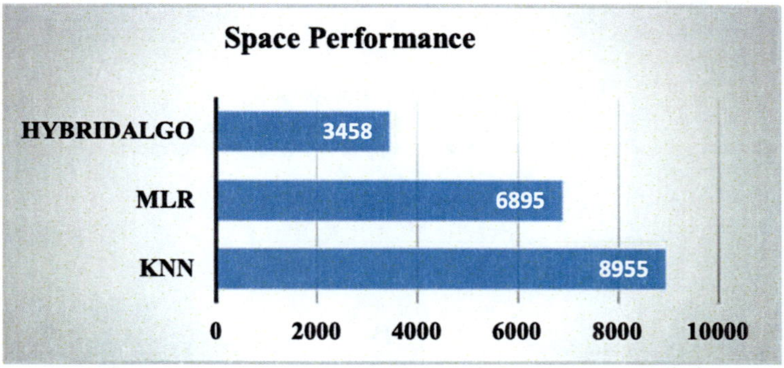

Fig. 6 Comparison graph showing the space complexities (kb) of existing and proposed algorithm

Table 5 Correct and incorrect classified instances of different algorithms

Algorithm	No of records	Correct instances	Incorrect instances	Prediction accuracy
kNN	267	245	22	91.27%
MLR	267	236	31	88.36
Hybrid algo	267	256	11	95.88

Figure 6 shows a comparison of three algorithms of kNN, MLR, and hybrid algorithms' correct and incorrect classifiers and prediction accuracy. In this, bar chart discusses about 267 records, and out of the records, blue bar is correctly classified, and red bars are incorrectly classifiers.

Figure 7 shows a comparison graph of the accurate prediction of proposed K-MLR algorithms. In this graph, blue bar shows about the no. of records, and red bar represents the accuracy of kNN 91.27%, MLR 88.36, and hybrid K-MLR 95.88% (Fig. 8).

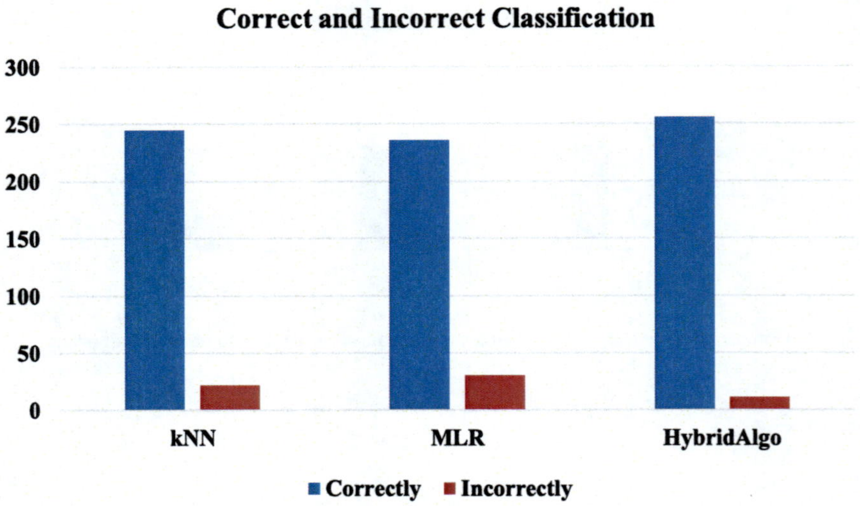

Fig. 7 Correct and incorrect classified instances by kNN, MLR with respect to K-MLR algorithm

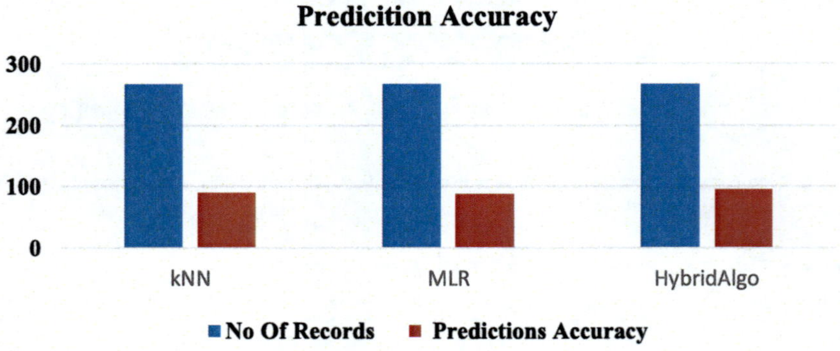

Fig. 8 Comparison graph showing the accurate prediction of proposed K-MLR algorithm

5 Conclusion

Today, one of the most deadly diseases in the world is cancer. In India, lung cancer is identified as the most fatal disease with highest probability of occurrence and death rate. It is very difficult to predict the occurrence of lung cancer as it bases on multiple diagnostic attributes which are very difficult to analyze [26]. This paper presents a novel method to discover the stage of malignant nodules. This paper concentrates on developing a new technique using which clusters can be generated with high accuracy and to assess the stage of the lung cancer patients by taking the diagnostic features from the CT scan reports [27] into consideration. The experimental results proved that this technique succeeded in generating highest performance ratio over other existing algorithms with respect to accuracy, complexity, and error rate.

References

1. Siegel, R. L., Miller, K. D., & Jemal, A. (2016). Cancer statistics, 2016. *CA: A Cancer Journal for Clinicians, 66*(1), 7–30.
2. MacMahon, H., Austin, J. H., Gamsu, G., Herold, C. J., Jett, J. R., Naidich, D. P., et al. (2005). Guidelines for management of small pulmonary nodules detected on CT scans: A statement from the Fleischner Society. *Radiology, 237*(2), 395–400.
3. Tsilimigras, D. I., Bakopoulos, A., Ntanasis-Stathopoulos, I., Gavriatopoulou, M., Moris, D., Karaolanis, G., et al. (2018). Clear cell "sugar tumor" of the lung: Diagnostic features of a rare pulmonary tumor. *Respiratory Medicine Case Reports, 23,* 52–54.
4. Rubin, G. D. (2015). Lung nodule and cancer detection in CT screening. *Journal of Thoracic Imaging, 30*(2), 130.
5. Wang, H., Guo, X. H., Jia, Z. W., Li, H. K., Liang, Z. G., Li, K. C., et al. (2010). Multilevel binomial logistic prediction model for malignant pulmonary nodules based on texture features of CT image. *European Journal of Radiology, 74,* 124–129.
6. Dennie, C., Thornhill, R., Sethi-Virmani, V., Souza, C. A., Bayanati, H., Gupta, A., et al. (2016). Role of quantitative computed tomography texture analysis in the differentiation of primary lung cancer and granulomatous nodules. *Quantitative Imaging in Medicine and Surgery, 6*(1), 6.
7. Gibbs, P., & Turnbull, L. W. (2003). Textural analysis of contrast-enhanced MR images of the breast. *Magnetic Resonance in Medicine, 50,* 92–98.
8. Cavouras, D., Prassopoulos, P., & Pantelidis, N. (1992). Image analysis methods for solitary pulmonary nodule characterization by computed tomography. *European Journal of Radiology, 14,* 169–172.
9. McNitt-Gray, M. F., Wyckoff, N., Sayre, J. W., Goldin, J. G., & Aberle, D. R. (1999). The effects of co-occurrence matrix based texture parameters on the classification of solitary pulmonary nodules imaged on computed tomography. *Computerized Medical Imaging and Graphics, 23,* 339–348.
10. Dujardin, M., Gibbs, P., Turnbull, L. W. (2014). Texture analysis of 3T high resolution T2 weighted images in ovarian cystadenoma versus borderline tumor. *Proc Intl Soc Magn Reson Med* 22, 2218. Available online: http://cds.ismrm.org/protected/14MPresentations/abstracts/2218.pdf.
11. Chae, H. D., Park, C. M., Park, S. J., Lee, S. M., Kim, K. G., & Goo, J. M. (2014). Computerized texture analysis of persistent part-solid ground-glass nodules: differentiation of preinvasive lesions from invasive pulmonary adenocarcinomas. *Radiology, 273,* 285–293.
12. Badic, B., Desseroit, M. C., Hatt, M., & Visvikis, D. (2018). Potential complementary value of noncontrast and contrast enhanced CT radiomics in colorectal cancers. *Academic radiology.*
13. Ganeshan, B., Abaleke, S., Young, R. C., Chatwin, C. R., & Miles, K. A. (2010). Texture analysis of non-small cell lung cancer on unenhanced computed tomography: initial evidence for a relationship with tumour glucose metabolism and stage. *Cancer Imaging, 10,* 137–143.
14. Ganeshan, B., Goh, V., Mandeville, H. C., Ng, Q. S., Hoskin, P. J., & Miles, K. A. (2013). Non-small cell lung cancer: Histopathologic correlates for texture parameters at CT. *Radiology, 266,* 326–336.
15. Krishnaiah, V. (2013). Diagnosis of lung cancer prediction system using data mining classification techniques. *International Journal of Computer Science and Information Technologies, 4*(1), 39–45. www.ijcsit.Com. ISSN: 0975-9646.

16. Zubi, Z. S. (2014). Improves treatment programs of lung cancer using data mining techniques. *Journal of Software Engineering and Applications, 7*, 69–77.
17. Balachandran, K. (2013). Classifiers based approach for pre-diagnosis of lung cancer disease. In *International Journal of Computer Applications® (IJCA) (0975–8887), Proceedings on National Conference on Emerging Trends in Information & Communication Technology (NCETICT 2013).*
18. Tariq, A., Akram, M. U., & Javed, M. Y. (2013). Lung nodule detection in CT images using neuro fuzzy classifier. In *Fourth International Workshop on Computational Intelligence in Medical Imaging (CIMI)*, pp. 49–53.
19. Wolfsen, A. R., & Odell, W. D. (1979). ProACTH: Use for early detection of lung cancer. *The American Journal of Medicine, 66*(5), 765–772.
20. Chen, D. (2009). Developing prognostic systems of cancer patients by ensemble clustering. *Journal of Biomedicine and Biotechnology*, 632786. Hindawi Publishing Corporation.
21. Al-Ameri, A., Malhotra, P., Thygesen, H., Plant, P. K., Vaidyanathan, S., Karthik, S., et al. (2015). Risk of malignancy in pulmonary nodules: A validation study of four prediction models. *Lung Cancer, 89*(1), 27–30.
22. Vesal, S., Ravikumar, N., Ellman, S., & Maier, A. (2018). Comparative analysis of unsupervised algorithms for breast MRI lesion segmentation. In *Bildverarbeitung für die Medizin 2018* (pp. 257–262). Berlin, Heidelberg: Springer Vieweg.
23. Travis, W. D., Asamura, H., Bankier, A. A., Beasley, M. B., Detterbeck, F., Flieder, D. B., … & Powell, C. A. (2016). The IASLC Lung Cancer Staging Project: Proposals for coding T categories for subsolid nodules and assessment of tumor size in part-solid tumors in the forthcoming eighth edition of the TNM classification of lung cancer. *Journal of Thoracic Oncology, 11*(8), 1204–1223.
24. Kennedy, G. T., Okusanya, O. T., Keating, J. J., Heitjan, D. F., Deshpande, C., Litzky, L. A., et al. (2015). The optical biopsy: A novel technique for rapid intraoperative diagnosis of primary pulmonary adenocarcinomas. *Annals of Surgery, 262*(4), 602–609.
25. Davis, J. N., Medbery, C., Sharma, S., Pablo, J., Kimsey, F., Perry, D., et al. (2015). Stereotactic body radiotherapy for centrally located early-stage non-small cell lung cancer or lung metastases from the RSSearch® patient registry. *Radiation Oncology, 10*(1), 113.
26. Gao, X., Chu, C., Li, Y., Lu, P., Wang, W., Liu, W., et al. (2015). The method and efficacy of support vector machine classifiers based on texture features and multi-resolution histogram from 18F-FDG PET-CT images for the evaluation of mediastinal lymph nodes in patients with lung cancer. *European Journal of Radiology, 84*(2), 312–317.
27. Kumar, D., Wong, A., & Clausi, D. A. (2015). Lung nodule classification using deep features in CT images. In *2015 12th Conference on Computer and Robot Vision (CRV)* (pp. 133–138). IEEE.

Core Performance Based Packet Priority Router for NoC-Based Heterogeneous Multicore Processor

K. Indragandhi and P. K. Jawahar

Abstract Nowadays, NoC plays a vital role in improving the performance of multicore processor. Router plays a major role in traversing a packet from source to destination in an NoC. In this paper, we designed a C3PR: Core Performance Based Packet Priority Router for NoC-based heterogeneous multicore processor. The main aim of the proposed router is to increase the utilization of high-performance processing core and reduce the starvation of low priority packets across the NoC. The performance of the C3PR is evaluated through Xilinx 14.4 ISE simulator with Spartan 3E family and compared with round robin router. The average number of deflections per packet across the network is termed as deflection rate. The experimental result reveals that deflection rate for high-performance core packet is reduced when compared with RR router design.

Keywords Heterogeneous multicore processor · Router · Network on chip · Flow control

1 Introduction

The advancement of IC technology leads to integrate more number of cores in a single chip [1]. System on Chip (SoC) is an integrated circuit that contains many processors, memory and peripherals based on the system. The integrated components of SoC are communicated conventionally through bus interface system [2]. There are two types of multicore SoC, namely homogeneous and heterogeneous. Homogeneous multicore SoC consists of identical processing cores and peripherals in a single chip whereas heterogeneous multicore SoC consists of non-identical processing cores and other peripherals on a single chip.

K. Indragandhi (✉) · P. K. Jawahar
B. S. Abdur Rahman Crescent Institute of Science & Technology, Chennai, India
e-mail: indra@crescent.education

P. K. Jawahar
e-mail: jawahar@crescent.education

S. C. Satapathy et al. (eds.), *Intelligent System Design*, Advances in Intelligent Systems and Computing 1171, https://doi.org/10.1007/978-981-15-5400-1_40

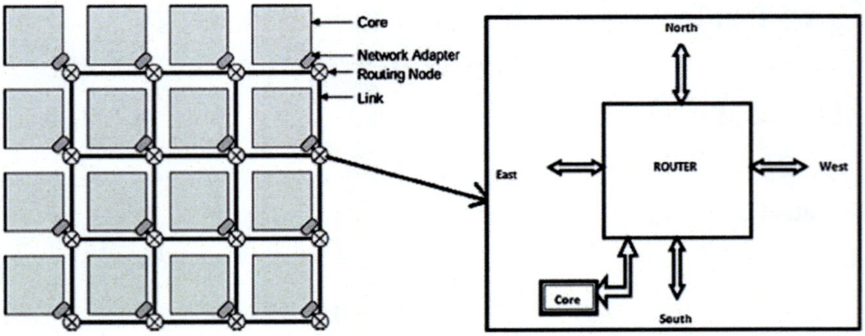

Fig. 1 NoC architecture

For the high-speed application, to satisfy the performance, more number of cores are integrated on an SoC. For such applications, the conventional method of on-chip communication is not scalable and does not support redesign. This paved way for a new paradigm Network on Chip (NoC). NoC architecture is shown in Fig. 1. It consists of processing core (PC), network adaptor, link and router. Each router has communication link from north, south, east, west and local core as shown in Fig. 1.

The building blocks of NoC are topology, routing algorithm, flow control and router architecture. Topology determines the connection pattern between the PC and the physical layout of the network [3]. The various types of topologies are ring, star, mesh and torus. Routing algorithm computes the packet delivery path from source to destination PC [4]. Flow control is how efficiently the resources are allocated for the successful delivery of packet across the network when two or more packets try to access the same output port in the router at the same time. Router plays a vital role in traversing a packet from source to destination processing core. There are buffered and bufferless at the input port of the router. The bufferless router is area and cost-efficient [5]. In this paper, Core Performance Based Packet Priority Router is designed and analyzed for NoC-based heterogeneous multicore processor.

2 Related Work

Mishra et al. [6] proposed a novel heterogeneous NoC which consists of power-efficient small router and high-performance big routers on single NoC. The resources are distributed among router based on the demand. The resources namely buffer and wider links are partitioned in an optimal manner. The hetero-NoC design evaluated in asymmetric CMP and showed a good power & performance compared the traditional method. Schoeberl et al. [7] proposed a minimal network interface for lightly weight NoC and consumes only 6% resources of RISC processor. Hossein et al. [8] proposed Minimally Buffered Router Infrastructure (Muffin) concentrate on static power consumption. Muffin designed using power gating techniques without powering the routers. Muffin improves 73.7% average packet latency and 95.4% static power consumption. Choudary and Dakhole [9] proposed a five-port router for Network on chip architecture. The packet flow control was store and forward method. FSM controller was used for deterministic routing of packet in mesh topology. If multiple packets try to access the same output port, round robin scheduling algorithm was used to traverse the packet from source to destination port. Shen et al. [10] discussed the software and hardware issues of reconfigurable NoC for heterogeneous multicore architecture. The adaptive routing method was used to deliver the packet to the processing core. Rakshe et al. [11] designed a 1X3 router for efficient resource sharing among the processing cores and to reduce the starvation of low priority packets due to high priority packets. Zhou et al. [5] designed a load balancing bufferless defection router: LBBDR to reduce defection rate during heavy traffic in the NoC. In LBBDR nearest-first algorithm is used to balance the load across the network.

3 Proposed C3PR Router

Router plays a major role in delivering a packet from source to destination PC across the network. Router is capable of receiving packets on all input ports, determining the packet destination PC based on the routing algorithm and then sending packet to the appropriate output port. The proposed router C3PR is a bufferless router as shown in Fig. 2.

This router designed with 4X4 mesh topology and it has 16 different performances PC. This router has five input and output ports: North input (Ni), South input (Si), East input (Ei), West input (Wi), local core input (Ci), North output (No), South output (So), East output (Eo), West output (Wo) & local core output (Co). The router is designed for heterogeneous multicore processor. The packet width assumed is 20 bit and a packet format is shown in Fig. 3.

For each input and output port, 4 bit is assigned for identification as shown in Table 1. Each PC has assigned a 4-bit value as shown in Table 2. Values are assigned based on the performance of the core. Least value represents low-performance and

Fig. 2 C3PR router

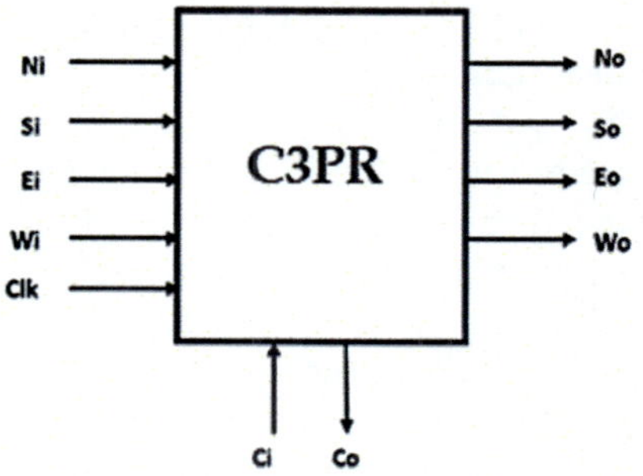

Destination port (4 bits)	Source port (4 bits)	Data (8 bits)	Processing Core(4 bit)

20 bit

Fig. 3 Packet format

Table 1 Port identification

Input ports	Assigned values	Output ports	Assigned values
Ni	0101	No	0000
Si	0110	So	0001
Wi	0111	Wo	0010
Ei	1000	Eo	0011
Ci	1001	Co	0100

Table 2 Priority of processing core based on performance

Processing core	Priority	Processing core	Priority
C0	0000	C8	1000
C1	0001	C9	1001
C2	0010	C10	1010
C3	0011	C11	1011
C4	0100	C12	1100
C5	0101	C13	1101
C6	0110	C14	1110
C7	0111	C15	1111

highest value represents high-performance. The PC value is considered as a priority when two packets compete for the same output port.

The packet allocated to the desired output port is termed as a productive port and the deflected packet to the neighboring router is termed as non-productive port. The main focus of the proposed C3PR router design is flow control. When two or more packets compete for the same output port then our proposed design provides a better solution.

In the proposed C3PR design, the deflection rate for high-performance core packet is reduced by assigning the productive port. The pseudo-code for proposed C3PR router design is shown in Fig. 4. The proposed router receives 20-bit packet in all the five ports at every clock cycle. At raising edge of clock, packets are received then the router checks the destination output port of each packet. If all the packets are requesting different output ports then all the packets successfully get the productive port. C3PR router plays a vital role if two or more packets try to send the packets to the same output port at the same time. Each processing core performance is different and always we want to utilize core efficiently by directing the packets on time. Since all the packets are having the deadline, delivering a packet after the deadline may degrade the overall performance of the application. The incoming packets from

```
C3PR Router algorithm:
1. Input ,IP={Ni,Si,Ei,Wi,Ci}
2. Output,OP={No,So,Eo,Wo,Co}
3. Assign priority based on the performance of core in
   NOC
4. Priority,P={0-15}:Largest value → highest priority
5. Assumptions:D=Direction,C=counter,ROP=Requested
   output port,AOP=available output port,LPP=Low
   Priority Packet, HPP=high priority
   packet,T=threshold value
6. At rising edge of clock
7. If ( IP- D unequal)
        OP=ROP;
8. else if (2 or more ports= >same ROP)
        then compare P
        If (P== high)&&(C==T)
            LPP→R OP
        else
        {  HPP→ OP & LPP→AOP;
            C++;}
9. else if (2 or more packets == same P)
    OP← random among HPP & remaining packet → AOP.
10. go to step 6
```

Fig. 4 Pseudo code of C3PR router algorithm

various PC may be allocated either productive port or non-productive port based on router design.

In the proposed C3PR design, All the packets have 4 bit for destination processing core and this value is considered as a priority during multiple packet compete for the same output port. The high priority packet is allocated to the productive port and low priority packets are assigned to the available port on that clock cycle. If the priority also the same for both the packets then randomly packet is sent to a productive port. By comparing the priority, a packet that has high-performance core as a destination traversed to productive port and a low priority packet is deflected across the network.

If the router flow control continues to be in the above scenario, the packet which has low-performance core as a destination leads to starvation. To prevent the starvation of low priority packet, in our proposed C3PR router design, a counter variable is introduced.

Whenever the low priority packet gets deflected, counter variable gets incremented. A threshold value is assigned for counter variable. If the counter variable reached the threshold value and multiple packets try to access the same output port then highest priority is considered for low priority packet and allotted a productive port. By this method, the proposed C3PR router design reduces the low priority packet starvation and performs the load balancing between high-performance and low-performance cores.

4 Experimental Results

The proposed C3PR router design is compared with the round robin (RR) based router. In round robin based router, whenever the multiple packets try to access the same output port, then packets will be allotted based on round robin scheme. The C3PR and RR routers were written in verilog hardware description language.

Simulation and synthesis of C3PR and RR router were performed in XILINX 14.4 ISE with Spartan 3E FPGA family. Verification of router was done using test bench. Different sets of test cases were simulated to verify the router design. At rising edge of clock, the input 20-bit packets are received through ports and 16-bit value is sent to output port. The 16-bit value consists of 8-bit data, 4-bit source input port and 4-bit processing core. Table 3 shows the four different test cases for C3PR and RR router. In test case 1, all the input packets need different output ports, so both C3PR and RR router design allocate the productive port for all the packets. Figure 5a shows the simulation result of test case 1 of RR and C3PR router design.

By considering the test case 2, the packets received from north and east port both need the same south output port. In RR router the packet is allotted based on round robin, so a packet from north input which is having low-performance PC has the destination wins the race and allotted to the productive port and a packet from east input port which is having high-performance PC has the destination packet, allotted to non-productive port.

In C3PR router, since a packet from north and east input seeks same output port, it checks PC value as a priority and a packet from east input port which is having high-performance PC has the destination packet allotted to productive port and a packet from north input which is having low-performance PC has the destination allotted to the productive port.

From Table 3, LPP represents low priority packet to productive port, LPD represents deflected low priority packet, HPP represents high priority packet to productive port and HPD represents deflected high priority packet. Figure 5b, c shows test case 2 simulation result of C3PR and RR router design. RTL schematic of C3PR router is shown in Fig. 5d.

5 Conclusion

The proposed C3PR router is a core performance based packet priority router designed for heterogeneous multicore processor which efficiently utilizes the high-performance core by assigning high priority for the packet, when multiple accesses to

Table 3 Test cases of RR and C3PR router

Source port	Destination		Allotted output port	
	Output port	Processing core (PC)	RR	C3PR
1				
Ni	So	C3	So	So
Si	Eo	C0	Eo	Eo
Wi	Co	C2	Co	Co
Ei	No	C4	No	No
Ci	Wo	C8	Wo	Wo
2				
Ni	So	C1	**So-LPP**	**No-LPD**
Si	Eo	C4	Eo	Eo
Wi	Co	C6	Co	Co
Ei	So	C3	**No-HPD**	**So-HPP**
Ci	Wo	C7	Wo	Wo
3				
Ni	Wo	C7	**Wo-LPP**	**Eo-LPD**
Si	Wo	C14	**Eo-HPD**	**Wo-HPP**
Wi	Co	C9	Co	Co
Ei	So	C3	So	So
Ci	No	C1	No	No
4				
Ni	Eo	C0	**Eo-LPP**	**Co-LPD**
Si	No	C3	No	No
Wi	Eo	C13	**Co-HPD**	**Eo-HPP**
Ei	So	C5	So	So
Ci	Wo	C7	Wo	Wo

same output port of router occurred. The proposed router design reduces the deflection rate of high priority packet and increases the utilization of high-performance core compared to round robin based router.

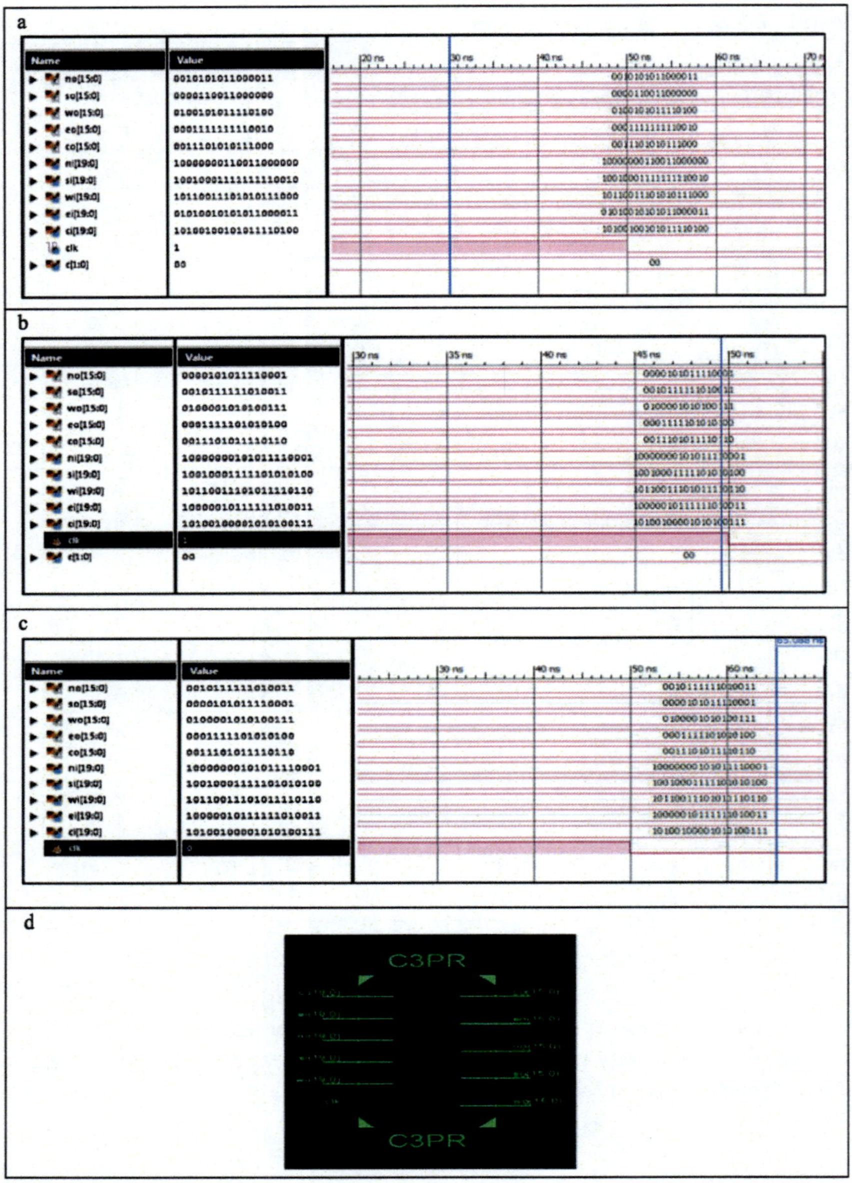

Fig. 5 **a** Simulated result of test case 1-RR and C3PR router, **b** Simulated result of test case 2-C3PR router, **c** Simulated result of test case 2-RR router, **d** RTL schematic of C3PR router

References

1. Dally, W., & Towles, B. (2004). Principles and practices of interconnection networks. San Francisco: Morgan Kaufmann Pub.
2. Saponara, S., & Fanucci, L. (2012). Homogeneous and heterogeneous MPSoC architectures with network-on-chip connectivity for low-power and real-time multimedia signal processing. Hindawi Publishing Corporation, VLSI Design.
3. Swapna, S., Swain, A., & Mahapatra, K. (2012). Design and analysis of five port router for network on chip, 51–55. https://doi.org/10.1109/primeasia.2012.6458626.
4. George, M., Alagarsamy, A., & Lakshminarayanan, G. (2017). Design of five port priority based router with port selection logic for NoC. *ICTACT Journal on Microelectronics, 02*, 293–299. https://doi.org/10.21917/ijme.2017.0051.
5. Zhou, X., Zhu, Z., & Zhou, D. (2016). A load balancing bufferless deflection router for network-on-chip. *Journal of Semiconductors, 37*(7).
6. Mishra, A. K., Narayanan, V., & Das, C. R. (2011). A case for heterogeneous on-chip interconnects for CMPs. In *38th Annual International Symposium on Computer Architecture (ISCA)*. IEEE.
7. Schoeberl, M., Pezzarossa, L., & Spars, J. (2019). A minimal network interface for a simple network-on-chip. In: M. Schoeberl, C. Hochberger, S. Uhrig, J. Brehm, & T. Pionteck (Eds.), *Architecture of Computing Systems—ARCS*.
8. Farrokhbakht, H., Kamali, H. M., & Jerger, N. E. (2019). Muffin: Minimally-buffered zero-delay power-gating technique in on-chip routers. In *ACM/IEEE International Symposium on Low Power Electronics and Design*.
9. Choudhari, E. M., & Dakhole, P. K. (2014). Design and verification of five port router for network on chip. *IEEE International Conference on Communication and Signal Processing*, Melmaruvathur, pp. 607–611.
10. Shen, J. -S., Hsiung, P. -A., & Lu, J. -M. (2014). Reconfigurable network-on-chip design for heterogeneous multi-core system architecture. In *IEEE International Conference on High Performance Computing & Simulation*, pp. 523–526.
11. Rakshe, P., & Prajapati, P. (2018). Router RTL design for network on chip (NoC) with common resource sharing scenario by arbitration technique. *IJCEM International Journal of Computational Engineering & Management, 21*(3).

Sequential Pattern Mining for the U.S. Presidential Elections Using Google Cloud Platform (GCP)

M. Varaprasad Rao, B. Kavitha Rani, K. Srinivas, and G. Madhukar

Abstract All traditional growth-based pattern methods for sequential pattern mining can result recursively with length ($k + 1$) models on the basis of the given long-k pattern databases. In log2 ($k + 1$) recursion rates at best, you can detect a length-k pattern which leads to fewer recursion rates and quicker pattern development. Here we suggested a cloud-based strategy in the minimum moment and precision to obtain patterns. On the basis of the U.S. presidential candidate donations, the suggested method is implemented. These patterns are easily identified using Cloud Dataprep application from Google Cloud Platform. The time taken to generate the patterns is linear i.e., O(n).

Keywords Cloud computing · Pattern matching · Accuracy · Time complexity · Google CloudPlatform

1 Introduction

Sequential pattern mining is a data mining subject concerned with discovering statistically appropriate trends between data instances where values are supplied in a sequence. The values are generally assumed to be discrete and therefore time series mining is strongly linked, but generally considered to be a distinct activity. Within this sector, there are several important traditional computing issues discussed. These include constructing effective sequence data databases and indexes, extracting

M. Varaprasad Rao (✉) · B. Kavitha Rani · K. Srinivas · G. Madhukar
CMR Technical Campus, Hyderabad, India
e-mail: varam78@gmail.com

B. Kavitha Rani
e-mail: phdknr1@gmail.com

K. Srinivas
e-mail: phdknr@gmail.com

G. Madhukar
e-mail: madhu.mani536@gmail.com

S. C. Satapathy et al. (eds.), *Intelligent System Design*, Advances in Intelligent Systems and Computing 1171, https://doi.org/10.1007/978-981-15-5400-1_41

frequently occurring patterns, comparing similarity sequences, and restoring incomplete sequence representatives. Sequence mining issues can generally be categorized as string mining, which usually is based on string processing algorithms and item sets, which usually is based on the teaching of association rules.

TRIFACTA ® Cloud Dataprep [1] is a smart information service designed to visually explore, clean and prepare organized and unstructured information for assessment.

Cloud Dataprep is serverless and functions on any scale. On top of the strong cloud dataflow provider, Cloud Dataprep is constructed. There is no deployment or management infrastructure. Click-and-no-code easy information preparation. It consists of six steps such as Cloud Data preparation, Data preparation of objects, workflows, tasks, identity and access management, and finally cross data access. To use Dataprep in Machine Learning applications, a six-step procedure is applied like, in step1 the selection of variables parameterization; the dataset uploading, and flow of data and object selection is performed in step2; whereas in step3 the workflow of data and as well as normalization is computed; in step4 parsing of data and correspondingly the data will be stored in cloud storage; the user control and management happens at step5, and finally the accessing of data across the GCP.

2 Literature Review

The subject of information mining is sequence pattern mining, where statistically appropriate patterns are to be found between information instances when values are given sequence [2]. The values are generally assumed to be discrete, and time-series mining is therefore strongly linked, but generally regarded as an activity differently. Specific instance of structured data mining is sequential pattern mining. Sequential pattern mining is an important data mining problem, which detects frequent subsequences in a sequence database [3]. Integrate frequency sequence mining with the frequency pattern and use sequence databases for the purpose of confining the search to sequence fragments development. FreeSpan decreases the whole set of models but significantly decreases the efforts of the post-generation candidate [4]. Deep-first search strategy integrating profound first crossing of the search area with efficient cutting processes [5]. Multi-dimensional sequential pattern mining theme that integrates the analysis and sequential data mining [1]. TRIFACTA ® Cloud Dataprep from the Google Cloud Platform [6] is used to generate valid patterns for our data.

3 Methodology

The architecture of Cloud Dataprep is shown in the above Fig. 1; the input is either a file or a query from the storage location submitted as raw data to Cloud Data preparation, in which generating of data by cleansing and blending procedures. By

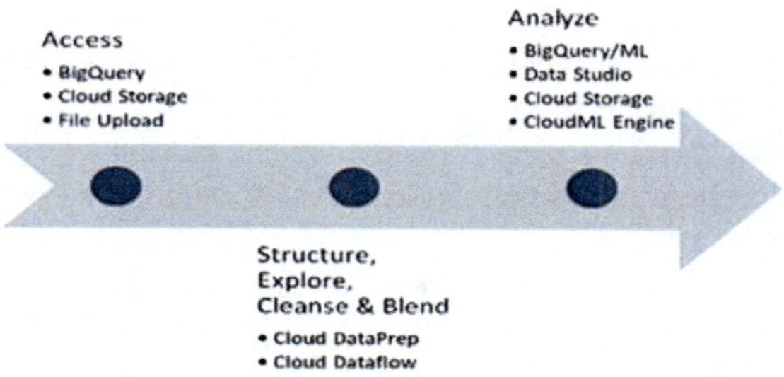

Fig. 1 The architecture of Cloud Dataprep at GCP *(from GCP)*

applying dataflow of this cleaned data, can be generated for required data and is stored in bigtable of GCP. Through which the data can easily be analyzed and interpret by using Machine Learning (ML) engine or data studio in the required format of end user.

4 Experimental Results

Experimental setup has been done on Google Cloud Platform (GCP) with Dataprep API by considering the dataset *U.S. presidential candidate donation*. The workflow has been created and made sequence of query steps executed to obtain results. The results are showing 97% accuracy in the data set, 3% of data may be missing and there are no mismatch cases. Total number of rows are 97 with six different columns that have been displayed as result and are stored at storage bucket. Entire process of this approach is obtained in linear time and is executed on GCP in only the duration of 17 min of time, $O(n)$.

The following are the screenshots of the above approach.

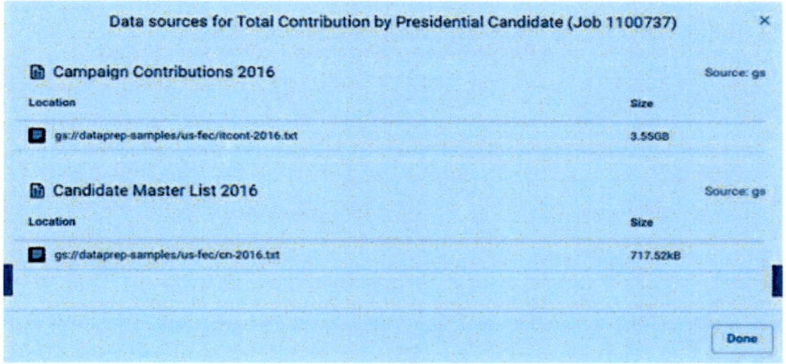

Fig. 2 Data set the U.S. president election contribution 2016

In Fig. 2, Cloud Dataprep enables a parameterization variable to select—to manage executions in the same recipes over serialized data sets of variable manageable managed datapaths; to match patterns, to identify column-specific data patterns and create recipes; predictive transformation—to return data from one format to another; Data sampling—produces one or more specimens of information for display and modification in the customer implementation; planning the implementation of recipes in your streams; sharing—various people working on the same assets; targeting—the number of rows, their order and their sizes to which you are trying to wrangle your information collection; and visual profiling—visual profiling provides real-time interactive visualizations.

Figure 3 will deal with flow structure and objects—organizing the work flow and its objects; imported data set—is the source of original data; recipe—is a functional task to transform a data set; outputs and publishing destinations—outputs.

Contain one or more publishing destinations, and publishing destinations specify the output format, location, and other publishing actions.

Workflow has the key capabilities like Import from a flat file, databases, or distributed storage systems; Locate and remove or modify missing or mismatched data; Unnest complex data structures; Identify statistical outliers in your data for review and management; Perform lookup from one data set into another reference data set; Aggregate columnar data using a variety of aggregation functions; Normalize column values for more consistent usage and statistical modeling; Merge datasets with joins; Appendone data set to another through union operations, is shown in Fig. 4.

Step5 is focused on parsing data set, reshaping data from rows to columns or vice versa; enriching data; write the query using big query where the data is stored in bigtable or cloud storage is shown in Fig. 5.

Figure 6 allows to control user and group access to your project's resources.

Figure 7 shows the number of valid pattern data within the Google Cloud Platform project from which Cloud Dataprep is run. It shows about the top 20 candidates list

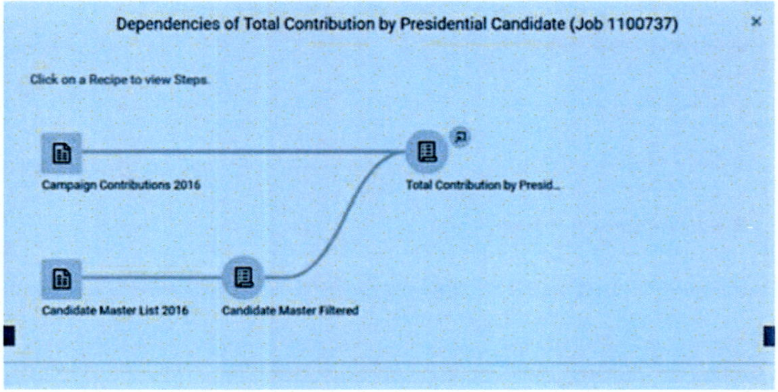

Fig. 3 Functional workflow with filters

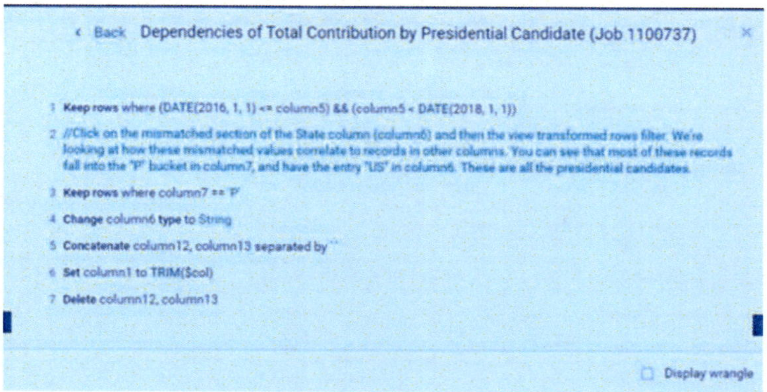

Fig. 4 Sequence of query steps on functional dependency in pass1

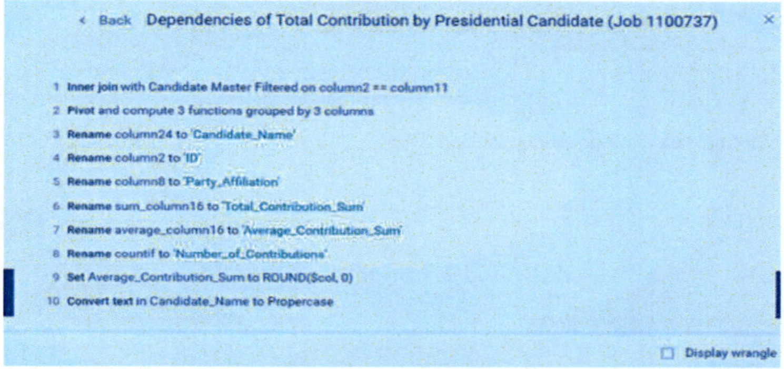

Fig. 5 Sequence of query steps on functional dependency in pass2

have been given its values based on the donations collected; along with their party from which the candidate contested. The total contribution collected from various sources around 292 million dollars and average contribution sum is about 25 thousand dollars. There are a total of 97 candidates selection rows and whose donations have analyzed and shown in visualized patterns with 97% valid patterns have generated and 3% of values have not identified, therefore who are called missing values.

5 Conclusion

The presidential candidate donations collected from various sources for total contributions by presidential candidates in the 2016 elections in the United States. The findings show 97% precision in the dataset, 3% information may be lacking, and there are no instances of mismatch. As a consequence, the total amount of lines is

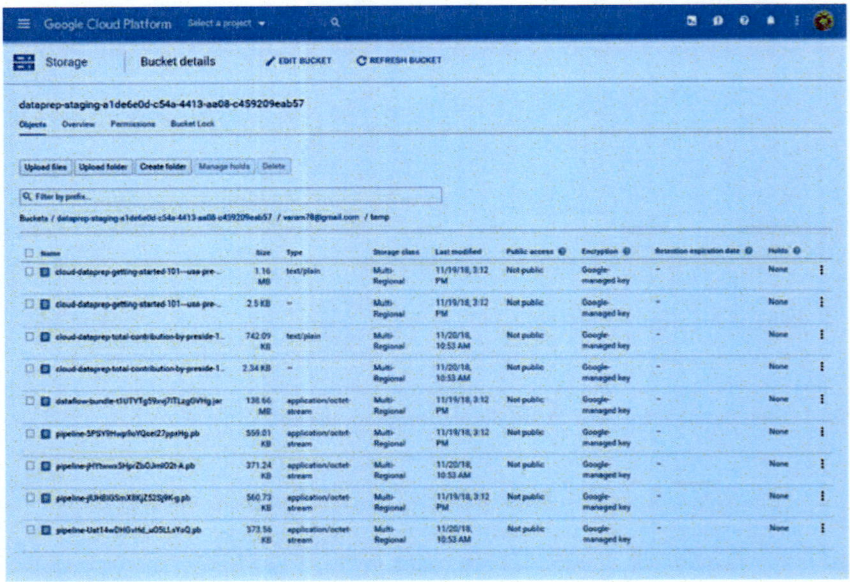

Fig. 6 Storing data in cloud storage/bucket

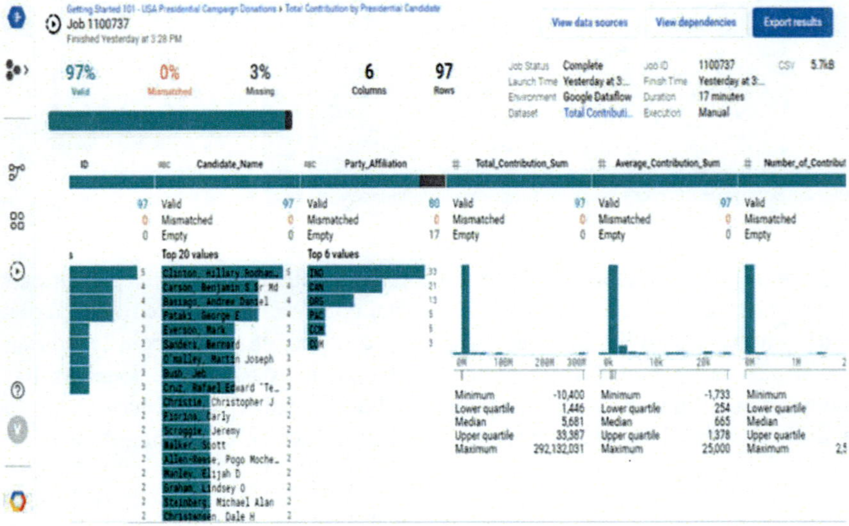

Fig. 7 Valid patterns for total contribution by presidential candidate

97 with 6 distinct columns presented and deposited at the storage tank. The entire method of this strategy is acquired in higher dimensions and is performed on GCP in linear time of 17 min, $O(n)$.

References

1. Pinto, H., et al. (2001). Multi-dimensional sequential pattern mining. In *International Conference on Information and Knowledge Management*, pp. 81–88.
2. Mabroukeh, N. R., & Ezeife, C. I. (2010). A taxonomy of sequential pattern mining algorithms. *ACM Computing Surveys, 43*, 1–41. CiteSeerX 10.1.1.332.4745. https://doi.org/10.1145/1824795.1824798.
3. Chen, J. (2010). An updown directed acyclic graph approach for sequential pattern mining. *IEEE Transactions on Knowledge and Data Engineering, 22*(7).
4. Han, et al. (2000). FreeSpan: Frequent pattern-projected sequential pattern mining. In *International Conference on Knowledge Discovery and Data Mining*, pp. 355–359.
5. Ayres, J., et al. (2002). Sequential pattern mining using a bitmap representation. CiteSeerX doi: 10.1.1.12.5575.
6. Google Cloud Platform. (2019). https://cloud.google.com/dataprep/taken.

Efficient Lossy Audio Compression Using Vector Quantization (ELAC-VQ)

Vinayak Jagtap, K. Srujan Raju, M. V. Rathnamma, and J. Sasi Kiran

Abstract Compression is the technique for effective utilization of space in servers as well as in personal computers. Most significantly, being multimedia compression. In this paper, the focus is on the audio compression method. Audio compression method has two types: lossy and lossless compression. Vector quantization is an effective way of lossy compression technique. The important tasks in vector quantization are codebook generation and searching. Simple Codebook generation algorithm is used which enhances the compression process. The proposed method is Efficient Lossy Audio Compression using vector quantization (ELAC-VQ). A centroid-based compression reduces the operation of the comparison with the codebook and helps to improve the performance. At the time of decompression is simple to audio compression. The experimental results show that ELAC-VQ approach reduces the computational complexity, increases the compression percentage and speeds up the vector quantization process. The universal codebook generation is the key which will reduce the overhead of vector finding. These can be preprocessed to reduce runtime processing in compression. To achieve better results as well as for validation, clustering is used for generalized codebook generation.

V. Jagtap
Department of Computer Engineering and Information Technology, College of Engineering Pune, Pune, Maharashtra, India
e-mail: vinayak5jagtap@gmail.com

K. Srujan Raju (✉)
Department of Computer Science Engineering, CMR Technical Campus, Hyderabad, Telangana, India
e-mail: ksrujanraju@gmail.com

M. V. Rathnamma
Department of Computer Science Engineering, KSRM College of Engineering, YSR Kadapa, Andhra Pradesh, India
e-mail: rathnamma@ksrmce.ac.in

J. Sasi Kiran
Lords Institute of Engineering & Technology, Hyderabad, Telangana, India
e-mail: sasikiranjangala@gmail.com

S. C. Satapathy et al. (eds.), *Intelligent System Design*, Advances in Intelligent Systems and Computing 1171, https://doi.org/10.1007/978-981-15-5400-1_42

Keywords Lossy compression · Vector quantization · LBG · Codebook · Clustering

1 Introduction

For efficient transmission of digital data over low channel bandwidth, there is a need for compression. The basic goal of digital compression is to represent data with a minimum number of bits with acceptable quality. Compression is achieved by removing redundancy. The multimedia consists of high redundancy. For high compression with acceptable distortion, lossy compression is used.

Many methods are available for lossless compression such as Run-length encoding, LZW, Huffman coding and so on. There are also techniques for lossy compression. In this paper, the main area of discussion is the lossy audio compression. Depending on the compression ratio, vector quantization method produces high compression with acceptable distortion.

In quantization, a range of values is reduced to a single value [1]. Vector quantization is a widely used quantization technique in multimedia compression, pattern recognition, and so on [2], assuming LBG algorithm is used. Vector quantization maps k-dimensional vectors into finite set called as codebook. In vector quantization there are two phases namely, encoding method and decoding method.

At the encoder, an input image is divided into blocks called input vectors which go into compression method. The lowest distorted code vector in the codebook is found for each input vector. The closest code vector is found by using square of Euclidean distance. The code vector which has minimum square of Euclidean distance with input vector is selected. The corresponding index associated with the searched code vector is transmitted to the decoder. Compression is achieved as input vectors are replaced with index of the closest code vector [2] [3]. The procedure is shown in Fig. 1.

The codebook generation and vectors matching are tedious task and will require a lot of efforts so we tried to implement a small method which reduce these operations and can compress any audio file with high compression ratio.

The image-based codebook generation is designed in the paper published earlier by the author, here a similar codebook generation approach is taken for audio lossy compression. The digitized audio file typically wave file is having data in the range of $[-1, 1]$ and there are two types of the wave file mono and stereo audio wave files. These types are based on the channel, after rendering data single or two-column audio node data is available based on the types, respectively. These codebooks are generated as per density-based clustering to find its mean and based on training audio file codebook generation is done. Optimization is done based on the distortion added in the reconstruction.

The idea came from research work done on audio and image file, where are both the methods can be merged from references [15–18].

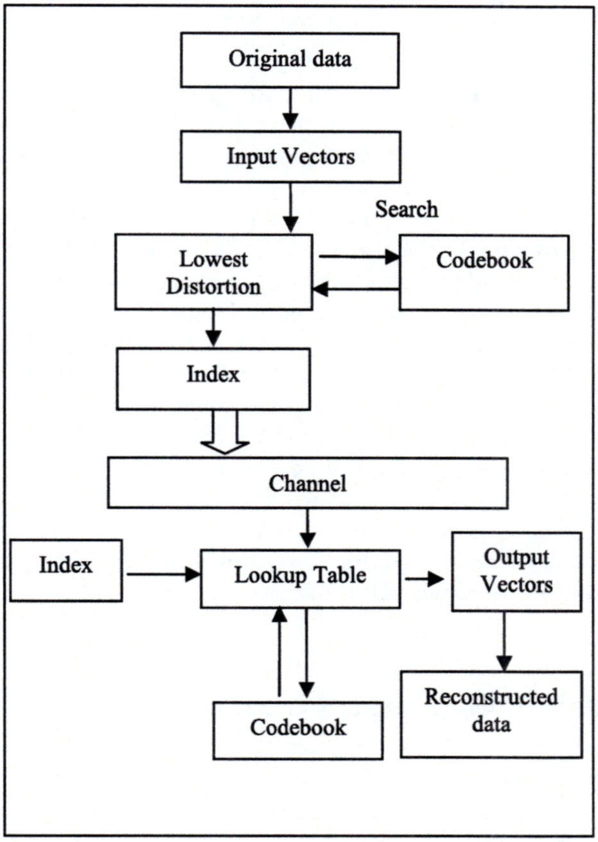

Fig. 1 Vector quantization method

2 Related Work

The important task in vector quantization is to generate an optimal codebook. To achieve it, generalized Lloyd's algorithm is proposed by Linde, Buzo, Gray referred to as LBG algorithm [4–6].

In LBG, an exhaustive search is used. The drawback of full exhaustive search is that it increases computational complexity as each input vector is compared with all code vectors in codebook [7]. Another drawback of LBG is that the codebook generation is dependent on initial codebook chosen. Poor results can be obtained, if initial codebook selection is wrong. Thus, standard LBG does not provide globally optimal results [8]. The search complexity increase as the number of code vectors in the codebook increases. The performance of vector quantization is directly dependent on codebook size and vector size [9, 10].

3 LBG Image VQ Based Compression

The standard LBG method with exhaustive search has more computational complexity. For example, let input image is divided into 4*4 blocks so that the code vector has same 4*4 block size. If number of code vectors in the codebook are 1024 then each code vector index is 10 bit (2^10 = 1024).

Consider, for good compression ratio with minimum distortion codebook of size 10000, having each codebook of 4*4 code vectors, needs to compare with 256*256 pixel images of having, 4*4 block segmentation means, 4096 input code vectors. These 4096 code vectors need to be compared with codebook, then the minimum number of comparisons in this compression by using LBG approach will be 10000*16*4096*16. This uses very high computational power and yet there is no surety of finding the exact codebook index from codebook. There is a need for reduction in computation along with standardization of global codebook. Hence, centroid-based codebook generation is proposed earlier [11, 12, 15].

In ELIC-VQ approach, distance metrics like Euclidean distance are not used for calculating a similarity between input vector and code vector. The centroid of input vectors is used as an index. Thus, ELIC-VQ method increases the performance by reducing the number of comparisons [15].

3.1 Standard Global Codebook Generation

Global codebook is generated using several images as training set. The main benefit of using the image as training image, is it gives proper codebook, which might not be the case in the random selection of training vectors. Because the values of the random training set can vary from 0–255 (28) for 8-bit grayscale image, which might be redundant and complex. Thus, the local optimum problem of LBG is removed by global codebook.

The standard global codebook is generated by taking 256 indices, having centroid mapping with 0–255 (28) which reduces the computation for a generation of codebook as digital image has the values from 0 to 255. Once the codebook is generated having unique values between 0 and 255, then the codebook is sorted using quicksort, which reduces the heavy computation of searching codebook further. The centroid-based sorted codebook is mapped with the index of codebook [12–14].

3.2 Centroid-Based Comparison

To reduce the number of operations per vector the centroid of input vector is calculated. The centroid of each input vector is stored as centroid value + 1 as index, which is similar to index of lowest distorted code vector. This method reduces operational complexity and thus increases the speed per operation.

4 Elac-Vq

Audio file is time-series dataset. So ELAC-VQ includes division of audio files depending on the timeline of 34 ms each which are the input vectors. The 34 ms are considered as human ear cannot detect the audio change in 34 ms data. Then the mean of input vectors are found and those are considered as centroid and as index of vector too. Computational complexity is reduced by 34 each time. Computational complexity reduced shown below n is no of sample inputs. ELAC-VQ considers only sample data instead of processing for vector computation, the complexity is in relation to sample points rather than in *f(n)* (Table 1).

Flow of compression and decompression in ELAC-VQ is shown in Figs. 2 and 3, respectively.

Table 1 Complexity comparison

	LBG	*ELAC-VQ*
Complexity	$O(n^2)$	O(n)

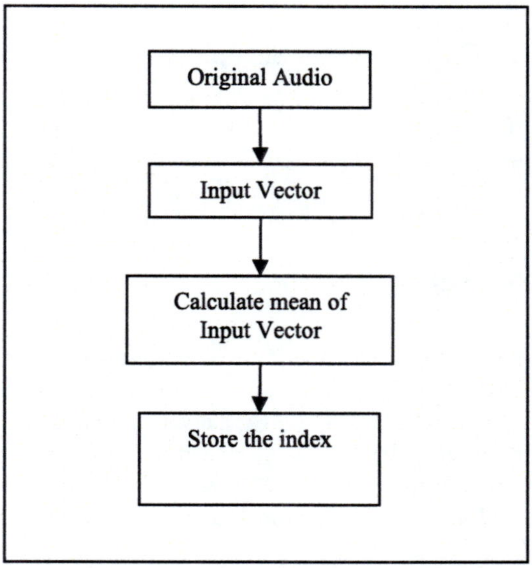

Fig. 2 ELAC approach compression

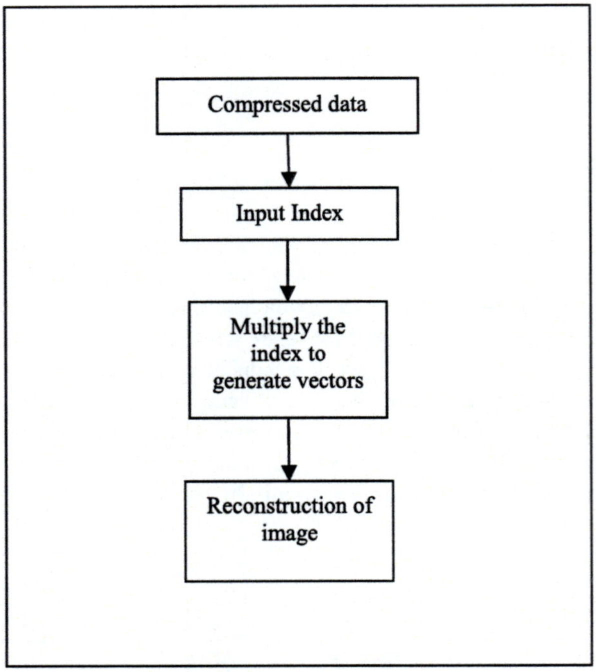

Fig. 3 ELAC approach decompression

5 Experimental Setup

Windows 7 home basic, Matlab R2016a are used for experimentation. Experiments were carried out on MP3 as well as wave files.

6 Results

Figures 4 and 5 show time comparison between LBG with global codebook and ELAC-VQ. The following are the results of ELAC-VQ, which shows that the compression ratio is 1:34 as a division is applied. 50 sample files are taken from both wave and MP3 format and each having a compression ratio the same. Ex Sample file size reduced from 1.47 MB to 49.27 KB and when the file is decompressed size is 1.47 MB with acceptable distortion. Figures 6 and 7 show the results.

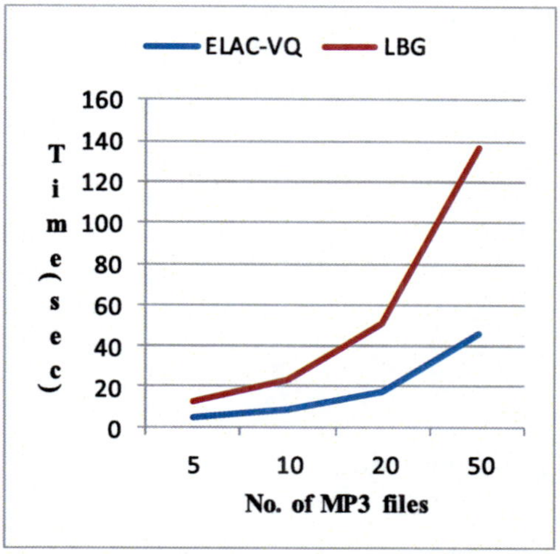

Fig. 4 MP3 Compression Time versus no. of MP3 files

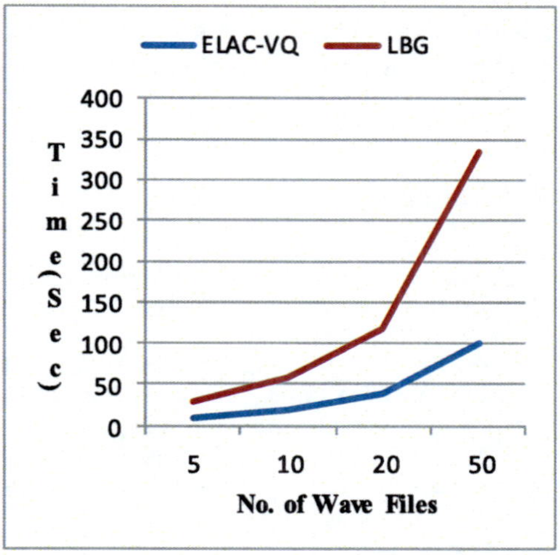

Fig. 5 Wave compression time versus no. of wave files

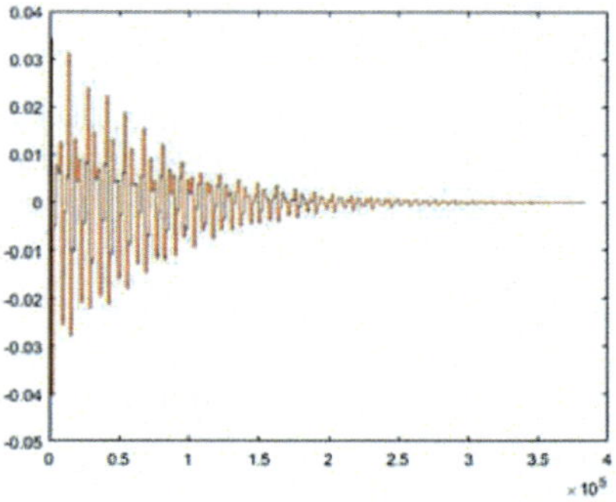

Fig. 6 Original audio

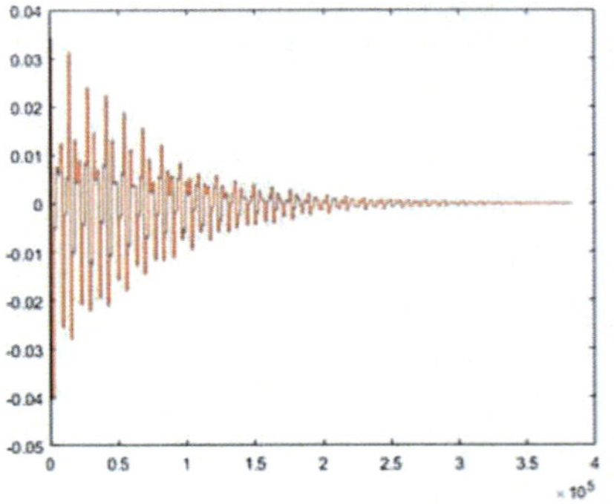

Fig. 7 Decompressed audio

7 Conclusion

Results show that ELAC-VQ method enhances the compression ratio and computation complexity. PSNR and MSE are higher sides than other compression method but it's still acceptable as humans can't detect those changes This technique can be combined with encryption to reduce the transmission time.

References

1. Huang, B., & Xie, L. (2010). An improved LBG algorithm for image vector quantization. In *2010 3rd International Conference on Computer Science and Information Technology* (Vol. 6, pp. 467–471). IEEE.
2. Gray, R. (1984). Vector quantization. *IEEE Assp Magazine., 1*(2), 4–29.
3. Linde, Y., Buzo, A., & Gray, R. (1980). An algorithm for vector quantizer design. *IEEE Transactions on Communications, 28*(1), 84–95.
4. Chang, C. C., Chen, T. S., & Xiao, G. X. (2003). An efficient and effective method for VQ codebook design. In *Fourth International Conference on Information, Communications and Signal Processing, 2003 and the Fourth Pacific Rim Conference on Multimedia. Proceedings of the 2003 Joint* (Vol. 2, pp. 782–786). IEEE.
5. Bardekar, A. A., & Tijare, P. A. (2011). Implementation of LBG algorithm for image compression. *International Journal of Computer Trends and Technology, 2*(2).
6. Chen, C. Q. (2004). An enhanced generalized Lloyd algorithm. *IEEE Signal Processing Letters, 11*(2), 167–170.
7. Kekre HB, Sarode TK. (2010). New Clustering algorithm for vector quantization using rotation of error vector. arXiv preprint arXiv:1004.1686.
8. Pal, A. K., & Sar, A. (2011). An efficient codebook initialization approach for LBG algorithm. arXiv:1109.0090.
9. Shen, G., & Liou, M. L. (2000). An efficient codebook post-processing technique and a window-based fast-search algorithm for image vector quantization. *IEEE Transactions on Circuits and Systems for Video Technology, 10*(6), 990–997.
10. Ying, L., Hui, Z., & Wen-Fang, Y. (2003). Image vector quantization coding based on genetic algorithm. In *IEEE International Conference on Robotics, Intelligent Systems and Signal Processing, 2003. Proceedings. 2003* (Vol. 2, pp. 773–777). IEEE.
11. Lu, T. C., & Chang, C. Y. (2010). A survey of VQ codebook generation. *Journal of Information Hiding and Multimedia Signal Processing., 1*(3), 190–203.

12. Kekre, H. B., & Sarode, T. K. (2009). Vector quantized codebook optimization using k-means. *International Journal on Computer Science and Engineering (IJCSE), 1*, 283–290.
13. Kekre, H. B., & Sarode, T. K. (2009, January). Fast codebook search algorithm for vector quantization using sorting technique. In *Proceedings of the international conference on advances in computing, communication and control* (pp. 317–325). ACM.
14. Sathappan, S., & Pannirselvam, S. (2011). An Enhanced Vector Quantization Method For Image Compression With Modified Fuzzy Possibilistic C-Means Using Repulsion. *International Journal of Computer Applications., 975*, 8887.
15. Jagtap, Vinayak, Reddy, Sudhakar, & Joshi, Pallavi. (2018). Efficient Lossy Image Compression Using Vector Quantization (Leic-Vs). *CIIT International Journal of Digital Image Processing, 10*(9), 165–170.
16. Jagtap, V. G., & Pande, S. S. (2012). Noisy Node Detection in Wave File by using Iterative Silhouette Clustering and Vivaldi. *International Journal of Computer Applications, 59*(4).
17. Jagtap V. G., Pande S. S., Kulkarni P. (2013). Intelligent Silhouette Wave Steganography (I -SiWaS), *CIIT International Journal of Artificial Intelligence Systems and Machine Learning: Issue*.
18. Jagtap V., Rafee., Raju S., Kulkarni P., Ravikanth M. (2019). Image Recognition and Content Retrival to Build Context: A Survey, *Jour of Adv Research in Dynamical & Control Systems, 11*(01)-Special Issue, 1656–1666.

Multi-objective Evolutionary Algorithms for Data Mining: A Survey

D. Raghava Lavanya, S. Niharika, and A. S. A. L. G. Gopala Gupta

Abstract Real-world optimization problems usually involve multiple objectives to be optimized at the same time under multiple constraints and with reference to many variables, whereas multi-objective optimization itself will be a difficult task, equally tough is that the ability to form a sense of the obtained solutions. This article presents various data mining techniques that can be applied to extract information regarding problems of multi-objective optimization. Further, this knowledge provides a deeper insight into prediction for the decision-makers.

Keywords Data mining · Multi-objective optimization · Classification · Machine learning · Clustering · Feature selection

1 Introduction

Data Mining involves discovering interesting, novel, and useful patterns from large databases. The main objective is to build an efficient descriptive or predictive model. Optimization of model parameters plays a vital role for successful applications of any data mining approach. Frequently such issues, because of their intricate nature, can't be unraveled utilizing standard numerical strategies. Also, because of the enormous size of the information, the issues here and there become recalcitrant.

In this way, planning effective deterministic calculations is regularly not attainable. Uses of transformative calculations have been observed to be hypothetically helpful for preparing huge information for setting ideal parameters and to find

D. Raghava Lavanya (✉) · A. S. A. L. G. G. Gupta
Department of Computer Science and Engineering, Koneru Lakshmaiah Education Foundation, Vaddeswaram, Guntur, Andhra Pradesh 522502, India
e-mail: draghavalavanya@gmail.com

A. S. A. L. G. G. Gupta
e-mail: amara_gupta@yahoo.com

S. Niharika
V.R.Siddhartha Engineering College, Vijayawada, India
e-mail: sakrutiniharika@gmail.com

S. C. Satapathy et al. (eds.), *Intelligent System Design*, Advances in Intelligent Systems and Computing 1171, https://doi.org/10.1007/978-981-15-5400-1_43

generous and significant data [1, 2]. Traditionally, evolutionary algorithms (EAs) [3] were used to solve single-objective problems. Be that as it may, some genuine issues concoct numerous destinations, which must be upgraded simultaneously to accomplish an exchange off. Consequently, a considerable lot of the advancement issues including various goals, while amplifying or limiting one target work numerous prompts clashes with the simultaneous augmentation or minimization of any of the other target's capacities, as an exchange off exists in one arrangement and, thus, nobody arrangement can give an improvement to every one of the destinations. Or maybe, numerous arrangements are potential, every one of which is superior to all the others, in any event, one of the destinations. The decent variety containing in these arrangements is called as ideal Pareto-front and named as Pareto-ideal arrangements. Throughout the previous couple of years, in information mining space multi-objective transformative calculations (MOEAs) [4, 5] have turned out to be progressively mainstream. Regular information mining errands incorporate element determination, order, bunching/biclustering, affiliation standard mining, deviation location, and so on. An assortment of MOEAs for settling such information mining errands can be found in the writing. This article audits such techniques in an efficient manner. This overview centers around the essential information mining errands, to be specific component determination, arrangement, grouping, and, since the vast majority of the multi-target calculations that are connected to information mining have managed these assignments.

Genetic Algorithms have also been shown to be an effective method in data mining and machine learning [11, 12]. Genetic algorithms also show a great aspect in a learning context in pattern recognition. There are two different approaches to applying GA:

1. We can apply a GA directly as a classifier.
2. GA can be utilized as an advanced instrument for the parameters resetting in different classifiers. Most utilization of GAs in information mining streamlines a few parameters in the grouping procedure. For achieving more exploration GA is the choice. It has likewise been connected to locate an ideal arrangement of highlight loads to improve grouping exactness.

Hereditary calculations (GAs) rely upon hypothesis of advancement and depend on organic applications. At the point when GAs are utilized for critical thinking, the arrangement has three particular stages:

The arrangements of the issue are encoded into portrayals that help the vital variety and choice tasks; these portrayals, are called chromosomes, are as basic as bit strings.

A wellness capacity passes judgment on which arrangements are the "best" living things, that is, most fitting for the arrangement of a specific issue.

Hybrid and change produce other people qualities created by recombining highlights of their folks. In the long run, an age of people will be deciphered back to the first issue area and the fit individual speaks to the arrangement.

2 Fundamental of Data Mining

Information mining includes finding intriguing and conceivably helpful examples of various kinds, for example, rundown, characterization, affiliations, grouping, and anomaly's location. As a rule, information mining procedures incorporate: Kind of information, a model to be utilized and inclination measure. The most well-known systems in current information mining approach are arrangement, affiliation standard mining, grouping, relapse, succession and connection examination, and reliance displaying are some of. Model portrayal decides both the adaptability of the model for speaking to the basic information and the interpretability of the model in human terms. Information mining undertakings can comprehensively be grouped into two classes: prescient or managed and unmistakable or unaided [3, 6]. The prescient procedures gain from the present information so as to make forecasts about the conduct of new datasets. Then again, the clear systems give a rundown of the information. The most normally utilized assignments in the area of information mining incorporate element determination, characterization, relapse, grouping, affiliation principle mining, deviation recognition, and so forth. In this paper, we have for the most part centered around the four assignments, i.e., include

determination, characterization, bunching, and affiliation guideline mining. In the writing, the greater part of the multi-target calculations connected to information mining has managed these assignments. MOEAs have completely been connected in these four essential fields of information mining four primary fields of data mining (Fig. 1).

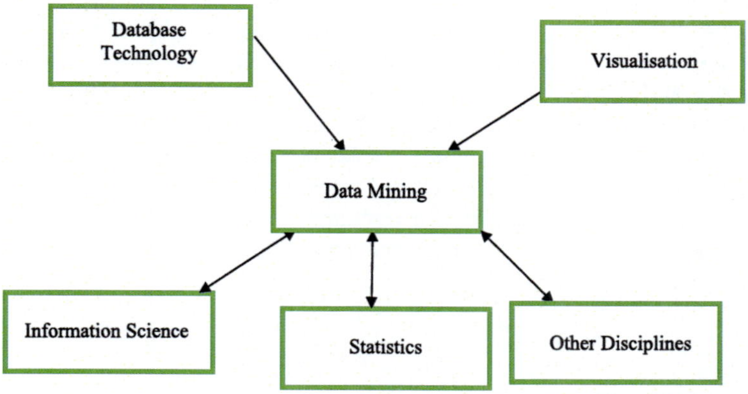

Fig. 1 Data mining as an interdisciplinary filed

3 Feature Selection

Highlight determination issue manages the choice of significant arrangement of highlights or traits for the given informational index. It is vital for the order or grouping process. It is predominantly used to play out the dimensionality decrease to limit the space. The objective of highlight choice is essential of the triple. To begin with, computationally hard to work with enormous number of highlights. Second, most of this present reality informational indexes that may have uproarious, repetitive, and immaterial highlights that may diminish the exhibition of the arrangement or bunching. At last, it is an issue when the quantity of highlights turns out to be a lot bigger than the quantity of information. For such cases, dimensionality decrease is required to change over the information into important information investigation.

The component choice issue can without much of a stretch be demonstrated as an improvement issue. The objective is to choose a subset of highlights for which some component subset assessment foundation is streamlined. Thus, developmental calculations have been broadly utilized for the component determination issue [7, 8]. Transformative calculations, for the most part, adopt a wrapper strategy for highlight determination strategies wherein the subset of highlights is encoded in a chromosome and an element assessment foundation is utilized as the wellness work. The component subsets are assessed depending on how well the chosen highlights are performing as far as managed learning (order)/unaided getting the hang of (bunching) for a given dataset. Be that as it may, assessment of choice includes a solitary basis that does not work similarly well for all datasets. Subsequently, the need for various advancing will emerge for such criteria. Multi-target include determination improves the heartiness of the element choice strategies. In the ongoing past, various MOEAs, both in administered and solo zones, have been proposed. Hereditary Algorithm (GA) can be utilized to discover the subset of highlights wherein the chromosome bits portrayal if the element is incorporated or not. The worldwide most extreme for the target capacity can be discovered which gives the best problematic subset. Here again, the target capacity is the indicator execution. The GA parameters and administrators can be changed inside the general thought of a developmental calculation to suit the information or the application to acquire the best execution or the best query item. An adjusted variant of the GA called the CHCGA can be utilized for highlight choice.

4 Classification

Classification is called administered learning and it essentially segments the element space into locales, one area for preparing the model and another for testing [9]. Along these lines, each datum point in the whole element space is mapped to one of the potential classes (K classes). Classifiers are related to marked information, in which case these issues are in some cases alluded to as regulated grouping. Regulated

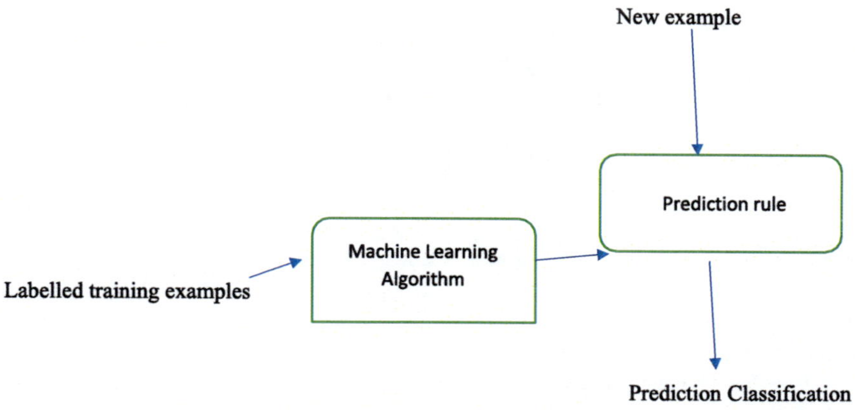

Fig. 2 Classification

classifiers accept that a lot of preparing information is accessible. The preparation dataset comprises of a lot of cases that are appropriately marked with the right class names. A learning calculation at that point creates a model that endeavors to limit the forecast blunder on the preparation occasions, and furthermore sum up beyond what many would consider possible to anticipate the new information. In writing, different arrangement calculations are accessible, for example, Decision Tree(DT), closest neighbor (NN) rule, the Bayes classifier, bolster vector machines (SVM), and neural systems.

MOEAs have been broadly utilized for the order. There are principally three unique approaches. I. Building up a decent arrangement of characterization rules. II. Characterizing the hyperplanes for class limits in the preparation informational collection. III. For the development of a decent classifier. In the past, scientists connected solo malignant growth order procedure dependent on multi-objective hereditary fluffy grouping for tissue information tests. In such manner, organize of the bunch focuses have been encoded in the chromosomes and three fluffy group legitimacy lists are at the same time streamlined. Every arrangement of the resultant Pareto-ideal set has been helped by a novel method dependent on Support Vector Machine (SVM) grouping (Fig. 2).

5 Clustering

Clustering is a significant unaided arrangement strategy [10]. In bunch investigation, the informational indexes are mapped in a multidimensional, are gathered into groups so that information object in a similar group is like one another, and information questions in various bunches are not at all like one another. The grouping

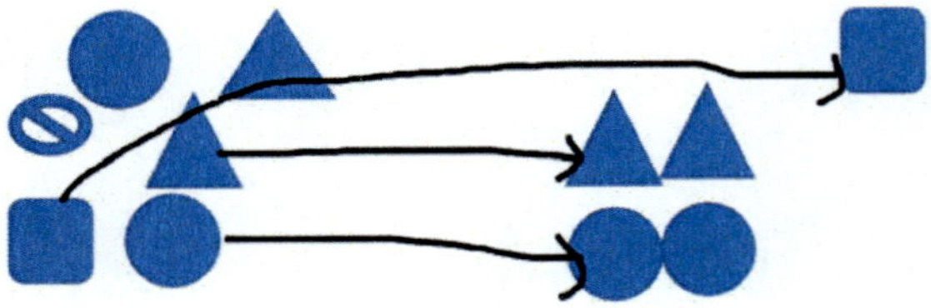

Fig. 3 Clustering

systems check for more intra likeness and less bury similitude among the information objects. In the writing, they are different bunching calculations, for example, K-mean calculations, K-middle calculations, DBSCAN calculations, Density-based calculations.

There have been a fewer works identified with the advancement of transformative calculations for multi-objective solo calculations. For the unaided case, the calculations don't expect the informational indexes have earlier class marks, and, in this manner, there is no preparation set. For this case, typically, a bunching calculation is utilized to assess dependent on how well these highlights can recognize the grouping structure of the dataset. In such a manner, to assess the integrity of the grouping arrangement bunch legitimacy record has been utilized to create the element subset. An epic multi-objective developmental grouping approach has been proposed utilizing Variable-length Real Jumping Genes Genetic Algorithms (VRJGGA). The proposed calculation utilized hereditary methodology called Jumping Genes Genetic Algorithm (JGGA) to develop close ideal grouping arrangements utilizing numerous bunching criteria, without the earlier information of the real number of groups (Fig. 3).

6 Conclusion

As most data mining algorithms need to have model parameters optimization with multiple performance criteria and multi-objective algorithms are the best choice for dealing with such tasks. Over the past decade, MOEAs have become very popular because of their flexibility in solving a data mining problem within the data mining community. Therefore, over the years, a number of MOEAs have been proposed for solving a variation of data mining problems. This article introduced some basic concepts related to multi-objective optimization and basics of data mining. Then, we discussed how MOEAs employed for solving data mining tasks, such as feature selection, classification and clustering. This article invites future research to work in this field.

References

1. Maulik, U., Bandyopadhyay, S., & Mukhopadhyay, A. (2011). *Multiobjective genetic algorithms for clustering—applications in data mining and bioinformatics*. Berlin, Germany: Springer.
2. Han, J., & Kamber, M. (2000). *Data mining: Concepts and techniques*. San Francisco, CA, USA: Morgan Kaufmann.
3. Goldberg, D. E. (1989). *Genetic algorithms in search, optimization and machine learning*. New York, NY, USA: Addison-Wesley.
4. Deb, K. (2001). *Multi-Objective optimization using evolutionary algorithms*. London, U.K.: Wiley.
5. Coello, C. A. C., Lamont, G. B., & Van Veldhuizen, D. A. (2007). *Evolutionary algorithms for solving multi-objective problems* (Vol. 5, pp. 79-104). New York: Springer.
6. Maulik, U., Holder, L. B., & Cook, D. J. (Eds.). (2005). *Advanced methods for knowledge discovery from complex data*. (Advanced Information and Knowledge Processing). London, U.K.: Springer-Verlag.
7. Tseng, L. Y., & Yang, S. B. (1997). Genetic algorithms for clustering, feature selection and classification. In *Proceedings of International Conference on Neural Networks (ICNN'97)* (Vol. 3, pp. 1612-1616). IEEE.
8. Karzynski, M., Mateos, A., Herrero, J., & Dopazo, J. (2003). Using a genetic algorithm and a perceptron for feature selection and supervised class learning in DNA microarray data. *Artificial Intelligence Review, 20*(1–2), 39–51.
9. Bandyopadhyay, S., & Pal, S. K. (2007). *Classification and learning using genetic algorithms: applications in bioinformatics and web intelligence*. Natural Computing Series. Berlin.
10. Jain, A. K., & Dubes, R. C. (1988). *Algorithms for clustering data*. Englewood Cliffs, NJ, USA: Prentice-Hall.
11. Falkenauer, E. (1998). *Genetic algorithms and grouping problems*. John Wiley & Sons, Inc..
12. Freitas, A.A., (2002). A survey of evolutionary algorithms for data mining and knowledge discovery. www.pgia.pucpr.br/~alex/papers. A chapter of: A. Ghosh and S. Tsutsui. (Eds.) Advances in Evolutionary Computation. Springer-Verlag.

A Multi-Criteria Decision Approach for Movie Recommendation Using Machine Learning Techniques

D. Anji Reddy and G. Narasimha

Abstract The challenges in decision making can be viewed as selecting an appropriate method from multi-criteria decision making (MCDM). The MCDM methods do not offer similar answers to decision makers. In such cases, selection of the best answer becomes an issue. This is possible when there may be a better choice which was not considered or the unavailability of the right information at that time. MCDM contains a limited number of alternatives for every problem. In the initial process for the solution, all these solutions are explicitly identified. In this work, we have used machine learning techniques to fulfill the objective of MCDM also known as multi-criteria decision making (MCDM). The objective is to rank the movies by using various alternatives based on the weights of several criterions. In this proposed work, different kinds of movies like science fiction, cartoon, and adventures are used to generate the ranking. Other criterions that are used such as various factors like price, movie reviews and duration are used for movie ranking. The movie ranking is generated based on the weightage.

Keywords MCDM · Ranking · Reviews · Decision making · And weightage

1 Introduction

This paper exclusively focuses on the process to apply the multi-criteria decision making (MCDM). approach using machine learning techniques. The approach of MCDA is appropriate when an intuitive method is not suitable. This will happen when the researchers feel that the decision is complex and too large to deal intuitively. This situation may arise because of a number of inconsistent objectives or it also involves numerous participants with dissimilar views [5, 6, 7]. This kind of recommendation

D. Anji Reddy (✉)
Vaageshwari College of Engineering, Karimnagar, Telangana, India
e-mail: anjireddy.knr@gmail.com

G. Narasimha
Computer Science and Engineering, JNTUHCE, Sultanpar, Telangana, India

S. C. Satapathy et al. (eds.), *Intelligent System Design*, Advances in Intelligent Systems and Computing 1171, https://doi.org/10.1007/978-981-15-5400-1_44

systems can be developed using various approaches to obtain ranking-based results or to obtain expert views [1, 2, 3, 4].

MCDA provides a functional and logical framework to define the best overall result. MCDA methods are developed and used to support the decision-makers in their distinctive and personal decision process. The knowledge gaining through MCDA includes informatics, economics, mathematics, management, social science, and psychology. Its applications are used in wide areas to solve any problem wherever a substantial decision needs to be made. The decisions made can be strategic or it can be tactical, depending on the perception of the time and its significance [9]. There are a large number of approaches that are developed to solve multi-criteria problems [8, 9].

The following table shows MCDA problems and methods (Table 1).

Various softwares are available to address the MCDM methods. Few of them are Decision Lab, Frontier Analyst D-Sight, Visual Promethee, Smart Picker Pro, Efficiency Measurement System, DEA Solver Online, Electre III–IV Right Choice, DEAFrontier, and DEA-Solver PRO.

1.1 Machine Learning

Machine learning is known as a popular application of artificial intelligence (AI). Machine learning provides the ability of automatic learning to the systems. This learning will improve from experience of the system when it is used, without being explicitly programmed. Applications of machine learning focus on developing computer programs to access data and to use these models to learn for themselves.

Table 1 MCDA problem classification and solutions

S. No.	Choice problems	Sorting problems	Ranking problems	Description problems
1	AHP	AHP	AHPSort	
2	ANP	ANP		
3	MAUT/UAT	MAUT/UAT	UTADIS	
4	MACBEATH	MACBEATH		
5	PROMOTHEE	PROMOTHEE	FlowSort	GAIA, FS-Gaia
6	ELECTRE-I	ELECTRE-III	ELECTRE-Tri	
7	TOPSIS	TOPSIS		
8	GOAL PROGRAMMING			
9	DEA	DEA		
10	HYBRID METHODS			

The rest of the paper is organized as follows. Section 2 describes the literature review, Sect. 3 discusses methodology and Sect. 4 gives an insight into experimental results and finally, conclusion about the paper is given in Sect. 5.

2 Literature Review

In 1981, Roy has identified four types of decision-making systems. These are described as the description problem, the choice problem, the sorting problem, and ranking problem [10]. In the year 1996, Bana e Costa has proposed the elimination problem as a particular branch of the sorting problem [11]. The design problem is defined as the goal to create or identify a new action. To meet the goals and objectives of the decision maker is expressed by Keeney in 1992 [12]. In the year 1999, Guitouni et al. have proposed a framework for choosing an appropriate multi-criteria procedure using an initial investigative method [13]. Since 45 years, there have been almost 70 techniques for MCDM have been explored [14]. MADM associates the selection of the best alternatives that are pre-specified and described in terms of multiple attributes [15]. MCDM can be applied in many areas of research and is used for selection, evaluation, and ranking in these areas [16, 17, 18, 19]. Multi-attribute decision making provides a trivial number of alternatives that are to be evaluated against a set of attributes that are often hard to quantify [20].

3 Methodology

In the University of Minnesota, a research team by name GroupLens Research Project has collected MovieLens data sets that contain 100,000 ratings ranges from 1 to 5. They have collected these reviews from 1682 movies by 943 users. Every user has rated nearly 20 movies and provided with simple demographic information for the users like age, gender, occupation, and zip code [21].

The data file *u.item* gives information about the items (movies). This data file consists a list of parameters such as movie-id, movie title, release date, video release date, IMDb URL, unknown, animation, action, comedy, Adventure, Children's, Crime, Drama, Documentary, Fantasy, horror, film-noir, musical, romance, mystery, science fiction, thriller movies, war, and western. From the above, 19 fields are the genres, and here, *1* indicates the movie is of that genre and *0* indicates as it is not a genre. Movies can be in one genre or in several genres. The other data file used here is *u.genre* which is a list of the genres.

Initially, we applied a statistical prediction to calculate precision and recall measurements. Later, logistic regression which is known as a famous machine learning algorithm is used to build a model. Then, finally, decision tree algorithm is also used on the same data set for finding out precision and recall values.

4 Experimental Results

Here, we have used statistical prediction decision tree and logistic regression algorithms for obtaining ranking. Data loading and pre-processing have been done for the given data. Here, the authors have added supporting columns *price* in a range of 5 levels ($-2, -1, 0, 1, 2$) and *buy probability* in two levels (cheaper and costlier) for rank prediction. The calculation of rank prediction is as follows.

4.1 Price

Starting date of the movies in data set (considered as 'oldest_date')

Ending date of the movies in data set (considered as 'most_recent_date')

$$normalized\_age = \frac{most\_recent\_date - movie\_releasedate}{most\_recent\_date - oldest\_date}$$

$$normalized\_rating = \frac{5 - movie\_data['ratings\_average']}{5 - 1}$$

$$price = \left(1 - normalized_{rating}\right) * \left(1 - normalized_{age}\right) * 10$$

4.2 Buy Probability

$$movie\_data['buy\_probability'] = 1 - movie\_data['price'] * 0.1$$

$$Buy\_Probability = 1 - price * 0.1$$

Here, the authors have a build model which shows as, if rank is Lower(Lower *the Rank)* then it will be treated as the highest priority(*Higher the Priority).* In the graph, below 0 denotes the high rank *(0- First rank).* In the below graphs, X-AXIS represents the feature, and Y-AXIS represents the count of MOVIES for marking data analysis. Figure 1 explains data distribution among various features of the data set.

Figure 2 represents the price vs buy_probablity which is a general structure of movie ranking.

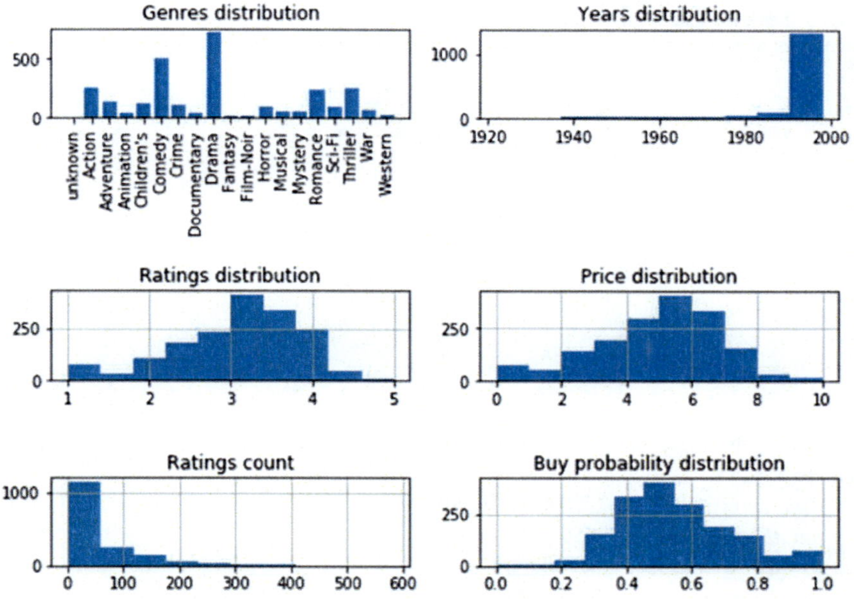

Fig. 1 Data genres distribution

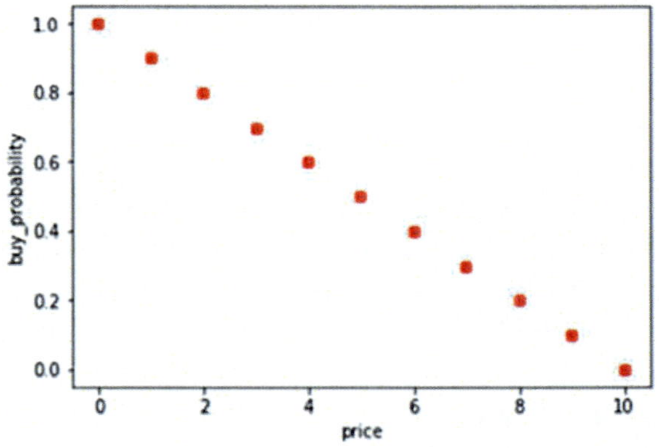

Fig. 2 Price versus buy_probablity which is a general structure of ranking

4.3 Perfect Predictor (Statistical Prediction)

The data set used is applied with scaling the data from minimum to maximum and then rankings are plotted which results with a minimum cost. The measurements, precision recall, are calculated by training the model with 75% of the data and testing the model with 25% of the data which consist of 23 various features as shown in Fig. 3.

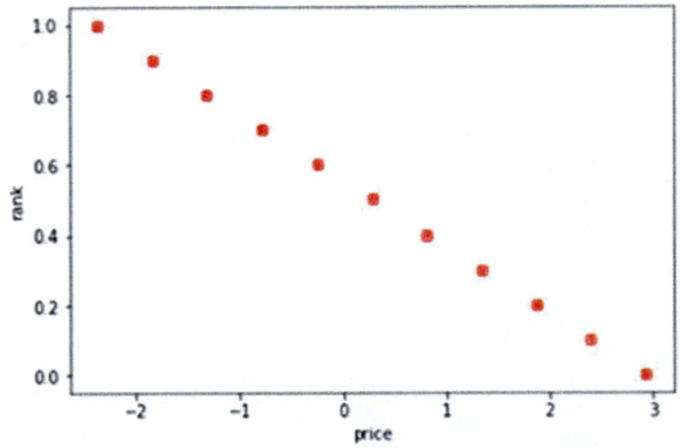

Fig. 3 Scale the data from minimum to maximum

After submitting the data into the model, the obtained precision and recall values are as follows.

Precision(training): 0.737659068662
Recall(training): 0.586567330146
Accuracy(training): 0.650025025025
Precision(testing): 0.714364336819
Recall(testing): 0.588128681468
Accuracy(testing): 0.642642642643
overall input shape: (1681, 23).

4.4 Logistic Regression Ranking

Logistic regression is used to get the probability along with raw events and run the regression. We will get the coefficient to represent probability generated artificially as shown in Fig. 4. It also explains that the price is not only a factor to rank a movie, where the movie of y-axis is trained on 23 features.

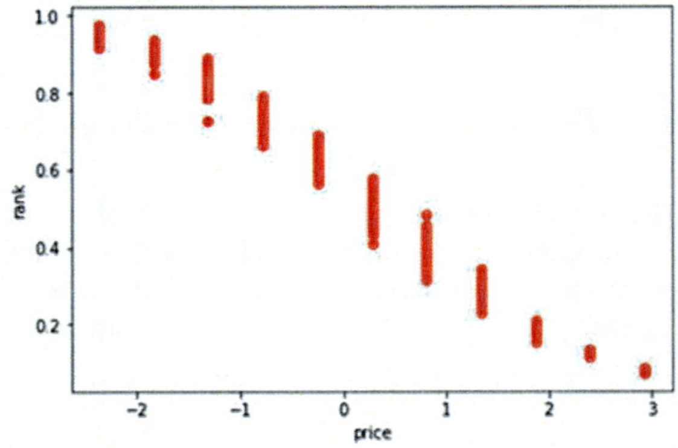

Fig. 4 Price along with movie

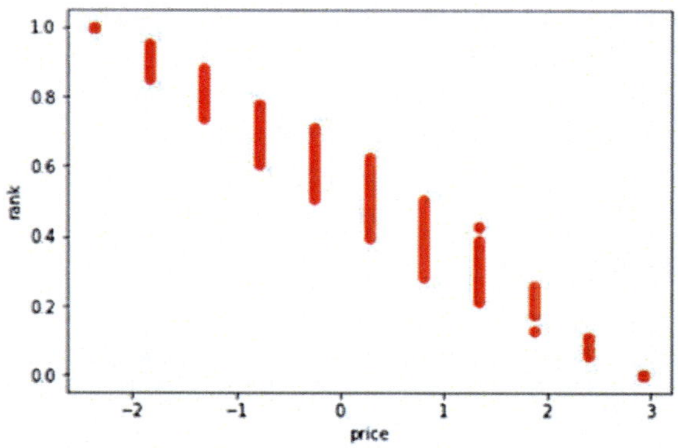

Fig. 5 Price along with movies

The precision, recall, and accuracy values are as follows for the logistic regression.
Precision(training): 0.674018877298
Recall(training): 0.754364505727
Accuracy(training): 0.656531531532
Precision(testing): 0.649592549476
Recall(testing): 0.758495695514
Accuracy(testing): 0.640640640641.

4.5 Decision Tree Ranking

The decision tree ranking method also trained using 23 features. The following graph in Fig. 5 explains the price along with the movies.

The precision, recall, and accuracy values are as follows for the decision tree
Precision(training): 0.680947848951
Recall(training): 0.711256135779
Accuracy(training): 0.653892069603
precision(testing): 0.668242778542
recall(testing): 0.704538759602
accuracy(testing): 0.644044702235

After analyzing the results, the comparative study shows that logistic regression performs better than other algorithms used in this work with 65% accuracy. The other algorithms used are decision tree and statistical prediction to find movie ranking.

5 Conclusion

In this work, statistical and machine learning methods are used to predict the movie rankings based on the multiple criterions. The results obtained by statistical methods are compared with the results obtained using popular machine learning techniques, logistic regression and decision tree. Though the statistical method developed satisfactory results, logistic regression has outperformed by producing the accuracy of 65% in the prediction of ranking. This confirms that machine learning can work effectively in multi-criterion decision making.

References

1. Lakiotaki, K., Matsatsinis, N. F., & Tsoukias, A. (2011). Multicriteria user modeling in recommender systems. *IEEE Intelligent Systems, 26*(2), 64–76.https://doi.org/10.1109/mis.2011.33.
2. Sridevi, M., et al. (2016). A survey on recommender system. *International Journal of Computer Science and Information Security, 14*(5), 265–272.
3. Khakbaz, S. B., & Maryam Karimi Davijani, M. K. (2015). Ranking Multi Criteria Decision Making Methods for a Problem by Area Under Receiver Operating Characteristic. *Journal of Investment and Management, 4*(5), 210–215. https://doi.org/10.11648/j.jim.20150405.21.
4. Mardani, Abbas, Jusoh, Ahmad, Nor, Khalil M. D., Khalifah, Zainab, Zakwan, Norhayati, & Valipour, Alireza. (2015). Multiple criteria decision-making techniques and their applications—a review of the literature from 2000 to 2014. *Economic Research-Ekonomska Istraživanja, 28*(1), 516–571. https://doi.org/10.1080/1331677X.2015.1075139.
5. Mabin, V., & Beattie, M. (2006). A practical guide to multi-criteria decision analysis. *Victoria University of Wellington.*
6. Belton, V. (1990). Multiple Criteria Decision Analysis: Practically the only way to choose. In L. C. Hendry & R. W. Eglese (Eds.), *Operational Research Tutorial Papers*. Birmingham: Operational Research Society.
7. Visual Thinking International Limited VISA http://www.simul8.com/products/visa.htm.
8. http://www-bd.cricket.org/.
9. Ishizaka, A. (2013). *Multi-criteria decision analysis: methods and software*. John Wiley & Sons. ISBN 978-1-119-97407-9.
10. Roy, B. (1981). The optimisation problem formulation: Criticism and overstepping. *Journal of the Operational Research Sociey, 32*(6), 427–436.
11. Costa, B. E., & Carlos, A. (1996). Les problématiques de l'aide à la décision: vers l'enrichissement de la trilogie choix-tri-rangement. *RAIRO-Operations Research-Recherche Opérationnelle, 30*(2), 191–216.
12. Keeney, R. (1992). *Value-Focused Thinking: A Path to Creative Decision Making*. Cambridge, MA: Harvard University Press.
13. Guitouni, A., Martel, J. M., Vincke, P. (1998). A framework to choose a discrete multicriterion aggregation procedure. Technical Report.
14. Alias, M. A., Hashim, S. Z. M., & Samsudin, S. (2008). Multi criteria decision making and its applications: a literature review. *Jurnal Teknologi Maklumat, 20*(2).
15. Rao, R. V. (2007). Introduction to multiple attribute decision-making (MADM) methods. *Decision Making in the Manufacturing Environment: Springer Series in Advanced Manufacturing*, 27–41.
16. Karmperis, A. C., Aravossis, K., Tatsiopoulos, I. P., & Sotirchos, A. (2013). Decision support models for solid waste management: Review and game-theoretic approaches. *Waste management*, ELSEVIER.
17. Aruldoss, M., Lakshmi, T. M., & Venkatesan, V. P. (2013). A survey on multi criteria decision making methods and its applications. *American Journal of Information Systems, 1*(1), 31–43.
18. Bernroider, E. W., & Mitlohner, J. (2005). Characteristics of the multiple attribute decision making methodology in enterprise resource planning software decisions. *Communications of the IIMA, 5*(1), 49–58.

19. Agarwal, Prince, Sahai, Manjari, Mishra, Vaibhav, Bag, Monark, & Singh, Vrijendra. (2011). A review of multi-criteria decision making techniques for supplier evaluation and selection. *International Journal of Industrial Engineering Computations, 2,* 801–810.
20. Hwang, C. L., & Yoon, K. (1981). *Multiple attribute decision making: methods and applications*. Berlin: Springer.
21. Harper, F. M., & Konstan, J. A. (2015). The movielens datasets: History and context. *ACM Transactions on Interactive Intelligent Systems (Tiis)*, *5*(4), 19. https://doi.org/10.1145/2827872.

Scope of Visual-Based Similarity Approach Using Convolutional Neural Network on Phishing Website Detection

Check for updates

J. Rajaram and M. Dhasaratham

Abstract Phishing website is an illegitimate website that is designed by dishonest people to mimic a real website. Those who are entering such a website may expose their sensitive information to the attacker who might use this information for financial and criminal activities. In this technological world, phishing websites are created using new techniques allows them to escape from most anti-phishing tool. So, the white list and blacklist based techniques are less effective when compared with the recent phishing trends. Advanced to that, there exist some tools using machine learning and deep learning approaches by examining webpage content in order to detect phishing websites. Along with the rapid growth of phishing technologies, it is needed to improve the effectiveness and efficiency of phishing website detection. This work reviewed many papers that proposed different real-time as well as non-real-time techniques. As the result, this study suggests a Convolutional Neural Network (CNN) framework with 18 layers and scope of transfer learning in Alex Net for the classification of websites using screenshot images and URLs of phishing and legitimate websites. CNN is a class of deep, feed-forward artificial neural networks (where connections between nodes do not form a cycle) & use a variation of multilayer perceptions designed to require minimal preprocessing.

Keywords Phishing · Character level CNN · Deep learning · Transfer learning · APWG · Alex net

1 Introduction

In recent years, the most spreading online fraud activities are from the result of phishing and malware-based attacks. A recent study from RSA Security, a Dell Technologies business reported that the US, the Netherlands and Canada are the top three countries targeted by phishing attacks. It is shocking that India serves a fourth place in that report. The phishing concept came into existence in the early 1990s via

J. Rajaram (✉) · M. Dhasaratham
Department of Computer Science & Engineering, Teegala Krishna Reddy Engineering College, Hyderabad, India
e-mail: drjrajaram81@gmail.com

S. C. Satapathy et al. (eds.), *Intelligent System Design*, Advances in Intelligent Systems and Computing 1171, https://doi.org/10.1007/978-981-15-5400-1_45

America Online or AOL. A group of hackers and pirates together named the warez community is the first "phishers." They developed an algorithm to generate random credit card numbers, which they would then attempt to use to make fraudulent AOL accounts in trial and error manner. If there was a match found with any real card, they were able to create an account and spam others in AOL's community. By 1995, AOL halted those generators but contradicts to that the warez group moved on to other methods, specifically behaved as AOL employees and messaging people using AOL Messenger for their information. On January 2, 1996, the word "phishing" was first broken away in a Usenet group dedicated to American Online. Then AOL eventually included warnings on all its email and messaging software to alert users of potential phishing abuse. But phishing attacks moved on the different approaches [1].

Phishing is an attack that involves both social engineering research and deception to steal user's personal and financial information. Techniques targeted at using spoofed emails purporting will be from businesses and legal entities, to obtain sensitive information from the victim through third-party fraud. Technical deception schemes work directly on computers on steep credentials, which often use user-friendly user names and password tracking systems—and to corrupt local navigation, infrastructure to attract false websites to consumers [2]. Attacks using phishing websites growing drastically in the internet world since web surfing became inevitable to individuals nowadays. Phishing websites are a similar web site for legitimate websites to swindle web users in order to obtain their personal and financial information. Detection and protection of phishing websites must be carefully done by predicting the trends of attackers for doing so. According to quarterly based APWG Phishing Activity Trends Report 2018, the total number of phishing websites was 138,328, reduced around 50% within one year. Phishing that targeted SaaS/webmail services (29.8%) doubled in Q4 and the most targeted sector is the payment service with 33% of phishing attacks. The total count of phishing attacks hosted on web sites that have HTTPS and SSL certificates decreased in that period [3].

There are different forms of phishing attacks in theory [1] as listed below:

- Phishing Attack by Fraud
- Phishing Attack by Infectious software
- Phishing Attack by MITM approach
- Phishing Attack by DNS spoofing
- Phishing Attack by Inserting harmful content
- Phishing Attack by Search Engine indexing.

In phishing attack by fraud, the user is fooled by fraudulent emails to disclose personal or confidential information. While phishing attack by infectious software such as key loggers and screen loggers, the attacker succeeds in running dangerous software on user's computer. The DNS spoofing is used by phishers compromises the domain lookup process so that the user's click would lead him or her to a fake website. Another way of phishing is the attacker puts malicious content into a normal website in order to extract personal or sensitive information. In phishing attack by MITM approach, the attacker stays in between the user and the legitimate site to taps sensitive information. The fake web pages created by attacker give attractive offers

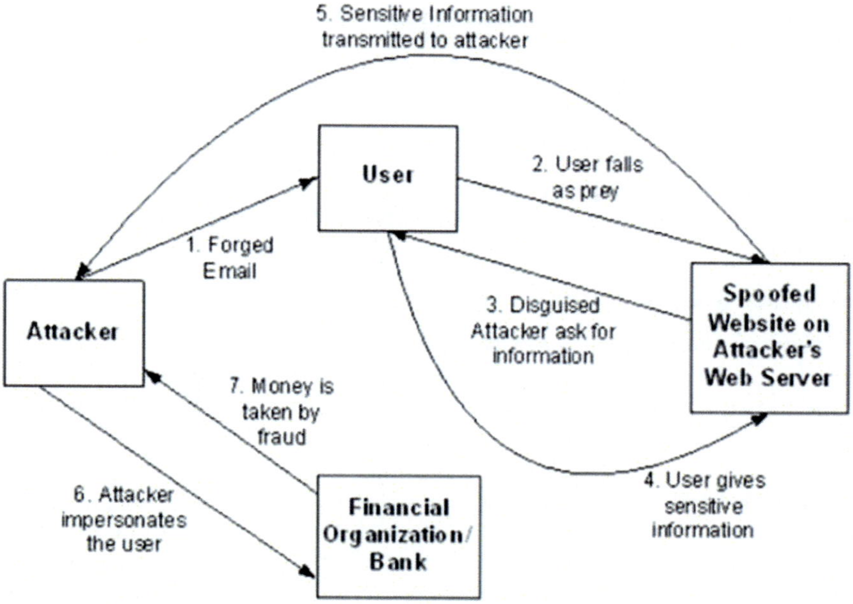

Fig. 1 Stages of phishing [1]

to be top indexed by a search engine so that a user would easily fall in it. Similarly, there are many ways for phishing attack detection and prevention. The stages of a general phishing attack are summarized in Fig. 1. Many researchers found different techniques against phishing but counter wards phishing experts developed many other practical ways of spoofing. Researchers in the advanced technology domain worked hard to prevent phishing in real-time by inventing many tools for detection and prevention of websites. Out of many domains, upcoming technology called Deep Learning has a significant role in handling phishing attacks.

1.1 Deep Learning

Deep learning is an advancement to the machine learning concept; a model learns itself to perform classification tasks directly from images, text, or sound. Deep learning is usually implemented using neural network architecture [4]. Traditional neural networks contain solely a pair of or three layers, while deep networks can have hundreds. Computer programs that use deep learning bear a lot of an equivalent method. Each algorithm in the hierarchy applies a nonlinear transformation on its input and uses what it learns to create a statistical model as output. The characteristics of deep learning are

- Easy access to large volumes of named text files:

Data sets are freely available and are useful for training on many different types of objects.

- Increased computing power:

High-performance GPUs in deep learning speed up the training of the massive amounts of data needed for deep learning, which reduces training time from weeks to hours.

- Pretrained models built by experts:

Pretrained models such as AlexNet can be used to perform new recognition tasks using a technique called *transfer learning* with the highest accuracy.

A deep neural network combines multiple nonlinear process layers, using simple elements operating in parallel and inspired by biological nervous systems. It consists of the associate input layer, several hidden layers, and an output layer. The layers are interconnected over nodes, or neurons, with each hidden layer using the output of the previous layer as its input. With the use of these features, real-time phishing detection became more accurate as compared with the traditional approach [4].

Phishing website detection tools can be developed by using the features of websites such as text, frames, and images [25]. Many researchers introduced phishing detection scheme based on text features only, some introduced detection scheme with the use of image features only, some based on text and frame features only, and based on text and image features only. Author Adebowale et al. 2019 introduced a scheme called intelligent phishing website detection and protection scheme (IPDPS) with the use of ANFIS architecture by looking into integrated features of images, frames and text [5]. Further details about this scheme are included in the later section.

2 Review

Detecting and preventing the phishing attack is an important step towards securing phishing attacks on websites. There are some approaches to detecting these attacks. There are several phishing-related review papers available currently. From those papers, it is noted that there are different techniques for detection phishing attacks. Generally, they are classified as follows [6]:

- Attribute-Based Anti-phishing Techniques
- Genetic Algorithm Based Anti-phishing Techniques
- Identity-Based Anti-phishing Techniques
- Content-Based Anti-phishing Approach
- Character-Based Anti-phishing Approach
- Visual Similarity Based Anti-phishing Approach.

2.1 Attribute-Based Anti-phishing Techniques

Attribute-based anti-phishing strategy applies both reactive and proactive anti-phishing defenses. This technique has been implemented in the Phish Bouncer tool [7]. An important aspect of the PhishBouncer approach is the plug-in framework, which provides a flexible way of applying personal adaptive and reactive logic to HTTP (S) streams. All anti-phishing logic was implemented as a set of plug-ins. They divided the plug-ins into three broad classes based on their role in the overall control flow and threading logic. The three classes were as follows: Dataplugins; called on every HTTP request and associated response to perform analysis on header and payload data. Checks; sequentially execute when HTTP requests entering the proxy and decide whether to accept the request, reject the request, or set a numeric value to indicate the confidence and choice selection. Probes; allow embedding proactive behavior into the proxy. PhishBouncer contains a set of nine anti-phishing checks and their supporting data plug-ins including Suspected by Image Attribution check, Domain Too Young check, HTML Crosslink check, False Info Feeder check, Certificate Suspicious check, URL Suspicious check, Leaking User Data check, Phish Signature Match check, and Referred by Webmail check. PhishBouncer combines the outcome of individual checks to detect phishing attacks.

Advantages

- More phished sites were detected than traditional approaches
- Unknown attacks also are detected.

Disadvantage

- Slow response time due to multiple checking of authentication.

2.2 Genetic Algorithm Based Anti-phishing Techniques

In this approach [8], the genetic algorithm (GA) is applied to develop rules that discern the phishing link from the legal link. In order to assess the parameters, such as the evaluation function, crossing and mutation, GA generates a set of rules that only adapts to phishing links. Those rules are stored in a database and a hyperlink activates as a phishing link if it matches any of these system rules and, therefore, is kept safe from the false hackers. The genetic algorithm not only useful for detecting phishing attacks but also protects users from malicious or unwanted links to web pages. The rules preserved in the rule base are in the following general form [9]:

If {condition}

Then {act} [9]

This rule can be explained with an example as follows: if there exists an IP address of the URL in the received email and it does not match the defined Rule Set for White List then the received mail is reported as phishing mail and denoted as follows:

If {IP address of the URL in the received email matches the Rule set}
Then {E-mail is Phishing mail}.

Advantages

- Before the user enters into the mail, it notifies the feature of malicious status
- Not only phishing detection but also malicious web link detection.

Disadvantages

- Need multiple rules set for each URL type to detect phishing
- It is to be needed to write new rules set which leads to a more complex algorithm for other parameters.

2.3 *Identity-Based Anti-phishing Techniques*

The identity-based anti-phishing exploits the non-technical/inexperienced users who cannot identify spoofed emails or web sites and the relative ease of masqueraders. Authors Tout and Hafner [10] present Phishpin, an approach that uses mutual authentication concepts, requires online elements to prove their identities. This anti-phishing approach that combines client-based filtering and domain-based identity techniques and integrates partial credentials and filtering to enforce bi-directional authentication without revealing sensitive information. By doing mutual authentication, users do not need to re-enter their credentials. Thus, the passwords between users and individuals through the internet have changed only through the initial account setup method.

Advantages

- Provide mutual authentication for both client-side and server-side
- Client password disclosure is not needed for each session.

Disadvantages

- If a hacker disabled the browser plug-in after gaining access permission to the client computer and then the method will not detect phishing attack.
- When the session is initialized for the first time, it is compulsory to disclose the password.

2.4 *Content-Based Anti-phishing Approach*

This approach identifies the similarity between the legitimate web page and suspicious/phishing web pages with respect to textual as well as visual contents. A content-based approach to detecting phishing sites CANTINA [11], examines the content of a webpage to determine whether or not it is legitimate, rather than looking at the surface characteristics such as URL and domain name of a web page. CANTINA uses the well-known TF-IDF (document term/inverse frequency) algorithm used to retrieve information. Experiments result that CANTINA was good at detecting phishing sites with accuracy approximate to 95%. CANTINA was implemented as a Microsoft internet Explorer extension. Similarly, another tool named GoldPhish [12] implemented this approach and used Google as its search engine. This tool protects against zero-day phishing attacks with high accuracy. The procedure is as follows: First, it captures an image of a page, and then uses optical character recognition to convert the image to text, and grasps the Google PageRank algorithm to reach decision on the validity of the site.

Advantages

- Both CANTINA and GoldPhish provides zero-day phishing and very low false-positive results.

Disadvantages

- Time lag entangled in querying Google degrades the performance of CANTINA
- CANTINA: No dictionary for languages other than English
- GoldPhish delays webpage login
- GoldPhish may the webpage vulnerable to attacks on Google's PageRank algorithm and Google's search service.

2.5 *Character-Based Anti-phishing Approach*

Phishers always try to steal information of users by misleading them by clicking on the hyperlink that embedded into phishing emails. The format of a hyperlink is as follows:

< ahref="URL" > Anchor text <\a>

The "Universal Resource Locator" (URL) a unique identifier that provides the web link where the user will be re-directed and "Anchor text" is the hypertext represents the visual link [6]. The unique characteristics or features of hyperlink were used in the character-based anti-phishing technique especially to detect phishing links. LinkGuard [13] tool implements this technique which extracts the DNS names from the actual and the visual links and then compares with each other, if these names have matched, then said to be a legitimate otherwise phishing. Authors [13] considered the different possibilities of mismatch and rectified with respect to it. If IP address in the actual DNS is in the form of dot-decimal, then suspicious. If the actual link or the visual link is in encoded form, then the link is decoded first and then analyzed. When there is no destination information (DNS name or dotted IP address) in the visual link then the hyperlink is analyzed. During the period of analysis, LinkGuard searches the DNS name in blacklist as well as whitelist, if it is present in either of the list then classify accordingly. If it is not contained in either whitelist or blacklist, pattern matching will be done. During pattern matching, the sender email address is extracted, and then it is searched in a list of address is maintained that is manually visited by the user. URLNet is the deep learning technique proposed to character-based anti-phishing technique. This network uses character level CNN as well as word-level CNN to URL classification.

Advantage

- More Effective; not only for known attacks but also to the unknown ones
- LinkGuard can detect up to 96% unknown phishing attacks in real-time.

Disadvantage

- The dotted-decimal IP addresses may be relevant in some special situations even though LinkGuard results it as phishing (False Positives).

2.6 *Visual Similarity Based Approach*

Phishers always try to design a website that looks exactly as a corresponding legitimate website, but the reality is they always keep a minor difference between them. So the visual similarity cannot be omitted when thinking about anti-phishing method. The two key points of visual similarity based approaches are as follows: [31]

(1) Attackers usually insert ActiveX, images, Java Applet, and Flashes in place of HTML text to get hidden from anti-phishing detection. Visual similarity based detection approaches easily detect such embedded objects.

(2) Techniques based on visual similarity use a signature to identify phishing web pages. The signature is created with common site-wide features rather than a single web page. Therefore, a signature is needed enough to detect multiple web pages targeted from a single site or from different versions of a site.

Author Dhamija et al. [26] conducted a survey on the visual similarity based approach on phishing detection, as the result they found the phishing intelligence of attackers using the visual similarity. In their study, it was noted that about 23% of the experienced users missed to verify the URLs of the phishing websites. So it is evident that even experienced users may fooled by phishing attackers. Sadia Afroz et al. [27] presented a new approach called PhishZoo to handle this attack with similar accuracy to blacklisting approaches. It was used the profiles of trusted website's appearances to detect phishing. Chen et al. used screenshot of web pages to detect phishing sites [28]. They used Contrast Context Histogram (CCH) to describe the images of web pages and k-mean algorithm to cluster nearest key points. Finally Euclidean distance between two descriptors is used to find matching between two sites. Their approach has 95–99% accuracy with 0.1% false-positive. Fu et al. [29] used Earth Mover's Distance (EMD) to compare low-resolution screen capture of a webpage. Images of web pages are represented using color of a pixel in the image (alpha, red, green, and blue) and the centroid of its position distribution in the image. They used machine learning to select different thresholds suitable for different web pages. Shuichiro Haruta et al. [30] proposed visual similarity based phishing detection scheme using image and CSS with target website finder which results in an accuracy of 72.1%. Since CSS contains the website visual contents, they used this feature to detect the website which plagiarizes appearance or CSS of legitimate website.

2.7 *Intelligent Phishing Detection System*

All the above anti-phishing techniques have many features as well as limitations. Those techniques can be integrated into many ways to phishing detection. Intelligent phishing detection can be developed in different ways; Using Artificial Neural Network (ANN), Neuro-Fuzzy logic, machine learning algorithms, and Convolutional Neural Network of Deep Learning approach.

2.7.1 Artificial Neural Network Architecture

An artificial neural network is an interconnected group of nodes, inspired by a simplification of neurons in a brain [15]. In other words, neural networks are a set of algorithms, modeled loosely after the human brain, that is designed to recognize patterns. Author Ramy Mohammad et al. [16] introduced an Artificial Neural Network model with backpropagation to phishing detection. They studied many NNs to determine required NN parameters, such as "the number of hidden layers, the number of hidden neurons, learning speed, etc."; which results in better prediction accuracy. This technique in predicting phishing websites with reduced training time gives out as the automation of the process of NN building. They show the result on different number of neurons 8,5,4,3,2 but the better result is obtained when number of hidden neurons are set to 2, used 18 features to classify and trained the neural network. Many researchers made advancements in this ANN to improve accuracy. Reference [17] presented an ANN-MLP based phishing website classification model instead of single-layered ANNs. This results in an accuracy of 98.23% at the test phase. Reference [18] presented another variant ANN with Particle Swarm Optimization (PSO) algorithm instead of Back Propagation to classify the Uniform Resource Locator (URL) into Phishing URL or Non-phishing URL. Dataset with 31 attributes was used for PSO-ANN modeling and achieved better accuracy with a different number of hidden layer and output layer than backpropagation neural network. Most of the research on this was based on text features only which gives the new scope.

2.7.2 Neuro-Fuzzy System

Neural networks have strong learning capabilities at a numerical level, but users have difficulty understanding it. On the other hand, Fuzzy systems have good ability to explain vague arguments and integrate the knowledge of skills. Both paradigms provide the ability to learn hybridization, include good interpretation and insight knowledge. Thus the concept of Neuro-Fuzzy systems evolved. This model was developed by inputting the fuzzification layers on neural networks for learning practice. Neuro-fuzzy models describe the systems by using fuzzy *if-then* rules, such as "*If x is small then y is large*" represented in a nested manner, to which learning algorithms known from the area of artificial neural networks can be applied [19]. A paper on Neuro-Fuzzy Methods for Modeling and Identification refers that this model can be named as a graybox technique on the boundary between neural networks and qualitative fuzzy models [19]. Many anti-phishing tools were developed using this model or system so far. Author Barrowclough et al. [20] presents a solution for inadequacy faced by online transactions. They claim that the inputs used for their study were not considered before in a single protection platform as hybrid way. They extracted a total of 288 features from these five input sources. Neuro-Fuzzy systems with five inputs offer better accuracy and also be effective in detecting phishing sites in real-time.

A drawback of Neuro-Fuzzy modeling is that the current techniques for constructing and tuning fuzzy models are rather complex, and their use requires specific skills and knowledge. Modeling of complex systems will always remain an interactive approach [19].

2.7.3 Machine Learning Approach

Machine learning is the concept comes after artificial intelligence as subset of itself. This technology provides an ability of self-learning and self-improving without any external instructions but with explicit feature engineering. Machine learning (ML) algorithm allows software applications to become more accurate in predicting outcomes without being explicitly programmed. This type of approach focuses on applying machine learning and data mining techniques to phishing detection. These related techniques are classified into three main categories: Classification, Clustering, and Anomaly Detection [21]. Classification techniques try to map inputs (features or variables) to desired outputs (response) using a specific function. In the case of classifying phishing emails, a model is created to categorize an email into phishing or legitimate by learning certain characteristics of the email. Phishing classifiers can be grouped into three main categories: Classifiers based on URL features, Classifiers based on textual features, Classifiers based on hybrid features. Clustering-based countermeasures partition a set of instances into phishing and legitimate clusters. The objective of clustering is to group objects based on their similarities. If each object is represented as a node, and the similarities between different objects are measured based on their shared common features, then a clustering algorithm can be used to identify groups (of nodes) of similar observations. The anomaly-based approaches to phishing detection essentially treat phishing attempts as outliers. Every website claims a unique identity in the cyberspace either explicitly or implicitly. Anomaly detection methods assign a score to the suspicious material under analysis by comparing the features of phishing material with those of one or more nearest neighbors. If the anomaly score goes above a cut-off point, the webpage would be classified as phishing [21].

A classification model based on Extreme Learning Machine [22] for phishing website features classification achieved better performance than other machine learning methods (Support Vector Machine (SVM), Naive Bayes (NB)). ELM model can be defined as the feed-forward Artificial Neural Network (ANN) model with a single hidden layer [22]. To activate cells in a hidden ELM layer, a linear function was used, as well as nonlinear (sigmoid, sinus, Gaussian), non-derivable or discrete activation functions. and achieved highest accuracy of 95.34% [22]. But this classification model restricted to text features only and needs more effectiveness in its performance.

2.7.4 Deep Learning Approach

Deep learning (also known as deep structured learning or hierarchical learning) is an advanced concept of machine learning research that gives the capability of learning from unsupervised data that is unstructured or unlabeled to the networks. That is any designed model itself learns to perform classification tasks directly from images, text, or sound, usually implemented using neural network architecture. The term "deep" refers to the number of layers in the network—the more layers, the deeper the network. Traditional neural networks contain only 2 or 3 layers, while deep networks can have hundreds [23]. Online Malicious URL and DNS Detection Scheme proposed by Jiang, Lin [24] considered two common attack vectors URL and DNS in malicious activities for phishing detection with the help of character-based Convolutional Neural Network framework. CNN framework automatically extracts the malicious features hidden within the URL strings and trains the classifying model. CNN network has been widely used in image recognition due to its ability to directly perform some convolution operations on the original pixel binary data to find hidden features hidden between pixels. CNN can also be used in word sequence feature mining with Neural Language Processing (NLP) [23]. In Reference [24] evaluated this approach using real-world datasets to demonstrate that this approach is both accurate and efficient. Figure 2 shows that the online malicious URL detection approach proposed in it [14].

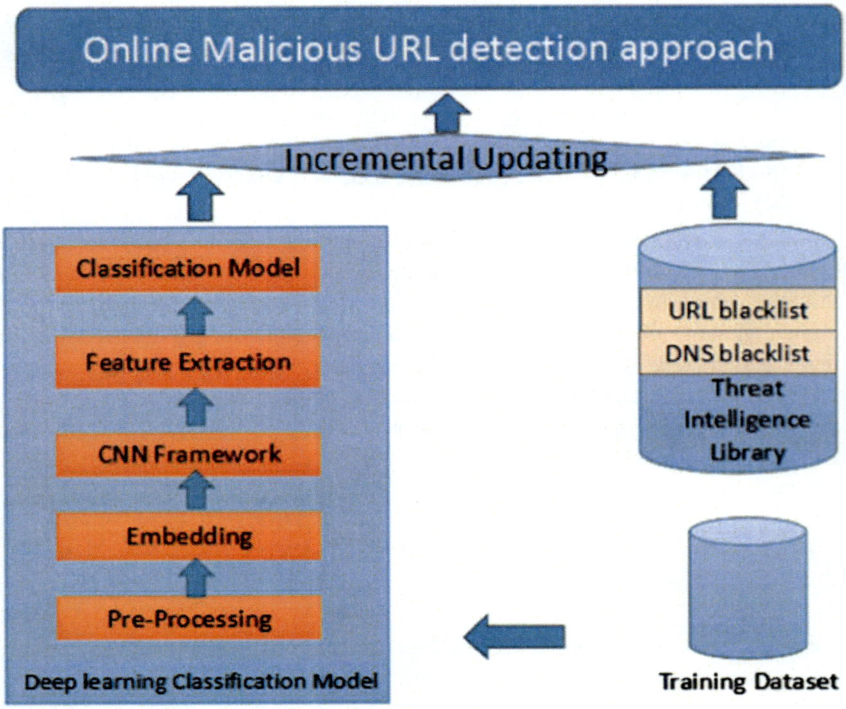

Fig. 2 Online malicious URL detection approach [24]

3 Results and Discussions

Deep learning was developed to understand and analyze text using a multi-layered deep neural network, limiting natural language processing (NLP) capabilities to traditional machine learning algorithms using raw machines. The approach described here uses a deep learning network called the Convolutional Neural Network to train the classification model. Over the network, every output of the previous layer turns to the next level of input. Particularly deep learning techniques have the potential for language analysis, and distributed vectors trained under text-based vector representation for words are widely used in linguistic analysis systems.

In our proposed approach (see Fig. 3) consists of two main components, as follows:

(1) Dataset: The real-world dataset of phishing websites with URLs, screenshots, and webpage content from PhishTank, 2017. Another dataset was Phish-IRIS with screenshots of phishing and legitimate websites.
(2) Deep learning classification model, which consists of three processes such as preprocessing the input data, embedding, feature extraction and training a classification model using deep learning method.

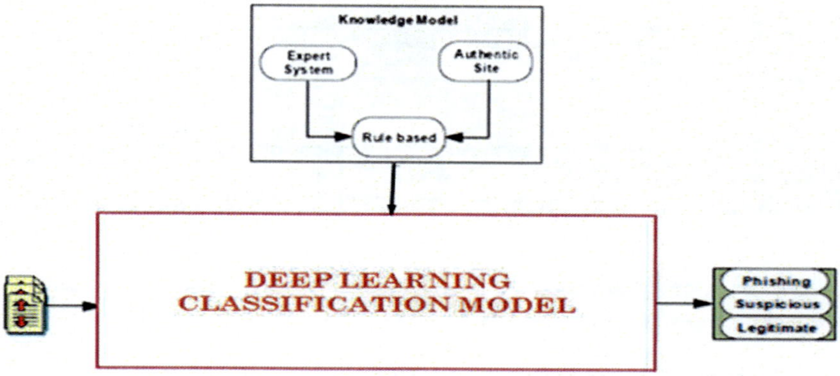

Fig. 3 Conceptual diagram of smart phishing detection system using deep learning

A Convolutional neural network (CNN or ConvNet) is one of the most popular algorithms for deep learning with images as well as texts. Like other neural networks, a CNN is composed of an input layer, an output layer, and many hidden layers in between. The CNN framework mainly divided into two layers.

3.1 Feature Detection Layers

These layers perform one of three types of operations on the data: convolution, pooling, or rectified linear unit (ReLU). Convolution puts the input data through a set of Convolutional filters, each of which activates certain features from the input datasets. Pooling simplifies the output by performing nonlinear downsampling, reducing the number of parameters that the network needs to learn about. Rectified linear unit (ReLU) allows for faster and more effective training by mapping negative values to zero and maintaining positive values. These three operations are repeated over tens of layers, with each layer learning to detect different features.

3.2 Classification Layers

After feature detection, the architecture of a CNN shifts to classification. The next-to-last layer is a fully connected layer (FC) that outputs a vector of K dimensions where K is the number of classes that the network will be able to predict. This vector contains the probabilities for each class of any image being classified. The final layer of the CNN architecture uses a softmax function to provide the classification output (Fig. 4).

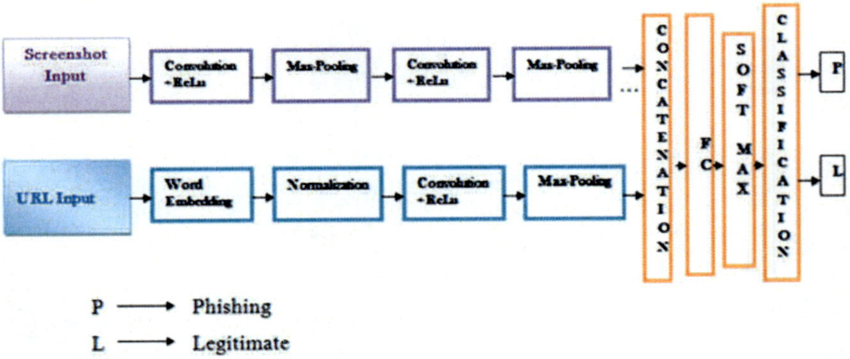

Fig. 4 Proposed CNN architecture

The above architecture takes the input as screenshot of a website along with its corresponding URL and then passes this data to respective layers. Screenshot image passes through eight layers of convolution, ReLu and max pooling. The URL input passes through embedding layer which quantizes the URL into unique words and then batch normalization applies to convert it into images. Then these can be easily passed through the convolution layers to make it as CNN model. The outputs from both the max pooling layers are to be concatenated and then send to classification layers. The detailed description is follows.

3.3 Convolutional Layer

In the Convolutional layer, the first argument is filter size, which is the height and width of the filters the training function uses while scanning along with the images. In this example, the number 3 indicates that the filter size is 3-by-3. You can specify different sizes for the height and width of the filter. The second argument is the number of filters, num filters, which is the number of neurons that connect to the same region of the input. This parameter determines the number of feature maps. Use the "Padding" name-value pair to add padding to the input feature map. For a Convolutional layer with a default stride of 1, "same" padding ensures that the spatial output size is the same as the input size. You can also define the stride and learning rates for this layer using name-value pair arguments of convolution 2d layer.

3.4 Batch Normalization Layer

Batch normalization layers normalize the activations and gradients propagating through a network, making network training an easier optimization problem. Use batch normalization layers between Convolutional layers and nonlinearities, such as Re LU layers, to speed up network training and reduce the sensitivity to network initialization. Use batch normalization layer to create a batch normalization layer.

3.5 *ReLU Layer*

The batch normalization layer is followed by a nonlinear activation function. The most common activation function is the rectified linear unit (ReLU). Use ReLU Layer to create a ReLU Layer.

3.6 *Max Pooling Layer*

Convolutional layers (with activation functions) are sometimes followed by a downsampling operation that reduces the spatial size of the feature map and removes redundant spatial information. Downsampling makes it possible to increase the number of filters in deeper Convolutional layers without increasing the required amount of computation per layer. One way of downsampling is using a max pooling, which you create using max Pooling 2d layer. The max pooling layer returns the maximum values of rectangular regions of inputs, specified by the first argument, poolsize. The "Stride" name-value pair argument specifies the step size that the training function takes as it scans along the input.

3.7 *Fully Connected Layer*

The Convolutional and downsampling layers are followed by one or more fully connected layers. As its name suggests, a fully connected layer is a layer in which the neurons connect to all the neurons in the preceding layer. This layer combines all the features learned by the previous layers across the image to identify the larger patterns. The last fully connected layer combines the features to classify the images. Therefore, the output size parameter in the last fully connected layer is equal to the number of classes in the target data. The output size is 2, corresponding to the two classes (phishing and legitimate). Use fully connected layer to create a fully connected layer.

3.8 Softmax Layer

The softmax activation function normalizes the output of the fully connected layer. The output of the softmax layer consists of positive numbers that sum to one, which can then be used as classification probabilities by the classification layer. Create a softmax layer using the softmax layer function after the last fully connected layer.

3.9 Classification Layer

The final layer is the classification layer. This layer uses the probabilities returned by the softmax activation function for each input to assign the input to one of the mutually exclusive classes and compute the loss. To create a classification layer, use classification layer.

4 Conclusion

Phishing is a significant problem involving fraud email and web sites that mislead unsuspecting users into disclosing private information. Among the different anti-phishing approaches, integration of visual similarity based approach and character-based anti-phishing approach was not yet considered. The objective of the work was to identify the scope of developing such phishing website detection scheme to detect phishing websites more accurately with the help of deep learning algorithm called Convolutional Neural Network (CNN) classification model. The CNN framework was suggested to automatically extract malicious features and train the classifying model. Thus it avoids manual handcrafted feature engineering method needed for real-world phishing detection schemes. The integration of hybrid features extracted from URL texts and screenshot images may lead to the betterment of existing phishing detection schemes.

References

1. Issac, B., Chiong, R., & Jacob, S. M. (2014). Analysis of Phishing Attacks and Countermeasures. IBIMA, Bonn, Germany, ISBN 0-9753393-5-4, 339–346.
2. Ali, W. (2017). Phishing Website Detection based on Supervised Machine Learning with Wrapper Features Selection. *International Journal of Advanced Computer Science and Applications, 8*(9), 72–78.
3. APWG report [Online] Available at: https://docs.apwg.org/reports/apwg_trends_report_q4_2018.pdf.
4. https://www.mathworks.com/content/dam/mathworks/tagteam/Objects/d/80879v00_Deep_Learningebook. pdf.

5. Adebowale, M. A., Lwin, K. T., Sanchez, E., & Hossain, M. A. (2019). Intelligent web-phishing detection and protection scheme using integrated features of Images, frames and text. *Expert Systems with Applications, 115*, 300–313.
6. Deshmukh, M., Popat, S. K., & Student, U. (2017). Different Techniques for Detection of Phishing Attack. *International Journal of Engineering Science, 10201.*
7. Atighetchi, M., & Pal, P. (2009). Attribute-based prevention of phishing attacks. In *2009 Eighth IEEE International Symposium on Network Computing and Applications.*
8. Shreeram, V., Suban, M., Shanthi, P., & Manjula, K. (2010). Anti-phishing detection of phishing attacks using genetic algorithm. In *2010 International Conference on Communication Control and Computing Technologies* (pp. 447–450), Ramanathapuram.
9. Almomani, A., Gupta, B. B., Atawneh, S., Meulenberg, A., & Almomani, E. (2013). A survey of phishing email filtering techniques. *IEEE communications surveys & tutorials, 15*(4), 2070-2090.
10. Tout, H., & Hafner, W. (2009). Phishpin: An identity-based anti-phishing approach. In *2009 international conference on computational science and engineering* (Vol. 3, pp. 347–352).
11. Zhang, Y., Hong, J. I., & Cranor, L. F. (2007). Cantina: a content-based approach to detecting phishing web sites. In *Proceedings of the 16th international conference on World Wide Web* (pp. 639–648), ACM.
12. Dunlop, M., Groat, S., & Shelly, D. (2010). Goldphish: Using images for content-based phishing analysis. In *2010 Fifth international conference on internet monitoring and protection* (pp. 123–128).
13. Deshmukh, M., Popat, S. K., & Student, U. (2017). Link Guard Algorithm Approach on Phishing Detection and Control. *International Journal of Advance Foundation and Research in Computer (IJAFRC).*
14. Jiang, J., Lin, X., Ghorbani, A., Ren, K., Zhu, S., Zhang, A. (2017). A deep learning based online malicious URL and DNS detection scheme. In *International Conference on Security and Privacy in Communication Systems* (pp. 438–448). Springer, Cham.
15. www.wikipedia.org.
16. Mohammad, R., McCluskey, T. L., & Thabtah, F. A. (2013). Predicting phishing websites using neural network trained with back-propagation. In: *Proceedings of the 2013 World Congress in Computer Science, Computer Engineering, and Applied Computing. WORLDCOMP 2013 . World Congress in Computer Science, Computer Engineering, and Applied Computing*, Las Vegas, Nevada, USA, (pp. 682–686). ISBN 1601322461.
17. Ferreira, R., Martiniano, A., Napolitano, D., Romero, M., De Oliveira Gatto, D., Farias, E., et al. (2018). Artificial Neural Network for Websites Classification with Phishing Characteristics. *Social Networking, 7*, 97–109.
18. Gupta, S., & Singhal, A. (2017, August). Phishing URL detection by using artificial neural network with PSO. In *2017 2nd International Conference on Telecommunication and Networks (TEL-NET)*, Noida, (pp. 1–6).
19. Babuška, R. (2003). Neuro-fuzzy methods for modeling and identification. In Abraham A., Jain L.C., Kacprzyk J (Eds.)*Recent advances in intelligent paradigms and applications* (pp. 161-186). Physica, Heidelberg.
20. Barraclough, P. A., Hossain, M. A., Tahir, M. A., Sexton, G., & Aslam, N. (2013). Intelligent phishing detection and protection scheme for online transactions (Report). *Expert Systems with Applications, 40*(11), 4697.
21. Aleroud, A., & Zhou, L. (2017). Phishing environments, techniques, and countermeasures: A survey. *Computers & Security.*
22. Sönmez, Y., Tuncer, T., Gökal, H., & Avcı, E. (2018). Phishing web sites features classification based on extreme learning machine. In *2018 6th International Symposium on Digital Forensic and Security (ISDFS)* (pp. 1–5).
23. https://www.mathworks.com/content/dam/mathworks/tagteam/Objects/d/80879v03_Deep_Learning_ebook.pdf.
24. Jiang, J. (2017). A deep learning based online malicious URL and DNS detection scheme. *Security and Privacy in Communication Networks: SecureComm* 2017. *Lecture Notes of the*

Institute for Computer Sciences, Social Informatics and Telecommunications Engineering, (Vol. 239). Springer, Cham.

25. Adebowale, M. A., Lwin, K. T., Sanchez, E., & Hossain, M. A. (2019). Intelligent web-phishing detection and protection scheme using integrated features of Images, frames and text. *Expert Systems with Applications*.
26. Dhamija, R., & Tygar, J. D. (2005). The battle against phishing: Dynamic security skins. In *Proceedings of the 2005 Symposium on Usable privacy and security* (pp. 77–88).
27. Afroz, S., & Greenstadt, R. (2011). Phishzoo: Detecting phishing websites by looking at them. In *2011 IEEE fifth International Conference on Semantic Computing* (pp. 368–375).
28. Chen, K.-T., Chen, J.-Y., Huang, C.-R., & Chen, C.-S. (2009). Fighting phishing with discriminative keypoint features. *IEEE Internet Computing, 13*(3), 56–63.
29. Fu, A. Y., Wenyin, L., & Deng, X. (2006). Detecting phishing web pages with visual similarity assessment based on earth mover's distance (EMD). *IEEE transactions on dependable and secure computing, 3*(4), 301–311.
30. Haruta, S., Asahina, H., & Sasase, I. (2017). Visual similarity-based phishing detection scheme using image and CSS with target website finder. In *GLOBECOM 2017-2017 IEEE Global Communications Conference* (pp. 1–6), Singapore.
31. Jain, A. K., & Gupta, B. B. (2017). Phishing detection: analysis of visual similarity based approaches. *Security and Communication Networks, 2017*. 5421046:1-5421046:20.

Survey on Massive MIMO System with Underlaid D2D Communication

Suresh Penchala, Deepak Kumar Nayak, and B. Ramadevi

Abstract The massive MIMO (mMIMO) provides high spectral efficiency (SE) by exploiting the advantage of hundreds of antennas. The Device-to-Device (D2D) technology is proven as a promising communication for 5G cellular networks. The maximum benefits of these two technologies can be explored by combining mMIMO transmission with D2D communication. The combined technology can enhance the SE by exploiting the proximity of the devices via direct transmission. The D2D transmission increases the data rate, energy efficiency (EE) of the mMIMO system, and reduces the delay with the short-distance transmission. In this paper, a brief review of D2D communication, advantages, applications, and open issues are discussed. Later, Mmimo, the coexistence of mMIMO and D2D communication, and literature review are discussed in brief.

Keywords D2D communication · 5G · Massive MIMO · Resource allocation · Interference · Spectral efficiency

1 Introduction

The modern 5G cellular communication needs D2D communication to support high data speed services with less delay, where conventional cellular networks fail in modern services like video conferencing, interactive gaming, etc. The D2D communication explores the physical vicinity of nodes [1, 2] and supports high-speed data

S. Penchala (✉) · D. K. Nayak · B. Ramadevi
Koneru Laxmaiah Education Foundation, Guntur, India
e-mail: penchalasuresh@gmail.com

D. K. Nayak
e-mail: nayak@kluniversity.in

B. Ramadevi
e-mail: ramadevikitsw@gmail.com

D. K. Nayak · B. Ramadevi
Kakatiya Institute of Technology & Science, Warangal, India

S. C. Satapathy et al. (eds.), *Intelligent System Design*, Advances in Intelligent Systems and Computing 1171, https://doi.org/10.1007/978-981-15-5400-1_46

transmission with low latency via direct link by bypassing base station (BS) or access point (AP).

The massive mMIMO and D2D technology are two prominent technologies in 5G. The mMIMO employs a huge antenna array at each BS and provides high SE with multiplexed gain. The advantages of mMIMO are high faithfulness; greater EE and SE, less power, cost-effective, low latency, robust to array element failures, accurate user tracking, and less inter-user interference. The disadvantages of mMIMO are pilot contamination, complicate processing techniques, and sensitive to beam alignment.

The mMIMO system with D2D communication is a blend of these two technologies which takes all the potential advantages of each and strives to improve the overall network performance. The advantages of D2D communication are high data rates with low delay, superior link coverage, low power consumption, high SE and EE, optimum utilization of radio resources, robust towards safety, and infrastructure failures.

The massive MIMO system together with the D2D underlay network will boost the system performance and SE [3]. The direct transmission in short-range D2D paired nodes provides high SE, reduces the delay, and improves EE of the network. The D2D communication helps to offload massive MIMO network traffic. It reduces the computational complexity at BS, boosts up offloading services with the high data rate, less power, and enhances the network capacity. The interference between inter cells and cellular to D2D are the major issues in mMIMO system with D2D communication. These interferences will decay the performance of the system if they are not addressed properly. A few key challenges in mMIMO system with D2D communication are interference mitigation and coexistence, nodes proximity discovery and mobility issues, device mode selection, link establishment, resource allocation, and security.

The main aim of this paper is to provide an overview of D2D communication, mMIMO, and coexistence. As per our knowledge, there is no review paper published describing D2D with massive MIMO. Due to the space limitation of this review paper, only D2D and mMIMO with D2D are explained in the next section. In this paper, a survey of D2D communication, advantages, challenges in diversified aspects and applications are studied in Sect. 2. The mMIMO, mMIMO with D2D communication, and challenges are given in Sect. 3. The conclusions are drawn in Sect. 4.

2 D2D Communication

The D2D technology allows user equipment (UE) which is proximity to exchange information via direct transmission without the inference of network infrastructures like BS or AP. This UE centric communication expected to solve the 5G network complications like low latency, reliability, high data rate, EE, etc. The D2D communication use licensed or unlicensed ISM band, i.e., inband or outband. Few standards support D2D communication are 3GPP LTE-advanced, 20-20 information society,

5G-PPP association, Networld 2020 platform, etc. The different D2D scenarios in cellular communication are given in Fig. 1.

The D2D communication is of two types, inband and outband, as shown in Fig. 2. The inband D2D is further divided into (i) Underlay and (ii) Overlay modes. The outband D2D is further classified as (i) Controlled, and (ii) Autonomous D2D. In the first case, D2D and cellular networks share the spectrum, and interference mitigation

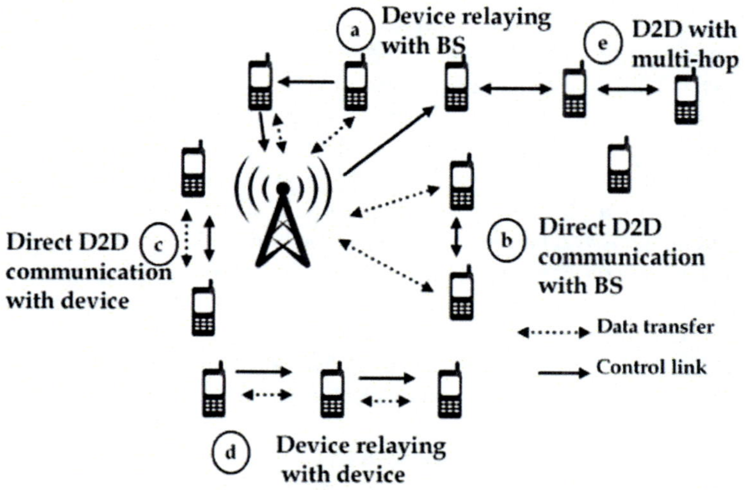

Fig. 1 D2D scenarios

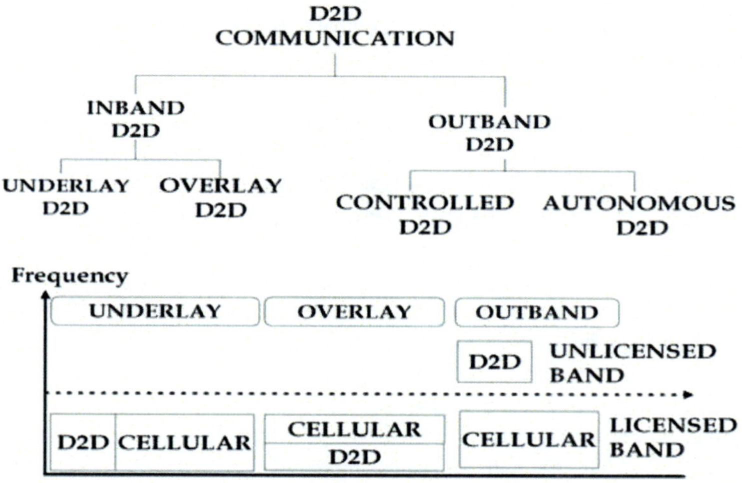

Fig. 2 Inband and outband D2D communication

is a big challenge. In contrast to it, a part of the licensed spectrum is devoted to D2D which simplifies resource allocation (RA) in overlay inband D2D mode.

Applications of D2D: Few applications of D2D communications are local service, emergency communication, Internet of Things enhancement, virtual MIMO, multiuser MIMO enhancement, cooperative relaying, M2M communications, content distribution, gaming, wearables, cellular offloading, advertisements, vehicular communication, and social networks.

Challenges and open issue in D2D: The challenges and open issue in D2D communication are discovery of proximity nodes, D2D synchronization, device mode selection, efficient utilization of spectrum, interference mitigation techniques to coexist with D2D UEs and cellular network, new architectures and device design issues, link failures due to mobility of the nodes, resource management, D2D RA to guarantee QoS, and D2D MIMO and mMIMO transmission.

Neighbor discovery: The discovery of peer nodes in the proximity for D2D communication with low power consumption is important, and these issues were addressed in [1]. A D2D discovery scheme based on random access procedure for proximity-based services in the LTE-Advance system by exploiting location information of user equipment (UE) was proposed in [2]. A joint neighbor discovery using a sound reference signal for cellular uplink (UL) transmission in D2D LTE-Adv SC-FDMA System was proposed in [4]. The peer to peer discovery and session setup; RA and QOS guaranteeing, mode selection, and interference mitigation in HETNETs were discussed in [5].

Mode selection: The different mode switching strategies for D2D communication in both single-cell and multi-cell scenarios to maximize ergodic achievable sum-rate was given in [6]. A queuing model was developed in [7] for mode selection (MS) to route mode for every D2D connection. The UL transmission in cellular networks using D2D with truncated channel inversion power control (PC) using MS was given in [8]. The modeling of MS and power control for D2D was given in [9].

Resource allocation (RA): The RA issues for network-assisted D2D were given in [10]. The SE of D2D and cellular underlying communication enhanced by sharing the resources. The SE can be obtained by multiplexing resources by using mixed mode transmission, i.e. D2D mode or cellular mode. The maximum weighted D2D sum-rate can be obtained by optimizing transmission modes via sub-channel allocation and power control [11]. Frequency reuse-based resource block allocation by maintaining enough distance between D2D and cellular users, macrocell BS, and D2D users were proposed in [12]. Joint power and block RA based on the sparse group for LTE-advanced networks to maximize the sum-rate was proposed in [13]. Resource utilization can be enhanced by the physical proximity of communication devices. Neighbor user's discovery, RA, and power control with channel quality and their design aspects were discussed in [14]. Resource sharing in D2D underlying cellular network (CN) causes significant interference, which affects the performance of the network. The D2D user's distribution was modeled as a homogenous spatial Poisson point process to overcome the above problem [15]. A cross-layer model to obtain a maximal achievable packet rate by sharing resources of D2D pairs in D2D underlay cellular link was given in [16].

Power control: In D2D underlying CN, the power can be saved by transferring the load to D2D devices with BS shut down [17]. Frequency reuse in D2D underlying CNs for multi-cell uplink system using location-based power allocation (PA) to improve outage performance with EE was described in [16]. Localization and optimal PA by decomposing infrastructure and cooperation phases using sparsity property in cooperative networks were explained in [18]. The stochastic geometry-based reciprocal impact of D2D and CNs using power control was given in [19]. An optimal transmission period with power control for underlay D2D to relay-assisted CNs to maximize ergodic capacity was proposed in [20].

Interference control and coexistence: A survey on D2D anti-interference was given in [21]. A joint transmit power and rate control scheme in D2D to the cellular network was optimized to mitigate the interference that was proposed in [22]. Multiuser D2D exploits MIMO using bucket based degree of freedom assignment to reduce interference was proposed in [23]. Cooperative interference cancellation (CIC) and network management with interference correlation to enhance downlink throughput were proposed in [24]. The distributed MAC for interference aware transmission for unicast and broadcast scenarios for D2D communication in 3GPP was discussed in [25]. The reusing cell resources under BS in single and multi-cellular communication networks and anti-interference analysis was given in [26].

3 Massive MIMO

The mMIMO system contains huge number of BS antennas that cover more users simultaneously with precise spatial focusing. It offers high SE via spatial multiplexing, accompanies massive users in the corresponding radio channel, and suppress interference. The energy consumption increases due to the circuit power consumption of a large number of antenna. The advantages of mMIMO system are less latency, cost-effective, high data rate, robust to interference, simplifies the MAC layer, and high SNR ratio for a link.

The mMIMO system in combination with D2D supports to overcome the limitations in D2D and enhances the EE of the system. The mMIMO system with D2D communication is shown in Fig. 3.

The massive MIMO cellular users use different sets of orthogonal pilots at each cell with reuse factor one with D2D underlay pairs to improve SE was explained in [27]. The coexistence of massive MIMO systems with underlaid D2D with two metrics average sum-rate and EE and a tradeoff between EE and sum-rate was given in [3]. The interference mitigation and coexistence of mMIMO systems with uplink D2D underlaid network using a spatially dynamic power control to enhance SE and EE was discussed in [28]. A distributive learning algorithm with rate adaptation for a MIMO with underlaid D2D links using the reuse of downlink (DL) subframes and

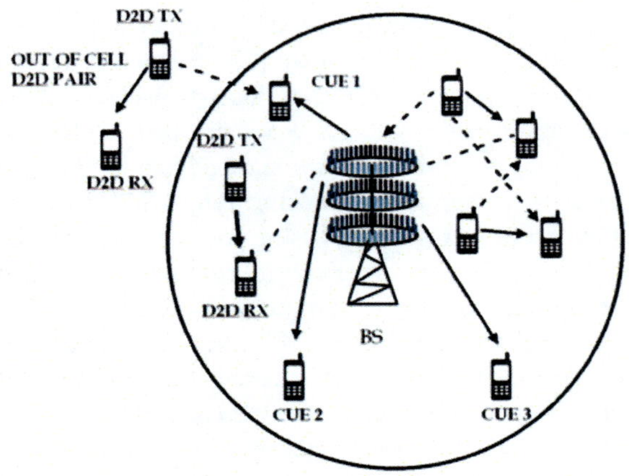

Fig. 3 mMIMO with D2D communication

uplink channel estimation in associated cells was proposed in [29]. The cooperative precoder feedback with adaptive CSI for FDD mMIMO with the D2D system was proposed in [30] to enhance the system efficiency. The inversion power control strategy based energy harvesting technique for mMIMO with D2D to enhance EE was proposed in [31]. The interference among users was considered to optimize the EE using pilot reuse with power control in an uplink mMIMO with a D2D underlay system [32]. An energy-efficient scheme with orthogonal pilot set reusing using graph coloring based pilot allocation for mMIMO with D2D was proposed in [33]. The BS precoder design and PA using MAXSUM algorithm for optimum data rates for mMIMO with D2D were discussed in [34] (Table 1).

4 Conclusions

In this paper, D2D communication, its advantages, opportunities, challenges, a review on different issues like neighborhood node discovery, resource allocation, power control, mode selection, interference control, and coexistence are discussed in brief. Further, massive MIMO, the coexistence of massive MIMO with D2D communication, issues, scope, and literature review discussed in brief. From the literature survey, it is identified that merging these technologies will help in achieving future requirements of 5G. Also, the 5G network or system performance can be improved drastically in massive MIMO with D2D communication.

Table 1 Comparision on massive MIMO system with D2D communication

Author	Adopted system	Results	Description	Challenges
Zhou et al. [26]	Power control, Pilot allocation, MR and ZF processing	SE increased, eliminates interference, controlled data, and power	Reuse of orthogonal pilots for cellular communication, and allocating a set of pairs to the D2D communication	Interference
Shalmashi et al. [3]	Uniformly distributed CUs DUs distributed according to Poisson point process	Tradeoff between EE and SE	DL resources also used for D2D communication	Applicable for low-density D2D users
He et al. [28]	Open-loop power control method, PC scheme	SE and EE increased	Two different power control schemes for CUs and DUs	Interference
Zhang et al. [29]	Distributive learning algorithm with rate adaptation	Improve the D2D link SINR, better SE	Controlling the BSs and D2D devices based on statistical means and variances with packet outage probability evaluation	D2D links only reuse DL resources
Chen et al. [30]	Cooperative precoder feedback strategy	Interference limited	User exchange their CSI using D2D communication then determine the precoder by themselves and feedback the precoder to BS	Interference
Jia et al. [31]	massive MIMO with hybrid network	Enhanced SE, improved secrecy of CU and DU	The D2D transmitters have harvested the power from dedicated power beacons	Interference
Xu et al. [32]	pilot reuse scheme	Power consumption reduced, EE improved	EE optimization for D2D communication using pilot reuse and PC using UL resources	Interference among users and power utilization

(continued)

Table 1 (continued)

Author	Adopted system	Results	Description	Challenges
Xu et al. [33]	Pilot reuse and data transmit power control (GCPA algorithm)	Reduce the length of pilots, pilot resources can be saved greatly	Pilots allocation using GCPA algorithm. LMMSE filters are used to detect the signal. lower bound of D2D links derived based on average SINR	Interference
Amin et al. [34]	The BS precoder design and PA using MAXSUM algorithm	Enhance SE and EE, the secrecy of CU and DU links	BS with huge antennas multicasting an individual data packet to the CUEs	To manage the interference between the BS and D2D communicating devices

References

1. Zou, K. J., Wang, M., Yang, K. W., Zhang, J., Sheng, W., Chen, Q., & You, X. (2014 June). Proximity discovery for device-to-device communications over a cellular network. *IEEE Communications Magazine* (pp. 98–107).
2. Choi, K. W., & Han, Z. (2015 January). Device-to-device discovery for proximity-based service in LTE-advanced system. *IEEE Journal on Selected Areas in Communications, 33*, 55–66.
3. Shalmashi, S., Björnson, E., Kountouris, M., Sung, K. W., & Debbah, M. (2016). Energy efficiency and sum rate tradeoffs for massive MIMO systems with underlaid device-to-device communications. *EURASIP Journal on wireless Communications and Networking, 2016*(175), 1–18. https://doi.org/10.1186/s13638-016-0678-1.
4. Tang, H., Ding, Z., & Levy, B. C. (2014 October1). Enabling D2D communications through neighbor discovery in LTE cellular networks. *IEEE Transactions on Signal Processing, 62*(19), 5157–5170.
5. Feng, D., Lu, L., Yuan-Wu, Y., Li, G. Y., Li, S., & Feng, G. (2014 April). Device-to-device communications in cellular networks. *IEEE Communications Magazine*, 49–55.
6. Ni, Y., Qiao, D., Li, X., Jin, S., & Zhu, H. (2014). Transmission modes witching for device-to-device communication aided by relay node. *EURASIP Journal on Advances in Signal Processing, 13.*
7. Lei, L., Dohler, M., Lin, C., & Zhong, Z. (2014 December). Queuing models with applications to mode selection in device-to-device communications underlaying cellular networks. *IEEE Transactions on Wireless Communications, 13*(12), 6697–6715.
8. ElSawy, H., Hossain, E., & Alouini, M.-S. (2014 November). Analytical modeling of mode selection and power control for underlay D2D communication in cellular networks. *IEEE Transactions on Communications. 62*(11).
9. Wang, X., Sheng, Z., Yang, S., & Leung, V. C. M. (2016 August). Tag-assisted social-aware opportunistic device-to-device sharing for traffic offloading in mobile social networks. *IEEE Wireless Communications*, 60–67.
10. Lei, L., Kuang, Y., (Sherman) Shen, X., Lin, C., & Zhong, Z. (2014 June). Resource control in network assisted device-to-device communications: Solutions and challenges. *IEEE Communications Magazine*, 108–117.
11. Tang, H., & Ding, Z. (2016 January). Mixed mode transmission and resource allocation for D2D communication. *IEEE Transactions on Wireless Communications, 15*(1).

12. Sobhi-Givi, S., Khazali, A., Kalbkhani, H., Shayesteh, M. G., & Solouk, V. (2017). Resource allocation and power control for underlay device-to-device communication in fractional frequency reuse cellular networks. *Telecommunication Systems, 65*, 677–697.
13. Li, X.-Y., Li, J., Liu, W., Zhang, Y., & Shan, H.-S. (2016 January). Group-sparse-based joint power and resource block allocation design of hybrid device-to-device and LTE-advanced networks. *IEEE Journal on Selected Areas in Communications, 34*(1).
14. Fodor, G., Dahlman, E., Mildh, G., Parkvall, S., Reider, N., Miklos, G., & Turanyi, Z. (2012 March). Ericsson research, "design aspects of network assisted device-to-device communications". *IEEE Communications Magazine*, 170–177.
15. Sheng, M., Li, Y., Wang, X., Li, J., & Shi, Y. (2016 January). Energy efficiency and delay tradeoff in device-to-device communications underlaying cellular networks. *IEEE Journal on Selected Areas in Communications, 34*(1).
16. Lu, H., Wang, Y., Chen, Y., & Ray Liu, K. J. (2016 April). Stable throughput region and admission control for device-to-device cellular coexisting networks. IEEE *Transactions on Wireless Communications, 15*(4), 2809–2824.
17. Li, Y., Jin, D., Hui, P., & Han, Z. (2016). Optimal base station scheduling for device-to-device communication underlaying cellular networks. *IEEE Journal on Selected Areas in Communications, 34*(1), 27–40.
18. Dai, W., Shen, Y., & Win, M. Z. (2015 January). Distributed power allocation for cooperative wireless network localization. *IEEE Journal on Selected Areas in Communications, 33*(1).
19. Al-Rimawi, A., & Dardari, D. Analytical characterization of device-to-device and cellular networks coexistence. *IEEE Transactions on Wireless Communications*. https://doi.org/10.1109/twc.2017.2712640.
20. Lee, D., Kim, S.-I., Lee, J., & Heo, J. (2014). Power allocation and transmission period selection for device-to-device communication as an underlay to cellular networks. *Wireless Personal Communications, 79*, 1–20. https://doi.org/10.1007/s11277-014-1837-5.
21. Zhou, X., Zhang, Y., Sheng, Q., & Wu, D. The anti-interference study on D2D communications. In *International Conference on Mechatronics, Electronic, Industrial and Control Engineering (MEIC 2015)*, pp. 1370–1373.
22. Xiao, Y., Niyato, D., Chen, K.-C., & Han, Z. Enhance device-to-device communication with social awareness: A belief-based stable marriage game framework. *IEEE Wireless Communications*.
23. Chiu, S.-L., Lin, K. C.-J., Lin, G.-X., & Wei, H.-Y. (2017 April). Empowering device-to-device networks with cross-link interference management. *IEEE Transactions on Mobile Computing, 16*(4), 950–963.
24. Osama, M. F., Abu-Sharkh, E. A., & Hasan, O. M. (2017). Adaptive device-to-device communication using Wi-Fi direct in smart cities. *Wireless Networks, 23*, 2197–2213. https://doi.org/10.1007/s11276-016-1278-z.
25. Chun, Y. J., Cotton, S. L., Dhillon, H. S., Ghrayeb, A., & Hasna, M. O. A stochastic geometric analysis of device-to-device communications operating over generalized fading channels. *IEEE Transactions on Wireless Communications*. https://doi.org/10.1109/twc.2017.2689759.
26. Zhou, X., Zhang, Y., Sheng, Q., & Wu, D."The anti-interference study on D2D communications. In *International Conference on Mechatronics, Electronic, Industrial and Control Engineering (MEIC 2015)*, pp. 1370–1373.
27. Ghazanfari, A., Bjornson, E., & Larsson, E. G. (2019). Optimized power control for massive MIMO with underlaid D2D communications. *IEEE Transactions on Communications, 67*(4), 2763–2778. https://doi.org/10.1109/TCOMM.2018.2890240.
28. He, A., Wang, L., Chen, Y., Wong, K.-K., & Elkashlan, M. (2017). Spectral and energy efficiency of uplink D2D underlaid massive MIMO cellular networks. *IEEE Transactions on Communications, 65*(9), 3780–3793.
29. Zhang, Z., Li, Y., Wang, R., & Huang, K. (2019). Rate adaptation for downlink massive MIMO networks and underlaid D2D links: A learning approach. *IEEE Transactions on Wireless Communications, 18*(3), 1819–1833.

30. Chen, J., & Yin, H. (2017). Laura Cottatellucci and David Gesbert "feedback mechanisms for FDD massive MIMO with D2D-based limited CSI sharing". *IEEE Transactions on Wireless Communications, 16*(8), 5162–5175.
31. Jia, X., Xie, M., Zhou, M., Zhu, H., & Yang, L. (2017 September). D2D underlay massive MIMO hybrid networks with improved physical layer secrecy and energy efficiency. *International Journal of Communication Systems* (wileyonlinelibrary.com/journal/dac) *30*(3), 10.
32. Shenghao, X. U., Zhang, H., Tian, J., & Takis Mathiopoulos, P. (2017). Pilot reuse and power control of D2D underlaying massive MIMO systems for energy efficiency optimization. *Science China Information Sciences, 60*(10), 100303. https://doi.org/10.1007/s11432-017-9194-y.
33. Xu, H., Yang, Z., Wu, B., Shi, J., & Chen, M. (2016 May 15–18). Power control in D2D underlay massive MIMO systems with pilot reuse. In *2016 IEEE 83rd Vehicular Technology Conference (VTC Spring).*
34. Amin, B. S., Ramadan Y. R., Ibrahim, A. S., & Ismail, M. H. (2015 March 9–12). Power allocation for device-to-device communication underlaying massive MIMO multicasting networks. In *2015 IEEE Wireless Communications and Networking Conference (WCNC).*

Prevention of DDoS Attacks and Detection on Cloud Environment—A Review and Survey

Sharath Kumar Allam and Gudapati Syam Prasad

Abstract Cloud computing has raised a revolution in IT technology that offers accessible, virtualized on-demand resources to the clients with more flexibility, less need for maintenance, and decreased infrastructure expenses. These resources are administered by diverse management firms and delivered through the Internet using acknowledged networking practices, standards, and layouts. The fundamental technologies and legacy practices have bugs and susceptibilities that can bring opportunities for interference by the attackers. The distributed denial-of-service (DDoS) attacks are quite frequent that impose severe harm and shake the cloud's performance. This paper will provide an overview of the DDoS attacks and discourse on the prevention and detection of DDoS attacks with respect to cloud environment.

Keywords Cloud environment · DDoS attack · Botnet · DST

1 Introduction

Cloud computing is considered as the computing services like the servers, databases, software, infrastructure, analytics, storage services, and other resources that are available on the cloud. Cloud computing provides access to flexible resources and helps the companies to achieve economies of scale by saving operating costs, by using the cloud computing services [1]. Although cloud computing services were introduced to provide faster services with better security, it faces some security concerns.

S. K. Allam (✉)
Department of CSE, K L Deemed to Be University, Guntur, India
e-mail: sarath.allam@gmail.com

G. S. Prasad
VKR, VNB &AGK College of Engineering, Gudivada, India
e-mail: syamprasad.gudapati@gmail.com

S. K. Allam
Department of CSE, G. Narayanamma Institute of Technology and Science (for Women), Hyderabad, India

S. C. Satapathy et al. (eds.), *Intelligent System Design*, Advances in Intelligent Systems and Computing 1171, https://doi.org/10.1007/978-981-15-5400-1_47

Resting on the nature of the security concerns, the prevention and mitigation strategies have been designed. It is significant to ensure effective security from various security threats and malicious attacks. Distributed denial-of-service (DDoS) attacks are considered as foremost security threat in cloud environment [2].

In this paper, we discussed as follows: Sect. 2 reviews types of problems in DDoS attacks and different methods proposed to prevent and detect DDoS attacks, Sect. 3 gives some recommendations for prevention and detection of DDoS attacks, and Sect. 4 concludes our work.

2 Literature Review

The DDoS attacks are aimed at disturbing a particular site on the Internet, also other sites that are connected to that particular site. DDoS attacks may attack vectors like TCP, HTTP, and HTTPS and SSL [2]. DDoS attacks are one of the foremost security threats in cloud computing. A scenario of DDoS attack is shown in Fig. 1, wherein various bots targeting a particular virtual machine [3].

The survey will provide an outline of the nature of DDoS attacks. This literature review will also converse the existing and prospective methods for the prevention of DDoS attacks and detection of the DDoS attacks.

2.1 Problems in DDoS Attacks

Cloud computing is in peril pertaining to various security threats coming up. DDoS is a major security attack. DDoS is defined as ‘an attempt to make operational services unavailable by overwhelming it with traffic from multiple sources. They target the wide variety of important resources from bank to news Web sites and present a major

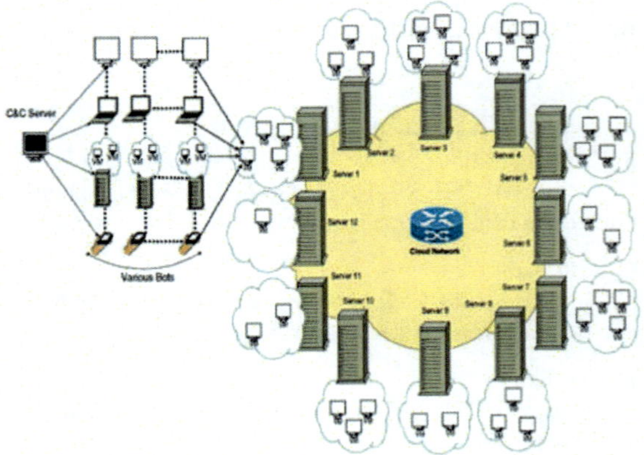

Fig. 1 Scenario of DDoS attack in the cloud (*Source* CC, 107, pp. 32)

challenge to making sure individuals can publish and access important information' [4]. Both denial of Service (DoS) and distributed denial-of-service (DDoS) attacks disturb and critically affect the availability of services on the cloud environment. Flood attack is the most occurred DDoS attack in which the attacker directs the victim numerous ping packets. In the ping of death attacks, the attackers send oversized Internet control message protocol (ICMP) packets which results into deafening of the site [5]. The aim of the DDoS attack is to disturb the usual functioning of the Web site and crashing the popular Web site that leads to huge losses. Such attacks severely affect the ecommerce businesses and Web sites. Growth is witnessed in DDoS attacks in the recent years.

The first DOS attack occurred in 1974, and it was designed by David Dennis with the help of CERL's PLATO terminal [4]. The DDoS attacks first occurred in 1999, and hackers used the tool 'Trinoo' which crashed the University of Minnesota Web site for around two days. The occurrence of DDoS attacks increased since the year 2000, and different famous Web sites like Yahoo, Amazon, eBay, and CNN were severely affected by these attacks. DDoS attacks represent almost 30% gaming traffic while they occur, and almost 5% gaming activities. By the year 2015, the quantity of DDoS attacks grew by almost 25% [4].

DDoS attack is the main issue for online organizations, since most e-business organizations are connected to Internet and they use cloud services for data storage [6]. With time, a significant increase has been witnessed in the magnitude as well as in the destruction caused by the DDoS attacks. A major DDoS attack occurred during the October sites blackout which severely affected Dyn Organization. This attack was successfully used in the scandalous Mirai malware which powered the broad botnet. Hence, it is evident that severity of the DDoS attack has increased over the past decade [7].

In case of DDoS attacks, a legion of the malevolent hosts which is also known as 'zombies' is harmonized which leads to further aggravation for sending data to the target server. The network nodes that rest at the edge of the target server may develop resource crunches leading to further vulnerability. There is an inverse proportional relation between the resource overruns and the distance from the edge. There are two major reasons that lead to such impact. First is that the node which is nearer to the server has less capacity, and hence, they only handle fewer users [7]. Another reason is that such nodes that are closer to the edge will agonize from more accumulated attacks, and there is a possibility that the attacks would compound inside the network. In such situations, the entire system becomes vulnerable to disruption. Considering the severity of the DDoS attacks, it is difficult to design moderation and anticipation policies. Identifying the nature of attack will further help in prevention of these attacks.

2.2 Prevention of DDoS Attack and Detection on Cloud Environment

Numerous techniques were designed for mitigation of the DDoS attacks. The different detection and defense mechanisms designed by the famous authors will be discussed in this section.

Feature Extraction/Identification for Classification. The two mechanisms for prevention and detection of DDoS attacks in cloud environment are intrusion detection system (IDS) and intrusion prevention system (IPS). IDS is correlated to hardware and software that help in detecting and recoding anomalous activities. The IPS functions are similar to IDS; though it is more sophisticated, it is designed to take necessary actions for preventing the malicious activities.

Intrusion Detection System. IDS includes a sensor subsystem which is used for collecting the events that are related to the security of the protected network. It also includes a subsystem analyzer which is intended for detecting the web attacks and additional suspicious activities. It also includes storage component which helps in collecting primary events and also for analyzing the results [7]. It also includes a management console which will allow the user for configuring the IDS and monitor the status of the protected system and also the IDS. It also helps in viewing the subsystem analysis of the identified events.

The anomalous activity helps in effective identification of the attacks. This method helps the administrator to start as well as adjust the security measures, regardless of their knowledge in the field of security. The anomaly method is supportive in identifying the irregular behavior arising on the cloud network. The detector anomalies used in the method of intrusion detection which helps in creating the profiles that denote the usual behavior of the users and network connections. The profiles so created are based on the history data collected while normal operations. Event data is collected, and different metrics is used for determining and analyzing if the data is different from the normal operation [2]. The anomaly method is one of the finest methods since it is useful in detecting the attack without having any specific details [2].

Intrusion Prevention System. IPS is the typical tool which deploys NIDS and NIPS systems. The function of IPS is similar to IDS. The only difference between the systems is that they operate in real time as well as they involuntarily block the network. It is important to ensure the correct placement of the IDS and the IPS systems to attain better protection of the system. IPS is capable of stalling the attacks that will affect the network, and it prevents intrusion of attacks with help of probing the different databases and the attack pattern recognition sensors. When the attack is identified, the IPS system will block the attack and then logs the felonious data. The IPS will also perform host detection within inbound packets and outbound packets and then attempts to block the attacks earlier causing any indemnities.

The host-based IPS can be deployed with the help of installing a resident application program in the host which functions in the system level. This type of application is known as the monitoring agent. The monitoring agents will check for any suspicious actions on the portion of the host and reported to the central monitoring station [2]. These monitoring agents will thereby generate some alerts depending on the activity, and they will be characterized on the basis of priorities which will be later customized. The monitoring agents are involved in completing various levels of monitoring like the file system monitoring, kernel-based intrusion detection, and others [4]. A major drawback of the system is that in case if any malevolent intruder is successful in making any alterations in the system, the only way in which the attacks can be dissatisfied by deploying adequate controls on the part of the security administrator [2].

The network-based approach helps in effectually providing the complete security of the network and the servers. A major benefit of the network-based approach is that there is no need for installing any monitoring software on the host machines [7]. A new technology 'Trap' for enabling the Internet security so that it will help to detect the attacks and prevent further attacks.

Dempster–Shafer Theory. DST is a useful way for surveying the probability of DDoS attacks, which was exhibited by a few research papers with respect to arrange intermission discovery frameworks. Dissanayake [8] introduced an overview on interruption location utilizing DST. DST is utilized to break down the outcomes got from every sensor. Information utilized in trials utilizing DST differs: Yu and Frincke [9] utilized DARPA DDoS interruption recognition assessment datasets; Chou et al. [10] utilized DARPA KDD99 interruption discovery assessment dataset; Chen and Aickelin utilized the Wisconsin Breast Cancer dataset and IRIS plant information, while others researchers created their own information.

Siaterlis et al. [11] and Siaterlis and Maglaris [12] carried out a comparable investigation of recognizing DDoS utilizing data combination, and their field was an active campus network; although in a couple of arrangements, the DDoS attacks are proposed to be identified and examined in private cloud computing environment [13].

Architecture. To invigorate assets against a DDoS attack, it is imperative to make the architecture as strong as feasible. Invigorating network architecture is a critical advancement in DDoS network defense as well as in ensuring business congruity and security from any sort of blackout or catastrophe circumstance [14].

Hardware. It is essential to make use of proper hardware that can deal with realized attack types and utilize the substitutions that are in the hardware that would guarantee that resources are arranged [15]. Also, while supporting resources won't keep a DDoS attack from occurring, doing as such will diminish the effect of an attack.

Moreover, hardware upgrading is successful against SYN flood attacks. Largely the present-day hardware, network firewalls, web application firewalls, and load balancers will usually have a setting that allows a network operator to begin concluding off TCP links once they achieve a specific limit [16].

Challenge–Response Method (CAPTCHA entering). A CAPTCHA is a kind of challenge–response framework envisioned to discriminate people from automated software programs. CAPTCHAs are utilized as safety authorizations to prevent spammers and programmers from utilizing forms on site pages to embed noxious or paltry coding [17].

CAPTCHAs are a sort of Turing test [7]. Simply, end clients are approached to play out some assignments that both cannot do.

Tests regularly include JPEG or GIF pictures, as though bots can distinguish the presence of a picture by reading the source code, they cannot tell what the picture delineates. Since certain CAPTCHA pictures are hard to translate, clients are generally given the choice to ask for another test.

- The highly known kind of CAPTCHA is the text CAPTCHA, which needs the user to see a misshaped string of alphanumeric letterings in a picture and enter the letterings in an appended shape. Content CAPTCHAS are additionally rendered as MP3 sound chronicles to address the concerns of those with visual impairment. Similarly, bots can distinguish the existence of a sound record; however, just a human can tune in and know the data the document has.
- Picture acknowledgment CAPTCHAs solicit users to distinguish a subsection from pictures inside a bigger arrangement of pictures. For example, the client might be given a lot of pictures and requested to tap on every one of the ones that have vehicles in them.
- Math CAPTCHAs—need the client to tackle a simple math sum, for example, addition or subtraction of two numbers.
- 3D Super CAPTCHAs—involve the client in distinguishing a picture extracted in 3D.
- I am not a robot CAPTCHA—needs the client to check a box.
- Promotional CAPTCHAs—need the client to type a specific word or expression linked with the patron's brand.

CAPTCHAs are currently practical standard security systems for guarding from unwanted and noxious boot programs on the Internet. It is alternatively called as human interaction proofs (HIPs).

A great CAPTCHA should be human friendly as well as sufficiently strong to oppose PC programs that attackers write to naturally waft through CAPTCHA tests. In any case, making CAPTCHAs that display both great control and ease of use is a lot tougher than it appears [18].

There are essentially two key steps linked with creating a solid CAPTCHA arrangement:

- The base for the riddle or challenge should be something that is really troublesome for PCs to solve.
- The manner in which riddles and replies are handled should simple for human clients.

The proposed technique has been produced to differentiate human and PC programs from one another by a similar certainty that human wants to give information in the wake of settling the inquiry related with CAPTCHA execution [19]. The inquiry must be troublesome for PCs to resolve and moderately simple for people.

3 Recommendations

The paper has assessed the nature of DDoS attacks that pose a serious threat to network security on the cloud environment. Considering the severity of the DDoS attacks, it is critical to design mitigation and prevention strategies. The paper has described the important mechanisms that are used for detection and prevention of DDoS attacks. The paper recommends IDS and IPS methods that will be helpful in prevention and detection of DDoS attacks. A major advantage of network-based IPS approach is that there is no need for deploying any monitoring program on the host machines. The IDS-based anomaly method is one of the best methods since it is useful in detecting the attack without having any specific details. Hence, the paper supports the deployment of IDS and IPS as the most secure methods for dealing with DDoS attacks.

One more method is a new generation of the CAPTCHA method that utilizes query linked to CAPTCHA rather than simple CAPTCHA. It can be termed as a THREE-TIER CAPTCHA; as in this technique, CLAD node must perform three acts: firstly, the alphanumeric CAPTCHA coding linked to picture. Secondly, the query linked to that CAPTCHA coding. In this procedure, human/user can give input as per the query that is not simple for software bots to solve. Thirdly, picture-based CAPTCHA icon image is positioned on background picture; first user needs to click on the picture in the nested pictures and then the CAPTCHA test gets cleared. The algorithm of this technique makes it tough for boot programs denoting that it is highly safe.

4 Conclusion

DDoS is one of the most destructive attacks on the Internet [4]. Considering the severity of the DDoS attacks, it is difficult to design mitigation and prevention policies. The identification of the nature of the attack will further help in the prevention of these attacks. IDS is related to hardware and software that help in detecting and recoding anomalous activities. The IPS functions are similar to IDS; however, it is classier, because it is designed to take necessary actions for preventing and reducing the malicious activities. The paper supports the deployment of IDS and IPS as the most secure methods for dealing with DDoS attacks.

References

1. Jaber, A., Zolkkipli, M., Anwar, S., & Majid, M. (2016). Methods for preventing DDoS attacks in cloud computing. *Advanced Science Letters, 23.*
2. Fakeeh, K. (2016). An overview of DDoS attacks detection and prevention in the cloud. *International Journal of Applied Information Systems (IJAIS), 11.*

3. Somani, G., Gaur, M. S., Sanghi, D., Conti, M., & Buyya, R. (2017). DDoS attacks in cloud computing: Issues, taxonomy, and future directions. In *Computer communications* (Vol. 107, pp. 30–48). Elsevier.
4. Verma, A., Arif, M., & Hussain, M. (2018). Analysis of DDOS attack detection and prevention in cloud environment: A review. *International Journal of Advanced Research in Computer Science, 9.*
5. Khadka, B., Withana, C., Alsadoon, A., & Elchouemi, A. Distributed denial of service attack on cloud: Detection and prevention. In *International Conference and Workshop on Computing and Communication (IEMCON)* (Vancouver, BC, IEEE, Canada, 2015).
6. Yang, L., Wang, S., Jhang, T., & Song, K. (2012). Defense of DDoS attack for cloud computing. In *IEEE International Conference on Computer Science and Automation Engineering (CSAE)*, IEEE.
7. Ambrose Thomas, V., & Kaur, K. (2013). Cursor CAPTCHA captcha mechanism using mouse cursor. *International Journal Of Computer Applications, 67,* 13–17. https://doi.org/10.5120/11526-7253.
8. Dissanayake, A. (2008). *Intrusion detection using the Dempster-Shafer theory—Literature review and survey.* School of Computer Science: University of Windsor.
9. Yu, D., & Frincke, D. (2004). A novel framework for alert correlation and understanding. In *International Conference on Applied Cryptography and Network Security (ACNS) 2004* (Vol. 3089, pp. 452–466). Springer's LNCS series.
10. Chou, T., Yen, K. K., & Luo, J. (2008). Network intrusion detection design using feature selection of soft computing paradigms. *International Journal of Computational Intelligence, 4,* 102–105.
11. Siaterlis, C., Maglaris, B., & Roris, P. (2003). *A novel approach for a distributed denial of service detection engine.* Greece: National Technical University of Athens.
12. Siaterlis, C., & Maglaris, B. (2005). One step ahead to multisensor data fusion for DDoS detection. *Journal of Computer Security, 13,* 779–806.
13. Batham, G., & Sejwar, V. (2016). Implementation of Dempster-Shafer theory for trust based communication in MANET. *International Journal of Computer Applications, 150,* 27–32.
14. Luo, H., Lin, Y., Zhang, H., & Zukerman, M. (2013). Preventing DDoS attacks by identifier/locator separation. *IEEE Network, 27,* 60–65.
15. Sheeba, R., & Rajkumar, K.: Enhancing Security by Preventing DoS and DDoS Attack using Hybrid Approach. Indian Journal Of Science And Technology, 9 (2016).
16. Parno, B. Trust extension as a mechanism for secure code execution on commodity computers.
17. Kumar, P. H. (2012). Assuage Bandwidth Utilization DDoS Attacks by Using Prototype Analyzer and Transfer Scheduling Scheme. *IOSR Journal Of Engineering, 2,* 187–189. https://doi.org/10.9790/3021-0281187189.
18. Ming, L. (2011). CAPTCHA in security ECIS: Depress phishing by CAPTCHA with OTP. *HKIE Transactions, 18,* 18–31. https://doi.org/10.1080/1023697x.2011.10668241.
19. Hubbard, B. (2010). *Code breakers.* Tunbridge Wells: TickTock.

A Review on Massive MIMO for 5G Systems: Its Challenges on Multiple Phases

Lalitha Nagapuri

Abstract The massive multiple input—multiple output (mMIMO) is focused on considering the different current strategies in the framework plan for 5G technology. There is a lot of scope for research to enhance the performance of 5G up-gradation regarding data rate, latency, throughput, energy, and spectral efficiency. Although there exist, latest philosophies, procedures, and structures including D2D communication for planning and executing 5G systems, the mMIMO adopts spatial multiplexing and enhances the system capacity. It provides higher spectral efficiency (SE), energy efficiency (EE), robust to individual element failures, and jamming. It plays a key in 5G advancement and the effective plan of mMIMO stays an open issue in the latest. Different papers are checked for EE and SE execution in mMIMO technology through various algorithms and parameter examinations are made to distinguish the better algorithms regarding SE and EE to accomplish the higher data rates, BER, and mitigating the inter-noise interference.

Keywords mMIMO · D2D · 5G · Energy efficiency · Spectral efficiency

1 Introduction

The prominent technological developments and current demands rapidly paves steps toward 5G technologies in the modern wireless communication. 5G technology should provide higher data rates, low battery consumption, high-energy efficiency, better coverage, and high spectral efficiency [1–3].

5G approaches include non-orthogonal multiple access (NOMA), MIMO, mMIMO; cooperative, D2D, mmWave, green communications; and cognitive radio (CR), etc. The major challenges before 5G technology implementation are:

- Design of new scheduling and access control mechanisms with less control overhead to support massive connections in the wireless network

L. Nagapuri (✉)
Kamala Institute of Technology & Science, Huzurabad, India
e-mail: penchalalalitha@gmail.com

S. C. Satapathy et al. (eds.), *Intelligent System Design*, Advances in Intelligent Systems and Computing 1171, https://doi.org/10.1007/978-981-15-5400-1_48

- High end-to-end data capacity to support high screen resolution and higher data rates
- High data capacity with optimum operating cost
- Fast and flexible architectures for easy deployment
- High reliability to support real-time-critical services like emergency services, medical monitoring, etc.
- Low latency to support real-time implementations like virtual reality services
- Integration of heterogeneous networks to support communication among different types of nodes
- Radio resource management for better coverage, higher energy, and spectral efficiencies.

Guaranteed Quality of Service (QoS) in various conditions like high density, high traffic, and high mobility scenarios.

The 5G interchanges targets supporting unbounded capacity of networking, enormous information transmission capacity alongside broad sign inclusion to give a decent scope of top-notch customized administrations to the end-customers. To achieve this, 5G interchanges will incorporate various existing cutting edge innovations with inventive new systems. The different methodologies and procedures for 5G systems were discussed in.

In this paper, technologies involved in 5G, advantages, and applications are studied in Sect. 2. Section 3 describes massive MIMO system. A comparative study on massive MIMO systems in terms of SE and EE is discussed in Sect. 4. Finally, the conclusions and future scope are discussed in Sect. 5.

2 Technologies Involved in 5G

2.1 Software-Defined Networking (SDN)

These models segment the system control capacities and information-sending capacities, and in SDN, system control capacities are programmable. These models comprise of three modules as (i) programming controller, which carries capacities, for example, organize the executives, network OS. (ii) An interface in the midst of the controller, SDN-empower foundation (iii) an interface crosswise over different SDN.

Specific advantages of SDN are centralized network provisioning, comprehensive enterprise management, security, economical, less complex, cloud abstraction, and assured content dispatch, etc.

2.2 Cognitive Radio Network (CRN)

Cognitive radio is the lead technology that empowers a CRN to utilize range in a unique way. In wireless communication, the management of radio resources is very crucial, and CRN employs dynamic spectrum sharing techniques. The CRN can self-configure its infrastructure, it can precisely accommodate to outside circumstances and maximize its performance. Intelligent and self-management services are essential to work with a cognitive paradigm. CRNs are being advanced to solve current remote system issues, resulting from the bounded accessible range and inefficiency in the range usage, by exploiting the current wireless spectrum opportunistically. CRN requires specific framework, explicit model, new protocol stack or multilayered protocols. So, these challenges swiftly demand for cross-layer design. Challenges in CRNs are spectrum sensing, management, mobility, and sharing. Issues in CRNs are implementation, regulatory, QoS, security, and cross-layer design.

CRNs can be applied to the following cases: utilities/smart grid, leased network, monitoring, control, telemetry, and automation, municipal government, public safety, emergency network, military network, rural deployments, wireless regional area network, location-based services, entertainment and multimedia, security.

2.3 Device-to-Device (D2D) Communication

The D2D transmission in a cell system is characterized as a point-to-point radio innovation where correspondence between two gadgets is managed without the use of BS. The D2D communication, for the most part, works on both licensed and unlicensed spectrum ranges. The motivation to pick inband range is to make the interface controllable while speak with one another.

The D2D communication framework can be characterized by imagining a two-level 5G cell system and viewed as device level and macrocell level. The first case has a universal cell structure for correspondence accomplished via base station (BS). Second case, ON the OFF chance of the device is associated legitimately with different devices. In the device level, corresponding BS have full or fractional authority over transferring device and asset assignment of source and goal [4].

In 5G cell systems, two kinds of D2D interchanges can be set up, neighborhood and worldwide D2D correspondences. In the neighborhood, D2D interchanges either away between two devices associated with a similar BS, straightforwardly or by handing-off data through different devices. They found the nearest devices and reduce the correspondence cost. Then again, in worldwide D2D interchanges, two devices identified with various BSs by bobbing through different systems. They abridge both BS-to-BS (B2B) interchanges and device-to-BS (D2B) correspondences.

2.4 Millimeter Waves (mmWave) Communication

The frequency range of mmWave is from 30 to 300 GHz and can be adopted for fast remote interchanges. This frequency band has wavelength going from 10 to 1 mm and along these lines called millimeter waves. The mmWaves accommodate higher data rates and support 5G requirements. Millimeter waves can supplant customary fiber optic transmission lines associating versatile BSs. Ordinary higher information rate transmission requires fiber optic link establishment. It experiences issues for execution, upkeep and it is not affordable. Any harm to the delicate fiber strand could cause full disturbance of the transmission framework. Millimeter-wave development can without a very remarkable stretch achieve 10 Gbps data rate for correspondence and supports high-security transmission.

Millimeter-wave [2] innovation is one of the quickest developing advancements in this decade. More popularity for rapid information, ultra top-notch interactive media, HD gaming, security, and observation and so on will drive millimeter-wave innovation to the next level. It will ceaselessly create and offer a wide range of utilizations later on. Because of a lot littler wavelength of mm-wave, it might abuse the polarization and new spatial handling systems, for example, mMIMO and adaptive beamforming [4].

2.5 Massive MIMO (MMIMO)

The MIMO capabilities can be enhanced by adopting mMIMO technology. A simple massive MIMO communication system is shown in Fig. 1. The mMIMO system employs 100 s of antennas, all of them serving in a similar time-frequency resource tens of terminals. The mMIMO system enhances all the gains accomplished by using MIMO in a larger degree. The mMIMO is the way to empower the advancement of 5G networks (fixed, mobile), which provides high EE and SE, robust and secured [5].

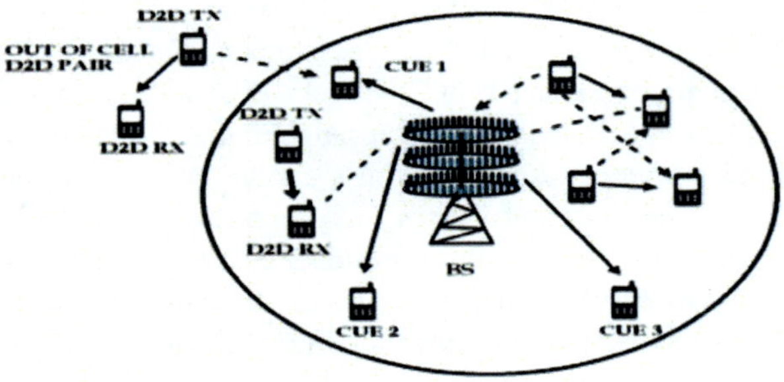

Fig. 1 Massive MIMO system

Massive MIMO Advantages: The massive MIMO uses large antenna arrays, which provide high antenna gain, high gain due to multiplexing, and results in greater SE. In massive MIMO system, the energy is radiated from antenna arrays to the mobile station or user equipment via a focused beamforming with high concentration, which results in greater energy efficiency. The massive MIMO via large diversity gain using narrow beam supports to accomplish high reliable and secured communication with least inter-user interference. The massive MIMO system is robust to failures of individual array elements, cost-effective, consumes low power, and reduces the latency on air interface.

3 MMIMO-Associated Work

In mMIMO systems, more number of antennas can provide more amount of degrees of opportunity to facilitate efficient wireless communication signals, thus enhancing spectral efficiency (SE). Huge amount of degrees of opportunity in spatial domain will facilitate power allocation, contributing to energy efficiency (EE). SE defines the data rate that can be sent over a given bandwidth in a particular communication system. EE defines the number of bits that can be sent over a unit of power consumption.

In 2017, Zhao et al. [6] has presented a calculation which manages obstruction arrangement and power parting engineering for Information Decoding and Energy Harvesting at every beneficiary by thinking about the Quality of Service prerequisites, The raised non-convex EE enhancement issue illuminated by structuring mix of Power Splitters, Transmitter Beam formers, Transmitter Power and Receiver Filters, it is contrasted and TZF arrangement where it ready to wipe out the Interference issue just on Transmitter side not on Receiver side and it finished up with better EE advancement for MIMO impedance issues.

In 2016, Jing et al. [7] had built up a plan for down -link multi-users to expand the energy efficiency in massive MIMO applications through millimeter-wave innovation and it additionally targets better utilization of order gains with simple beamforming calculation and Energy Efficiency with Digital beamforming calculations, It likewise demonstrated that better BER (Bit Error Rate) contrasted with all customary Beamforming calculations when number of clients expanded.

In 2015, Zhao et al. [8] have assessed mMIMO-OFDM with zero forcing and maximum ratio combination to upgrade the energy and spectral efficiency in uplink applications; he likewise proposed a trade-off among EE and SE when creating the base station with number of reception apparatuses, and found that the exhibition of the framework execution not upgraded just by expanding the spatial or recurrence assets which additionally required to improve the reasonable parameters as number of multiplexing clients, various identifiers, number of radio wires at base station with transmitted power.

In 2017, Liu et al. [9] have examined a trade-off among EE and SE utilizing linear pre-coding and transmit antenna selection (TAS) in massive MIMO frameworks; this exchange off simply settled the Complexity about multi-objective optimization (MOO) issue and demonstrated that this issue neither convex or concave, optimization of SE and EE done dependent on transmit antennas and transmit power by considering circuit power utilization and large-scale fading.

In 2018, Gao et al. [10] had presented a full-array and sub-Array exchanging systems for the antenna selection in massive MIMO channels, he broke down branch and Bound algorithms for antenna selection, this examination created a powerful results for Finite-Dimensional MIMO.

In 2016, Patcharamaneepakorn et al. [11] has presented a generalized modulation strategy for fifth-era remote systems in massive MIMO frameworks to improve EE and SE, and this proposition entirely reasonable to homogeneous systems and littler-dimensional antenna arrays, and explored a trade-off among EE and SE, and this work is quite certain to the MF and ZF linear receivers.

In 2014, Jiang et al. [12] has recommended an improvement strategy in virtual MIMO frameworks to upgrade the energy efficiency; under this article the EE done dependent on allotment of capacity to the transmitter and the relays, designation of transmission capacity to the information and participation channels, a littler measurements 2×2 had been considered for the MIMO framework, EE correlation made among MIMO and MISO, completed outcomes demonstrated MIMO have better EE when relays given additional power.

In 2016, Le et al. [13] had presented antenna selection procedures for MIMO-OFDM frameworks in the point of view of energy efficiency and furthermore made a trade-off between EE-SE; he demonstrated that The GA-based antenna selection give better EE contrasted with regular MIMO frameworks exhibit loss in EE.

4 MMIMO Analysis

Regardless of reality, a couple of procedures and strategies are available for 5G system structure as given in Fig. 2. The mMIMO is a pleasant advancement to ensure the various essentials required for 5G systems.

To build an efficient mMIMO system, the performance is refined in terms of latency, throughput, SE/EE, pilot contamination, and peak data rate as given in Fig. 3.

This paper examines the different structure techniques for building up a mMIMO system to meet the challenges of 5G. Few existing works are compared and given in Table 1. Table 1 demonstrates adopted methodology, highlights, and difficulties of regular algorithms intended for enhancing energy and spectral efficiencies in MIMO and massive MIMO frameworks.

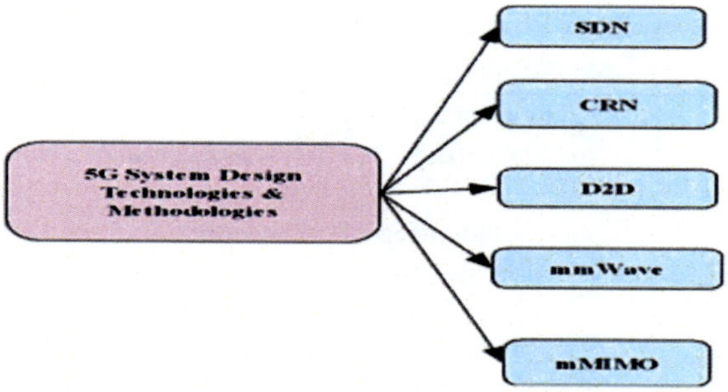

Fig. 2 Technologies used in 5G system design

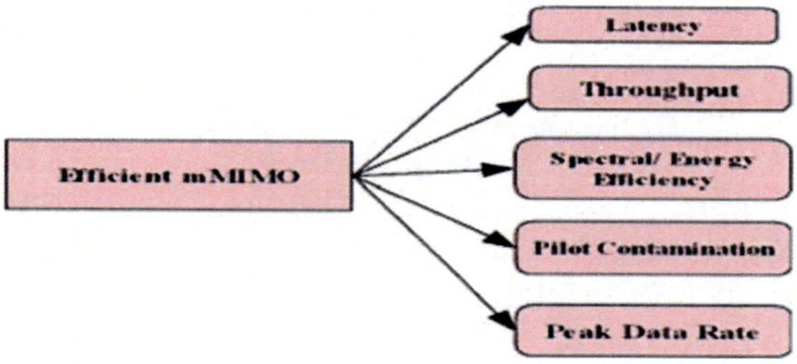

Fig. 3 Efficient massive MIMO system design

5 Conclusion and Future Scope

This paper gives a comprehensive overview of the current EE mMIMO framework plan with their starters, condition of-craftsmanship to investigate a vitality effective MIMO framework for 5G correspondences. Furthermore, it illuminates open difficulties in the advancement procedure. Later on, a comparative study of energy efficiency in multiple-user and multiple-cell framework is done. Concerning the propelled strategies that will be utilized in the remote correspondences, for example, OFDMA, MIMO, and relay, current exploration has demonstrated that bigger EE can be achieved through EE design. In any case, the research is going on and higher exertion expected to examine future themes. The energy efficiency issues in MIMO in multiple-user and multiple-cell scenarios are discussed in brief. The prime challenge is to boost the energy efficiency using the spatial resource by reducing the interference. Reduction of complexity plays an important role in the development of any system. It is adequate but basic methods are needed to get a decent agreement between intricacy and performance.

Table 1 Result of conventional EE and SE techniques in massive MIMO systems

Author	Adopted system	Results	Challenges
Zhao [6]	The MAX-SINR ID algorithm utilizing SWIPT	Improved energy efficiency, eliminate interference	Power efficiency, non-linear and concave problems
Jing [7]	Hybrid beamforming using ZERO-gradient	Better energy efficiency, mitigate inter-user interference	Baseband processing, low power efficiency
Zhao [8]	ZF detection, maximum ratio combination	Trade-off between EE and SE provides the desired amount of antennas chosen	In one-cell condition, inter-cell interference problem
Liu [9]	Particle swarm optimization (NBI, WS)	Excellent trade-off between EE, SE	EE decreases with decrease in SE
Gao [10]	Greedy search based antenna selection	Minimized complexity, better spectrum efficiency	RF fully switching networks are less power efficient with larger MIMO
Patcharamaneepakorn [11]	Generalized spatial modulation	Improved EE, moderate SE	Gen-SM for heterogeneous network CC for massive antenna arrays
Jiang [12]	Compress and forward cooperation	Enhance EE for MIMO over MISO	Shared band cooperation channels, extra power for relays
Le [13]	GA-based antenna selection	Better EE compared to conventional systems, lower complexity	Multi-user MIMO-OFDMA, larger antenna arrays

References

1. Panwar, N., Sharma, S., & Singh, A. K. (2016). A survey on 5G: The next generation of mobile communication". *Physical Communication, 18,* 64–84.
2. Gupta, A., & Jha, R. K. (2015 July). A survey of 5G network: Architecture and emerging technologies. *IEEE Access, 3,* 1206–1232.
3. Samsung Electronics 5G Vision, White Paper. (2015 February). https://images.samsung.com/is/content/samsung/p5/global/business/networks/insights/white-paper/5g-vision/global-networks-insight-samsung-5g-vision-2.pdf.
4. Qiao, J., Shen, X., Mark, J. W., Shen, Q., He, Y., & Lei, L. (2015 January). Enabling device-to-device communications in millimeter-wave 5G cellular networks. *IEEE Communications Magazine, 53*(1), 209–215.
5. Arfaoui, M.-A., Ltaief, H., Rezki, Z., Alouini, M.-S., & Keyes, D. (2016). Efficient sphere detector algorithm for massive MIMO using GPU hardware accelerator, Vol. 80.(pp. 2169–2180).
6. Zhao, H., Liu, Z., & Sun, Y. (2018). Energy efficiency optimization for SWIPT in K-user MIMO interference channels. *Physical Communication, 27,* 197–202.

7. Jing, J., Xiaoxue, C., & Yongbin, X. (2016). Energy-efficiency based downlink multi-user hybrid beamforming for millimeter wave massive MIMO system. *The Journal of China Universities of Posts and Telecommunications, 23*(4), 53–62.
8. Zhao, L., Li, K., & Zheng, K. (2015 March). An analysis of the trade off between the energy & spectrum effeciences in an uplink massive MIMO-OFDM system. *IEEE Transactions on Circuits and Systems, 62*(3).
9. Liu, Z., Du, W., & Sun, D. (2017). Energy and spectral efficiency tradeoff for massive MIMO systems with transmit antenna selection. *IEEE Transactions on Vehicular Technology, 66*(5), 4453–4457.
10. Gao, Y., Vinck, H., & Kaiser, T. (2018 March 1). Massive MIMO antenna selection: Switching architectures, capacity bounds, and optimal antenna selection algorithms. *IEEE Transactions on Signal Processing, 66*(5), 1346–1360.
11. Patcharamaneepakorn, P., Wu, S., & Wang, C.-X. (2016 December). Spectral, energy, and economic efficiency of 5G multicell massive MIMO systems with generalized spatial modulation. *IEEE Transactions on Vehicular Technology, 65*(12).
12. Jiang, J., Dianati, M., Imran, M. A., Tafazolli, R., & Zhang, S. (2014 June). Energy-efficiency analysis and optimization for virtual-MIMO systems. *IEEE Transactions on Vehicular Technology, 63*(5).
13. Le, N. P., Safaei, F., & Tran, L. C. (2016 April). Antenna selection strategies for MIMO-OFDM wireless systems: An energy efficiency perspective. *IEEE Transactions on Vehicular Technology, 65*(4).

A New Ensemble Technique for Recognize the Long and Shortest Text

M. Malyadri, Najeema Afrin, and M. Anusha Reddy

Abstract In data mining, shorter-text analysis is performed more widely for many applications. Based on the syntax of the language, it is very difficult to analyze the short text with several traditional tools of natural language processing, and this is not applied correctly either. In short text, it is known that there are rare and insufficient data available with this text, and it is difficult to identify semantic knowledge. With the great noise and ambiguity of short texts, it is very difficult to find semantic knowledge. In this paper, it was proposed to replace the coefficient of similarity of cosine with the measure of similarity of Jaro–Winkler to obtain the coincidence of similarity between pairs of text (source text and target text). Jaro–Winkler does a much better job of determining the similarity of the strings because it takes order into account when using positional indices to estimate relevance. It is presumed that the performance of CACT driven by Jaro–Winkler with respect to one-to-many data links offers optimized performance compared to the operation of CACT driven by cosine. An evaluation of our proposed concept is sufficient as validation.

Keywords Text mining · Cosine's similarity coefficient · Jaro–Winkler similarity

1 Introduction

In data mining, text mining is the domain most commonly used to extract knowledge from semantic data. Text mining is the method of extracting data from similar varied information and has the necessary relationship between the entities extracted. Within the text mining space, the classification of texts is one of each technique. It is one of many difficult data mining problems, since it manages high-dimensional data indexes

M. Malyadri (✉) · N. Afrin · M. Anusha Reddy
CMR Technical Camus, Hyderabad, India
e-mail: malyadri.mtech@gmail.com

N. Afrin
e-mail: najeema.afrin@gmail.com

M. Anusha Reddy
e-mail: anushamandadi@ymail.com

S. C. Satapathy et al. (eds.), *Intelligent System Design*, Advances in Intelligent Systems and Computing 1171, https://doi.org/10.1007/978-981-15-5400-1_49

with subjective samples of missing information. The address of the problem in this document was to classify text documents without a label. The problem would be described by taking a large group of tagged text documents and designing the data mining classifier.

The wide accessibility of Internet records in electronic structures requires a programmed strategy to mark documents with a predefined set of topics, which is known as automatic short-text categorization (ASTC). Since the last few years, a large number of machine learning algorithms have been observed to handle this test task. In designing the TC journey as a meeting problem, several current learning methodologies are connected, but its unit of constraint area was discovered once the basic content is small.

There is an extensive type of use to handle short messages. Short messages present new problems with connected content assignments, as well as knowledge retrieval (IR), order and clustering. Unlike long files, two short messages that have a comparable meaning do not share very different words. For example, the implications of "uploading and returning Macintosh items" and the "new iPhone and iPad" area unit are firmly connected; however, they do not share traditional words. The absence of adequate factual knowledge causes successful challenges in the estimation of similarity, and therefore, the calculations of varied current content examinations do not specifically distinguish short messages.

Stopping too short messages specifically. All this much, the lack of objective information still means issues that will be overlooked after we treat long papers. Take lexical ambiguity for example. "Apple" offers ascend to varied implications in "apple item" and "apple tree". Due to the shortage of relevant knowledge, these ambiguous words build short messages arduous to understand by machines. In this paper, the proposed system replaces cosine's similarity coefficient with Jaro–Winkler similarity measure to obtain the similarity matching of text pairs (source text and destination text). Jaro–Winkler does a much better job at determining the similarity of strings because it takes order into account using positional indexes to estimate relevancy. It is presumed that Jaro-Winkler-driven CACT's performance with respect to one-to-many data linkages offers an optimized performance compared to cosine-driven CACT's workings. An evaluation of our proposed concept suffices as validation.

This instruction file for word users (there is a separate instruction file for LaTeX users) may be used as a template. Kindly send the final and checked word and PDF files of your paper to the contact volume editor. This is usually one of the organizers of the conference. You should make sure that the word and the PDF files are identical and correct and that only one version of your paper is sent. It is not possible to update files at a later stage. Please note that we do not need the printed paper.

We would like to draw your attention to the fact that it is not possible to modify a paper in any way, once it has been published. This applies to both the printed book and the online version of the publication. Every detail, including the order of the names of the authors, should be checked before the paper is sent to the volume editors.

2 Related Work

2.1 Various Text Mining Techniques

1. Wang and Wang [1] explain the concept I in three different ways, first is the syntax of a written language does not observe the short tests, second is short texts do not contain sufficient statistical information to support and approaches for text mining, and third, they are more ambiguous and noisy, which further increases the difficulty to handle them.
2. McCallum and Li [2] from the tagged data, it is known that this method obtains the seeds for the lexicons and this is called Web list, according to the HTML rules and the service as the search. For example, based on the appearance of Arnold Palmer in the tagged data, we collect from the Web an extensive list of other golf players, including Tiger Woods (a phrase that is difficult to detect as a name without a good lexicon).
3. G. Zhou and J. Su, "Entity recognition named using a fragment tagger based on HMM". This article proposes a hidden Markov model (HMM) [3] and a fragment tagger based on HMM, from which an entity recognition (NE) system (NER) is constructed to recognize and classify names, times and quantities numerical.
4. D. M. Blei, A. Y. Ng and M. I. Jordan, "Assignment of Latent Dirichlet". This article describes the allocation of latent Dirichlet [4] (LDA), a generative probabilistic paradigm for discrete data groups, such as text bodies. LDA is a three-level, hierarchical, Bayesian model, in which each element of a meeting is recorded as a measurable mix over an underlying set of topics. Each theme, in turn, is modeled as an infinite mix over an underlying set of subject probabilities.

3 Ensemble Semantic Knowledge Approach

The challenging problem of inducing a taxonomy from a set of keyword phrases instead of large text corpus is the current project context. Conceptualization is the most widely used multiple inferencing mechanisms to work under various contexts [5]. All these contain a huge number of probabilistic concepts, interconnected and fine-grained, and it is also called as Probase API. To express the semantics explicitly, the concept based on the data is most powerful in comprehending the understanding of the short text [6]. For comparing the two short texts and also for classification, it is also alone to be not compatible for the tasks. Consider the same two short texts: "upcoming apple products" and "new iphone and ipad".

In this system, for every short text, it is known that there are no common terms. To improve the performance such as similarity of their semantics by using the inferencing mechanism on Probase to retrieve the most popular similar terms such as noun, these are checked for new contexts for that short text. The first process is clustering the Probase terms based on their similar relationship, and with this, it is known that this belongs to the same cluster [7]. For instance, If we have a keyword "dogs", our Probase driven extracts other polysemy words like "mutt", "canines", "mongrel", etc., which definitely forms a cluster group, and we shall repeat the process for other nouns in the short text and from the obtained results we shall identify the most commonest matching entities using a three-layer stacked auto-encoders for hashing terms to reduce processing complexity.

1. Cosine similarity coefficient, a measure that is commonly used in semantic text classifications which measures the similarity between two texts and determines the probable measure.
2. CACT's approach to use cosine's similarity coefficient increases time complexity exponentially [8].
3. So we propose to replace cosine's similarity coefficient with Jaro–Winkler similarity measure to obtain the similarity matching of text pairs(source text and destination text).
4. Jaro–Winkler does a much better job at determining the similarity of strings because it takes order into account using positional indexes to estimate relevancy [9].
5. It is presumed that Jaro-Winkler-driven CACT's performance with respect to one-to-many data linkages offers an optimized performance compared to cosine-driven CACT's workings [10].
6. An evaluation of our proposed concept suffices as validation.

3.1 Algorithm

Step 1: Start
Step 2: Load datasets
Step 3: Preprocessing of data
Step 4: CACTS + SAE + Cosine Similarity + Short Texts
Step 5: Calculate time
Step 6: Display time
Step 7: Results.

4 Evolution Results

By using the Netbeans 8.0.2 and install jdk 1.8 with 4 GB RAM to process, the huge datasets load the articles by browsing from datasets. The analyzing of results is done in four phases. In the first phase, TF-IDF + LongTexts where the only LONG TEXTS are analyzed it was in a existing system, and the articles are be categorized and article classes also generated by running algorithm. The time complexity for TF-IDF Classification Build is completed in 20.214587612 s for 15 files.

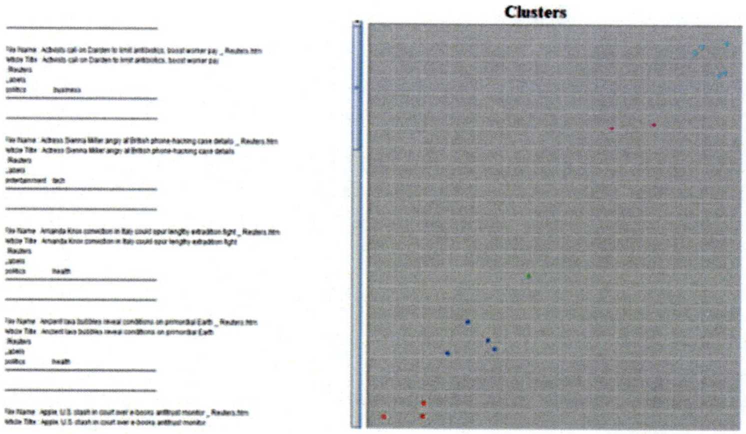

Fig. 1 Semantic knowledge results

In the second phase, CACTS + SAE + Cosine Similarity + ShortTexts where only SHORTTEXT can be analyzed, mostly used algorithm, when the algorithm is applied then the articles are categorized as each and individual articles classified and given the result. The time complexity of TF-IDF Classification Build is completed in 20.214587612 s for 15 files and for CACTS + SAE + Cosine Similarity + ShortTexts algorithm the time complexity is 13.500744525 s for 15 files.

In the third phase, CACTS + SAE + Jaro–Winkler Similarity + ShortTexts where only SHORTTEXT can be analyzed, mostly used algorithms, when the algorithm is applied then the articles are categorized as each and individual articles classified and given the result. The time complexity of CACTS + SAE + Cosine Similarity + ShortTexts Classification Build is completed in 13.500744525 s for 15 files and for CACTS + SAE + Jaro–Winkler Similarity + ShortTexts algorithm the time complexity is 8.576099063 s for 15 files (Fig. 1).

5 Conclusion

In this paper, the ensemble algorithm CACTS and SAE are adopted with Jaro–Winkler similarity approach is utilized for short text for better results. Many existing systems performed semantic knowledge to understand the shortest text but those existing system does not work properly. Finally, the proposed system works better and reduces the computation time to get better results. A huge time gap to have the program proposed.

References

1. Hua, W., Wang, Z., Wang, H., Zheng, K., & Zhou, X. (2017 March). Understand short texts by harvesting and analyzing semantic knowledge. *IEEE Transactions on Knowledge and Data Engineering, 29*(3).
2. McCallum, A., & Li, W. (2003). Early results for named entity recognition with conditional random fields, feature induction, and web enhanced lexicons. In *Proceedings of the Seventh Conference on Natural Language Learning at HLT-NAACL 2003* (Vol. 4, ser. CONLL '03, pp. 188–191), Stroudsburg, PA, USA.
3. Zhou, G., & Su, J. (2002). Named entity recognition using an hmm-based chunk tagger. In *Proceedings of the 40th Annual Meeting on Association for Computational Linguistics, ser. ACL '02* (pp. 473–480). Stroudsburg, PA, USA.
4. Blei, D. M., Ng, A. Y., & Jordan, M. I. (2003). Latent Dirichlet allocation. *Journal of Machine Learning Research, 3,* 993–1022.
5. Rosen-Zvi, M., Griths, T., Steyvers, M., & Smyth, P. (2004). The author-topic model for authors and documents. In *Proceedings of the 20th Conference on Uncertainty in Artificial Intelligence, ser. UAI '04* (pp. 487–494). Arlington, Virginia, United States.
6. Mihalcea, R., & Csomai, A. (2007). Wikify! Linking documents to encyclopedic knowledge. In *Proceedings of the Sixteenth ACM Conference on Conference on Information and Knowledge Management, ser. CIKM '07* (pp. 233–242). New York, NY, USA.
7. Milne, D., & Witten, I. H. (2008). Learning to link with Wikipedia. In *Proceedings of the 17th ACM Conference on Information and Knowledge Management, ser. CIKM '08* (pp. 509–518). New York, NY, USA.
8. Kulkarni, S., Singh, A., Ramakrishnan, G., & Chakrabarti, S. (2009). Collective annotation of Wikipedia entities in web text. In *Proceedings of the 15th ACM SIGKDD International Conference on Knowledge Discovery and Data Mining, ser. KDD '09* (pp. 457–466). New York, NY, USA.
9. Han, X., & Zhao, J. (2009). Named entity disambiguation by leveraging Wikipedia semantic knowledge. In *Proceedings of the 18th ACM Conference on Information and Knowledge Management, ser. CIKM '09* (pp. 215–224). New York, NY, USA.
10. Structural semantic relatedness: A knowledge-based method to named entity disambiguation. In *Proceedings of the 48th Annual Meeting of the Association for Computational Linguistics, ser. ACL '10* (pp. 50–59). Stroudsburg, PA, USA (2010).

Image Denoising Using NLM Filtering and Convolutional Neural Network

K. Srinivasa Babu, K. Rameshwaraiah, A. Naveen, and T. Madhu

Abstract Image denoising technology is one of the forelands during the field of computer graphic along with computer vision. By comparing different filtering techniques which technique is giving the best results of filtering. Non-local means technique is one of the great performing techniques which arouse tremendous research. During this article, an improved weighted non-local means algorithm designed for image denoising is planned. The non-local means denoising technique replaces each pixel through the weighted average of pixels by the surrounding neighborhoods. Furthermore, explanation of the convolutional neural network also gives the greatest results compared to further filtering techniques. The planned method evaluates under testing images among various levels noise. Experimental results are shown used for different filtering techniques by using PSNR.

Keywords Image denoising · Non-local mean filtering · CNN

1 Introduction

Picture denoising [1] is a hot research issue in the field of advanced picture preparing. Picture denoising is significant under ensuring the viability as well as vigor of other picture handling calculations within the business picture process systems, for example, picture enrollment, also picture division. So picture denoising has pulled into an ever increasing number of considerations as of numerous specialists along with numerous denoising techniques have been approved, for example, PDE-based methodologies [2–4], change area strategies [5], the Gaussian smoothing model, the area separating [6], the observational Wiener channels [7] along with the wavelet thresholding strategy [8].

K. Srinivasa Babu (✉) · K. Rameshwaraiah · A. Naveen · T. Madhu
Computer Science & Engineering, Nalla Narasimha Reddy Education Society's Group of Institutions, Hyderabad, India
e-mail: srinivasbabu.k@nnrg.edu.in

K. Rameshwaraiah
e-mail: rameshwaraiah.k@nnrg.edu.in

S. C. Satapathy et al. (eds.), *Intelligent System Design*, Advances in Intelligent Systems and Computing 1171, https://doi.org/10.1007/978-981-15-5400-1_50

It has demonstrated that the spatial space smoothing is successful toward evacuating the added substance Gaussian commotion within the uproarious picture. The key thought is toward supplanting the power estimation of every pixel through a weighted normal of all force estimations of its neighborhood. The weight can be figured by means of the Gaussian channel otherwise the case channel. The essential thought of the Gaussian channel is that the estimation of the pixels of its neighborhood is given distinctive weighting which is characterized through a spatial Gaussian dissemination. In 1998, Manduchi and Tomasi arranged a two-sided channel and also utilized it toward register the weighting capacity.

The non-local means denoising technique replaced each pixel during the noisy picture with the weighted average of pixels by related surrounding neighborhoods. The weighting function is determined via the similarity between neighborhoods. Many articles have introduced an extra detailed analysis under the non-local means algorithm. The most important problem is toward determining the weighting function.

In this article, it proposes a different weighting function also gets an improved non-local means algorithm used for picture denoising. This article uses a dissimilar weighting function toward computing the weight also makes some experiments toward comparing the different weighting function against the original function, also it is shown that the better non-local algorithm outperforms the original non-local means technique.

2 Previous Works

Buades A, Coll B, Morel J M. “A non-local algorithm for image denoising” [9] We suggest another measure, the technique commotion, toward assessing as well as to look at the presentation of advanced picture denoising strategies. We initially figured that this technique commotion is used for a wide class of denoising calculations, and to be specific, the neighborhood smoothing channels. Second, we suggest another calculation, the nonneighborhood implies (NL-implies), in view of a non-nearby averaging of all pixels within the picture. The human eye is just a single ready toward choosing if the nature of the picture has been improved through the denoising strategy. We explain some denoising encounters contrasting the NL-implies calculation along with nearby smoothing channels.

Goossens B, Luong H Q. A fast non-local image denoising algorithm [10] designed for the non-local denoising advance presented through Buades et al., remarkable denoising results are obtained at high expense of computational cost. During this article, a novel algorithm that reduces the computational cost used for calculating the similarity of neighborhood windows is planned. We first introduced an approximate measure about the similarity of neighborhood windows, also then we utilize an efficient summed square image (SSI) scheme along with fast Fourier transform (FFT) toward accelerating the calculation of this determine. Our algorithm is concerning fifty times faster than the original non-local algorithm both theoretically as well as experimentally, yet produces comparable results in terms of mean-squared error (MSE) as well as perceptual image quality.

Exploiting Summed Square Image (SSI) along with Fast Fourier Transform (FFT), we planned a fast non-local denoising algorithm within this article. Theoretically as well as experimentally, the efficiency of our accelerated algorithm is concerning fifty times of the original algorithm, along with the denoising results are still comparable toward the results of the novel algorithm both within MSE as well as perception. Thus, our accelerated algorithm is feasible toward tackling through practical problems.

3 Proposed Work

3.1 *The Non-local Means Filter*

In this article, an adaptation of the non-local (NL) means filter is planned for speckle reduction into ultrasound (US) imagery. At first, developed on behalf of additive white Gaussian noise, we propose toward use a Bayesian framework to get a NL-means filter adapted toward a relevant ultrasound noise model.

Within the standard formulation of the NL-means filter, the restored intensity $NL(u)(x_i)$ of the pixel x_i, is a weighted average of the pixels intensities $u(x_i)$ within the "search volume" V_i. Let us denote:

$$NL(u)(x_i) = \sum_{x_i \in V_i} w(x_i, x_j)u(x_j) \tag{1}$$

where $w(x_i, x_j)$ is the weight assigned toward intensity value $u(x_j)$ used for restoration of pixel x_i. For each pixel x_j within V_i, the L$_2$-norm $\|.\|_2^2$ is computed among $u(N_j)$ (the neighborhood of x_j) as well as $u(N_i)$ (the neighborhood of x_i). Then, these distances are weighted through the weighting function defined as follows:

$$w(x_i, x_j) = \frac{1}{Z_i} e^{-\frac{\|u(N_i) - u(N_j)\|_2^2}{h^2}} \tag{2}$$

where Z_i is the normalization constant ensuring that $\sum_{x_j \in V_i} w(x_i, x_j) = 1$, along with h acts like a smoothing parameter. This filter produces high quality denoising although is computationally expensive.

3.2 Block-Wise Approach

Toward decreasing the computational difficulty of the algorithm, we begin a block-wise approach. During our block-wise NL-mean filter, a weighted average of patches is performed during its place of weighted average of pixel intensities (Fig. 1). The block-wise advance consists in

a. Separating the volume into blocks with overlapping supports;
b. Performing NL-means like restoration of these blocks;
c. Restoring the pixels values based on the restored intensities of the blocks they belong toward:

A partition of the picture Ω into overlapping blocks B_{i_k} of size $P = (2\alpha + 1)^d$ is performed (d is the dimensionality of the picture: 2 or 3), such as $\Omega = \sum_k B_{i_k}$, under the constraint that the intersections among the blocks B_{i_k} are non-empty (i.e., $2\alpha \geq n$). These blocks are centered under pixels x_{i_k} which represent a subset of Ω. The x_{i_k} are equally distributed at positions $i_k = (k_1 n, k_2 n, k_3 n)$, $(k_1, k_2, k_3) \in \mathbb{N}^d$ where n represents the distance among the centers of B_{i_k}. The restoration of a block B_{i_k} is based under a NL-means scheme defined as follows:

$$NL(u)(B_{i_k}) = \sum_{B_j \in V_{i_k}} w(x_{i_k}, x_j) u(B_j) \tag{3}$$

With

$$w(x_{i_k}, x_j) = \frac{1}{Z_{i_k}} e^{-\frac{\|u(B_{i_k}) - u(B_j)\|_2^2}{h^2}} \tag{4}$$

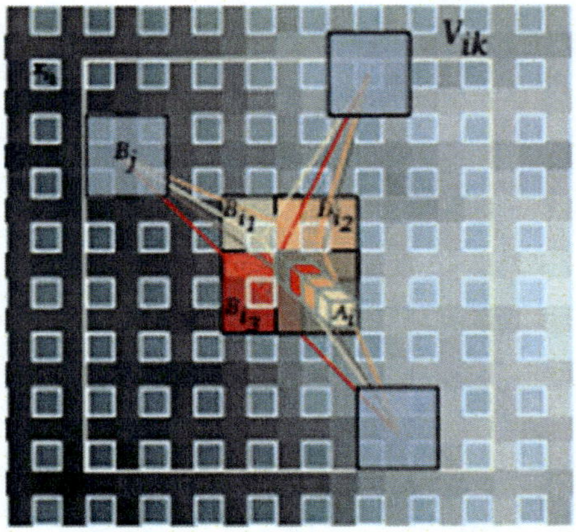

Fig. 1 Block-wise NL-means filter

where $u(B_i) = \left(u^{(1)}(B_i), \ldots u^{(P)}(B_i)\right)^T$ is an picture area containing the intensities of the block B_i, Z_{i_k} is a normalization constant ensuring so as to $\sum_j w\left(x_{i_k}, x_j\right) = 1$ in addition toward

$$\|u(B_i) - u\left(B_j\right)\|_2^2 = \sum_{p=1}^{P}\left(u^{(p)}(B_i) - u^{(p)}\left(B_j\right)\right)^2 \tag{5}$$

In favor of a pixel x_i incorporated within numerous blocks B_{i_k}, more than a few estimations of the identical pixel x_i as of different $NL(u)\left(B_{i_k}\right)$ are computed during addition toward stored inside a vector A_i (Fig. 1). The previous restored intensity of pixel x_i is then defined as:

$$NL(u)(x_i) = \frac{1}{|A_i|}\sum_{l \in A_i} A_i(l) \tag{6}$$

This advance allows toward drastically reducing the complexity of the algorithm. On behalf of instance, if we place $n = 2$, the complexity is divided through a factor 4 inside 2D as well as 8 inside 3D.

3.3 Convolutional Neural Network (CNN)

In this article, we utilize convolutional layers of a linear-CNN model on the way toward execute image filtering. The enhancement of using the CNN model is so as to it always optimizes the weights of convolution kernel throughout network training. We too suggest the linear-CNN model toward evaluating among the traditional linear with nonlinear filtering methods. Behind network training, we obtain two models which communicate toward filter Gaussian noise during addition toward salt-and-pepper noise. Beginning the efficiency of filtering operations, the denoising during the planned linear-CNN model has exposed its better performance compared among those traditional ones.

In this article, we will use CNN [18] for image denoising; we suggest a linear-CNN model in addition to utilize the convolutional layer inside this model to simulate average filtering. Allow the initial size of the convolution kernel is 11×11, throughout the training procedure, if a smaller kernel be capable of get better results, next the CNN model resolve to optimize the convolution kernel through letting the outside fundamentals of convolution kernel be 0; later, a novel convolution kernel be capable of be obtained which is equal toward the smaller kernel.

Input layer: The input picture is represented as a matrix, with the intensity of each pixel during the matrices is an integer within [0, 255]. During organize toward construct the computing easier, we regularize the pixel values of range within [0, 1] of Layer 1.

Convolutional layer: This layer is the core part within the CNN model. The parameter optimizations as well as filtering operations are conducted inside this layer. The first size of this convolution kernel is 11 × 11 which is self-adaptive meant for better picture denoising based under network training.

Compared through the traditional average filter, the planned CNN model uses 9 convolution kernels along with each output color component is collected through the convolutional outputs of every input color components along with multiple convolution kernels. The concentrate of this filtering procedure is in the direction of utilize offered imagery while the training dataset toward rebuild the denoised picture.

Predict layer: The pixel values of output picture beginning the convolutional layer are not essentially in the interval [0, 1]; consequently, in the predict layer, we resolve to normalize the output picture. Designed for the intensities of individual's pixels $I(x, y)$ which are take away than 0, we construct them equal to 0, meant for those pixel values are better than 1.0, we take 1.0.

Validation layer: During this layer, the MSE consequences among the normalized output imagery along with the normalized labeled imagery will be considered.

Throughout the training procedure, we compute the MSE among the output picture earlier than normalization with its normalized labeled picture. We will utilize an optimizer toward to optimize the parameters of this CNN model accordingly since on the way toward to decrease the MSE. We make use of the stochastic gradient descent (SGD) for the optimization.

$$W^{(i+1)} = W^{(i)} - \lambda \cdot \partial L/\partial W, i = 1, 2, \ldots. \quad (7)$$

where $W^{(i)} = \left(w_1^{(i)}, w_2^{(i)}, \ldots, w_n^{(i)}\right)$ is the parameters of ith iteration of our linear-CNN model, λ is the learning rate as well as L is the loss function. On behalf of the output consequences on or after the convolutional layers, if we desire the outputs, after that, we utilize an optimizer to optimize the parameter $W^{(i)}$ moderately than choose $W^{(0)}$ as well as compute the MSE directly. For this cause, let us think that x_{ij} is the pixel x under the input picture, (i, j) is its position, and y_{ij} is the estimated output of the pixel x. Behind the model training, the real output of x_{ij} is $f\ (W_{ij}; x_{ij})$. The equivalent loss function is represented via using Eq. (8).

$$L(W; x, y) = h\big(g\big(f\big(W_x, x_{ij}\big)\big), y_{ij}\big)$$

$$L(W) = h(g(f(W_x))) \quad (8)$$

Wherever g is the normalized function, h is the function used for calculating MSE. We utilize Eq. (8) on the way toward characterizing the normalized function.

$$g(x) = \begin{cases} 1x > 1 \\ 0x < 0 \\ x0 \leq x \leq 1 \end{cases} \quad (9)$$

With observe toward pixel x, the actual output $f\left(W_x, x_{ij}\right) < 0$, and next, we obtain $g(f(W) = 0$. We utilize Eq. (8) on the way toward characterizing $\partial L/\partial W$ in Eq. (10)

$$\partial L/\partial W_x = \partial L/\partial h \cdot \partial h/\partial g \cdot \partial g/\partial f \cdot \partial f/\partial W_x \tag{10}$$

In Eq. (8), we observe, if $g(f(W) = 0$, next $\partial g/\partial f = 0$; thus, $\partial L/\partial W = 0$, which means the SGD algorithm cannot optimize W_x. During this study, we compute the MSE one more all through the training procedure.

The quantitative assessments are additionally made among all the tried strategies. Two understood assessment files as peak signal-to-noise ratio (PSNR) also structural similarity index metrics (SSIM) are utilized for execution gratefulness, which are characterized as:

$$\text{PSNR} = 10\log_{10}\left(\frac{255^2}{\frac{1}{W\times H}\sum_{i=1}^{W}\sum_{j=1}^{H}\left(\hat{u}(i,j) - u(i,j)\right)^2}\right) \tag{11}$$

4 Experimental Results

See Figs. 2, 3 and Table 1.

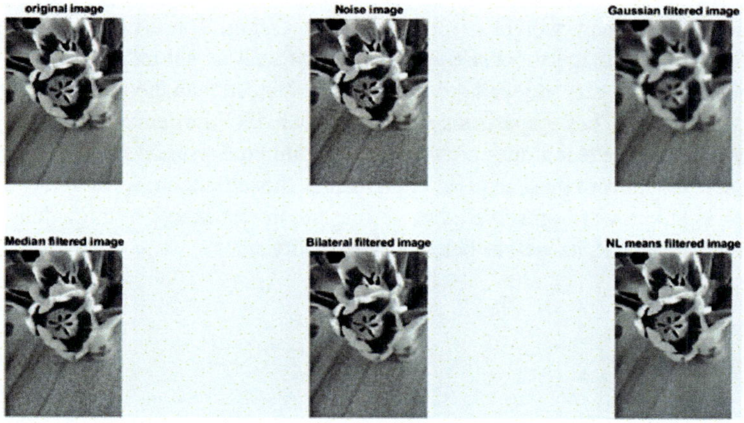

Fig. 2 Different filter techniques (Gaussian, median, and bilateral along with NL-means filtering)

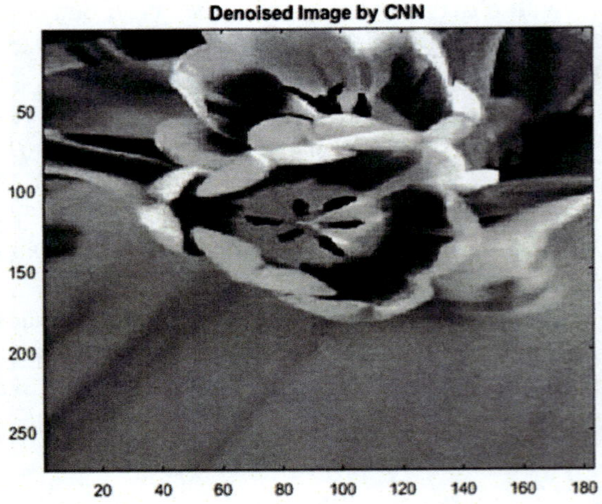

Fig. 3 Denoised image by CNN

Table 1 Denoising result of filtering using PSNR

Parameter/Methods	Gaussian	Median	Bilateral	NLM	CNN
PSNR	24.3209	27.6080	28.3062	29.7626	**30.833**

5 Conclusion

This article suggested a better non-local means method as well as CNN. The NLM technique uses a new weight kernel for preprocessing. The article has tested our algorithm under imagery consisting of edges as well as smooth regions. Various features of the imagery are well preserved still enhanced than the classical non-local means technique. The comparisons prepared on behalf of the denoising imagery as of different denoising techniques are presented. Compared with NLM, CNN gives the greatest results. From these experimental results, this article observes to the planned method is able toward preserve edges during the flower image by high details also gets a higher PSNR, as well as better visual quality.

References

1. Liu, C., Szeliski, R., Kang, S. B., et al. (2008). Freeman, Automatic estimation and removal of noise from a single image. *IEEE Transactions on Pattern Analysis and Machine Intelligence, 30,* 299–314. https://doi.org/10.1109/TPAMI.2007.1176.
2. Rudin, L., Osher, S., & Fatemi, E. (1992). Nonlinear total variation based noise removal algorithms. *Physica D: Nonlinear Phenomena, 60,* 259–268. https://doi.org/10.1016/0167-2789(92)90242-F.
3. Perona, P., & Mlik, J. (1990). Scale-space and edge detection using anisotropic diffusion. *IEEE Transactions on Pattern Analysis and Machine Intelligence, 12,* 629–639. https://doi.org/10.

1109/34.56205.
4. Bayram, I., & Kamasak, M. E. (2012). Directional total variation. *IEEE Signal Processing Letters, 12,* 781–784. https://doi.org/10.1109/LSP.2012.2220349.
5. Portilla, J., Strela, V., Wainwright, M. J., & Simoncelli, E. P. (2003). Imagedenoising using scale mixtures of gaussians in the wavelet domain. *IEEE Transactions on Image Processing, 12,* 1338–1351. https://doi.org/10.1109/TIP.2003.818640.
6. Lindenbaum, M., Fischer, M., & Bruckstein, A. M. (1994). On Gabor's contribution to image-enhancement. *Pattern Recognition, 27,* 1–8. https://doi.org/10.1016/0031-3203(94)90013-2.
7. Ghael, S., Sayeed, A. M., & Baraniuk, R. G. (1997). Improved wavelet denoising via empirical wiener filtering. *Proceedings of SPIE, Wavelet Applications in Signal and Image Processing V, 3169,* 389–399. https://doi.org/10.1117/12.292799.
8. Donoho, D. L. (1995). Denoising by soft-thresholding. *IEEE Transactions on Information Theory, 41,* 613–627. https://doi.org/10.1109/18.382009.
9. Buades, A., Coll, B., & Morel, J. M. (2005). A non-local algorithm for image denoising. In *IEEE Computer Society Conference on Computer Vision and Pattern Recognition, 2005. CVPR 2005* (Vol. 2, pp. 60–65). IEEE.
10. Goossens, B., & Luong, H. Q. (2008). A fast non-local image denoising algorithm (Vol. 6812, pp. 81210–81210).

A Survey on Diverse Chronic Disease Prediction Models and Implementation Techniques

Nitin Chopde and Rohit Miri

Abstract In today's world, chronic diseases are a crucial reason for death. The chronic disease is gradually taking the patient into control and then take over. Chronic start slowly and continues for a long time. There is a need to predict chronic disease at early stages before it reaches an uncontrolled situation so timely treatments can resist it. Prediction system effectively controls chronic disease at early stages. Our study aims to cover various prediction models for chronic disease and techniques for the development of prediction models. This review gives a comprehensive overview of the predictions system and implemented techniques for basic chronic disease. We go through prediction models are developed for basic chronic diseases like heart disease, cancer, diabetes and kidney disease with a different set of techniques. The survey paper discusses an overview of different chronic disease prediction models and its implementation techniques. The survey shows that machine learning approach is efficient to design a prediction system for chronic diseases in the welfare of health organizations and ultimate benefit to patients. This paper reviews basic chronic disease prediction models and suggested that to achieve accurate results of chronic disease prediction system machine learning is promising.

Keywords Chronic disease · Prediction model · Classifier

1 Introduction

They proposed hybrid method and filtering method classifications for a chronic system. The decision tree algorithms based on classification are used to develop a predictive model that predicts risk diagnosis using the observed things [1]. This paper starts from handling extreme values to the use of predictive analytics for selecting

N. Chopde (✉) · R. Miri
Department of Computer Science & Engineering, Dr. C. V. Raman University Kota, Bilaspur (C.G.), India
e-mail: nitin.rchopde@gmail.com

R. Miri
e-mail: rohitmiri@gmail.com

S. C. Satapathy et al. (eds.), *Intelligent System Design*, Advances in Intelligent Systems and Computing 1171, https://doi.org/10.1007/978-981-15-5400-1_51

optimal subset of parameters. It selects the best parameter for prediction and applied strategy to begin from acute values for the prediction [2]. To predict disease they studied genetic issues of polygenic diseases [3] Author forced on accuracy for chronic kidney disease dataset in which various parameters processed using two critical [4]. They studied multiple machine learning algorithms to predict the risk of chronic disease diabetes disease, heart disease, hepatitis disease and liver disease; also they evaluate machine learning algorithms and tools for decision making and prediction of disease [5].

They concentrate on claim data in which they studied clinical and claim data of chronic disease arthritis, kidney disease, osteoporosis [6]. Authors proposed neural network models for CKD detection may effectively and feasibly equip medical staff with the ability to make precise diagnosis and treatment to the patient [7]. They developed neural network models to diagnose chronic kidney disease and treatment to patient which improve the accuracy of prediction. To improve the performance in terms of accuracy of chronic disease prediction they applied adaptive and parallel classification systems [8]. Machine learning effectively classifies healthcare data, as well as diagnosis efficiently and generates accurate results [9].

2 Literature Survey

To provide security to handle big data of the health industry, they introduced a novel model using advanced security techniques along with machine learning algorithms. The innovation lies in the incorporation of optimal storage and data security layer used to maintain security and privacy [10]. They performed comparative analysis of different machine learning algorithms and suggested that ANN is good for prediction but it is slow as compared to other algorithms. Trees algorithm useful for prediction but its complexity is more and to maximum output, you should use RS theory [5].

To get deep information about Medicare data, codebook from CMS is available [11]. In [12] proposed model they used different classification algorithm of machine learning like K-Nearest Neighbour (KNN), tree-based decision tree, LDA classifier and lazy learner for decision support system also they analyzed probability-based Naive Bayes neural network based back propagation (BPN) [13]. These algorithms have been compared with classification accuracy to each other on the basis of correctly classified instances, mean absolute error, Kappa statistics and RMSE metric. The results show that MLP classifier outperforms Naïve Bayes classifier in all sectors with respect to the parameters specified [14].

Then Machine Learning techniques were applied to derive the correlation between the 11 chronic diseases and the ICD9 diagnostic codes. Reduced set of diagnostic codes were obtained for each chronic disease in the training phase. These codes were tested on a test data sample using accuracy and confusion matrix as the performance metric. It was understood that these reduced set of diagnostic codes are relevant and important for each chronic disease [12]. They proposed accuracy point of view for extraction rule best choice is SQR ex-SVM and eclectic approaches for SVM and experimental result shows that the best approach is C5 [15]. They investigate prediction by machine learning over big data which comes from healthcare and archives accurate result [16].

They proposed chronic disease prediction model, where they evaluated different algorithms like decision tree (J48) and RF algorithm ensemble classifier on various parameters like ROC performance measures, accuracy, precision, recall and F-measure and experimental results both algorithm performance is superior [17].

Authors constructed a system which used UCI repository dataset having 400 data records and 25 attributes and concluded highest accuracy 99.1% from Multiclass Decision forest algorithm [18]. The author implemented AdaBoost machine learning ensemble technique for diabetic diagnostic and result shows in line with J48 and decision tree it performs well [19]. They designed a system for diabetes prediction, focused on the diagnosis of the patient at a particular age; to achieve this, they have implemented a system using machine learning algorithm decision tree they found higher accuracy with decision tree [20, 21]. The author applied classification with a rule for mining for healthcare data [22]. The author has experimented prediction system for diabetic patients in two parts—one is artificial neural network and second one is fast blood sugar [23]. The author classifies biological data into different classes for this they applied rule-based classifier due to this it filters out many problems. And provide a solution for noisy instances and over fitting problems [24].

3 Analysis of Various Approaches and Techniques for Prediction Model

Decision tree classifications applied to predict chronic disease risk diagnosis [1]. The authors have developed prediction model using advances in machine learning to recognize intelligent solutions that enhance prediction in chronic kidney disease [2].

They investigate K-Nearest Neighbour, J48 tree Bayesian Network and Random tree in context with error rate, accuracy and learning time and suggested that on classification Bayesian algorithm is healthy. They had compared the analysis of classification function techniques for heart disease prediction [1]. The heart disease prediction is done by using classification and applied Sequential Minimal Optimization algorithm, LMT algorithm, Multilayer Perception algorithm [4].

This paper focused on clinical and claims data for studying 11 chronic diseases such as kidney disease, osteoporosis, arthritis, etc. using the claims data. The correlation between chronic diseases and the corresponding diagnostic tests is analyzed, by using ML techniques [6]. Hadoop/Map Reduce techniques was used in which predictive analysis algorithm predict the diabetes types prevalent, complications associated with it and further analysis, this system provides an efficient way to cure and care the patients [25]. They used a special model SVM named black box model which gives beneficial information for SVM's pinpointing decision. SVM is the best solution for diagnosing real-life diabetic problems [15].

Authors predict kidney diseases by applying basic data mining techniques such as clustering, regression, time series and sequence analyses, classification, association analysis, and if the proper technique of data mining is applied it gives promising results [26]. Authors developed a prediction model that diagnosed chronic kidney failure disease using c4.5 learning algorithm which is machine learning algorithms astutely exploit information and extract the best knowledge. Authors designed predictive model for Egyptian patients' using Artificial Neural Networks is machine learning approach where responses are based on their clinical and biochemical data [27] (Table 1).

4 Conclusion

In this survey, it is found that machine learning gives promising accuracy of prediction model of chronic disease as compared to other techniques like data mining. This paper gives an analysis of various machine learning techniques for different chronic diseases; we covered chronic disease as diabetes disease, kidney disease, heart disease and hepatitis disease. Many authors have experimented with different sets of machine learning algorithm which result in an acceptable level of accuracy for healthcare.

Many efforts have been taken by researchers to develop an efficient prediction system for chronic diseases and researcher used different sets of data mining techniques and machine learning techniques. The study previous research shows that for kidney disease ensemble Random forest classifier gives 100% accuracy with less set of attribute and maximum instances from given dataset and J48 decision tree gives an accuracy of 99%. In the case of diabetic disease, SVM gives 94% accuracy and 93% sensitivity. For heart disease SVM provides good accuracy rate up to 95%. In case of chronic Hepatitis Neural network with back propagation shows highest accuracy of 98% although it takes more time to show maximum output. In case of heart disease SVM gives maximum accuracy with 95.2%.

We contributed by reviewing various recent and past research of chronic disease prediction system and it is suggested that machine learning is a promising approach to design efficient and effective prediction systems in which researcher can filter out various classifiers as per the need of prediction system of chronic disease and achieve a promising result.

Table 1 Comparative analysis of basic chronic disease prediction models with its implemented technique

Types of chronic disease	Implemented techniques and algorithm for prediction model	Outcome
Heart disease and kidney disease [28]	k-nearest neighbours	Standard accuracy of 90%
Heart disease and diabetic [11]	Naive Bayes and support vector machine (SVM)	SVM gives highest accuracy rate of 95.56% as compared to Naïve Bayes classifier has 73.58%
Heart disease [9]	Data mining techniques Naïve Bayes followed by neural network and decision trees	Neural network gives more accurate predictions (49.34%) Naïve Bayes (47.58%) and decision trees (41.85)
Heart disease [27]	Data mining, support vector machine (SVM) and ANN	SVM accuracy 95.2% and artificial neural network accuracy 94.27%
Diabetic disease [29]	Weka tool used along with Naïve Bayes, support vector machine, and functional trees	Support vector machine gives 88.3% accuracy
Diabetic [25]	predictive analysis algorithm in Hadoop/MapReduce	It is an efficient way to cure and care the patients with affordability and availability
Diabetes [15]	Data mining and machine learning approach support vector machines	Accuracy is 94%, sensitivity of 93%, and specificity of 94%
Diabetic [27]	Data mining techniques Naive Bias, SVM, decision tree and artificial neural network	Highest accuracy decision tree 86.47% and SVM 87.32%
Pima Indians diabetes database [30]	Tree, SVM and Naïve Bayes	Naive Bayes outperforms with the highest accuracy of 76.30% comparatively other algorithms
kidney [26]	Naive Bayes(NB), J48, and random forest (RF), bagging, AdaBoost	J48 decision tree 99% and random forest an average accuracy 100%
Chronic kidney failure and heart disease [29]	k-nearest neighbours	Accuracy of 90% with error rate 5%
Chronic kidney disease [31]	Multiclass decision jungle, forest, multiclass neural network and logistic regression	Multiclass decision forest algorithm gives 99.17% accuracy
Kidney disease [32]	Naïve base classifier and ANN	Naïve base classifier gives 100% accuracy and ANN gives 72.73%
Chronic kidney disease [17]	Decision tree, J48 and SMO classifier and ensemble classifier random forest	Random forest gives 100% accuracy

(continued)

Table 1 (continued)

Types of chronic disease	Implemented techniques and algorithm for prediction model	Outcome
Chronic Hepatitis [27]	Machine learning algorithm Artificial Neural Networks (ANN) and decision trees (DT)	ANN and prediction accuracy 0.76 and 0.80
Chronic Hepatitis [33]	Data mining algorithms	Neural network with back propagation shows highest accuracy of 98%
Chronic Hepatitis [33]	Data mining techniques CART ID3 Algorithm C4.5 algorithm and binary decision tree algorithm	CART algorithm accuracy of 83.2% and C4.5 gives 71.14%

References

1. Hussein, A. S., Omar, W. M., & Xue, L. (2012). Efficient chronic disease diagnosis prediction and recommendation system. In *Biomedical Engineering and Sciences (IECBES), IEEE EMBS Conference* (pp. 209–214).
2. Aljaaf, A. J., Al-Jumeily, D., Haglan, H. M., Alloghani, M., Baker, T., Hussain, A. J., & Mustafina, J. (2018). Early prediction of chronic kidney disease using machine learning supported by predictive analytics. In *IEEE Congress on Evolutionary Computation (CEC)* (pp. 1–9).
3. Grasso, M. A., Dalvi, D., Das, S., & Gately, M. (2011). Genetic information for chronic disease prediction. In *IEEE International Conference on Bioinformatics and Biomedicine Workshops* (p. 997).
4. Jena, L., & Swain, R. (2017). Work-in-progress: Chronic disease risk prediction using distributed machine learning classifiers. In *International Conference on Information Technology (IEEE).*
5. Fatima, M., & Pasha, M. (2017). Survey of Machine learning algorithms for disease diagnostic. *Journal of Intelligent Learning Systems and Applications, Scientific Research Publishing* (pp. 1–16).
6. Gupta, D., Khare, S., & Aggarwal, A. (2016). A method to predict diagnostic codes for chronic diseases using machine learning techniques. In *International Conference on Computing, Communication and Automation (ICCCA2016)* (pp. 281–287). IEEE.
7. Kei Chiu, R., Chen, R. Y., Wang, S., & Jian, S. (2012). Intelligent systems on the cloud for the early detection of chronic kidney disease. In *Machine learning and cybernetics* (pp. 1737–1742). IEEE.
8. Jain, D., & Singh, V. (2018). Feature selection and classification systems for chronic disease prediction. *Egyptian Informatics Journal, 19,* 179–189.
9. Tayeb, S., Pirouz, M., Sun, J., Hall, K., Chang, A., Li, J., et al. (2017). Toward predicting medical conditions using k-nearest neighbors. In *IEEE International Conference on Big Data (BIGDATA)* (pp. 3897–3903).
10. Kaur, P., Sharma, M., & Mittal, M. (2018). Big data and machine learning based secure healthcare framework. In *International Conference on Computational Intelligence and Data Science (ICCIDS)* (pp. 1049–1059). Elsevier.
11. Palaniappan, S. P., & Awang, R. (2008). Intelligent heart prediction system using data mining techniques, IEEE (pp. 108–115).
12. Ruiz-Arenas, R. (2017). A summary of worldwide national activities in chronic kidney disease (CKD) testing. *The Electronic Journal of the International Federation of Clinical Chemistry and Laboratory Medicine, 28*(4), 302–314.

13. Ani, R., Sasi, G., Sankar, U. R., & Deepa, O. S. (2016). Decision support system for diagnosis and prediction of chronic renal failure using random subspace classification. In *Advances in Computing, Communications and Informatics (ICACCI), International Conference IEEE* (pp. 1287–1292).
14. Lin, R. H., Ye, Z. Z., Wang, H., & WuI, B. (2018). Chronic diseases and health monitoring big data: A survey. *IEEE Transactions Journal*, 1 –15.
15. Mellitus, Barakat, N. H., & Bradley, A. P. (2010). Intelligible support vector machines for diagnosis of diabetes senior member. *IEEE Transactions on Information Technology in Biomedicine, 14*(4).
16. Chen, M., Hao, Y., Hwang, K., Wang, L., & Wang, L. (2017). Disease prediction by machine learning over big data from healthcare communities. *IEEE Access*, 1–9.
17. Sisodia, D. S., & Verma, A. (2017). Prediction performance of individual and ensemble learners for chronic kidney disease. In *Proceedings of the International Conference on Inventive Computing and Informatics IEEE Xplore* (pp. 1027–1031).
18. Gunarathne W. H. S. D, Perera, K. D. M, & Kahandawaarachchi, K. A. D. C. P. (2017). Performance evaluation on machine learning classification techniques for classification and forecasting through data analytics for chronic kidney disease. In *IEEE International Conference on Bioinformatics and Bioengineering.*
19. Perveen, S., Shahbaz, M., Guergachi, A., & Keshavjee, K. (2016). Performance analysis of data mining classification techniques to predict diabetes. *Procedia Computer Science.*
20. Orabi, K. M., Kamal, Y. M., & Rabah, T. M. (2016). Early predictive system for diabetes mellitus disease. In *Industrial Conference on Data Mining* (pp. 420–427).Springer.
21. Rashid, T. A., & Abdullah, R. M. (2016). An intelligent approach for diabetes classification, prediction and description. *Advances in Intelligent Systems and Computing*, 323–335.
22. Song, S., Warren, J., & Riddle, P. (2014). Developing high risk clusters for chronic disease events with classification association rule mining. In *CRPIT—Health Informatics and Knowledge Management* (Vol. 153, pp. 69–78).
23. NaiArun, N., & Moungmai, R. (2015). Comparison of classifiers for the risk of diabetes prediction. *Procedia Computer Science, 69*, 132–142.
24. Farid, D. M., Al-Mamun, M. A., Manderick, B., & Nowe, A. (2016). An adaptive rule-based classifier for mining big biological data. *International Journal of Expert Systems with Applications, 64*, 305–316.
25. Saravana Kumar, N. M., Eswari, T., Sampath, P., & Lavanya (2015). Predictive methodology for diabetic data analysis in big data. In *2nd International Symposium on Big Data and Cloud Computing (ISBCC'15).*
26. ElHefnawi, M., Abdalla, M., Ahmed, S., Elakel, W., Esmat, G., Elraziky, M., et al. (2012). Accurate prediction of response to Interferon-based therapy in Egyptian patients with chronic hepatitis C using machine-learning approaches. In *IEEE/ACM International Conference on Advances in Social Networks Analysis and Mining* (pp. 772–778).
27. Deepika, K., & Seem, S. (2016). Predictive analytics to prevent and control chronic diseases. In *IEEE 2nd International Conference on App Applied and Theoretical Computing and Communication Technology* (pp. 381–386).
28. Wickramasinghe, M. P. N. M., Perera, D. M., & Kahandawaarachchi, K. A. D. C. P. (2017). Dietary prediction for patients with chronic kidney disease (CKD) by considering blood potassium level using machine learning algorithms (pp. 300–303). IEEE.
29. Suresh Kumar, P., & Pranavi, S. (2017). Performance analysis of machine learning algorithms on diabetes dataset using big data analytics. In *International Conference on Infocom Technologies and Unmanned Systems* (pp. 508–513).
30. Sisodiaa, D., & Sisodiab, D. S. (2018). Prediction of diabetes using classification algorithms. In *International Conference on Computational Intelligence and Data Science (ICCIDS 2018.*

31. Dilli Arasu, S. (2017). Review of chronic kidney disease based on data mining techniques. *International Journal of Applied Engineering Research, 12*(23), 13498–13505.
32. Jena, L., & Kamila, N. K. (2015). Distributed data mining classification algorithms for prediction of chronic kidney-disease. *International Journal of Emerging Research in Management & Technology, 4*(11). ISSN: 2278–9359.
33. Sathyadevi, G. (2011). Application of CART algorithm in hepatitis disease diagnosis. In *IEEE International Conference on Recent Trends in Information Technology (ICRTIT)* (pp. 1283–1287).

An Efficient Text-Based Image Retrieval Using Natural Language Processing (NLP) Techniques

P. M. Ashok Kumar, T. Subha Mastan Rao, L. Arun Raj, and E. Pugazhendi

Abstract The image retrieval system is a computer system for browsing, searching and retrieving images from a large database of digital images or text. Most traditional and common methods of image retrieval utilize some method of adding metadata such as captioning, keywords or descriptions to the images so that retrieval can be performed over the annotation words. Manual image annotation is time-consuming, laborious and expensive. To address this, there has been a large amount of research done on automatic image annotation. With the rapid development of information technology, the number of electronic documents and digital content within documents exceeds the capacity of manual control and management. The usage of images is increased in real time. So, the proposed system concentrates on retrieving image by using the text-based image retrieval system. Text documents are given as input to the preprocessing stage, and features are extracted using TF-IDF. Finally, document clustering method can be used to automatically group the retrieved documents into list of meaningful categories. Document clustering clusters the document of different domains and latent Dirichlet allocation (LDA), each document may be viewed as a mixture of various topics where each document is considered to have a set of topics, and after that relevant documents are retrieved and then images of those relevant documents are retrieved.

Keywords Document clustering · K-means clustering · Latent Dirichlet allocation · TF-IDF · Topic modelling

---

P. M. Ashok Kumar (✉)
Department of Computer Science Engineering, K.L. University, Vaddeswaram 522502, India
e-mail: profpmashok@gmail.com

T. Subha Mastan Rao
Department of Computer Science Engineering, CMR Technical Campus, Hyderabad 501401, India

L. Arun Raj
Department of Computer Science Engineering, B.S. Abdur Rahman Crescent Institute of Science and Technology, Chennai 600048, India

E. Pugazhendi
Department of Information Technology, Anna University, MIT Campus, Chennai 600048, India

S. C. Satapathy et al. (eds.), *Intelligent System Design*, Advances in Intelligent Systems and Computing 1171, https://doi.org/10.1007/978-981-15-5400-1_52

1 Introduction

Information retrieval refers to retrieving related information from databases. It deals with the user's access to large amounts of information which is stored in repository. Users have to give input query [1] to the IR system which retrieves relevant information from the database which satisfies user needs. Information overload is reduced by automated information retrieval systems. Applications of information retrieval are integrated solutions, distributed IR, efficient, flexible indexing and retrieval, semantic matching, routing and filtering, effective retrieval, multimedia retrieval, information extraction, relevance feedback and Web Search Engine. Among those applications, the most viable application is Web Search Engine. IR operates on three prominent scales: Web search, personal information retrieval and enterprise, institutional and domain—specific search. Top five leading IR research groups are Centre for Intelligent Information Retrieval (CIIR) at the University of Massachusetts Amherst; Information Retrieval Group at the University of Glasgow; Information and Language Processing Systems (ILPS) at the University of Amsterdam; Information Retrieval Group (THUIR) at Tsinghua University; and Information Storage, Analysis and Retrieval Group (ISAR) at RMIT University.

Image retrieval can be done in two ways—text-based image retrieval and content-based image retrieval. In text-based approach, input query is given as text to retrieve images relevant to input query. This is traditional approach. Each image is annotated with textual description which is matched with input text query to retrieve relevant images. Some of the text-based approaches are bag of words, natural language processing and Boolean retrieval. In content-based approach, input query is given as images to get similar images. This is one of the applications of computer vision. This approach solves the problem of searching digital images in large databases by analysing the actual content of the image, i.e. texture, shape and colour of the images in the database with the query image. Image retrieval can be improved by relevance feedback method in which the user is allowed to with IR system by giving feedback about the results which are relevant to the given query which makes IR to understand the query and to improve the performance. The evaluation criteria for IR are accuracy, speed, consistency and ease of use. These metrics are used to check whether user's need is satisfied.

The main objective of this project is to retrieve images from documents using text-based image retrieval in an unsupervised manner. This can be done by collecting Web documents and images by scraping the websites, making unstructured documents into structured documents, extracting features from those documents and clusters are formed by applying clustering algorithm on extracted feature. Similarly, features of text query are extracted, and then, document is retrieved by predicting to which cluster the query belongs to. After that topic modelling is applied to retrieve topic from the retrieved documents and input query, and then, similarity measure is applied to get more relevant documents. Finally, images of relevant document are retrieved.

2 Literature Survey

2.1 Image Retrieval

A survey on text- and content-based image retrieval system for image mining was done by Karthikeyan et al. [2]. Image retrieval is performed by matching the features of a query image with those in the image database. The collection of images in the Web is growing larger and becoming more diverse. Retrieving images from such large collection is a challenging problem. The research communities study about image retrieval from various angles on text-based and content-based. The traditional text retrieval techniques [3, 4] are used for image annotations in text-based image retrieval. This is based on annotations that were manually added for disclosing the images (keywords, descriptions), or on collateral text which is available with an image (captions, subtitles, nearby text). The image-processing techniques are used to first extract image features and then retrieve relevant images based on the match of these features in content-based image retrieval. Similarity measures are also used to determine how similar or dissimilar in the given query image or image database collections.

Text-based approach for indexing and retrieval of image and video was proposed by Bhute et al. [5]; this paper discussed the different techniques for text extraction from images and videos and then reviewed the techniques for indexing and retrieval of image and videos by using extracted text. Text extraction is the stage where the text components are segmented from the background. Extracted text components are usually low resolution and are susceptible to noise, and it is required to enhancement. By using OCR technology, the extracted text images are transformed into plaintext. The different information sources like colour, texture, motion, shape, geometry, etc., are used for text. By merging the different sources of information, they enhance the performance of a text extraction system and text-based video retrieval systems.

Image retrieval using multiple evidence ranking was proposed by Coelho et al. [6]. The World Wide Web is the largest publicly available image repository. The consequence is that searching for images on the Web has become a current and important task. To search for images of different user interest, the most direct approach used is keyword-based searching. The keyword-based image searching techniques [7] frequently yield poor results because images on the Web are poorly labelled. In this approach, multiple sources of evidence related to the images are considered. To allow combining these distinct sources of evidence, they introduce an image retrieval model based on Bayesian belief networks. This is an interesting result because current image search engines in the Web usually do not take text passages into consideration with relative gains in average precision figures of roughly 50%, when compared to the results obtained by the use of each source of evidence in isolation.

Towards privacy-preserving content-based image retrieval (CBIR) in cloud computing was presented by Xia et al. [8]. In this paper, without revealing the actual content of the database to the cloud server, they propose a privacy-preserving content-based image retrieval scheme, which allows the data owner to outsource the CBIR service and image database to the cloud. To evaluate the similarity of images, local features are utilized to represent the images, and earth mover's distance (EMD) is employed. In order to improve the search efficiency, they design a two-stage structure with LSH. In the first stage, dissimilar images are filtered out by prefilter tables to shrink the search scope. In the second stage, the remaining images are compared under EMD metric one by one for refined search results. The experiments and security analysis show the security and efficiency of the proposed scheme.

Image retrieval based on its contents using features extraction was presented by Shinde et al. [9]. Image retrieval is based on its contents using features extraction, and an image retrieval system returns a set of images from a collection of images or digital text in the database to meet users' demand with similarity matching evaluations such as image content similarity, edge pattern similarity and colour similarity. An image retrieval system offers an efficient way to search, access, browse, identify and retrieve a set of similar images in the real-time applications. Some attempts have been addressed to describe the visual image content, and several approaches have been developed to capture the information of image contents by directly computing the image features earlier.

2.2 Document Clustering

Efficient information retrieval using document clustering was proposed by Bansal et al. [10]. The purpose of information retrieval is to store document electronically and assist user to effectively navigate, trace and organize the available Web documents. The IR system accepts a query from the user and responds with the set of documents. The system returns both relevant and non-relevant materials, and a document organization approach is applied to assist the user in finding the relevant information in the retrieved sets. Clustering is an approach to improve the effectiveness of IR. The documents are clustered [11, 12] either before or after retrieval in clustering. The motivation of this paper is to explain the need of clustering in retrieving efficient information that closely associates documents which are relevant to the same query. Here IR framework has been defined which consists of four steps: (1) IR system, (2) similarity measure, (3) document clustering and (4) ranking of clusters. The goal of clustering is to split or separate relevant documents from non-relevant documents. Hence, a cluster is a collection of objects which are similar to each other but dissimilar to the objects of other clusters.

A survey on semantic document clustering was made by Naik et al. [13]. Clustering is considered as one of the most important unsupervised learning problems. In the clustering process, objects are organized into group of similar members. Hence, a cluster is a collection of objects which are similar to each other but dissimilar to the objects of other clusters. The goal of clustering is to divide a collection of text documents into different category groups so that documents in the same category group describe the same topic. A text clustering is the process of partitioning a set of data objects into subsets. The grouping is an unstructured collection of documents into semantically related groups. A document is considered as a bag of words in traditional document clustering methods; however, semantic meaning of word is not considered. Thus, more informative features like concept weight are important to achieve accurate document clustering, and this can be achieved through semantic document clustering because it takes meaningful relationship into account. Moreover, it is observed that semantic-based approach for document clustering provide better accuracy, result and quality of cluster than traditional approach.

2.3 *Topic Modelling*

Pattern-based topic models for information filtering, topic modelling, such as latent Dirichlet allocation were proposed by Gao et al. [14]. This is to generate statistical models to represent multiple topics in a collection of text documents, which have been widely utilized in the fields of information retrieval and machine learning. Information filtering is a system to remove document stream based on document representations which represent users' interest or unwanted information from an information or redundant information. The information filtering models were developed based on a term-based approach, whose advantage is efficient computational performance. However, effectiveness in information filtering is rarely known. Patterns are always more representative than single terms for representing documents. In this paper, pattern-based topic model (PBTM), which a novel information filtering model, is proposed to represent the text documents not only using the topic distributions at general level but also using semantic pattern representations at detailed specific level, both of which contribute to the document relevance ranking and accurate document representation.

The performance of content-based image retrieval (CBIR) system was presented by Rubini et al. [15]. Generally, there exist two approaches for searching and to retrieving images. The first one is based on textual information done manually by a human. This is called concept-based or text-based image indexing. A human describes and evaluates the images according to the image content, the caption or the background information. However, the representation of an image with text requires significant effort and can be expensive, tedious, time-consuming. To overcome the limitations of the text-based approach, the second approach known as content-based image retrieval (CBIR) techniques is used. In a CBIR system, images are automatically indexed by visualizing their valued features such as colour, texture and shape. It is observed that colour features provide approximately similar results with very large processing time. The proposed method of text-based image retrieval retrieves images with very small processing time.

3 Implementation

3.1 Anaconda

Anaconda is a free and open-source software used by the data scientist to perform Python programs and machine learning on Windows, Linux and macOS. Anaconda 3.5.3 version is used as Anaconda tool, and Python 3.6.4 version is used for Python programming language. The latest version of Python is Python 3.7.2, and latest version of Anaconda is Anaconda 5.3.0. In Anaconda, Jupyter Notebook is an IPython platform which is used to code Python programs. Jupyter Notebook allows to create and share documents that contain live code (Python code), visualization and explanatory text. Its use cases are data processing/transformation, numerical simulation, statistical modelling and machine learning. Jupyter notebook can be installed using two different ways. One is using Python's package manager using pip, and another is using Anaconda distribution. The user interface of anaconda is split into files, running and clusters. File tab is the default view where notebooks are created and operated. The extension for Jupyter Notebook is .ipynb.

4 Proposed Work

Multiple text documents are collected from news websites such as Times of India, Financial Express and Indian Express along with its associated images and stored in a separate folder with same names for both text documents and images but with different extension. Here, domains of collected documents are sports, education, entertainment, politics and technology. The features of text documents are extracted by using natural language processing (NLP), term frequency-inverse document frequency (TF-IDF) [16, 17]. Again, features are extracted from fresh input query and then using the features of text document collection, document clustering is done, which separates documents according to different clusters, input document is predicted to which cluster [18] does it belongs to, and documents of particular clusters are retrieved. This is the first stage of document retrieval. Then, the topic modelling is done by using latent Dirichlet allocation which is used to classify text in a document to a particular topic. It builds a word-per-topic model and topic-per-document model, modelled as Dirichlet distributions. After that similarity measures are used to retrieve a more related document. From that more relevant images are retrieved. The detailed architecture of the proposed system is shown in Fig. 1.

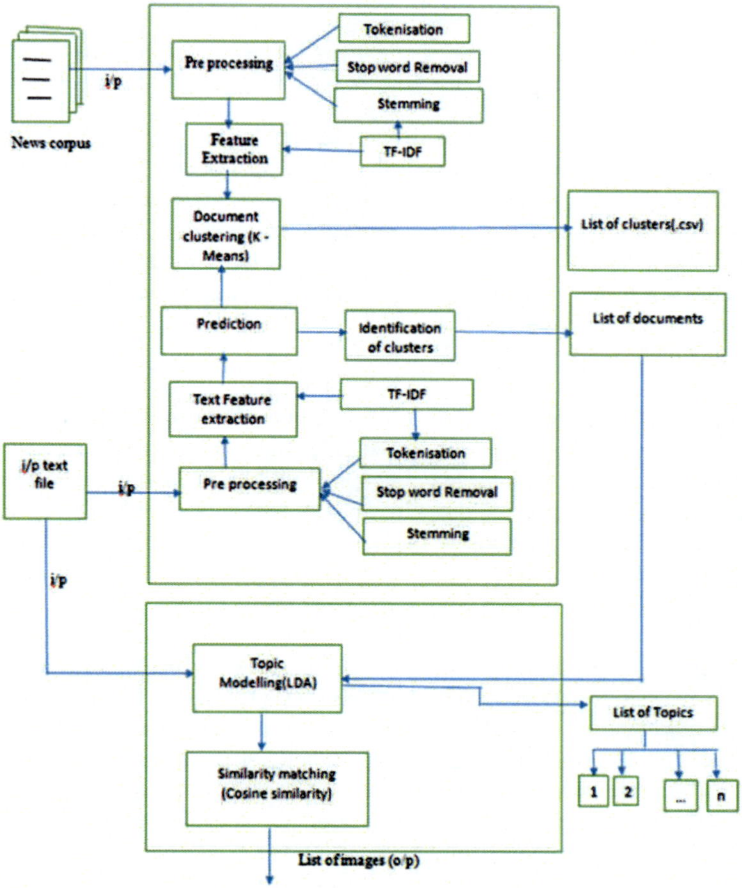

Fig. 1 Architecture of the proposed system

4.1 Dataset Description

Initially, the text documents [19] and their associated images are collected from newspaper websites such as Times of India, Indian Express and Financial Express. The domains taken are Politics, Sports, Entertainment, Education and Technology. The total number of documents collected is 1085. Politics contains 246 documents, Sports contains 220 documents, Education contains 266, Entertainment contains 200, and Technology contains 153 documents.

4.2 Feature Extraction

Tokenization

Tokenization is the common process in natural language processing. It is the process of splitting larger text into sentences and sentences into tokens. The sent_tokenize uses nltk.tokenize.punkt module for splitting text into sentences and word_tokenize uses TreebankWordTokenizer for splitting sentences into words. Other than English, many different languages can be tokenized using these modules. punkwordtokenizer does not separate punctuations from words, but wordpuncttokenizer separate punctuations from words.

Stop Word Removal

The process of stop word removal is to filter out unwanted tokens (stop words) from the text. Stop words are removed using Natural Language Toolkit (NLTK) in Python which is a Python programming and used to make programs, and its latest version is NLTK3.3. It has list of stop words in 16 different languages and stored in nltk_data directory as a text file. Stop words can be modified by adding additional words in the stop word text file.

Stemming

Stemming is the process of converting a word into its stem root form. For example, study, studying, studied are converted into its root word study. There are two types of errors in stemming. Over-stemming occurs when two words are rooted to same root when they have different roots. Under-stemming occurs when two words are rooted to same root when they do not have different roots. Stemming algorithm is categorized into truncating methods (affix removal), statistical methods and mixed methods. Truncating method is further divided into Lovins, Porter, Paice/Husk and Dawson. Statistical method is further divided into N-Gram, HMM and YASS. Mixed method is further divided into inflectional and derivational, corpus based and context sensitive. Inflectional and derivational is then divided into Krovetz and Xerox.

Here, Snowball stemmer is used which is detailed framework of Porter stemmer which is available in NLTK toolkit. It gets language as a parameter when object is initialized. Then, stem() function is called that performs stemming using different clauses.

TF-IDF

TF-IDF [20] is a score which is used to evaluate importance of a word in a document based on the product of TF and IDF score of that document. It is one of the information retrieval technique. If a word appears many times in a document, then it is important but it is appearing in many documents, and then it is not an important word. The equation of TF-IDF is shown in Eq. 1.

For a term t in a document d, the weight $W_{t,d}$ of term t in document d is given by Eq. 1:

$$W_{t,d} = TF_{t,d} \cdot \log(N/DF_t) \tag{1}$$

where

$TF_{t,d}$ is the number of occurrences of t in document d.
DF_t is the number of documents with the term t.
N is the total number of documents in the corpus.

TF * IDF is an information retrieval technique that weighs a term's frequency (TF) and its inverse document frequency (IDF). Each word or term has its respective TF and IDF score. The product of the TF and IDF scores of a term is called the TF * IDF weight of that term.

Term frequency calculates number of times each word appears in single document. Term frequency is not found for stop words, and before finding term frequency, all words are converted into lower cases. The formula for TF is given in the Eq. 2

$$TF(t) = \frac{(\text{Number of times term } t \text{ appears in a document})}{(\text{Total number of terms in the document})} \tag{2}$$

The inverse document frequency (IDF) of a word is the measure of how important that term is in the whole corpus. The formula for IDF is given in the Eq. 3

$$IDF(t) = \log_e \frac{(\text{Total number of documents})}{(\text{Number of documents with term } t \text{ in it})} \tag{3}$$

In Python, tdidfvectorizer is used to calculate TF-IDF score for the whole text document collected. Initially, tfidfvectorizer object is created, fit() is called which is used to learn vocabulary using word frequencies, and the vocabulary is wrapped in a list. Finally, transform() is called to encode as a vector.

TF-IDF can also be implemented using CountVectorizer and HashingVectorizer. Their algorithm and implementations are similar to tfidfVectorizer.

4.3 Document Clustering (K-Means Clustering)

K-Means is one of the centroid-based clustering techniques which is used to group similar documents into clusters. It is an iterative algorithm where we have to define the fixed number of clusters. Here, the number of clusters is 5. By calculating the similarity between the data points and the centroid of the cluster using Euclidean distance, the data points are grouped into clusters. The data points are the extracted features using TF-IDF.

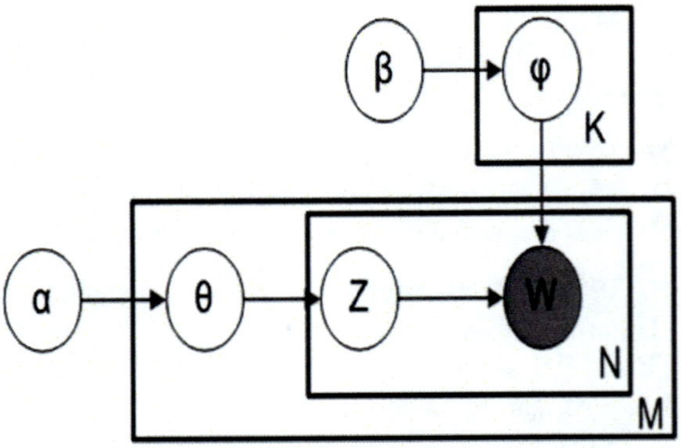

Fig. 2 Plate diagram of LDA

4.4 *Topic Modelling (LDA)*

In LDA [21], each document may be viewed as a mixture of various topics where each document is considered to have a set of topics that are assigned to it via Dirichlet distribution, where the topic is a mixture of various words. Dirichlet distribution is a generalization of beta distribution and is a multivariate distribution. LDA builds two matrices—one is topic per document and another is words per topic.

Plate Diagram of LDA

The documents which are retrieved after predicted are taken for LDA. First, create dictionary and corpus using gensim package, which are the main inputs for LDA. Each word has its unique ID in the document created by gensim. The number of topics specified is 2. α and φ are set to default as 1.0/num_topics and affect the sparsity of the topics as they are hyperparameters. The number of documents used in training chunk tells by chunksize. How the model parameter is updated and total number of passees is determined by update_every and passes. Then, build the LDA model. Perplexity and coherence scores are calculated to measure the topic model. The topics are visualized using pyLDAvis package. It is an interactive chart package which works well with Jupyter Notebook. The plate diagram for LDA is shown in Fig. 2.

5 Results and Analysis

5.1 *K-Means Clustering*

Table 1 shows as the number of documents available in each domain such as Politics, Sports, Education, Entertainment and Technology and total number of documents in the corpus.

Table 1 Documents before clustering

S. No.	Domain name	Number of documents before clustering
1	Politics (.txt)	246
2	Sports (.txt)	220
3	Education (.txt)	266
4	Entertainment (.txt)	200
5	Technology (.txt)	153
6	Total number of documents (.txt)	1085

Table 2 Documents after clustering

S. No.	Domain name	Number of documents after clustering
1	Politics (.txt)	165
2	Sports (.txt)	92
3	Education (.txt)	486
4	Entertainment (.txt)	234
5	Technology (.txt)	108

Table 3 Documents after LDA

S. No.	LDA for politics	Number of documents
1	Before LDA	165
2	After LDA	72

Table 2 shows number of documents available in each domain after clustering.

Latent Dirichlet Allocation

Table 3 discusses number of documents retrieved after performing LDA and similarity matching with text query in Politics domain.

5.2 *Performance Evaluation of Proposed Work*

Table 4 shows the elbow method evaluation which is used to find the appropriate number of clusters in a dataset. In this, when the number of clusters is 5, the sum of the squared distance between each member of the cluster and its centroid (SSE value) is less and the score is high. The formula for SSE is given in Eq. 4:

$$\mathrm{SSE} = \sum_{i=1}^{n} (X_i - \bar{X})^2 \tag{4}$$

Table 4 Elbow method evaluation

Number of cluster	SSE	Score
1	1051.22	0.54
2	1038.33	0.65
3	1026.28	0.72
4	1014.68	0.75
5	1005.43	0.81
6	1006.24	0.69
7	1006.1	0.61
8	1005	0.63
9	1005.49	0.57

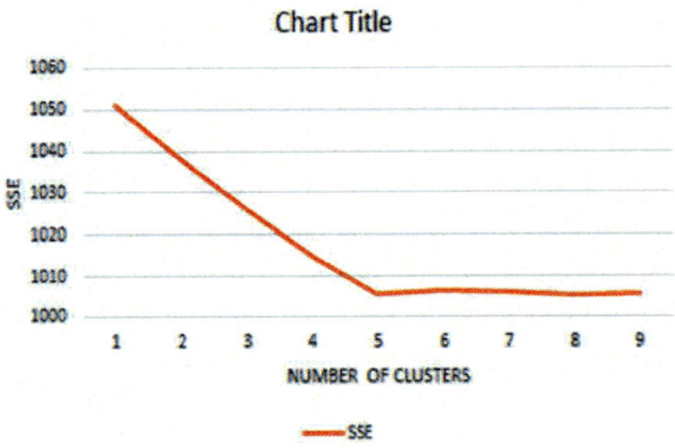

Fig. 3 Elbow method evaluation

Figure 3 shows the graph for the elbow method in which the K-means algorithm runs from range of k values from 1 to 9 on an unlabelled dataset. The elbow is the best value for k, if the line looks like an arm. From graph, the best value for k is 5.

Figure 4 shows the graph for the score of K-means clustering in which the algorithm runs from range of k values from 1 to 9 on an unlabelled dataset. The graph tells when the number of clusters is 5, and the score of K-means algorithm is high.

Figure 5 shows the graph for the coherence score of LDA in which the algorithm runs from range of list of topics values from 1 to 40. From the graph, when the number of topics is 2, the coherence score is high.

Fig. 4 Score for K-means clustering

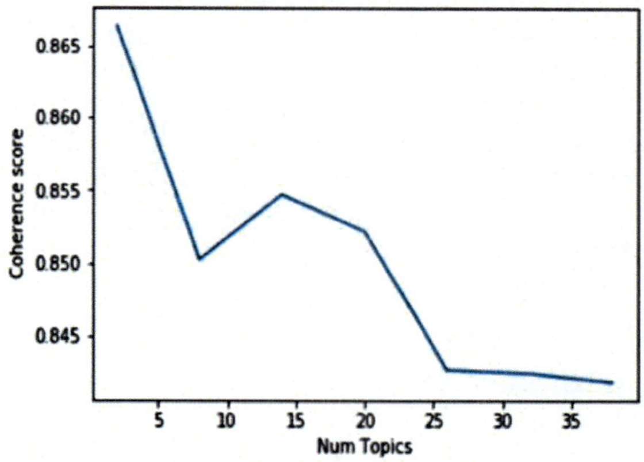

Fig. 5 Coherence score for LDA

6 Conclusions

Multiple text document and its associated images are collected from news websites. To the text documents, term frequency-inverse document frequency is applied which involves tokenization, stop word removal and stemming which are applied where text documents converted into tokens in tokenization, stop words are removed from the tokens in stop word removal and tokens are converted into root words in stemming, and these tokens form vocabulary for TF-IDF using word frequencies. Using TF-IDF, important words are extracted from the text documents by products of TF and IDF scores. TF is the frequency of each word in a document, and IDF is the frequency of each word in whole corpus. This process is called feature extraction. Features extracted from TF-IDF are given as input to K-means clustering which forms group of clusters.

An input file is given as input query and is predicted to which cluster this input file belongs to. Then, the documents in the predicted cluster are retrieved and used in topic modelling. Principal component analysis (PCA) is used for 2D visualization of clusters. Finally, LDA is used for topic modelling which builds list of topics, and topics contain list of words along with words weightage for the documents retrieved from clustering and input file. Document similarity is done which will retrieve documents similar to input file. Finally, images of the similar documents are retrieved.

7 Future Work

In future, we extend the proposed work by retrieving the documents from web using web scraping packages like Beautiful Soup, request and htmllib. All the above processes will be done using semi-supervised method. This semi-supervised method can be done by unsupervised document clustering and supervised ensemble learning method. The future work of the proposed system involves documents retrieved from the Web using Web scraping packages such as BeautifulSoup, request and htmllib and do it in semi-supervised method (i.e.) doing document clustering as unsupervised method and label them according to the clusters and then doing ensemble learning as supervised method in which boosting and stacking algorithms are combined so that accuracy can be improved.

References

1. Mahmud, T., Hasan, K. M. A., Ahmed, M., & Chak. T. H. C. (2015). A rule based approach for NLP based query processing. In *2nd International Conference on Electrical Information and Communication Technologies (EICT)*.
2. Karthikeyan, T. (2014 March). A survey on text and content based image retrieval system for image mining. *International Journal of Engineering Research & Technology (IJERT), 3*(3).
3. Jaiswal, A., Janwe, N. (2011). Hierarchical document clustering: A review. *International Journal of Computer Applications (IJCA)*.
4. Chen C. L, Tseng F. S. C., & Liang, T. (2009). An integration of fuzzy association rules and WordNet for document clustering. In *Proceedings of the 13th Pacific-Asia Conference on Knowledge Discovery and Data Mining* (pp. 147–159).
5. Bhute, A. N. (2014 March). Text based approach for indexing and retrieval of image and video. *An International Journal (AVC), 1*(1).
6. Coelho, T. A. S. (April 2004). Image retrieval using multiple evidence ranking. *IEEE Transactions on Knowledge and Data Engineering, 16*(4).
7. Dinakaran, B., Annapurna, J., & Aswani Kumar, C. (2010). Interactive image retrieval using text and image content. *Cybernetics and Information Technologies, 10*.
8. Xia, Z. (2018 January–March). Towards privacy-preserving content-based image retrieval (CBIR) in cloud computing. *IEEE Transactions on Cloud Computing, 6*(1).
9. Shinde, P. (2016 January). Image retrieval based on its contents using features extraction. *International Research Journal of Engineering and Technology (IRJET, 03*(01).

10. Bansal, S. (2010). Efficient information retrieval using document clustering. *International Journal of Advanced Research in Computer Science, 1*(03).
11. Isa, D., Kallimani, V. P., & Lee, L. H. (2009 July). Using the self organizing map for clustering of text documents. *A Journal Paper of Elsevier, 36*(5).
12. Karthikeyan, M., & Aruna, P. (2013). Probability based document clustering and image clustering using content-based image retrieval. *A Journal Paper of Elsevier Applied Soft Computing, 13*, 959–966.
13. Naik, M. P. (2015). A survey on semantic document clustering. In *IEEE International Conference on Electrical, Computer and Communication Technologies (ICECCT).*
14. Gao, Y. Pattern-based topic models for information filtering. In *2013 IEEE 13th International Conference on Data Mining Workshops.*
15. Rubini, S. (2018 March). Content-based image retrieval (CBIR) system. *International Research Journal of Engineering and Technology (IRJET), 05*(03).
16. Bafna, P., & Pramod, D. (2016). Document clustering: TF-IDF approach. In *Published in International Conference on Electrical… .*
17. Qiu, J., & Tang, C. (2007). Topic oriented semi-supervised document clustering. In *Workshop on Innovative Database Research, Proceedings of the SIGMOD.*
18. Jang, M., Choi, J. D., & Allan, J. Improving document clustering by eliminating unnatural language. In *Conference Paper 17 March 2017.*
19. Yeshambel, T., Assabie, Y. Amharic document image retrieval using morphological coding. In *Conference Paper October 2012.*
20. Renukadevi, D., & Sumathi, S. (2014 April). Term based similarity measure for text classification and clustering using Fuzzy C-means algorithm. *International Journal of Science, Engineering and Technology Research (IJSETR), 3*(4).
21. Blei, D. M., Ng, A. Y., & Jordan, M. I. (2003). Latent Dirichlet allocation (LDA). *Journal of Machine Learning Research, 3*, 993–1022.

An Efficient Scene Content-Based Indexing and Retrieval on Video Lectures

P. M. Ashok Kumar, Rami Reddy Ambati, and L. Arun Raj

Abstract Recently, the popularity for massive online open course (MOOC) learning among student community is increasing day by day. In most of the MOOC learning Web sites, indexing and retrieval techniques are based on the keywords like names, titles and Web addresses of the videos. As a result, the response for content-based queries is not good in most of the scenarios. In this paper, we propose an efficient scene content-based indexing and retrieval (ESCIR) framework for lecture videos. The proposed ESCIR framework consists of two phases: offline phase and online phase. In offline phase, we apply novel block-level key frame extraction (BLKFE) technique to segment video frames into shots and choose the right frame. The optical character recognition (OCR) tool is applied to the generated key frames in the videos for extracting the text. The jaccard similarity coefficient measure is used to eliminate the duplicate text frames, and then use of stop word removal and stemming algorithms are applied to get meaningful keywords from the scene. We used single-pass in-memory indexing (SPIMI) technique for building the indexing with the help of extracted keywords. In online phase, search and matching algorithms will map input queries given by user to the corresponding videos. We applied the okapi BestMatch25 (BM25) ranking function to rank those matched videos for best relevance results. The proposed ESCIR framework has been validated on standard video lectures and found to give better results than the existing state of art algorithms in terms of precision and accuracy. We compared our results with existing methods of lecture video techniques reflects improvising robust performance on lecture videos.

Keywords Block-level key frame extraction (BLKFE) · Indexing · Lecture videos · Multimedia · Okapi BM25

P. M. Ashok Kumar (✉)
Department of Computer Science Engineering, K.L. University, Vaddeswaram 522502, India
e-mail: profpmashok@gmail.com

R. R. Ambati
Department of Computer Science Engineering, RVR & JC College of Engineering, Guntur 522019, India

L. Arun Raj
Department of Computer Science Engineering, B.S. Abdur Rahman Crescent Institute of Science and Technology, Chennai 600048, India

S. C. Satapathy et al. (eds.), *Intelligent System Design*, Advances in Intelligent Systems and Computing 1171, https://doi.org/10.1007/978-981-15-5400-1_53

1 Introduction

Nowadays, the demand for e-education is growing. Universities are facing a new competitor in the form of massive open online courses (MOOC) to meet this demand. The big advantages of digitally delivered courses, which teach students via the browser or tablet apps through Internet that are, learn courses for anywhere, anytime. The online courses dramatically lesser the price of learning and broaden access to it, with low startup costs and powerful economies of scale by removing the need for students to be taught at set times or places.

To access the significant information from these video databases, efficient techniques were needed. The search and retrieval technique plays a vital role in MOOC systems. As a result, lot of research works got started in the area of image retrieval and video retrieval [1] during the last decade. The retrieval of videos from large video collections is becoming increasingly common due to their wide variety of application related to business, social media and entertainment areas.

Conventional methods for video indexing are based on video characterization of content using a set of computational features. These techniques do not fit for the commonly available text in lecture videos. So the retrieval of relevant videos from large video collection databases is an era applications in various domains. The effective information access depends on the retrieval from lecture video database. The main two important issues in content-based video access: (a) a human friendly query/interface; (b) a indexing and retrieval scheme for the content.

Our work mainly focuses on those lecture videos formed with the help of screen clutching technique. Since, in recording process, two videos are synchronized. Hence, the sequential possibility of a whole unique slide can be measured as a lecture segment. This way, slicing two-scene lecture videos can be attained by only processing video streams, which comprise more visual metadata. To provide a pictorial guideline for video content navigation, the extracted slide frames can be used.

We extract metadata from audio and visual resources of lecture videos automatically by applying OCR analysis techniques. In a large lecture video portal, various automatic indexing functionalities were developed for evaluation purposes; these functionalities can guide both text-oriented and visual users to navigate within lecture video. So we extract keywords from the raw results. Keywords [2] are generally used for information retrieval in digital libraries and also be used to summarize the document.

In a large database lecture video collection, one can query for relevant videos based on popular content present in video lectures. The user interface textual query is a promising approach for querying and retrieval of corresponding videos from large storage of lecture video databases, since it offers a more natural interface. Users are interested to retrieve the best matched relevant video based on content present in the slide, e.g., title, subtitle and keyword instead of video name and author name. The problem observed in the traditional models is efficient key frame extraction, which on further analysis provides indexing for slides in the lecture video.

Our novelty lies in two aspects:

1. A block-level key frame extraction (BLKFE) technique for extracting frames in each video shot.
2. Application of okapi BestMatch25 (BM25) ranking function to retrieve relevant videos.

The rest of the paper is organized as follows. Section 2 gives an overview of the recent works. Section 3 gives details on novel scene content-based indexing and retrieval (ESCIR) framework. In this section, the importance of OCR analysis, stop word removal and stemming, indexing is presented. Finally, Sect. 4 describes the experiments and analysis on the standard video lectures. Conclusions are drawn in Sect. 5.

2 Related Works

During the last decade, several works were reported for indexing and retrieval of video lectures collections. They considered as spatial and temporal characteristics of the video for content representation in the video lectures. The feature vectors in spatial domain are computed by evaluating different pixels of the frames and the encoded text descriptors in videos. Temporal analysis was done by dividing the video into basic elements like frames, shots, scenes or video-segments. The appearance and dynamism of the video content present in each of the video segments are then characterized by content. The information content of the video clip assumed that features like motion vectors, moments, texture and histograms. In a database of lecture videos, one can query for relevant videos with examples text in images. Extend this approach for lecture video indexing retrieval system.

Yang and Meinel [3] used two kinds of techniques to retrieve the relevant videos optical character recognition (OCR) technology for key frames, with the help of TFID score and document frequency indexed the terms in the lecture videos. Wanga et al. [4] proposed a novel idea for capturing the documents using topic indexing and querying of documents by structuring the lecturing videos. The two stage matching algorithm used to retrieve the videos from data collection. Tuna et al. [5] partitioned the lecture videos into key frames by using worldwide frame differencing metrics and background distributions. With the help of segmentation and text detection procedures, they attained better results rather than applying image transformations.

Jeong et al. [6] used scale invariant feature transform (SIFT) for extracting features on lecture video, with the help of segmentation method and adaptive threshold selection algorithm. Moritz et al. [4] proposed lecture video retrieval and video search by apply tagging data. Beyond the keyword-based tagging, Yu et al. [7] proposed an method of associated data to interpret lecture video resources using linked data to further repeatedly interpret the extracted textual metadata (Fig. 1).

The text content in key frames is used as keywords for indexing. The content-based video retrieval using visual features techniques are partially successful in retrieval

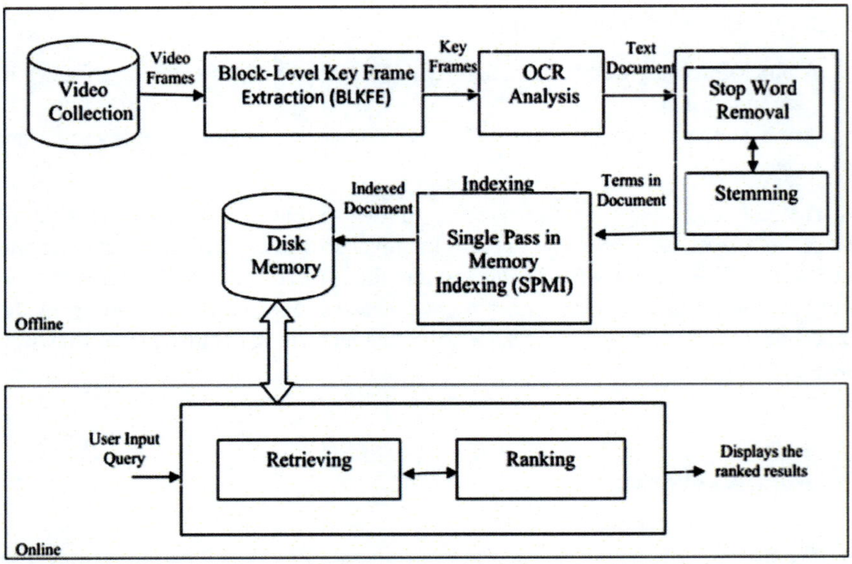

Fig. 1 Efficient scene content-based indexing and retrieval (ESCIR) framework

based on semantic content [8]. To search for the related videos (where the content is present) with the help of given a textual query, based on attributes like size of the text and duration of its appearance rank those videos. To detect the textural properties of a text region in an image using the techniques based on Gabor filters, Wavelet, FFT, spatial variance, etc. Many of these methods are also suitable for processing in the domain compression [7]. The content recognition in video lectures, the difficulty appears in the low resolution and visual quality of text slides. The single image is available in the traditional image super-resolution problem [9]. Lecture video super-resolution can get advantage of multiple frames that contain the same objects or scenes.

In [10], parallelize FCM and T2FCM using pure GPU implementation [11] and we parallelize IT2FCM, PCM and IT2MFPCM algorithms using hybrid CPU-GPU implementation. The images will be processed in a fast manner.

In [12], compressive sensing of medical images with confidentially homomorphism aggregations, the proposed mainly on memory point of view and reduced the size of the images. Most of the existing works fails on the extraction of key frame, which contains more useful information describing about the video. So, we proposed an efficient technique for extracting the key frame and character to create index and retrieval purposes.

3 Proposed Efficient Scene Content-Based Indexing and Retrieval (ESCIR) Framework

The proposed work is mainly on key frame extraction and indexing the lecture videos based on lecture content present in lecture videos, and hence retrieval will be done based on the videos using the title, subtitle, keywords, etc., present in the slides. The existing systems are retrieved based on the video name, author name, etc., used as input user query.

The novelty of proposed system is we will index the video using single-pass in-memory indexing and then store it into disk of the system, so that it can helps to retrieve the videos in a fast manner than existing methods in the lecture videos. The real-world examples of this work is used in MOOCs courses retrieval, videos are retried based on the text present in the scenes, etc.

3.1 System Design

The system design mainly deals with the two kinds of modes: 1. Offline and 2. Online. In offline mode, video frames are extracted and then histogram difference mechanism is applied at each block level to find shot boundaries and then last frame is selected as key frame. For this frame, optical character recognition (OCR) analysis is performed to extract the text. As a part of post processing step, stop word removal and stemming algorithms are applied to identify important contents in the lecture videos and place it into the candidate documents. From these documents, index is created using single-pass in-memory indexing (SPIMI). The key ideas are used to create distinct dictionaries for each block, no need to maintain term-term ID mapping across blocks and do not sort the terms accumulate the postings in posting list as they occur. So, the complete inverted index for each block can created with the help of these two ideas. The separate indexing used to merge into one big index. Finally, the indexed documents stored in disk memory.

3.2 Block-Level Key Frame Extraction (BLKFE) Technique

This module extracts the frames from a video and identifies the key frames from sequence of frames. A key frame is a frame that differs from its previous frame in a considerable amount. Frame difference can be measured using pixel or histogram representation. But pixel representation is sensitive to illumination, camera movements; we used histogram for representing frames. Normal histogram-based approaches for entire frame are simple and but they not tolerant to small object changes. So, we used histogram representation at block level, i.e., each frame '*i*' is divided into '*b*' blocks and are compared with their corresponding blocks in frame

'$i + 1$'. Typically, the histogram difference $\lambda_{k,i}$ of block 'k' between ith frame and $(i + 1)$th frame is measured by Eq. (1).

$$\lambda_{k,i} = \sum_j |H_i(j,\ k) - H_{i+1}(j,\ k)|, \tag{1}$$

where $H_i(j,\ k)$ represents histogram value at gray level j for the block k and n is the total number of the blocks.

After block-comparison, we calculated the cumulative differences between blocks of the consecutive frames, to detect the gradual transition in the lecture video. A high threshold T_h is selected for detecting cuts and low threshold T_l is selected to detect the potential starting frame f_s of a gradual transition. To this selected starting frame $f_{s,}$ a cumulative histogram difference is calculated using Eq. (2)

$$\eta_{k,i} = \sum_{i=f_s}^{f_e} \lambda_{k,i}, \text{ for } i = f_s \text{ to} f_e \tag{2}$$

The end frame block f_e of the video transition is detected when the histogram difference between consecutive frames decreases to less than T_l, while the cumulative histogram difference has increased to a value higher than $T_{h.}$ Similarly, start frame block f_s is detected, if consecutive histogram difference falls below T_l before the cumulative histogram difference exceeds T_h, and then the potential start frame f_s is dropped and the search will continue for the rest of the video transitions. This is process is shown in Eqs. (3) and (4).

$$\delta_{k,i} = 1, \text{ if } \eta_{k,i} \geq T_h \text{ and } \lambda_{k,i} < T \tag{3}$$

$$\delta_{k,i} = 0, \text{ if } \eta_{k,i} \leq T_h \text{ and } \lambda_{k,i} > T_l \tag{4}$$

The key frame selection is done on the basis of large number of significant changes in all blocks of the image. This is specified in the form of parameter T_k in Eq. (5)

$$D(i,\ i + 1) = \begin{cases} 1, \text{ if } \sum_{k=1}^{n} \delta_{k,i} \geq T_k, \\ 0, \quad \text{Otherwise} \end{cases} \tag{5}$$

Table 1 Procedure for OCR analysis

Step 1: Load the key frame image into OCR engine
Step 2: Use Tesseract OCR tool to extract text from key frame image
Step 3: Run the command <Input image path> <output path>
Step 4: Save the output text files into a folder

3.3 OCR Analysis

Optical character recognition (OCR) is the mechanism to retrieve the text from the images. In this paper, we used Tesseract optical character recognition engine open-source software to extract text [13] from the key frames. The process of performing OCR analysis is shown in Table 1.

3.4 Jaccard Index

The jaccard similarity coefficient (as shown in Eq. 6) is a number used for comparing the diversity and similarity of sample sets text documents extracted from OCR tool to eliminate the duplicates frames.

$$J(A, B) = \frac{|A \cup B|}{|A \cap B|} \quad (6)$$

If A and B are both empty, we define. Clearly, $0 < J(A, B) < 1$ so we set the measure $J(A, B) = 0.7$ for compare the text documents.

3.5 Stop Word Removal and Stemming

This module uses the stop word removal and stemming algorithms to reduce the text in the documents and it is effectively useful index the documents. In computing world, stop words does not convey any useful information about the content and hence they are filtered out during processing of text. There is no single universal list of stop words used by all processing tools, and indeed not all tools even use such a list. Any group of words can be chosen as the stop words for a given purpose. For example is, at, which, the and so on.

Stemming [8] is the term used in language information retrieval and morphology and to describe the process for reducing inflected (or sometimes derived) words to their word stem, base or root form—usually a written word. The stem needs not to be equal to the morphological origin of the term or word; it is typically enough that associated words map to the equal stem, though this stem is not in itself a valid root.

Table 2 Single-pass in-memory indexing algorithm

```
Input:   Filtered Text Files
Output: Indexed Text Files
Steps:
1.  SPIMI_Invert (token_stream)
2.   file_output= NEWFILE ()
3.  dictionary= NEWHASH ()
4.  while (free memory available)
5.    do token <--- next (token_stream)
6.    if the term(token) not belongs to existing dictionary then
7.      POS_LIST = ADDTODICTIONARY (dictionary,
                            term(token))
8.    else
9.       postings_list = GETPOST_LIST ((dictionary, term
                         (token))
10.   if full(postings_list)then
11.      POST_LIST= DOUBLEPOST_LIST ((dictionary, term
                         (token))
12.      ADDTOPOST_LIST (post_list, docID (token))
13.      sorted_terms <--- SORTTERMS (dictionary)
14.      WRITEBLOCKTODISK (sorted_terms, dictionary,
                                  file_output)
15.   return file_output
```

3.6 Indexing

Sort-based blocking indexing has excellent scaling properties it always needs a data structure for mapping terms to term IDs. For very large lecture video collections, this data structure does not fit into memory. The scalable approach is single-pass in-memory indexing (SPIMI). SPIMI uses terms instead of term IDs, writes each block's dictionary to disk, and it starts creating a new dictionary for the next new block. The SPIMI approach (as shown in Table 2) can be able to create index for large collections of any size as long as there is enough free disk space available.

3.7 Retrieving

The transformation of a string of characters is usually done by hashing for its advantage of shorter fixed-length values or keys was mapped to the string. The technique hashing is used to search and index the text items in a database because of this feature it is faster to find the item using the shorter hashed key than to find it using the original value. It is also still using in many encryption algorithms. A search for any name would first consist of computing the hash value (using the same hash function to store the item) and then comparing for a match using that value. It would, in common, be much quicker to discover a match across the documents. The hash function is also called the hashing algorithm; the term is probably derived from the idea that the results hash value of 'mixed up' version of the represented value.

3.8 *Ranking*

The ranking function okapi BM25 (BM stands for best matching) is a used for search engines to rank matched documents according to their relevance to a given search queries.

The ranking function BM25 works as follows:
The bag-of-words retrieval function using BM25 that ranks a set of documents based on the query terms appeared in each document. The relationship between the query terms present in a document (e.g., their relative proximity). The whole families of scoring functions are not singleton functions. There are different components and parameters possible in ranking functions. One of the most prominent instantiations of the function is as follows.

Given a query and keywords, the BM25 score of a document is:

$$\text{Score}(D,\ Q) = \sum\nolimits_{i=1}^{n} IDF(q_i) \cdot \frac{f(q_i, D) \cdot (K_l + 1)}{f(q_i, D) + K_l(1 - b + b \cdot \frac{|D|}{avgdl})} \tag{7}$$

where $f(q_i, D)$ is q_i term frequency in the document D, $|D|$ indicates the length of the document in words, and *avgdl* shows the average document length in the text collection from which documents are drawn. k_1 and b are free parameters chosen in absence of an advanced optimization, as $k_1 = [1.2, 2.0]$ and $b = 0.75$. The inverse document frequency (IDF) weight of the query term q_i is usually computed as:

$$IDF(q_i) = \log \frac{N - n(q_i) + 0.5}{n(q_i) + 0.5} \tag{8}$$

where N represents the total number of documents in the storage collection.

4 Experimental Results and Discussion

The data set we considered was the NPTEL lecture video available in the government Web site. The data set contains the standard professional formats for every lecture videos used the recording of lecture videos using latest technologies like capture the whole scene of class room using Web cam and the grab the speakers lecture slides using tools.

The lecture slides will be divided into 4:3 ratio, and one part of the lecture slide contains the title part and another part contains the content in the lecture videos. We are going to test the different style formats of lecture slides because to increase the best relevance results of the proposed system.

4.1 Tools Used

The tools we are mainly used are as follows.

4.2 Net Beans

The Net Beans IDE for writing java code for the algorithms used in the works are stop word removal, stemming, indexing, search, matching and ranking algorithms.

4.3 MATLAB

The MATLAB simulation tool is used for extracting the frames and key frames from the lecture videos and the key frames stored in separate folder.

4.4 Performance Results

The performance metrics are considers as precision, recall and F-measure to compare the results with existing models (Figs. 2 and 3).

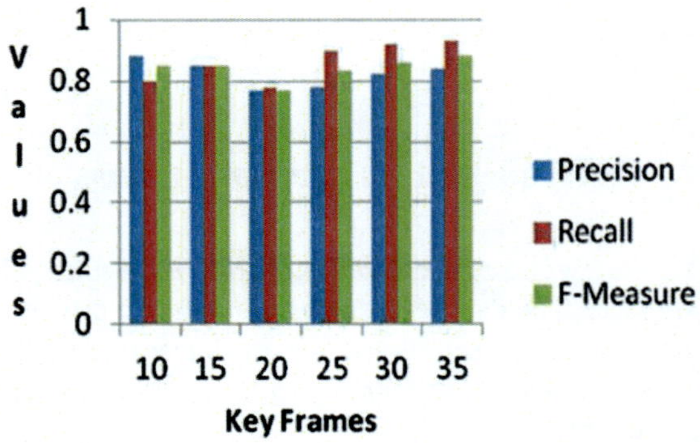

Fig. 2 Precision, recall and F-measure values

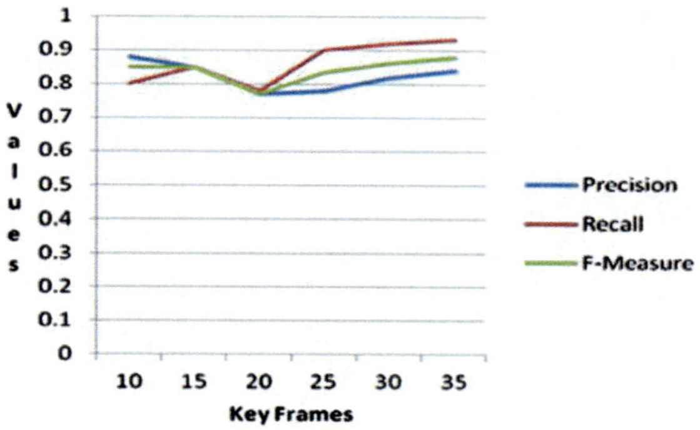

Fig. 3 Precision, recall and F-measure values

$$\text{precision} = \frac{\#(\text{relevant items retrieved})}{\#(\text{retrieved items})} \tag{9}$$

$$\text{precision} = \frac{\#(\text{relevant items retrieved})}{\#(\text{relevant items})} \tag{10}$$

$$\text{F - Measure} = \frac{2 * \text{Precision} * \text{Recall}}{\text{Precision} + \text{Recall}} \tag{11}$$

We considered the different video available in the NPTEL and YouTube. The original data set will be available on request. Table 3, 4 and 5 represents the size of the videos, total number of frames and key frames used in our experiment.

Figure 4 shows the comparison of precision values with the help of existing work, i.e., TF-IDF score and our proposed work, i.e., okapi BM25 score. Hence, it gives the better performance. Similarly, graphs for recall and F-measures gives the better performances (as shown in Tables 6 and 7).

Table 3 Summary of videos

Videos	Size of the video in mega bytes (MB)	Total number of frames	Key frames in videos
Test1	138	140,998	20,027
Test2	13.5	8143	280
Test3	23.2	20,049	3677
Test4	138	140,971	22,123
Test5	76.8	78,317	15,638
Test6	80.8	82,437	19,270
Test7	103	89,821	19,880
Test8	124	96,747	18,026

Table 4 Values of precision, recall and F-measure

Random frames in videos	No. of relevant items in collection	Total no of items retrieved	Relevant item in retrieved	Precision	Recall	F-measure
10	10	9	8	0.88	0.8	0.85
15	15	14	12	0.85	0.85	0.85
20	19	18	15	0.77	0.78	0.79
25	20	23	18	0.78	0.9	0.834
30	25	28	23	0.82	0.92	0.86
35	30	33	28	0.84	0.93	0.88

Table 5 Precision comparison

Methods	Title	Key point
TF-IDF score	0.95	0.77
okapi BM25 score	0.97	0.89

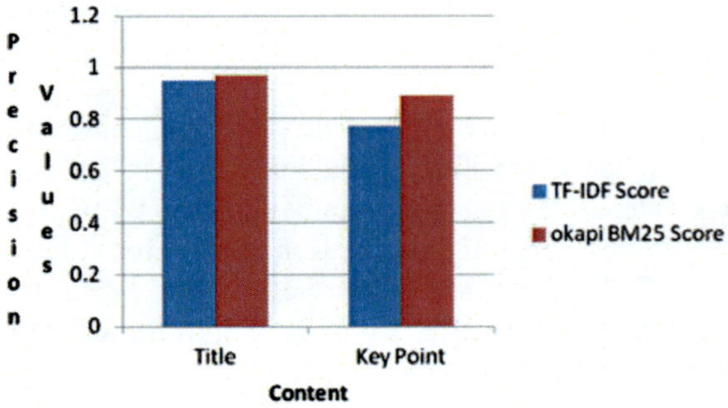

Fig. 4 Precision comparison using different input query content

Table 6 Recall comparison

Methods	Title	Key point
TF-IDF score	0.95	0.77
Proposed okapi BM25 score	0.97	0.89

Table 7 F-measure comparison

Methods	Title	Key point
TF-IDF score	0.9	0.68
Proposed okapi BM25 score	0.94	0.87

5 Conclusion and Future Work

In this paper, we presented an approach for content-based indexing and retrieval based on variant video lectures. The main focus on text slides in the videos and the input queries were content present in the slides. The indexing technique used was single-pass in-memory indexing (SPIMI) because no need to maintain the term-term ID mapping across blocks. For search and ranking, the okapi Best Match25 (BM25) ranking function used to retrieve best relevance video lectures results based on the score.

In the future, we are going to retrieve the lecture video based on the different kinds of the input queries, i.e., audio, image and short video clips. To extract the key frames, we used existing OCR tool to extract text from key frames, for improving the performance, we need to propose novel algorithm in the future. We tried standard professional lecture slides for extracting; indexing and retrieval want to test all kinds of lecture slides.

References

1. Yang, H., & Meinel, C. (2014 April–June). Content based lecture video retrieval using speech and video text information. *IEEE Transactions on Learning Technologies, 7*(2). https://doi.org/10.1109/tlt.2014.2307305.
2. Rathod, G. I., Nikam, D. A. An algorithm for shot boundary detection and key frame extraction using histogram difference. *International Journal of Emerging Technology and Advanced Engineering, 3*(8). https://doi.org/10.1109/icip.2008.4712476.
3. Pong, T.-C., Wang, F., & Ngo, C.-W. (2008). Structuring low-quality videotaped lectures for cross reference browsing by video text analysis. *Journal of Pattern Recognition, 41*(10), 3257–3269. https://doi.org/10.1016/j.patcog.2008.03.024.
4. Jeong, H. J., Kim, T.-E., & Kim, M. H. (2012). An accurate lecture video segmentation method by using sift and adaptive threshold. In *Proceedings in 10th International Conference on Advances in Mobile Computing* (pp. 285–288). http://doi.acm.org/10.1145/2428955.2429011.
5. Tuna, T., Subhlok, J., Barker, L., Varghese, V., Johnson, O., & Shah, S. (2012). Development and evaluation of indexed captioned searchable videos for stem coursework. In *Proceedings of 43rd ACM Technical Symposium on Computer Science Education* (pp. 129–134). http://doi.acm.org/10.1145/2157136.2157177.
6. Sivic, J., & Zisserman, A. (2003 October). Video google: A text retrieval approach to object watching in videos. In *Proceedings of the 9th International Conference on Computer Vision (ICCV)*, Nice, France. https://doi.org/10.1109/iccv.2003.1238663.
7. Meinel, C., Moritz, F, & Siebert, M. (2011). Community tagging in tele-teaching environments. In *Proceedings of 2nd International Conference e-Education, e-Bussiness, e-Management and e-Learning*. https://doi.org/10.1109/icis.2010.141.
8. Baker, S., Kanade, T. (2002). Limits on super-resolution and how to break them. *IEEE Transactions on Pattern Analysis and Machine Intelligence, 24* (9). https://doi.org/10.1109/cvpr.2000.854852.
9. Kay, A. (2007 July). Tesseract: An open-source optical character recognition engine. *Linux Journal*. https://doi.org/10.5120/8794-2784.
10. Accelerating compute intensive medical imaging segmentation algorithms using hybrid CPU-GPU implementations. *Multimedia Tools and Applications, 76*(3), 3537–3555 (2017). https://doi.org/10.1007/s11042-016-3884-2.

11. Accelerating compute intensive medical imaging segmentation algorithms using hybrid CPU-GPU implementations. *Multimedia Tools and Applications, 76*(3), 3537–3555 (2017). https://doi.org/10.1109/JIOT.2018.2844727.
12. Zhong, Y., Zhang, H. J., & Jain, A. (2000). K. Automatic caption localization in compressed video. *IEEE Transactions on Pattern Analysis and Machine Intelligence*, 385–392. https://doi.org/10.1109/icip.1999.822862.
13. Porter, M. F. (1980). An algorithm for suffix stripping. *Program, 14*(3), 130–137. https://doi.org/10.1108/eb046814.

Text Summarization Using Natural Language Processing

Kota Prudhvi, A. Bharath Chowdary, P. Subba Rami Reddy, and P. Lakshmi Prasanna

Abstract With the advancements in the technology, most of the things in this world have become automated. The concept of text summarization came into limelight as summarization of text manually has become a tough and time-consuming task. So, the main purpose of text summarization is to overcome the difficulties faced during manual summarization of text documents or other information from various sources. Text summarization is the process of extracting the main idea of the context or the text and briefly explaining about the context. This process is not only to extract key idea and phrases from the text sources but also generating meaningful summary in a concise and crisp way. The demand for text summarization is raising nowadays because of the large amounts of data from multiple sources like Internet, Twitter, Facebook, Instagram, research papers, and other news articles. Text summarization can be efficiently implemented using NLP as it has many packages and methods in Python or R. Text summarization is also related to text mining as summary is generated based on classifying the given input text. There are different approaches for text summarization and some algorithms are identified to implement these approaches. In this paper, unsupervised learning approach is implemented and cosine similarity technique is used to find the similarity between sentences. To generate rank based on similarity, text rank algorithm is used and sentences with top rank are placed in summarized text.

Keywords Text summarization · Text mining · Natural Language Processing · Cosine similarity · Text rank algorithm

K. Prudhvi (✉) · A. Bharath Chowdary · P. Subba Rami Reddy
Computer Science and Engineering, KL University, Vaddeswaram, India
e-mail: prudhvi.kvnsn@gmail.com

A. Bharath Chowdary
e-mail: bharathchowdary2909@gmail.com

P. Subba Rami Reddy
e-mail: padigapatiram@gmail.com

P. Lakshmi Prasanna
KL University, Vaddeswaram, India
e-mail: pprasanna@kluniversity.in

S. C. Satapathy et al. (eds.), *Intelligent System Design*, Advances in Intelligent Systems and Computing 1171, https://doi.org/10.1007/978-981-15-5400-1_54

1 Introduction

Data and information have become more important in our day-to-day life. This can be explained with the fact given by International Data Corporation (IDC) that the total amount of digital data travelling all around the world in the Internet every year would increase from 33 zettabytes in 2018 to 175 zettabytes in 2025. With increase of online information and other resources, text summarization has become an important part in our fast-paced lives. Text summarization is a part of text analytics which deals with analysing textual data in order to understand customer opinions, reviews, activities, and their feedback. This text analytics has become an important part for data scientists and engineers and they do not face much difficulties in dealing with structured or relational data. But they face difficulties in understanding and analysing online information like articles, blogs, and social media data as these mediums generate unstructured data in form of audio, video, and image.

Text summarization can be said as a subdomain in domain of Natural Language Processing (NLP). Automatic text summarization can be said as a task which is used to generate brief and coherent version of larger text or documents. Researches these days show interest in summarizing the text due to presence of large volume of data in the one hand and time constraint in the other hand. Furthermore, automatic text summarization reduces the reading time of the users, speeds up the selection process of research among documents, and automatic text summarization provides more effective and efficient summaries than human or manual summarization.

Automatic text summarization is a common problem in both the fields of Natural Language Processing (NLP) and deep learning (DL). But it can be better implemented in NLP as several methods and packages like nltk are available in Python and R. The main types of automatic text summarization are extractive-based summarization and abstraction-based summarization. In extractive-based summarization, algorithm identifies for key phrases and important sentences from the text or document and generates a summary using the same key phrases and sentences from the text without any change. In abstraction-based summary, algorithm generates a summary which consists of brief explanation of the text in few lines. In this paper, we will be using extractive text summarization procedure to implement automatic text summarization.

2 Literature Review

2.1 An Overview of Extractive Text Summarization

Rahimi et al. [1] put in the picture that NLP researches are showing more interest in summarizing textual information. In this paper, authors defined text summarization as process of automatically creating and reducing form of given document and retaining its information content source into shorter version with correct meaning. In this paper, the authors also explained about the relationship between text mining and text summarization. The authors of this paper explained about different processes in text summarization like Case Folding, Bag of Words, Tokenization, N-grams, Stemming. This paper also discusses about different criteria in text summarization systems designing such as summarization based on output summary, based on details, based on contents, based on limitation, based on number of input texts, based on language acceptance. Finally, this paper discusses about various approaches to text summarization such as statistical approaches, Lexical chain-based approaches, graph-based approaches, cluster-based approaches, and fuzzy logic-based approaches.

2.2 A Review Paper on Text Summarization

Gaikwad et al. [2] explains that in text summarization, the information that is important is extracted and collected from original text and presented in the form of summary. In this paper, the authors discussed about the two main approaches of text summarization which are extractive-based summarization and abstractive-based summarization. This paper explains about various features of text summarization like term frequency, location, cue method, sentence length, similarity, it also explains about various approaches present in abstractive-based summarization and extractive-based summarization. This paper is also an attempt to present the view of text summarizers for Indian languages and comparison of their performances.

2.3 An Overview of Text Summarization Techniques

Andhale et al. [3] in this paper presented comprehensive survey of both the abstractive-based and extractive-based summarizations. The authors explain about the automatic text summarization becomes important in finding relevant information from long texts in short period of time with less efforts. The authors in this paper mainly focused on extractive-based text summarization methods such as term frequency-inverse document frequency method, cluster-based method, text summarization using neural networks, graph-based method, text summarization with fuzzy logic, latent semantic analysis method, machine learning approach, query-based summarization. The authors also discussed about final observations of summarization methods.

2.4 Document Summarization Using Sentence-Based Topic Modelling and Clustering

George et al. [4] discuss about various approaches to identify significant portion of each document and determine that topic representation and modelling are intermediate representation of text that captures topics discussed in the input and aids the automatic summarization. This article attempts to provide comprehensive summary that includes sentence extraction, tokenization of extracted sentences. The authors discuss about sentence-based structural topic modeling to determine important content for each domain in integrated document and sentences are grouped using k-means clustering. Finally, this paper discusses about lexical ranking score in summarized text.

2.5 A Novel Technique for Efficient Text Document Summarization as a Service

Bagalkotkar et al. [5] discuss the need of automatic text summarizer and its effect in reducing human efforts. The authors also explain about text summarization as the process of generating summary of input document by extracting important sentences from it. In this paper, authors proposed a unique technique for generating summarization of domain specific text from single document from Web by using statistical NLP techniques. The system proposed by authors have four distinct phases which are generation of list of frequent words, sentence generation, update details in database, setup Web service, also proposed an algorithm for single document summarizer. Finally, authors discussed about experimental results, parameters, and performance of proposed system.

2.6 Automatic Text Summarization Using Neural Networks

Kaikhah [6] proposed a different technique for summarizing news articles using neural networks. In this article, author discussed about a neural network that is trained to learn important features of sentences that should be included in summary of article, after that the neural network is modified to generalize and combine important features possible in summary sentences and the neural network that is modified is used as filter to summarize news articles. In this paper, the author discussed about the three phases in text summarization process in neural network that are neural network training, feature fusion, and sentence selection and analyzed about their results.

2.7 *Automatic Text Summarization Using Natural Language Processing*

Devihosur et al. [7] proposed a system of automatic text summarization using unsupervised learning. The authors used a technique of simplified Lesk calculation to assess the significance and importance of sentence in information, an online semantic lexicon wordnet is utilized. This paper discussed about the advantages of proposed system and system architecture of proposed system. The system architecture consists of three stages that are data preprocessing, evaluation of weights, summarization, and the overall representation of automatic text summarization using Natural Language Processing consists of input document, preprocessing, Lesk algorithm, generation of summary, and lesk algorithm is connected to WordNet. Finally, authors discussed about the outputs of the lesk algorithm and its calculation and future scope of the proposed system.

2.8 *Automatic Text Summarization and It's Methods—A Review*

Bhatia et al. [8] in this survey paper investigated the popular and important work done in the field of single and multiple document summarizations. The authors discussed about the method-based approaches for text summarization. These method-based approaches include term-based frequency method, graph-based method, time-based method, separation and merging-based method, semantic dependency method, topic-based approaches, discourse-based approaches, Latent Semantic-based approaches, approaches based on lexical chain, and approaches based on fuzzy logic.

2.9 *NLP-Based Text Summarization Using Semantic Analysis*

Moiyadi et al. [9] presented a unique technique for generating domain specific text by using semantic analysis for text summarization and it is the subset of Natural Language Processing. In this paper, authors proposed a system that uses latent semantic analysis to summarize the documents from the user. According to this paper, the product evaluation process is done in four distinct phases that are preparation, criteria establishment, characterization, testing and the stages in the proposed system are parsing, generation of terms by singular value decomposition (SVD) matrix, analysis of SVD matrix, and arranging of sentences from the final SVD matrix. The authors also suggested a code for implementation of latent semantic analysis using Python library.

2.10 Automatic Amharic Text Summarization Using NLP Parser

Mekuria et al. [10] proposed a system that investigates the problem of building the domain-based single and multiple document Amharic text summarization. In this paper, authors suggested that multi-document summarization targets to compress the most important information from a set of documents to produce a short summary and also suggested that text summarization can be performed based on input, purpose, and output. In this paper, the authors proposed a system that solves the existing problem by developing the combinations of extractive and abstractive-based approach on single and also multiple document input from the user. The authors discussed about the importance of page rank algorithm as it is used in finding out sentence score and weights of sentence in document. In this paper, authors also concluded that the proposed model summarizes only text documents but in future a system would be proposed to develop text summarization for all types of documents including images, graphs, pictures, and videos.

2.11 Text Summarization Using Latent Semantic Analysis Model in Mobile Android Platform

Foong et al. [11] in this paper, presented about latent semantic analysis in automatic text summarization on single English document in android platform of mobiles. The author's objective in this paper is to investigate the LSA model by examining the semantic relationship between terms and sentences in a document for text summarization. In this paper, authors proposed a model for text summarization that consists of three steps that are pre-processing which includes stop words removal, word stemming, POS tagging; latent semantic analysis which includes input matrix, singular value decomposition (SVD); generation of summary.

2.12 Text Summarization Using Latent Semantic Analysis

Ozsoy et al. [12] explain about text summarization as it solves the problems of presenting the information needed by a user in compact form. The authors of the paper explained different latent semantic analysis-based summarization algorithms and two algorithms were proposed by the authors. In this paper, authors used ROGUE scores to evaluate the performance of the algorithms. The authors mainly discussed about latent semantic analysis approach and sentence selection approach for text summarization. One of the algorithms proposed by authors produced best scores and both the algorithms equally performed well on summarizing English and Turkish document texts.

2.13 Text Summarization Within the Latent Semantic Analysis Framework: Comparative Study

Badry et al. [13] explained that text summarization is used to provide shorter version of original text by retaining the overall meaning. In this paper, authors discussed about various approaches of text summarization using latent semantic analysis that are Gong and Liu's approach, Steinberger and Jezek's approach, Murray, Renals and Carletta's approach, Ozsoy's approach which consists of crossmethod and topic method. Finally, in this paper, author presented a table that explains about the comparative summary among the LSA-based summarization algorithms.

2.14 A Survey on Extractive Text Summarization

Moratanch et al. [14] explained about the technique of text summarization is that extracted information is obtained as summary report and presented as a short summary to the user. In this paper, authors discussed about word-level features, sentence-level features, and various extractive text summarization methods. Word-level features consist of content word feature, title word feature, cue phrase feature, biased word feature, and upper case word feature. Sentence-level features consist of sentence location feature, sentence length feature, paragraph length feature, and sentence-to-sentence cohesion. Various summarization methods are categorized based on supervised and unsupervised learning methods. In this paper, unsupervised learning methods discussed are graph-based approach, fuzzy logic-based approach, concept-based approach and supervised learning methods are machine learning approach based on Bayes rule, neural network-based approach, and conditional random fields. The authors in this paper suggested some of the evaluation metrics such as human evaluation, Rogue score, recall, precision, F-measure, compression ratio.

2.15 Literature Review on Extractive Text Summarization Approaches

Saziyabegum et al. [15] discussed about the basis on which types of summarization are classified. The main types of summarization are classified based on preprocessing techniques that include abstractive, extractive; based on purpose of preprocessing that include indicative summaries, informative summaries, critical or evaluative summaries, update summaries; based on audience that include generic summaries, query-based summaries, user-specific or topic-specific summaries; based on number of documents that include single document summarization, multi-document summarization; and based on language that include single, multi-language summarization. The authors in this paper discussed about problems of text summarization, discussed about features of extractive text summarization. This paper also explained about approaches for summarization of single document that include term frequency-inverse document frequency-based approach, cluster-based approach, naïve-bayes-based approaches, rich-feature and decision trees-based approaches, hidden Markov model-based approach, log-linear model approach, neural network-based approach, graph theoretic approach, latent semantic-based approach, concept obtained approach, fuzzy logic-based approach, genetic algorithm and mathematical regression model-based approach, query-biased and structure-preserving-based approaches, lexical chain-based approach, ranking-based sentence clustering approach, and hybrid approaches.

3 Existing System

The above research papers in literature survey suggest that many researchers have done their research in text summarization using latent semantic analysis and fuzzy logic. The proposed system or model in many research papers consists of latent semantic analysis which mostly included singular value decomposition (SVD) and some researchers proposed neural networks to implement text summarization. Most of the researches used python libraries for implementation of text summarization using LSA. Fuzzy logic-based approaches were also used by some researches. The evaluation criteria were mostly done by using parameter called ROGUE score to evaluate performance of various algorithms.

4 Proposed Method

In this paper, a model was proposed by combining two techniques that are used for implementing automatic text summarization. The techniques used to summarize the text are cosine similarity and textrank method. Cosine similarity can be defined as measurement of similarity between documents that consists of group of sentences or vectors of single sentence (non-zero vector) and this cosine similarity approach measures the cosine angle between them. In this paper, we represented each sentence in a vector, so we can measure the similarity between the sentences using cosine similarity. If the angle measured using cosine similarity technique is equal to zero, then the sentences present in the vectors are similar. Another technique used in this paper is text rank algorithm which is commonly used algorithm for text summarization in NLP. Textrank algorithm is similar to PageRank algorithm but in pagerank algorithm, online Web pages are ranked in online search results using pagerank score. But, in text rank algorithm, sentences are used in place of Web pages and rank is generated.

Textrank is unsupervised learning method and used for extractive text summarization. Generally, text rank algorithm alone is used for generating text summaries. But, in this paper, text rank algorithm is combined with cosine similarity to extract summaries.

5 Flowchart

In this architecture, initially, a document which consists of the user's text is passed as an input and then the group of sentences in the paragraph are divided into vectors of sentences where each vector holds each sentence. In the next step, pre-processing techniques like removal of punctuations and removal of stopwords are applied. Then, a technique called cosine similarity is used to build a similarity matrix between the vectors that consists of sentences. Then, weigh the sentences using an algorithm called textrank algorithm to generate ranks in the matrix. Then, the proposed summarizer generates summary that includes on top ranked sentences (Fig. 1).

6 Procedure

- Import required packages like nltk (as it includes stopwords, cosine_distance), numpy (for using text rank algorithm).
- Define a function to read the document as input and divide the document into sentences.
- Define a function to measure similarity between the sentences using cosine similarity that is consider two sentences at each time and convert these sentences into vectors, and then return the value of cosine similarity.
- Define a function to build a matrix basing on the cosine similarity value which is measured in the above function.
- Rank the sentences in the similarity matrix using text rank algorithm imported from numpy, networkx packages, and sort the sentences using the rank and place top ranked sentences in the summary.

7 Results

See Figs. 2, 3 and 4.

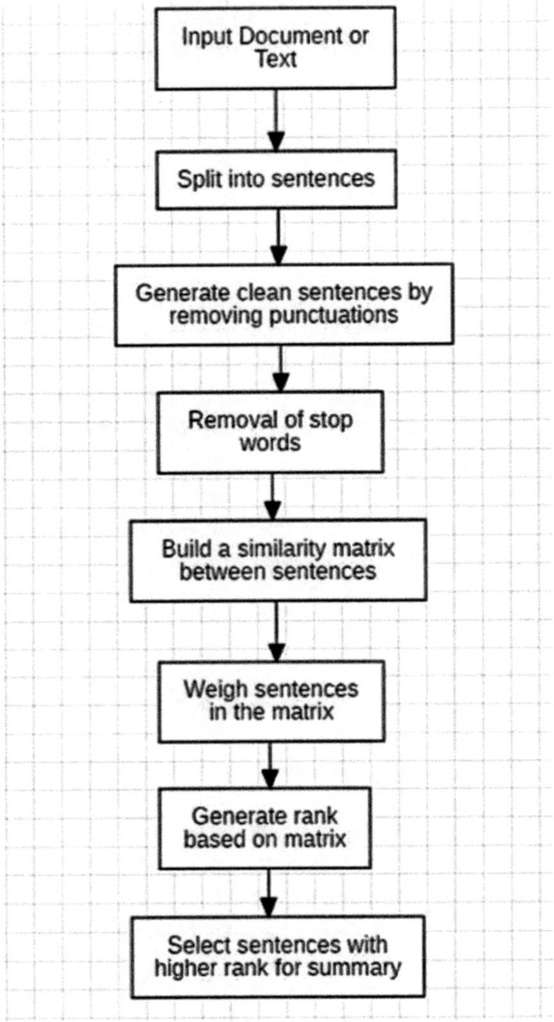

Fig. 1 Flowchart for the implementation of text summarization

```
==================== RESTART: C:\Users\prudh\Desktop\2.py ====================
While complex in the extreme, Derridas work has proven to be a particularly infl
uential approach to the analysis of the ways in which language structures our un
derstanding of ourselves and the world we inhabit, an approach he termed deconst
ruction
In its simplest formulation, deconstruction can be taken to refer to a methodolo
gical strategy which seeks to uncover layers of hidden meaning in a text that ha
ve been denied or suppressed
The term text, in this respect, does not refer simply to a written form of commu
nication, however
Rather, texts are something we all produce and reproduce constantly in our every
 day social relations, be they spoken, written or embedded in the construction o
f material artifacts
At the heart of Derridas deconstructive approach is his critique of what he perc
eives to be the totalitarian impulse of the Enlightenment pursuit to bring all t
hat exists in the world under the domain of representative language, a pursuit h
e refers to as logocentrism
Logocentrism is the search for a rational language that is able to know and repr
esent the world and all its aspects perfectly and accurately
Its totalitarian dimension, for Derridas at least, lies primarily in its tendenc
y to marginalize or dismiss all that does not neatly comply with its particular
linguistic representations, a tendency that, throughout history, has all too fre
quently been manifested in the form of authoritarian institutions
Thus logocentrism has, in its search for the truth of absolute representation, s
ubsumed difference and oppressed that which it designates as its alien other
For Derridas, western civilization has been built upon such a systematic assault
 on alien cultures and ways of life, typically in the name of reason and progres
s.In response to logocentrism, deconstruction posits the idea that the mechanism
 by which this process of marginalization and the ordering of truth occurs is th
rough establishing systems of binary opposition
Oppositional linguistic dualisms, such as rational or irrational, culture or nat
ure and good or bad are not, however, construed as equal partners as they are in
, say, the semiological structuralism of Saussure
Rather, they exist, for Derridas, in a series of hierarchical relationships with
 the first term normally occupying a superior position
Derridas defines the relationship between such oppositional terms using the neol
```

Fig. 2 Screenshot 1 of the text present in the file

8 Conclusion

Text summarization which is one of the important problems in Natural Language Processing has become an interesting research area for many data scientists and researchers. There are various methods and several approaches in implementing automatic text summarization and still more new techniques and approaches are being identified, experimented, and implemented to derive automatic text summarization. In this paper, initially, literature survey is done on various research papers and existing model for text summarization is explained. Later, a model is proposed with slight extensions to improve the performance of text summarization. The methods that are introduced in this paper are slight extension to previous identified methods but these extensions improve the performance of generating summaries.

```
Derridas defines the relationship between such oppositional terms using the neol
ogism difference
This refers to the realization that in any statement, oppositional terms differ
from each other for instance, the difference between rationality and irrationali
ty is constructed through oppositional usage, and at the same time, a hierarchic
al relationship is maintained by the deference of one term to the other in the p
ositing of rationality over irrationality, for instance
It is this latter point which is perhaps the key to understanding Derridas appro
ach to deconstruction.For the fact at any given time one term must defer to its
oppositional other, means that the two terms are constantly in a state of interd
ependence
The presence of one is dependent upon the absence or absentpresence of the other
, such as in the case of good and evil, whereby to understand the nature of one,
 we must constantly relate it to the absent term in order to grasp its meaning
That is, to do good, we must understand that our act is not evil, for without th
at comparison the term becomes meaningless
Put simply, deconstruction represents an attempt to demonstrate the absent-prese
nce of this oppositional other, to show that what we say or write is in itself n
ot expressive simply of what is present, but also of what is absent
Thus, deconstruction seeks to reveal the interdependence of apparently dichotomo
us terms and their meanings relative to their textual context that is, within th
e linguistic power relations which structure dichotomous terms hierarchically
In Derridas own words, a deconstructive reading must always aim at a certain rel
ationship, unperceived by the writer, between what he commands and what he does
not command of the patterns of a language that he uses.It attempts to make the n
ot-seen accessible to sight.Meaning, then, is never fixed or stable, whatever th
e intention of the author of a text
For Derridas, language is a system of relations that are dynamic, in that all me
anings we ascribe to the world are dependent not only on what we believe to be p
resent but also on what is absent
Thus, any act of interpretation must refer not only to what the author of a text
 intends, but also to what is absent from his or her intention
This insight leads, once again, to Derridas further rejection of the idea of the
 definitive authority of the intentional agent or subject
The subject is decentred, it is conceived as the outcome of relations of differe
nce
As author of its own biography, the subject thus becomes the ideological fiction
 of modernity and its logocentric philosophy, one that depends upon the formatio
n of hierarchical dualisms, which repress and deny the presence of the absent ot
her
```

Fig. 3 Screenshot 2 of the text present in the file

```
As author of its own biography, the subject thus becomes the ideological fiction
 of modernity and its logocentric philosophy, one that depends upon the formatio
n of hierarchical dualisms, which repress and deny the presence of the absent ot
her
No meaning can, therefore, even be definitive, but is merely an outcome of a par
ticular interpretation.

Summarized Text:
 It is this latter point which is perhaps the key to understanding Derridas appr
oach to deconstruction.For the fact at any given time one term must defer to its
 oppositional other, means that the two terms are constantly in a state of inter
dependence. The presence of one is dependent upon the absence or absentpresence
of the other, such as in the case of good and evil, whereby to understand the na
ture of one, we must constantly relate it to the absent term in order to grasp i
ts meaning. Derridas defines the relationship between such oppositional terms us
ing the neologism difference
>>> |
```

Fig. 4 Screenshot of the output that consists of the summarized text

References

1. Rahimi, S. R., Mozhdehi, A. T., & Abdolahi, M. An overview on extractive text summarization. In *2017 IEEE 4th International Conference on Knowledge-Based Engineering and Innovation (KBEI) Dec. 22nd, 2017.* Iran University of Science and Technology, Tehran, Iran.
2. Gaikwad, D. K., & Namrata Mahender, C. (2016 March). A review paper on text summarization. *International Journal of Advanced Research in Computer and Communication Engineering, 5*(3).
3. Andhale, N., & Bewoor, L. A. *An overview of text summarization techniques.*
4. George, A., & Hanumanthappa. Document summarization using sentence based topic modelling and clustering. *International Journal of Advanced Research (IJAR).* Received: 04 March 2018, Final Accepted: 06 April 2018, Published: May 2018.
5. Bagalkotkar, A., Khandelwal, A., Pandey, S., Sowmya Kamath, S. A novel technique for efficient text document summarization as a service. In *2013 Third International Conference on Advances in Computing and Communications.*
6. Kaikhah, K. (2004 June). Automatic text summarization with neural networks. In *Second IEEE International Conference on Intelligent Systems.*
7. Devihosur, P., & Naseer, R. (2017). Automatic text summarization using natural language processing. *International Research Journal of Engineering and Technology (IRJET), 04*(08).
8. Bhatia, N., & Jaiswal, A. Automatic text summarization and its methods—A review. In *2016 6th InternationalConference—Cloud System and Big Data Engineering (Confluence).*
9. Moiyadi, H. S., Desai, H., Pawar, D., Agrawal, G., & Patil, N. M. (2016). NLP based text summarization using semantic analysis. *International Journal of Advanced Engineering, Management and Science (IJAEMS)* Infogain Publication (Infogainpublication.com) *2*(10).
10. Mekuria, G. T., Jagtap A. S. (2017). Automatic amharic text summarization using NLP parser. *International Journal of Engineering Trends and Technology (IJETT), 53*(1).
11. Foong, O.-M, Yong, S.-P., Jaid, F.-A. Text summarization using latent semantic analysis model in mobile android platform. In *2015 9th Asia Modelling Symposium.*
12. Ozsoy, M. G., & Alpaslan, F. N., & Cicekli, I. (2011). Text summarization using latent semantic analysis. *Article in Journal of Information Science.*
13. Badry, R. M., Eldin, A. S., & Elzanfally, D. S. (2013). Text summarization within the latent semantic analysis framework: Comparative study. *International Journal of Computer Applications (0975–8887), 81*(11).
14. Moratanch, N., & Gopalan, C. A survey on extractive text summarization. In *IEEE International Conference on Computer, Communication, Signal Processing (ICCCSP-2017).*
15. Saziyabegum, S., & Sajja, P. S. (2016). Literature review on extractive text summarization approaches. *International Journal of Computer Applications (0975–8887), 156*(12).

Light-Weight Key Establishment Mechanism for Secure Communication Between IoT Devices and Cloud

Syam Prasad Gudapati and Vidya Gaikwad

Abstract Nowadays, the security of communication between cloud server and Internet of things (IoT) devices helps in the development, improve standards of application, and related software tools to smoothen the security of connected IoT devices and the cloud environments. Data is very sensitive, whether this data belong to entity's privacy or it can be any private business information. It is profitable chance for the attackers as data is easily commoditized in our society. Connecting insecure IoT devices like smart devices, botnets, smart routers, etc., to cloud server is causing problems in communication between IoT devices and cloud server. In the mob of conceivable platforms and security algorithms, selecting the most suitable to implement a specific service is quite a difficult task, particularly from a security perspective. This paper focuses on light-weight key establishment mechanism for secure communication between IoT device and cloud server. The proposed algorithm provides vital security attributes, also prevention of popular attacks like denial of service (DoS), resist replay attack, known key attack.

1 Introduction

IoT is the booming technology nowadays, and every real-life problem is solved using IoT smart devices. IoT is a connection of devices to the Internet. IoT devices collect data using smart sensors and direct the data to cloud servers. This has resulted in large size of data. The data is deployed on cloud environment. Cloud computing makes IoT-connected devices work hand in hand. Actually, cloud computing and

S. P. Gudapati (✉) · V. Gaikwad
VKR, VNB and AGK College of Engineering, Gudivada, India
e-mail: syamprasad.gudapati@gmail.com

V. Gaikwad
e-mail: gaikwad.vidya30@gmail.com

V. Gaikwad
Department of Computer Science and Engineering Koneru Lakshmaiah Education Foundation, Vaddeswaram, AP, India

S. C. Satapathy et al. (eds.), *Intelligent System Design*, Advances in Intelligent Systems and Computing 1171, https://doi.org/10.1007/978-981-15-5400-1_55

IoT are quite different technologies, they both handle respective data in own way. Cloud computing and IoT both are useful in increasing efficiency in our day-to-day tasks. The IoT devices generate larger amount of data and cloud computing helps data to travel its respective destination.

Some few principles should be considered for IoT solution architect.

(1) Symmetric encryption—In this encryption, communication is done with shared key.
    AES128 is most favorable for IoT devices.
(2) Asymmetric encryption—These keys are public or private key. A private key decrypts data by public key. In this elliptic curve, Diffie Hellman is used to generate a shared key using asymmetric elliptic curve keys. These keys are smaller in size and efficient.
(3) Datagram transport layer security (**DTLS**) is better in secured communication.
(4) Hash algorithms—These algorithms are cryptographically secured and used for integrity of documents. Hash values are encrypted by private key, and for verification, public key is needed to know that the document has not been tampered.
(5) Cloud server security is also important requirement.

IoT devices including smart TVs, smart speakers, toys, wearables and smart appliances, smart meters, smart routers, IP cameras and even smart locks and connected doors are being targeted by attackers to feat them as a entry for hacking and other attacks. Attackers are always trying to change the firmware of these devices.

Typically, a architecture of communication between IoT and cloud server consists of a number of varied smart devices, such as, low-cost sensor, actuator, smart light, smart window shutter, smart thermostat and surveillance camera or other type of smart devices that are combined with *intelligence*, as shown in Fig. 1.

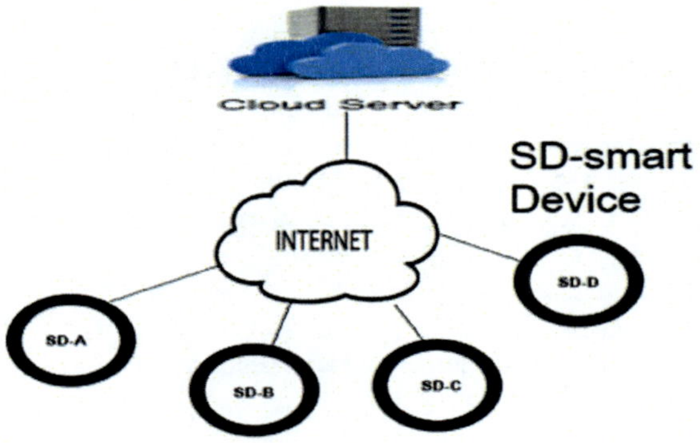

Fig. 1 Architecture of communication between IoT-end devices and cloud platform

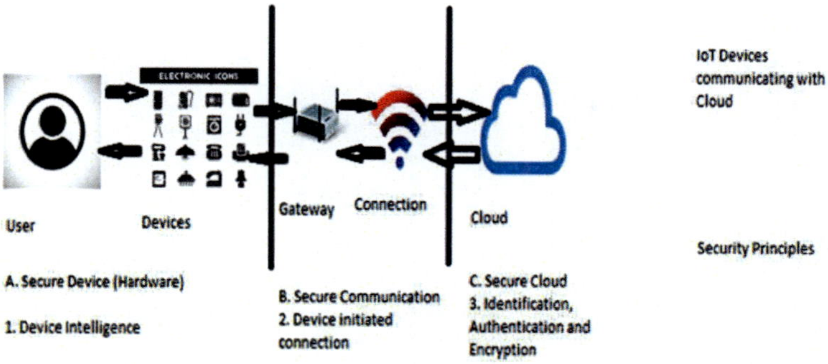

Fig. 2 Security principles for IoT devices communicating with cloud

Nowadays, IoT systems strappingly rely on the cloud platform. IoT-end devices are progressively used as trivial light-weight devices that gather data and connect to influential cloud servers responsible for all the application intelligence and data analytics [1–4].

In this new setting, IoT cannot indifference Cloud computing, as the Cloud is becoming essential component of the overall IoT architecture, rather than an outdoor entity.

Considering through security viewpoint, this vision of cloud computing as a key factor of the IoT architecture infers that all safekeeping issues that the cloud drags on need to be studied and addressed when signifying to IoT security. The result, represented in Fig. 1, is a symbolic shower of cyber-security issues hypothetically disturbing every context of the current and future society. This is the aim; in Fig. 2 we strongly trust that a clear and detailed study of the security issues of the "clouds" is essential to progress the security on the "ground".

2 Literature Survey

S. No.	Name of journal	Title of paper	Authors	Comments/Remarks
1	IEEE Sensors Journal	Lightweight and secure session-key establishment scheme in smart home environments	Pardeep Kumar, Member, IEEE, Andrei Gurtov, Senior Member, IEEE, Jari Iinatti, Senior Member, IEEE, Mika Ylianttila, Senior Member, IEEE, and Mangal Sain	Kumar et al. proposed a session-key establishment mechanism for smart home environments using a three-party-protocol based on symmetric key cryptography. In their algorithm, a trusted service provider stores certain security parameters, including a symmetric key, offline in the home gateway and each of the smart devices. Before a secure session is established between the home gateway and a smart device, the two of them first establish a session-key and authenticate each other based on the parameters they previously received. ISSUE: here is that each time a new device needs to be installed and it must be registered offline with the service provider, to have the security
2	ELSEVIER	Automated security protocol analysis with the AVISPA tool 1	Luca Vigan'o2	For a better implementation, formal verification is required to prove or disprove the correctness of a protocol. This section presents a formal verification of the proposed scheme using SPAN. The AVISPA tool provides a suite of applications for building and analyzing formal models of security protocols

(continued)

(continued)

S. No.	Name of journal	Title of paper	Authors	Comments/Remarks
3	Personal and Ubiquitous Computing	A novel secure key paring protocol for RF4CE ubiquitous smart home systems	Kyusuk Han Jangseong Kim Taeshik Shon Daesik Ko	Author Han et al. presented a secure key-pairing protocol for "radio frequency for consumer electronics" smart home systems, where several heterogeneous consumer electronics devices (e.g., IEEE 802.15.4) form a smart home network. Each device sends its own authentication information to a mobile operator. After the first level of authentication, the mobile operator sends the device information to the device manufacturers to be authenticated again. The proposed scheme is also based on symmetric key cryptography. The main requirement of this scheme is that the manufacturers must always be online, which may not always be practical. Also, the communication costs of their proposed scheme would be high for the resource-constrained devices
4	IEEE Transactions on Information Theory	New directions in cryptography	W. Diffie M. Hellman	Diffie and Hellman introduced the first key exchange protocol that is also the first public key system. However, the Diffie–Hellman key exchange is vulnerable to man-in-the-middle attack (MIMA) due to the lack of authentication

(continued)

(continued)

S. No.	Name of journal	Title of paper	Authors	Comments/Remarks
5	International Journal of Communication Systems	Lightweight and provably secure user authentication with anonymity for the global mobility network	Chun Chen Daojing He Sammy Chan Jiajun Bu Yi Gao Rong Fan	The trust model of FA, HA, and MN is described in. The user authentication needs to provide the followings: first, the location of a user may be traced by an adversary because a user accesses services provided from service providers in anytime and in anywhere
6	Computer Networks Journal: ELSEVIER	The internet of things: a survey	Atzori, Luigi, Iera, Antonio, Morabito, Giacomo	
7	Journal of Cyber Security Technology	Lightweight cryptography methods	Buchanan, William J. Li, Shancang Asif, Rameez	Light-weight cryptography methods are proposed to overcome many of the problems of conventional cryptography. This includes constraints related to physical size, processing requirements, memory limitation, and energy drain. This paper outlines many of the techniques that are defined as replacements for conventional cryptography within an Internet of things space and discuss some trends in the design of lightweight algorithms
8	IEEE Communications Surveys and Tutorials	Context aware computing for the internet of things: a survey	Perera, Charith, Zaslavsky, Arkady, Christen, Peter, Georgakopoulos, Dimitrios	Authors objectives were in revisiting the literature are three-fold: (1) to learn how context-aware computing techniques have helped to develop solutions in the past, (2) how can we apply those techniques to solve problems in the future in different paradigms such as the IoT, and (3) to highlight open challenges and to discuss future research directions

(continued)

(continued)

S. No.	Name of journal	Title of paper	Authors	Comments/Remarks
9	CRYPTREC: Lightweight Cryptography Working Group	CRYPTREC cryptographic technology guideline (lightweight cryptography)		Light-weight cryptography, which can be implemented at lower costs and with lower power consumption compared to the conventional cryptography, is expected to have many applications including on-vehicle devices and medical equipment. It is often not easy for engineers without expertise to exploit this technology by choosing a proper light-weight cryptographic algorithm
10	Sony Corporation	Lightweight cryptography for the Internet of Things	Katagi, Masanobu, Moriai, Shiho	This paper gives an overview of the state-of-the-art technology and standardization status of light-weight cryptography, which can be implemented efficiently in constrained devices. This technology enables secure and efficient communication between networked smart objects

(continued)

(continued)

S. No.	Name of journal	Title of paper	Authors	Comments/Remarks
11	Digital Communications and Networks	A roadmap for security challenges in the internet of things	Riahi Sfar, Arbia, Natalizio, Enrico, Challal, Yacine, Chtourou, Zied	A case study is presented to highlight the components and interactions of the systemic and cognitive approach. Then, security questions about privacy, trust, identification, and access control are discussed. According to the novel taxonomy of the IoT framework, different research challenges are highlighted, important solutions and research activities are revealed, and interesting research directions are proposed. In addition, current standardization activities are surveyed and discussed to the ensure the security of IoT components and applications
12	Security and Communication Networks	Authentication protocols for internet of things: a comprehensive survey	Ferrag, Mohamed Amine, Maglaras, Leandros A., Janicke, Helge, Jiang, Jianmin, Shu, Lei	In this paper, a comprehensive survey of authentication protocols for Internet of things (IoT) is presented. Specifically, more than forty authentication protocols developed for or applied in the context of the IoT are selected and in-detail examined. These protocols are categorized based on the target environment: (1) Machine-to-machine communications (M2M), (2) Internet of vehicles (IoV), (3) Internet of energy (IoE), and (4) Internet of sensors (IoS). Threat models, countermeasures, and formal security verification techniques used in authentication protocols for the IoT are presented

(continued)

(continued)

S. No.	Name of journal	Title of paper	Authors	Comments/Remarks
13	Journal of Ambient Intelligence and Humanized Computing	Advanced lightweight encryption algorithms for IoT devices: survey, challenges and solutions	Singh, Saurabh, Sharma, Pradip Kumar, Moon, Seo Yeon, Park, Jong Hyuk	Authors have analyzed many light-weight cryptographic algorithms based on their key size, block size, number of rounds, and structures. In addition, we discuss the security architecture in IoT for constrained device environment, and focus on research challenges, issues, and solutions. Finally, a proposed security scheme with a service scenario for an improvement of resource constrained IoT environment and open issues are discussed
14	IEEE Internet of Things Journal	Secure range search over encrypted uncertain IoT outsourced data	Guo, Cheng, Zhuang, Ruhan, Jie, Yingmo, Choo, Kim-kwang Raymond, Member, Senior	IoT could have broader characterization where diverse data or information could come from ubiquitous and persistent sources. Influenced by surrounding environment, digitization process, measurement errors, network transmission, and other factors, the data collected from IoT devices are often in the distribution-based form instead of the point-valued form, such as in the applications of wireless sensor network, RFID, online travel agents and so on. All these forms of information are important for IoT to accomplish a mission. Therefore, the study of classification for distribution-based data is very valuable in the field of IoT

(continued)

(continued)

S. No.	Name of journal	Title of paper	Authors	Comments/Remarks
15	International Journal of Computer Applications	Privacy and security in mobile cloud computing: review	Malik, Sapna, M Chaturvedi, M.	Authors have mentioned that mobile cloud computing is an answer to intensive processing and storage demand of real-time and high-end applications. Being in nascent stage, mobile cloud computing has privacy and security issues which deter the users from adopting this technology. This review paper throws light on privacy and security issues of mobile cloud computing
16	Computers in Biology and Medicine	Security and privacy preserving approaches in the eHealth clouds with disaster recovery plan	Sahi, Aqeel, Lai, David, Li, Yan	Authors mentioned that third-party authenticator is required which authenticates the cloud data from the side of users or holders of the data. Security of sensitive data is of questionable nature due to the presence of various entities. Cloud service providers along with data users are equally in charge for putting the security of the pivotal confidential data at risk. Surveys have suggested that huge storage systems cannot be trusted completely; and all of them can be hacked

2.1 Security Challenges in IoT with Cloud

At a distance from sensors and actuators, IoT often joins legacy and new systems together to enable interoperability. These systems connect through gateways to the cloud where sensors' data are being processed, in order to make a choice. Sporadically, this interoperability issue applies in environments where critical operations take place, such as healthcare, energy, aviation, wearables, smart home applications, smart cities, agriculture, industrial automation.

There is a rising anxiety about security for the IoT ecosystem in its entireness. Attacks to IoT devices can possibly ease treading pebble attacks and access to network or cloud resources and vice versa.

Nowaday's new attack surfaces for both cloud and end-devices of the IoT ecosystem are being exposed. Deprived security on the level of the IoT devices or the IoT gateways can hypothetically outcome in insecure connectivity and data flow toward the cloud with significances to the overall security of the ecosystem [5, 6, 9].

European Union Agency for cybersecurity (ENISA) identified the following security challenges dwindling under each of the three main classes derived from the ENISA IoT high-level reference model. An analysis of these challenges is available in this section [3].

Connectivity	1. Heterogeneous protocols for communication 2. Insecure data flow from the edge to the cloud
Analysis	1. Real-time processing at the edge overshadows security 2. Impact of cloud decentralisation on security
Integration	1. Security depends on the vertical that cloud is serving 2. Security relies much on the implementation from IoT developers

3 Proposed System Architecture

A. **Security at Device (Hardware Layer)**

4 Intelligent Devices Is Needed for Complex Security Task

Smart devices, toys, smart tools, and wearables have the ability to communicate with cloud server via Ethernet or WIFI. Some of the devices are equipped with microprocessor; these devices are week to handle the complications of Internet connectivity.

IoT smart devices should be effective and secured. These devices should be able to follow all security principles like encryption, authentication, proxies, buffering, firewalls, and disconnectivity. Also, these devices should be strong enough to operate in every environment with limited support.

B. **Security at Communication Layer**

The communication layer is used for connectivity between networks and IoT devices. Sensitive data is in transit over the physical layer, networking layer, or application layer and unsecure communication channels can be susceptible to intrusions such as man-in-the-middle attacks.

The IoT device should start the connection to the cloud. A connection to the cloud can also facilitate a bi-directional channel, and it allows the IoT device to be remotely controlled.

5 Our Proposed Algorithm

Notation used:

EK_A = Encryption Key for SD-A

CI_A = Cloud Identifier for SD-A

DI_A = Device Identifier for SD-A

$E_A[m]$ = Encrypt Message m for SD-A

$D_A[m]$ = Decrypt Message m for SD-A

h() = Hash Function

|| = Concatenation Operation

DST_A = Device Session token for SD-A

AT_A = Authentication Token for SD-A

SD-A(Smart Device - A)	**Cloud Server**
• Generate NONCE=[S]	
• Generate a random number "r1"	
• Calculate Authentication Token $AT_A = h(CI_A \| DI_A \| r1)$	
• Calculate Session Token	
$DST_A = E(EK_A, S \| r1)$	
. Send DST_A, AT_A	
• Calculate Session Key	
$SK = h(CI_A \| DI_A \| S)$	
	. Receive DST_A, AT_A

- Decrypt DST_A

$S', r1' = D(EK_A, DST_A)$

- Calculate

$AT_A' = h(CI_A || DI_A || r1)$

- If $AT_A' = AT_A$ then proceed else exit
- Calculate r2

Device not authenticated

- Calculate Session Key

$SK = h(CI_A || DI_A || S')$

- Calculate $CST = E(SK, r1 || r2)$
- Send CST

- Receive CST
- Decrypt CST with SK

$r2', r1' = D(SK, CST)$

if $r1' == r1$ then proceed with Data Transfer
else Not Authenticated Cloud Server

Smart Device-A

- Generate NONCE [S] at SD-A, also generate a random number r1 for randomness in the data.
- Calculate the Authentication token $AT_A = h(CI_A \parallel DI_A \parallel r1)$, using hash by Cloud Identifier of SD-A to which it is connected, Device Identifier of SD-A and random number r1. By adding a random number r1 into the data of the hash function, we are creating randomness in the output of hash, thus it becomes difficult for an attacker to crack the hash function or perform man-in-the-middle attack.
- Calculate Device Session Token $DST_A = E(EK_A, S \parallel r1)$, using Encryption Key EK_A for SD-A with data S and random number r1. By adding the random number into the data, it becomes difficult to an attacker to crack since he does not know the key and the data is also random.
- Send this generated DST_A and AT_A to cloud server.
- Calculate Session Key $SK = h(CI_A \parallel DI_A \parallel S)$ using hash by Cloud Identifier of SD-A with Device Identifier and data S. Once the handshaking between the cloud server and smart device occurs, a Session Key is generated for secure communication. A new Session Key will be generated for every time new handshake takes place between cloud server and smart device.

Cloud Server

- Cloud server receives DST_A, AT_A generated by SD-A.
- Cloud server decrypts the DST_A by $S', r1' = D(EK_A, DST_A)$.

- Calculate $AT_A' = h(CI_A \| DI_A \| r1)$; here, if $AT_A' = AT_A$ then proceed, this is used to check Masquerade Attack, if it does not match, then the smart device is not an authentic device.
- Now, generate random number r2, calculate Session Key $SK = h(CI_A \| DI_A \| S')$ using hash with Cloud Identifier, Device Identifier with data S'.
- Calculate Cloud Server Token $CST = E(SK, r1 \| r2)$, using encryption on generated session key with r1 and r2.
- Send this CST to Smart Device-A.

Smart Device-A

- Receive CST sent from Cloud Server.
- Decrypt this CST with SK
  $r1', r2' = D(SK, CST)$
  if $r1' == r1$, then proceed with data transfer by encrypting the data using generated session key, and if not matching, then cloud server is not authenticated [7].

6 Contribution and Outline of the Paper

This paper's objectives are to provide an up-to-date and well-structured solution of the security issues of cloud computing in the era of the IoT revolution [8–11].

For this purpose, we use a structured approach; subsequently, there is no IoT without cloud, and we cannot secure IoT without securing the cloud.

In short, the contribution of the paper is to focus on communication security between cloud and IoT-end devices.

- We provide a new cloud-centered perspective of IoT security. Since, cloud computing has become of highest standing for Internet of things. Since, most of the works related to IoT security focus on the security of end devices. In this paper, we fill this gap providing an analysis of communication security between cloud and IoT devices.

7 Conclusion

In this paper, we proposed a light-weight key establishment mechanism for secure communication between IoT devices and cloud.

In the future, we would like to demonstrate the proof of concept of security analysis for

(1) Fight Masquerade Attack
(2) Fight Man-in-the-middle Attack
(3) Fight Message-Forgery Attack.

References

1. Guan, Z., Li, J., Wu, L., Zhang, Y., Wu, J., & Du, X. (2017). Achieving efficient and secure data acquisition for cloud-supported internet of things in smart grid. *IEEE Internet of Things Journal, 4,* 1934–1944.
2. Mihovska, A., & Sarkar, M. (2018). Smart connectivity for internet of things (IoT) applications. In R. R. Yager & J. Pascual Espada (Eds.), *New advances in the internet of things* (pp. 105–118). Springer International Publishing: Cham, Switzerland.
3. Desnitsky, V. A., Kotenko, I. V., & Nogin, S. B. (2015). Detection of anomalies in data for monitoring of security components in the internet of things. In *2015 XVIII International Conference on Soft Computing and Measurements (SCM)*. https://doi.org/10.1109/scm.2015.7190452.
4. Malik, A., & Om, H. (2018). Cloud computing and internet of things integration: Architecture, applications, issues, and challenges. In W. Rivera (Ed.), *Sustainable cloud and energy services: Principles and practice* (pp. 1–24). Cham, Switzerland: Springer International Publishing.
5. Díaz López, D., Blanco Uribe, M., Santiago Cely, C., Tarquino Murgueitio, D., Garcia Garcia, E., Nespoli, P., et al. (2018). Developing secure IoT services: A security-oriented review of IoT platforms. *Symmetry, 10,* 669.
6. Biggest security challenges for IoT. (2019, April 29). Retrieved from https://www.peerbits.com/blog/biggest-iot-security-challenges.html.
7. Abood, O. G., & Guirguis, S. K. (2018). A survey on cryptography algorithms. *International Journal of Scientific and Research Publications (IJSRP), 8*(7) (7978). https://doi.org/10.29322/ijsrp.8.7.2018.
8. Chen, Y., & Luo, B. (2012). S2A. In *Proceedings of the Second ACM Conference on Data and Application Security and Privacy—CODASKY '12*. https://doi.org/10.1145/2133601.2133628.
9. Coppolino, L., D'Antonio, S., Mazzeo, G., & Romano, L. (2019). A comprehensive survey of hardware-assisted security: From the edge to the cloud. *Internet of Things, 6,* 100055. https://doi.org/10.1016/j.iot.2019.100055.
10. Darwish, S., Nouretdinov, I., & Wolthusen, S. D. (2017). Towards composable threat assessment for medical IoT (MIoT). *Procedia Computer Science, 113,* 627–632. https://doi.org/10.1016/j.procs.2017.08.314.
11. Towards secure convergence of Cloud and IoT. (n.d.). Retrieved from https://www.enisa.europa.eu/news/enisa-news/towards-secure-convergence-of-cloud-and-iot.

BUGVILLA: Calibrating Bug Reports with Correlated Developers, Tracking Bug Reports, and Performance Analysis

Gudapati Syam Prasad, D. Kiran, and G. Sreeram

Abstract A bug is defined as the error in code because of which a system cannot produce the desired output. Whenever a bug occurs, a bug report is created. To find the cause of a bug, developers will execute the whole code continuously and analyze the code depending upon the report they received. So the bug reports should be accurate as the developers depend on the bug report received to solve a particular bug. Bug reports are the documents which provide total information of a particular bug with different accuracy. We need to manage the bugs which arise while working on a software project. Developers usually take the help of spreadsheets and fill them with new bugs which are raised. But when the list of bugs increases to a large number and all developers working on a single spreadsheet to update bugs may create ambiguity. Every software project reaches this point, especially during testing and deployment when users tend to find an application's bugs. So, a tool for tracking bug reports will be useful instead of a spreadsheet. Where tracking bug reports is an important feature which helps to observe the activities of software developers. Moreover, it is important to observe that a bug will be assigned to the developer who is an expert in solving that particular type of developer. This paper introduces a new tool for tracking bug reports, using which we can assign bugs to their respective developers, view bug status, check bug history, check the total performance of developers, and also communicate with developers by means of a group chat.

Keywords Bug reports · Bug repository · Tracking bug · Timestamp · Spreadsheets

---

G. S. Prasad
Department of CSE, V.K.R, V.N.B & A.G.K College of Engineering, Krishna District, Gudivada, Andhra Pradesh, India
e-mail: syamprasad.gudapati@gmail.com

D. Kiran
Department of CSE, Daripally Anantha Ramulu College of Engineering and Technology, Khammam, Telangana, India
e-mail: kiran.daripally@gmail.com

G. Sreeram (✉)
Koneru Lakshmaiah Education Foundation, Vaddeswaram, Guntur, Andhra Pradesh, India
e-mail: gsmnaidu@gmail.com

S. C. Satapathy et al. (eds.), *Intelligent System Design*, Advances in Intelligent Systems and Computing 1171, https://doi.org/10.1007/978-981-15-5400-1_56

1 Introduction

An unexpected output is called a bug whose cause is a tedious task to find in any software project. Everyday many companies go through a huge number of bugs. A bug cannot be avoided, and it is necessary to fix them because the system may not produce expected output unless a bug is solved [1]. Usually, software developers preserve the bugs and update the solutions in a single spreadsheet when there is less number of bugs. Every bug report is a textual document which contains the name of the bug with its description [1]. Sometimes inaccurate data can be present in bug report which makes the bug solving task more difficult [2]. But when the number of bugs becomes large, then ambiguity gets created if a single spreadsheet is used. To avoid this problem, bug tracking systems are developed to handle the bugs. In the process of handling bugs, a lot of resources are being spent by software companies. Maintaining a bug repository is important to manage the bugs. All the bugs raised with an updated solution can be preserved in the bug repository.

A bug repository preserves the bugs and its updates for the future purpose. These bug repositories are used by large companies to help developers in the process of handling bugs [3]. The data preserved in a software repository is valuable and confidential. This data which is preserved helps to manage the future bugs in the same project. These bug repositories can also be called a bug tracking system (BTS). Keeping track of bugs that occurred under a particular software project can also be called as a BTS. All the bug reports will be saved in BTS. Some bug tracking systems lack the ability to preserve the previous bugs [4].

There are two problems in these bug tracking systems. They are a large number of bugs and inaccurate bug reports. On one hand, while working on a project, the number of bugs increases day-to-day. On the other hand, there will be inaccuracy in the bug reports. When it comes to bugs, a particular bug should be assigned to only that developer who can solve it or who has experience in solving it. Or else, it may take a long time to the developer to study the bug and then to solve it. Overall, a large number of inaccurate bugs and lack of expertise about the bug may make bug solving more difficult and time taking.

In this paper, we introduce two modules, namely the manager module and employee module. A manager assigns bugs to the respective developer and the developer updates the bug with a solution. Once the bug is updated, the information of the bug is preserved in the repository. The bug repository gives information about the bugs which are raised solved and also the solution. We can also fetch the bugs occurred under a particular period of time from the bug repository. We can also analyze the performance of software developers and also communicate with the developers by means of group chat.

2 Literature Survey

In 2004, Bugzilla was developed as the tool for the bugs of Firefox and Eclipse projects [5]. In this, one can form a bug report and the report will be sent to the email of the developer of that particular project. It does not keep a count of bugs that are solved for a particular project (Table 1).

In 2007, BTS was developed for MNC's projects [6]. Here a few developers will be selected to solve the bugs and also keep a record of the errors that can happen in the project.

In 2008, Zimmerman introduced a card sort technique [7]. It shows suggestions for the better performance of the bug tracking system. It just shows suggestions but does not provide any tool like a bug tracking system. Later, Infozilla was developed for the maintenance of large projects [8]. Here all the duplicates are considered as useful information for future purposes. It totally depends on survey results for a particular bug report. Cuezilla was developed for product developing and manufacturing [9]. Cuezilla indicates the time taken by the developer while solving a particular bug. It depends on the result which is mostly suggested by a large number of developers. But accurate relevant information isn't provided.

In 2010, automatic bug triage was implemented which uses a classifier to differentiate bugs among the bugs [10]. It does not deal with the bug handling which is a major drawback. In 2014, Weka was introduced for the large-sized android projects [11]. It determines the grouping of bugs and their duplicates.

In 2015, instance selection approach is introduced where bugs will be selected based on token or keywords [11]. But here too many repositories are needed to be handled.

In 2019, we are introducing BUGVILLA with which both the manager and the developer can communicate with each other. Bugs will be assigned to the relevant developer and developer updates it with a relevant solution. We can also make an instance selection of bugs that occurred under particular days. All the bugs are preserved in the repository for the future purpose. At the end of the project, we can also analyze how many bugs are assigned and how many are solved by means of a graph.

Table 1 Literature survey table

Paper	Year	Tool
Bugzilla, iTracker, and other bug trackers	2004	Bugzilla
Defect tracking systems in global software development—a work practice study	2007	BTS
Toward the next generation of bug tracking systems	2008	Card sort technique
Duplicate bug report considered harmful really?	2008	Infozilla
What makes a good bug report?	2008	Cuezilla
Automatic bug triage using semi-supervised text classification	2010	Automatic bug triage
New feature for duplicate bug detection	2014	Weka

3 Implementation

The purpose of the proposed work is as follows:

A. Creating an environment of manager and developers.
B. The manager can assign bugs to the relevant developers.
C. Developers can update the bugs.
D. Saving the updated bugs.
E. Instance the selection of bugs.
F. Chat between the manager and developers.
G. Performance analysis.

Bugvilla is developed in JAVA which is platform-independent and can run on any browser. MySQL is used to store all the data. Figure 1 illustrates the realistic architecture of "BUGVILLA".

Here, a manager assigns bugs to their relevant developers. When the developer updates the bugs, they get preserved in the software repository. With the help of the saved bug history, we can analyze the number of bugs raised and the number of bugs solved in the form of performance analysis. Here the developers and manager can communicate with each other by means of group chat where five people can chat at once. This chat may help for easy communication and solving the bugs in time.

A. *Creating an environment of manager and developers*: An environment for the communication of manager and developer is created as shown in Fig. 2.

A manager can login by clicking on to the manager login. Once clicked on the manager login, manager will be directed to the login page where the manager has to provide username and password as shown in Fig. 3.

A developer can register/login by clicking on to the employee login. When the developer registers, he/she will be given an id and then the developer should login with that particular id.

Once a manager logins, he can see the tasks that he can do as shown in Fig. 4, i.e., assigning bugs, bug status checking, maintaining repository, timestamp (instance selection of bugs) and do performance analysis.

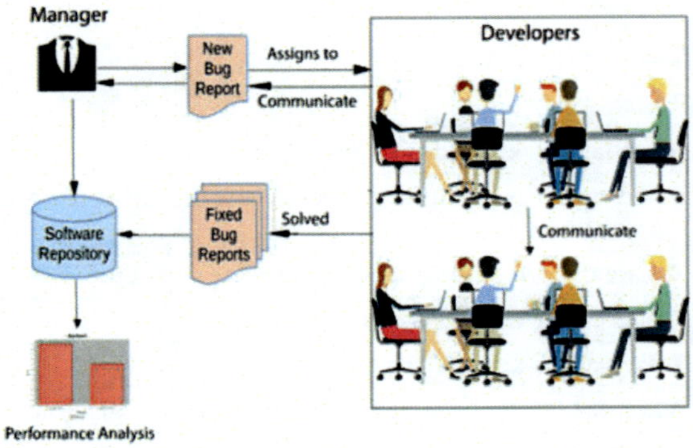

Fig. 1 System architecture

Fig. 2 Environment for login

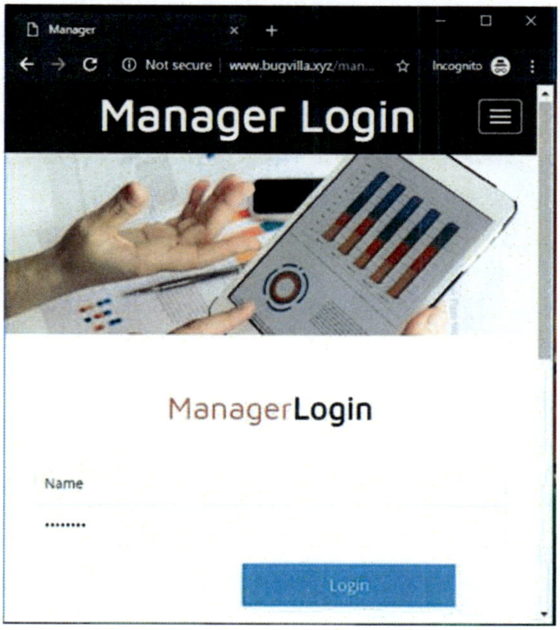

Fig. 3 Manager login

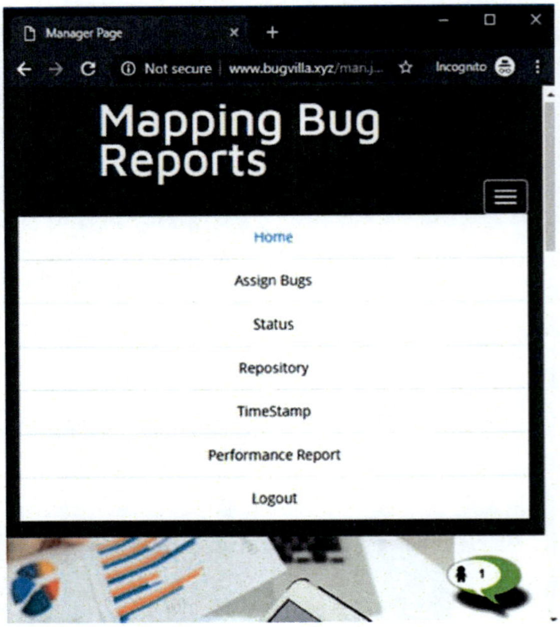

Fig. 4 Manager tasks

B. *The manager can assign bugs to the relevant developers*: A manager firstly assigns a bug to a developer based on his/her expertise. To assign a bug, manager may need to enter bug name, bug description, and date. Then a manager will set the bug triage, i.e., high, mid, or low and select a developer who can solve the respective bug. After this, the manager can raise the bug as shown in Fig. 5.

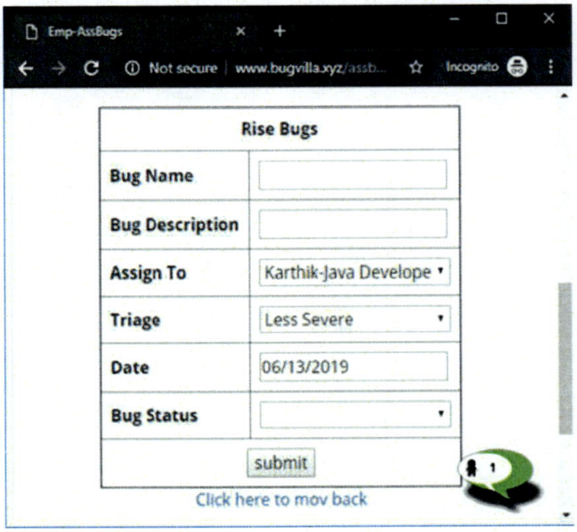

Fig. 5 Assigning bugs

Table 2 Assigned bugs table

Bug name	Bug desc.	Triage e	Assigned d to	Status
Navigation error	User cannot navigate from profile to inbox	Low	Kalyani	Unresolved
Display picture does not update	Database does not store the new display picture	Mid	Laya	Unresolved
Error in sending messages	User cannot see two ticks	High	Santoshini	Unresolved

Fig. 6 Bug status

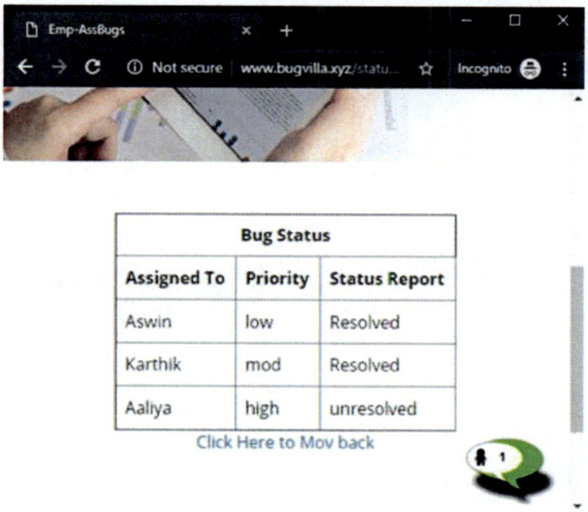

Once the bug is raised, the bug name and bug description get preserved in the repository with the triage under unresolved status bugs as shown in Table 2. This raised bug will be shown to the respective developer for updates when he/she logins.

The manager can also check bug status as shown in Fig. 6 for the bugs which are assigned to particular manager can understand that how and whom to assign the future bugs. If a developer has high bug triage bug assigned and it is unresolved, then the manager will assign another high triage bug to another developer who has low triage bugs. In this way, it is easy to solve the bugs in time to complete projects on time.

C. *Developers can update bugs*: If no bugs are assigned to a developer, then it shows that "NO BUGS ASSIGNED" as shown in Fig. 7.

If a bug is assigned to a developer, then the bug name, description, and triage are shown to the developer as shown in Fig. 8, where the developer can update the bug with its solution developers. The bug status shows the developer name, bug triage, and bug status. By this, whenever the developer updates the bugs, they get saved into the repository of bugs with the solution and with the date when the bug is raised.

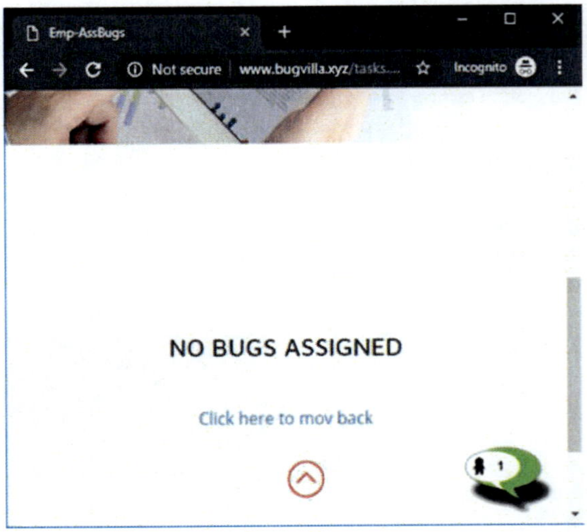

Fig. 7 No bugs assigned

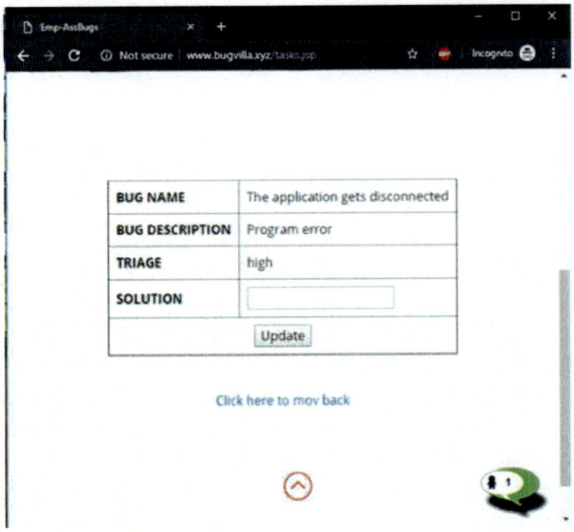

Fig. 8 Bug update

Saving the date helps to fetch the bugs when bugs are fetched under a particular range of dates.

D. *Saving the updated bugs*: Whenever the bugs are assigned to a developer, they get saved into the bug repository with bug info and their solution as "WAITING" as shown in Fig. 9.

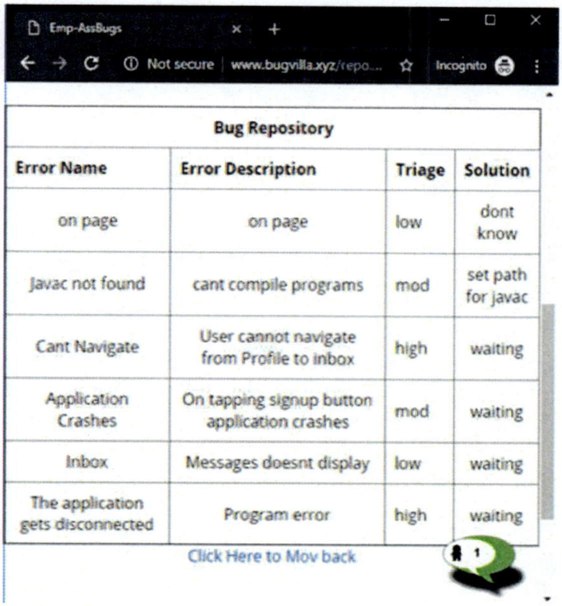

Bug Repository			
Error Name	**Error Description**	**Triage**	**Solution**
on page	on page	low	dont know
Javac not found	cant compile programs	mod	set path for javac
Cant Navigate	User cannot navigate from Profile to inbox	high	waiting
Application Crashes	On tapping signup button application crashes	mod	waiting
Inbox	Messages doesnt display	low	waiting
The application gets disconnected	Program error	high	waiting

Fig. 9 Bug repository before updating

When the bugs are updated by the developers, the solution gets saved in the bug repository as shown in Fig. 10. When the manager wants to check the bug repository, the bug repository will be displayed in the form of a table with columns, namely error name, error description, bug triage, bug solution.

If the bug solution has "WAITING", then it means that the bug is yet to be solved.

E. *Instance Selection Of Bugs*: The bugs which are saved in the repository can be fetched by using timestamp; i.e., a particular date range should be given as shown in Fig. 11 and the bugs that occurred under a particular period of time can be shown to the manager as shown in Fig. 12.

When the time range is selected as shown in Fig. 11, the bugs occurred between Jan 1st, 2017, and Jan 1st, 2019, are fetched and shown as a table with columns, namely bug name, description, triage, developer, and date.

F. *Chat between manager and developer*: A chat communication is developed between manager and developer. To keep the chat secure every user is asked to login to the group chat before chatting as shown in Fig. 13.
G. *Performance Analysis*: Here the number of bugs raised and the number of bugs are solved which is shown in the form of bar chart. So that the manager can know the total number of bugs that are yet to be solved as shown in Fig. 14.

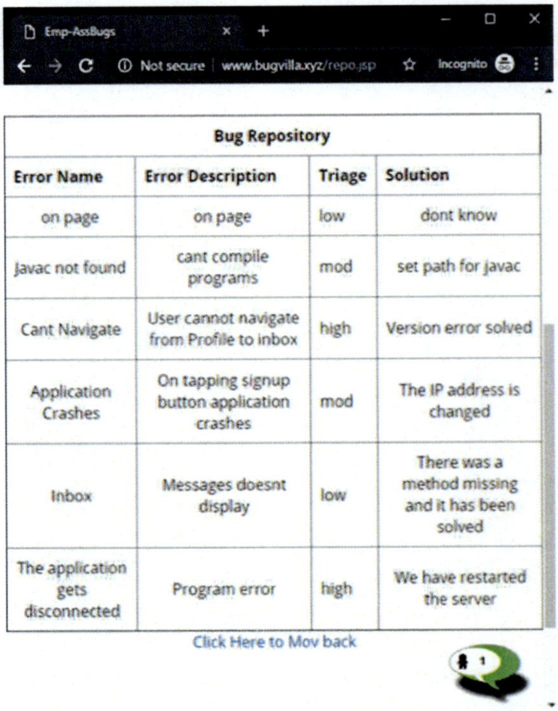

Bug Repository			
Error Name	Error Description	Triage	Solution
on page	on page	low	dont know
Javac not found	cant compile programs	mod	set path for javac
Cant Navigate	User cannot navigate from Profile to inbox	high	Version error solved
Application Crashes	On tapping signup button application crashes	mod	The IP address is changed
Inbox	Messages doesnt display	low	There was a method missing and it has been solved
The application gets disconnected	Program error	high	We have restarted the server

Fig. 10 Bug repository after updating

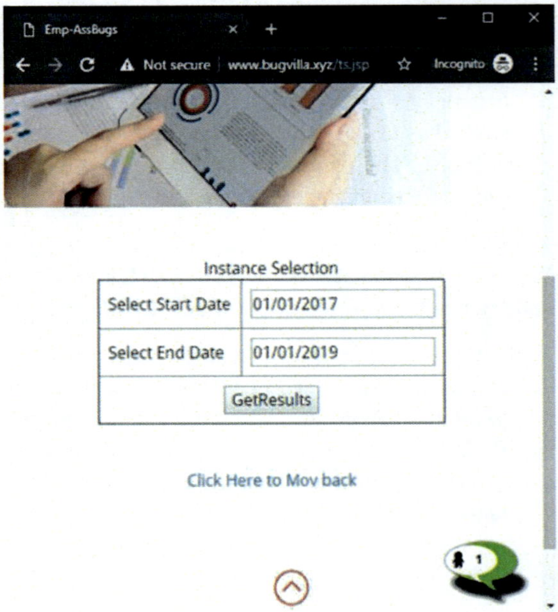

Fig. 11 Timestamp

Fig. 12 Instance selection

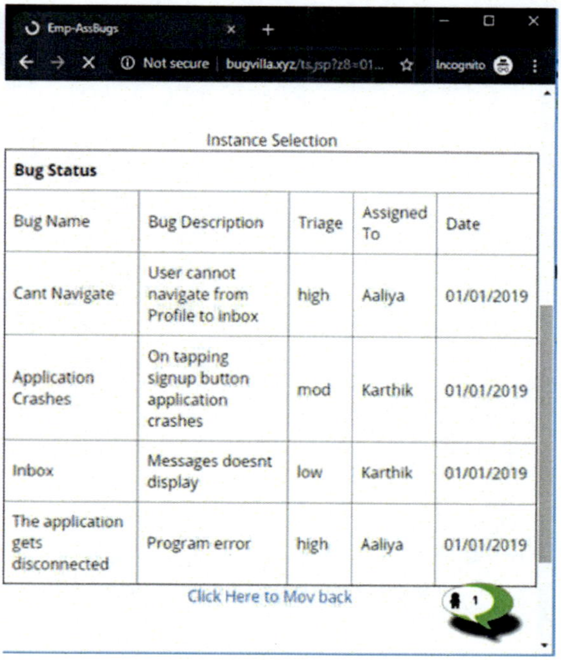

Bug Status				
Bug Name	Bug Description	Triage	Assigned To	Date
Cant Navigate	User cannot navigate from Profile to inbox	high	Aaliya	01/01/2019
Application Crashes	On tapping signup button application crashes	mod	Karthik	01/01/2019
Inbox	Messages doesnt display	low	Karthik	01/01/2019
The application gets disconnected	Program error	high	Aaliya	01/01/2019

Fig. 13 Chat login

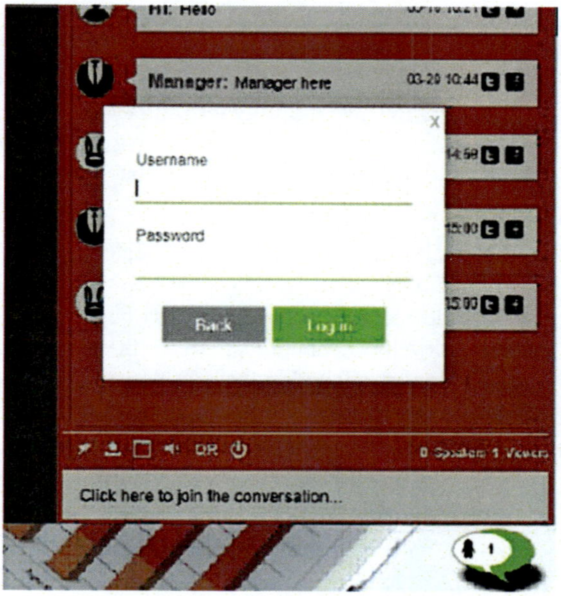

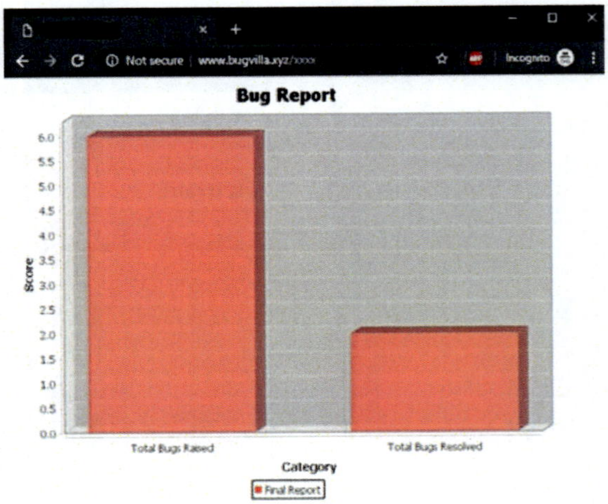

Fig. 14 Performance analysis

This bar chart works on the algorithm of instance selection algorithm (Algorithm 1), i.e. whenever a bug is raised the raised count increases to 1 and whenever a bug is resolved, the update count increases to 1.

Algorithm 1 : Instance Selection

```
Raised=0;    Update=0;

Resolved=0; Bug = 0;
If (bug>0)        //If any bug is raised Raised++;
If (Update>0)     //If any bug is resolved Resolved++;
```

These values get saved in the bug update table, as shown in Table 3.

WORKFLOW: The workflow of the proposed work is shown in Fig. 15. Firstly, a developer logins and assigns bugs to the relevant developers. Then an employee/developer should login to see the bugs assigned to him/her. After solving bugs, developers will update the bugs with their solutions. Meanwhile, manager and developers can chat by means of a group chat. After the bugs are updated, they get saved in the bug repository. The manager can check bug status, maintain a repository, and do performance analysis. Later, spreadsheets can be exported with bug info when all the bugs are solved or when performance analysis shows equal results.

Table 3 Bug update table

Raised	Resolved
6	2

Fig. 15 Workflow

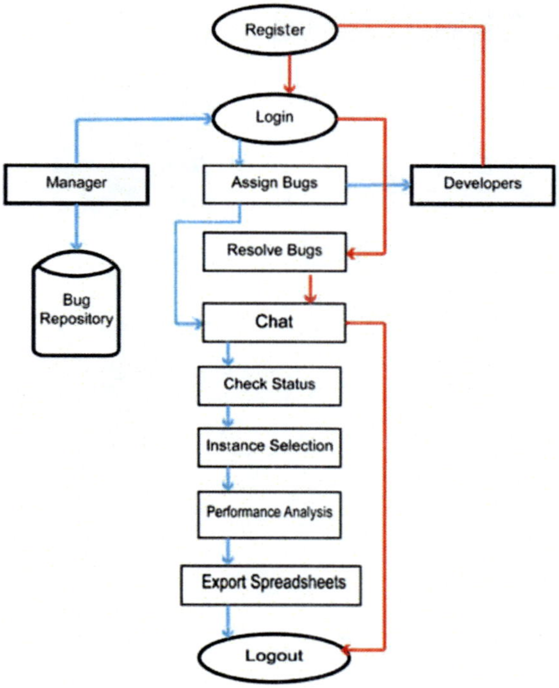

4 Results and Discussion

In our proposed work, bug reports are mapped to their relevant developers, tracked, and analyzed as shown in Fig. 16.

By mapping bug reports to their relevant developers, the developers can easily solve the bugs as they are expertise in that particular bug solving. For example, *android bug will be assigned to the android developer*. This helps in reducing the time taken for solving bugs. By tracking bugs, we can easily identify which type of bugs is occurred in a particular period of time from timestamp. With performance analysis, we can analyze that the bugs are fully solved or not. If fully solved, then we can export the final spreadsheets with bug info.

There is also group chat facility where a single manager can chat with four different developers at a time in one chat group as shown in Fig. 17. The number of persons in the chat can also be shown in the left status bar of the chat with their names.

Here the manager acts as a support to the developers to give instant solutions, and the developers can also chat with each other to solve a particular bug. This helps the project to get completed in time.

Here we can also generate two types of spreadsheets by exporting the database. The spreadsheet can be generated in a single click with all the updated bugs and it won't create any ambiguity. One spreadsheet will contain bug info and developer info

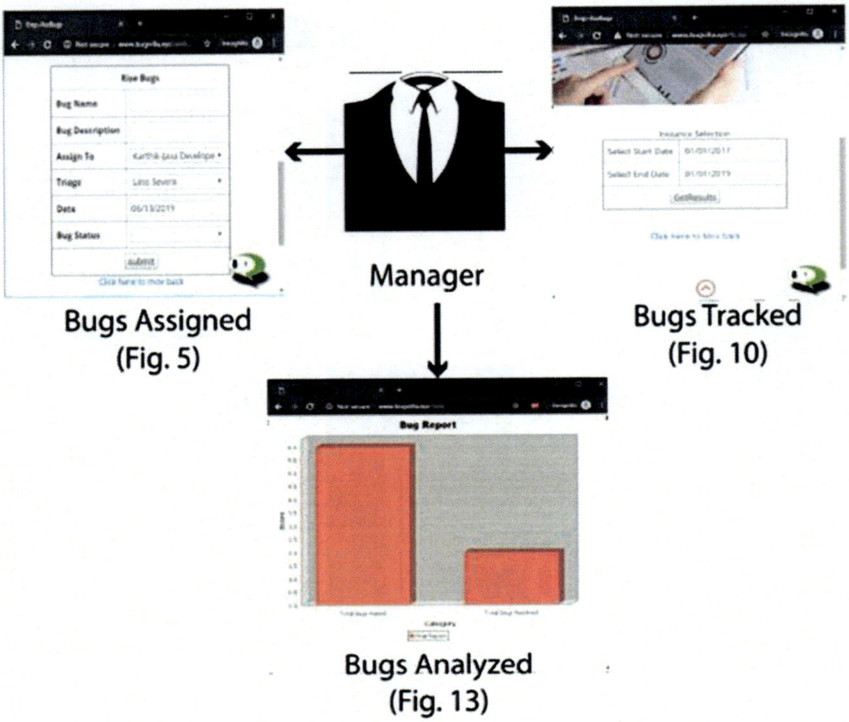

Fig. 16 Bug reports assigned (mapped), tracked, and analyzed

Fig. 17 Chat results

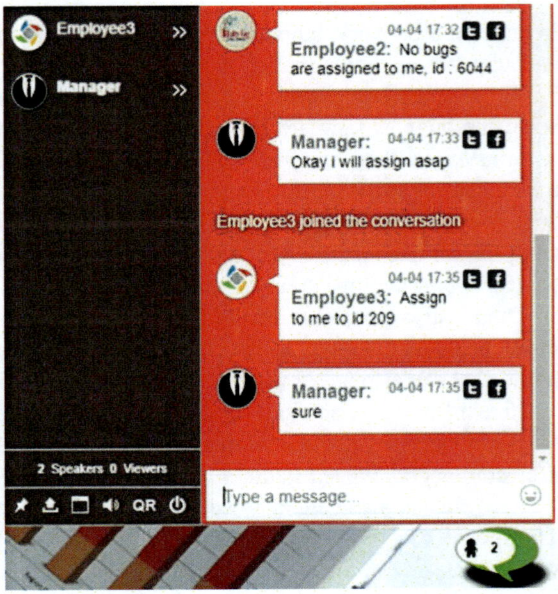

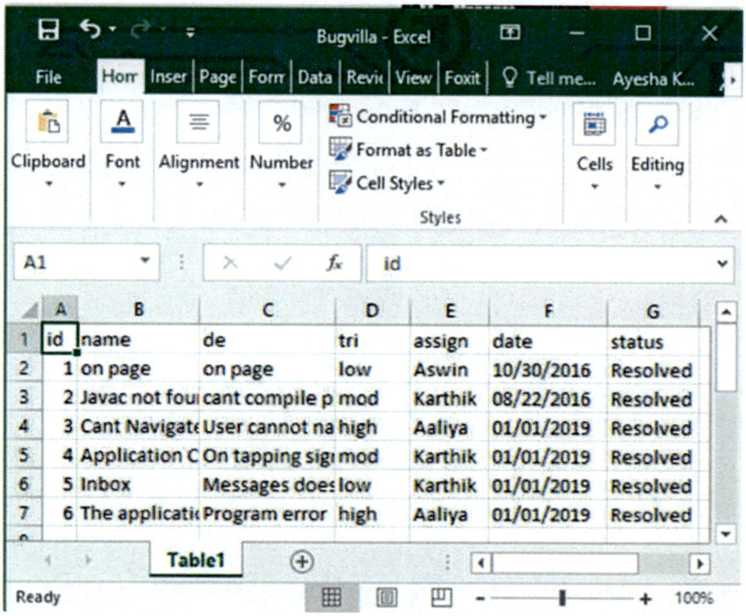

Fig. 18 Generated spreadsheet with bug info and developer info

as shown in Fig. 18. This spreadsheet contains bug id, bug name, bug description, bug triage, developer name, date, and status of the bug whether it is "RESOLVED" or "UNRESOLVED". With this easily, a report of bug reports can be preserved.

Another spreadsheet will contain only bug info with a solution as shown in Fig. 19. The bugs assigned to developers will be updated by the developers with a correct solution. And the updated bugs get saved in the repository. From the repository, we can generate/export the data of bugs in the form of spreadsheets. These spreadsheets will be 100% accurate without any ambiguity.

While working on a software project, the developers usually fill the bugs in a spreadsheet. By using bugvilla, there will be no ambiguity in the generated spreadsheets while compared to human-made spreadsheets. The information related to bugs will be accurate. The difference between the accuracy of human-made spreadsheets and bugvilla spreadsheets is shown in Fig. 20. So, bugvilla is a handy web tool for developers who are working on large-scale projects. The bugs with the most accurate info can be assigned to the developers.

Advantages of the proposed system are as follows:

a. Assigning bug reports to relevant developers makes it easy to solve bugs.
b. Bug expertise developers can solve the bugs in time.
c. This tool is easily accessible from anywhere.
d. Accurate spreadsheets can be exported.
e. Bug tracking becomes easy.

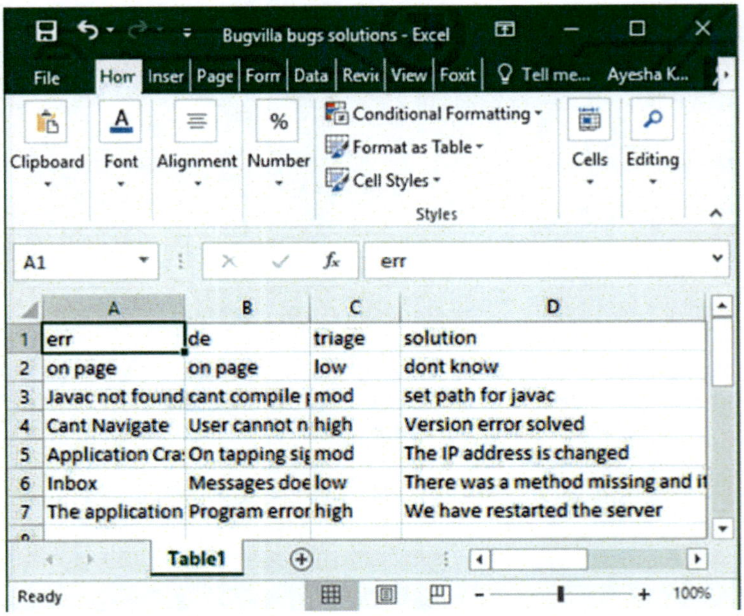

Fig. 19 Generated spreadsheet with bug info and solution

f. The manager can act as instant support for the developers by means of a group chat.
g. Inexpensive, companies just need to deploy it on to their domain.

5 Conclusion

The proposed system aims to map the bug reports to their relevant developers in order to reduce the time taken for solving a particular type of bug. Every developer gets assigned bugs in which he/she is expertise. So that, they can solve bugs in time and update on time. Tracking bug reports help to know the type of bugs that occurred in a particular time period from the timestamp. Analyzing bugs help to export spreadsheets with accurate info of the bugs. Since there is a way of communication by means of group chat, the manager acts as a support and can have a conversation with developers and solve the problems of login/other. Four developers can chat once and solve the bugs in time. This helps if the developers are located in different places. This tool is platform-independent and runs on any browser. The software companies just need to deploy the tool on to their Web site, and they can use it easily. This reduces the cost and time which the software companies spend huge in number.

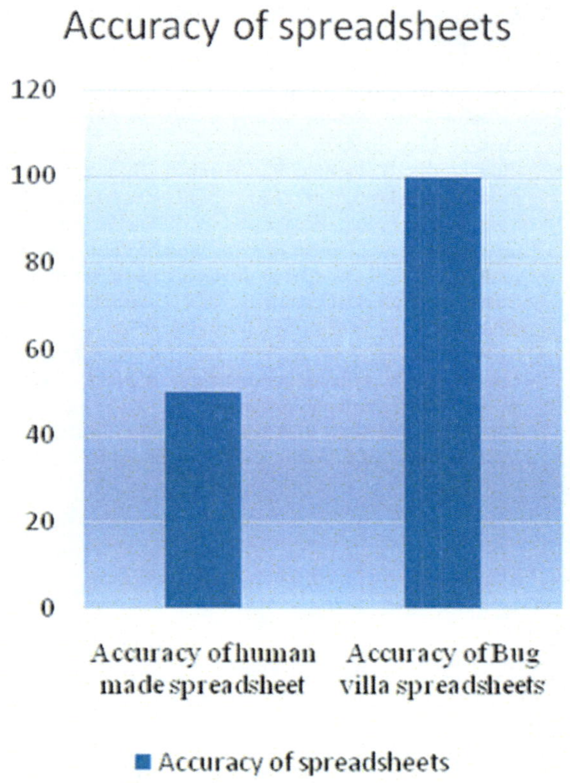

Fig. 20 Bar chart of accuracy

Future Work In future scope, additional features will be leveraged like companies can have their own profile with a manager and group of developers. Also, an option of creating different group chats will be developed.

References

1. Musale, K., & Gagare, G. (2015). Towards effective troubleshooting with data truncation. *International Journal of Advanced Research in Computer and Communication Engineering, 4*(11).
2. Xuan, B. J., Jiang, H., Hu, Y., Ren, Z., Zou, W., Luo, Z., & Wu, X. (2015). Towards effective bug triage with software data reduction techniques. *IEEE Transactions on Knowledge and Data Engineering, 27*(1).
3. Jadhava, A., & Jadhava, K. (2015). A survey on software data reduction techniques for effective bug triaging. *International Journal of Computer Science and Information Technologies, 6*(5), 4611–4612.
4. Pradeep, S., & Sharma, Y. K. (2019). A pragmatic evaluation of stress and performance testing for web applications. In *IEEE Proceedings of AICAI* (pp. 399–403), doi: 978-1-5386-9347-6.

5. Serrano, N., & Ciordia, I. (2005). Bugzilla, ITracker, and Other Bug Trackers. In *Published in IEEE Software*.
6. Avram, G., Sheehan, A., & Sullivan, D. K. (2008). Defect tracking systems in global software development-a work practice study. In *Proceedings of the 2008 International Workshop on Cooperative and Human Aspects of Software Engineering*, CHASE 2008, Leipzig, Germany, Tuesday, May 13, 2008.
7. Zimmermann, T., Premraj, R., Bettenburg, N., Just, S., Schroter, A., & Weiss, C. (2010). What makes a good bug report? *IEEE Transactions on Software Engineering, 36*(5).
8. Xuan, J., He, J., Ren, Z., & Luo, J. Y. Z. (2019). Automatic bug triage using semisupervised text classification. *Journal of Systems and Software,* March 2019.
9. John, A., & Gail, M. (2011). Reducing the effort of bug report triage: Recommenders for development-oriented decisions. *ACM Transactions on Software Engineering and Methodology, 20*(3).
10. Krisztian, B. Leif, A., & Maarten, D. R. (2006). Formal models for expert finding in enterprise corpora. In *29th Annual International ACM SIGIR Conference on Research and Development in Information Retrieval*, Washington.
11. Zou, W., Hu, Y., Xuan, J., & Jiang, H. (2011). Towards training set reduction for bug triage. In *IEEE 35th Annual Computer Software and Applications Conference*, NJ.

Annihilate Unsighted Dots in Operating and Aviation Using Artificial Intelligence

P. S. Rajakumar, S. Pradeep Sunkari, and G. Sreeram

Abstract With the new trends in technology, artificial intelligence is playing a major role in the advancement of machines and software's. Remote sensing, robot control, medical diagnosis, and electronic trading are the various activities that are performed with the use of structured algorithms using this technology. One such main demand in today's world is its use in the field of automated driving and aviation. In automated driving, we need to cover a lot of aspects and evaluate each situation to improve the performance if we are completely depending on this innovation. In the long term, automated driving has the potential to improve traffic safety and follow rules and regulations to the extent, thus, minimizing accident deaths and loss of public property. In the aviation scenario, a number of tasks are initiated using this technology such as customer queries, facial recognition of passengers, baggage check-in, autopilot mode, and aircraft fuel optimization. Also, a huge amount of data is collected and carefully examined so that certain steps in the future can be taken to ensure safe journey and quality travel. The flight's scheduling on various routes is also determined, and our idea here is to perform end-to-end passenger identification such as biometric impression is made essential after booking of the flight, and during check-in, it is again checked. Proper techniques and filtered data will be used to enhance this process, and sorted information will be provided to the pilots regarding the present condition of the aircraft and the weather conditions at certain heights or regions. AI demonstrates many unwitting consequences for the foreseeable future in the context of driving. For many years, this process is in the testing phase to overcome blind spots and human errors recognized from the data collected. Many of the driver assistance mechanics have reached a tipping point and are poised to take control in most situations, and researchers are excited regarding its function

P. S. Rajakumar
Department of CSE, Dr. M.G.R Educational and Research Institute, Chennai, TamilNadu, India

S. P. Sunkari (✉)
Associate Professor, Dept. of CSE, Bhoj Reddy Engineering College for Women, Saidabad, Telangana, India
e-mail: pradeep.sunkari87@gmail.com

G. Sreeram
Department of CSE, Koneru Lakshmaiah Education Foundation, Guntur, Andhra Pradesh, India

S. C. Satapathy et al. (eds.), *Intelligent System Design*, Advances in Intelligent Systems and Computing 1171, https://doi.org/10.1007/978-981-15-5400-1_57

to improve transportation and huge human benefits. The following theory will take assumptions and evaluate different calculations with computer software to upgrade the system and look thoroughly into restrictions and necessary steps to overcome these barriers will be initiated. In this research paper, we will elaborate and define various algorithms and diagrams through which we can achieve and reach a next level in automated driving and aviation.

Keywords Aviation · Artificial intelligence · Automated driving · Safety · Automation · Lidar

1 Introduction

1.1 Defining Artificial Intelligence

The field of computer science that improvises and demonstrates activities and tasks using machines defines the application of artificial intelligence. Any machine that takes certain steps to maximize its chance of successfully achieving a goal is termed as 'intelligent agent.' Many scientists approve of the fact that it is basically a system's ability to interpret data from different sources and apply techniques to resolve issues with minimum human interference. They have designed such tools which have the ability to understand human emotions and consider many aspects like social intelligence to help them in successful decision making. This term was first explained in 1956 but it had many limitations in its research such as reasoning, knowledge, representation, planning, learning, natural language processing, perception, and its ability to move and manipulate elements. This is a very procedural approach which includes computed intelligence, traditional symbolic AI, and statistical methods to perform a task. Artificial neural networks, search and mathematical optimization, probability and economics are the various tools used in AI. This field was founded to trigger human intelligence in machine form but it raised many philosophical arguments about the ethics and nature of the mind in the machine developed as this may cause a lot of problems and may prove to be a dangerous situation for the humans based on myths, fictions and may even create a risk of mass unemployment. In this era, AI methodology has experienced a recovery phase following consecutive advances in computer function, large amounts of data and detailed understanding, and it has become an essential part of the technology field; it helps in solving many difficult problems in computer science, software engineering, and operational research. AI involves a set of algorithms and structural procedure for its use and applications. An algorithm consists of a set of instructions which a machine can execute on demand. Many of these algorithms have the capability to organize data and learn from it using strategies and planned framework. For example, decision trees, nearest neighbor, and Bayesian network can evaluate and describe any amount of data provided to them in a definite span of time and will provide a set of calculations that are relevant and accurate. However, as stated earlier, many problems do arise that inhibits the overall research goal of the AI. Earlier, humans used a step-by-step algorithm to solve a puzzle or while making logical decision but with the increasing amount of data and incomplete information provided, and these algorithms were found insufficient, gave complex results, and were slow in function. After this, knowledge representation was also unconventional. It became very difficult to understand the result displayed, and the method followed by the machine is to solve a problem. It was now time-consuming and even created a sense of disbelief among the users to use such platforms. Objects, their properties, categories, and relation among elements were hard to evaluate. These limitations and backlogs in this technology were detected and then researchers worked on these problems to recover AI from the bad phase and developed such intelligent methods that were reliable and easy to understand. A proper hierarchical system was initiated and constructed to normalize the concept of AI, and it has now several applications in our day-to-day life.

1.2 Let Us Take a Quick Look at Some of the Basic Applications of AI in Today's World

1.2.1 Automated Driving (AD) and Aviation

Scientists explain the concept of driverless cars, drones, and air carriers by understanding and conducting many demonstrations, test drives, etc., and have suggested that AI could save many million dollars in this industry.

1.2.2 Healthcare Sector

This technology has proved to be a boon by assisting doctors as AI determines the accurate dosage of the immunity drug that should be given to the organ patients, and Microsoft has developed a system for the right treatment of cancer because more than 500 types of vaccines are involved in cancer treatment.

1.2.3 Finance and Economics, Kasisto and Moneystream

These were the programs developed in the 1980s to counter the banking fraud problems that include debit card cloning and use of unauthorized credit cards. Today, many banks use AI to operate, maintain passbook, and attend customer queries. Also, people get the benefits of Internet banking by just sitting at their homes.

1.2.4 Defense and Security

Governments in various nations have allowed the use of AI to improve their military and other defense operations. Robots, drones, and other devices are placed to keep track of the illegal movements of the suspicious elements in different regions of the world. This helps in maintaining peace and security in unstable regions.

So, we can say that AI in the long run will play a significant role in the upliftment of the society and overall development of the world. Let us discuss our main objective of the research paper and highlight all the issues involved.

2 Automated Driving

Over the years, the necessity to have a safe travel and relaxing journey has led to the advancement in the technology. As a result, AI has paved a path that takes us to a next level of driving with automatic control and handling. Automated driving (AD) is a brand new concept that emphasizes on the use of driverless cars and supervision of each component inside the vehicle. Due to the advancement in sensors, information systems, and microchips, it is also termed as dynamic driving. The community has also shown great interest when they were introduced to such cars at the auto expo held in Bangalore in 2015. Many international companies such as Tesla, Google, and Apple have designed their own driverless car that has triggered the need of such vehicles on a huge scale.

There are many elements involved in the functioning of self-driven cars, and it includes systematic developments in the regulation of brake system, lane changing, navigation, mapping, and detection of speed breakers and prevention of collision by maintaining an adequate gap from the vehicles around. This whole system is integrated with high-performance computer to form one such complex vehicle that undertakes the job. Many nations are now planning out to work on this technology and laying out such networks in which these vehicles can be accessed with ease. In the UK, the government has begun self-driving truck platoons in 2018 in which a single truck leads all the other non-driver trucks to their destination. Apart from this, in Germany, an institute is carrying out trials for firefighting operations based on AI.

Many experts are speculating and analyzing the subject of the evolution of AD, and for how long it will last but now it seems that it will soon be a part of our routine life. It is very exciting to note that companies having primarily specialization in web-based technologies and digital devices are now entering the automobile industry. Google named its driverless car as Waymo LLC, and it started its trial operation for taxi service in the USA in 2017. Waymo One was launched in 2018, and this was a commercial self-driven car service which uses eight types of sensors and microchips which are less expensive to manufacture and the prototype car uses several cameras connectivity installed inside the vehicle to perform the mapping function. Also, the information of the highways is already stored in the system regarding position of speed breakers and traffic lights so that there is minimum risk to life and property.

The company has designed such an effective system in which all the traffic rules that are to be strictly followed and adjustment of temperature inside the vehicle is completely automated. Company has issued various guidelines for customers before purchasing their vehicle. The components giving 360° views and gear shifting panel has long-life durability, and the hardware installed meets the requirements of the customer (Figs. 1, 2, and 3).

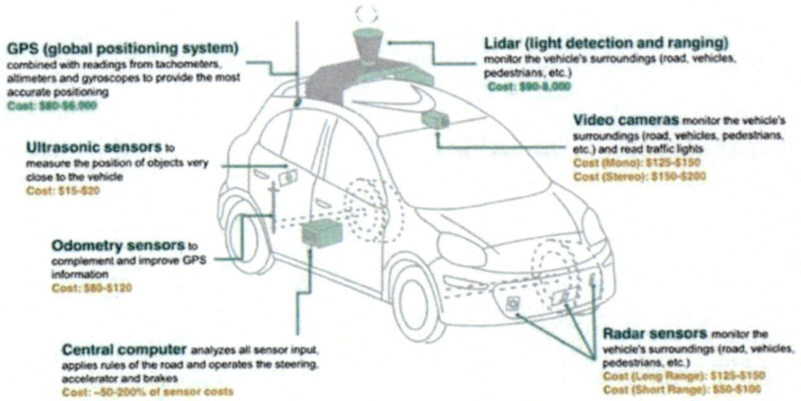

Fig. 1 Prototype car of Google

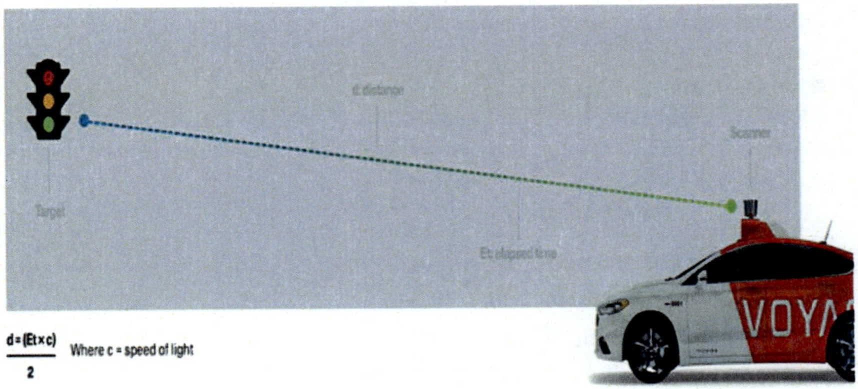

Fig. 2 Visualization of light

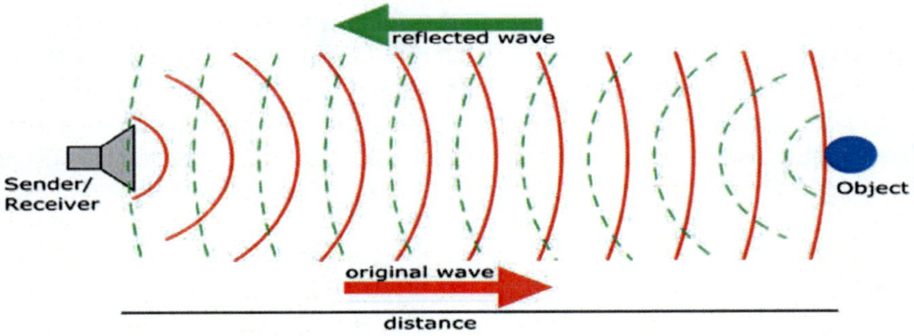

Fig. 3 Sender or receiver

2.1 SAE Levels of Automation

Society of Automotive Drivers (SAE) has considered six different levels of driving automation.

Level 0: No Driving Automation—It includes the performance of the dynamic driving test (DDT) in which all the real-time operations required to operate a vehicle are executed.

Level 1: Driver Assistance—Here, operational design domain (ODD) which includes the specific features in the system is controlled and assisted.

Level 2: Partial Driving Assistance—Lateral and longitudinal vehicle motion it sustained with the expectation that the driver completes object and event detection and response (OEDR) subtask.

Level 3: Conditional Driving—In any bad situation occurs, the human driver will be informed so that he can intervene and respond to a request when issued.

Level 4: High Driving—This level is an integration of all the above-mentioned levels of driving.

Level 5: Full Driving—All the practical work done by the automated driving system based on DDT is grouped and turned into data and then display the statistics.

2.2 Two-Way Matrix to Explain AD

Figure 4 describes the extent of the technological development of automated driving. Quadrant 1 depicts matureness but may not engage at all speeds and lane-changing patterns. Quadrant 2 shows developing strides like electronic stability and emergency braking. Quadrant 3 includes the rapidly evolving vehicles that operate on certain highways and under specific weather conditions. Quadrant 4 represents full automation in all of the time and is colored black to emphasize and shows that it is completely driverless. This diagram shows scheme for categorizing levels of automated driving from a human-centric perspective, ADAS, i.e., advanced driver assistance systems.

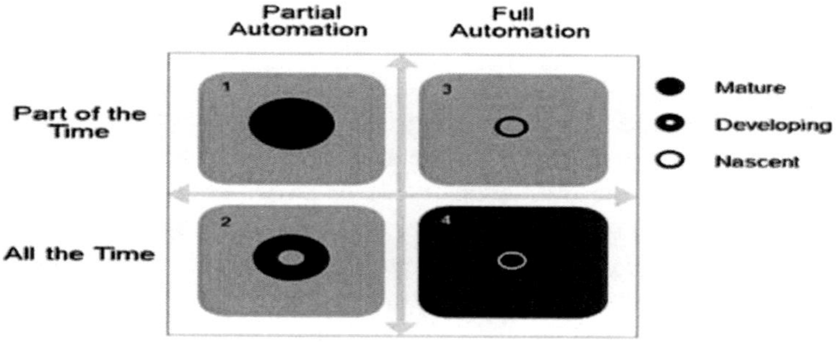

Fig. 4 Scheme for categorizing levels of AD from a human-centric perspective

2.3 *Examining Blind Spots and Their Prevention*

2.3.1 Human Error

Indiana University stated that human error was the definite cause of nearly 90% accidental cases based on a tri-level study conducted on 'accidental causes.' Other studies by different organizations reached similar conclusions. Now scholars involved in AD illustrated that this problem can be solved by using AI components in the vehicle such as replacing the driver by robotic control technologies, communication, computing, and sensors and thus reducing the damage by 90%. Computer Science and Artificial Intelligence Laboratory (CSAIL) has developed corner cameras for a better imaging system and to keep away the driving blind spots. This object can detect very tiny details of light and shadows in order to track the position of surrounding objects and then determining the projected speed and trajectory. This new system works in combination with various driver assistance hardware, and it suggests a future where AD will be able to react more quickly than a human being in a dander situation.

2.3.2 Investigation Error

Sometimes, during a crash, the on-duty officer files a false statement against the driver because of the less evidence found at the accidental spot. It is again a sub-part of human error. In such cases, AD components could serve better as they are capable of providing each and every minute information regarding the accidental spot. Aspects like road condition, rotational degree of the road curve, traffic lights timing, pothole's position are recorded in the system, and this data helps in the better investigation of the case. In long drive journeys, it is often noticed that driver gets into a sleepy mode and loses his consciousness that results in a vehicle crash. To remove this kind of blind spot, auto-mode is an option that comes into function where the driver can relax and leave the driving to the AI system to ensure a relaxing journey ahead.

2.3.3 Complications and Exceptions

As we all know, drunk driving and reckless driving can prove fatal but still people oversee their consequences. This technology is to assist people after they are returning to their homes from a party or event. The auto-mode could be turned ON for the passenger safety, and even instructions could be passed vocally to the system to maintain a definite limit of speed for a safer travel. We know that machines are fast in calculations, precise, and well-programmed to avoid risky driving, and the AI network provides a wide range of information to the system to estimate accurate goals of driving. So, this technique is a powerful source to help millions and billions of people.

3 Aviation Sector

The air transport and travel system of a nation comprise of the aviation sector, and it is the aircraft industry that is responsible for the movement of goods and services and people from one region to another. Over the decades, this sector has seen an enormous boost in its operations and huge number of passengers opts for this traveling service as it is fast and time-saving. This has led the aviation companies to increase the number of flights on certain important routes throughout the country. This industry has seen huge changes and has been revolutionized with the trending new technology. Earlier, aircrafts could carry nearly 50 people but now certain aircrafts can accommodate more than 500 people on a single journey. Let us talk about India's aviation industry so basically it is divided into two categories—military aviation and civil aviation. Our topic focuses and revolves around civil aviation and its operational activities. In India, there are various government-owned and privately based companies such as Air India, IndiGo, SpiceJet, and GoAir, and these are major carriers which provide air travel facilities throughout the world. Various domestic and international airports are set up across the country with the help of government aid and foreign investment.

3.1 How Technology Improves This Industry?

Air Operations Division (AOD) uses artificial intelligence to navigate the airline system, and it also conducts training sessions for new professionals based on this technology. Various management ideas, technical decisions, and support systems are taken into consideration that is essential for the proper functioning of the aircraft. All over the world, there is a huge network of air carriers and this leads to the generation of big database. This data is collected, restored, then evaluated, and analyzed using AI technology so that appropriate steps could be performed in the coming future to overcome difficult situations. These computer systems are programmed to regulate the air traffic and give them proper instructions to manage and perform operations inside the cockpit. Other services like passenger check-in, luggage handling, barcodes are also assisted by AI.

3.2 Applications of AI in the Aviation Field and How to Eliminate the Errors

3.2.1 Intelligent Processor

Earlier, a lot of paperwork was done in designing, maintaining, and planning the movement of the aircraft. It took long hours to come up with a solid plan in order to regulate a flight session. Now, with the introduction of AI, an expert system is installed in the aircraft which assists the pilot and other crew members of the plane. An interactive fault diagnosis and isolation system are set up which provide sorted information and expert advice to the pilot in case any mishap occurs. Also, a strong communication link gets established between the pilot and air traffic regulators at the airport and takeoff and landing of the aircraft gets an assistance with proper timing. Customer satisfaction has reached a next level on the positive side, and the business is now growing with the implementation of AI systems.

3.2.2 Passenger and Baggage

Identification of the passenger and his luggage were a concern of security in earlier days because it was easy to smuggle illegal things and travel on false identity. Now new devices such as barcodes, speech recognition, biometrics are installed at the airports to ensure better safety norms. Entry in the airport premises is only permitted after verification by the machines, and specific barcodes are attached to the luggage and then the luggage is moved to surveillance for security purposes. At the airport, there are different sections and departments where only the authorized people can enter. For this entry purpose, scanning eye retina machines and other kind of biometric machines are used to avoid any of breakage in the system.

3.2.3 Autopilot Mode

Busy schedule and long journeys in adverse weather conditions make the travel worse and dangerous for the pilots to carry on with their work. There are certain other factors that affect the safe passage of an aircraft. AI system demonstrated the use of autopilot mode in its experiment sessions. Later on, this mode came into use and was adopted by many air companies to assist the pilot and departmental people. This mode is designed to regulate and control the trajectory of an aircraft by assisting a human operator and not completely replacing him. Information regarding the bad weather conditions or any glitch in the hardware engines is directly detected by the AI system, and then suitable guideline is provided to the pilot so that he can control the situation in a better way. In the long hour journey, human expert can relax for some time and he can rely on the autopilot mode if the weather condition is extremely fine. The AI technology has reduced the workload on human beings and has allowed the industry to gain more profit in several ways.

4 Conclusion

Artificial intelligence is evolving at an accelerated pace, and this research paper carefully examines all the advantages and disadvantages that are involved in this concept. Many obstacles and hindrances have shown up their immense effect on life and economy. To keep them out of the way, several experiments and tests are carried out in different situations so that safety which is a major concern in the traveling part can be highlighted. Also, the government has issued various guidelines and notices to the companies that manufacture automated cars to follow strict rules, and they are not allowed to compromise with human safety at any cost.

References

1. Google. 2016. Google self-driving car project monthly report. February, 2016. Online at https://static.googleusercontent.com/media/www.google.com/en//selfdrivingcar/files/reports/report-0216.pdf.
2. IEEE. 2014. You won't need a driver's license by 2040. IEEE.org, online at http://sitesieee.org/itss/2014/09/15/you-wont-need-a-drivers-license-by-2040/.
3. SAE. 2016. *Taxonomy and definitions for terms related to driving automation systems for on-road motor vehicles* (p. 30). SAE International. http://standards.sae.org/j3016_201609.
4. Pradeep, S., & Sharma, Y. K. (2019). Storing live sensor data to the platforms of internet of things (IoT) using Arduino and associated microchips. In *Proceedings of 3rd International Conference on Computational Intelligence and Informatics*, December 2018.
5. VDA. (2015). *Automation: from driver assistance systems to automated driving*. Berlin: VDA Magazine. Verband der Automobiliinsdustrie e.V.
6. Schoettle, B., & Sivak, M. (2014). *Public opinion about self-driving vehicles in China, India, Japan, the US, the UK, and Australia*. Ann Arbor, Michigan: The University of Michigan, Transportation Research Institute.
7. Schoettle, B., & Sivak, M. (2014). *A survey of public opinion about autonomous and selfdriving vehicles in the US, the UK, and Australia*. Ann Arbor, Michigan: University of Michigan, Transportation Research Institute.
8. Salmon, P. M., Walker, G. H., & Stanton, N. A. (2016). Pilot error versus sociotechnical systems failure: a distributed situation awareness analysis of Air France 447. *Theoretical Issues in Ergonomics Science, 17*(1), 64–79.
9. Pradeep, S., & Sharma, Y. K. (2019). Effectual secured approach for internet of things with fog computing and mobile cloud architecture using IFogSim, WEC. In *IAENG International Conference*.
10. Noy, I. Y. et al. A (2018). Automated driving: Safety blind spots. *Safety Science, 102*, 68–78. www.elsevier.com/locate/safety.

Miss Rate Estimation (MRE) an Novel Approach Toward L2 Cache Partitioning Algorithm's for Multicore System

Pallavi Joshi, M. V. Rathnamma, K. Srujan Raju, and Urmila Pawar

Abstract The concept of cache memory partitioning is handled and implemented few times earlier. Partitioning is essential in cache memory to attain still higher speed while searching and transaction of data, as L1 cache is comparatively small in size, memory interference in L1 cache is very low data, and instruction were partitioned easily. For effective use of multicore processor, cache partitioning or cache sharing is important. However, there are some limitations identified from many studies, like increase in simulation time, because of inter thread memory interfering, and inaccuracies in simulation. This paper explains a cache partitioning method for multicore systems using miss rate estimation. It reduces the interference of cache among concurrently executing process threads. The cache partitioning algorithm, using our novel approach, calculates the miss rate characteristics of individual threads at the run time, then dynamically partitions and distributes the cache memory among the threads, which are running concurrently, and thus improves the performance of multicore systems. The paper is divided into several sections; first, it explains all the basics related to the domain; secondly, it gives the details about all the papers which have been referred with a comparison. Finally, it focuses on future scope.

Keywords Cache · Performance · Benchmark · Multiprocessor · Fairness · Interference · Miss rate

---

P. Joshi · U. Pawar
Department of Computer Engineering, Pune Institute of Computer Technology, Pune, India
e-mail: pallavee.j3@gmail.com

U. Pawar
e-mail: pawarurmila123@gmail.com

M. V. Rathnamma
Department of Computer Science Engineering, KSRM College of Engineering, YSR Kadapa, Andhra Pradesh, India
e-mail: rathnamma@ksrmce.ac.in

K. Srujan Raju (✉)
Department of Computer Science Engineering, CMR Technical Campus, Hyderabad, Telangana, India
e-mail: ksrujanraju@gmail.com

S. C. Satapathy et al. (eds.), *Intelligent System Design*, Advances in Intelligent Systems and Computing 1171, https://doi.org/10.1007/978-981-15-5400-1_58

1 Introduction

1.1 Multicore System

A multicore processor is an individual single computing entity with (multiple) more than two independent actual central processing units (CPUs) which are required for reading and executing program instructions. These instructions are normal CPU instructions like addition, subtraction, moving data, or branching instructions, but many cores run multiple instructions simultaneously, thus increasing the overall speed of the program enabling parallel computing. Manufacturers typically integrate multiple processing cores on to single integrated die (called as a chip multiprocessor or CMP) or multiple dies in one single chip package. Originally, processors were developed with one core (single core). A dual-core processors work with two cores (e.g., AMD Phenom II X2, Intel Core Duo), a quad-core processor has four processing cores within (e.g., AMD Phenom II X4, Intel's quad-core processors) [1, 2].

Designers may couple cores in a multicore device loosely or tightly. For instance, all present cores are not bound to share the cache memory, but they might implement message passing (MP) algorithms or shared memory inter core communication methods. Common network topologies to interconnect cores out of many include bus topology, ring topology, mesh topology (two dimensional), and crossbar, wherein homogeneous multicore systems include only identical cores, while in heterogeneous multicore systems, cores are not identical. Multicore processors have wide range of application domains which include embedded, general purpose, network, graphics, and digital signal processing. [3–5]. Figure 1 represents generic dual core processor with CPU levels.

1.2 Real-Time System

Real-time systems are the systems which need to execute the instruction and must give the response within predefined time. Real time includes several domains of computer sciences like defense, space system, network multimedia system, embedded automotive electronics, etc. Here, the correctness of system behavior is not based on logical results of the computation, but also is dependent on the physical instant at which the results are generated [6].

Classification of real-time systems:

- Hard real time versus soft real time
- Fail-safe versus fail-operational
- Guaranteed timeliness versus best effort
- Resource adequate versus resource inadequate
- Event triggered versus time triggered.

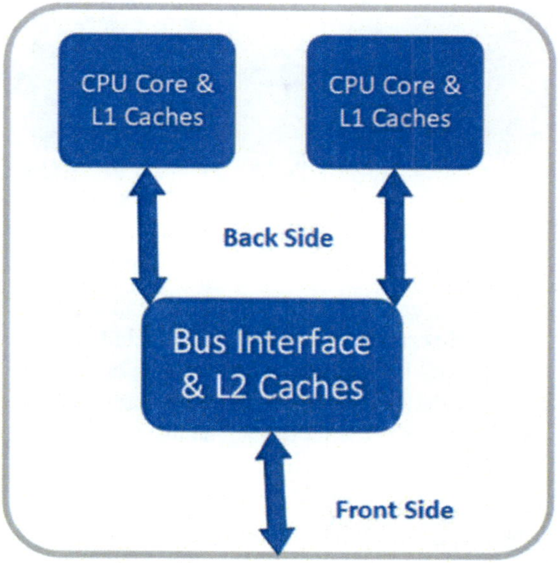

Fig. 1 A generic dual core processor with CPU local level cache and shared on die level cache

A hard RTS is the one in which response should be obtained before the deadline is actually meet [7, 8].

1.3 Cache Memory

Cache memory is also called as random-access memory, made up of SRAM (type of RAM) computer microprocessor can access SRAM memory more quickly than any other regular RAM memory. While program executes, at run time when the microprocessor executes the instruction, and is in need of the data, it first looks into the cache memory and checks for the presence of the data there. If it finds the data there (from the previous readings), it saves lot of time-consuming efforts to read it from larger memory. Cache is very small, but a fast storage device that stores the data (operands and instructions) that might be required at the time of execution by the CPU. Figure 2 shows transfer data in hardware memories

- It is a very small amount of the fast memory
- It is present between normal, main memory, and the CPU.

Because of the increasing distance between the CPU (CPU registers) and the main memory, a small SRAM memory, called as L1 cache memory, is inserted there. We can use L1 cache memory as fast as we can access the CPU registers, i.e., in 1 or 2 clock cycles. Because of even more increasing distance between CPU and main memory, additional cache memory, i.e., L2 cache memory, is inserted between L1 cache and the main memory, which requires few more clock cycles than L1 cache

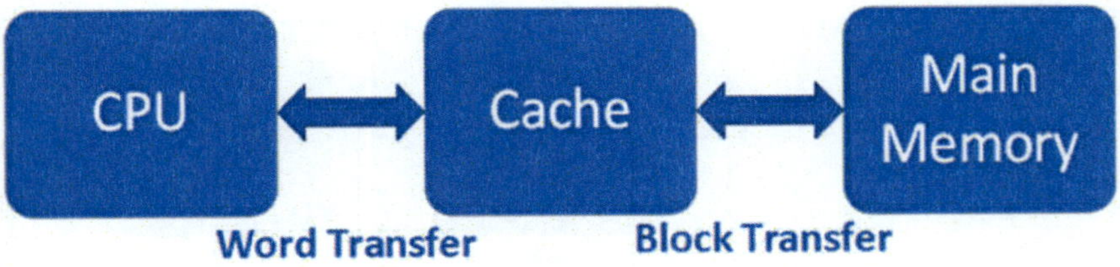

Fig. 2 Data transfer in memory

memory. L2 cache memory is linked with memory bus or to its own cache bus. Some high-performance systems also include extra (additional) L3 cache memory, which resides between the L2 cache memory and the main memory. It is arranged differently, having same principle. The cache is placed both physically closer and logically closer to the CPU than the main memory [9, 10]. Figure 3 shows memory hierarchy of early computers.

- CPU registers
- DRAM memory
- Disk storage.

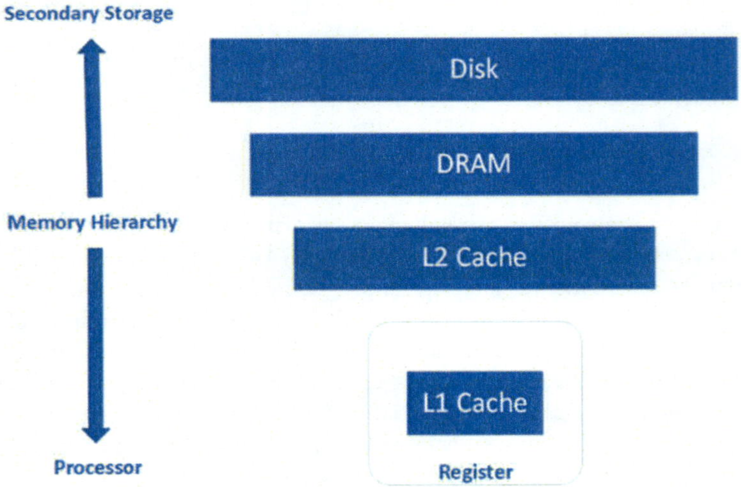

Fig. 3 A generic memory hierarchy

2 Objectives

The objective of proposed paper is to develop such an algorithm which will overcome the drawbacks of the earlier project and cope up with the challenges faced with the L2 layer cache partitioning.

The proposed system aims at:

a. To study different cache partitioning systems for multicore systems.
b. Development of efficient cache partitioning algorithm for L2 layer.
c. Checking the performance of the algorithm.
d. Relative study of the performance of available algorithm with the new one.

3 Comparative Study of Different Partitioning Algorithm

3.1 Kpart Present

- KPart also called as a hybrid cache partitioning and sharing technique overcomes the drawbacks of way-partitioning algorithm and allows significant performance on the current systems.
- First step of KPart is to group all the applications into clusters, and after that, partition the cache memory among these clusters.
- To build clusters, KPart depends on a new approach for estimating the performance loss of an application which suffers when sharing a partition.
- KPart then spontaneously picks the number of clusters, and KPart automatically chooses the number of clusters, keeping intact the advantages of isolation of way-partitioning technique with its potential performance effect.
- Kpart makes use complete summarizing information to make this decision. This information can be assembled offline or online with minimum efforts using a novel profiling (summarizing) mechanism [11].

3.2 Fair Cache Partitioning Algorithm

- The algorithm is developed to deal with three problems.
- Out of the three, the first and the important problem is thread starvation, and it happens if one thread is unsuccessful in getting sufficient cache space required for making satisfactory forward progress.
- However, the second problem is inversing the priority, where a less priority thread gets higher forward progress then the higher priority thread should get, irrespective of the attempt by the OS to provide more time-slices to the higher priority thread.

- The last problem is that the forward progress rate of any thread is hugely dependent on threads mix in co-schedule. Thus, we cannot predict or characteristic the forward progress rate, and also, it makes the system behavior unpredictable.
- To solve this problem, ideally the hardware should provide fair caching, which means the technique which promises that the impact (effect) of sharing cache memory is unvarying for all the co-scheduled threads. With fairness in cache sharing, the operating system can mostly expect priority-based time-slice assignment to work as effectively as in time sharing uniprocessor system [12].

The fair cache algorithm offers the solution

- Initially, it proposes and assesses five cache memory fairness metrics, which measures the grade (degree) of how fair the cache sharing is and execution-time fairness can be termed as how evenly (unvaryingly) the execution time of co-scheduled threads are changed.
- Secondly, we have both, static and dynamic L2 cache partitioning algorithm which works on optimization of fairness. Among them, the dynamic partitioning algorithm is very easy to understand and implement, requires very little or negligible profiling, also has low overhead, but at the same time, does not restrict the cache replacement algorithm to LRU technique. The static algorithm, although requires the cache memory to preserve the information of LRU stack, can help the thread scheduler of operating system to neglect cache thrashing [12, 13].

3.3 The Dynamic Cache Partitioning Algorithm

- In the dynamic cache partitioning, algorithm can be used for applying set-associative cache at any partition granularity.
- Furthermore, in this scheme, threads are allowed to have overlapping partitions that will provide greater degree of liberty when partitioning caches with very low associativity. Simulation results show following observations.
- Comparative improvement in the L2 hit rate of up to 40.05% over normally generated by standard least recently used (LRU) replacement policy, and IPC improvements were observed of up to 17%.
- Later, a data-centered method for partitioning the cache in embedded real-time database system is followed where it allows various tasks to have shared locking partition in cache memory.
  - The hard real tasks will have their separate partitions, and in turn, they can perform higher predictability.
  - However, on the similar grounds, a shared non-locking partition is reserved for some soft real-time tasks. Like this, performance improves based on the data that is frequently used by many tasks in the system [14].

3.4 Static and Dynamic Algorithm for L1 Cache Partitioning

- In this scheme, first, the static cache partitioning algorithm is presented. This algorithm is used for effective partition assignment of cache. It supports many types of memory models, which includes multiple layers of caches partitions. To exemplify the effectiveness of the static cache partitioning algorithm, we have used few ranges of benchmarking units.
- Dynamic cache partitioning is an algorithm to measure the effectiveness of cache partitioning for the given application without the need of time-consuming simulation. There, performance estimates are in comparison with the previous algorithms' simulation results. Experiments show that performance of this algorithm is better as compared to the previous algorithm.
- This algorithm reduces the cache interference between simultaneously executing benchmark application.
- The same algorithm is inefficient to reduce the cache interference in L2 level [15, 16].

3.5 Survey Techniques for Cache Partitioning

- This paper discusses in details on the background of cache partitioning basics.
- Then, it classifies cache partitioning techniques (CPT's):
  - CP based on characteristic and goal
  - CP based on (cache partitioning) CP algorithms and implementation
  - CP based on evaluation, platform, and approach
  - CP based on other features
- Discusses granularities of CP:
  - Way-based cache partitioning technique
  - Set-based cache partitioning technique
  - Block-based cache partitioning technique
- Also discusses various strategies for cache partitioning:
  - Static cache partitioning technique
  - Pseudo partitioning (CPT)
  - Cache statistics-based cache partitioning technique

    IPC based cache partitioning technique
    Bandwidth-based cache partitioning technique [17, 18].

4 Simulators

4.1 MARSSx86 Micro-Architectural and System Simulator for x-86 Based System

- MARSS is a tool for cycle accurate full system simulation of the x86-64 architecture, specifically for multicore implementations.

4.2 Simics

- This is multiprocessor full system simulator.
- Here, processes are simulated at the instruction-set level.
- Supports various ISAs for, e.g., SPARC,x-86,PPC,ARM, etc.

4.3 Sniper Multicore Simulator

- Sniper multicore simulator is a next-generation, parallel, high-speed, and accurate x86 simulator. This kind of multicore simulator is based on interval core model and the graphite simulation infrastructure, and allows for fast and accurate simulation.

4.4 SESC

- Super E-scalar simulator is a microarchitecture level simulator developed primarily by i-acoma research group at UIUC and various groups at other universities.
- It is based on MIPS ISA, and SESC is an event-driven simulator.
- This is highly customizable and modular—includes ability to simulate multi-core/multiprocessor systems.
- SESC is written in C++ programming language.
- It is claimed to be very fast.
- SPEC and SPLASH2 binaries compiled for SESC should be available within CSL.
- To compile your own programs, you need to download sescutils from Web site.
- Setup to compile your own programs is a little more complicated.

The algorithm to be developed for the proposed project would be in C language and will make use of SESC simulator [4].

5 Miss Rate Estimation (MRE)-Based L2 Cache Partition: Proposed System

The proposed system is designed based on miss rate estimation (MRE)-based cache partitioning. Figure 4 shows the proposed diagram. MRE system will be computed by different cache partitioning algorithm, and on the basis of miss rate, cache is partitioned for the computation. We are working in Hadoop environment for experimental analysis of the proposed system.

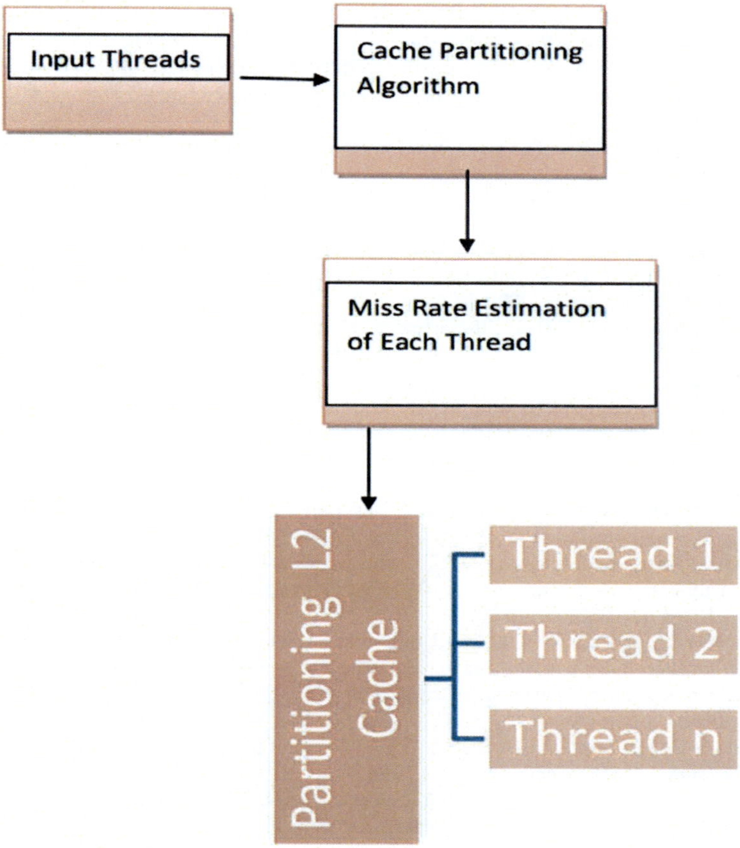

Fig. 4 Proposed architecture MRE

6 Conclusion

This paper will be working on the idea of designing an algorithm which will be simulated on SESC simulator for L2 cache partitioning, and a comparative study will be made between already existing algorithm to eliminate their drawbacks.

References

1. Xie, Y., & Loh, G. H. (2009). PIPP: promotion/insertion pseudo-partitioning of multi-core shared caches. *ACM SIGARCH Computer Architecture News, 37*(3), 174–183.
2. Tam, D., Azimi, R., Soares, L., & Stumm, M. (2007). Managing shared L2 caches on multicore systems in software. In *Workshop on the Interaction Between Operating Systems and Computer Architecture* (pp. 26–33), June 2007.
3. Wang, W., Mishra, P., & Ranka, S. (2011). Dynamic cache reconfiguration and partitioning for energy optimization in real-time multi-core systems. In *2011 48th ACM/EDAC/IEEE Design Automation Conference (DAC)* (pp. 948–953). IEEE, June 5, 2011.
4. Lin, J., Lu, Q., Ding, X., Zhang, Z., Zhang, X., Sadayappan, P. (2008). Gaining insights into multicore cache partitioning: Bridging the gap between simulation and real systems. In *2008 IEEE 14th International Symposium on High Performance Computer Architecture* (pp. 367–378). IEEE, February 16, 2008.
5. Yu, C., & Petrov, P. (2010). Off-chip memory bandwidth minimization through cache partitioning for multi-core platforms. In *Design Automation Conference* (pp. 132–137). IEEE, June 13, 2010.
6. Plazar, S., Lokuciejewski, P., & Marwedel, P. (2009). WCET-aware software based cache partitioning for multi-task real-time systems. In *9th International Workshop on Worst-Case Execution Time Analysis (WCET'09)*. Schloss Dagstuhl-Leibniz-ZentrumfürInformatik.
7. Liedtke, J., Hartig, H., & Hohmuth, M. (1997). OS-controlled cache predictability for real-time systems. In *Proceedings Third IEEE Real-Time Technology and Applications Symposium* (pp. 213–224). IEEE, June 9, 1997.
8. Fu, X., Kabir, K., Wang, X. (2011). Cache-aware utilization control for energy efficiency in multi-core real-time systems. In *2011 23rd Euromicro Conference on Real-Time Systems* (pp. 102–111). IEEE, July 5, 2011.
9. Qureshi, M. K, Patt, Y. N. (2006). Utility-based cache partitioning: A low-overhead, high-performance, runtime mechanism to partition shared caches. In *2006 39th Annual IEEE/ACM International Symposium on Microarchitecture (MICRO'06)* (pp. 423–432). IEEE, December 9, 2006.
10. Suh, G. E., Rudolph, L., & Devadas, S. (2004). Dynamic partitioning of shared cache memory. *The Journal of Supercomputing, 28*(1), 7–26.
11. El-Sayed, N., Mukkara, A., Tsai, P. A., Kasture, H., Ma, X., & Sanchez, D. (2018). KPart: A hybrid cache partitioning-sharing technique for commodity multicores. In *2018 IEEE International Symposium on High Performance Computer Architecture (HPCA)* (pp. 104–117). IEEE, February 24, 2018.
12. Nesbit, K. J, Aggarwal, N., Laudon, J., & Smith, J. E. (2006). Fair queuing memory systems. In *Proceedings of the 39th Annual IEEE/ACM international Symposium on Microarchitecture* (pp. 208–222). IEEE Computer Society, December 9, 2006.
13. Kim, S., Chandra, D., & Solihin, Y. (2004). Fair cache sharing and partitioning in a chip multiprocessor architecture. In *Proceedings. 13th International Conference on Parallel Architecture and Compilation Techniques, PACT* (pp. 111–122). IEEE, October 3, 2004.

14. Kang, S., & Dean, A. G. (2010). DARTS: Techniques and tools for predictably fast memory using integrated data allocation and real-time task scheduling. In *2010 16th IEEE Real-Time and Embedded Technology and Applications Symposium* (pp. 333–342). IEEE, April 12, 2010.
15. Chang, J., & Sohi, G. S. (2014). Cooperative cache partitioning for chip multiprocessors. In *ACM International Conference on Supercomputing 25th Anniversary* (pp. 402–412). ACM, June 10, 2014.
16. Wei, H., Tianzhou, C., Qingsong, S., & Ning, J. (2008). A data centered approach for cache partitioning in embedded real-time database system. *WSEAS Transactions on Computers, 7*(4), 140–146.
17. Pei-Xiang, Y., Jiang, J., Xian-Ju, Y., & Min-Xuan, Z. Partitioning mechanism based on dynamic Allocation of Data entries for chip multiprocessors. In *5th International Conference on Computer Sciences and Convergence Information Technology*.
18. Mittal, S. (2017). A survey of techniques for cache partitioning in multicore processors. *ACM Computing Surveys (CSUR), 50*(2), 27.

Hierarchical Agglomerative Based Iterative Fuzzy Clustering to Impute Missing Values in Health Datasets

Ravindar Mogili and G. Narsimha

Abstract As most of the real-world datasets contain missing values (MVs) and prediction accuracy is influenced by them, they need to be handled. The popular approaches to deal with MVs are imputation methods. In this paper, a new imputation approach called Agglomerative based Iterative Fuzzy Clustering is proposed. This method imputes the MVs by systematic calculations of fuzzy c-means clusters. The initial seed points for fuzzy c-means clusters are chosen from Agglomerative clusters and number of fuzzy clusters is decided by Fuzzy Silhouette Index. The accuracy of the proposed method with other widely used cluster-based imputation methods on health data with different percentage of missing entries is estimated in terms of Root Means Square error. The experiment results show that the proposed method provides better performance than other cluster-based imputation methods.

Keywords Imputation · Missing values · Fuzzy C-mean cluster · Agglomerative cluster · Fuzzy silhouette index · Health dataset

1 Introduction

Incorrect handling of missing values (MVs) in dataset may mislead the results of data mining tasks, thereby resulting in false analysis and incorrect decision-making. Specifically, missing values in health data sets can lead to the wrong diagnosis and thus to incorrect medical treatment, which is a very serious potential hazard. So, handling missing values is a very crucial step in data mining. There exist several techniques to handle missing values; one of the best among them is the imputation technique. An imputation technique imputes MVs in a data sample with the best estimate from the most similar patterns derived from other data samples of the dataset.

R. Mogili (✉)
Department of CSE, Jyothishmathi Institute of Technology & Science, Karimnagar, India
e-mail: ravindarmogili@live.com

G. Narsimha
Department of CSE, Jawaharlal Nehru Technological University, Hyderabad, India

S. C. Satapathy et al. (eds.), *Intelligent System Design*, Advances in Intelligent Systems and Computing 1171, https://doi.org/10.1007/978-981-15-5400-1_59

Among these imputation methods, cluster-based imputation methods are considered as better tools because of their ability to find similar patterns.

When a person is suffering from disease, his/her health state may not be in 100% unhealthy state. But it maybe 90% unhealthy state and 10% health state. As fuzziness exists in the health states, fuzzy clustering is more appropriate for deriving similar patterns to impute missing values in health data. Even though fuzzy clustering is preferred as one of the best choices for imputation, its accuracy depends on right number of clusters to be formed and also on initial seed points to start fuzzy clustering. In order to handle this, we used Fuzzy Silhouette Index value to find number of clusters to be formed and Hierarchical Agglomerative Clustering is used to find initial seed points to start fuzzy clustering.

2 Related Work and Concepts

One of the simplest imputation methods is mean/mode imputation [1, 2]. In this method, an attribute with numerical missing value is imputed by mean of that attribute; categorical attribute value is imputed by mode of that attribute. One of the major disadvantages of this method is not considering any dependency in the dataset while imputing missing values.

K-Nearest Neighbor Imputation (KNNI) method [1, 3] considers k most similar neighbor values in dataset to impute every missing value. While imputation, 'k' neighbors mean value is chosen for numerical attributes, and for categorical attributes mode value is used. One of the major disadvantages of KNNI is high dependency on k value. Usually, k value is chosen between 5 and 10 and higher k leads to negative results on the imputation performance.

In the KNNI method, every k neighbor is considered equally important while imputation. But in reality, the first neighbor (closest) is more similar than last neighbor (farthest) of missing value. Therefore, Weighted K-Nearest Neighbor Imputation method (WKNNI) [4] is similar to KNNI but instead of considering mean, weighted mean is chosen for numerical data and the mode for nominal data. The weights are reciprocally proportional to the neighbor distance such that the first neighbor has the highest weight value and the kth neighbor has the lowest weight value. In reality, every missing value may not exactly depend on same k number of samples.

In KMI (K-means Clustering Imputation) [5], the data set is divided into k clusters and missing values are imputed with cluster mean it belongs. In Fuzzy C-Mean clustering Imputation method (FCMI) [5, 6], the missing values are imputed with the clustering centers of Fuzzy C-Mean and the membership matrix. The iterative fuzzy clustering algorithm (IFC) [7] imputes missing values using fuzzy clusters and updates imputed values through iterations. However, it improves the imputation accuracy of missing data iteratively, it is more time consuming due to choosing initial seed points randomly to generate fuzzy clusters.

2.1 *Hierarchical Agglomerative Clustering*

Agglomerative clustering is a type of hierarchical based clustering method in which it begins with considering each cluster contains only one data sample [8, 9]. Latter, at every round the two most similar clusters are merged to form a cluster. Repeat this process until the entire dataset becomes one cluster. The two most similar clusters to be merged at every round are determined by ward's method [8, 10]. The main objective of Ward's method is to minimize the total within-cluster variance and it is determined by the sum of the squared Euclidean (SSE) distance between data samples and their mean in each cluster. The formula for SSE is

$$\mathrm{SSE}_i = \sum_{k=1}^{p} (\bar{x} - x_k)^2 \tag{1}$$

where x_k is a data sample in ith cluster C_i containing 'p' number of data samples such that $x_k \in C_i$ where $1 \leq k \leq$ p and $\bar{x}$ is mean of all the data samples in that cluster. At each round, the sum of SSEs of each cluster in that round is to be minimized. It is given as

$$\mathrm{SSE}_{\mathrm{total}(m)} = \mathrm{SSE}_1 + \mathrm{SSE}_2 + \mathrm{SSE}_3 + \cdots + \mathrm{SSE}_m \tag{2}$$

where 'm' is the total number of clusters at that round. At the initial stage when each data sample is in one cluster, $\mathrm{SSE}_i = 0 \forall\, 1 \leq i \leq m$. At every round, pairwise squared euclidean distances between all the clusters are calculated and minimum distance pair (r, s) is chosen to merge clusters C_r and C_s to form new clusters $C_t = C_r \cup C_s$. This process is repeated until only one cluster is created for entire dataset. The pairwise squared Euclidean distance in the initial round between clusters r and s is calculated using

$$\mathrm{dist}(r, s) = |x_r - x_s|^2 \tag{3}$$

In the next round, the pairwise squared euclidean distances between clusters are the same as in the previous round except for the pairwise distance between new cluster $(r \cup s)$ and the remaining cluster, let say 'rem'. These new pairwise distances can be calculated recursively by using Lance–Williams formula [11]:

$$\mathrm{dist}(r \cup s, \mathrm{rem}) = \frac{(n_r + n_{\mathrm{rem}})\mathrm{dist}(\mathrm{r,rem}) + (n_s + n_{\mathrm{rem}})\mathrm{dist}(s, \mathrm{rem}) - (n_{\mathrm{rem}})\mathrm{dist}(r, s)}{(n_r + n_s + n_t)} \tag{4}$$

where n_r, n_s and n_t are number of data samples in the cluster r, s, and t, respectively.

2.2 Fuzzy Clustering

Fuzzy clustering is a soft portioning method, in which every data sample belongs to many clusters with different degrees of membership values [12]. In hard partition clustering methods, when a data sample is on the boundaries among many clusters is forced to belong to one of them. But in the case of fuzzy clustering, these data samples belong to all the boundary clusters with different membership values between 0 and 1 such that sum of the membership values of any data sample is equal to 1.

The dataset $D = \{x_i/1 \leq i \leq n, x_i$ is ith data sample, n is no. of data samples$\}$ is partitioned into k fuzzy clusters and produce membership matrix U_{ik} and centroid matrix $\mathrm{CM}_k = \{c_i/1 \leq k \leq c, c_k$ is centroid point of kth cluster$\}$ based on minimization of objective function

$$J_m = \sum_{i=1}^{n} \sum_{k=1}^{c} u_{ik}^{m} x_i - c_k^2 \tag{5}$$

where m is any positive real number greater than 1, u_{ik} is the membership value of x_i^{th} data sample in kth cluster, c_k is the centroid of the kth cluster and $x_i - c_k$ is the similarity between data sample x_i and its center c_k. The fuzzy partitioning is done by iteratively optimizing the above objective function, through updating membership u_{ik} and cluster center c_k given by:

$$u_{ik} = \frac{1}{\sum_{j=1}^{c} \left(\frac{x_i - x_k}{x_i - x_j} \right)^{\frac{2}{m-1}}} \tag{6}$$

The cluster center c_k is calculated with

$$C_k - \frac{\sum_{i=1}^{n} u_{ik}^{m} . x_i}{\sum_{i=1}^{n} u_{ik}^{m}} \tag{7}$$

2.3 Fuzzy Silhouette Index

The Fuzzy Silhouette Index (FSI) is a measure used to check how well fuzzy clustering is performed [13, 14]. The FSI for each cluster k is given by

$$\mathrm{FSI}(k) = \frac{\sum_{i=1}^{n} \left(u_{ig(i)} - u_{ig(i)'} \right)^{\alpha} s_i(k)}{\sum_{i=1}^{n} \left(u_{ig(i)} - u_{ig(i)'} \right)^{\alpha}} \tag{8}$$

where U_{nk} is fuzzy membership matrix with 'n' rows and 'k' columns such that row represents a data sample and column represents a cluster and $u_{ij} \in U_{nk}$. The $u_{ig(i)}$

is the first largest and $u'_{ig(i)}$ is the second-largest values in ith row, respectively; α is a weight coefficient (default value = 1); and $s_i(k)$ is the silhouette value of data sample i, and it is calculated as

$$s_i(k) = \frac{(a(x_i) - b(x_i))}{\max\{a(x_i), b(x_i)\}} \tag{9}$$

where $a(x_i)$ is the average similarity between data sample x_i and the remaining data samples from the clusters in which x_i is a member; $b(x_i)$ is the are the average dissimilarity between data sample x_i and all the data samples from the clusters in which x_i is not a member. The fuzzy silhouette index value is calculated for all the clusters and average of them is found. The range of average fuzzy silhouette index value is in between -1 and $+1$. If the average FSI is close to $+1$ then the fuzzy clusters are well-formed and if that value is close to -1, then they are not formed well.

3 Proposed Model

The fuzzy c-mean (FCM) method is best suited to impute missing values in health data. But the correctness of missing values is influenced by

1. How many fuzzy clusters to be formed?
2. What are the initial seed points to start clustering process?

Generally, number of fuzzy clusters to be formed is set by the user and the initial seed points are chosen randomly. This approach may lead to local optimal as FCM purely relies on the initial conditions and as a result, the imputed values are influenced. In order to overcome this, we used hierarchical agglomerative clustering to identify the initial seed points to start fuzzy clustering and the number of fuzzy clusters to be formed is found with fuzzy silhouette index [14]. Our approach to impute the MVs (missing values) is explained as follows:

Step 1 *Initial imputation is done by WKNN imputation*

Every missing value in the original dataset is initially imputed with WKNN imputation method, in which 'k' nearest neighbor's weighted average for numerical values and mode for nominal values is used. This step is essential to map every data sample with missing values into one of the clusters. We used $k = 5$ to perform initial imputation.

Step 2 *Generating Hierarchical agglomerative cluster (HAC) and round wise storing centroid points of each cluster*

Initially the dataset D with 'n' data samples is divided into 'n' clusters, such that each cluster contains only one data sample. Latter at every round, the two most similar clusters identified by Ward's method are merged and forms one bigger cluster. This process is repeated until all the 'n' data samples form as only one cluster. During the above process, at every round store each cluster centroid point round-wise from $\sqrt[3]{n}$ round to 2 clusters round. That is, at a round when there are 'k' clusters, find and store 'k' clusters centroids points. In the next immediate round these 'k' clusters becomes '$k - 1$' clusters and store those '$k - 1$' clusters centroid points for $(k - 1)$th clusters round. Repeat this process from $\sqrt[3]{n}$ clusters to 2 clusters. These cluster centroid points are used as initial seed points for generating fuzzy clusters. That is, when we need to generate 'c' fuzzy clusters, we use the round with 'c' centroid points are chosen as initial seed points. Moreover, we have chosen the process of storing centroid points in the range of $[2, \sqrt[3]{n}]$ because, the dataset may be divided into 2 fuzzy clusters in the best case and it may be divided into $\sqrt[3]{n}$ clusters in the worst case.

Step 3 *Generating Fuzzy clusters*

The dataset 'D' is partitioned into 'k' fuzzy clusters by considering the round with 'k' clusters centroid points as initial seed points identified in Hierarchical Agglomerative Clustering process. After generating 'k' fuzzy clusters, we get the membership degree Matrix U_{ik} and centroids matrix c_k. Store this data for the case of 'k' clusters. Repeat the process of generating 'k' fuzzy clusters from $k = 2$ to $\sqrt[3]{n}$.

Step 4 *Determining optimal number of fuzzy cluster*

Find average fuzzy silhouette index value for each k number of fuzzy clusters from $2 \le k \le \sqrt[3]{n}$ and the k with maximum average fuzzy silhouette index value is chosen as optimal number of clusters; let us say 'p'.

Step 5 *Imputing missing values using optimal number of fuzzy cluster values*

Previously initialized missing values as per step 1 are updated in this step iteratively. The membership degree Matrix U_{ip} and centroids matrix C_p associated with 'p' (Optimal no of Fuzzy Clusters) is considered to impute missing values. If the missing value in ith data sample at jth attribute to be imputed is represented with EV_{ij}, then each missing value is estimated using

$$\mathrm{EV}_{ij} = \sum_{k=1}^{p} \cdot C_{kj} * U_{ik} \tag{10}$$

where C_{kj} is the value of jth attribute in centroid of the kth cluster and U_{ik} is the membership degree of ith data sample in kth cluster.

Step 6 *Repeat the process to improve imputation values*

Repeat the above process from step (2) to step (5) m times. This improves imputation value. The default value of m is 5.

4 Experimental Results

The effectiveness of an imputation method can be determined by comparing the respective imputed values with their original values. If the imputed values are very close to their original values, then the imputation method used is the best method. The efficiency of the proposed method is determined by conducting experiments on datasets collected from UCI machine learning repositories [15]. We randomly introduced each time different percentages of missing values into the datasets. Then we applied our proposed method and other imputation methods like mean imputation, KNN imputation and fuzzy imputation methods to calculate imputed values. We used metric called Normalized Root Mean Square Error (NRMSE) for determining the correctness of these imputation methods. The NRMSE is given by

$$\text{NRMSE} = \frac{1}{x_{\text{max}} - x_{\text{min}}} \sqrt{\frac{1}{n} \sum_{i=1}^{n} (x_i - x_i')^2} \tag{11}$$

where x_i is original value and x_i' is imputed value; x_{max} and x_{min} are minimum and maximum values, respectively. The datasets used for evaluating imputation methods should not contain any missing values because; as original values are missing it is not possible to compare imputed values with original values. In order to use such datasets for evaluation purposes, the data samples with missing values in those datasets should be removed and modified datasets should be generated.

The Wisconsin breast cancer dataset (WBCD) is a well-known dataset consist of 699 instances with 10 attributes. As this dataset contains 16 instances with missing values, those data samples are removed and modified dataset is generated with 683 instances to evaluate imputation method.

Randomly Different percentages of missing values are introduced into the modified WBCD and initial imputation (Step 1) is implemented to fill these missing values. Hierarchical agglomerative clustering is implemented and round-wise cluster centroids are stored (Step 2). These cluster centroids are used as initial seed points and round-wise Fuzzy clusters are generated (Step 3). Average Fuzzy silhouette index value is calculated for fuzzy clusters of each round and the appropriate number of clusters found (Step 4). Table 1 shows the average Fuzzy silhouette index (FCI) values

From the above table, it is clear that WBCD can be soft partitioned into 2 fuzzy clusters ($k = 2$). As $k = 2$, the centroid matrix and membership matrix belongs to 2 fuzzy clusters are used to impute missing values. The performance evaluation of the proposed method is done by NRMSE and the results are summarized in Table 2.

Table 1 Average fuzzy silhouette index values for Wisconsin breast cancer dataset (WBCD)

K value	2	3	4	5	6	7	8	9
FSI	0.8354	0.8123	0.5906	0.6592	0.5764	0.5692	0.5341	0.5484

Table 2 Comparisons of imputation methods NRMSE values

Dataset	Missing ratio (%)	Mean	KNNI	WKNNI	KMI	FCMI	Proposed method
Wisconsin breast cancer	2	0.32	0.31	0.27	0.24	0.2	0.15
	5	0.36	0.33	0.31	0.3	0.23	0.17
	10	0.37	0.35	0.33	0.31	0.27	0.19
	15	0.39	0.36	0.35	0.33	0.28	0.22
Hepatitis	2	0.31	0.28	0.26	0.26	0.21	0.18
	5	0.34	0.32	0.29	0.28	0.25	0.21
	10	0.38	0.34	0.32	0.31	0.28	0.23
	15	0.41	0.37	0.36	0.35	0.31	0.24
Cleveland heart	2	0.32	0.32	0.3	0.29	0.21	0.15
	5	0.35	0.33	0.32	0.3	0.23	0.19
	10	0.38	0.36	0.34	0.32	0.26	0.21
	15	0.39	0.39	0.37	0.35	0.3	0.25

Similarly, the above process is repeated on the Hepatitis dataset and the Cleveland Heart dataset. Latter, 2, 5, 10, and 15% of missing values introduced into each of the datasets and these missing values are imputed with other methods (Mean, KMI, WKNNI, KNNI, FCMI) and our proposed method. The imputed values are compared with original values and NRMES scores are generated. The summary is given in below Table 2.

It is clear that our proposed method imputing the missing values very close to the original values when compared with other methods.

5 Conclusion

Missing values in the datasets not only cause computational challenges but also influence strategic decision-making. So, it became a big challenging issue in data science. The proposed method imputes missing values close to the original data without disturbing the existing relationship among the attributes. None of the imputation methods are best suitable for all types of datasets. Due to existence of fuzzy nature in medical datasets, our method is best suitable for imputing missing values in medical data set.

References

1. Batista, G. E., & Monard, M. C. (2003). An analysis of four missing data treatment methods for supervised learning. *Applied Artificial Intelligence, 17*(5–6), 519–533.
2. Zhang, Z. (2016). Missing data imputation: Focusing on single imputation. *Annals of Translational Medicine, 4*(1).
3. Batista, G., & Monard, M. C. (2002). A study of K-nearest neighbour as an imputation method. In *Proceedings of Second International Conference on Hybrid Intelligent Systems*, vol. 87 (pp. 251–260). IOS Press.
4. Troyanskaya, O., et al. (2001). Missing value estimation methods for DNA microarrays. *Bioinformatics, 17*(6), 520–525.
5. Li, D. et al. (2004). Towards missing data imputation: A study of fuzzy k-means clustering method. In *International Conference on Rough Sets and Current Trends in Computing*. Springer, Berlin, Heidelberg.
6. Hathaway, R. J., & Bezdek, J. C. (2001). Fuzzy c-means clustering of incomplete data. *IEEE Transactions on Systems, Man, and Cybernetics, Part B (Cybernetics), 31*(5), 735–744.
7. Nikfalazar, S. et al. (2017). A new iterative fuzzy clustering algorithm for multiple imputation of missing data. In *2017 IEEE International Conference on Fuzzy Systems (FUZZ-IEEE)*. IEEE.
8. Murtagh, F., & Legendre, P. (2014). Ward's hierarchical agglomerative clustering method: Which algorithms implement Ward's criterion? *Journal of Classification, 31*(3), 274–295.
9. Mendoza, Z., Lisandra, M., & Resendis-Antonio, O. (2013). Hierarchical agglomerative clustering. *Encyclopedia of Systems Biology*, 886–887.
10. Murtagh, F., & Legendre, P. (2011). Ward's hierarchical clustering method: Clustering criterion and agglomerative algorithm. arXiv:1111.6285.
11. Lance, G. N., & Williams, W. T. (1967). A general theory of classificatory sorting strategies: 1. Hierarchical systems. *The Computer Journal, 9*(4), 373–380.
12. Bezdek, J. C., Ehrlich, R., & Full, W. (1984). FCM: The fuzzy c-means clustering algorithm. *Computers & Geosciences, 10*(2-3), 191–203.
13. Roubens, M. (1982). Fuzzy clustering algorithms and their cluster validity. *European Journal of Operational Research, 10*(3), 294–301.
14. Campello, R. J. G. B., & Hruschka, E. R. (2006). A fuzzy extension of the silhouette width criterion for cluster analysis. *Fuzzy Sets and Systems, 157*, 2858–2875.
15. Bache, K., & Lichman, M. (2013). UCI machine learning repository.

Association Rules Mining for STO Dataset of HSES Knowledge Portal System

Santosh Kumar Miri, Neelam Sahu, Rohit Miri, and S. R. Tandan

Abstract HSES knowledge portal system is one of the web portals. In this portal, we collect information into student, teacher, and other datasets. Now we combine these datasets with common features with class descriptors. This is named as STO dataset which is handled using traditional tools and techniques. This large dataset is generally known as big data. Increasing the use of social media and new innovations generate a vast amount of data. The quantity of data makes HSES-KP very powerful and useful. This portal needs to manage and analyze big data for better association rules and outcomes. Association rule mining is the technique for finding frequent patterns, associations, correlations, or causal structures among sets of items in transaction databases. Lift, support, and confidence are user-defined measures of interestingness. In this paper, we review the background and the various techniques of association rules mining. This study grows up with a discussion of challenges and future directions.

Keywords Data preprocessing · Association rules mining · RStudio · Visualizations and correlations

1 Introduction

HSES Knowledge Portal System includes the student data, teacher data, other data, student feedback, teacher feedback, and other feedback. So, huge amount of HSES-KP data are available for data scientists. By understanding stencils and trends within

S. K. Miri (✉) · N. Sahu · R. Miri · S. R. Tandan
Dr. C.V. Raman University, Kota, Bilaspur, Chhattisgarh, India
e-mail: skmiri29@gmail.com

N. Sahu
e-mail: neelamsahu16@gmail.com

R. Miri
e-mail: rohitmiri@gmail.com

S. R. Tandan
e-mail: srtandan26@gmail.com

S. C. Satapathy et al. (eds.), *Intelligent System Design*, Advances in Intelligent Systems and Computing 1171, https://doi.org/10.1007/978-981-15-5400-1_60

the data, data science seems to improve care, save times, and reduce costs. RStudio in HSES-KP takes advantage of extracting rules from data for better confidence, support, and lift purposes. Association rule mining is the technique for finding frequent patterns, correlations, associations, or causal structures among item sets in transaction databases.

Antecedent → Consequent [support, confidence] (Confidence and support are user-defined measures of interestingness).

2 Literature Survey

Agrawal et al. [1] introduced association rules mining. A rule is defined as an implication of the form $X \Rightarrow Y$ where X and Y are itemsets of the database. X and Y are called antecedent (left-hand-side) and consequent (right-hand side) of the rule. Association rules surpass user-specified minimum confidence and minimum support threshold. An association rule $X \Rightarrow Y$ will satisfy $\text{supp}(X \cup Y) \geq \sigma$ and $\text{conf}(X \Rightarrow Y) \geq \delta$ where σ and δ are the minimum support and minimum confidence respectively. The support, confidence, and lift measures help with focusing on potentially more interesting rules. The lift measure was first introduced by Brin et al. [2]. They argued the need for extending the support-confidence framework and using correlation coefficient for tapping negative relationships between itemsets. The lift of a rule is defined as $\text{lift}(X \Rightarrow Y) = \text{supp}(X \cup Y)/(\text{supp}(X)\ \text{supp}(Y))$. Greater lift values ($\gg 1$) indicate positively or stronger associations. The lift value <1 indicates negatively correlation. Unwin et al. [3] introduced two-key scatter plot that includes support and confidence. It can be used for large sets of association rules. It gives an impression of the distribution of rules concerning large and small values for interest measures. Zaki [4] performed Mining Non-Redundant Association Rules. The authors presented a new framework for associations based on the concept of closed frequent itemsets. It produces number of non-redundant rules which are smaller than the rule set from the traditional approach. Antonie and Zaïane [5] extended the work of Brin et al [2]. They have used Pearson's correlation coefficient as a measure of negative association. They developed an algorithm to generate negative AR with a sliding correlation coefficient threshold. Ashrafi et al. [6] discussed redundant association rules reduction techniques. They have included support and confidence thresholds to find hidden correlations among different items within a dataset. However, association rule mining algorithms produce many redundant rules using them. The proposed methods eliminate a significant number of redundant rules. Graph-based visualization techniques [7] concentrate on the relationship between individual items reflecting their membership in different association rules. Bruzzese and Davino [8] discussed visual mining of association rules. Three aspects of visual data mining are tasks, visual representation (visualization), and the process. Items visualization, rules visualization, and Conjoint Visualization are included in this paper.

Hahsler and Chelluboina [9] discussed Visualizing Association Rules. Visualization techniques can be used to analyze a data set. They presented the R-extension package arulesViz. The arulesViz implements several known and novel visualization techniques to explore association rules. In this paper, scatter plots, Matrix-based visualization techniques, Grouped matrix-based visualization, Graph-based visualizations, Parallel coordinates plot, and Double Decker plots are practically discussed. Kumbhare and Chobe [10] discussed an overview of six association rule mining algorithms AIS, SETM, Apriori, Aprioritid, Apriorihybrid, and FP-growth. The performance of these algorithms is evaluated based upon various parameters like execution time and data support and accuracy. It is observed that each algorithm has some advantages and disadvantages. So the authors concluded that FP-growth performs better. Avcilar and Yakut [11] discussed Association Rules in Data Mining. They applied it to a clothing and accessory specialty store. In this research, association rules were estimated by using market basket analysis and taking confidence, support, and lift measures into consideration. Zhao[12] discussed association rule mining with Rprogramming, pruning redundant rules, visualizing, and interpreting association rules. Association rules are rules presenting or correlation or association between itemsets. Support($A \Rightarrow B$) = $P(A \cup B)$, confidence($A \Rightarrow B$) = $P(B|A)$ and lift($A \Rightarrow B$) =confidence($A \Rightarrow B$)/$P(B)$ are used. The Apriori algorithm employs level-wise search for frequent itemsets with default settings minimum support: supp = 0.1, minimum confidence: conf=0.8, and maximum length of rules: maxlen = 10. After removing redundant rules, the author interprets rules and visualizing through Scatter plot, grouped matrix, and graph. Karthikeyan et al. [16] performed a survey when applying association rule mining in Education management systems. Association rules mining discovers relationships among attributes in educational databases. It has been applied to e-learning systems for traditionally association analysis. The strength of association rule is measured by its support and confidence. Confidence and support factors can be used for obtaining interesting rules. These two parameters allow the pruning of many associations.

The grouped matrix-based visualization technique (Hahsler and Karpienko, [14] enhance matrix-based visualization by organizing the large set of different antecedents into a small set of groups through the clustering technique. Santhi and Lilly [15] implemented association rule mining algorithm for the recommender system of e-business. E-business connected the customer and company very close without their presence. It is necessary to uphold the websites to run a successful e-business. It is required to identify the frequently visited and associated pages by applying the Apriori association rule mining algorithms. Michael Hahsler [17] performed a research paper entitled arulesViz: Interactive Visualization of Association Rules with R. The R-extension package arulesViz provided most popular visualization techniques for association rules. In this paper, the author discussed three major interactive visualization techniques scatter plots, graph-based visualization, and matrix visualization. Sevri et al. [13] performed crime analysis based on association rules using apriori algorithm. It has been used to extract the relationships between features of criminal records. This research reveals the relationships between the attributes of different criminal records. It allows the authorities to establish relationships between new and old incidents.

3 Data Preprocessing

Data mining is a knowledge discovery process from raw data. It discovers interesting patterns (or rules) and knowledge. It involves data cleaning, data selection, data integration, data transformation, pattern evaluation, pattern discovery, and knowledge presentation.

Data Collection: Dataset can be used to predict the performance of a knowledge portal. Raw Data of student, teacher, and other datasets can be collected from the www.hseshiksha.in for nearly 5 months of period. The student dataset includes a total of 301 samples represented by 19 attributes. The teacher dataset includes a total of 93 samples represented by 19 attributes. The other dataset includes a total of 110 samples represented by 17 attributes.

Data Cleaning: Data quality is assessed based on the intended use of data which is defined in the terms of accuracy, completeness, consistency, timeliness, interpretability, and believability. Noise and inconsistent data remove during data cleaning. Data cleaning process requires filling missing values, smooth out noise, and correct inconsistencies in the data.

Data Integration: Multiple data sources may be combined during data integration. It integrates multiple databases, data cubes, or flat files. Now, we are combined student, teacher, and other datasets into STO dataset. In the STO dataset, one new attribute category is included. The STO dataset includes a total of 504 samples represented by 26 attributes.

Data Selection: In data selection step, data relevant to the analysis tasks are retrieved from the dataset. Data reduction technique techniques obtain a reduced representation of the data while minimizing the loss of information content. These include methods of numerosity reduction, dimensionality reduction, and data compression. Dimensionality reduction reduces the number of random variables or attributes under consideration. By using it, we are removed 13 attributes (Stuid/Tid/Oid, Name, Stream, Board, Medium, Gender, DOR, State, District, LM, TM, Parent, and Help) from the STO dataset. Finally, STO dataset has 13 attributes. One new field category is included for categorizing the records.

Data Transformation: Data transformation routines convert the data into appropriate forms for data mining. In this preprocessing step, the data are transformed or consolidated so that the patterns found may be easier to understand. For data transformation, smoothing, attribute construction, normalization, aggregation, etc. are used. Mining methods that require conversion of raw data into an understandable format called Data preprocessing. It is a proven method of resolving errors. The consequence of the data preparation stage is the final dataset which can be ready to analyze and identify patterns by using data mining tool. STO dataset has 14 attributes that are related to the responses about 13 questionnaires.

Extraction of Data Patterns: Data mining is an essential process where intelligent methods are applied to extract data patterns. Frequent patterns occur frequently in data which includes frequent item sets, frequent subsequences, and frequent substructure. Mining frequent patterns lead to the discovery of interesting associations and correlations within data. Apriori algorithm is used to extract data patterns from the normalized STO dataset.

Knowledge Presentation: Knowledge representation and data visualization techniques are used to present mined knowledge to users. Knowledge obtained from data mining in visual forms may include scatter plots and box plots as well as decision trees, clusters, outliers, association rules, and generalized rules. Visualization tools are used in the data mining process.

4 Association Rules Mining

A technique to detect and extract useful information from large scale transaction data is known as association rules mining. It was first introduced by Agrawal et al. [1] and can formally be defined as:

Let $I = \{i1, i2, i3, \ldots, in\}$ and $D = \{t1, t2, t3, \ldots\ tm\}$. A set of binary attributes is called items and set of transactions is called the database. Each transaction has a unique ID and contains a subset of the items in I. Rule is defined as an implication of the form:

$$A \Rightarrow B \text{ where } A, B \subseteq I \text{ and } A \cap B = \phi$$

The set of items A and B are called antecedent (left-hand-side or lhs) and consequent (right-hand-side or rhs) of the rule. In the consequent, rules are restricted to only a single item. Association rule mining is a procedure for finding the frequent patterns, correlations, and associations from datasets. The measures of the effectiveness of it are as follows:

- Support: It means how much historical data supports our rule.
- Confidence: It means how confident we that the rule holds.
- Lift: It is the ratio confidence is to support.
- Others: Affinity, Leverage.

Association rules surpass a user-specified minimum support and minimum confidence threshold. The support supp(A) of A is defined as the proportion of transactions in the dataset which contains the itemset.

$$\text{Support} = \frac{\mathbf{supp(A \cup B)}}{\mathbf{N}} = \frac{\mathbf{frq(A, B)}}{\mathbf{N}}$$

The confidence of a rule is defined as

$$\text{Confidence} = \text{conf}(A \Rightarrow B) = \frac{\mathbf{supp(A \cup B)}}{\mathbf{supp(A)}} = \frac{\mathbf{frq(A, B)}}{\mathbf{frq(A)}}$$

Table 1 Association rules satisfaction

Effectiveness measures	Association rules $A \Rightarrow B$
$supp(A \cup B) \geq \sigma$	Satisfy
$conf(A \Rightarrow B) \geq \delta$	*Satisfy*

Table 2 Association rules correlation

Lift	Correlated
Lift($A \Rightarrow B$) $\gg 1$	Positively/strongly
Lift($A \Rightarrow B$) < 1	Negatively
Lift($A \Rightarrow B$) $= 1$	Not

Therefore, an association rule $A \Rightarrow B$ will satisfy:

$$\text{supp}(A \cup B) \geq \sigma \text{ and conf}(A \Rightarrow B) \geq \delta$$

where and δ are the minimum support and minimum confidence, respectively. The Brin et al. [2] discussed that lift of rule is defined as

$$\text{Lift} = \text{Lift}(A \Rightarrow B) = \frac{\text{support}(A \cup B)}{\text{supp}(A) \times \text{supp}(B)}$$

Greater lift values ($\gg$1) indicates stronger associations. If the lift is less than 1 then A and B are negatively correlated else positively correlated and it is equal to 1 then it is not correlated (Tables 1 and 2).

According to Hahsler and Chelluboina (2017), association rules are typically generated in two-step process. Frequent itemsets are those itemsets that satisfy the minimum support. In first step process, minimum support is used to generate the set of all frequent itemsets for the dataset. In second step process, each frequent itemsets is used to generate all possible rules from it and all rules which do not satisfy the minimum confidence are removed. In the worst case, rules are generated in the order of $2n - (n + 1)$. In the least case, rules are generated in the order of $O(2n)$. It removes potentially interesting rules with less support.

5 Data Visualization Tools

The famous data visualization tools are JupiterNotebook, R, and RStudio, etc. To analyze association rules, many researchers introduced visualization techniques like scatter plots, grouped matrix, graphs, and mosaic plots, etc. We have chosen RStudio to analyze association rules.

RStudio: RStudio is a free and open-source integrated development environment for R which was founded by JJ Allaire in 2011. It was written in Java, C++, and JavaScript. R is a programming language for graphics and statistical computing.

It runs on the desktop (Windows, macOS, and Linux) or in a browser connected to RStudio Server. RStudio contributes to many R packages. They are Tidyverse, Shiny, RMarkdown, knitr, packrat, and devtools. Tidyverse is a package for data science, including ggplot2, dplyr, tidyr, and purr. Shiny includes interactive web technology. RMarkdown inserts R code into markdown documents. Knitr is used to generate dynamic reports combining R, TeX, Markdown, and HTML. The devtools is a Package development tool. RStudio provides a mechanism for executing R functions.

The bookdown package of Addins creates books. The color picker package is a graphical tool to pick colors for plots. Searching and loading of datasets is possible through Datasets.load package.

Tidyverse Package: The tidyverse package is an opinionated collection of RStudio packages designed for data science. The dplyr package is a tool for Splitting, Applying, and Combining Data. The ggplot2 package creates graphics and charts. Grouping of attributes can be done by ddply package.

arules Package: The arules [18] package provide the infrastructure for representing, manipulating, and analyzing transaction data and patterns (frequent itemsets and association rules). It handles and mines association rules. It is installed and loads by using library (arules). After that we have get this: >library (arules).

arulesViz package: The arulesViz package extends package 'arules' with various visualization techniques for association rules and item sets. It also includes several interactive visualizations for rule exploration. Loading of grid package is possible by using arulesViz. The arulesViz package is installed by using library(arulesViz). After that, we have got this: >library (arulesViz).

Scatter Plots: A scatter plot is a two-dimensional data visualization technique that uses Cartesian coordinates to display values for typically two variables for a set of data. It is also called scatter graph, scatter chart or scatter diagram. It graphs pairs of numerical data to look for a relationship between them. It is a set of points plotted on horizontal and vertical axes. It is useful data visualization tool for illustrating a trend. It is similar to the line graph and created using the function plot (x, y). The Scatter Plot in RStudio Programming is very useful to visualize the relationship between two sets of data. A straight-forward visualization of association rules is to use a scatter plot using plot() of the arulesViz package. It uses Support and Confidence on the axes. The third measure Lift is used by default to color (grey levels) of the points. In two-key plot, support and confidence are used on x and y-axis respectively. It uses order for coloring where order is the number of items in the rule. An amazing interactive plot can be used to present our rules that use arulesViz and plotly. It is showing support, confidence, and lift of rules. The plot mainly gives an overview of the distribution of support and confidence in the ruleset. It is run by>plot (rules). Now, scatter plot for selected association rules is produced.

Grouped matrix Visualization: Matrix-based visualization is limited in the number of rules. It can visualize effectively since large sets of rules are produced. It requires the analyst to repeatedly zoom in and out. Grouped matrix-based visualization enhances matrix-based visualization by grouping antecedents of rules via sorting and clustering rules. Grouped rules are presented as aggregate in a matrix that is visualized as a balloon plot. The color indicates lift and balloon size represents support. We have get RHS {attribute=data} items. The grouped matrix visualization allows the user to identify the rules making up groups. It is run by:

>plot (rules, method = "grouped", engine = "interactive").

Now, grouped matrix for selected association rules is produced.

Graph-based Visualization Technique: Graph-based techniques concentrate on the relationship between individual items in the rules. Graph-based visualization tends to become cluttered and thus is only viable for very small sets of rules. The warning message informs the user that only the top 100 rules are included in the visualization. The force-directed layout moves items that are included in many rules to share many items. Items that are in very few rules are pushed to the periphery of the plot. Higher lift rules are also on the outside of the plot. A big restriction of this technique for association rules is that they are only useful for a very small set of rules. Filtering, grouping, and coloring nodes feature is used to explore large sets of rules with graph.

6 Unified Interface of Arulesviz for STO Dataset

The first data frame STO contains 14 attributes. Use read.csv (E:/STO.csv) to read the STO dataset into RStudio. Before applying Association Rule mining, we need to convert data frame into transaction data. We can see that each record is in atomic form as in relational databases. This format is also called as the singles format. Next, as Stuid/Tid/Oid, Name, Gender, LM, TM, State, District, DOR, Parent, Help, Occupation will not be of any use in the rule mining, we can set them to NULL. Now, we have to store this transaction data into a STO.csv file. It is possible by using the write.csv(). After that, we have to load this transaction data into an object of the transaction class. This is done by using the Rstudio function (Fig. 1).

MyData ← read.csv(file = "E:/STO.csv", TRUE, ",")

read.transactions of the arules package. Now, we have taken transaction data file E:/STO.csv and convert it into an object of the transaction class. We can view the MyData transaction object (Fig. 2).

Fig. 1 Output for association rules

```
Apriori
Parameter specification:
confidence minval smax arem aval originalSupport
0.8 0.1 1 none FALSE TRUE
maxtime support minlen maxlen target ext
5 0.1 1 10 rules FALSE
Algorithmic control:
filter tree heap memopt load sort verbose
0.1 TRUE TRUE FALSE TRUE 2 TRUE
Absolute minimum support count: 50
set item appearances ...[0 item(s)] done [0.00s].
set transactions ...[34 item(s), 504 transaction(s)] done [0.03s].
sorting and recoding items ... [25 item(s)] done [0.00s].
creating transaction tree ... done [0.00s].
checking subsets of size 1 2 3 4 5 6 7 8 9 10 done [0.11s].
writing ... [830551 rule(s)] done [0.47s].
creating S4 object ... done [3.47s].
```

Fig. 2 Summary of association rules

```
set of 830551 rules
rule length distribution (lhs + rhs):sizes
1  2   3    4     5     6     7      8      9      10
13 258 2211 11056 36616 86051 149318 195917 196916 152195
Min.  1st Qu. Median  Mean  3rd Qu.  Max.
1.000 7.000   8.000   8.016 9.000    10.000

summary of quality measures:
support          confidence        lift              count
Min.   :0.1012   Min.   :0.8000    Min.   :0.8831    Min.   : 51.0
1st Qu.:0.1409   1st Qu.:0.9759    1st Qu.:1.0602    1st Qu.: 71.0
Median :0.2242   Median :0.9890    Median :1.0729    Median :113.0
Mean   :0.2795   Mean   :0.9820    Mean   :1.0770    Mean   :140.9
3rd Qu.:0.3849   3rd Qu.:1.0000    3rd Qu.:1.0909    3rd Qu.:194.0
Max.   :0.9444   Max.   :1.0000    Max.   :1.1533    Max.   :476.0

mining info:
data    ntransactions support confidence
MyData  504           0.1     0.8
```

Fig. 3 Summary of MyData

```
Category    Locale          Gender      IK      CLSP    QP      SY
Other:110   Rural:119       Female:245  No:42   No: 34  No:28   No:38
Student:301 Semi-urban: 93  Male :259   Yes:462 Yes:470 Yes:476 Yes:466
Teacher:93  Urban:292
Time    Region  HM      PExam   SMSearch Euse   AMCL    SMDL    ASM
No:31   No:56   No:63   No:67   No:58    No:39  No:35   No:47   No:38
Yes:473 Yes:448 Yes:441 Yes:437 Yes:446  Yes:465 Yes:469 Yes:457 Yes:466
```

Now, we install an arules package for finding the association rule. Again we run MyData to generate apriori rules. It mines the rules by using the APRIORI algorithm. The function apriori() is from package arules.

>library(arules)

>rules ← apriori(MyData).

The result is a set of 830,551 association rules for 34 items and 504 transactions. 54 Transaction objects are created and 25 items are sorted and recorded. Now we generate summary of rules.

>summary(rules)

The summary (MyData) is a very useful command that gives us information about our transaction

>summary(MyData) (Fig. 3).

7 Unified Interface of Arulesviz Forsto Dataset Whenlocale = Rural/Urban/Semi-urban

It is possible to generate association rules for the parameter Locale = Rural, Locale = Urban or Locale = Semi-urban.

>rules ← apriori(MyData, parameter = list(minlen = 1, maxlen = 10, conf = 0.2), appearance = list(rhs = c("Locale = Rural"), default = "lh s"))

It is generated a set of 14,913 association rules. We can display top most association rules. For example, top two rules with respect to the confidence measure are: top2rules ← head (rules, n = 2, by = "confidence")

>inspect(top2rules) (Fig. 4).

Fig. 4 Inspection of top two rules

```
lhs                                                              rhs
[1] {Gender=Male, QP=Yes, SY=Yes, Region=Yes,
     PExam=Yes, SMSearch=Yes, Euse=Yes}      => {Locale=Rural}
[2] {Gender=Male, QP=Yes, SY=Yes, Region=Yes, HM=Yes,
     PExam=Yes, SMSearch=Yes, Euse=Yes} => {Locale=Rural}

     support   confidence  lift   count
     0.12      0.31        1.31   59
     0.12      0.31        1.31   59
```

Now, we generate association rules for urban locale by using the given apriori function:

>rules ← apriori(MyData, parameter = list(minlen = 1, maxlen = 10, conf = 0.2), appearance = list(rhs = c("Locale=Urban"), default = "l hs"))

It is generated a set of 41,137 association rules. Now, we generate association rules for semi-urban locale by using the given apriori function:

>rules ← apriori(MyData, parameter = list(minlen = 1, maxlen = 10, conf = 0.2), appearance = list(rhs = c("Locale = Sem-urban"), default = "lhs"))

It is generated a set of 11,162 association rules. Now we are created scatter plots for 14,913, 41,137 and 11,162 association rules. It is shown in Fig. 5. In the scatter

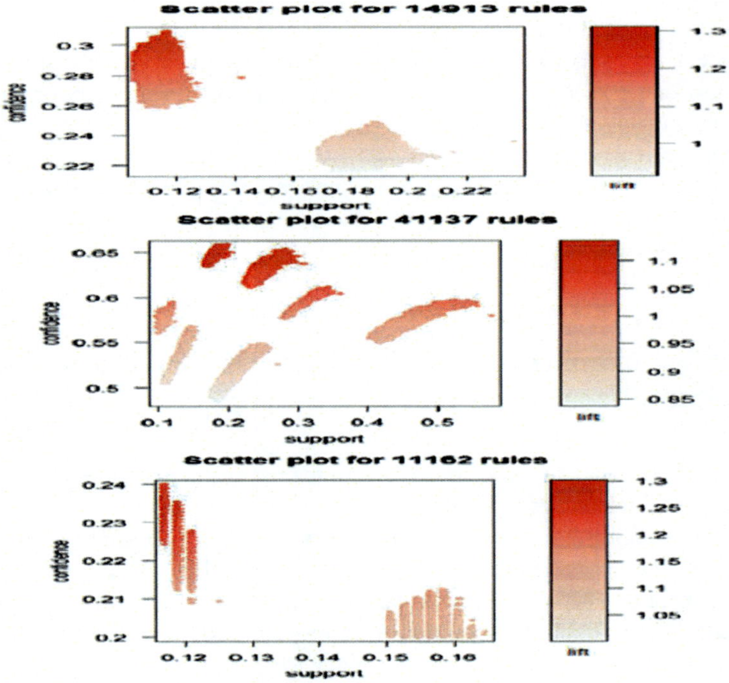

Fig. 5 Scatter plots for association rules with locale field of STO dataset

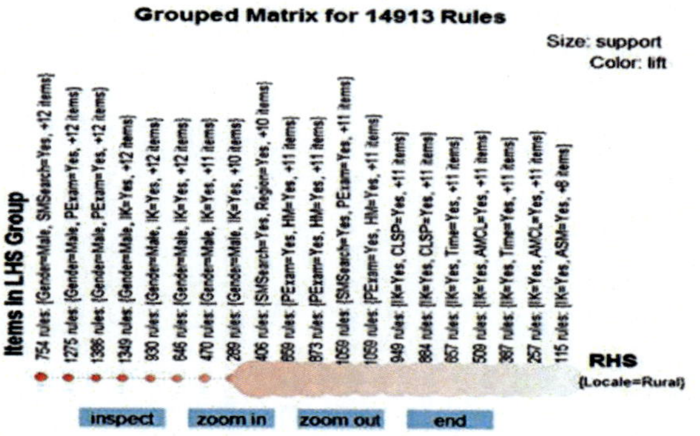

Fig. 6 Grouped matrix for 14,193 rules

plot for 14,913 rules, bulk of high confidence rules are plotted in the top-left corner and higher lift rules are located close to the minimum support threshold. In the scatter plot for 41,137 rules, a few high confidence rules are plotted in the top-left corner, some high confidence rules are plotted in the top-center corner and a few lift rules are located close to the minimum support threshold.

The parameter of 0.5 to 0.65 confidences, 0.1 to 0.5 supports and 0.85 to 1.1 lifts are seen in this scatter plot. In the scatter plot for 11,162, a few high confidence rules are plotted in the top-left corner, some high confidence rules are plotted in the middle-left corner and some lift rules are located close to the minimum support threshold. A bulk of low confidence rules are plotted bottom-right corner whose support parameter is high but lift is poor. The parameter of 0.2–0.24 confidences, 0.12–0.16 supports, and 1.05–1.3 lifts are seen in this scatter plot.

The highest interest group of grouped matrix for 14,193 rules are 754 rules which contains "Gender = Male and SMSearch = Yes +12 items" in the antecedent and "Locale=Rural" in the consequent. The resulting visualization of the set of 14,913 rules mined earlier is shown in Fig. 6. There are 20 LHS groups of rules are produced. Up to 12, 11, 10, or 8 items in LHS group are included. Lowest rules of LHS group are 115 in which up to 8 items are included. According to lift, the most interesting rules are shown in the top-left corner. The balloon sizes represent support which is small for top 7 groups of rules. The lift of balloon for 754 rules is maximum. The highest interest group of grouped matrix for 41,137 rules are 992 rules which contains "Gender = Female, Category = Student, +12 items" in the antecedent and "Locale = Urban" in the consequent. The resulting visualization of the set of 41,137 rules mined earlier is shown in Fig. 7. There are 20 LHS groups of rules are produced. Up to 11, 12, 13 or 14 items in LHS group are included. Lowest rules of LHS group are 457. According to lift, the most interesting rules are shown in the top-left corner. The balloon sizes represent support which is small for top 7 groups of rules. The lift of balloon for 992 rules is more.

The highest interest group of grouped matrix for 11,162 rules are 218 rules which contains "SY = Yes, SMSearch = Yes, +12 items" in the antecedent and "Locale = Semi-urban" in the consequent. The resulting visualization of the set of 11,162 rules mined earlier is shown in Fig. 8. There are 20 LHS groups of rules are produced.

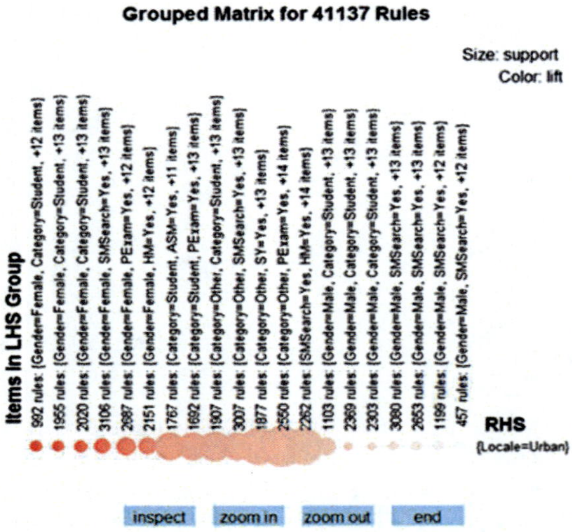

Fig. 7 Grouped matrix for 41,137 rules

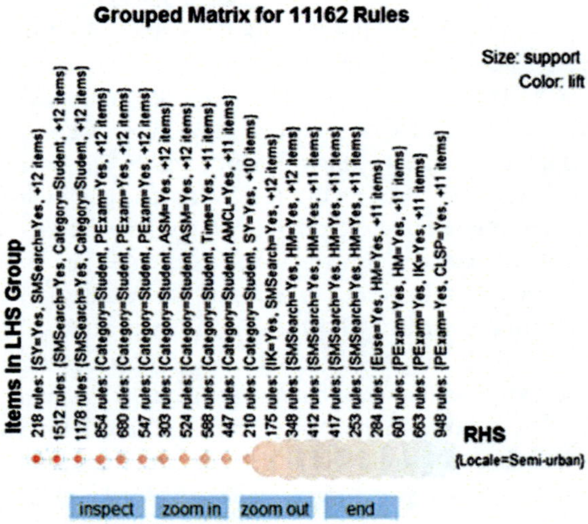

Fig. 8 Grouped matrix for 11,162 rules

Up to 10, 11 or 12 items in LHS group are included. Lowest rules of LHS group are 948. According to lift, the most interesting rules are shown in the top-left corner. The balloon sizes represent support which is small for top 11 groups of rules with higher lift. The lift of balloon for 219 rules is more.

In graph-based visualization for 100 rules of 14,913, lowest support size is 0.198 and highest support size is 0.236. The supports of association rules are increasing when size of bubbles are increasing. Gray levels are growing along with lift value. It is found here that A and B are positively correlated. It is shown in Fig. 9.

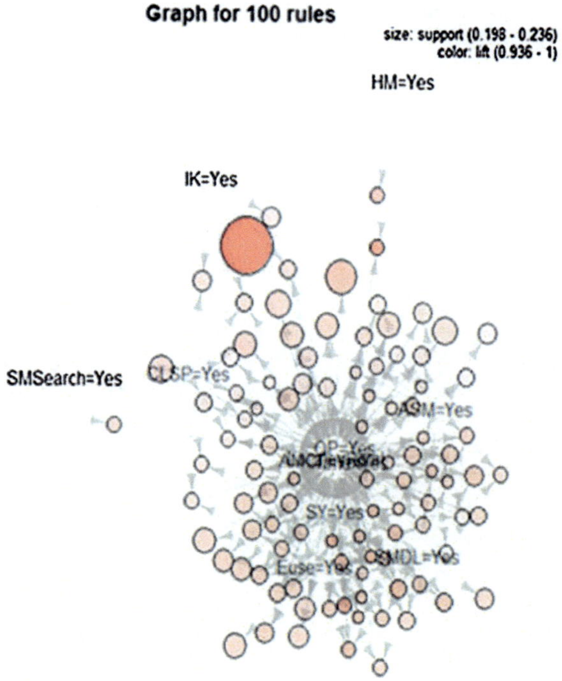

Fig. 9 Graph for 100 rules out of 14,913

In the graph-based visualization for 100 rules of 41,137, lowest support size is 0.522 and highest support size is 0.579. The range of lift is 1 to 1.029. The supports of association rules are increasing when size of bubbles is increasing. Gray levels are growing along with lift value. It is found here that A and B are positively correlated. It is shown in Fig. 10.

In the graph-based visualization for 100 rules of 11,162, lowest support size is 0.161 and highest support size is 0.165. The range of lift is 1.084 to 1.137. The supports of association rules are increasing when size of bubbles are increasing. Gray levels are growing along with lift value. It is found here that A and B are positively correlated. It is shown in Fig. 11.

8 Unified Interface of Arulesviz for STO Dataset Whencategory = Student/Teacher/Others

It is possible to generate association rules for the parameter Category = Student, Category = Teacher or Category = Others.

>rules ← apriori(MyData, parameter = list(minlen = 1, maxlen = 10, conf = 0.1),

Appearance = list(rhs = c("Category = Teacher"), default = "lhs"))

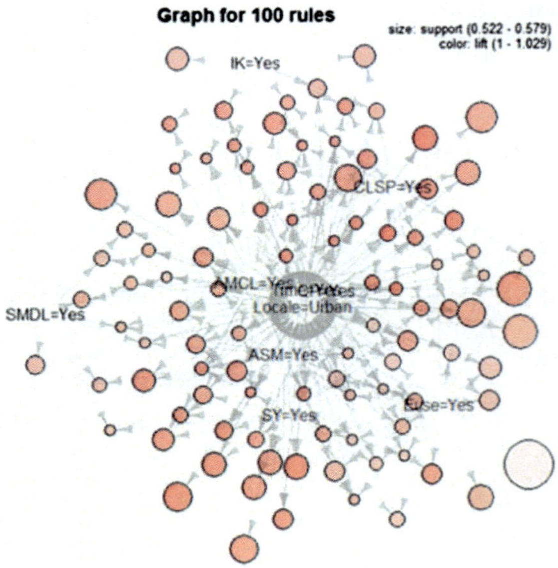

Fig. 10 Graph for 100 rules out of 41,137

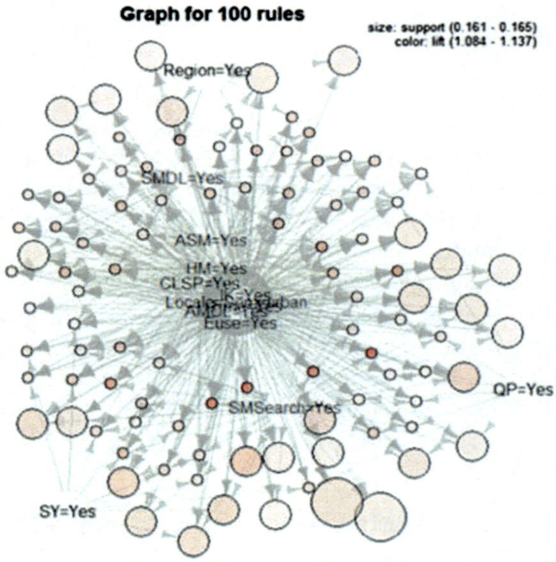

Fig. 11 Graph for 100 rules out of 11,162

It is generated a set of 7814 association rules. We can display the top most association rules. It is plotted in Fig. 12. In this scatter plot, a few high confidence rules are plotted in the top-center corner, some high confidence rules are plotted in the center to top-right corner direction and some lift rules are located close to the minimum support threshold. A bulk of low confidence rules are plotted bottom-left corner

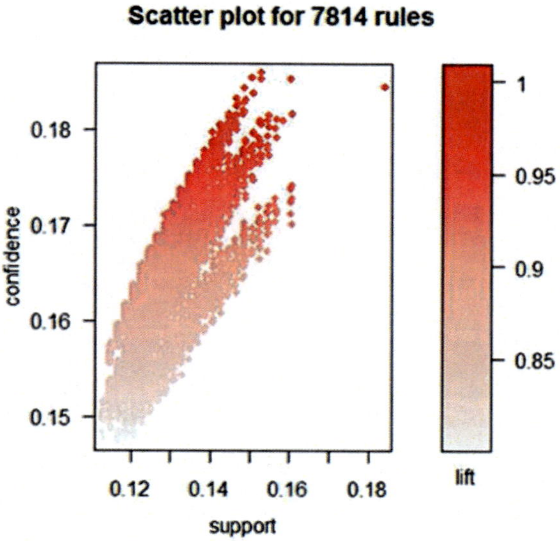

Fig. 12 Scatter plot for 7814 rules

whose support parameter is 0.12 or less but the lift is poor. The parameter of 0.15–0.185 confidences, 0.12–0.18 supports, and 0.08 to 1.0 lifts are seen in this scatter plot.

Now, we generate association rules for student locale by using the given apriori function:

>rules ← apriori(MyData, parameter = list(minlen = 1, maxlen = 10, conf = 0.1), appearance = list(rhs = c("Category = Student"), default = "lhs")

It is generated a set of 47,898 association rules which is shown in Fig. 13. In this scatter plot, a few high confidence rules are plotted in the top-center corner, some high confidence rules are plotted in the center and some lift rules are located close to

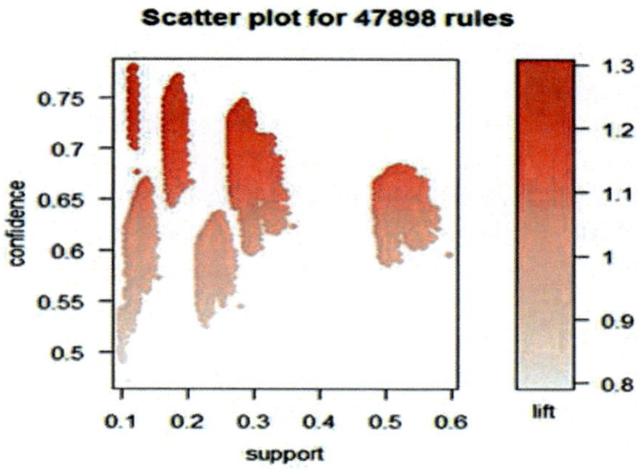

Fig. 13 Scatter plot for 47,898 rules

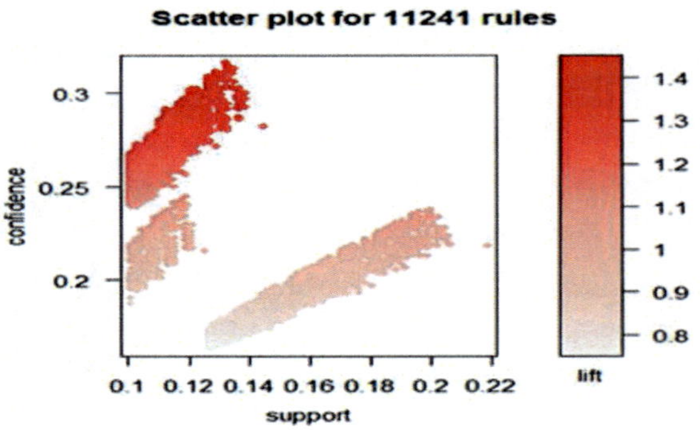

Fig. 14 Scatter plot for 11,241 rules

the minimum support threshold. A bulk of confidence rules are plotted middle-right side whose support parameter is 0.5 or higher but the lift is poor as compared to the high confidence rules. The parameter of 0.5 to 0.8 confidences, 0.1–0.6 supports, and 0.8–1.3 lifts are seen in this scatter plot. Now, we generate association rules for other categories by using the given apriori function:

>rules ← apriori(MyData, parameter = list(minlen = 1, maxlen = 10, conf = 0.1), appearance = list(rhs = c("Category = Other"), default = "lhs"))

It is generated a set of 11,241 association rules which is shown in Fig. 14. In this scatter plot, high confidence rules are plotted in the center to left-middle corner with decreasing confidence and support, some medium confidence rules are plotted in the center-left corner and some low confidence rules are plotted right-center to bottom-left side with decreasing order. The parameter of 0.1–0.3 confidences, 0.1–0.22 supports, and 0.8–1.4 lifts are seen in this scatter plot.

The resulting visualization of the set of 47,898 rules mined earlier is shown in Fig. 16. There are 20 LHS groups of rules are produced. Up to 12, 13, 14, or 15 items in LHS group are included. According to the lift, the most interesting rules are shown in the top-left corner. The balloon sizes represent supports which are increasing and start decreasing from the 12th LHS group for the left to right side corner. The lift of balloon for first group of 3867 rules is more but size of balloon is small.

The highest interest group of grouped matrix for 7815 rules are 79 rules which contains "SMSearch = Yes, PExam = Yes, +11 items" in the antecedent and "Category = Teacher" in the consequent. The lowest interest group of (bottom-right-hand corner) consists of 134 rules which contain "QP = Yes, IK = Yes, +10 items" in the antecedent and "Category = Teacher" in the consequent. The resulting visualization of the set of 7814 rules mined earlier is shown in Fig. 15. There are 20 LHS groups of rules are produced. Up to 10 or 11 items in LHS group are included. According to the lift, the most interesting rules are shown in the top-left corner. The balloon sizes represent supports which are decreasing for the left to right side corner. The lift of balloon for first group of 79 rules is more.

The highest interest group of grouped matrix for 47898 is 3867 rules which contains "Locale = Semi-urban, PExam = Yes, +14 items" for "Category = Student" which is shown in the (Fig. 16).

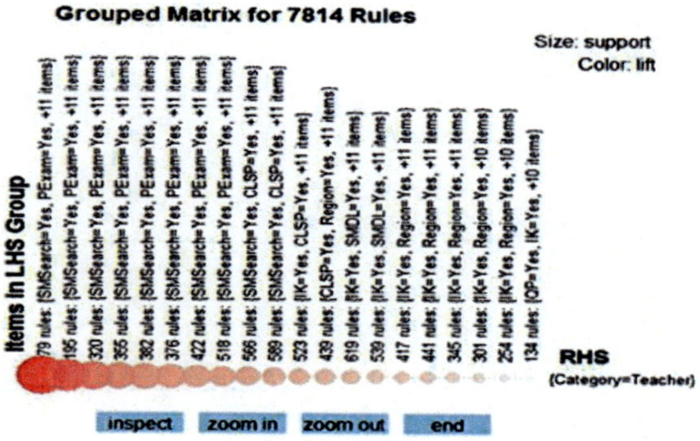

Fig. 15 Grouped matrix for 7814 rules

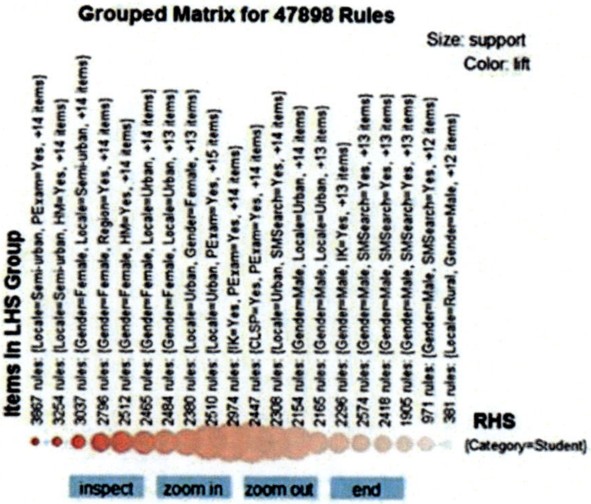

Fig. 16 Grouped matrix for 47,898 rules

The highest interest group of grouped matrix for 11241 rules is 121 rules which contains "Gender = Male, SMSearch = Yes, +9 items" in the antecedent and "Category=other" in the consequent. The lowest interest group of (bottom-right-hand corner) consists of 1021 rules which contain "IK = Yes, PExam = Yes, +11 items" in the antecedent and "Category = other" in the consequent. The resulting visualization of the set of 11,241 rules mined earlier is shown in Fig. 17. There are 20 LHS groups of rules are produced. Up to 9, 10, 11, 12, or 13 items in LHS group are included. According to the lift, the most interesting rules are shown in the top-left corner. The balloon sizes represent supports that are decreasing and start increasing from the 9th LHS group for left to right side corner. The lift of balloon for first group of 121 rules is more.

In Fig. 18 graph-based visualization, lowest support size is 0.149 and highest support size is 0.185. The range of lift is 0.901 to 1.008. The supports of association rules are increasing when size of bubbles is increasing. Gray levels are growing along with lift value. It is found here that A and B are positively correlated.

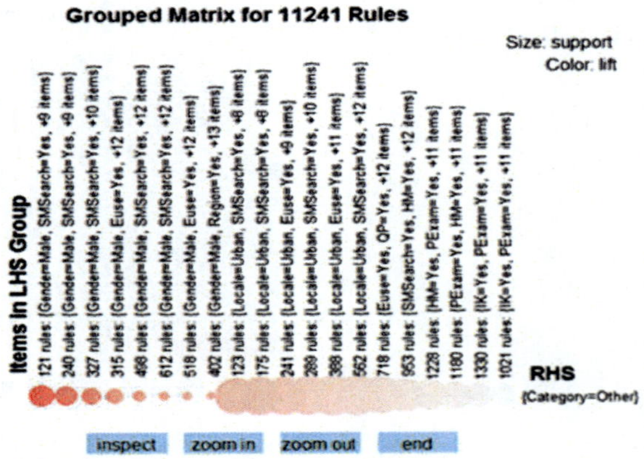

Fig. 17 Grouped matrix for 11,241 rules

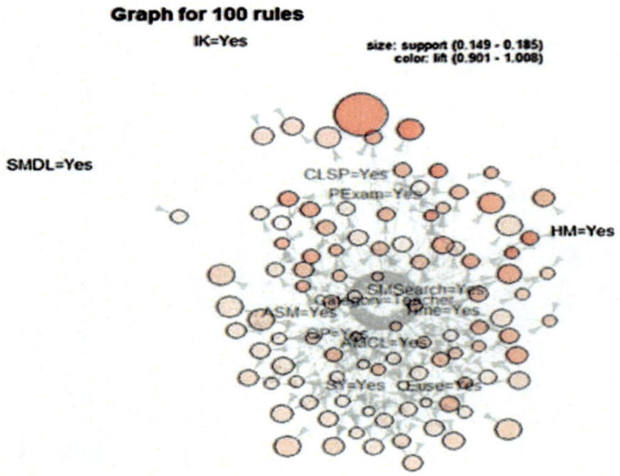

Fig. 18 Graph for 100 rules out of 7814

In Fig. 20 graph-based visualization, lowest support size is 0.556 and highest support size is 0.597. The range of lift is 1.0 to 1.122. The supports of association rules are increasing when size of bubbles are increasing. Gray levels are growing along with lift value. It is found here that A and B are positively correlated.

In Fig. 19 graph-based visualization, lowest support size is 0.192 and highest support size is 0.218. The range of lift is 0.977–1.089. The supports of association rules are increasing when size of bubbles are increasing. Gray levels are growing along with lift value. It is found here that A and B are positively correlated.

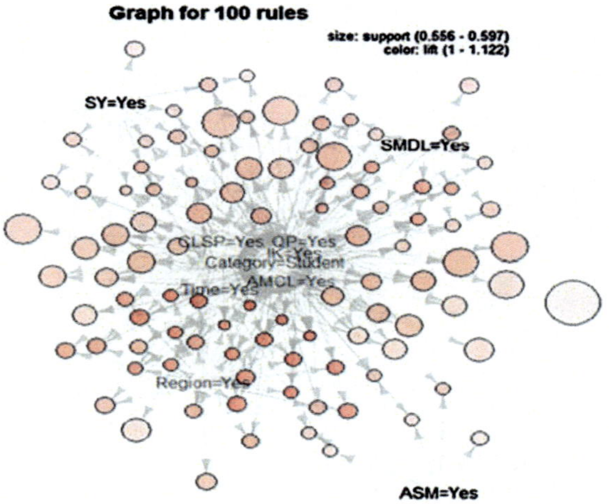

Fig. 19 Graph for 100 rules out of 47,898

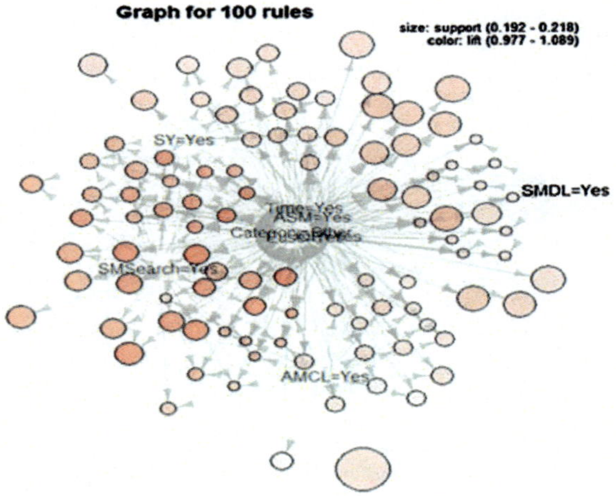

Fig. 20 Grouped matrix for 11,241 rules

9 Conclusion

We concluded the association rules A ⇒B of satisfaction and correlation. It is outcome from STO dataset which is a collection of all records of student, teacher, and other tables of HSES knowledge portal. It includes Locale, Category, IK, CLSP, SY, QP, Time, Region, Euse, PExam, Medium, SMDL, SMSearch, and ASM. In our experiment, supports ($A \Rightarrow B$) for all field descriptors are found more than minimum support 0.1. So, STO dataset satisfies support measure of association rules. Only the student category of STO dataset is satisfying confidence measure of association rules because of the item of the association rules. Based on the lift, experimental outcomes are discussed. A and B are positively or strongly correlated for field descriptors of dataset STO. Here all association rules are plotted with lift($A \Rightarrow B$) $\gg 1$. For semi-urban locale, A and B are not correlated with those association rules which are plotted with lift($A \Rightarrow B$) $= 1$. For Rural and urban locale, teacher, student, and others category, A and B are not correlated and negatively correlated also because some association rules are plotted with lift($A \Rightarrow B$) $= 1$ and lift($A \Rightarrow B$) < 1.

References

1. Agrawal, R., Imielinski, T., & Swami, A. (1993). Mining association rules between sets of items in large databases. In *Proceedings of the 1993 ACM SIGMOD International Conference on Management of Data* (pp. 207–216).
2. Brin, S., Motwani, R., Ullman, J. D., & Tsur, S. (1997). Dynamic itemset counting and implication rules for market basket data. In *Proceedings ACM SIGMOD 1997 International Conference on Management of Data* (pp. 255–264).
3. Unwin, A., Hofmann, H., & Bernt, K. (2001). The two key plots for multiple association rules control. In *Proceedings of the 5th European Conference on Principles of Data Mining and Knowledge Discovery* (pp. 472–483). Springer-Verlag.
4. Zaki, M. J. (2004). Mining non-redundant association rules. *Data Mining and Knowledge Discovery, 9*(3), 223–248.
5. Antonie, M. L., & Zaïane, O. R. (2004). Mining positive and negative association rules: An approach for confined rules. *Knowledge Discovery in Databases: PKDD 2004* (pp. 27–38). Berlin Heidelberg: Springer.
6. Zaman, A. M., David, T., & Kate, S. (2005). Redundant association rules reduction techniques. In S. Zhang, & R. Jarvis (Eds.), AI 2005, LNAI 3809, (pp. 254–263).
7. Ertek, G., & Demiriz, A. (2006). A framework for visualizing association mining results. In *ISCIS* (pp. 593–602).
8. Bruzzese, D., & Davino, C. (2008). Visual mining of association rules. *Visual data mining: Theory, techniques and tools for visual analytics* (pp. 103–122). Berlin: Springer-Verlag.
9. Hahsler, M., & Chelluboina, S. (2011). Visualizing association rules: Introduction to the R-extension package arulesViz. *Semantic Scholar*, 1–26.
10. Kumbhare, T. A., & Chobe, S. V. (2014). An overview of association rule mining algorithms. *International Journal of Computer Science and Information Technologies, 5*(1), 927–930.
11. Yüksel, A. M., & Emre, Y. (2014). Association rules in data mining: An application on a clothing and accessory specialty store. *Canadian Social Science, 10*(3), 75–83.
12. Zhao, Y. (2015). Association Rule Mining with R*. In *R and Data Mining Workshop* (pp. 1–30). Melbourne: Deakin University. http://www.RDataMining.com, May 28, 2005.
13. Sevri, M., Karacan, H., & Akcayol, M. A. (2017). Crime analysis based on association rules using apriori algorithm. *International Journal of Information and Electronics Engineering, 7*(3), 99–102.
14. Hahsler, M., & Karpienko, R. (2016). Visualizing association rules in hierarchical groups. *Journal of Business Economics*, 1–19.
15. Santhi, J. D., & Lilly, F. M. (2016). Implementation of the association rule mining algorithm for the recommender system of e-business. *International Journal of Computational Intelligence and Informatics, 6*(3), 198–205.
16. Karthikeyan, R., Venkatesan, K. G. S., Sundararajan, M., & Arulselvi, S. (2015). An applying association rule mining in education management systems. *International Journal of Innovative Research in Science, Engineering and Technology, 4*(5), 3771–3777.

17. Hahsler, M. Chelluboina, S. ,Hornik, K. and Buchta, C. (2010). The arules R-Package Ecosystem: Analyzing Interesting Patterns fromLarge Transaction Data Sets. *Journal of Machine Learning Research, 12*, 2021–2025.
18. Hahsler, M. (2017). ArulesViz: Interactive visualization of association rules with R. *The R Journal, 9*(2), 163–175 ISSN 2073-4859.

The Role of Heart Rate Variability in Atrial ECG Components of Normal Sinus Rhythm and Sinus Tachycardia Subjects

B. Dhananjay and J. Sivaraman

Abstract Heart Rate Variability (HRV) is a biological process that indicates any disturbances in the time duration between heartbeats. It is quantified by the number of disturbances in beat to beat interval and the contrast observed in heart rate may indicate various cardiac diseases. This study aims to delineate the role of HRV in Normal Sinus Rhythm (NSR) and Sinus Tachycardia (ST) subjects of atrial ECG components. Hundred young healthy individuals (20 women) of mean age (28 $\pm$ 5.36) were included in this study. ECGs were recorded in supine position using the EDAN SE-1010 PC ECG system. The mean atrial rate of NSR subjects is 87 $\pm$ 7.7 beats per minute (bpm) and 108 $\pm$ 4.2 bpm for ST subjects. The mean P wave duration for NSR and ST subjects is 96 $\pm$ 8.5 ms and 87 $\pm$ 18.6 ms, PR interval is 139 $\pm$ 17.4 ms and 107 $\pm$ 18.6 ms, and PP interval is 718 $\pm$ 56 ms and 534 $\pm$ 41 ms, respectively. Atrial sinus cycle length (PP interval) and HRV mainly affected the repolarization segment in ST subjects than the NSR subjects.

Keywords Electrocardiogram · Heart rate variability · Normal sinus rhythm · Sinus tachycardia

1 Introduction

Einthoven was the first to device standard limb lead electrode positioning to record the electrocardiogram (ECG) [1]. Later, unipolar leads were introduced by Wilson et al. [2], and subsequently, a standard protocol was introduced for chest lead positions [3]. In a few circumstances, where limbs of the person are not clinically accessible, modified limb electrode positions were used to resolve the issue [4]. For a complete analysis of atrial ECG morphology in tachycardia patients, modified limb electrode placements were used and besides it, few other alternative lead systems were also utilized [5]. Lewis et al. [6] designed a lead system to study the atrial ECG components during atrial fibrillation and noted that atrial fluctuations were maximum when

B. Dhananjay · J. Sivaraman (✉)
Bio-signals & Medical Instrumentation Laboratory, Department of Biotechnology and Medical Engineering, National Institute of Technology, Rourkela, Odisha, India
e-mail: jsiva@nitrkl.ac.in

S. C. Satapathy et al. (eds.), *Intelligent System Design*, Advances in Intelligent Systems and Computing 1171, https://doi.org/10.1007/978-981-15-5400-1_61

electrodes were positioned closer to the right atrium. Drury et al. [7] proved the existence of maximum atrial fluctuations throughout atrial fibrillation by placing leads on the sternum, anteriorly and posteriorly, complied with the study of Lewis.

P wave in healthy subjects represents atrial depolarization in the Standard Limb Lead (SLL) recording of ECG signals. However, atrial repolarization (Ta wave) transpires in PR segment, generally, it is overlooked in the SLL system because the amplitude of the Ta wave is between 10 and 60 μV, moreover, the activity of the ventricles occurs during atrial repolarization phase [8]. In third-degree AV block patients, the Ta wave can be visualized in SLL [9]. The SLL has its limitation in recording the Ta waves in NSR subjects and, to overcome it Sivaraman et al. [10–13] put forward an MLL ECG recording system for revealing the atrial repolarization wave in NSR subjects, Atrial Tachycardia (AT) and AV block patients. To boost the understanding of atrial ECG components in normal limits, MLL system was used [14, 15].

The irregular heartbeats caused at the atrial part of the heart is generally because of alterations in action potential duration (APD) which give rise to disturbances in repolarization. Disturbances in the repolarization phase are due to unstable changes of APD which occur at the cell level. In ECG, the QT interval represents the depolarization and repolarization in ventricles; similarly, the PTa interval represents the same in atria. Abnormal atrial repolarization is the key indicator for several classes of atrial disturbances. The alterations seen in the Ta wave and subsequently in the PTa interval during arrhythmias are clearly illustrated by Childers [16] and Roukoz [17]. Conditions like tachycardia and hyperthyroidism show an effect on atria by modifying the atrial repolarization phase and the variations observed are considered as principal changes in the atrial repolarization phase.

Generally, HRV is analyzed using R-R interval which in turn requires detection of R peaks in an ECG signal. However, in this study, the prime vision is on analyzing the role of HRV in atrial ECG components using PPInterval (PPI). Further, the study focuses on the analysis of PPI histogram and HRV trends between NSR and ST subjects in understanding the mechanisms of atrial waves under normal and tachycardia conditions.

2 Methodology

Subjects: The study comprised of 100 young healthy subjects (20 females) of mean age 28 ± 5.36 years. The volunteers interested in this study had given a written consent for their participation in the study. All the subjects were examined for any cardiovascular diseases. Smokers and drinkers were excluded from the study (Table 1).

Data Collection: The subjects considered in this study did not have any past medical history. Data processing was done with the help of EDAN software system. The noise reduction in the ECG signal was done by applying 25 Hz filter. The ECG was recorded in the supine position with EDAN SE-1010 PC ECG instrument. The

Table 1 Statistics of the subjects

Age statistics		
Healthy volunteers	80 (22 ± 1.4)	(20.0, 22.0, 24.0)
ST subjects	20 (30 ± 3.3)	(27.0, 29.5, 33.0)
Total subjects	100 (28 ± 2.4)	(20.0, 23.0, 33.0)

All values are expressed in *n* (mean ± SD), (minimum, median and maximum)

time duration of the recording of data for each subject was 60 sec. The sampling rate of the instrument is 1000 samples per second and the frequency response is 0.05–150 Hz. For a better understanding of the ECG signal, the signal could be printed in variable gain and with variable speed.

Statistical Analysis: All the data are expressed in mean ± standard deviation. Shapiro–Wilk W test was performed to test the normality of the given data. Pearson correlation was used for correlation analysis and all the data was statistically computed in Origin Pro-8 of Origin Lab Corporation, 2016 version.

3 Results

The duration of the ECG recording of each subject was 60 s. Figure 1a represents the PP histogram for an individual NSR subject. The x-axis is the duration of (PPI) and the y-axis represents the number of observations of PPIs. The PPI of NSR subject varied between 781 and 910 ms and the average PPI is 846 ms. Figure 1b describes the trend of the heart rate of the NSR subject. The trends of the heart rate of NSR subjects show an average of 71 beats per minute (bpm). Figure 2a depicts the PP

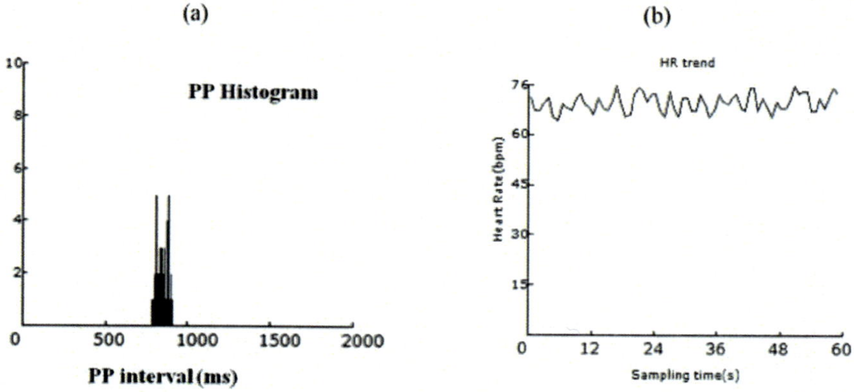

Fig. 1 Analysis of HRV in NSR subjects. **a** PP Histogram of NSR subject, **b** Heart rate trend of a NSR subject

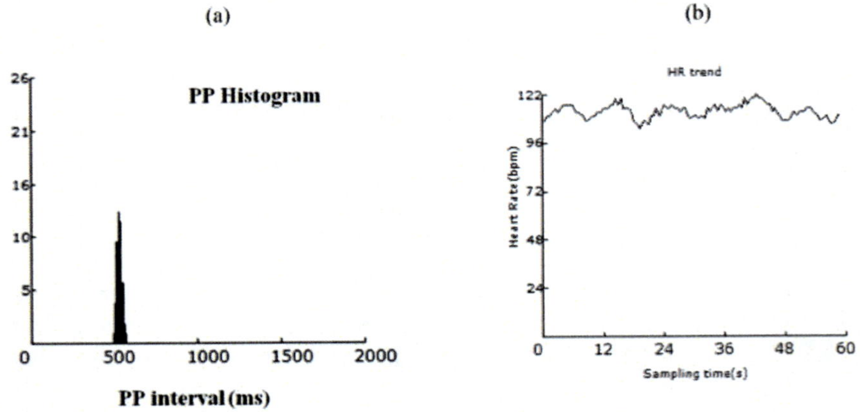

Fig. 2 Analysis of HRV in ST subjects. **a** PP Histogram of ST subject, **b** Heart rate trend of a ST subject

histogram for an individual ST subject. The PPI of ST subject varied between 491 and 567 ms and the average PPI is 527 ms. Figure 2b denotes the heart rate trend of a ST subject and the average heart rate of a ST subject is 113 bpm.

The box and whisker plot in Fig. 3 shows the duration of P wave in NSR subjects and ST patients. In ST subjects, the minimum value of P wave duration is about 66.5 ms. The lower quartile is at about 77.2 ms, median value is 84 ms, upper quartile is 93.7 ms and the maximum P wave value is 102.3 ms. In NSR subjects, the minimum value of P wave duration is 85.5 ms, lower quartile value is 92.7 ms, median value is 95.5 ms, upper quartile value 99.7 ms and the maximum value are 108.1 ms.

In Fig. 4, the whisker plot shows the variation of HRV in ST condition and NSR subjects. The minimum heart rate value of ST subjects is 100 (bpm), the lower quartile value is 102.25 (bpm), median value is 105 (bpm), upper quartile value is 108 (bpm) and the maximum value is 113 (bpm). In NSR subjects, the minimum heart rate value

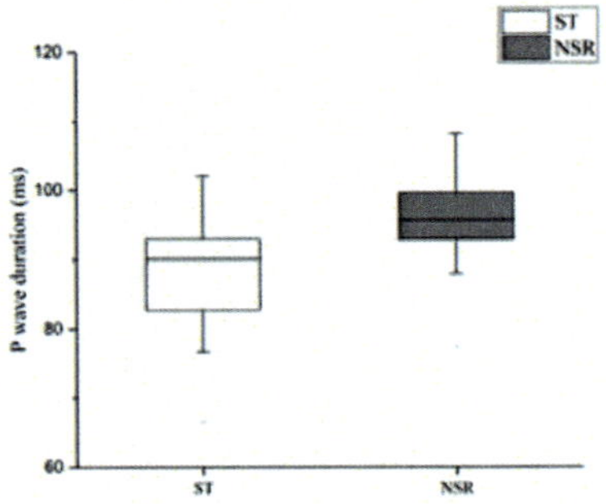

Fig. 3 Box plot of *P* wave duration in ST and NSR Subjects

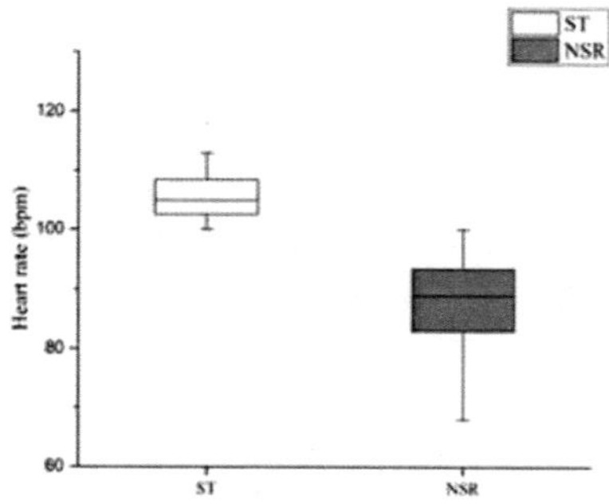

Fig. 4 Box plot of Heart rate in ST and NSR subjects

is 68 (bpm), lower quartile value is at 83 (bpm), median value is 89 (bpm), upper quartile value is 93.75 (bpm) and maximum value of heart rate is 100 (bpm).

The time-domain analysis and comparison of HRV between the two subjects of study are shown in Table 2. The mean and standard deviation of heart rate, average PP interval, max PP, min PP, and other statistical analysis of all the subjects are shown for comparative purposes.

Table 2 Comparison of HRV between NSR and ST subjects

Measurement	Units	NSR subjects		ST subjects	
		Mean	SD	Mean	SD
Atrial rate (PPI)	bpm	83.5	6.29	109.6	5.05
Average PP interval	ms	718.3	56.56	534.5	41.57
Max PP	ms	841.8	92.78	623.3	74.55
Min PP	ms	576.8	138.23	482.8	49.81
SDNN	ms	54.4	26.82	27.06	13.56
RMSSD	ms	40.5	14.27	17.85	7.69
NN50 (count)		12.5	8.57	2.7	3.77
PNN50	%	15.7	11.84	2.28	3.12
LF		5.6	5.95	4	6.55
HF		2.94	4.17	1.54	1.69
LF [norm]		57.6	15.12	42.9	20.97
HF [norm]		29.43	14.06	19.05	10
LF/HF		2.8	2.39	3.5	2.94
Total power	ms^2	18.4	20.82	16.4	19.68

4 Discussion

P wave duration in Atrial Tachycardia: Konieczny et al. recorded ECG using the SLL system in athletes using signal-averaged ECG (SAECG) on P wave parameters. In his study, the P wave duration showed a direct relation with the volume of the right atrium in the athletes. The length of the P wave duration, in this case, was unrelated to atrial arrhythmia in the athletes [18]. In interatrial block patients, the atrial rate is statistically high when compared with atrial fibrillation [19]. In this study, it is noted that the P wave duration decreased linearly for ST patients and it is not true for NSR group. The reason for decrease of P wave duration in ST patients is due to the higher atrial rate.

Relationship between HRV and Myocardial infarction (MI): Patients suffering from acute MI display a very low parasympathetic cardiac control and a very high proportion of sympathetic activity [20]. Ventricular fibrillation in MI patients is caused due to a decrease in sympathetic activity [21]. There is a direct association between sinus arrhythmia and parasympathetic cardiac control, which can be used as an analytical tool in MI patients. [22, 23]. It has been documented that, in MI patients HRV decreases [24]. Though MI affected patients who underwent exercise training, there was no improvement in HRV indexes [25]. In the present study, the relationship between HRV and ST subjects was studied and the results of this study showed that the HRV shows an increasing trend in ST when compared with NSR subjects. The P wave duration of ST subjects was less when compared with NSR subjects. Heart rate of ST subjects was more according to NSR subjects.

Statistical analysis of HRV: In a study [26] between congestive heart rate (CHF) patients and NSR subjects it was found that frequency domain, nonlinear, and time-domain parameters showed a variation. Sivaraman et al. [27] in the study of stability analysis between NSR subjects and AT patients, a comparison of HRV showed a statistical difference between them. In the present study, a few more statistical parameters such as NN50, pNN50, LF, HF were included to substantiate the role of HRV between the two groups. NN50 mean value of NSR subjects and ST patients is 12.5 $\pm$ 8.57 ms and 2.7 $\pm$ 3.77 ms respectively. PNN50 mean value of NSR subjects is 15.7 $\pm$ 11.84 and 2.28 $\pm$ 3.12 ms for ST patients.

5 Conclusion

The focus of the study is to clearly understand the role of HRV in NSR and ST subjects in atrial ECG components. In P wave duration analysis, it is found that there was a declining trend in ST subjects when compared with NSR subjects. Heart rate of ST subjects was higher than NSR subjects and this study also brings out statistical analysis of HRV in ST and NSR subjects. Temporal aspects in HRV are higher for NSR than ST subjects and the total power for HRV is higher in NSR subjects.

Acknowledgements The authors acknowledge the support from MHRD, Government of India, for sponsoring the Ph.D. Programme of the first author. The present study was supported by financial grants from Science Engineering Research Board (SERB), Department of Science and Technology, Government of India, (EEQ/2019/000148).

References

1. Einthoven, W. (1912). The different forms of the human electrocardiogram and their signification. *Lancet, 1,* 853–861.
2. Wilson, F. N., Johnston, F. D., MacLeod, A. G., & Barker, P. S. (1934). Electrocardiographic electrode Electrocardiograms that represent the potential variations of a single electrode. *American Heart Journal, 9,* 447–471.
3. Barnes, A., Pardee, H. E. B., White, P. D., Wilson, F. N., & Wolferth, C. C. (1938). Effect of Standardization of precordial leads—Supplementary report. *American Heart Journal, 15,* 235–239.
4. Diamond, D., Griffith, D. H., Greenberg, M. L., & Carleton, R. A. (1979). Torso mounted electrocardiographic electrodes for routine clinical electrocardiography. *Journal of Electrocardiology, 12,* 403–406.
5. Szekely, P. (1944). Chest leads for the demonstration of auricular activity. *British Heart Journal, 6,* 238–246.
6. Lewis, T. (1910). Auricular fibrillation and its relationship to clinical irregularity of the heartbeat. *Heart, 1,* 306–372.
7. Drury, A. N., & Iliescu, C. C. (1921). The electrocardiograms of clinical fibrillation. *Heart, 8,* 171–191.
8. Briggs, K. L. (1994). A digital approach to cardiac cycle. *IEEE Engineering in Medicine and Biology, 13,* 454–456.
9. Sprague, H. B., & White, P. D. (1925). Clinical observations on the T wave of the auricle appearing in the human electrocardiogram. *Journal of Clinical Investigation, 1,* 389–402.
10. Sivaraman, J., Uma, G., & Umapathy, M. (2012). A modified chest leads for minimization of ventricular activity in electrocardiograms. In *Proceedings of International Conference on Biomedical Engineering* (pp. 79–82).
11. Sivaraman, J., Uma, G., Venkatesan, S., Umapathy, M., & Dhandapani, V. E. (2013). A novel approach to determine atrial repolarization in electrocardiograms. *Journal of Electrocardiology, 46,* e1.
12. Sivaraman, J., Uma, G., Venkatesan, S., Umapathy, M., & Ravi, M. S. (2015). Unmasking of atrial repolarization waves using a simple modified limb lead system. *Anatolian Journal of Cardiology, 15,* 605–610.
13. Sivaraman, J., Uma, G., Venkatesan, S., Umapathy, M., & Keshav Kumar, N. (2014). A study on atrial Ta wave morphology in healthy subjects: an approach using P wave signal averaging method. *Journal of Medical Imaging and Health Informatics, 4,* 675–680.
14. Sivaraman, J., Uma, G., Venkatesan, S., Umapathy, M., & Dhandapani, V. E. (2015). Normal limits of ECG measurements related to atrial activity using a modified limb lead system. *Anatolian Journal of Cardiology, 15,* 2–6.
15. Sivaraman, J., Venkatesan, S., Periyasamy, R., Joseph, Justin, & Ravi, M. S. (2017). Modified limb lead ECG system effects on electrocardiographic wave amplitudes and frontal plane axis in sinus rhythm subjects. *Anatolian Journal of Cardiology, 17,* 46–54.
16. Childers, R. (2011). Atrial repolarization: its impact on electrocardiography. *Journal of Electrocardiology, 44,* 635–640.
17. Henri, R., & Kyuhyun, W. (2011). Response to letter to the editor. *The Annals of Noninvasive Electrocardiology, 16,* 416–417.

18. Konieczny, K., Banks, L., Osman, W., Glibbery, M., & Connely, K. A. (2019). Prolonged P wave duration is associated with right atrial dimensions, but not atrial arrhythmias, in middle-aged endurance athletes. *Journal of Electrocardiology, 56,* 115–120.
19. Tekkesin, A. I., Cinier, G., & Cakilli, Y. (2017). Interatrial block predicts atrial high rate episodes detected by cardiac implantable electronic devices. *Journal of Electrocardiology, 50,* 234–237.
20. Rothschild, M., Rothschild, A., & Pfeifer, M. (1988). Temporary decrease in cardiac parasympathetic tone after acute myocardial infarction. *Journal of Cardiology, 18,* 637–639.
21. Schwartz, P. J., Rovere, M. T., & Vanoli, E. (1992). Autonomic nervous system and sudden cardiac death. Experimental basis and clinical observations for post-myocardial infarction risk stratification. *Circulation, 85*(Suppl I), 177–191.
22. Katona, P. G., & Jih, F. (1975). Respiratory sinus arrhythmia: non-invasive measure of parasympathetic cardiac control. *Journal of Applied Physiology, 39,* 801–805.
23. Malliani, A., Pagani, M., Lombardi, F., & Cerutti, S. (1991). Cardiovascular neural regulation explored in the frequency domain. *Circulation, 84,* 482–492.
24. Carney, R. M., Blumenthal, J. A., Watkins, L. L., & Catellier, D. (2001). Depression, heart rate variability, and acute myocardial infarction. *Circulation, 104,* 20–24.
25. Duru, F., Candinas, R., Dzeikan, G., Goebbels, U., Myers, J., Dubach, P., et al. (2000). Effect of exercise training on heart rate variability in patients with new-onset left ventricular dysfunction after myocardial infarction. *The Journal of Heart, 140*(1), 157–161.
26. Wang, Y., Wei, S., Zhang, S., Zhang, Y., Zhao, L., Liu, C., et al. (2018). Comparison of time-domain, frequency-domain, and non-linear analysis for distinguishing congestive heart failure patients from normal sinus rhythm subjects. *Biomedical Signal Processing and Control, 42,* 30–36.
27. Sivaraman, J., Uma, G., Langley, P., Umapathy, M., Venkatesan, S., & Palanikumar, G. (2016). A study on stability analysis of atrial repolarization variability using ARX model in sinus rhythm and atrial tachycardia ECGs. *Computer Methods and Programs in Biomedicine, 137,* 341–351.

Road Traffic Counting and Analysis Using Video Processing

A. Kousar Nikhath, N. Venkata Sailaja, R. Vasavi, and R. Vijaya Saraswathi

Abstract The vehicle counting system can detect and count the number of vehicles that are moving on roads at any particular time period accurately using video processing. Highways, expressways, and roads are becoming overcrowded with the increase of a large number of vehicles. Many of the government operations to find a solution for the limitless traffic have been continuously working on this kind of data retrieval on traffic counting. The requirement to detect and count the moving vehicle is getting very high. Moving vehicle detection, tracking, and counting are very critical for traffic flow monitoring, planning, and controlling. The video-based solution is easy to install and use and does not disturb traffic flow. By analyzing the traffic sequence on roads at every instance of timing with video processing, the proposed method can detect and count moving vehicles accurately using video processing. As of now operations and research on how to find, the data for the further analysis of traffic counting are going on in the government sector to implement solutions for controlling the heavy traffic.

Keywords Vehicle count · Blobs · Video-based system · Traffic sequence

1 Introduction

The main objective of this project is to develop a video-based system that can be used to count the road traffic, and it does not disturb traffic flow as we only capture the video at that instance and we process it. By analyzing the traffic sequence on

A. Kousar Nikhath (✉) · N. V. Sailaja · R. Vasavi · R. Vijaya Saraswathi
Department of Computer Science and Engineering, VNR VJIET, Hyderabad, Telangana, India
e-mail: kousarnikhath@vnrvjiet.in

N. V. Sailaja
e-mail: sailaja_nv@vnrvjiet.in

R. Vasavi
e-mail: vasavi_r@vnrvjiet.in

R. Vijaya Saraswathi
e-mail: vijayasaraswathi_r@vnrvjiet.in

S. C. Satapathy et al. (eds.), *Intelligent System Design*, Advances in Intelligent Systems and Computing 1171, https://doi.org/10.1007/978-981-15-5400-1_62

roads at every instance of timing with computer vision, we count the number of vehicles traveling on that road or lane [1]. The video-based system can detect and count moving vehicles [2] accurately using computer vision.

The project is all about detecting and counting vehicles from a CCTV video footage and to finally present an idea of what the real-time scenario on-street situation is across the road network in any lane or area or location.

1.1 OpenCV

Open-source computer vision is the full form of OpenCV. It is a library of many programming functions mainly aimed at real-time computer vision [3]. Originally, OpenCV was developed by Intel, and it was later supported by Willow Garage then 'Itseez' (which was later acquired by Intel). The library is a cross-platform and is a free to use software under the open-source BSD license.

It was officially launched in 1999 and the OpenCV project was initially launched by the Intel Research initiative to advance all of the CPU-intensive applications, part of a series of projects including real-time ray tracing and display walls (3D) [4]. The main contributors to the project include a number of optimization experts in Intel Russia, as well as Intel's Performance Library Team.

1.2 Vehicle Counter Algorithm

In this algorithm, we have two stages. They are vehicle and vehicle counter. In vehicle stage, we identify the motion of the detected object and decide whether the object is vehicle or not and assign id if the object is vehicle. In vehicle counter stage, we take out the id and increment the count once the vehicle crosses the road [5]. We identify the motion of the object by plotting it over a graph and select the object, which lies under the region. We maintain separate counters for both left and right lanes.

2 Proposed System

Our project is a video-based solution that captures the video at that instance, and then the video is processed to evaluate the results. By analyzing the traffic sequence on roads at every instance of timing with computer vision [6, 7], we count the number of vehicles traveling on that road or lane. The proposed method can detect and count moving vehicles accurately using computer vision. The purpose of this project is to detect and count vehicles from a CCTV feed. The project work will result as a standalone system to retrieve and analyze the real-time on-street traffic situations across the road network. The work can be further considered as base for tracking pollution and accident analysis

Advantages

1. The proposed system can identify the vehicles that travel in nonlinear motion by changing into multiple lanes and increase the count.
2. The system provides individual lane counts so that the roads can be analyzed more precisely.
3. With the help of count, we can analyze the road traffic and based on that we can rectify the traffic problem by providing alternate routes.
4. The generated count can be used to analyze the traffic over bridges. Also based on the count of vehicles, we can recognize any structural damages prior to the event by relating the count with the capacity of the structure.
5. The proposed system can be used even with low-quality video feed.

3 Methodology

Initially, the lane has to be identified in order to get a background reference, as we need a base for the identification of the vehicles that are traveling on that particular lane. This lane identification is used to set certain borders for the identification and is helpful for identifying the objects traveling on that lane [6, 8].

After lane detection, we take the lane image as a reference and identify vehicle with the help of blob formation. Here the image is identified as a binary image, and whenever an object is identified passing through the lane as the pixel value changes, these variations are identified, and then the pixel values are highlighted and which leads to the creation of blobs [4, 9]. This finally results in the blob formation. The image is formed by dilating the initial blob. This process is done to increase the threshold value of the blobs so that we can easily identify the object [10]. Along with dilation, we fill out the gaps to make the blob falls under an object. In this process, we clear all the noises from the background so that we can easily differentiate the objects from the noise.

After dilating the blobs, we draw rectangular box around the blob. These rectangular boxes are drawn using the height and width of the blob. This is done to make us understand the object has been correctly identified. Along with that, we note down the centers of the blob, track the motion of the identified object, and then draw a line joining this centroid, and this line is considered as a vector [11, 12]. Now if the identified vector [13] is in a certain angle limit, then it is identified as a vehicle [5], whereas when it is not in the given angle limit, then it is not identified as a vehicle

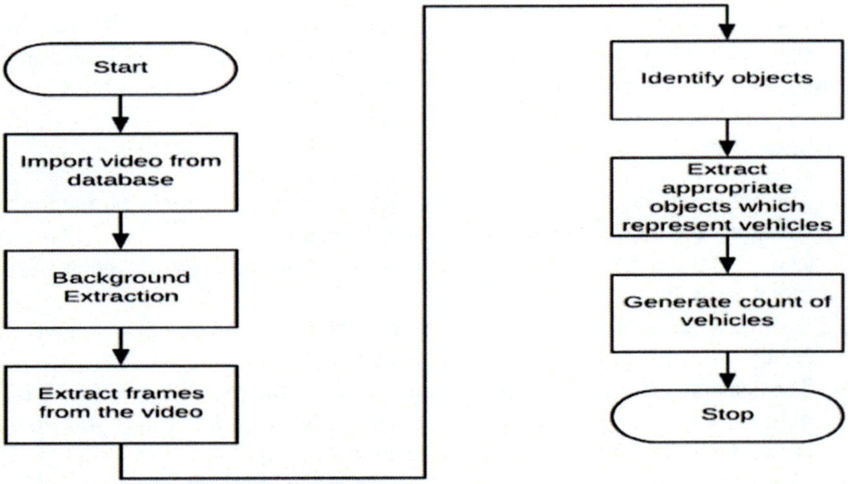

Fig. 1 Process flow of the entire project

[7, 14]. To uniquely identify each vehicle, we give id numbers for each vehicle individually. This helps us to identify whether the vehicle has been counted or not. If the vehicles are not counted, then we increment the vehicle count.

Process Flow

Figure 1 represents the workflow or process, showing the steps as boxes of various kinds of the modules, and their order by connecting them with all other arrows. Every symbol represents a part of code written in the program. The start/end symbol, represented by oval shape, can be used to represent either the beginning or ending of a program. The symbol for process, represented by rectangle shape box, allows you to show how the program is functioning. When you decide to enter data, show it on the screen, or print it to paper, you use the input/output symbol, represented by parallelogram. The decision symbol, represented by a rhombus shape, is used for things like 'if statements,' where you must choose an option based on a specified criterion. The flow line denotes the direction of logic flow in the program.

4 Result

The system is tested by giving the road traffic videos taken from TFL-jam-cams. Figures 2, 3, 4, 5, and 6 are pictures of the implementation at different steps.

Figure 2 is the lane image, which is used as reference in order to identify vehicles.

Figure 3 is the blob image, which is formed by the help of reference image. The blobs represent the vehicles. Figure 4 is the dilated image of Fig. 3. This is done in order to increase the threshold intensity of the blobs.

Figures 5 and 6 are the screenshots of initial and final count of the video given.

Fig. 2 Lane image

Fig. 3 Blob image

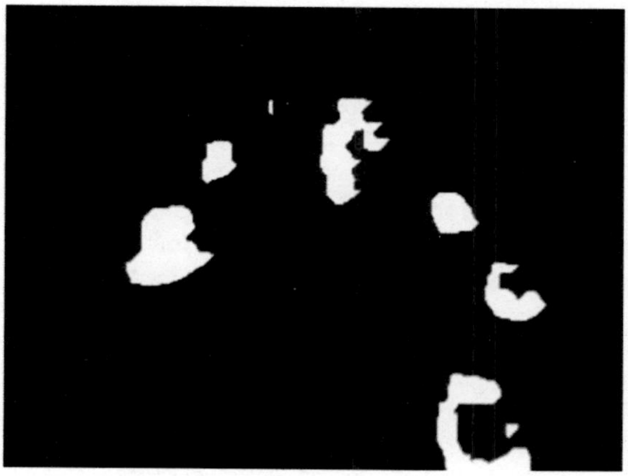

Fig. 4 Dilated blob image

Fig. 5 Initial count

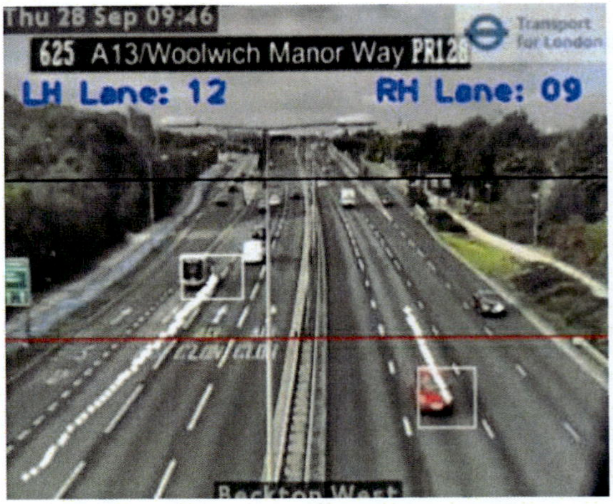

Fig. 6 Final count

5 Conclusion

We aim to create a system, which can count the vehicles that are traveling on roads by implementing a vehicle counting system [12]. It detects the lanes, then identifies the objects that are present on the lanes, and then detects the vehicles from the vehicles and give us the count of the vehicles traveling in that road. This data can be used to analyze the traffic system off the roads.

References

1. Coifman, B. (2006). Vehicle level evaluation of loop detectors and the remote traffic microwave sensor. *Journal of Transportation Engineering, 132*(3), 213–226.
2. Chigurupati, S., Polavarapu, S., Kancherla, Y., & Nikhath, A. K. (2012). Integrated computing system for measuring driver safety index. *International Journal of Emerging Technology in Advanced Engineering, 2*(6).

3. Collins, R. T., et al. (2000). *A system for video surveillance and monitoring* (pp. 1–68). VASM final Report: Robotics Institute, Carnegie Mellon University.
4. Yang, G. (2012). Video vehicle detection based on self adaptive background update. *Journal of Nanjing Institute of Technology (Natural Science Edition), 2,* 13.
5. Alpatov, B. A., Babayan, P. V., & Ershov, M. D. Vehicle detection and counting system for real -time traffic surveillance. In *Published by IEEE Conference.*
6. Beymer, D., McLauchlan, P., Coifman, B., & Malik, J. (1997). A real-time computer vision system for measuring traffic parameters. In *Proceedings of IEEE Conference on Computer Vision and Pattern Recognition* (pp. 496–501), Puerto Rico, June 1997.
7. Lowe, D. G. (1999). Object recognition from local scale-invariant features. In *Computer Vision, Seventh IEEE International Conference.*
8. Bas, E., Tekalp, A. M., & Salman, F. S. (2007). Automatic vehicle counting from video for traffic flow analysis. In *IEEE Intelligent Vehicles Symposium.*
9. Wu, K. et al. (2011). Overview of video-based vehicle detection technologies.
10. Friedman, N., & Russell, S. (1997). Image segmentation in video sequences: A probabilistic approach. In *Proceedings of the Thirteenth Conference on Uncertainty in artificial intelligence.* Morgan Kaufmann Publishers Inc.
11. Rabiu, H. (2013). Vehicle detection and classification for cluttered urban intersection. *International Journal of Computer Science, Engineering and Applications, 3*(1), 37.
12. Mithun, N. C., Rashid, N. U., & Rahman, S. M. (2012). Detection and classification of vehicles from video using multiple time-spatial images. *IEEE Transactions on Intelligent Transportation Systems, 13*(3), 1215–1225.
13. Li, D., Liang, B., Zhang, W. (2014). Real-time moving vehicle detection tracking and counting system implemented with OpenCV. *Information Science and Technology*, 631–634.
14. Seenouvong, N., Watchareeruetai, U., Nuthong, C., Khongsomboon, K., & Ohnishi, N. (2016). A computer vision based vehicle detection and counting system. *Knowledge and Smart Technology*, 224–227.

SCP Design of Situated Cognitive Processing Model to Assist Learning-Centric Approach for Higher Education in Smart Classrooms

A. Kousar Nikhath, S. Nagini, R. Vasavi, and S. Vasundra

Abstract With the use of a variety of instructional design approaches being used to let learners fit into real-world applications, the field of situated cognition will let the current-generation students to get suitable for industry endorsements. The use of blended learning methodologies and activities has a culture change for the learners to link with the real-world scenarios, wherein learners are ready with high-end engagement and industry readiness. The proposed situated cognitive processing (SCP) model will be a helpful tool in assisting the learner-centric approach of creativity and application of the current trends. The proposed model would like to address the advantages of the situated learning behavior, wherein student is involved in learning and adopting the knowledge based on situations and conditions.

Keywords Situated cognition · Situated learning · Speech recognition · Linguistics · Human–computer interaction · Face recognition · Psychology

1 Introduction

The current pedagogy teaching–learning methodologies being used have changed the pace of student-centric learning process. The techniques like Think-Pair-Share, learning by doing, flipped classrooms, online assessments, etc., are letting the student involvement and skilled enough for the course content in which students are eventually utilizing the resources to the maximum extent. As a result, instructors too are

A. Kousar Nikhath (✉) · S. Nagini · R. Vasavi
Department of Computer Science and Engineering, VNR VJIET, Hyderabad, Telangana, India
e-mail: kousarnikhath@vnrvjiet.in

S. Nagini
e-mail: nagini_s@vnrvjiet.in

R. Vasavi
e-mail: 3vasavi_r@vnrvjiet.in

S. Vasundra
Department of Computer Science and Engineering, JNTUA, Ananthapur, India
e-mail: vasundras.cse@jntua.ac.in

S. C. Satapathy et al. (eds.), *Intelligent System Design*, Advances in Intelligent Systems and Computing 1171, https://doi.org/10.1007/978-981-15-5400-1_63

feeling it comfortable for a complete satisfactory delivery process and full immersion of classroom encouragement, letting a scope for students to be skilled enough for sharing their ideas and knowledge [1]. The proposed model makes the study of various learning patterns in such forcible situation of immersion kind of learning demands [13]. The current pace of digital technology and the use of smart devices and gadgets have a tremendous growth in creating zeal and interest in learner [2, 3]. The use of online assessments, case study approach, and real-time problem solving has translated the incorporation of interactive involvement and simulations of real-life settings. The proposed model will be a helpful tool in assisting the learner-centric approach of creativity and application. The proposed model would like to address the advantages of the situated learning behavior over a traditional concept-based learning strategy [5, 14]. This model will let an instructor to integrate and practice contextual learning into their curriculum.

2 Background

The arise of technologies like artificial intelligence, machine learning, deep learning, and human–computer interaction has led to the inception of domains of new era like cognitive science, cognitive computing, cognitive linguistics, and so on [4]. Cognitive computing describes the use of computerized models to simulate the human cognition process and find solutions in complex situations where the answers may be ambiguous and uncertain. Cognitive computing is interchangeably being used with the terms AI, expert systems, neural networks, robotics, and virtual reality (VR) [5, 6, 15].

- **Working Process of Cognitive Computing**

Cognitive computing systems can synthesize data from various information sources, assign weights to the context, and provide suitable evidence in suggesting the best possible outcome or a result [16]. The system involves the use of self-learning technology interface, pattern recognition, and natural language processing to simulate the human working functionality. *Theoretically*, cognitive science and cognitive computing differ from some other fields, but *methodologically,* the tools for investigation will differ in terms of information processing, analytics, simulation, responsiveness, statistical, and descriptive modeling [7–9]. Features of cognitive computing are listed as shown in Figs. 1 and 2.

Figure 3 illustrates various kinds of learning styles in terms of behavioral, cognitive, and emotional engagement.

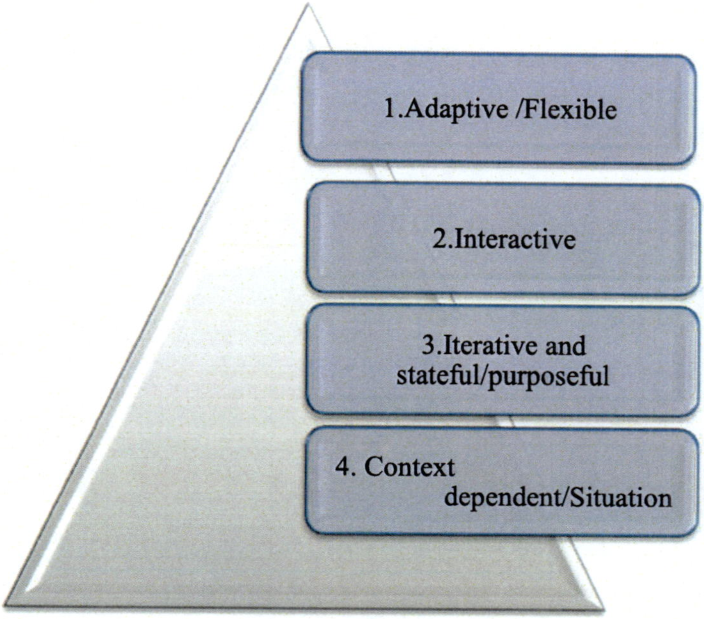

Fig. 1 Features of cognitive systems

3 Need for Situated Cognitive Processing Model

Sometimes, a student complete involvement may be negatively engaged if they report any dislike or if any anxiety distress toward their learning process is affected indirectly or directly. In addition, the use of mobile phones and ongoing social media apps is making students to be always associated to the instant responses or status updates from these social applications. In such situations, it is difficult to conclude the outcome or the actual learning strategy that took place in a classroom [9, 10]. The proposed model will result in rubrics and statistical dashboard resulting in interpreting social and emotional engagement of the students in the classrooms [11, 12]. The use of the methodologies like online quiz, teamwork, learning by doing, showcase the project, each one teach one, choice-based answering, group activities, design thinking, and problem analysis will result in improved and effective engagement of learners. In addition to that, the assessment and responsiveness toward the instructor will be increased, leading toward overall improvement for learner-centric needs.

Trowler (2010) identifies positive and negative elements of all three dimensions in terms of student engagement process (Table 1).

Research Questions required in processing the model
See Fig. 4, 5 and Table 2.

Fig. 2 Strategic teaching methodologies used in situated cognitive processing (SCP) model

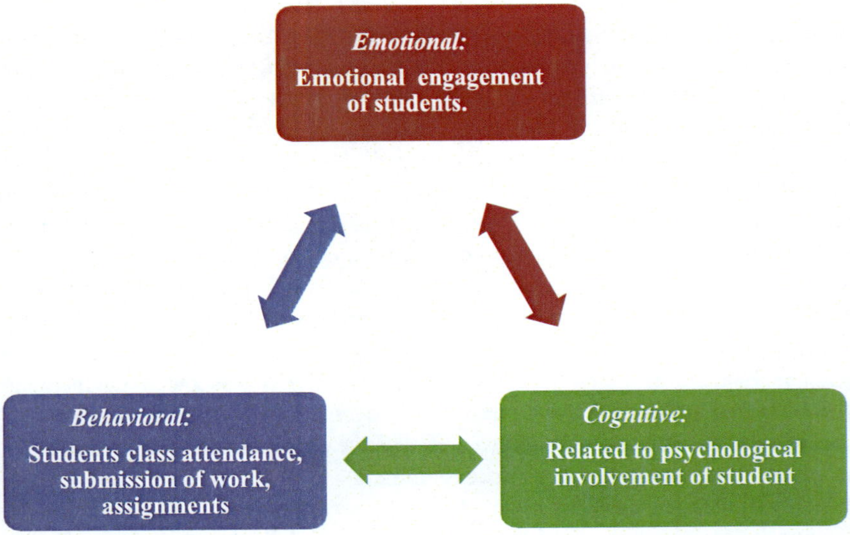

Fig. 3 Classification of learning patterns

Table 1 Engagement for learner-centric approach

Learner-centric			
Learning pattern	Positive engagement	Non-engagement	Negative engagement
Behavioral	Attends lectures, participates with enthusiasm, full attendance	Skips lectures, without excuse, postponements	Boycotts, pickets, or disrupts lectures, mobile engagement
Emotional	Shows interest	Feels boredom	Rejection
Cognitive	Meets or exceeds, assignment requirements	Rushed or absent to classes	Redefines parameters for assignments
Response will be observed, recorded, and given as an input to the model to assess from the instructor-centric level and then be evaluated in the model			

Fig. 4 Parameters that are involved in assessment of model

4 Results

See Fig. 6, 7, 8, 9, 10, 11, 12 and 13.

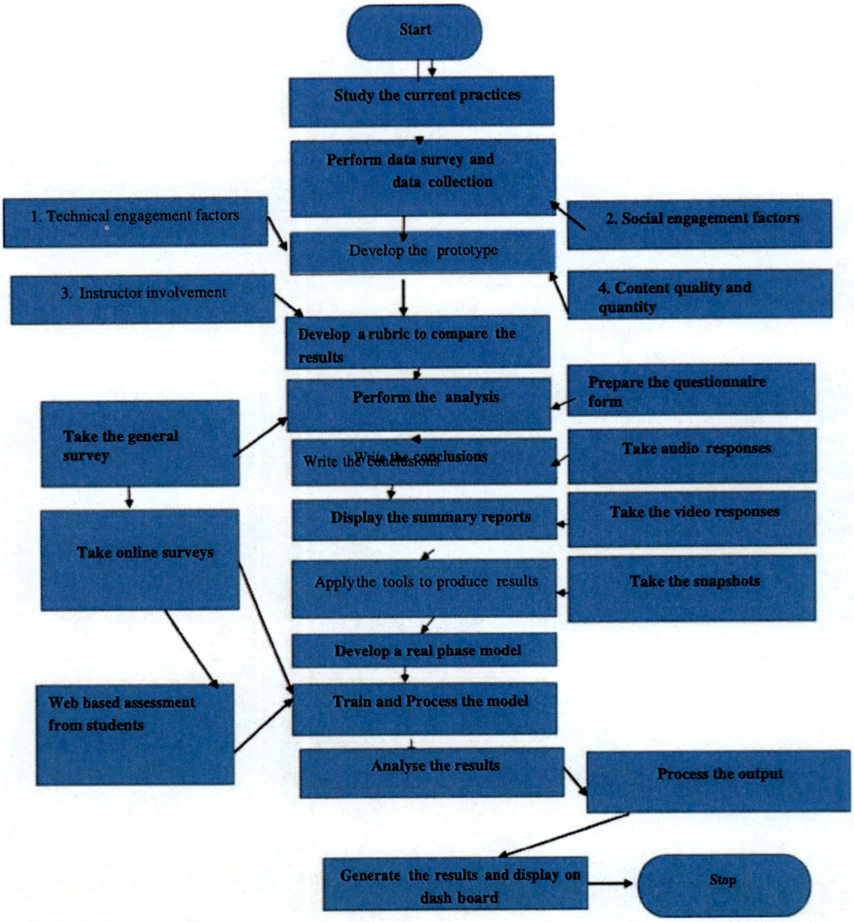

Fig. 5 Framework designed to evaluate the learning patterns of student's in different situations

Table 2 List of research questions designed to process model

	List of research questions designed to process the model	Rate score 0–10
1.	What are the instructional practices that are mostly used in smart classrooms?	
2.	In what ways does SCP support the development and assessment of cognitive skills in undergraduate education?	
3.	What is the correlation between course assessments, ratings, and SCP test scores?	
4.	What is the difference between the demonstration of cognitive skills before and after the assessment of SCP?	
5.	How effective is the SCP model at propagating change in assessment practices for a assigned course?	

Fig. 6 Branch-wise students considered for assessment

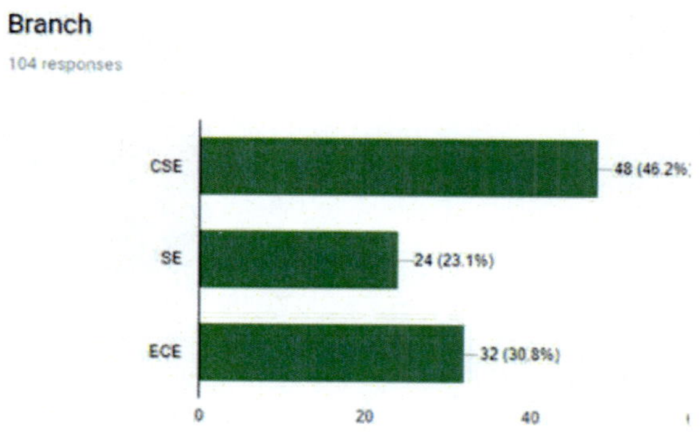

Fig. 7 Year-wise students considered for assessment

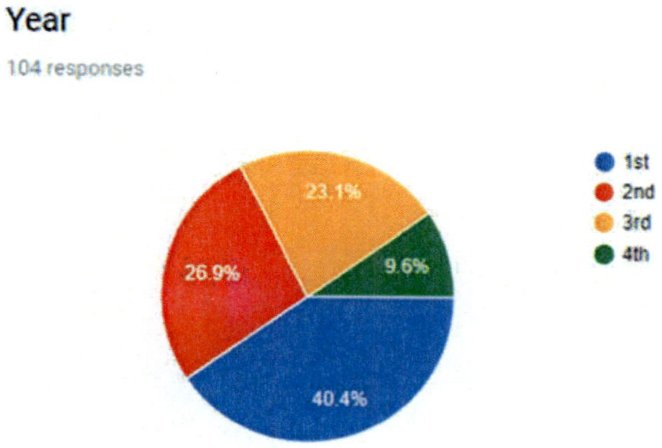

Fig. 8 Chart illustrating instructional practices

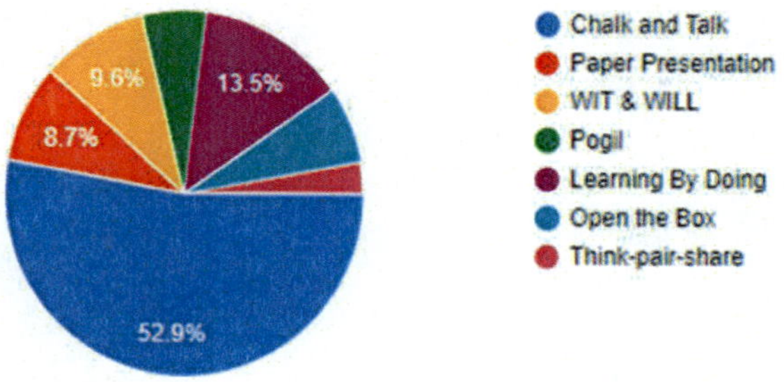

Fig. 9 Chart illustrating purpose of SCP

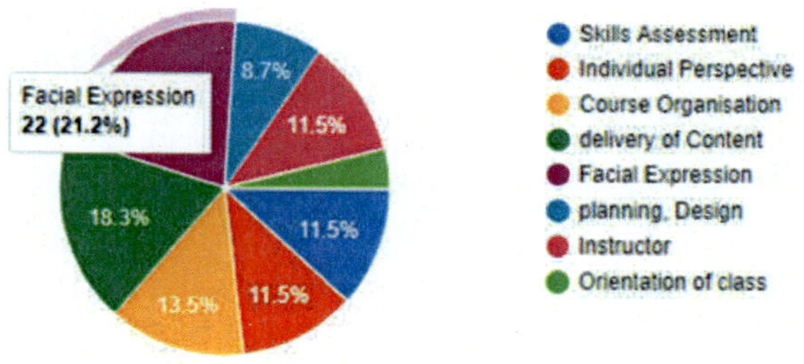

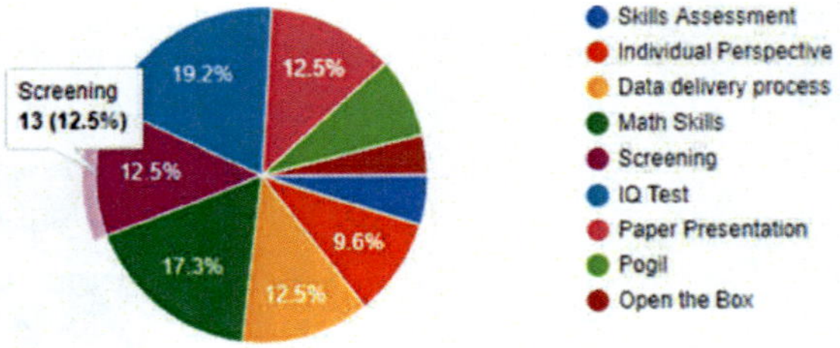

Fig. 10 Result of students performance before SCP

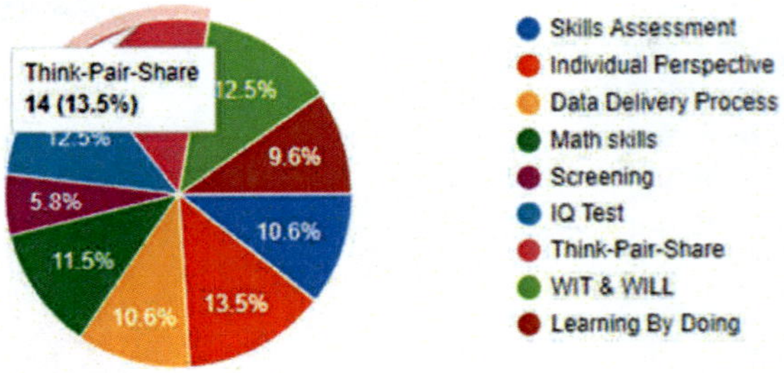

Fig. 11 Result of students performance after SCP

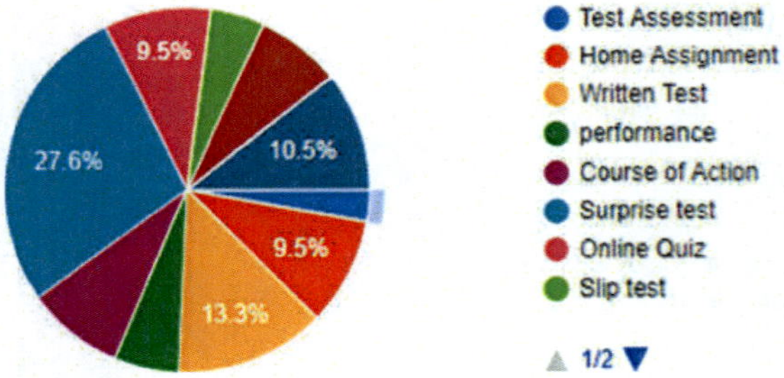

Fig. 12 Correlation of assessment metrics

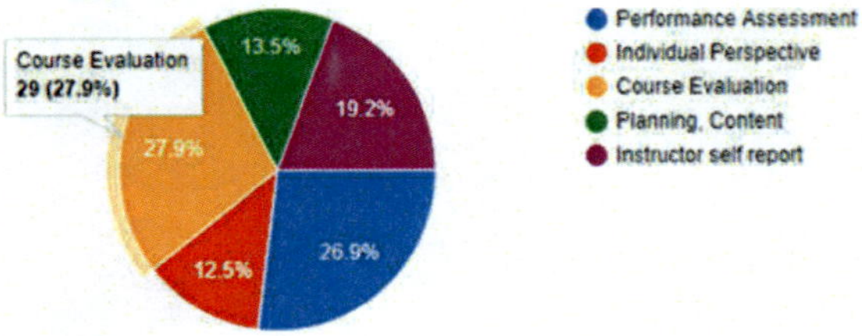

Fig. 13 Course-wise mapping considered for evaluation

5 Conclusion

Teaching is the God's given art of leading the learning practices that will form a base for a student's career. The use of so many advanced teaching methodologies; there is a need to observe the learning and understanding patterns of students, of how well he/she is able to absorb concepts and relate them accordingly for given scenarios. The teaching methods as field trips, internships, group activities, and quiz will strive the students and indirectly drag the students into the subject, wherein the student learns and gains knowledge effectively than the traditional theoretical kind of practices. The proposed model will result in evaluation of a learner performance and bring in the culture change of a student learning practices.

References

1. Miller, G. A. (2003). The cognitive revolution: a historical perspective. *Trends in Cognitive Sciences, 7*.
2. Sun, R. (Ed.), *Grounding Social Sciences in Cognitive Sciences*. Cambridge, MA: MIT Press.
3. Thagard, P. (2008). *Cognitive Science, The Stanford Encyclopedia of Philosophy (Fall 2008 Edition)*. In E. N. Zalta (Ed.).
4. Brown, J. S., Collins, A., & Duguid, P. (1989). Situated cognition and the culture of learning. *Educational Researcher, 18*(1), 32.
5. Greeno, J. G. (1998). The situativity of knowing, learning, and research. *American Psychologist, 53*(1), 5–26.
6. Lave, J. (1988). *Cognition in practice: Mind, mathematics and culture in everyday life*. Cambridge: Cambridge University Press.
7. Lave, J., & Wenger, E. (1991). *Situated learning: Legitimate peripheral participation*. Cambridge: Cambridge University Press.
8. Varela, F. J., Thompson, E., & Rosch, E. (1991). *The embodied mind: Cognitive science and human experience*. Cambridge, Mass.: MIT Press.
9. Marr, D. (1982). *Vision: A computational investigation into the human representation and processing of visual information*. W. H. Freeman.
10. Miller, G. A. (2003). The cognitive revolution: A historical perspective. *Trends in Cognitive Sciences, 7*, 141–144. https://doi.org/10.1016/s1364-6613(03)00029-9. PMID 12639696.
11. Sun, R. (Ed.). (2008). The cambridge handbook of computational psychology. New York: Cambridge University Press.
12. Isac, D., & Reiss C. (2013). *I-language: An introduction to linguistics as cognitive science* (2nd ed., p. 5). Oxford University Press. ISBN 978-0199660179.
13. Pinker, S., & Bloom P. (1990). Natural language and natural selection. *Behavioral and Brain Sciences, 13*(4), 707–784. CiteSeerX https://www.google.com/search?q=10.1.1.116.4044&ie=utf-8&oe=utf-8&client=firefox-b-ab, https://doi.org/10.1017/s0140525x00081061.
14. Lewandowski, G., & Strohmetz, D. (2009). Actions can speak as loud as words: Measuring behavior in psychological science. *Social and Personality Psychology Compass, 3*(6), 992–1002. https://doi.org/10.1111/j.1751-9004.2009.00229.
15. UCSD Cognitive Science—UCSD Cognitive Science. Retrieved July 8, 2015.
16. d'AvilaGarcez, A. S., Lamb, L. C., & Gabbay, D. M. (2008). *Neural-symbolic cognitive reasoning. Cognitive technologies*. Springer, ISBN 978-3-540- 73245-7.

A Simple Method for Speaker Recognition and Speaker Verification

Kanaka Durga Returi, Y. Radhika, and Vaka Murali Mohan

Abstract Speaker recognition is developing and is very much useful in the modern life. This paper clearly explained the speaker recognition as a technique of automatically identifying the persons liable upon the data incorporated in signals of speech, classification of the speaker recognition method, speaker identification and verification method. It also explained the recognition text-dependent and text-independent methods, different phases involved in speaker recognition systems such as speaker verification process and speaker identification process. This paper explained the speech production technique which involves voiced, unvoiced and perceived speech.

Keywords Speaker recognition · Speaker identification · Text dependent · Text independent · Speech signals

1 Introduction

Speaker recognition is a general technique and utilized in several areas like "telebanking, telephone shopping, voice dialing, services related to database access, information services, security control over confidential areas, voice mail, remote access to computers, forensic applications, etc." The development of instinctive speaker recognition technique is the combination of pattern recognition and speech science. Speaker recognition technique is the composition of "computer science, voice linguistics, signal processing, and intelligent systems." It is a significant area in common valuable application and also in research.

K. D. Returi (✉) · V. M. Mohan (✉)
Professor, Department of CSE, Malla Reddy College of Engineering for Women, Medchal, Hyderabad, Telangana, India
e-mail: durga1210@gmail.com

V. M. Mohan
e-mail: murali_vaka@yahoo.com

Y. Radhika
Director Academic Affairs & Professor of CSE, GITAM University, Visakhapatnam, AP, India

S. C. Satapathy et al. (eds.), *Intelligent System Design*, Advances in Intelligent Systems and Computing 1171, https://doi.org/10.1007/978-981-15-5400-1_64

The forming of person identification with respect to voice is known as speaker recognition. The most significant scientific issue remains appropriate and user pleasurable boundary on behalf of humanoid computer communication. The extensive computer input line is keyboard or mouse and graphical presentation for output. It is normal in lieu of persons to envisage such kind of interface through CPUs. Speech recognition methods authorize usual persons to communicate CPU in order to redeem information. The recognition of speaker and speech remains identical. The aim of speech recognition method is to identify the verbal confrontations, wherever the aim of recognition of speaker remains in the direction of behavior identification through extraction and categorization limited in signal of speech.

To attain the performance acceptance of speaker recognition method, several schemes are proposed in different areas by using artificial neural networks, i.e., pattern recognition, sentence recognition, image processing, discrete wavelet packet transform technique, genetic wavelet packets, wavelet analysis, text independent, back propagation, support vector machines, common vector, graph matching, kernel analysis, fuzzy inference system and wavelet filter banks.

2 Literature Review

The investigators projected so many works, such as Herbig et al. [1] reported the models for identification and recognition of speaker. Naito et al. [2] proposed models for identification speaker and speech by using vocal zone size. Vimala and Radha [3] presented an isolated system for recognition of speaker-independent speech for Tamil language by means of hidden Markov model (HMM). Jeong [4] presented a variation in hidden Markov model-based system for recognition of the speech to obtain speaker and its noise surroundings. Çetingül et al. [5] presented a model for recognition of speaker and speech to facilitate audio, texture and motion. Blomberg [6] described synthetic generation system for speech recognition through prototypes and symbolic transformation. Furui [7] introduced latest developments in the speaker recognition with VQ and HMM techniques. Talbot [8] reported the interface design of the speech recognizers through matching utterances to stored voice data. Furui [9] introduced a model for utilizing pitch information, adaptation techniques, HMM, neural networks training algorithms for speech recognition. Picone [10] introduced hidden Markov model-based model system in view of spectral and duration information.

Howard et al. [11] developed a model and analyzed the system performance by speech recognition devices through vocabulary, user's speech and algorithm. De Mori et al. [12] presented a network-based paradigm for speech recognition and description of speech properties by using Markov models. Hershey et al. [13] developed a model that recognizes the speech data which was recorded within a single channel. Weiss and Ellis [14] reported a model for segregation of single-channel combination of speech model with hidden Markov models and factorial HMM. Lee et al. [15] introduced a technique and described the improvements of the hidden Markov model used in SPHINX for speaker appreciation. Returi et al. [16] presented a method of speaker identification with respect to WA and SVM. Returi et al. [17] reported a relative method of changed lines for speaker appreciation. Returi and Radhika [18] developed an ANN model with WA.

3 Speaker Recognition

Speaker recognition is the progression to know an individual by using the data combined in speech signals. Biometric confirmation measures persons' communication features. The thumb impressions and images of retina remain surplus faithful in user credentials. Biometric confirmation takes certain important benefits over information and demonstration established verification methods. Speaker identification indorses the competency to renovate different appreciation numbers and PINs through thumb impressions that cannot be inappropriate.

3.1 Speaker Recognition Method

Speaker recognition is a technique of automatically identifying the persons liable upon the evidence combined in speaking signals. At the end, it deals through speaker identification shown in Fig. 1.

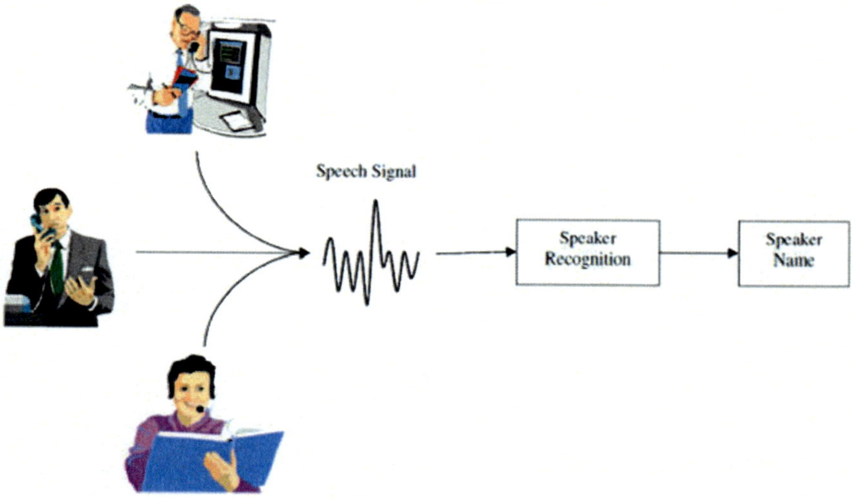

Fig. 1 Speaker recognition model

Speech recognition is dissimilar from verbal credit subsequently these ideas agreement through speech recognition, language recognition agreements through knowing the verbal in spoken sentences.

3.2 *Classification*

Recognition of the speaker is a method of estimating the performance of the speaker by a machine. The terms speaker identification and speaker recognition are used interchangeably. The classification of the Speaker Recognition is shown in Fig. 2. Speaker recognition is of two types: 1. verification and 2. identification.

In process of the speaker verification, the assignment is in the direction of verify person characteristics with his own expression. It includes decision with binary character and its identity. The speaker verification system is represented as Fig. 3.

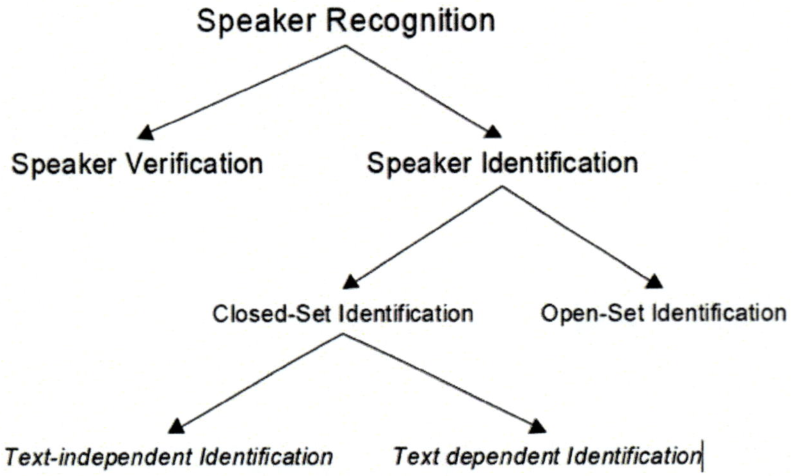

Fig. 2 Classification of the speaker recognition

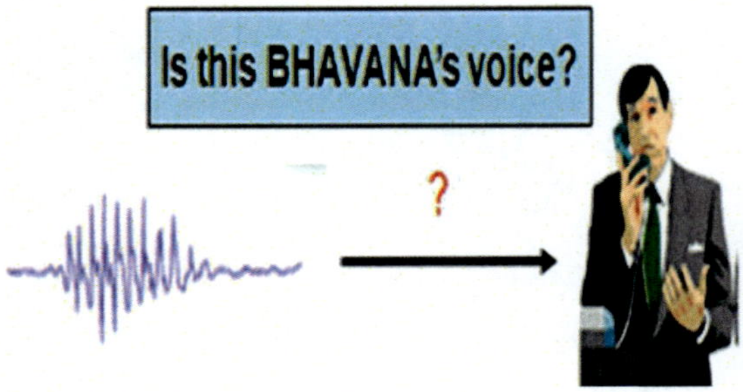

Fig. 3 Speaker verification system

Fig. 4 Speaker identification system

In the process of speaker identification, no features are claimed and method agrees the speaker identification The speaker identification system is represented as Fig. 4.

The closed set problem tries to establish the identity of a person among set of recognized voices. This is also referred to as closed identification since it is often assumed that the unidentified voices must come from a known set. Open set problem deals with deciding whether speaker of the particular investigation sound fits to a collection of recognized speakers. It is entitled usual problem because the unidentified voice could come from a large set of unknown speakers. In this case, the speaker makes an identity claim.

4 Phases of Speaker Recognition

Speaker recognition method essentially serves two phases—enrollment or training phase and recognition or testing phase. In enrollment phase, known speaker voice samples are collected using a microphone and stored in data acquisition card. The output values are compared with the original database obtained in enrollment phase. The testing process reveals the data approximately uniqueness of the speaker by its appreciation decision.

4.1 Speaker Verification Process

The Speaker Verification System with two distinct phases was shown in Fig. 5.

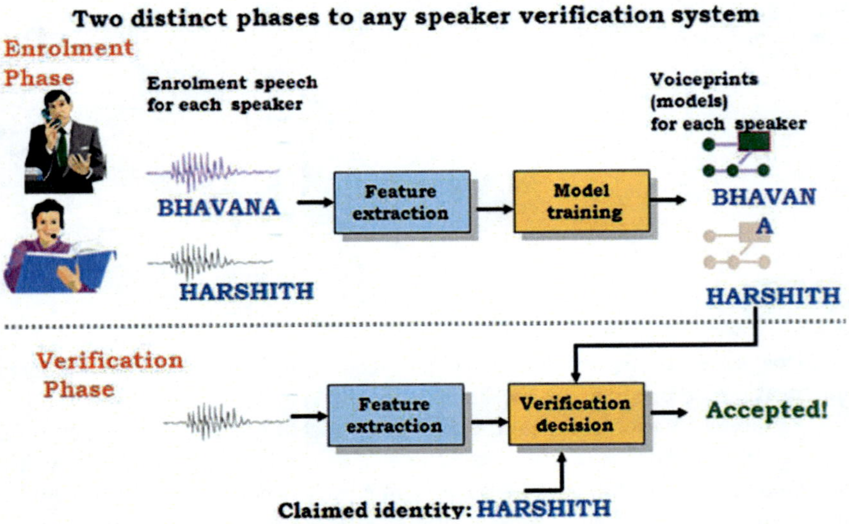

Fig. 5 Speaker verification system

4.2 *Speaker Identification Process*

The speech models are collected from the speaker, and the features were extracted. The features are extracted and compared with output of the speaker database. With this evaluation, final resolution identity of the speaker is finished and shown in Fig. 6.

5 Speech Production Technique

Speech was produced by the movements of various mechanisms of vocal region in different configurations creating speeches of voiced and unvoiced. As a result, pressure wave is generated in front of the lips. A speech indication is nothing than the tested form of force wave.

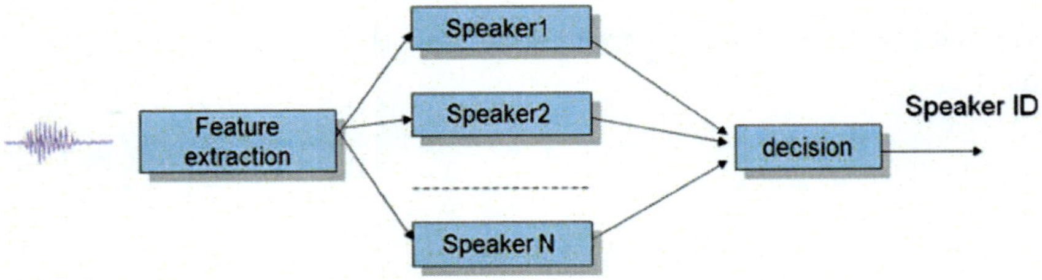

Fig. 6 Speaker identification process

Fig. 7 Human vocal system

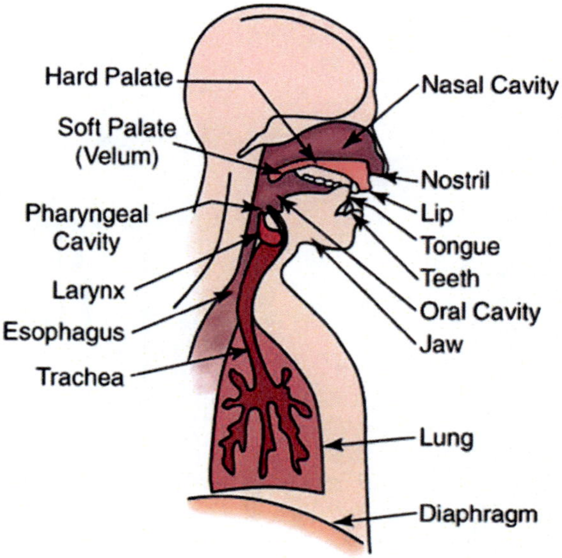

Vocal region consists of association from esophagus to mouth. The general form of vocal region diverges over period through the association of articulators consequently producing consistent differences in character and shown in Fig. 7.

6 Voiced, Unvoiced and Perceived Speech

Speech is the sound signal produced by the air and is perturbed through certain limitation at some moment in the vocal region. It is an all pole model capable of representing all sounds. Generally, nasal and fricative sounds consider poles and zeros, but once the order of the filter is very high it acts as an all pole model. This summarizes the fact that vocal tract response represents an all pole model and classified into three types:

1. Voiced
2. Unvoiced
3. Perceived

6.1 Voiced Speech

Vocal tract is creating quasi-periodic pulsations of the air (Fig. 8). Vowels remain regularly categorized by means of voiced noises. This sound requires high average energy levels and very distinct formant frequencies. Such noises remain produced

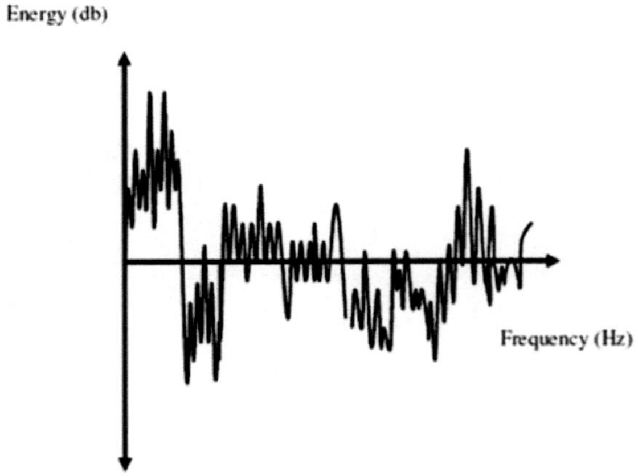

Fig. 8 Voiced speech

by imposing the air from lungs over vocal chords, and it produces series of air pulses termed as air pulses. Pitch of sound is time frequency with vocal cord vibration. Generally in women and children, due to a faster rate of vibration of the vocal chords while producing voiced speech, pitch is believed to be higher than in men.

6.2 Unvoiced Speech

The unvoiced speech sounds have a random behavior and are produced in the vocal tract by a restriction in various positions through the mouth and compelling the air at high velocity for generation of turbulence. Accordingly, sound is produced to stimulate the vocal region. Unvoiced speech referred to fricative communication. Consonants are categorized as unvoiced sounds. Unvoiced sounds have lower energy levels and high frequencies than voiced sounds (Fig. 9).

6.3 Perceived Speech

The perceived pitch fluctuates with the gender and age of speaker. Its range for humans lies between 50 and 500 Hz. Children have the highest pitch voices followed by females and then males with the lowest pitch. Pitch varies with time and tells about the prosody of utterance. With age, females tend to lower their pitch and male voices tend to rise in pitch. The acoustical equivalent of field remains important rate.

Fig. 9 Unvoiced speech

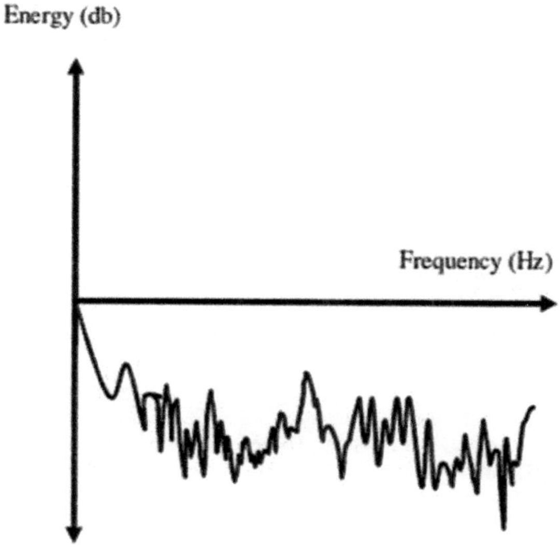

7 Conclusions

This paper clearly explained the speaker appreciation as a technique of automatically knowing persons liable upon the data combined in speaking signals, classification of the speaker recognition, verification and identification method. It also explained the recognition with text-dependent and text-independent methods, different phases involved in speaker recognition methods, such as speaker verification process and speaker identification process. This paper explained the speech production technique which involves voiced, unvoiced and perceived speech.

References

1. Herbig, T., & Gerl, W. Minker. (2012). Self-learning speaker identification for enhanced speech recognition. *Computer Speech & Language, 26*(3), 210–227.
2. Naito, M., Deng, L., & Sagisak, Y. (2002). Speaker clustering for speech recognition using vocal tract parameters. *Speech Communication, 36*(3–4), 305–315.
3. Vimala, C., & Radha, V. (2012). Speaker independent isolated speech recognition system for tamil language using HMM. *Procedia Engineering, 30,* 1097–1102.
4. Jeong, Y. (2014). Joint speaker and environment adaptation using TensorVoice for robust speech recognition. *Speech Communication, 58,* 1–10.
5. Çetingül, H. E., Erzin, E., Yemez, Y., & Tekalp, A. M. (2006). Multimodal speaker/speech recognition using lip motion, lip texture and audio. *Signal Processing, 86*(12), 3549–3558.
6. Blomberg, M. (1991). Adaptation to a speaker's voice in a speech recognition system based on synthetic phoneme references. *Speech Communication, 10*(5–6), 453–461.
7. Furui, S. (1997). Recent advances in speaker recognition. *Pattern Recognition Letters, 18*(9), 859–872.

8. Talbot, M. (1987). Adapting to the speaker in automatic speech recognition. *International Journal of Man-Machine Studies, 27*(4), 449–457.
9. Furui, S. (1992). Recent advances in speech recognition technology at NTT laboratories. *Speech Communication, 11*(2–3), 195–204.
10. Picone, J. (1990). Duration in context clustering for speech recognition. *Speech Communication, 9*(2), 119–128.
11. Nusbaum, H. C., & Pisoni, D. B. (1987). Automatic measurement of speech recognition performance: A comparison of six speaker-dependent recognition devices. *Computer Speech & Language, 2*(2), 87–108.
12. De Mori, R., Cardin, R., Merlo, E., Mathew, P., & Rouat, J. (1988). A network of actions for automatic speech recognition. *Speech Communication, 7*(4), 337–353.
13. Hershey, J. R., Rennie, S. J., Olsen, P. A., & Trausti, T. K. (2010). Super-human multi-talker speech recognition: A graphical modeling approach. *Computer Speech & Language, 24*(1), 45–66.
14. Weiss, R. J., & Ellis, D. P. W. (2010). Speech separation using speaker-adapted eigenvoice speech models. *Computer Speech & Language, 24*(1), 16–29.
15. Lee, K. F., Hon, H. W., Hwang, M. Y., & Huang, X. (1990). Speech recognition using hidden Markov models: A CMU perspective. *Speech Communication, 9*(5–6), 497–508.
16. Returi, K. D., Radhika, Y., & Mohan, V. M. (2015). A novel approach for speaker recognition by using wavelet analysis and support vector machines. In *2nd International Conference on Computer and Communication Technologies—IC3T 2015* (Vol. 1, pp. 163–174), July 24–26, 2015 at CMR Technical Campus, Hyderabad, Telangana, India (Technically co-sponsored by CSI Hyderabad Section) IC3T 2015, ISBN: 978–81–322–2517–1.
17. Returi, K. D., Mohan, V. M., & Praveen Kumar, L. (2016). A comparative study of different approaches for the speaker recognition. In *3rd International Conference on Information System Design and Intelligent Applications INDIA 2016* (Vol. 1, pp. 599–608), January 8–9, 2016 at ANIL NEERUKONDA Institute of Technology & Sciences, Visakhapatnam, AP, India Technically co-sponsored by CSI Visakhapatnam Section), INDIA 2016, ISBN: 978–81–322–2755–7.
18. Returi, K. D., & Radhika, Y. (2015). An artificial neural networks model by using wavelet analysis for speaker recognition" India 2015. In J. K. Mandal et al. (Eds.), *Information Systems Design and Intelligent Applications, Advances in Intelligent Systems and Computing* (Vol. 340, pp. 859–874). https://doi.org/10.1007/978-81-322-2247-7_87.

An Effective Model for Handling the Big Data Streams Based on the Optimization-Enabled Spark Framework

B. Srivani, N. Sandhya, and B. Padmaja Rani

Abstract The recent advancements in information technology tend to maximize the volume of the data used in day-to-day life. In big data era, analysis and extraction of knowledge from huge-scale data sets is a key issue in the research due to the complexity in the analysis of big data. Numerous techniques are devised for streaming big data. However, the techniques are highly dynamic, which may result in concept drift. Accordingly, this paper introduces a technique based on optimization for handling the concept drift effectively. The proposed dragonfly moth search (DMS) algorithm-based spark architecture is developed for handling the big data stream classification without any concept drift. Specifically, big data stream classification is done based on the proposed DMS-based stacked auto-encoder (SAE) in spark and the big data stream classification involves two phases, namely online phase and offline phase. The incremental data is tackled effectively in the online phase, where the concept drift is checked based on error, and in case of any concept drift, the rough set theory is employed, which is defined, based on DMS algorithm. The proposed DMS algorithm is the integration of dragonfly algorithm (DA) and moth search algorithm (MS), which aims at feature selection and optimal tuning of the SAE.

Keywords Big data streaming · Spark architecture · Incremental data · Entropy · Rough set theory

B. Srivani (✉)
Research Scholar, JNTUH, Hyderabad, India
e-mail: srivaanib@gmail.com

N. Sandhya
CSE Department, VNRVJIET, Hyderabad, Telangana, India
e-mail: sandhyanadela@gmail.com

B. Padmaja Rani
CSE Department, JNTUCEH, Hyderabad, India
e-mail: padmaja_jntuh@jntuh.ac.in

S. C. Satapathy et al. (eds.), *Intelligent System Design*, Advances in Intelligent Systems and Computing 1171, https://doi.org/10.1007/978-981-15-5400-1_65

1 Introduction

Due to advanced applications, there is an increasing scope for handling the massive data obtained from different sources. The huge volume of data is accumulated from various types of information sources. Thus, one can articulate that the data resides in era of big data. The massive data focuses on five 'V' principles, which involve variety, volume, veracity, value and velocity for defining the dynamic nature, different qualities, and effectiveness for human beings [1]. The high-speed data stream generates more data for the evaluation. The processing of massive data set is complex, and the entire processing task cannot be done in the main memory. In order to evaluate the data, commonly used machine learning techniques are clustering, prediction, and classification. The conventional regression methods are unable to analyse the massive data sets. Moreover, the complete data set should be accumulated in the memory for computing the tasks and to build the knowledge. The researchers devised several data stream mining methods for processing the incremental data in a sequential manner, thereby keeping the memory spacious [2]. In addition, the processing of loads in big data streaming may vary with respect to time by stipulating more processing resources [3]. The big data is massive data that has been captured from different fields and used to specify its untapped knowledge and connotation. However, the application of machine learning into the big data is a major challenge as the conventional machine learning systems failed to manage such huge amount of data with different velocity. The determination of proper programming model for the data analysis is confusing, while the platforms used for the data analysis pose the ability to handle the machine learning algorithms in huge scale [4].

The foremost learning problems deliberated in data stream mining involve classification algorithms, clustering techniques, and frequent pattern mining techniques [5]. The significant feature of big data is the massive data symbolized by heterogeneous and miscellaneous dimensionalities. This is due to the fact that different information collectors utilize their own schemata for recording the data and the characteristics of different applications result in various data representations. In bio-medical world, each human being is symbolized utilizing simple demographic information, like family disease, history, gender, and age [6]. Moreover, the big data is considered as a fascinating field especially in media governments and companies that tries to use the available information for processing [7]. The relentless and exceptional speed of big data streams raises issues in analysing real-time data. The cloud computing is utilized for tackling the issues, but due to massive data, the selection of suitable cloud resources for real-time analysis became a major issue. The recent techniques allocated the cloud nodes on the basis of GPU power, and user-defined memory size [8–11].

MapReduce is considered as the foremost programming paradigm developed for dealing with the big data events [1, 12]. A large-scale processing framework, named Apache Spark [1], has gained immense attention in the big data because of its improved performance in incremental steps. The goal of this paper is to devise a big data streaming model using an optimization algorithm. The big data streaming is progressed using the Apache Spark framework that uses the proposed optimization algorithm, named DMS algorithm. The proposed DMS algorithm is designed by combining DA with the MS algorithm. The big data generated from the distributed sources is subjected to the master node, wherein the obtained data is split into different slave nodes to perform feature selection based on proposed DMS algorithm and fitness function. The effective feature selection ensures the classification of the data such that the classification accuracy is enhanced. The selected features obtained from all the slave nodes are combined and then subjected to the master node for data classification, and the slave node utilizes the SAE, which is trained using the proposed DMS algorithm such that the data is classified as various classes.

The arrangement of the paper is done as follows: The introductory part depending on the big data streaming is deliberated in Sect. 1, and the review of eight existing works is elaborated in Sect. 2. In Sect. 3, the architecture of proposed method of big data streaming is presented. Section 4 elaborated the proposed algorithm steps and in Sects. 5 and 6 experiment as well as the results of the methods. At last, Sect. 7 illustrated the summary of the research work.

2 Motivation

The analysis of eight existing works based on big data streaming is briefly elaborated along with its merits and demerits. Ramirez Gallego et al. [1] designed an incremental and distributed classifier on the basis of popular nearest neighbour algorithm using a particular situation. The method was executed in Apache Spark and adapted distributed metric-space ordering for performing rapid searches. The method updated and eliminated the outdated instances from the case-bases. This method improved the high computational requirements of the classifier and makes it apposite for dealing with problems. However, the method failed to consider a condensation technique for handling the data with massive size. The method was unsuitable for refining huge imbalanced data streams. Wibisono and Sarwinda [2] developed an enhanced fast incremental model tree-drift detection (FIMT-DD) based on average restrain divider of evaluation value (ARDEV). The ARDEV utilized the Chernoff bound approach for evaluating the errors, and to compute the perceptron rule. However, the method was not applicable with different big data set for deeper analysis. Vicentini et al. [3] developed a multitenant-aware resource provisioning technique for performing big data streaming based on task scheduling and ongoing task rescheduling using virtual machines states. The method facilitates load balancing using different cloud-based clusters of virtual machine (VM) based on Software Defined Network (SDN). The method was executed with Apache Storm (big data). However, the method failed to consider security aspects of a given dynamic environment. Nair et al. [4] developed a real-time remote health status prediction system using the big data processing engine with Apache Spark, adapted in cloud environment. The system employed machine learning model for streaming the big data. This system poses the health records of the patients and uses those records for predicting the health status of the patient. This method failed to link the system with healthcare providers in order to give entire real-time health data. Puthal et al. [13] developed a method named dynamic prime number-based security verification (DPBSV) for streaming massive data. The method was based on common shared key, which was updated in a dynamic manner by producing synchronized prime numbers. The key was updated using source sensing devices and Data Stream Manager (DSM). The method was effective and increased the performance, but the method failed to detect the moving target defence. Yin et al. [14] designed a fog-assisted data streaming scenario for sharing the spare resources for processing the raw data in which the SDN controller regulated the application data for making effective decisions. The method was efficient in handling the communication overhead. Ruiz et al. [5] developed a fuzzy system named Fuzzy-CSar with adaptive fuzzy partitions (Fuzzy-CSar-AFP) for obtaining interesting fuzzy association rules using the data streams. The method failed to use advanced versions of Fuzzy-CSar-AFP for addressing the electroencephalogram problems. Fernandez-Basso et al. [15] developed a frequent item set mining method based on sliding windows. The method was used to extract the tendencies using the continuous data flows. However, the method did not consider association rule mining for studying the frequent items.

The challenges faced by the conventional methods are enlisted below,

- In [16], fuzzy associative classifier was designed for exploiting a new distributed discretizer based on fuzzy entropy for producing fuzzy partitions of the attributes. The method enhanced the classifier interpretability. However, the rules produced by the fuzzy distributed method are lower.
- In [17], dendritic cell algorithm (DCA) was devised for classifying the data using the MapReduce framework. The method was proficient, distributed, and scalable while distributing the computing elements, but showed poor performance.
- An effectual resource management system was designed for evaluating the data characteristics of massive data streams based on velocity, variability, volume, and variety. This method allocated the resources to stream on the basis of its data characteristics, but the resultant performance was poor [8].
- The deep learning methods are devised for rethinking the traffic flow prediction problem using the deep architecture models f or dealing with big traffic data. The method attained improved performance in predicting the traffic flow, but the method was not applicable with other data sets [18].
- In [19], the issues linked with the data stream mechanisms were about the capability for managing the high-speed data streams, incremental data, huge memory requirements, and classification accuracy.

2.1 Problem Statement

Conventionally, the processing of big data is based on store-then-process paradigm. Numerous big data applications are devised to transform the data between servers and nodes for real-time data processing for handling the faults. There exist several challenging aspects for performing big data streaming. The first challenge was to acknowledge the generated data set. It was complicated to sense the data with fewer features. Another challenge was devised in CNN-based classification techniques, wherein the method did not consider target nominal for classification and also failed to add linguistic prior knowledge into system. Another challenge was to make decision for selecting the machine learning algorithm that should be selected for addressing the streaming issues of big data. The aim was to balance the prediction accuracy and the speed for which the whole data could be trained on and classified. In NN-based methods, the consideration of massive data sets may lead to overhead. The consideration of neural networks may lead to performance degradation and overhead due to massive data size. The challenge was to build a classifier using any set of deep learning tools or software libraries such that it will learn based on the large data set and the optimization technique for training the classifier in order to classify an unlabelled set of data. However, the challenging issues of the optimization-based data classification techniques failed to use other kinds of data, like audio, video, images. However, the desired results are obtained by the best tool by employing suitable technique for designing a tool. The complications in the application and the usage of different techniques and algorithms to develop a tool became a major issue. Hence, this paper devises a technique for removing the above abnormalities by introducing a novel classifier named DMS-SAE, wherein the proposed DMS algorithm is utilized for training the SAE classifier. The proposed DMS is generated by altering the update equation of DA with the MS algorithm.

3 Structural Design of Big Data Streaming in Apache Spark Architecture

Figure 1 portrays the structural design of big data streaming using proposed DMS algorithm. The big data streaming is carried out using proposed DMS algorithm. The proposed DMS algorithm undergoes two phases, namely offline phase and online phase, respectively. In the offline phase, the input data from the master nodes is partitioned as subsets and is provided as input to the individual slave nodes, where the feature selection is performed using the newly designed fitness function, which is derived using entropy measure. The feature selection is performed using the proposed DMS, which is obtained by incorporating DA [20] algorithm on MS [21] algorithm. The proposed DMS selects the features from each subset of input data in the slave nodes based on fitness. The features obtained from the individual slaves are concatenated together and are fed to the master node for classification. In the master node, the classification is done based on the SAE classifier which is trained using the proposed DMS algorithm. In the online mode, several set of slave nodes are employed to process the data streams. The process involved in the online phase includes feature selection, classification, and remodelling of classifier. Initially, the incremental data (new data) from the master node is partitioned and forwarded to the slave set-1, where

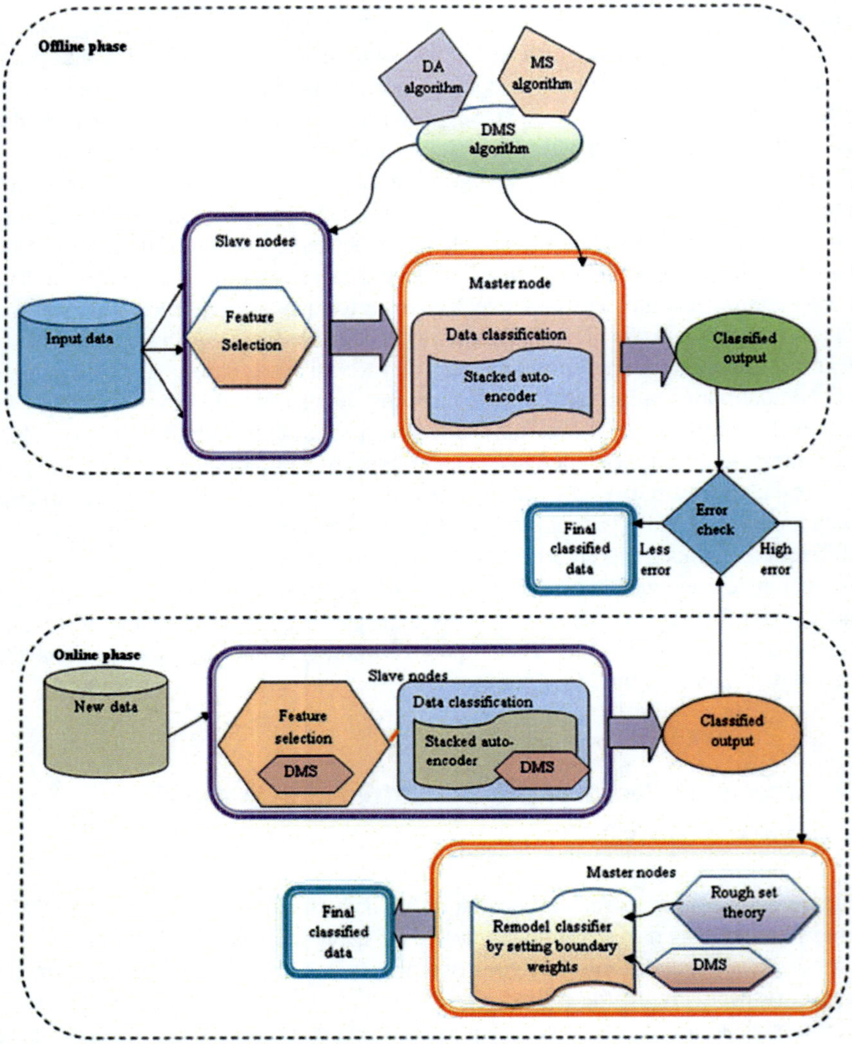

Fig. 1 Architecture of big data streaming using the proposed DMS algorithm

the feature selection is carried out using the index-based combined features obtained from the fitness function. Accordingly, the features selected from the slave set-1 in online phase are equivalent to the concatenated features in the offline phase. The resulted features obtained from the slave set-1 of online phase are further processed by the classification module in the slave set-2, which employs the SAE classifier for the classification. The classification is performed to acquire the output from the individual slave. Accordingly, the next slave set act as a decision-making phase, where the error for the classified output is calculated with respect to the target. If the error computed by the classification output varied much with a maximal error value, then the decision output becomes one, which signifies that the classifier remodelling is necessary. In case of the minimal error, the decision output from the individual slaves in slave set-2 is zero. Thus, the classifier remodelling in case of the maximal error is performed in the master node. The classifier remodelling is done using the RST, where the boundary is set for weight, and the weight is computed using the proposed DMS algorithm. Therefore, the output is attained in the master node.

3.1 *Offline Phase in Big Data Streaming*

The offline mode evaluates the stored time-consistent big data sets. Initially, the input data is uploaded and the inbuilt function reads the file in the spark context. The Spark architecture produces high processing power as the slave node servers poses the ability to run in parallel. The processing time and the flexibility of the big data are improved using spark in such a way that the input big data is split into numerous subsets of data and the individual slave nodes intake a subset and data on it to generate the desired output (Fig. 2).

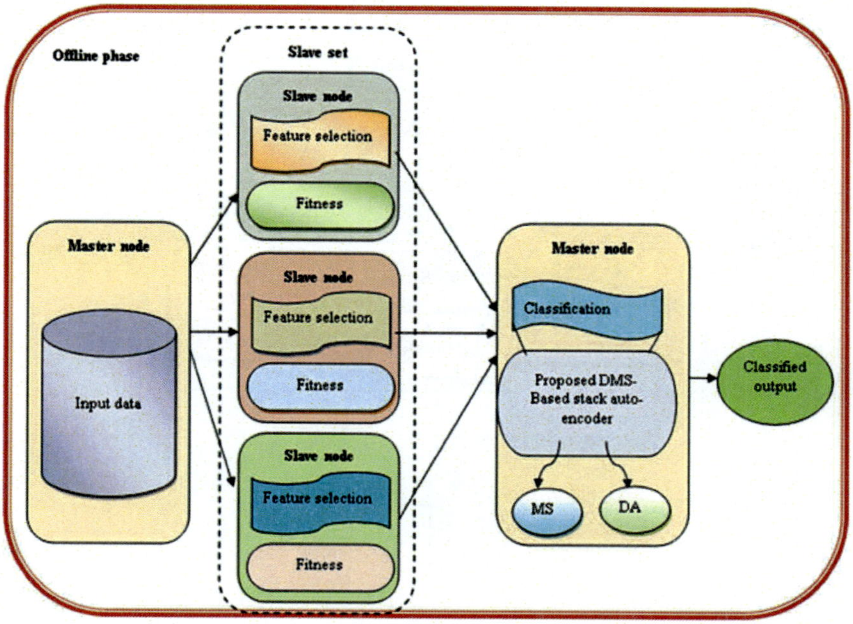

Fig. 2 Architecture of big data streaming in offline mode

3.2 *Online Phase in Big Data Stream Classification*

The online mode works similar to the offline mode, wherein the data streaming is performed in the master node. Figure 3 elaborates the schematic view of online for attaining effective big data classification using new data. Initially, the incremental data or new data is taken as an input in the master mode, which is then split to three different slave sets for performing feature selection, classification, and in this section, the concept drift phenomenon is executed. On slave set 1, the feature selection is carried out using indices obtained from the offline mode. The index of the feature computed on the offline mode is retrieved for selecting the features from the incremental data. Based on the obtained features, the classification is carried out on second slave set using proposed DMS-SAE. Finally, in slave set 3, the decision is taken by comparing the classified output obtained from the online phase and offline phase using an error function.

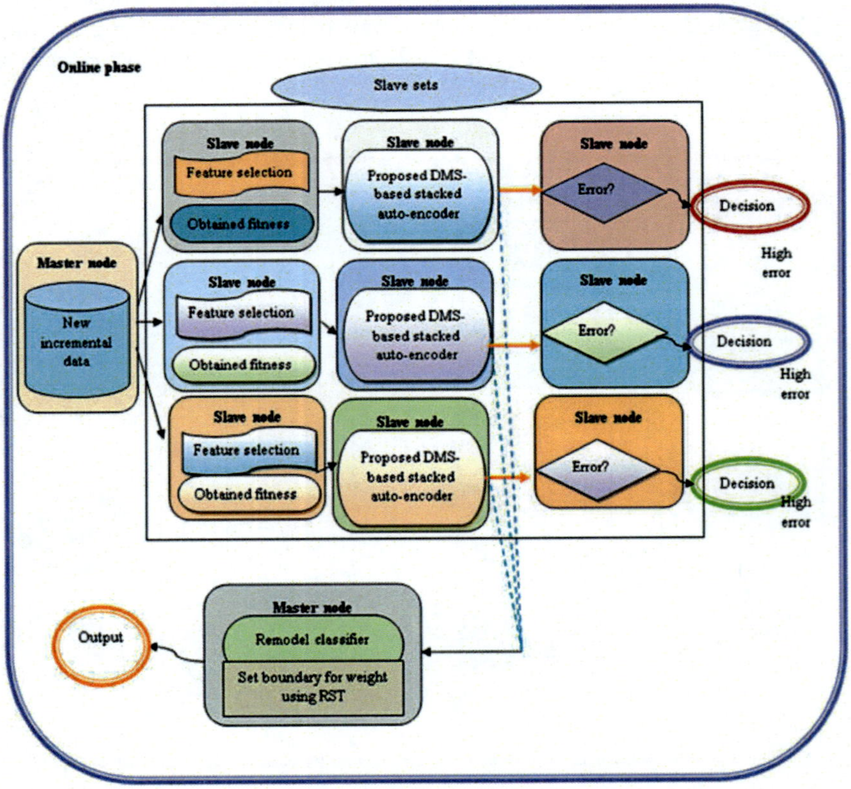

Fig. 3 Schematic view of online phase for effective big data classification

4 Proposed DMS-SAE Classifier for Big Data Stream Classification

Consider the input big data be given by I with different attributes, which is represented as,

$$I = \{D_{bc}\}; (1 \leq b \leq B); (1 \leq c \leq C) \tag{1}$$

where D_{bc} refers the data present in big data D indicating the cth attribute of the bth data. The total data points and attributes contained in the database are given as B and C.

4.1 Master Nodes in Spark

The master node is responsible for partitioning the obtained input data and sends the distributed data to different slave nodes. The slave nodes use the proposed DMS algorithm and the fitness function for the feature selection. The significance of feature selection is to select most significant features for assuring dimensional reduction, with maximized classification accuracy. At first, the big data is split into subsets of data represented as,

$$D_{bc} = \{S_m\}; (1 \leq m \leq Q) \tag{2}$$

where Q denotes total subsets generated utilizing a big data. The total subsets are equal to the total slave nodes, where the feature selection is done based on the proposed DMS algorithm. Once the significant features are selected in the slave nodes, the features are transferred to the master node at the output end in which the classification process is performed. The classification is performed using SAE-DMS algorithm, where the proposed DMS algorithm is used for training the SAE in order to gain optimal weights for the classification.

4.2 Slave Nodes in Spark

Each slave node acquires the subgroups of data from the master node and performs feature selection based on the fitness function in parallel. The fitness function is derived using the entropy, which is utilized for selecting the significant features and the feature indices are stored in the file. The input to the mth slave node is given as,

$$S_m = \{T_{p,q}\}; (1 \leq p \leq B); (1 \leq q \leq C) \tag{3}$$

F_1	F_2	...	F_x	...	F_A

Fig. 4 Solution encoding

where $T_{p,q}$ specifies mth subset data representing pth attribute of qth data. The feature selection is performed on each subset of data S_m. The slave nodes use the proposed DMS and fitness function to choose the optimal features, and the output obtained from all slave nodes is combined together at the master node to perform the classification. In feature selection, the significant features selected from the input data and the significance of feature selection is performed based on fitness function. The feature selection is performed for generating the significant features that facilitates the better classification of the input data. On the other hand, the complexity of analysing the data is minimized as the data is represented as the reduced set of features. Moreover, the accuracy associated with the classification is assured through the effective feature extraction and the features extracted using the input data.

(a) **Solution Encoding**

The solution encoding provides a symbolic solution representation of the obtained features using the proposed DMS algorithm. For feature selection, the solution vector consists of selected features in which the index of the feature is stored for the online analysis. Consider ℓ be the total number of features with dimension $[1 \times \ell.]$ From the total features, the features selected based on the fitness function is given as A with dimension $[1 \times y]$. Figure 4 illustrates the solution encoding using proposed DMS algorithm.

(b) **Fitness Function**

The fitness function-based entropy [22] and the fitness are intended to resolve the maximization function. The fitness function is represented as,

$$F = -\sum_{x=0}^{A-1} P_x \log P_x \tag{4}$$

where F indicates the fitness function derived using entropy, A is the total number of features, and P_x is the probability distribution of features.

(c) **Proposed DMS Algorithm**

The classification and feature selection are carried out employing proposed DMS algorithm, which is designed by integrating MS and DA algorithm. The proposed DMS algorithm inherits the advantages of both MS and DA and provides best performance of classification for big data streaming. MS algorithm [21] is developed based on the inspiration gained by the levy flights and the photo taxis of the moths. MS poses the ability to search the optimal solutions in an effective and accurate manner. This method does not contain complex operations thereby; the execution is simple as well as flexible. The DA algorithm [20] is a novel swarm intelligence optimization, originated from the swarming behaviours of the dragonflies. DA algorithm is benefitted from high exploration of dragonflies. The convergence rate of dragonflies produces global optimal solutions in multi-objective search spaces. Thus, the inclusion of MS in the update process of DA enhances the convergence and thereby improves the performance of the algorithm by generating the global optimal solution. The steps included in the proposed DMS algorithm are given as,

Step 1 **Initialization**

The initialization of the dragonflies' population (solutions) in the search space is given as,

$$U = \{U_1, U_2, \ldots, U_a, \ldots U_d\} \tag{5}$$

where U_a is the ath dragonfly in the solution, and d is the total dragonflies (solutions).

Step 2 **Computation of the fitness function**

The fitness of the solution is evaluated based on the formula shown in Eq. (4). The fitness of the individual solutions is computed, and the solution that attains the maximum value of fitness is selected as the optimal solution.

Step 3 **Determination of update velocity**

After computing the fitness, the solutions are updated based on the fitness and the solution update is followed using DA and MS. As per DA [20], the solution update is given as,

$$U_{g+1} = U_g + \text{Levy}(d) \times U_g \tag{6}$$

where g is current iteration, and d refers the dimension of position vector, $\text{Levy}(d)$ represents the levy flight, and U_g is the current solution.

MS algorithm [20] poses better convergence in the first stage of the algorithm and contains the ability to switch between the exploration and the exploitation phases for generating the optimal location. Thus, the update based on the MS algorithm is given as,

$$U_{g+1} = U_g + \rho \text{Levy}(d) \tag{7}$$

where U_g is the current solution, and ρ is the scale factor, and $\text{Levy}(d)$ is the step obtained from the levy flight.

The scale factor is used for controlling the convergence speed of the algorithm in order to solve the huge-scale global optimization issues. Thus, the scale factor ρ is formulated as,

$$\rho = \frac{l_{\max}}{g^2} \tag{8}$$

where $l_{\max}$ is the max walk step, and g is the current iteration.

Equation (7) is re-written as,

$$\text{Levy}(d) = \frac{U_{g+1} - U_g}{\rho} \tag{9}$$

After substituting Eq. (9) in Eq. (6), the update equation obtained is given as,

$$U_{g+1} = U_g + \left(\frac{U_{g+1} - U_g}{\rho}\right) \times U_g \tag{10}$$

$$U_{g+1} = \frac{\rho U_g + (U_{g+1} - U_g)U_g}{\rho} \tag{11}$$

$$\rho U_{g+1} = \rho U_g + (U_{g+1} - U_g)U_g \tag{12}$$

$$\rho U_{g+1} - U_{g+1}U_g = \rho U_g - U_g^2 \tag{13}$$

$$U_{g+1}(\rho - U_g) = \rho U_g - U_g^2 \tag{14}$$

The final update of the solution obtained based on the position of the dragonflies and moths in its preceding iteration is given as,

$$U_{g+1} = \frac{\rho U_g - U_g^2}{\rho - U_g} \tag{15}$$

The solution update obtained in the above equation is used for the weight update in order to train the SAE. Thus, the weights corresponding to the minimum value of error are employed for training SAE.

Step 4 **Determination of best solution**

The solutions are ranked based on the fitness, and the solution that ranked the highest forms the best solution.

Step 5 **Terminate**

The optimal solutions are derived in an iterative manner until the maximum number of iterations.

(d) **Formation of the significant feature vector**

The features from the individual slave nodes are carried to the master node, where the selected features from the slave nodes are combined together to perform the big data stream classification. The feature selection in the slave nodes supports dimensional reduction such that the selection of the highly significant features ensures the effective classification of big data. The selection of the features in the slave nodes removes the repetition of the features, thereby reducing the dimension of the data, and enables the accurate data classification. The selected features are given as,

$$F = \{F_1, F_2, \ldots, F_x, \ldots, F_y\} \tag{16}$$

where y is total selected features, and F_x is the xth feature.

4.3 *Proposed DMS-Based Stack Auto-Encoder for Big Data Stream Classification in the Master Node*

The big data classification is progressed in the master node using SAE based on the selected features. The SAE is highly fault-tolerant, possesses the ability to deal with noisy data, and is capable of dealing with the complex patterns. The DMS-based SAE is employed in the master nodes to form the required classification results. The following sections deliberate the architecture of SAE and the training of SAE.

4.3.1 Architecture of Stack Auto-Encoder

Auto-encoder [23] plays an important part in deep neural network (DNN). The auto-encoder takes the relevant features of the input using Principle Component Analysis (PCA). The single-layer auto-encoder does not contain directed loops. This auto-encoder comprises M input visible units, N hidden units, and O output visible units. Figure 4 explains the architecture of SAE. The auto-encoder is progressed by encoding the input vector into an advanced phase hidden version depicted as K. In encoder, the deterministic mapping X_θ transforms the input vector M into hidden vector given as,

$$K = \text{fun}(\omega_1 J + L_1 J) \tag{17}$$

where ω_1 indicates the weight matrix, and L_1 refers bias vector. After decoding the hidden version K back to reconstruction J is formulated as,

$$\hat{J} = \text{fun}(\omega_2 N + L_2 N) \tag{18}$$

where ω_2 denotes weight matrix, L_2 represents the bias vectors, and N represents hidden units. The back propagation has issues regarding the slow convergence and local minima. The local minima happen due to frequent alteration of the weights. Moreover, the back-propagation algorithm does not need normalization for input vectors. However, the normalization can increase the system performance and this algorithm does not pose the ability to determine the global minimum of the error function. For minimizing cost function and the squared reconstruction error, the auto-encoder is trained using proposed DMS instead of back-propagation algorithm and the reconstruction error is given as,

$$\mathrm{Jac}(\omega, L) = \frac{1}{2z}\sum_{u=1}^{z} ||\hat{J}_u^O - J_u^M||^2 \tag{19}$$

where ω denotes the weight set and L represents the set of bias vectors, z represents total layers where $1 \le u \le z, \hat{J}_u^O$ indicates the input reconstruction on uth layer, and J_u^M shows the output reconstruction on uth layer (Fig. 5).

In order to mitigate the over-fitting, a weight regularization term is adapted and the scarcity constraints are attained to produce a cost function given as,

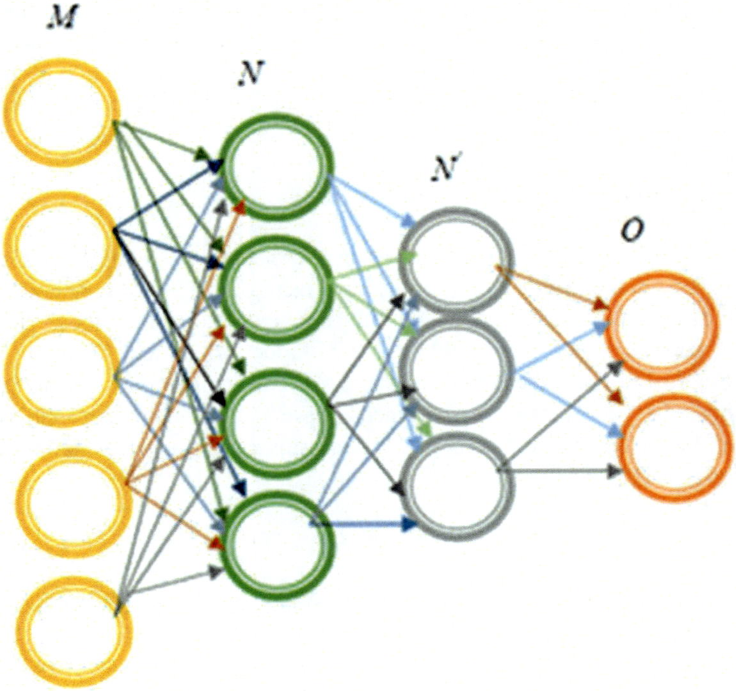

Fig. 5 Architecture of stack auto-encoder

$$\text{Jac}(\theta) = \frac{1}{z}\sum_{u=1}^{z} ||J_u^O - J_u^M||^2 + \alpha\left(||\omega_1||^2 + ||\omega_2||^2\right) + \beta\sum_{v=1}^{W} Y(J||J_v) \tag{20}$$

where J_u^O indicates the input reconstruction on uth layer, J_u^M shows the output reconstruction on uth layer, α indicates the weight for regulation condition, β refers the weight for sparse condition, $|\omega_1||^2 + ||\omega_2||^2$ represents the parametric conditions, J is the sparse parameter, J_v indicates the average activation of hidden unit v, $Y(J||J_v)$ refers the Kullback–Leibler divergence, and W is the number of hidden units such that $1 \leq v \leq W$. The average hidden unit is formulated as,

$$\hat{J}_v = \frac{1}{z}\sum_{u=1}^{z} k_{2,u}^v \tag{21}$$

where $k_{2,u}^v$ represents the hidden layer activation function of vth entry, and z represents total layers where $1 \leq u \leq z$. The Kullback–Leibler divergence is formulated as,

$$Y(J||J_v) = J\log\frac{J}{J_v} + (1-J)\log\frac{1-J}{1-J_v} \tag{22}$$

where J is the sparse parameter, and J_v indicates the average activation of hidden unit v. The auto-encoder contains several layers of sparse auto-encoder, which are the outputs generated by each layer that has been connected to the input of consecutive layer where N_1, N_2, and N_3 depict the hidden layers. The concatenation of auto-encoders in which the outputs of the auto-encoders stacked on the layer is linked to the successive layer inputs.

Finally, the auto-encoders are stacked in a hierarchical manner. The activation output of the zth layer is given as,

$$k_{tu}^M = nt\left(k_{t-1}^M \omega_{t-1} + r_{t-1}^M\right) \tag{23}$$

where $k_{1,u} = s_u$. By describing $h_{\omega,L}\left(s_u^M\right) = k_{o,u}$, the cost function is given as,

$$\text{jac}(\omega, L) = \frac{1}{2z}\sum_{u=1}^{z} ||h_{\omega,L}\left(s_u^M\right) - s_u||^2 \tag{24}$$

$$\text{jac}(\theta) = \text{jac}(\omega, L) + \frac{\alpha}{2}\sum_{t=1}^{i_t-1}\sum_{e=1}^{j_t}\sum_{f=1}^{j_{t+1}}\left(\omega_t^{e,f}\right)^2 + \sum_{t=z}^{i_t-1}\sum_{v=1}^{j_t}\gamma^t Y\left(J^t||\hat{J}_v^t\right) \tag{25}$$

where i_t represents the total layers in the network and γ^t and J^t represent the hyper parameters in the tth layer. Thus, the SAE [23] is processed with three steps. The first step is the initialization of parameters near a local minimum with each individual auto-encoder. The second step is to learn the hidden layer activations of the next auto-encoder hidden layer. In the third step, the fine-tuning is performed using the proposed DMS algorithm. In this model, each parameter varies in a simultaneous manner in order to enhance the classification result.

4.3.2 Training of SAE

The training of SAE [23] is performed using the proposed DMS optimization algorithm that aims at determining the optimal weights to tune the SAE for classifying the big data. The optimal weights derived from the proposed algorithm tune the SAE for deriving the optimal weights. The big data streaming using the proposed DMS-SAE is effective in classifying the data through deriving the optimal weights and is capable of dealing with the new data attributes arriving from the distributed sources.

When a new data is added to the model, the error is computed and the weights are updated without considering the weights of previous instance. If the error computed by the current instance is less than the error of previous instance, then the weights are updated based on the proposed DMS algorithm. On the other hand, if the error computed by the current instance is more than the error of previous instance, then the classifier is remodelled by setting boundary for weight using rough set theory (RST) [24] and then, select the optimal weight using proposed DMS algorithm. Thus, the weight bound is given as,

$$\omega^{g+1} = \omega^g \pm R^B \tag{26}$$

where R^B is the rough set bound, ω^{g+1}, and ω^g are the weights of next and current instance. The rough set bound is formulated as,

$$R^B = \frac{\eta^H(w)}{N_f} \tag{27}$$

where R^B is the roughest bound, $\eta^H(w)$ refers rough membership function, and N_f is the normalization function. The rough membership function as per RST [24] is given as,

$$\eta^H(w) = \frac{|B \cap H(w)|}{H(w)} \tag{28}$$

where B is the weight area ranging from (−10 to 10), and $H(w)$ is the rough sets, and $H(w) \in \omega^g$.

After updating the weight, the error is calculated for ensuring that the bounded weight is optimal. The process is iterated repeatedly until the new data is added to make sure that the feasible weights are selected for the classification. Finally, when the maximum number of iteration is reached, then the procedure is terminated.

5 Experiment

The results are attained by the proposed method using the three data sets, namely breast cancer data set, localization data set, and skin segmentation data set based on accuracy, specificity, and sensitivity.

5.1 *Database Description*

This method is responsible for performing the big data streaming in which the massive data is used for the experimentation, and they are taken from three standard databases, namely skin segmentation data set, breast cancer data set, and localization data for person activity data set, using UCI machine learning repository [25], which are elaborated below:

(i) **Skin Segmentation Data Set**

This data set [26] is donated by Rajen Bhatt, and Abhinav Dhall. The data set is obtained by collecting the random samples B, G, R values from the face images of different groups of ages. The size of the total learning sample size is 245,057 from which 50,589 are the samples of skin and 194,198 are non-skin samples. The number of attributes is 4, the number of Web hits attained by the data set is 161,252, and the data set is univariate in nature.

(ii) **Localization Data for Person Activity Data Set**

This data set [27] is donated by Bozidara Cvetkovic. The data set contains recordings of five people each acquiring different activities. Each person carries four sensors to perform the same scenario five times. The number of instances considered in the data set is 164,860, whereas the number of attributes is 8. The Web hits attained by the data set are 96,443.

(iii) **Breast Cancer Data Set**

The breast cancer data set [28] is donated by Ming Tan and Jeff Schlimmer for breast cancer classification. The data set consists of 286 instances from which 85 instances to one class and 201 instances belong to other class. Each instance is described by 9 attributes, which are either nominal or linear. The number of Web hits is 401,299, and the nature of the data set is multivariate.

5.2 *Performance Measures*

The results attained by the methods are evaluated using performance measures like accuracy, sensitivity, and specificity, respectively. The performance measures show the success of algorithm for big data streaming and are elaborated as follows:

(i) ***Accuracy***: The accuracy indicates the accurate detection of intrusions, which is computed as,

$$\text{Accuracy} = \frac{T_p + T_n}{T_p + T_n + F_p + F_n} \tag{29}$$

where T_p denotes the true positive T_n represents the true negative,F_p refers false positive, and F_n is the false negative.

(ii) ***Sensitivity***: It detects the correct portion of intrusions using the sample of inputs identified by the classification results. The sensitivity is also known as true positive rates and is formulated as,

$$\text{Sensitivity} = \frac{T_p}{T_p + F_n} \tag{30}$$

(iii) ***Specificity***: The specificity detects the wrong portion during the intrusion detection. The specificity is also known as false positive rate (FPR) and is formulated as,

$$\text{Specificity} = \frac{T_n}{T_n + F_p} \tag{31}$$

The proposed DMS-SAE method is compared with four existing techniques like linear regression [29], neural network (NN) [30], DS-RNGE-KNN [1], and SAE [31].

6 Results

6.1 Analysis Using Skin Segmentation Data set

Figure 6 illustrates the analysis of methods based on accuracy, sensitivity, and specificity parameters using skin segmentation data set. The analysis based on accuracy parameter is portrayed in Fig. 6a. For 2 chunks, the corresponding accuracy values computed by existing linear regression, NN, DS-RNGE-KNN, SAE, and proposed DMS-SAE are 94.9%, 95%, 95%, 95%, and 95%, respectively. Likewise, for 6 chunks, the corresponding accuracy values computed by existing linear regression, NN, DS-RNGE-KNN, SAE, and proposed DMS-SAE are 88.30%, 88.85%, 89.39%, 89.94%, and 90.48%, respectively. The analysis based on sensitivity parameter is

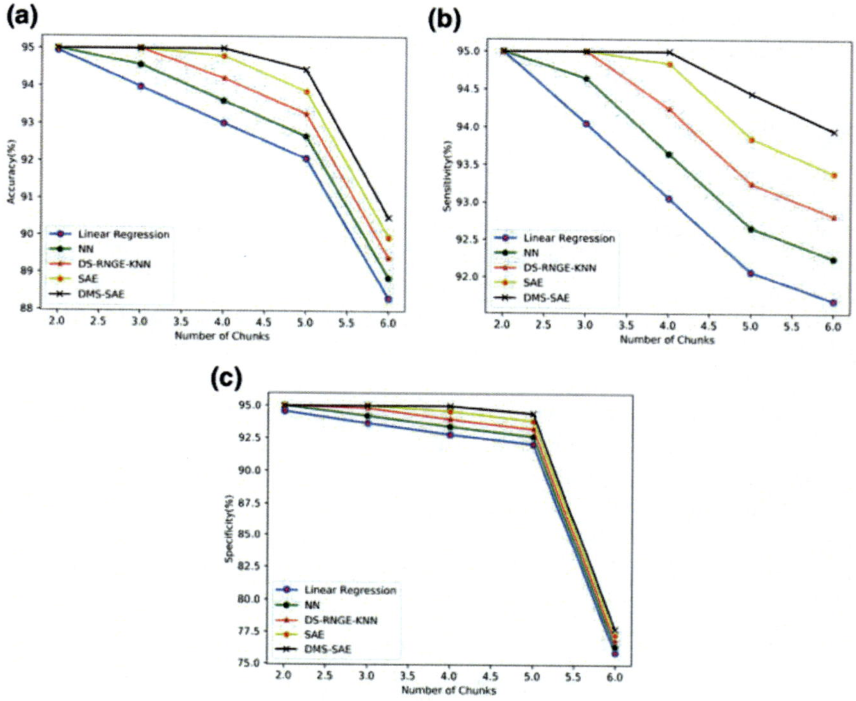

Fig. 6 Analysis of methods based on skin segmentation data set using metrics like **a** accuracy, **b** sensitivity, **c** specificity

portrayed in Fig. 6b. When the number of chunk is 2, then the corresponding sensitivity values computed by existing linear regression, NN, DS-RNGE-KNN, SAE, and proposed DMS-SAE are 95%, 95%, 95%, 95%, and 95%, respectively. Likewise, when the number of chunk is 6, the corresponding sensitivity values computed by existing linear regression, NN, DS-RNGE-KNN, SAE, and proposed DMS-SAE are 91.68%, 92.25%, 92.82%, 93.38%, and 93.95%, respectively. The analysis based on specificity parameter is portrayed in Fig. 6c. When the number of chunk is 2, then the corresponding specificity values computed by existing linear regression, NN, DS-RNGE-KNN, SAE, and proposed DMS-SAE are 94.56%, 95%, 95%, 95%, and 95%, respectively. Likewise, when the number of chunk is 6, the corresponding specificity values computed by existing linear regression, NN, DS-RNGE-KNN, SAE, and proposed DMS-SAE are 75.88%, 76.35%, 76.82%, 77.29%, and 77.76%, respectively.

6.2 Analysis Using Localization Data for Person Activity Data Set

Figure 7 illustrates the analysis of methods based on accuracy, sensitivity, and specificity parameters using localization data for person activity data set. The analysis based on accuracy parameter is portrayed in Fig. 7a. When the number of chunk is 2, then the corresponding accuracy values computed by existing linear regression, NN, DS-RNGE-KNN, SAE, and proposed DMS-SAE are 92.06%, 92.66%, 93.25%, 93.85%, and 94.45%, respectively. Likewise, when the number of chunk is 6, the corresponding accuracy values computed by existing linear regression, NN, DS-RNGE-KNN, SAE, and proposed DMS-SAE are 85.94%, 86.57%, 87%, 87.53%, and 88.06%, respectively.

The analysis based on sensitivity parameter is portrayed in Fig. 7b. When the number of chunk is 2, then the corresponding sensitivity values computed by existing linear regression, NN, DS-RNGE-KNN, SAE, and proposed DMS-SAE are 95%, 95%, 95%, 95%, and 95%, respectively. Likewise, when the number of chunk is 6, the corresponding sensitivity values computed by existing linear regression, NN, DS-RNGE-KNN, SAE, and proposed DMS-SAE are 85.27%, 85.80%, 86.33%, 86.85%,

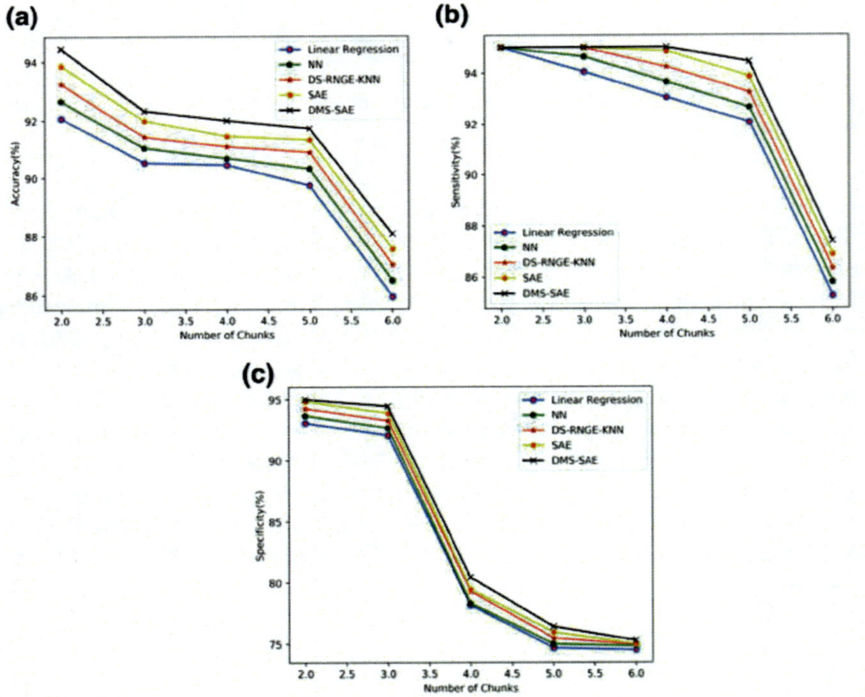

Fig. 7 Analysis of methods based on localization data for person activity data set using metrics like **a** accuracy, **b** sensitivity, **c** specificity

and 87.38%, respectively. The analysis based on specificity parameter is portrayed in Fig. 7c. When the number of chunk is 2, then the corresponding specificity values computed by existing linear regression, NN, DS-RNGE-KNN, SAE, and proposed DMS-SAE are 93.06%, 93.65%, 94.24%, 94.84%, and 95%, respectively. Likewise, when the number of chunk is 6, the corresponding specificity values computed by existing linear regression, NN, DS-RNGE-KNN, SAE, and proposed DMS-SAE are 74.53%, 76.35%, 76.82%, 77.29%, and 77.76%, respectively.

6.3 Analysis Using Breast Cancer Data Set

Figure 8 illustrates the analysis of methods based on accuracy, sensitivity, and specificity parameters using breast cancer data set. The analysis based on accuracy parameter is portrayed in Fig. 8a. For 2 chunks, then the corresponding accuracy values computed by existing linear regression are 92.06%, NN is 92.66%, DS-RNGE-KNN is 93.25%, SAE is 93.85% and proposed DMS-SAE is 94.45%, respectively. Likewise, for 6 chunks, the corresponding accuracy values computed by existing linear regression are 75%, NN is 75%, DS-RNGE-KNN is 75%, SAE is 75% and

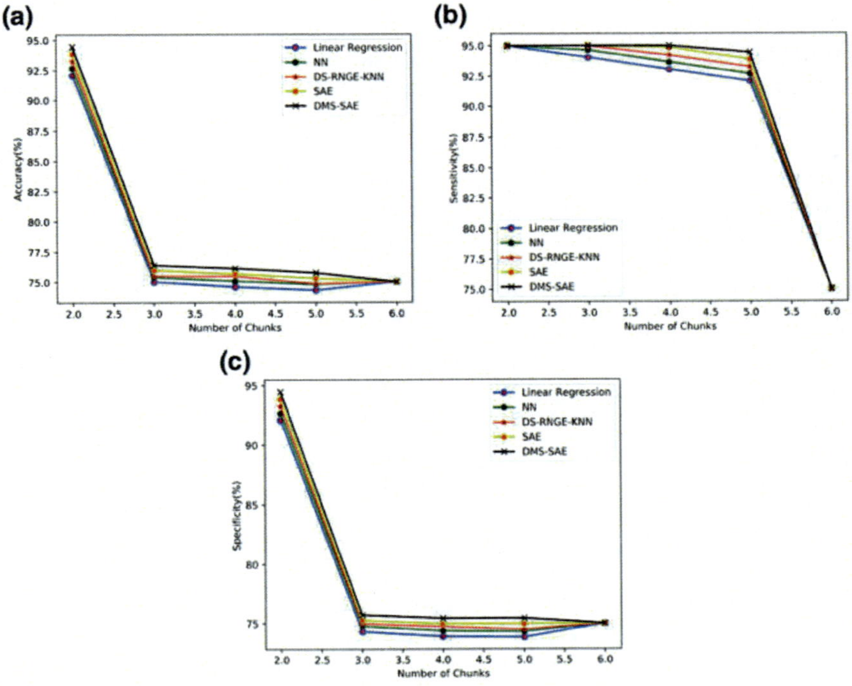

Fig. 8 Analysis of methods based on Breast cancer data set using metrics like **a** accuracy, **b** sensitivity, **c** specificity

proposed DMS-SAE is 75%, respectively. The analysis based on sensitivity parameter is portrayed in Fig. 8b. When the number of chunk is 2, then the corresponding sensitivity values computed by existing linear regression, NN, DS-RNGE-KNN, SAE and proposed DMS-SAE are 95%, 95%, 95%, 95%, and 95% respectively. Likewise, when the number of chunk is 6, the corresponding sensitivity values computed by existing linear regression, NN, DS-RNGE-KNN, SAE and proposed DMS-SAE are 75%, 75%, 75%, 75%, and 75%, respectively. The analysis based on specificity parameter is portrayed in Fig. 8c. When the number of chunk is 2, then the corresponding specificity values computed by existing linear regression, NN, DS-RNGE-KNN, SAE, and proposed DMS-SAE are 92.06%, 92.66%, 93.25%, 93.85%, and 94.44%, respectively. Likewise, when the number of chunk is 6, the corresponding specificity values computed by existing linear regression, NN, DS-RNGE-KNN, SAE and proposed DMS-SAE are 75%, 75%, 75%, 75%, and 75%, respectively.

6.4 Comparative Analysis

Table 1 illustrates the analysis of methods based on accuracy, sensitivity, and specificity measures using three data sets. The maximal accuracy is acquired by proposed DMS-SAE with accuracy value of 95%, whereas the accuracy value attained by the existing methods like linear regression is 95%, NN is 95%, DS-RNGE-KNN is 95%, and SAE is 95%. The maximal sensitivity is measured by the proposed DMS-SAE with sensitivity value as 94.56%, whereas the sensitivity value of existing methods like linear regression is 95%, NN is 95%, DS-RNGE-KNN is 95%, and SAE is 95%, respectively. The maximal specificity is computed by proposed DMS-SAE with specificity value as 95%, whereas that of the existing methods like linear regression is 95%, NN is 95%, DS-RNGE-KNN is 95%, and SAE is 95%, respectively.

Table 1 Comparative analysis

Data set	Metric	NN	DS-RNGE-KNN	Linear regression	SAE	Proposed DMS-SAE
Using skin segmentation data set	Accuracy	94.94	95	95	95	95
	Sensitivity	94.56	95	95	95	95
	Specificity	95	95	95	95	95
Using localization data for person activity data set	Accuracy	92.06	92.66	93.25	93.85	94.44
	Sensitivity	93.06	93.65	94.24	94.84	95
	Specificity	95	95	95	95	95
Using breast cancer data set	Accuracy	92.06	92.66	93.25	93.85	94.44
	Sensitivity	92.06	92.66	93.25	93.85	94.44
	Specificity	95	95	95	95	95

7 Conclusion

The paper deals with the proposed big data streaming that aimed at meeting the raising demands of high volume, high velocity, high value, high veracity, and huge variety. The big data streaming is performed using the Apache Spark framework such that the data from the distributed sources is handled parallel at the same time. The big data is evaluated using Apache Spark framework, wherein the processing is carried out in two phases. The first phase is offline phase; in this, data is split into different slave nodes for feature selection based on fitness that extracts the optimal features from the data using the proposed DMS algorithm. Then, the classification is carried out in the master nodes that are provided with the staked auto-encoders. The optimal tuning of the weights of staked auto-encoders is processed using the proposed DMS algorithm. The second phase is the online phase, in which the incremental data is considered for the evaluation. The new data obtained is split into several slave nodes for feature selection, classification, and for making effective decisions. Once the classification of new data is done, then the classified results generated from the online phase and offline phase are compared and adjusted based on the RST using weight bounds. The final output from the Apache Spark framework is the streamed data. The analysis of the classification methods confirm that the proposed method outperformed the existing methods with the accuracy, sensitivity, and specificity with values 95%, 95%, and 95%, respectively.

References

1. Ramirez, Gallego S., Krawczyk, B., García, S., Woźniak, M., Benítez, J. M., & Herrera, F. (2017). Nearest neighbor classification for high-speed big data streams using spark. *IEEE Transactions on Systems, Man, and Cybernetics: Systems, 47*(10), 2727–2739.
2. Wibisono, A., & Sarwinda D. (2019). Average restrain divider of evaluation value (ARDEV) in data stream algorithm for big data prediction. In *Knowledge-Based Systems.*
3. Vicentini, C., Santin, A., Viegas, E., & Abreu, V. (2019). SDN-based and multitenant-aware resource provisioning mechanism for cloud-based big data streaming. *Journal of Network and Computer Applications, 126,* 133–149.
4. Nair, L. R., Shetty, S. D., & Shetty, S. D. (2018). Applying spark based machine learning model on streaming big data for health status prediction. *Computers & Electrical Engineering, 65,* 393–399.
5. Ruiz, E., & Casillas, J. (2018). Adaptive fuzzy partitions for evolving association rules in big data stream. *International Journal of Approximate Reasoning, 93,* 463–486.
6. Wu, X., Zhu, X., Wu, G. Q., & Ding, W. (2014). Data mining with big data. *IEEE Transactions on Knowledge and Data Engineering, 26*(1), 97–107.
7. Torrecilla, J. L., & Romo, J. (2018). Data learning from big data. *Statistics & Probability Letters, 136,* 15–19.
8. Kaur, N., & Sood, S. K. (2017). Efficient resource management system based on 4vs of big data streams. *Big data Research, 9,* 98–106.
9. Hashem, I. A. T., Yaqoob, I., Anuar, N. B., Mokhtar, S., Gani, A., & Khan, S. U. (2015). The rise of "big data" on cloud computing: Review and open research issues. *Information Systems, 47,* 98–115.

10. Batista, B. G., Ferreira, C. H. G., Segura, D. C. M., Leite Filho, D. M., & Peixoto, M. L. M. (2017). A QoS-driven approach for cloud computing addressing attributes of performance and security. *Future Generation Computer Systems, 68,* 260–274.
11. Zheng, Z., Wu, X., Zhang, Y., Lyu, M. R., & Wang, J. (2013). QoS ranking prediction for cloud services. *IEEE Transactions on Parallel and Distributed Systems, 24*(6), 1213–1222.
12. Dean, J., & Ghemawat, S. (2008). MapReduce: simplified data processing on large clusters. *Communications of the ACM, 51*(1), 107–113.
13. Puthal, D., Nepal, S., Ranjan, R., & Chen, J. (2017). A dynamic prime number based efficient security mechanism for big sensing data streams. *Journal of Computer and System Sciences, 83*(1), 22–42.
14. Yin, B., Shen, W., Cheng, Y., Cai, L. X., & Li, Q. (2017). Distributed resource sharing in fog-assisted big data streaming. In *IEEE International Conference on Communications (ICC)* (pp. 1–6), May 2017.
15. Fernandez-Basso, C., Francisco-Agra, A. J., Martin-Bautista, M. J., & Ruiz, M. D. (2019). Finding tendencies in streaming data using big data frequent itemset mining. *Knowledge-Based Systems, 163,* 666–674.
16. Segatori, A., Bechini, A., Ducange, P., & Marcelloni, F. (2018). A distributed fuzzy associative classifier for big data. *IEEE transactions on cybernetics, 48*(9), 2656–2669.
17. Dagdia, Z. C. (2018). A scalable and distributed dendritic cell algorithm for big data classification. *Swarm and Evolutionary Computation.*
18. Lv, Y., Duan, Y., Kang, W., Li, Z., & Wang, F. Y. (2015). Traffic flow prediction with big data: a deep learning approach. *IEEE Transactions on Intelligent Transportation Systems, 16*(2), 865–873.
19. Almalki, E. H., & Abdullah, M. (5 March 2018). *A survey on Big Data Stream Mining.*
20. Mirjalili, S. (2016). Dragonfly algorithm: a new meta-heuristic optimization technique for solving single-objective, discrete, and multi-objective problems. *Neural Computing and Applications, 27*(4), 1053–1073.
21. Wang G. G. (2018). Moth search algorithm: A bio-inspired metaheuristic algorithm for global optimization problems. *Memetic Computing,* 1–14.
22. Wang, P., Fu, H., & Zhang, K. (2018). A pixel-level entropy-weighted image fusion algorithm based on bidimensional ensemble empirical mode decomposition. *International Journal of Distributed Sensor Networks, 14*(12), 1550147718818755.
23. Jayapriya, K., & Mary N. A. B. (2019). Employing a novel 2-gram subgroup intra pattern (2GSIP) with stacked auto encoder for membrane protein classification. *Molecular Biology Reports,* 1–14.
24. Pawlak, Z. (1995). Rough sets. In *Institute of Theoretical and Applied Informatics, Polish Academy of Sciences.*
25. UCI machine learning dataset. (2019). https://archive.ics.uci.edu/ml/datasets.php, Accessed on April 2019.
26. Skin Segmentation Data Set. (2019). https://archive.ics.uci.edu/ml/datasets/Skin+Segmentation, Accessed on August 2019.
27. Localization Data for Person Activity Data Set. (2019). https://archive.ics.uci.edu/ml/datasets/Localization+Data+for+Person+Activity, Accessed on August 2019.
28. Breast Cancer Data Set. (2019). https://archive.ics.uci.edu/ml/datasets/Breast+Cancer, Accessed on August 2019.
29. Morariu, O., Morariu, C., Borangiu, T., & Răileanu, S. (2018) Manufacturing systems at scale with big data streaming and online machine learning. In *In Service Orientation in Holonic and Multi-Agent Manufacturing* (pp. 253–264). Springer.
30. Hajar, A. A. S., Fukase, K., & Ozawa, S. (2013) A neural network model for large-scale stream data learning using locally sensitive hashing. In *In proceedings of International Conference on Neural Information Processing* (pp. 369–376). Springer.
31. Budiman, A., Fanany, M. I., & Basaruddin, C. (2015). Online marginalized linear stacked denoising autoencoders for learning from big data stream. In *In proceedings of International Conference on Advanced Computer Science and Information Systems,* IEEE (pp. 227–235).

Event Organization System for Turnout Estimation with User Group Analysis Model

P. Subhash, N. Venkata Sailaja, and A. Brahmananda Reddy

Abstract It is commonly observed that most of the events are organized through Internet social network services. Also, estimating the attendance of an event plays a major role in successfully organizing the event. Until now, major focus has been put on a user's personal profile in predicting whether he would attend the event or not. But there has been a minimal concentration on other user behaviors who are in the same group. Therefore, there is a fair chance that a user can be influenced by other users in the same group. In this paper, we develop a social network service which allows a group of users to get notified about the events organized by the event manager, and we also implement a novel framework for estimating or predicting the attendance of each user in a respective event. In addition, this methodology has been designed with a specially designed tool, and the performance is tested with an existing dataset. This experimental study yielded good results.

Keywords Dynamic social influence · Social event · Social network · User behavior

1 Introduction

Social media is playing an important role in various aspects in today's life. Especially in organizing the events, social network sites are being created and utilized effectively. For example, in meetups.com, tons of programs are conducted at different locations and lakhs of users are participating in the events every day. Like this, there are many other systems on large scale and small scale which are also providing platform for organizing events and managing it. This makes difficult to estimate whether a user would like to attend certain event or not, which in turn makes it harder to estimate the attendance of a particular event [1].

Event can be anything. It can be a party, or it can be a TED talk, or it can be some political gathering. Estimating the attendance of an event is very important. Event

P. Subhash (✉) · N. V. Sailaja · A. Brahmananda Reddy
Department of Computer Science and Engineering, VNR Vignana Jyothi Institute of Engineering and Technology Bachupally, Hyderabad, Telangana, India
e-mail: subhash_p@vnrvjiet.in

S. C. Satapathy et al. (eds.), *Intelligent System Design*, Advances in Intelligent Systems and Computing 1171, https://doi.org/10.1007/978-981-15-5400-1_66

involves a main thing, i.e., the gathering of the people together at a place. A main factor which can decide whether an event is successful or a failure is the attendance of the event. Suppose say, we have estimated 100 people to grace an event, but only 30 people turn up. This event is definitely a failure as the arrangements have been made for 100 people but very less number of people attends the event, and there is a lot of money and effort wasted for the event, and if the event involves distribution of food, there has been also a lot of wastage of food.

In another scenario, let us say there are 200 people who received invitations for an event. And the event organizers estimated the event attendance as 100. But number of people who attended the event is 150. This event will also be a failure because the arrangements and programs are made only for 100 people. There would be shortage of place, shortage of seating and also there would be shortage of food in the event. This will ultimately result in negative talk about the event, and the event will be a huge failure.

So, the problem most event managers face is the estimate of attendance of their event. If they estimate it less or if they estimate it more, it will result in a failure definitely. So, the estimation must be accurate. This means a lot of work for event managers. They have to first check the profiles of the people they are inviting, and they have check whether they are active or not in attending and participating in the social events.

Previously, a traditional method existed. It still exists now but as the things have become faster and easier through Internet, this traditional method is not used that often. So, the traditional method is that whenever an event is conducted, it involves a lot of people to attend it and make it a huge success; the invitations are sent to all the people. The people in response to that they will send an RSVP to the event manager or organizer, which means that the person would attend an event. In this way, by considering the number of RSVPs, the estimates were done in the past. Suppose say the event organizers sent invitations to 200 people and there are 80 RSVPs in response. So, the event attendance estimate is between 70 and 100. The arrangements were made to the maximum number in the estimate in case. So as we can see this is not a very good estimate, but still it does the job. This used to happen when there were no computers and there are no online systems.

However, because of the human factors and their changing minds, the people who submitted the RSVPs would not attend the events. But there were no other alternatives before. There was a high dependence on the RSVPs to estimate the attendance of the event, and it is not that reliable.

When the Internet came into existence, this process has transferred online. Event organizers are organizing the events and making people aware about those events. Then, online forms are circulated to the people. The people fill those online forms and give their details. Then, the event manager could personally call them or get into contact with them to know whether a specific person would attend an event or not. In this way, we can almost predict whether a person would attend the event or not. But, however, because we are humans, even though we say that we would definitely attend an event, but because of some sudden changes in our daily lives and our sudden changes in the minds, we decide not to attend the event. We are breaking the estimate rule here.

Facebook is a popular social media used by hundreds of millions of people. There are people who are addicted to Facebook and follow every development in the system. Facebook has a feature where people can create any event and describe about the events. There is also an extra feature where people have an option to select whether a person can attend the event or not. Since Facebook has collected the details of the person, the person who has added the event can have the details of the persons who are interested to attend the event. But due to the privacy issues, some people opt not to expose their data. At present, even though the details of the persons who are about to attend the event are not displayed, still an estimate of the event attendance can be there by seeing the number of people interested in the event. Like Facebook, there are many other systems which provide a feature for users to interact with others. Meetup.com is one such system which allows its users to organize events.

While predicting, there are two types of events that need to be considered. One of the types is the events which do not require any ticket to participate in the event. Another type is the events which would require a ticket to participate in the event. This is also an important factor in consideration to estimate the ground attendance at that event.

There are a lot of events which require people to buy a ticket to participate in that event. These are commercial events like sports matches (e.g., cricket, soccer, hockey), movie premieres, concerts, special events like celebrations of festivals, etc. These are also the events which involve a lot of promotions, sales and indirectly a lot of money. For example, if there is a music concert held by a famous musician, a lot of people would buy tickets for that event in advance. As the tickets require money to buy them, the number of advance bookings may give an approximate estimate of the people attending that event. However, the number of people who book tickets in advance and the number of people who would attend the event would never tally as the same. This is because the people might get caught up in other works. In these events, promotions are required to bring awareness in people that a certain event is happening. This might include sending text messages and emails to the persons who are interested in the topics on which the event is going to be held. There are also people who do not make advance bookings, but they buy the tickets on the spot when the event is going on. This factor is also important in estimating the attendance of a particular event. And knowing the number of people who would attend this event is very important as a lot of side businesses also happen like selling food items, clothes and toys. If it is involved in a large scale, providing security to the event site requires the attendance estimates as the security involves a lot of planning and effort. For example, if the capacity of the place where the event is held is 10000 people and the place capacity is filled with the people, then there is a fair chance that mishappening like stampedes or terrorist attacks can happen. The security has to be planned for the estimated number of people accordingly like how much time before the people should be allowed into the event site and where the security guards should be placed at the crowd places. So, if we overestimate the attendance in such situations, then surely it would be a waste of cost which may eventually lead to failure of the event.

If we underestimate the attendance, then definitely it would be very difficult and requires a lot of work to control the crowd.

Numerous investigations, for as far back as quite a long while, have thought about the social impacts as limitations or highlights in their examinations, from which the expectation results can be improved successfully. These explorers are assuming with the fundamental thought of, i.e., clients who hold comparable inclinations attempt to become a close acquaintance with one another, and because of their comparable inclinations, companions will in general carry on likewise. This event may be balanced as proceeding with associations continuously influence tendency. Be that as it may, in interpersonal organization which is occasion arranged, transient get-togethers are the main conceivable circumstances in which digital outsiders are associated, and these affiliations develop routinely, in this way impact probably will not be sufficiently relentless. Besides, it is traditional to see an individual who has interests in different subjects and contacts which just mirror incompletely regular interests ought not to deal with worldwide requirements. For instance, an individual may lean toward open-air recreations, yet his kinship with his individual mates does not compulsorily mean they additionally like outside amusements. To put it plainly, legitimately regarding social factors as fixed highlights or imperatives could be too unpleasant to even think about assessing the procedure of making a decision [1].

Especially, the arrangement procedure of a get-together is repeated, and we comprehend that social impacts should not be treated as static, rather it ought to be treated as a changing factor, i.e., when choices are made inside a gathering, individuals may tune into and can be convinced by certain sidekicks, and they may additionally influence different individuals. For the most part, if two partners hold a similar supposition, their penchant will be reinforced commonly; then again, opposing considerations leads to handicap certainty.

Till now, there is no integration of any method that estimates a user's attendance in an event. In a research work [1], the researchers have proposed a framework to include the social influence of other users on a user along with the user's personal interest. By involving the social influence of the other users, they have calculated a threshold value (h) and compared it with the personal preference function value (f) iteratively, and if $f > h$ of a user, the user is likely to attend the event. Otherwise the user is unlikely to attend an event. In this work, a lot of ground work is done to propose this algorithm and they have achieved the maximum precision of 96.15%.

The framework in the existing research is claimed to be correct, yet it is implemented only on one system's dataset (meetup.com). There are many event organizing systems which run on completely different datasets and different parameters. But the framework is suited only for one system. In this work, we try to implement the existing framework on a completely different event organization system built from scratch to prove that the existing system works properly on the new users also.

2 Literature Review

Even though human decisions and thinking patterns have a high degree of fluctuation and variation, they also exhibit structural patterns due to geographic, social constraints also other influences. Some of the works in the literature have given an event recommendation models [2–4]. With creating discussions among each other, making group decisions involving social networks, as well as influence within a group, in this paper [1] they proposed a system to understand what basic laws govern human presence and thinking patterns. There are few major factors which help in estimating these laws. They found out that humans experience a combination of periodic movement that is limited to an extent and seemingly random correlated with their social networks as well as its influence. Few events are both spatially and temporally and not affected by the social network structure, while other important events are more influenced by social network ties. They show that social relationships can explain about most of human influence which identifies an approximate estimation. Based on their findings, they developed a framework to estimate the attendance in an event that combines involvement of user group and analysis by their influence. Their model reliably predicts the number of users with the help of their activeness and gives an order of magnitude better performance than present models of event turnouts.

Gathering suggestion, which makes proposals to a gathering of clients rather than people, has turned out to be progressively imperative in both the workspace and individuals' social exercises, for example, meetings to generate new ideas for colleagues and social TV for relatives or companions. Gathering proposal is a difficult issue because of the elements of gathering enrollments and decent variety of gathering individuals. Past work concentrated principally on the substance interests of gathering individuals and overlooked the social qualities inside a gathering, bringing about problematic gathering suggestion execution. In this work [5], they propose a gathering suggestion strategy that uses both social and substance interests of gathering individuals. They examined the key attributes of gatherings and propose (1) a gathering agreement work that catches the social, skill and interest difference among different gathering individuals; and (2) a nonexclusive structure that naturally investigates a bunch of peoples' qualities and builds the relating bunch accord work. Nitty-gritty client investigations of differing bunches show the adequacy of the proposed procedures and the significance of consolidating both social and substance premiums in gathering recommender frameworks.

Despite the fact that "human movement and mobility patterns" have a high level of opportunity and variety, they additionally display basic examples because of geographic and social limitations. Utilizing phone area information, just as information from two online area-based interpersonal organizations, they planned to comprehend what fundamental laws administer human movement and elements. In this work [6], they found that people experience a mix of occasional development that is geologically restricted and apparently irregular bounces associated with their informal communities. Short-went travel is intermittent both spatially and transiently and not affected by the interpersonal organization structure, while long-separate travel is more impacted by informal community ties. They demonstrated that social connections can disclose about ten to 30% of all human development, while occasional conduct discloses half to seventy percent. In light of their discoveries, they built a model of "human mobility" that joins occasional short-range developments with movement because of the informal organization structure. They demonstrate that their model dependably predicts the areas and elements of future human development and gives a request of size preferable execution over present models of human portability.

There are situations where because of some peoples' actions in a group, the whole group gets affected. Finding the people who is the most influential in a group is a classical problem termed as "influence maximization" problem. For solving this problem, previously most of the works have ignored the past activities of the users. In this work [7], they have proposed that the influence maximization problem in NP-hard, and they have proposed an algorithm that is more accurate than the previous ones to solve the maximization of the influence problem.

3 Estimating Event Turnout

To estimate the attendance of the event, until now only the user's personal preference is considered. But to the best of our understanding, no popular event organization system in the literature has a working model of the estimate or predict module that can eventually predict the attendance for the events depending upon the attendance of the other users in the same group.

This work [1] proposes a framework based upon a concept called as independent cascade model [8]. The base of the concept is that there are a set of vertices in a network. These are also called as nodes. These nodes have two states: (i) active and (ii) inactive. These nodes are connected like a graph with the edges between them. Initially, all the nodes are not active. So, there are both active and inactive nodes at a time. An active node maybe connected to an inactive node via an edge. This active node can activate the inactive node with a probability. This is the concept of independent cascade model.

In calculating the preference function value of the users, we use other similarities instead of cosine similarity. For example, Levenshtein's distance algorithm can be used to calculate the similarity between two texts. This basic idea of the algorithm is that if there are two texts $t1$ and $t2$ which contain different strings, the number of edits required to convert $t2$ as $t1$ or $t2$ as $t1$ is calculated. This value is also called as edit distance. Based on this edit distance and the length of two texts, we can calculate the similarity between them. This algorithm is easy to implement and has a documentation support and can be implemented in any programming language. Similarly, there are other ways to calculate the similarity between two texts. For example, Jaccard's similarity coefficient, dice's coefficient, overlap similarity, etc. Based on the type of user attributes, event attributes and the method of storing them we can use these algorithms in the preference function to calculate the similarity (Figs 1 and 2).

If there are two nodes *A* and *A*| of which *A* is active and the other node *A*| is inactive, these two nodes are connected by an edge *E*. The node *A* gets only one

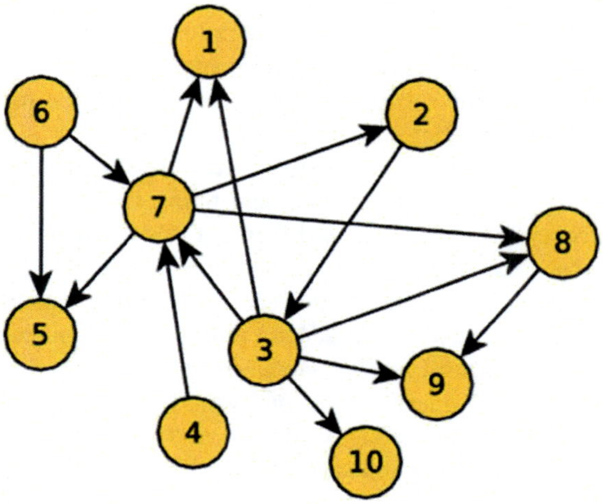

Fig. 1 Example of event network graph

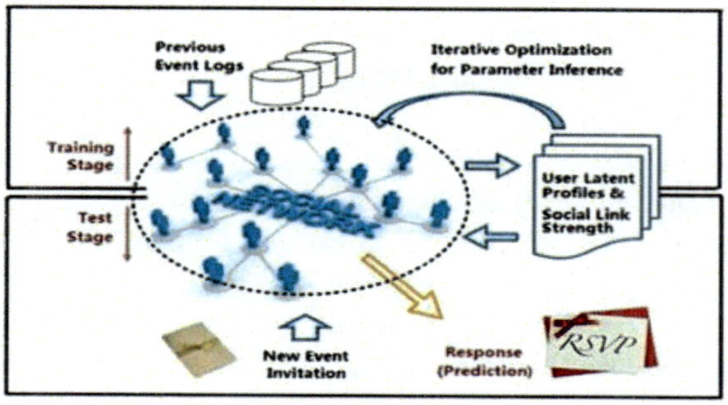

Fig. 2 System architectural model

chance to activate the *A*I. This event is considered as success. The success depends upon the probability with which the node *A* can activate the inactive node.

If necessary, instead of this independent cascade model concept, we can use other concepts like linear threshold model [9] to find the impact between nodes but for calculating the impact made by a user on other users in social media network, the former concept is used widely, and it has a very high accuracy.

To understand how the estimate is being done, first we have to know the features of this system. The event organization system mainly consists of two groups of people: (a) users and (b) event organizers. The users have to register in the system to see any events happening. At the time, we collect the basic details of users like their name, age, date of birth, phone number, their preferences of events, etc., while registering. A strong input validation is done to minimize the entry of invalid or unintelligible data. When the user logs in, he can view the events that are going to happen in the upcoming days and see the specifics of the events. In addition to it, the user can participate under the discussions related to specific events.

The event organizers can register into the system by giving their basic details like their name, email, phone number, etc. After registering, the event manager can login to the event manager module and can add events. The event manager has to provide details of the events like the event location, event description and time of the event. So, to describe the working of this system shortly, the users register and login into the user module, they see the events added by the event organizers and can make discussions under the events. However, we have created a log for some users which contains the details of their attendances of the events they attended previously. In the system, there are two types of users: (a) old users who are using this system and have records of previous attendances and (b) new users who would be registering into the system in future or who have registered recently [10].

As the attendance estimation/prediction is the problem at this point of time, we can define the problem statement as follows: There are a set of users who are in a same group, and we have to estimate for every person in the group whether the person would come to the event or not. The attendance estimate has to be a yes (positive) or a *no* (negative). To make it to happen, we use the values of two functions: threshold function and preference function. Threshold has a range between 0 and 1, it is basically a cut-off value for the user to attend for an event, and this threshold value calculation part involves the threshold values of other users and the preference function values of the other users to attend an event. In this way, the influence parameter of other users is incorporated into the system/model to make the event-wise timely attendance prediction. However, this threshold value for every user is calculated iteratively. In the preference function value, we find the cosine similarity between the users' preferences and event attributes. To make the estimation of the event, the framework proposed in the existing system is as mentioned. The existing framework consists of two stages: (i) training and testing stages.

Training stage: In this stage, the user's participation in the system as well as the previous attendance logs are given as the input. Based on the Eq. (1) and the algorithm shown in the figure below, a threshold value is calculated for every user in the system with respect to an event.

Here hi, 0 is the activeness of the user in the system. It fluctuates between 0 and 1. The higher the activeness of the user, the lower the hi, 0 value. The parameters taken into consideration for measuring the user's activeness are the number of logins of the user, number of previous attendances of the user for the similar events. For a new user, the attendances factor is not considered but for the old users, it plays an important role in deciding the threshold value.

For the preference function f value, the user preferences pk and the event attributes ak are taken into consideration and cosine similarity is calculated between those two. Finding cosine similarity between two records is nothing but taking the strings in both the records which make two vectors by recording the number of occurrences of the string in both the documents.

For example, consider two records "Hi there" and "Hello there". So, the strings in both the documents are "Hi", "there" and "Hello". For the record 1, the vector becomes [1,1,0] and for the record 2, the vector becomes [0,1,1]. Now, the cosine similarity between the two vectors is given by the cosine angle between these two vectors.

$$\cos\theta = \frac{(\vec{a})\cdot(\vec{b})}{|(\vec{a})||(\vec{b})|}$$

Using the above equation, for the example given above the cosine similarity would be calculated as:

$$\cos\theta = \frac{([1,1,0])\cdot([0,1,1])}{|([1,1,0])|\cdot|([0,1,1])|}$$

The algorithm for the training stage is as follows:

$$\begin{aligned} h(u_i, e_k) &= h_{i,o}\cdot\prod_{j\in N_i}\left[1 - I(f_{j,k} - h_{j,k})\right] \\ f(u_i, e_k) &= \text{cosine}(a_k^T, \text{p}_i^T) \\ I(X) &= \frac{1}{1+e^{-x}} \end{aligned} \tag{1}$$

So initially, to find the threshold value of a particular user in the first step, as we do not know the threshold values of the other users, we take hi, 0 value as the threshold value of that user. Suppose say there are four users—$u1$, $u2$, $u3$ and $u4$. Now if we want to calculate the threshold value for the user $u1$ and say there are no other threshold values available. So we calculate the threshold of the user as h1, 0. Using $u1$'s threshold value, we can find $u2$'s threshold value and so on. Using the algorithm above, we update the threshold value of each user until the required condition is met.

Input: target user group $U = \{u_i\}$, event set $E = \{e_k\}$ and attendance record
Store: event attributes a_k for each $e_k \in E$;
Output: user's threshold values $h = \{h_{i,k}\}$

1: Iteration = *True*;
2: **while** (Iteration)
3: Iteration = *False*;
4: **for** $u_i \in U$, $e_k \in E$
5: update $f_{i,k}$, $h_{i,k}$ based on Equation 1;
7: update $s_{i,k}$ as $I(f_{i,k} - h_{i,k})$;
8: **if** $s_{i,k}$ changed **then** Iteration = *True*;
9: **end if**
10: **end for**
11: **end while**

Testing stage:

In this stage, for every user, the algorithm, the f value is updated again and again until a stable value is achieved. For every user u_i and event e_k, if threshold value is less than preference function value, the user may likely attend the event e_k. Otherwise, the user may not attend the event. In this stage, for every user, the preference value of a user is calculated based on the cosine similarity between user preferences and event attributes. Then, using the algorithm, the f value is updated again and again until a stable value is achieved. For every user u_i and event e_k, if threshold value is less than preference function value, the user may likely attend the event e_k. Otherwise, the user may not attend the event. For finding preference function value initially, the event attributes and the user preferences can be mined using appropriate queries and writing an appropriate function to find the similarity between them. The preference value of a user is calculated based on the cosine similarity between user preferences and event attributes.

Input: target user group $U = \{ u_i \}$, event set $E = \{e_k\}$ and attendance records $\{s^0_{i,k}\}$;
Store: event attributes a_k for each $e_k \in E$;
Output: user's threshold values $h = \{h_{i,k}\}$

for $u_i \in U$
calculate $f_{i,k}$ for each u_i;
end for
Iteration = *True*;
while (Iteration)
Iteration = *False*;
for $u_i \in U$
update $h_{i,k}$ based on Equation 1;
update $s_{i,k}$ as $I(f_{i,k} - h_{i,k})$;
if $s_{i,k}$ changed **then** Iteration = *True*;
end if
end for
end while
return $S = \{s_{i,k}\}$;

4 Results

A specially designed tool has been developed for the event organization system, and it is created with the user, event organizer and admin modules whose functionalities are described in the above section. But it was never integrated with any social network service. We are implementing the framework on a new dataset with changed parameters and check the compatibility of the framework on a new system (Figs 3 and 4).

As the tool is opened, it offers various features out of which user selects any of service according to the requirement. The event manager can login and add events as on when it is required, this makes system update and on par with the societal needs. After adding an event, the user can view the events and make discussions under them as below.

After all these things done, the event manager can predict the attendance of a specific event and the report is submitted to the event manager thus can improve the way of organizing the events and many more timely decisions can be taken (Fig. 5).

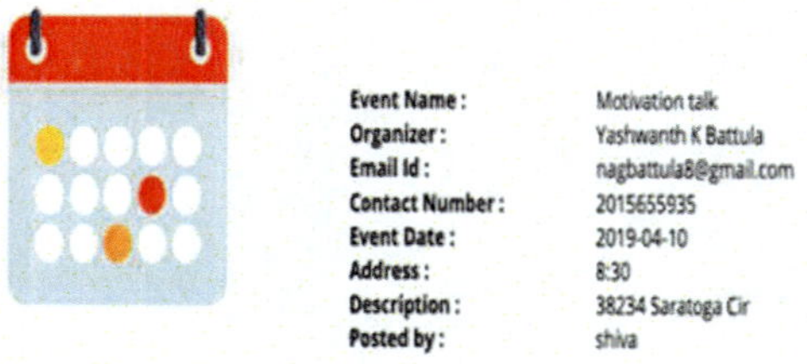

Fig. 3 Discussions by user

Fig. 4 Viewing the events

Here, the estimate for an old user (akhil) who is active in participation of the similar events and active in the system by frequently logging in and making discussions have the positive estimate. Also a user (tony) who is new and has no previous attendance records, but the participation of the user in the system is active, a positive estimate is shown.

5 Conclusions and Future Work

In this system, we designed a new tool with fresh database for event organizers and users and also implemented the existing framework proposed for estimating the attendances of an event. Estimating each person's presence in the event is again another crucial factor as it gives us a possibility to do a background check on them and

User Name	Type
shiva	Positive
krish	Negative
akhil	Positive
johndoe	Negative
janedoe	Negative
patrick	Negative
bauer	Negative
tony	Positive
chloe	Negative
lisa	Negative
yash	Negative
shivaqwer	Negative
neelesh	Negative

Fig. 5 Estimate results of an event

detect any security issues. It will also be beneficial for an event manager if the event involves distributing food and arranging seating for the events as the manager will now have an approximate idea of how many people would attend the event. As there will be an increase in the usage of Internet and increase in the population of internet users, there will be many more events that would be conducted through Internet portals, which in turn will lead to increase in the number of online portals organizing the events. This will lead to different types of datasets for the same purpose. In all these datasets, there are different types of people attending different type of events. Some might attend the events alone, and some people may attend the events with their friends. This work has given equal focus on both the aspects leading to a better estimation of the event attendance. However, we can make this framework even better if we use natural language processing techniques to detect positive discussions and negative discussions for estimating the user's attendance. In addition, if we get some more reliable parameters, this prediction model would be even more accurate.

References

1. Xu, T., Zhu, H., Zhong, H., Liu, G., Xiong, H., & Chen, E. (2018). Exploiting the dynamic mutual influence for predicting social event participation. *IEEE Transactions on Knowledge and Data Engineering, 31*(6), 1122–1135. https://doi.org/10.1109/tkde.2018.2851222.
2. Mynatt, E., & Tullio, J. (2001, January). Inferring calendar event attendance. In *Proceedings of the 6th international conference on Intelligent user interfaces* (pp. 121–128). ACM.
3. Khrouf, H., & Troncy, R. (2013, October). Hybrid event recommendation using linked data and user diversity. In *Proceedings of the 7th ACM conference on Recommender systems* (pp. 185–192). ACM.
4. Quercia, D., et al. (2010) Recommending social events from mobile phone location data. In *2010 IEEE international conference on data mining*. IEEE.
5. Gartrell, M., Xing, X., Lv, Q., Beach, A., Han, R., Mishra, S., & Seada, K. (2010, November). Enhancing group recommendation by incorporating social relationship interactions. In *Proceedings of the 16th ACM international conference on Supporting group work* (pp. 97–106). https://doi.org/10.1145/1880071.1880087.
6. Cho, E., Myers, S. A., & Leskovec, J. (2011, August). Friendship and mobility: user movement in location-based social networks. In *Proceedings of the 17th ACM SIGKDD international conference on Knowledge discovery and data mining* (pp. 1082-1090). https://doi.org/10.1145/2020408.2020579.
7. Goyal, A., Bonchi, F., & Lakshmanan, L. V. (2011). A data-based approach to social influence maximization. *Proceedings of the VLDB Endowment, 5*(1), 73–84. https://doi.org/10.14778/2047485.2047492m.
8. Nazemian, A., & Taghiyareh, F. (2012, November). Influence maximization in independent cascade model with positive and negative word of mouth. In *6th International Symposium on Telecommunications (IST)* (pp. 854–860). IEEE. https://doi.org/10.1109/istel.2012.6483105.
9. Goyal, A., Lu, W., & Lakshmanan, L. V. (2011, December). Simpath: An efficient algorithm for influence maximization under the linear threshold model. In *2011 IEEE 11th international conference on data mining* (pp. 211-220). IEEE. https://doi.org/10.1109/icdm.2011.132.
10. http://www.sumankundu.info/articles/detail/How-To-CodeIndependent-Cascade-Model-of-Information-Diffusion.

A Method of Speech Signal Analysis Using Multi-level Wavelet Transform

Kanaka Durga Returi, Y. Radhika, Vaka Murali Mohan, and K. Srujan Raju

Abstract Speech signal analysis using multi-level wavelet transform and signal decomposition is presented in this paper. This research work carried out by recording the voices of the 40 different speakers. In this work, same set of words were spoken by the speakers, those 4 words are 'Bhavana,' 'How,' 'Are,' and 'You.' From these words, features were extracted by using wavelet analysis. The results presented are based on the 800 data signals for the analysis. The study was carried out to measure the performance of the speech signal through the wavelet analysis and is reported.

Keywords Speech signals · Wavelet analysis · Discrete · Multilevel · Wavelet transform

1 Introduction

Wavelet analysis is an attractive and familiar tool used for several models in geophysics together with tropical convection. Wavelet analysis is an electrifying model used for solving complex problems of arithmetic, general science, engineering, and technology. The recent utilization of this analysis is for wave propagation, signal processing, pattern identification, data compression, computer graphics, image processing, recognition of airplanes, submarines, and medical technology. This analysis of the wavelets also allows difficult data like images, signal, speech,

K. D. Returi (✉) · V. M. Mohan
Department of CSE, Malla Reddy College of Engineering for Women, Medchal, Hyderabad, Telangana, India
e-mail: durga1210@gmail.com

V. M. Mohan
e-mail: murali_vaka@yahoo.com

Y. Radhika
GITAM University, Vizag, AP, India

K. Srujan Raju
Department of CSE, CMR Technical Campus, Kandlakoya, Medchal, Hyderabad, Telangana, India

S. C. Satapathy et al. (eds.), *Intelligent System Design*, Advances in Intelligent Systems and Computing 1171, https://doi.org/10.1007/978-981-15-5400-1_67

and patterns for decomposition of the basic structures at various situations, levels, and then reformed by means of elevated accuracy. Wavelet representation of the function is the enhanced description of Fourier transform and it is a potential tool designed to investigate the workings of the signal. This work describes the method of wavelets including exchange of the various functions of wavelets and provides particulars used to investigate the power spectrum of wavelets.

Wavelet analysis is a most popular tool for feature extraction which is utilized for signal analysis. The main goal of the work is to design a structure that can recognize the speaker depending upon the voice variations. They try to estimate the predictable output values with known historic data. A group of work also involves wavelet and other transformations. In those procedures, they tried to identify the speaker with the help of wavelet transformations and analysis. The wavelet analysis provides us valuable features of speech signals, and they are strong enough against noise as well as variations.

2 Literature Review

Several researchers proposed several research works comprises of wavelet analysis some of them are Tabibian et al. [1] presented a new method for the estimation of threshold value of the noisiness of the speech and noise coefficients of the wavelets. Lung [2] derived wavelet packets features for speaker identification by introducing the evaluation index. Bojan and Zdravko [3] presented an algorithm for feature extraction through the decomposition of the wavelet packets and signal model. Sahu et al. [4] reported a new formation by using wavelet packet analysis for English phoneme recognition which contains frequency bandwidth.

Pavez and Silva [5] proposed speech recognition systems through Cepstral coefficients of wavelet packets. The performances were calculated with Mel-Frequency Cepstral coefficients in database. Daqrouq [6] developed a model for wavelet transform by utilizing neural network for the recognition of the speaker. Khald and Khalooq [7] presented the technique for speaker identification using linear prediction coding with genetic algorithm. Vong and Wong [8] reported the wavelet packet transform for the extraction of features of the pattern through multi-layer perceptron and support vector machines. Fabrício [9] presented wavelet-based algorithm for speech and speaker recognition systems. Lung [10] proposed a model for wavelet transform-based speaker recognition system by means of clustering.

Szu et al. [11] developed a method to construct artificial neural networks-based recognition system. Shivnarayan and Ram Bilas [12] presented a new method for categorization of wavelet transforms using least squares support vector machine by means of different kernel functions. Lung [13] described a method of wavelet packet by utilizing neural network for the identification of the speaker with back propagation method. Shung [14] developed a thresholding method for noise reduction by using gradient-based adaptive learning algorithms. Lung [15] presented a new method of speaker recognition from wavelet decomposition within the significant computational advantages.

Hamid et al. [16] presented speech representations by using wavelet filter-banks with discrete wavelet transform and the performance is analyzed with neural networks. Leandro et al. [17] reported a genetic algorithm for the representation of the speech by using wavelet decomposition and classification. Avci and Zuhtu [18] reported a speech recognition system with the classification of the feature extraction and wavelet packet-based fuzzy system implication. Mohammed and Jean [19] presented a method for enhancement of speech by means of time and scale variation of wavelet thresholds. Szu et al. [11] presented a method for the utilization of the discrete and continuous wavelet transforms simultaneously by using artificial neural networks for the recognition of the speaker.

Returi and Radhika [20] presented an artificial neural networks model by using wavelet analysis for speaker recognition. Returi et al. [21] presented an approach for speaker recognition by using wavelet analysis and support vector machines. Returi et al. [22] compared and reported different approaches for the speaker recognition.

3 Wavelet Transform

This transform can be utilized for evaluation of the time series data to facilitate signals at various frequencies. The acceptable wavelet function should contain zero mean and it is limited to a small area in both frequency and time. Gaussian proposed the plane wave function and it is shown in Eq. (1).

$$\Psi_0(\eta) = \pi^{-1/4} e^{i\omega_0\eta} e^{-\eta^2/2} \tag{1}$$

where $\psi_0(\eta)$ is the wavelet function, η is the time variable, and ω_0 is the frequency.

The wavelet transform of the continuous function of the isolated progression x_n can be defined as the complexity of x_n through a scaled and transformed description of $\psi_0(\eta)$ as shown in Eq. (2).

$$W_n(s) = \sum_{n'=0}^{N-1} x_{n'} \Psi^* \left[\frac{(n' - n)\delta t}{s} \right] \tag{2}$$

where δt is the equal time spacing, x_n is the time series. Here, (*) represents the conjugate of the complex function through changeable scale of the wavelet s.

To estimate the wavelet transform, the convolution of each scale should be done N times; here, N is number of points. The theorem of convolution agrees to do all N simultaneous convolutions in Fourier space by using a discrete Fourier transform (DFT) and it is shown in Eq. (3).

$$\widehat{x}_k = \frac{1}{N}\sum_{n=0}^{N-1} x_n e^{-2\Pi ikn/N} \quad (3)$$

where i is the number iterations, k is frequency, x_k is the time series function, and N is the number of samples.

3.1 Discrete Wavelet Transform

Signal is represented in a different form by signal transformation, without changing the information of the content present in the original signal. The important region in many signals is the low-frequency region which represents the characteristics of the signal. The audio signal contains a variety of information about the speaker. This comprises 'high-level' properties like language, context, emotional state, speaking style, and many others. These properties are very complex and difficult in measuring the speaker identity. The most useful information of the speaker is retrieved by the 'xlow-level' properties of the audio signal; i.e., pitch, frequencies and bandwidths, intensity, short-time spectrum, spectral correlations, and others. That is why telephone and radio-circuits can be bandwidth-limited to low frequencies (up to about 3.5 kHz, even though our hearing extends much higher than this).

However, for the purpose of listening only, the high-frequency part of the speech without the low-frequency components might find it entirely unintelligible. To illustrate this idea of the discrete signal, it passes through the filters and then divides the frequency domains. If the signal passes through the filters of low and high pass, it gives two signals as a result: one with low frequency of signal and other is signal with high-frequency content. To recognize the distinction between the qualities of the low- and high-pass information, the result of low-pass output is represented as '*A* (Approximation)' and high-pass output as '*D* (Detail).' Sub-band coding is a technique of decomposing the source signal into constituent parts and decoding the parts separately.

Given the original signal with N samples, the decomposition would contain a total of 2N samples, because both A and D versions of the signal would each have N samples. To retain the same amount of information at each step of the decomposition of the signal, the A and D signals need to be down sampled by a factor of 2. The original signal is decomposed as approximation and detail with the low and high-pass filters. By applying down sampling, process continues by taking the A component and putting it through the mirror filters again. This would then result in another level of decomposition $A1$ and $D1$; the same process is applied for this level $A1$ and the resulted output is $A2$ and $D2$. The result of this successive filtering of the signal would be the signals that contain information in frequency bands successively reduced by a factor of 2. This is called multi-level wavelet transform. Complete coding of the signal would consist of the signals $A5$, $D5$, $D4$, $D3$, $D2$, $D1$ and it is represented in Fig. 1.

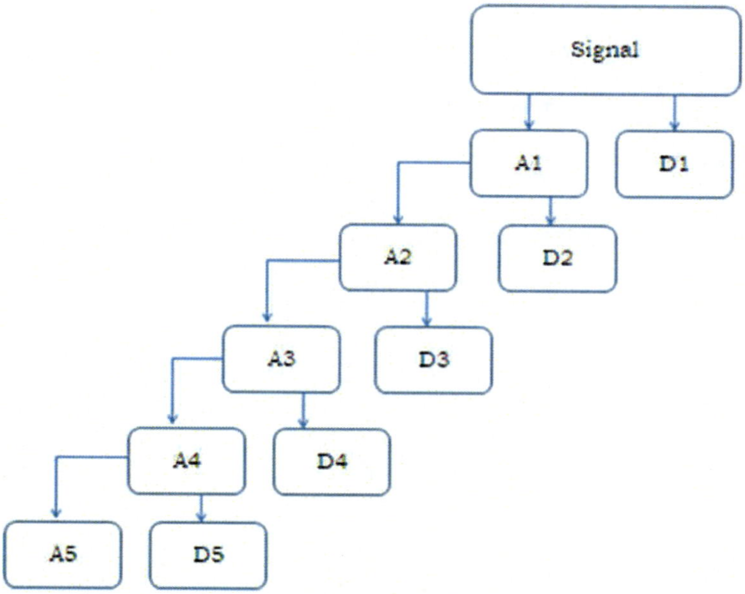

Fig. 1 Multi-level discrete wavelet transform

By moving upwards in the tree, and using mirror filters, the signal can be reconstructed with the filter structure shown in Fig. 2. The mirror filters are used for perfect reconstruction of the given level from the below level by upsampling the *A* and *D* components by the factor of 2.

$$\begin{aligned} S &= A_1 + D_1 \\ A &= A_{j+1} + D_{j+1} \end{aligned} \tag{4}$$

$$\begin{aligned} S &= A_1 + D \\ &= A_2 + D_2 + D_1 \\ &= A_3 + D_3 + D_2 + D_1 \\ &= A_4 + D_4 + D_3 + D_2 + D_1 \\ &= A_5 + D_5 + D_4 + D_3 + D_2 + D_1 \end{aligned}$$

In this, *S* represents the signal, *A* represents the approximation, *D* represents the detail, and *j* is the decomposition level. Equation 4 is that at any level *j* of analysis, and an approximation to the signal can be constructed from the next lower approximation and details coefficients.

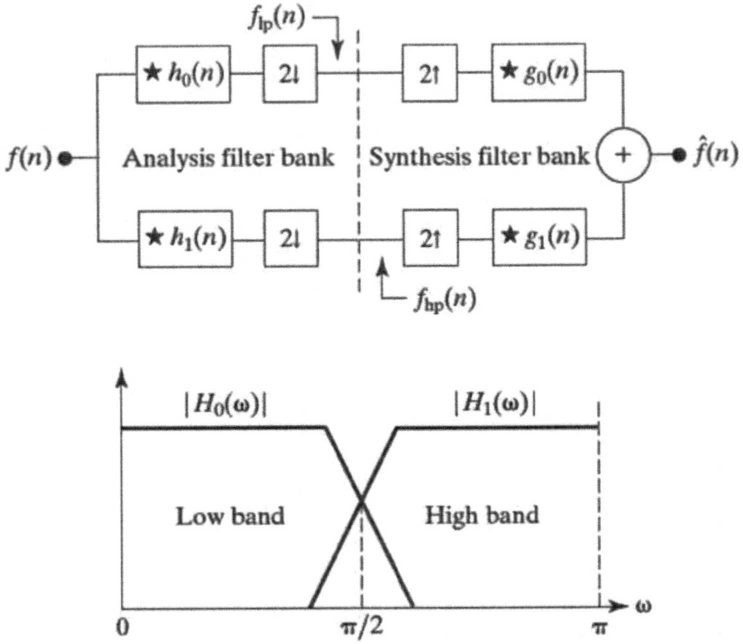

Fig. 2 Structure of the mirror filters

4 Problem Statement

This research work represents that wavelet analysis was used to extract the features. During this phase, speaker voices are recorded and stored. The stored voices are then converted into the form of digital signals. Later, the digital signal undergoes pre-process stage, where this is decomposed into several frequency bands. Features are extracted by utilizing the signal. The voices of 40 different speakers were recorded 20 male and 20 female speakers in the range of 20–40 were asked to spell out 4 words each 'Bhavana,' 'How,' 'Are,' and 'You.' Each speaker says again these words and the same were recorded. For the analysis of the wavelets, the kit of MATLAB 7.7.0 is applied. The number of levels précised as 5, which resulted in five different values from wavelet analysis.

To illustrate this idea, initially, the discrete signal is passed through the filters that divide the frequency domain. After passing the signal through the filters of low pass and high pass, two signals are obtained as a result: one signal with the content of low frequency and other is signal with high-frequency content. To recognize the distinction between the qualities of the low and high-pass information, the original signal is decomposed into two parts such as approximation and detail with the low- and high-pass filters. By applying down sampling, process continues by taking the

A component and putting it through the mirror filters again. Since the same decomposition and reconstruction filters are used at each level, and signal information is down sampled by 2 at each level, the above equation implies a very efficient method of coding a signal with wavelets.

5 Results of the Wavelet Analysis

The results presented are based on the 800 data signals for the analysis. The study was carried out to measure the performance of the speech signal through the wavelet analysis. The analysis of the wavelets is utilized for the known input signal. In this signal, it represents a specific word spoken by a specific speaker. Analysis of the wavelets is represented in the form of detail and approximation. The approximation is then split into a series of levels. These levels are original audio signal (shown as Fig. 3), symmetric transform output of the waveform (shown as Fig. 4), transform output of the waveform (shown as Fig. 5), signal train optimization (shown as Fig. 6), and regenerated output of the waveform (shown as Fig. 7). This analysis represents the degree of details. In this analysis, the number of details are fixed and taken as 5 and represented as $D1$, $D2$, $D3$, $D4$, $D5$, and $A5$.

Figure 3 shows the original audio signal of the DCT transform and is drawn between number of samples vs amplitude of the signal. Prior to the step of feature extraction, the signal needs to be processed with the help of silence removing algorithm and it is followed by the process of normalization of the speech signals to formulate the similar signals in spite of the differences in magnitude.

Fig. 3 Original audio signal

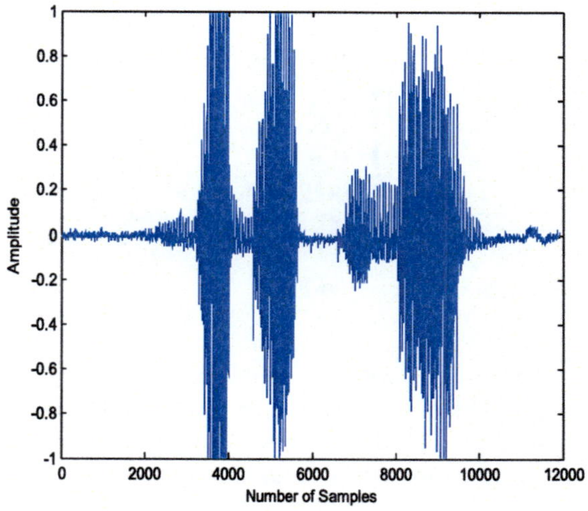

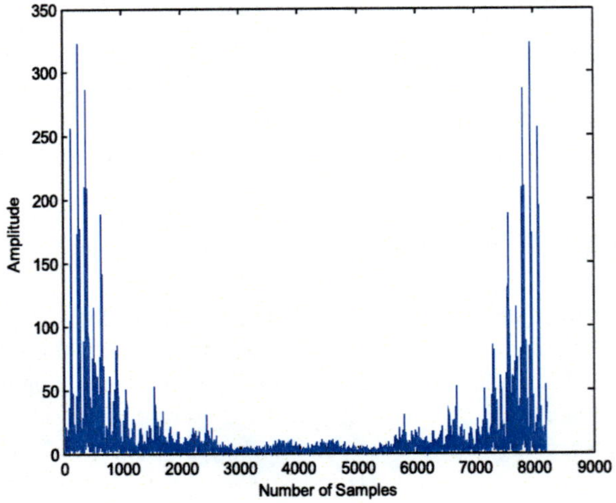

Fig. 4 Symmetric transform output waveform

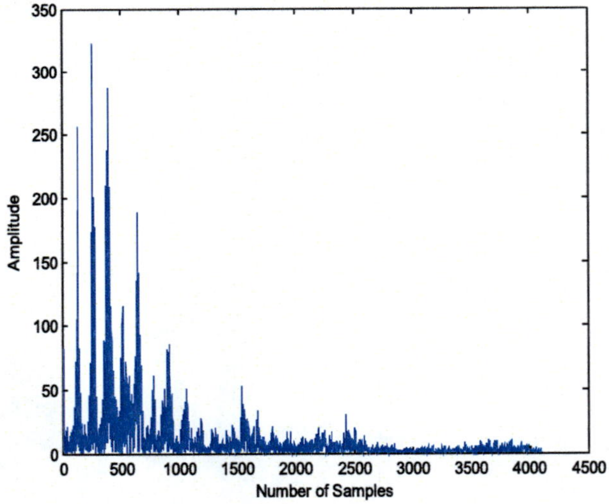

Fig. 5 Transform output waveform

Figure 4 shows the symmetric transform output waveform and is drawn against the number of samples versus amplitude of the signal. This figure reveals the transformation of the different frequencies from high to low.

Figure 5 shows the transform output waveform and is drawn against the number of samples versus amplitude of the signal. This figure reveals the transformation of the signal at different high-frequency values.

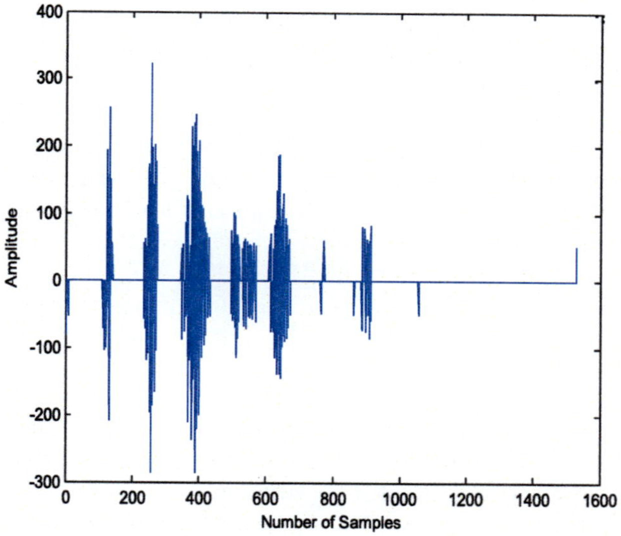

Fig. 6 Signal train optimization

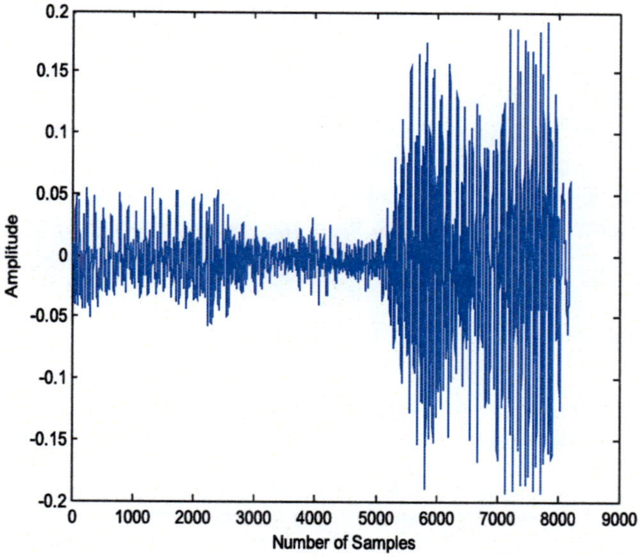

Fig. 7 Regenerated output waveform

Figure 6 shows the signal train optimization and is drawn against the number of samples versus amplitude of the signal. This figure reveals the transformation of the signal at different low-frequency values and its optimization.

Figure 7 shows the regenerated output waveform and is drawn against the number of samples versus amplitude of the signal. This figure reveals that the original signal can be reconstructed from the wavelet decomposition with the help of mirror filters. It also requires the identification of the components which are having noise. By removing those components, signal can be reconstructed.

6 Conclusion

This research work is carried out by recording the voices of the 40 different speakers (20–40 Years), which include 20 male and 20 female voices. In this work, same set of words were spoken by the speakers. They were made to say 4 words each 'Bhavana,' 'How,' 'Are,' and 'You.' Each speaker says again these words and the same were recorded. For the analysis of wavelet, the kit of MATLAB 7.7.0 is applied. The number of levels was précised as 5, which resulted in five different values from the wavelet analysis. From these words, features were extracted by using wavelet analysis.

The results presented are based on the 800 data signals for the analysis. The study was carried out to measure the performance of the speech signal through the wavelet analysis. Analysis of the wavelets is represented in the form of detail and approximation. The approximation is then split into a series of levels. These levels are original audio signal, symmetric transform output of the waveform, transform output of the waveform, signal train optimization, and regenerated output of the waveform.

References

1. Tabibian, S., Akbari, A., & Nasersharif, B. (2015). Speech enhancement using a wavelet thresholding method based on symmetric Kullback–Leibler divergence. *Signal Processing, 106*, 184–197.
2. Lung, S.-Y. (2006). Wavelet feature selection based neural networks with application to the text independent speaker identification. *Pattern Recognition, 39*(8), 1518–1521.
3. Kotnik, B., & Kačič, Z. (2007). A noise robust feature extraction algorithm using joint wavelet packet subband decomposition and AR modeling of speech signals. *Signal Processing, 87*(6), 1202–1223.
4. Sahu, P. K., Biswas, A., Bhowmick, A., & Chandra, M. (2014). Auditory ERB like admissible wavelet packet features for TIMIT phoneme recognition. *Engineering Science and Technology, an International Journal, 17*(3), 145–151.
5. Pavez, E., & Silva, J. F. (2012). Analysis and design of wavelet-packet cepstral coefficients for automatic speech recognition. *Speech Communication, 54*(6), 814–835.
6. Daqrouq, K. (2011). Wavelet entropy and neural network for text-independent speaker identification. *Engineering Applications of Artificial Intelligence, 24*(5), 796–802.

7. Daqrouq, K., & Al Azzawi, K. Y. (2012). Average framing linear prediction coding with wavelet transform for text-independent speaker identification system. *Computers & Electrical Engineering, 38*(6), 1467–1479.
8. Vong, C. M., & Wong, P. K. (2011). Engine ignition signal diagnosis with wavelet packet transform and multi-class least squares support vector machines. *Expert Systems with Applications, 38*(7), 8563–8570.
9. Sanchez, F. L., Júnior, S. B., Vieira, L. S., Guido, R. C., Fonseca, E. S., Scalassara, P. R., ... & Chen, S. H. (2009). Wavelet-based cepstrum calculation. *Journal of Computational and Applied Mathematics, 227*(2), 288–293.
10. Lung, S.-Y. (2004). Further reduced form of wavelet feature for text independent speaker recognition. *Pattern Recognition, 37*(7), 1565–1566.
11. Szu, H., Telfer, B., & Garcia, J. (1996). Wavelet transforms and neural networks for compression and recognition. *Neural networks, 9*(4), 695–708.
12. Patidar, S., & Pachori, R. B. (2014). Classification of cardiac sound signals using constrained tunable-Q wavelet transform. *Expert Systems with Applications, 41*(16), 7161–7170.
13. Lung, S. Y. (2007). Efficient text independent speaker recognition with wavelet feature selection based multilayered neural network using supervised learning algorithm. *Pattern Recognition, 40*(12), 3616–3620.
14. Lung, S. Y. (2007). Wavelet feature domain adaptive noise reduction using learning algorithm for text-independent speaker recognition. *Pattern recognition, 40*(9), 2603–2606.
15. Lung, S. Y. (2008). Feature extracted from wavelet decomposition using biorthogonal Riesz basis for text-independent speaker recognition. *Pattern recognition, 41*(10), 3068–3070.
16. Tohidypour, H. R., Seyyedsalehi, S. A., Behbood, H., & Roshandel, H. (2012). A new representation for speech frame recognition based on redundant wavelet filter banks. *Speech Communication, 54*(2), 256–271.
17. Vignolo, L. D., Milone, D. H., & Rufiner, H. L. (2013). Genetic wavelet packets for speech recognition. *Expert Systems with Applications, 40*(6), 2350–2359.
18. Avci, E., & Akpolat, Z. H. (2006). Speech recognition using a wavelet packet adaptive network based fuzzy inference system. *Expert Systems with Applications, 31*(3), 495–503.
19. Bahoura, M., & Rouat, J. (2006). Wavelet speech enhancement based on time–scale adaptation. *Speech Communication, 48*(12), 1620–1637.
20. Returi, K. D., & Radhika, Y. (2015) An artificial neural networks model by using wavelet analysis for speaker recognition. In *Proceedings of information systems design and intelligent applications. Second international conference on information systems design and intelligent applications (INDIA–2015).* Organized by Faculty of Engineering, Technology and Management University of Kalyani, Kalyani-741235, West Bengal, India Technically co-sponsored by IEEE Kolkata Section and IEEE Computational Intelligence Society Kolkata Chapter, 340, Vol. 2, pp 859–874.
21. Returi, K. D., Radhika, Y., & Mohan, V. M. (2015). A novel approach for speaker recognition by using wavelet analysis and support vector machines. In *2nd international conference on computer and communication technologies.* IC3T-2015 will be held during July 24–26, 2015 at CMR Technical Campus, Hyderabad, Telangana, India (Technically co-sponsored by CSI Hyderabad), 379, Vol. 1, pp 163–174.
22. Returi, K. D., Radhika, Y., & Mohan, V. M. (2016). A comparative study of different approaches for the speaker recognition. In: *3rd international conference on information system design and intelligent applications.* INDIA 2016 will be held during January 8–9, 2016 at ANIL NEERUKONDA Institute of Technology & Sciences, Visakhapatnam, AP, India (Technically co-sponsored by CSI Visakhapatnam Section), INDIA 2016, 433, Vol. 1, pp. 599–608.

A Systematic Survey on IoT Security Issues, Vulnerability and Open Challenges

Ranjit Patnaik, Neelamadhab Padhy, and K. Srujan Raju

Abstract The Internet of things has changed the technology to a new era, which enhanced our lifestyle. Because of the popularity of IoT in real world, the implementation of IoT in different areas created mislead of malicious access and security issue. Many researchers have studied, explored and opened the problems in security. The IoT is a concept which is an interconnection of our day-to-day devices. Devices are now elegantly communicating to human or each other. Each connected devices are attached to sensors and actuator senses and understands what is going on and performs the task accordingly. This paper discusses the different levels of IoT architecture and vulnerabilities in each level. The IoT devices are securely hackable and negotiable to use as per the requirement of malicious users. In this paper, we discuss the significant issues and challenges in the IoT environment and its communication, security, open problems, and their probable solutions. As the IoT devices are exposed to the vulnerabilities which need to be protected. To protect, we reviewed feasible solutions like use of blockchain, software define network (SDN), by using private and public keys and machine learning and deep learning algorithms to protect data in the cloud and IoT.

1 Introduction

Nowadays, IoT has been accepted very rapidly, and it has expanded widely, acknowledged, and famous as the leading low-power consuming networks standard is having constrained resources. In [1], authors have described the security issues, challenges

R. Patnaik · N. Padhy (✉)
Research Scholar, School of Engineering and Technology, CSE Department, GIET University, Gunupur, India
e-mail: dr.neelamadhab@giet.edu

R. Patnaik
e-mail: ranjitpatnaik@giet.edu

K. Srujan Raju
CMR Technical Campus, Secundrabad, India
e-mail: ksrujanraju@gmail.com

S. C. Satapathy et al. (eds.), *Intelligent System Design*, Advances in Intelligent Systems and Computing 1171, https://doi.org/10.1007/978-981-15-5400-1_68

and open issues in IoT. It indicates a network wherein the devices are interconnected in the network, which may be a private or public. These are embedded with a sensor [2, 3]. The methods in IoT are operable from remote interfaces. The devices are sharing information among themselves in a network. It gears the typically used protocols in communication [4]. These devices are embedded with sensor chips. The home appliance, like air-conditioner, washing machine, and refrigerators, can be controlled by the IoT devices.

Khan, A.et al. [5] have discussed the overview of CR-based IoT systems in this article. The author has highlighted the potential CR-based IoT systems and their applications. They have done the survey based on the CR-based IoT systems architectures and frameworks. Recent years, the faster development of technology and development of IoT is versatility in the application domain. In [5] cited the summary of IoT security attacks and developed classification based on the application domain and principal of system architecture. The vulnerabilities in IoT are exposed especially in the network where the devices are connected. Data are transmitted through the wireless network from device to device. The data must transfer to the authorised method.

2 Literature Survey

The IoT model contains extensive devices and gear from small to more significant parts (chipset) with a high-end server with security issue at a different level mostly address. It is categorising threats in security based on the architecture of IoT defined below. The papers discussed different security vulnerability in IoT and related publications. Here, it is classified based on above-cited categories.

- Security issues in physical level
- Security issues in middle level
- Security issues in logical level

2.1 Security Issues in Physical Level

The issues in data link and physical layer of communication are associated with this kind of security issue category. The problems that are discussed below are classified in this type of security issues:

Jamming adversaries: This kind of attack is done by decreasing of network sent frequency signal without following the specific protocols [6, 7]. The radio interference mostly affects the network operation like sending and receiving of data in IoT.

Insecure initialisation: To ensure a proper and secure network service in IoT, we must initialise and configure IoT devices in the physical layer without deviating secrecy and obstacle to the network [8, 9].

2.2 *Security Issues in Middle Level*

The communication, session management and routing taking place at "network and transport layers of IoT" concerned in middle-level security are explained.

2.3 *Security Issues in Logical Level*

The issues in logical level provide through the applications executing on IoT as follows.

Constrained Application Protocol security using the Internet: Application layer is consisting of various types of applications for different purposes, which are sometimes vulnerable. It leads to the attack by the unauthorised attacker.

In [10], the authors have proposed a technique on the security of IoT called Elliptic Galois Cryptography and Steganography Protocol. Elliptic Galois Cryptography (EGC) protocol is for protection against data infiltration during transmission over the IoT network. In the proposed work, different devices in the IoT network transmit data through the proposed protocol as a part of the controller. The encrypted algorithm within the controller encrypts the data using the EGC protocol, and then the encrypted and secured message is hidden in layers of the image, with the help of steganography technique. The model can then be easily transferred throughout the Internet such that an intruder cannot extract the message hidden inside the image. Initially, the EGC technique encrypts confidential data. Subsequently, the encoded secret message is inserted within the image by the XOR steganography technique.

3 Architectural Behaviour of IoT

In Fig. 1 describes the base-architecture of IoT. In the first layer, the sensors are connected to every individual component of IoT and sense the surrounding, then send the collected data to the base station in second level through the network. In the third level, data collected at the base station is stored at cloud storage and then it is given to the next level for analysis and prediction of data.

The IoT is described in four layers Fig. 2; 1. Application Layer; 2. Middleware layer; 3. Network Layer; and 4. Perception Layer. In Fig. 2, it describes the four-layered architecture.

In [11], the authors have described the genesis of IoT, phases of Internet and IoT application. IoT is the combination of information technology and operational technology through the network.

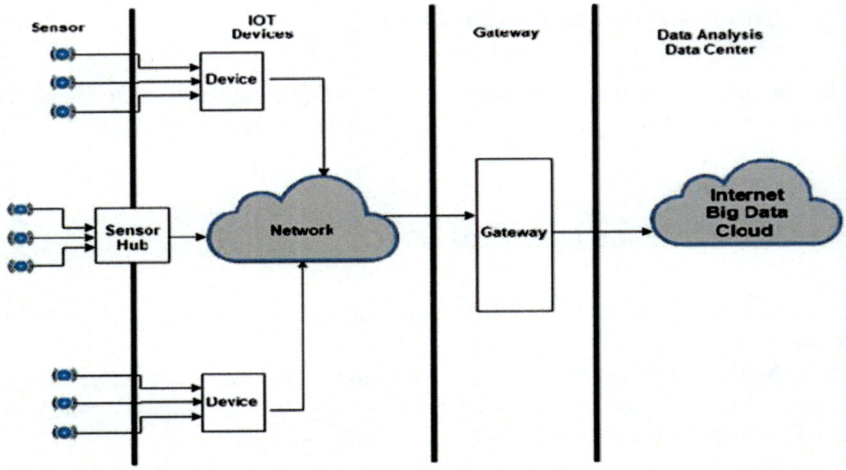

Fig. 1 Base IoT architecture

Fig. 2 Four-layered IoT architecture

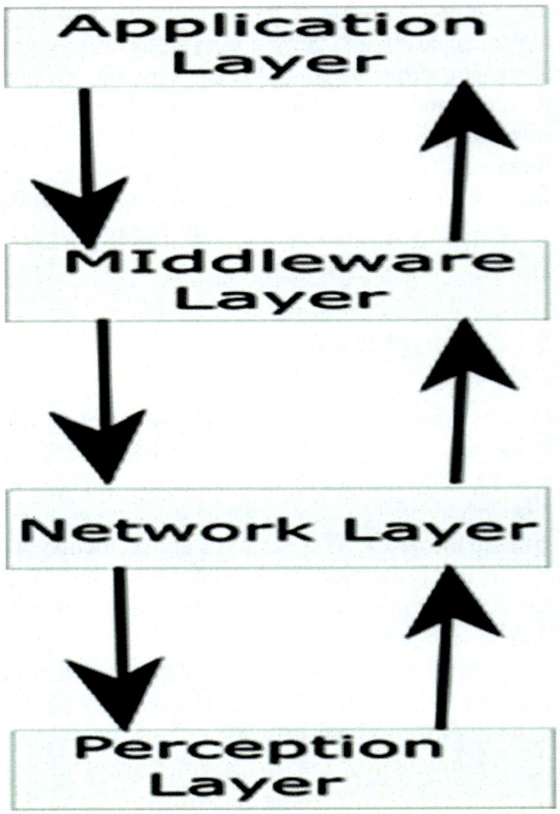

4 Probable Security Solution for IoT

4.1 Blockchain

A blockchain is a technique which uses data structure of distributed configuration that can be shared among the components of the network. The blockchain consists of a distributed ledger which is having all transactions within the system, enhanced with cryptography and carried out through the peer-to-peer nodes. This is an innovative and effective solution for the security challenges of centralised tracking, monitoring and security of data because of its decentralisation computing nature. Figure 3 shows the architecture of blockchain. The structure of blockchain is represented by list of blocks with transactions in a particular order. This list stores in the form of flat files (.txt format) which are the simple database. Blockchain has two primary data structure: 1. Pointer 2. Linked list. The blockchain uses the concepts of unspent transaction output (UTXO) model; the blockchain can able to provide the security in the network. The UTXO model mainly preserves the unspent electronic money in the account of the authorised customer.

In [10], the authors enlighten the blockchain and steganography for securing the data in the IoT-based network. Steganography is a technique used to hide secret data in an ordinary, non-secret, file, or message, to avoid detection, and then to extract confidential data at its destination. The usage of Steganography can be combined with encryption to hide or protect data. It has two different mechanisms to safeguard the data, i.e., encryption and decryption.

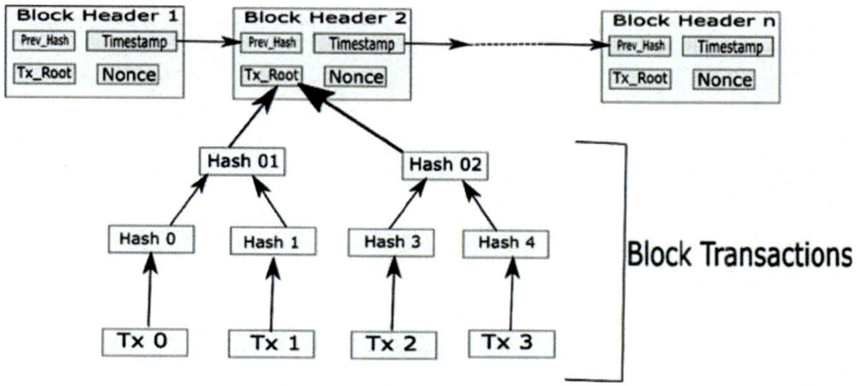

Fig. 3 Architecture of blockchain

4.2 Machine Learning and Deep Learning

In [12], the authors discussed the use of machine learning and deep learning algorithms like regression, classification and density estimation. The various ML and DL algorithms provide the intelligence to the IoT network and devices to tackle the vulnerability and then provide the security. There are multiple algorithms in ML and DL which offers protection in the different levels of IoT. Here, the authors have illustrated some exclusive characteristics of IoT, which leads to challenges in security while deploying IoT. Machine learning provides a technique of learning from past records or data. Google uses ML to analyse threats against mobile endpoints and applications running on android. Amazon has launched a service Macie that uses ML to sort and classify data stored in its cloud storage service. Deep learning is a new variant of machine learning which is a self-service version of machine learning for classification and prediction task in innovative IoT applications.

4.3 EdgeSec Technique

In [13], the authors proposed a technique of securing data in the IoT environment called EdgeSec: Edge Layer Security Service, which includes the security options of edge device in IoT environment. The IoT devices are moreover similar to the edge devices. Edges contain several components like Security Profile Management (SPM), Protocol Mapping (PM), Security Analysis Module (SAM), Protocol Mapping (PM), User Interface (UI) and Interface Manager (IM). The IoT devices are registered into EdgeSec module by SPM, where a security profile is created. N. Padhy et al. [14] developed threshold estimation algorithms for software metrics. They have used evolutionary intelligence algorithms for estimation purposes. They have taken a sample of object-oriented examples and derive software metrics Panigrahi et al. [15] N. Padhy discussed the software reusability metrics algorithms in terms of the software refactor perspective. They have used the bench-marked data set (social media) and extracted software metrics.

5 Conclusion

In this paper, we discussed the most imperative aspect of security issue of the IoT environment with highlighted further research on this area of security in IoT as a survey. In our research, we will find the overall IoT architecture and architecture, security issues related to and produced by different types of IoT devices. We also performed meticulous enquiry of issues on security in IoT system and the limitations of energy, lightweight protocol of cryptographic and resources. Furthermore, we cited the circumstances of real life wherein lacking security could pose various threats in

IoT. In this paper, the authors reviewed the different security techniques for the IoT data security and authentication. Here, also discussed some aspects of probable solution for security in IoT environment using algorithms of ML and DL. By using EdgeSec, we can strengthen the security in IoT. Not only the cryptography but also the combination of steganography with cryptography provides the security for IoT data.

6 Future Scope

In recent years, there has been rapid advancement in IoT in every sector. It is being seen that most examiners trust on the platforms like intelligent transportation, logistic monitoring, etc. The security challenges identified in the IoT are managed and accomplish to the development.

According to the survey, the most problematic thing in IoT is the security layer feeding The data is being stolen or leaked through layers. The cryptographic algorithm and blockchain algorithm help to secure layer from malicious attack.

Authentication Protection

As most of IoT gadgets prefer to have authentication from server-side. We apply some algorithms like SDN which authenticates the clients during entry such that data packets are secured in the application layer of SDN as the ML and DL have capability of self-driving policy which may help for protection of data in IoT.

As a part of future work, we will enhance and implement the probable solution for the issues and challenges of the IoT-based educational system. We will analyse various machine learning algorithms on several datasets, which are generated data from the model of IoT-based educational system. Various investigations will be conducted to propose an accurate, efficient and secured IoT-based educational system.

References

1. Hossain, M. M., Fotouhi, M., & Hasan, R. (2015). Towards an analysis of security issues, challenges, and open problems in the internet of things, In *IEEE World Congress on Services* (pp. 21–28).
2. Atzori, L., Iera, A., & Morabito, G. (2010). The internet of things: A survey. *Computer Networks, 54*(15), 2787–2805.
3. Giusto, D., Iera, A., Morabito, G., & Atzori, L. (Eds.). (2010). *The internet of things: 20th Tyrrhenian workshop on digital communications*. Springer Science & Business Media.
4. Khan, M. A., & Salah, K. (2018). IoT security: Review, blockchain solutions, and open challenges. *Future Generation Computer Systems, 82,* 395–411.
5. Khan, A. A., Rehmani, M. H., & Rachedi, A. (2017). Cognitive-radio-based internet of things: Applications, architectures, spectrum-related functionalities, and future research directions. *IEEE Wireless Communications, 24*(3), 17–25.

6. Xu, W., Trappe, W., Zhang, Y. & Wood, T. (2005). The feasibility of launching and detecting jamming attacks in wireless networks. In *Proceedings of the 6th ACM international symposium on Mobile ad hoc networking and computing* (pp. 46–57). ACM
7. Noubir, G., & Lin, G. (2003). Low-power DoS attacks in data wireless LANs and countermeasures. *ACM SIGMOBILE Mobile Computing and Communications Review, 7*(3), 29–30.
8. Chae, S. H., Choi, W., Lee, J. H., & Quek, T. Q. (2014). Enhanced secrecy in stochastic wireless networks: Artificial noise with secrecy protected zone. *IEEE Transactions on Information Forensics and Security, 9*(10), 1617–1628.
9. Hong, Y. W. P., Lan, P. C., & Kuo, C. C. J. (2013). Enhancing physical-layer secrecy in multiantenna wireless systems: An overview of signal processing approaches. *IEEE Signal Processing Magazine, 30*(5), 29–40.
10. Khari, M., Garg, A. K., Gandomi, A. H., Gupta, R., Patan, R., & Balusamy, B. (2019). Securing data in internet of things (IoT) using cryptography and steganography techniques. *IEEE Transactions on Systems, Man, and Cybernetics: Systems.*
11. Jha, S., Kumar, R., Chatterjee, J. M. & Khari, M (2019). Collaborative handshaking approaches between the internet of computing and internet of things towards a smart world: A review from 2009–2017. *Telecommunication Systems, 70*(4), 617–634.
12. Hussain, F., Hussain, R., Hassan, S. A., & Hossain, E. (2019). Machine learning in IoT Security: Current solutions and future challenges. arXiv preprint arXiv:1904.05735
13. Sha, K., Errabelly, R., Wei, W., Yang, T. A., & Wang, Z. (2017). Edges: Design of an edge layer security service to enhance IoT security. In *2017 IEEE 1st International Conference on Fog and Edge Computing (ICFEC)* (pp. 81–88).
14. Padhy, N., Panigrahi, R. & Neeraja, K. (2019). Threshold estimation from software metrics by using evolutionary techniques and its proposed algorithms, models. *Evolutionary Intelligence.* https://doi.org/10.1007/s12065-019-00201-0
15. Panigrahi, R., Padhy, N., Satapathy, S. C. (2019) Software reusability metrics estimation from the social media by using evolutionary algorithms: Refactoring perspective. *International Journal of Open Source Software and Processes (IJOSSP), 10*(2) , 21–36, IGI Global.

Prevention and Analysing on Cross Site Scripting

L. Jagajeevan Rao, S. K. Nazeer Basha, and V. Rama Krishna

Abstract Nowadays web application is highly demandable in different criteria based on their e-commerce setup which give assurance in security to lots of customers. If an applications contains small flaw of susceptibilities on it, Attackers had accidental to steal their credential password accidentally same way, Cross site scripting is a procedures where malicious code are to be exploit by an attackers to Grab the credential information of a user's it may within your web application or within a server side database its allow us by injecting malicious JavaScript when code wasn't sanitized for a web applications This study discusses about Cross site attack and its taxonomy, In addition to that the paper presents the XSS attack apparatuses, Analysis and preclude from XSS forgery.

Keywords Hacker directory · Malicious · Cross site scripting · Zed attacks · XSS attacks · Victem server

1 Introduction

Cross site scripting is a vulnerability in web application that's caused due to un sanitized code where malicious script are to execute to theft credential authentication cookies which allow to gain authorized administration from delimited resources. Cross site scripting classified into three groups Reflected (Non-persistent), stored (persistent) and DOM-based XSS attacks [1].

L. Jagajeevan Rao (✉)
Department of Computer Science and Engineering, KL University, Guntur, Andhra Pradesh, India
e-mail: jeevan@kluniversity.in

S. K. Nazeer Basha
SRK University, Bhopal, Madhya Pradesh, India

V. Rama Krishna
GITAM University, Hyderabad, Telangana, India

S. C. Satapathy et al. (eds.), *Intelligent System Design*, Advances in Intelligent Systems and Computing 1171, https://doi.org/10.1007/978-981-15-5400-1_69

The rest of this paper is organized as following sectors Sect. 2 represents procedures of XSS attacks prevention techniques of XSS attacks Sect. 4 comprehends about conclusions.

2 Procedures of XSS Attack

We can divide cross site attacks into different sorts [2]. Hacker inject a malicious script code into a web page, when client try to stopovers that website again the malicious scripts get activated to get credentials information of an user by hijacking session ID into his own hacker directory which he had created Through which he can manipulate all webpage and authorized database which may leads to forfeiture if hacker procure some vulnerabilities page he goanna assigned his own session id by allowing scripts to get exploit on it [3].

SCRIPT>location.href="HACKERDIRECTORY.PHP?data="+document. Cookies</SCRIPT>
<SCRIPT>Document. Cookie="PHP SESSID=12@9";</SCRIPT>

Second rally is to restrict the feedback of a client to a database whenever client try to defer to his forum in a input field it's show a alert dialog which a immense trouble for a website through which Attackers can influences website in his own elegance [4] (Fig. 1). This Diagram give us a indistinct illustration with a malicious code. to apparatuses.

Now instead of writing a text form As a hackers I try to manipulate his website by writing some scripts as follow:

<script>alert("xss")</script>

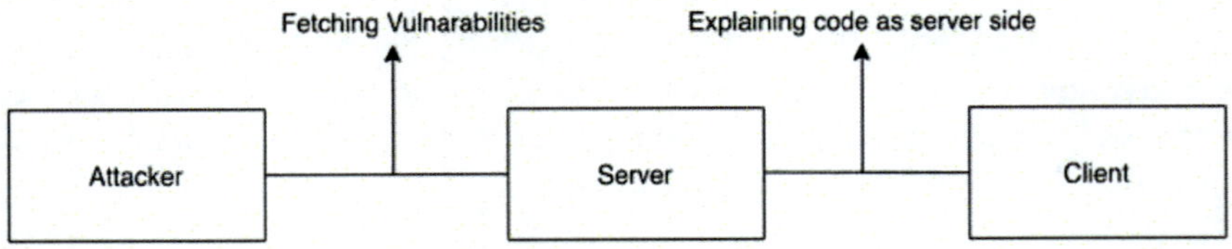

Fig. 1 Diagrammatical views of xss attack

These script are to be written in Message forum, which execute the flaws' in the page source of a particular website, until the input wasn't sanitized [4] (Figs. 2 and 3).

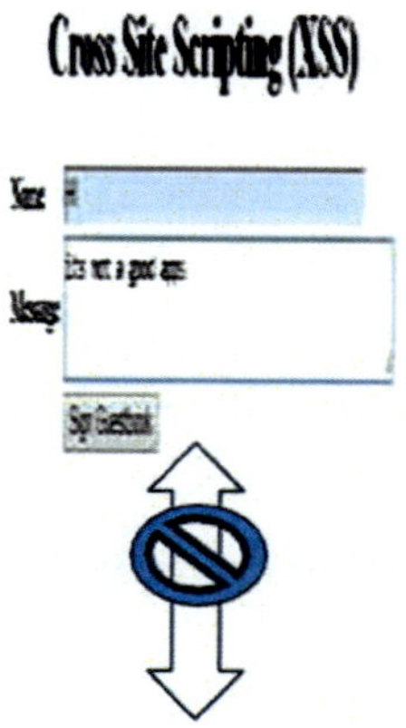

Fig. 2 Cross scripting

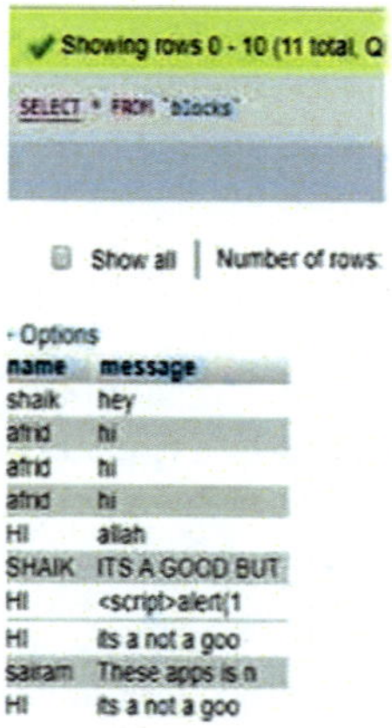

Fig. 3 cross scripting

Algorithms

```
<?php
include_once("b.php");

if(isset($_POST['Submit']))
{
$na=$_POST['Q'];
$ma=$_POST['R'];

file_put_contents("R.txt", $_POST["R"], FILE_APPEND);
echo file_get_contents("R.txt");


$sql= "INSERT INTO blocks(name,message) VALUES ('$na','$ma');";

$result=mysqli_query($con , $sql);

header("J.PHP");

}

?>
```

NOTE:
But still's code wasn't sanitized which leads in disrupted the page by allowing different script (Fig. 4).

This figure revealed us How attacker try to deploy by embedding script on it to stop the redirection of a database, whenever a haphazard try to open it off its illustrates alert option whenever client try to open the feedback forum again.

We can implement's different types of malicious scripts as follows in

The taxonomy of XSS attacks: [5]
XSS is classified into three main classes as follows:

1. **Reflected Xss attack**: Reflected XSS attack (Non-persistent attack) The purpose of this attack is to steal the session cookie of the user. This type of XSS attack requires a more interaction between the victim and the attacker in Fig. 2. The steps are as follows: [5]

1. The attacker sends associate malicious script code to the user via electronic mail or any analogous web page.
2. When the user clicks on this link, the malicious code is sent to the server without being become aware of web application.
3. The server sends the HTTP response to the user with the malicious script code.
4. The attacker's domain receives the user's cookies after executing the script contained in the response (Fig. 5).

Fig. 4 cross site scripting

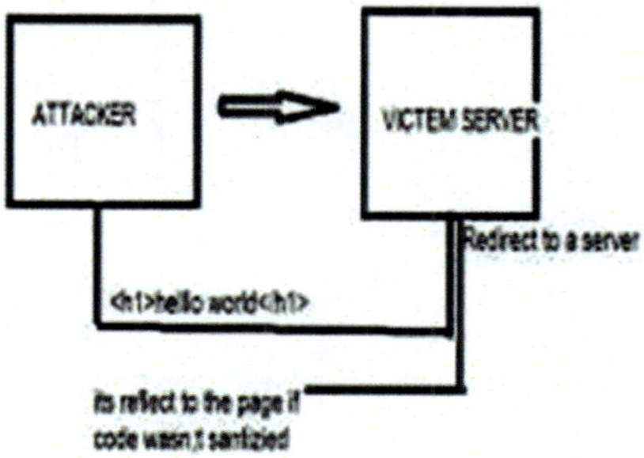

Fig. 5 Reflcted xss attack (mechanism)

2. **Stored XSS:** While browsing a E-commerce website, Attackers try to consent to embedded script tag in a comment section, these embedded tag are permanent execute on it Fig. 2

1. The Attackers try to add script in a comment tag as **<scriptsrc="http//hackerdirectory.js</script>** by assigning reference page a **CLICK ME!** [5]
2. Whenever client try to open it off its activates the code when he/she try to open again
   Attacker can compromised the visitor account, granting him to easy access to his personal information And credit card.
3. Unlike a reflected xss, where the script is activated after linked is clicked, stored attack only requires When victim visit the comprised page.
4. Stored xss relatively harder to execute because of the difficulties in enabling of permanent script Embedding (Fig. 6).

3. **DOM-based xss attack**: DOM Based XSS is an attack wherein the attack payload is executed as a result of modifying the DOM "environment" in the victim's browser used by the original client side script, so that the client side code runs in an "unexpected" manner. That is, the page itself (the HTTP response that is) does not change, but the client side code contained in the page executes differently due to the malicious modifications that have occurred in the DOM environment [5].

Suppose the following code is used to create a form to let the user choose his/her preferred language. A default language is also provided in the query string, as the parameter "default".

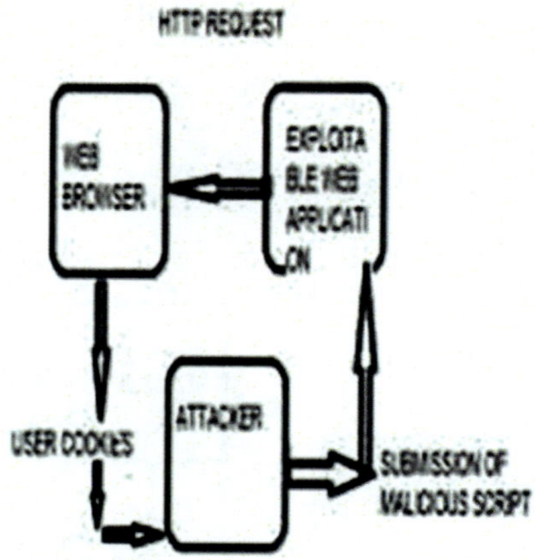

Fig. 6 Stored xss attack (mechanism)

Lets come upon to script how to exploit: [5]

```
<select><script>
Document.Write("optionvalue=1>)"+document.location.href.substring
(document.location.href.indexof ("default=")+8+"</option>");
```

The page invokes as:
http://www.afridpage.html/?default=Hindi
When Attackers try to manipulate the original code its look like
http://www.afridpage.html/?default=%3Cscript%3Ealert(document.cookies)%3C/script%3E

3 Preventation of Xss Attack

We can prevents xss attack by implementing proper code to get exceedingly secure which it's difficulties 'by attackers to inject to finite script on it, through which give a clear pictures how developers can secure their perspective web pages.

The manual testing can be performed by pen testing which uses proxy tools such as Burp suite, Zed attack proxy (ZAP) its validate the user input data and the "boundary injection" of html tags and JavaScript code [1].

The following algorithm we can prevents from xss attacks: [6]

PREVENTATION FROM EFLECTION CROSS SITE SCRPTING:

```
<?PHP
$name=htmlspecialchars($_POST['S']);
ECHO $NAME."THANKS FOR SUBMETTING";
?>
```

PREVENTATION FROM STORED CROSS SITE SCRPTING:-

```
if(isset($_POST['Submit']))
{
$na= trim($_POST['Q']);
$ma= trim($_POST['R']);

$ma = strip_tags($ma);

file_put_contents("R.txt",
$_POST["R"], FILE_APPEND);
$ma = file_get_contents("R.txt");
echo htmlspecialchars($ma);
```

The automated testing is performed with a commercial scanner tools such as Acunetix App scan [7]

1. **Html special chars** () functions converts some predefined character to html entities as [8]

- < (less than) becomes <
- >(greater than) becomes >
- '(single quotes) become &quote;
- " "(double quotes) become &quote;

2. **Strip tags()** functions mainly to srtrip out the tags (<>) **eg: </script>,</html>3.Trim** () functions remove whitespaces and other predefined character.
3. **Trim ()** functions remove whitespaces and other predefined character.

While xss attack continue to be a popular and successful vector of attack, a bit of thoughtful design and testing can go long way in keeping your website or web application from vulnerable and keeping your user protected.

NOTE: But these alone are not going to help you. there are many way on which they can bypass this filter functions .so we have to think about the all the possibilities

We need to escape css and xml data to prevents xss [9].

SOME OPEN SOURCE LIBARIES FOR PREVENTING XSS ATTACK: [9]

- Html purifier
- Xss_clean.php filter
- Php Antics xss

1. **Html purifier:** it's a html filtering library which are to be written in php. its also available

   As a plug-in for most p frame work, which help in prevents from xss attacks.

2. **Xss_clean.php filter:** its cleans various UTF encodings and nested exploits.[9]
3. **PhpAnticsxss:** this library automatically detects the encoding of the data that must be filtered.

   Eg:
   $ma =AntiXSS:: setEncoding($ma,"UTF-8");

4 Conclusion

Xss attack considered as a top 10 vulnerabilities which have lots of chance by an attackers to take hold of authorized information which its much decisive about their productivity, But still we can roughly find out 50% of a website are embraced due to un sanitized of a code, which may leads to disaster. We can figure it out by giving sanitized code to prevent from malfunction which was being mention in third modules [III. PREVANTION ON XSS ATTACK].

We being assign different flaws by which attackers can exploits and by manipulating out the pages by giving some script on it, end ways we had some some algorithm to prevents from these type of exploits.

These paper illustrate about analysing and preventions of xss attacks. As how we can aware by giving considerably secure to our web pages.

References

1. https://www.owasp.org/index.php/Cross-site_Scripting_(XSS).
2. https://www.incapsula.com/web-application-security.
3. https://www.owasp.org/index.php/Testing_for_Stored_Cross_site_scripting_(OTG-INPVAL-002).
4. Sighting of JavaScript code injection attacks in a web applications International Conference on Information Security & Privacy (ICISP).
5. https://www.owasp.org/Cross-site_Scripting_(XSS).
6. Priming tags by removing out the vulnerability flaws Proceedings of 11th International Conference on Ubiquitous Information Management and Communication.
7. https://www.acunetix.com/vulnerability-scanner/.
8. https://www.geeksforgeeks.org/php-strip_tags-function/.
9. https://resources.infosecinstitute.com/how-to-prevent-cross-site-scripting-attacks/.

First-Hand Information from the Spot of Accident and Resolution of Claims

G. Venkatram Reddy, B. Veera Mallu, B. Sunil Srinivas, and CH. Naga Lakshmi

Abstract There ought to be an application for congregating the first-hand information from the spot of the road accidents. The details gathered ought to consist of photographs of the accident spot, statements recorded from the bystanders, reports related to sufferings and casualty, reasons for the mishap, momentum of the vehicles, road track situation on considerable ground, etc. The data collected from various sources and means can be sent to a central database server. The authority of congregating the information can be assigned to concerned jurisdiction police officer, transport authority, ambulance, or even common men who step forward for the equivalent. The same application should also support the submission/swapping of information related to insurance policy or other details so as to settle the debate if any emerging out of mishap.

Keywords Mobile application · Insurance policy

1 Introduction

This mobile (android)-based application will provide a platform for the authorized users to collect the on the spot information regarding the accident, and this verified information is available to the authorities, victims, and mainly to the insurance agents. The insurance company can use this application to know the details of the accident in order to settle the dispute if any arising out of accident.

G. Venkatram Reddy (✉) · B. Veera Mallu · B. Sunil Srinivas · CH. Naga Lakshmi
TKR College of Engineering & Technology (TKRCET), Hyderabad, Telangana, India
e-mail: venkatram2003@tkrcet.com

B. Veera Mallu
e-mail: bvmallu@tkrcet.com

B. Sunil Srinivas
e-mail: sunilsrinivasb@tkrcet.com

CH. Naga Lakshmi
e-mail: ch.nagalakshmi18@tkrcet.com

S. C. Satapathy et al. (eds.), *Intelligent System Design*, Advances in Intelligent Systems and Computing 1171, https://doi.org/10.1007/978-981-15-5400-1_70

A. *Scope*

- The application facilitates the authorities and the naive users for easy reporting and management of the incident-related data at any point of time right from the incident place.
- The scope of this application can be extended toward the verification and authentication of the incident-related data by the concerned authorities. Thus, it leads to the elimination of counterfeit and fraudulent claims in medical legal cases.
- All the data right from the date of initial reporting can be accessed by the authorities of various departments, viz transport, government, health, insurance, victims if any from this single application.
- Overall, this integrated application serves the authorities and naive users by integrating data from distinct departments onto a single platform.

B. *Objective*

The objective is to have first-hand information of the road accidents and its status of report to the concerned authorities and victims through a mobile app. This is a mobile (android)-based application which provides a platform for the authorized users to collect the on the spot information regarding the accident, and this verified information is available to the authorities, victims, and mainly to the insurance agents. So that, the insurance company can use this application to know the details of the accident so as to settle the debate if any emerging out of mishap.

C. *Existing System*

In the existing system, all the accident records are written manually by the police only when the victim wants to case a file, otherwise the accident will not be recorded. It provides insufficient data since it is maintained manually.

And for claiming the insurance in the existing system, the victim of the accident will first call or meet with the insurance agents, and then, the insurance agents will investigate about the accident, i.e., he/she needs all the details about the accident so as to settle the debate if any emerging out of mishap. So, for gathering the details of the accident, they will visit the accident location and meet the police for more information. Thus, it will take more time for collecting the data.

D. *Disadvantages*

- The existing system may provide insufficient data since it is maintained manually.
- It may take lot of time for the police to collect the data and to prepare an accident report since it is done manually.
- It may take time for insurance company to the gather the information of the accident from the police to process the claims.

2 Proposed System

In the proposed system, all the accident records are maintained online by the authorized user such as police, transport authority, ambulance, and press. In this system, not only the police but also any authorized user can upload the data online.

And insurance claiming process in the proposed system is done in less time when compared to that in the existing system. Using this application, the insurance agents are provided with the full accident details of their customer such as insurance number, reasons for the mishap, momentum of the vehicles, and road track situation on considerable ground. And along with the photos of the site, interviews with eyewitnesses, information on injuries, and fatalities in order to settle the dispute if any arising out of accident. Thus, it will not take extra days for gathering the details of accident since the data is provided by the authorized person.

A. *Advantages of Proposed System*

- All the data are maintained in a central database.
- Reduce human effort.
- It takes less time for the police to prepare the accident report.
- The insurance company can use this application to know the details of the accident in order to process the claims.

3 Modules

This application consists of following modules.

1. Registration module
2. Login module
3. Data collection and upload to database
4. Retrieve of data from database.

A. *Registration Module*

If anyone wants to use this application, he/she need to register first, and after that, he/she can login. Based on authority, following types of user can register.

- Authorized user
- Non-authorized user

Authorized User

The authorized users such as the police, RTA officer, ambulance, and press can register by filling their respective details such as employ ID, email, phone number, station/region (for police and RTA officer), ambulance no/hospital name (for ambulance), and password (for providing security). The authorized users are responsible for gathering the information, and he/she can only upload the collected data in the application. The authorized user can upload the data and can view the information related to accident.

Non-Authorized User
The non-authorized users such as victims can register using email and password. The non-authorized users can only view the information related to the accident. He/she cannot upload the data related to the accident.

B. *Login Module*

The details of the authorized person will be validated with the details in the authentication tables, and if he/she is authorized person, they are allowed to access the application. Based on the authorization type, access to the different modules will be enabled/disabled. Only registered users can login to the application using their respective login ID (i.e., email ID) and password.

C. *Data Collection and Upload to Database*

The users who are authorized such as police, ambulance, transport authority, and press are responsible for gathering the information. To upload the data in the application, the authorized user need to register first after that he/she can login and then he can upload the collected data in the application.

Prepare the report by uploading photographs of the accident spot, statements recorded from the bystanders, reports related to sufferings and casualty, reasons for the mishap, momentum of the vehicles, road track situation on considerable ground, etc. And generate the unique file number for the report that is prepared, so that it is uniquely identified. While preparing the reports, fill the insurance policy details of the victims if any.

D. *Retrieval of Data From Database*

The insurance agents will select their respective insurance company. After selecting their company, the insurance agents are provided with the complete accident details of their customer such as insurance number, photographs of the accident spot, statements recorded from the bystanders, reports related to sufferings and casualty, reasons for the mishap, momentum of the vehicles, and road track situation on considerable ground, so as to settle the debate if any emerging out of mishap.

Along with the insurance agents, all users can retrieve or view the data by signing in this application.

4 Software Design

Software design is the process wherein the software architect or project manager creates a specification of a software artifact, intended to achieve, using a set of primitive components subject to constraints.

Software layout may additionally confer with either "all of the hobby involved in conceptualizing, framing, implementing, commissioning,, and ultimately reforming complicated systems" or "the activity following requirements specification and before programming, as in a software engineering process." Software design usually involves problem solving and planning a software solution. This includes both a low-level component design and a high-level architecture design.

5 Implementation and Testing

a. *Implementation*

Implementation is the most significant phase in attaining a successful system and giving the user's confidence that the new system is potential, powerful, and effective. Implementation is the reformed application to replace the current one. This type of conversation is comparatively easy to address, provided there are minor changes within the system.

Each program module is tested independently at the time of development using the data and also verified after integration with other modules whether it is working according to the specifications or not and the computer system and its environment is tested to the fulfillment of the user requirement. The system that has been developed is accepted and proved to be satisfactory for the user. And so, the system is going to be implemented very soon. A simple demo video is included, so that the user can understand the different functionalities clearly and quickly.

Primarily, the executable form of the application is to be created and loaded in the common server system which is reachable to all the users, and the server is to be linked to a network. The final stage is to document the complete system which provides components and the operating procedures of the system (Tables 1, 2 and 3).

Table 1 Test Case I

Test aspect	Try to login with wrong email/password
Initial condition	Empty login fields
Steps to run	Open application and enter wrong login details in login page
Expected output	It should show login error message
Actual output	It is shown in below image
Test aspect	Try to login with wrong email/password
Initial condition	Empty login fields
Steps to run	Open application and enter wrong login details in login page
Expected output	It should show login error message
Actual output	It is shown in below image

Table 2 Test Case II

Test aspect	Try to login with correct email/password
Initial condition	Empty login fields
Steps to run	Open application and enter correct login details in login page
Expected output	It should show login successful message and move to home page
Actual output	It is shown in below image

Table 3 Test Case III

Test aspect	Retrieve data from database
Initial condition	Access files
Steps to run	Open application and choose Access file option after login then select any file from the list of files
Expected output	It should show the file details
Actual output	It is shown in below image

Screenshot: 1

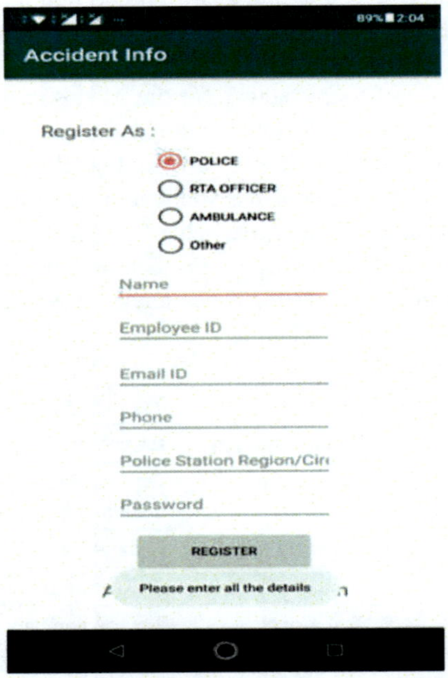

SCREEN SHOT:2

Screen shot: 3

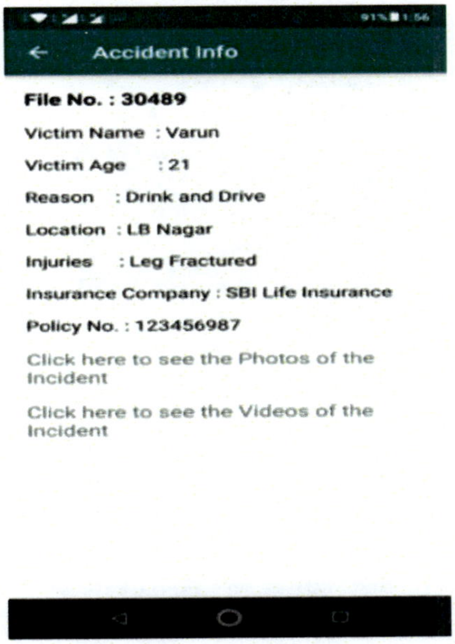

Screen shot: 4

screen Shot: 5

6 Conclusion

This project helps the victim of the road accident to settle down the insurance dispute in less time since the details of the accident is uploaded on the spot by the concerned police officer, road transport authority, ambulance or even common men who step forward for the equivalent. So that by using this application, the insurance agents are provided with the complete accident details of their customer such as insurance number, photographs of the accident spot, statements recorded from the bystanders, reports related to sufferings and casualty, reasons for the mishap, momentum of the vehicles, and road track situation on considerable ground, to process the claims. Thus, it will not take much time for gathering the details of accident since the data is provided by the authorized person.

The project can be expanded to not only for maintaining the accident information online but can also be integrated into an application available to general public that notifies a user and insurance company about the details of an accident and makes the insurance claiming process easier.

References

1. Schildt, H. (2008). *Java complete reference* (7th ed.). Tata McGraw-Hill.
2. Brooch, G., & Rambaugh, J. (1998). *Unified modeling language user guide, chapter 8–31*. Addison Wesley Publishing.
3. Roth, B. J., Wulff, R. W., & Cooper, C. A. *Alternative dispute resolution practice guide*. Lawyers Cooperative Publishing.
4. Zimmerman, D. E., & Muraski, M. L. (1995). *The elements of information gathering*. Oryx Press publisher.

Websites

5. www.android.com.
6. www.firebase.com.

Referred Urls

7. https://developer.android.com/guide
8. https://firebase.google.com/docs/android/setup

Survey on Biological Environmental Sequence Analysis Using Association Technique

G. Sivanageswara Rao, U. Vignesh, Bhukya Jabber, T. Srinivasarao, and D. Babu Rao

Abstract In this study, robust biological filters are used to combine the datasets for drug discovery. The two protein datasets are combined into a single form using a filtering instance to identify the proper association algorithm for protein analysis. Big data paves the way for identification of better protein association algorithm to avoid congestion in the analysis of large volume of data and to save time for the successful results. The proposed methodology used for the conversion of unstructured protein into a structured protein.

Keywords Big data · Prediction · Protein sequence · Data mining

1 Introduction

Data mining is defined as useful information from the database also said to be KDD (Knowledge Discovery in Databases). It is the process of extracting meaningful information from a large database. One can get useful patterns to form the database using the techniques of data mining which are efficient and accurate. This data mining has a very large scope to grow and which in hand can be used for society. The use of the database is required for any sector whether it is a government or non-government organization. Considering the example of DNA bank sectors, the data of each and

G. Sivanageswara Rao (✉) · U. Vignesh · B. Jabber · T. Srinivasarao · D. Babu Rao
KLEF, Vaddeswaram, Guntur (Dt), AP, India
e-mail: sivanags@kluniversity.in

U. Vignesh
e-mail: vigneshbun@gmail.com

B. Jabber
e-mail: jabber@kluniversity.in

T. Srinivasarao
e-mail: tsrinu@kluniversity.in

D. Babu Rao
e-mail: dbaburao@kluniversiity.in

S. C. Satapathy et al. (eds.), *Intelligent System Design*, Advances in Intelligent Systems and Computing 1171, https://doi.org/10.1007/978-981-15-5400-1_71

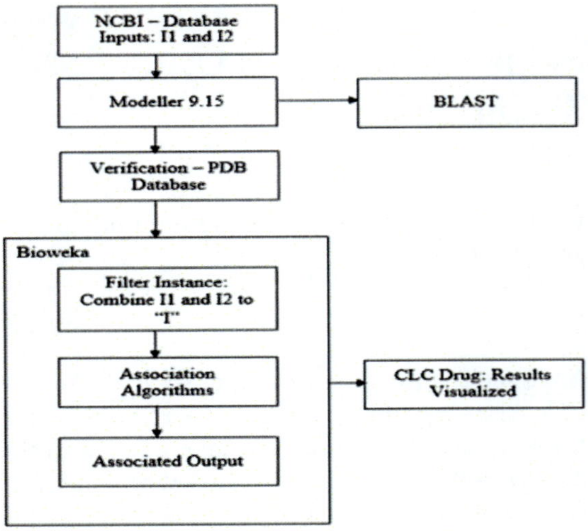

Fig. 1 Data model

every individual should be maintained properly. Efficient data storage is done through a database where a KDD process can be applied for the retrieval of data from it. The data process deals with the different stages of data access as shown in Fig. 1.

1.1 *Biological Data Mining and Association*

Data mining is an interdisciplinary area which includes many domain areas interconnected into it such as prediction, information retrieval, database systems, statistical calculations, and machine learning concepts. The applications of data mining are '*n*' in numbers which are uncountable and unpredictable; some of them are medicine, biology, pharmacology, etc. In biology, the microarray concepts have to lead to an increase in new databases and different strategies involved in management systems. Thus, the gen bank such as NIH (National Institute of Health), NCBI (National Centre of Biotechnology Information) started in 2004, etc., have become thousands in numbers and increased substantially compared with the past twenty years. These prospects make the bioinformatics as an important field integrated with computer science and also with association technique as shown in Fig. 2 [1–11].

Data mining deals with the action of data fusion in the majority, enabling different types of data to be organized in a consistent manner and efficiently with accurate information obtained from the single source or multiple data source. Sampling also has its part in the data mining domain. In the process of sample analysis in the digital form, if it is not in the digital format, then the conversion process takes place. The analysis can also be provided using multi-resolution processes for finding the occurrence of redundancy in data from the database. In the preprocessing aspect, the concept of de-noising removes the noise present in the data and accesses efficiently.

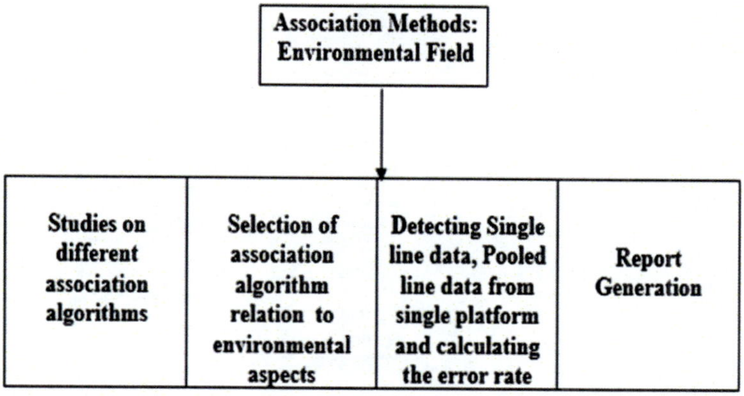

Fig. 2 Components of association data mining

Since data mining is the extraction process, it starts with the extraction of feature information [12–16].

Bioinformatics allows researchers and inventors to refer the existing data and submit new data into the database entries. It provides a way to develop biological software analysis tools, which in turn results in new and varied biological insights for reference. There are many applications directly connected with the bioinformatics in real time such as agriculture, the farming and organized study of biomaterials, etc [17–20].

2 Association Technique

Association technique plays a major role in bioinformatics for the coded region analysis. Association prospects the data in different ways framing the rules for its access. In this methodology, various parameters are considered for perception to deal with the biological data. There are various association algorithms in trends to deal with the Withania somnifera data, crab data, and egg data, such as Apriori, FP–Growth, Eclat, etc (Fig. 3).

2.1 *DNA and Amino Acids Data*

Data mining processes find wide use in education-related management systems with challenges involving quality, privacy, and data streaming. Security is also one of the major challenges in data mining. Ford company, Sony company, etc., are examples where it has failed to ensure that data are not hacked and accessed by third parties. Though there are many applications for data mining, bioinformatics being an ocean

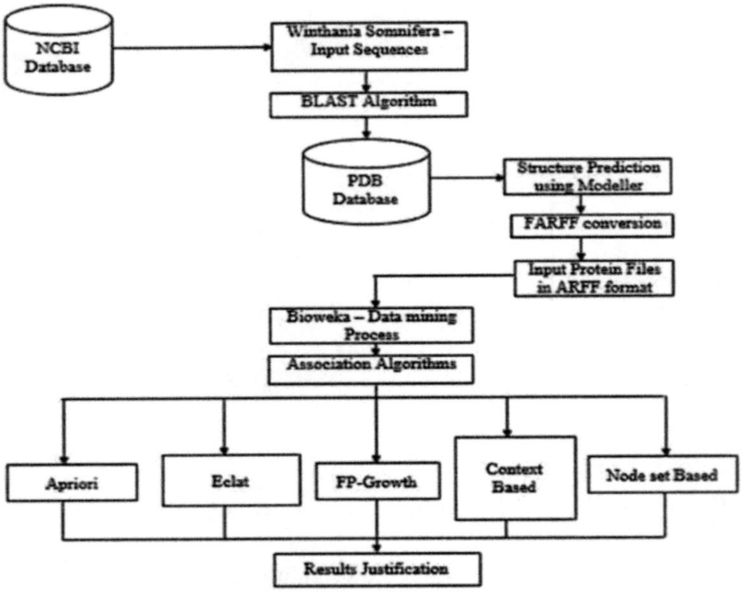

Fig. 3 Biological data mining architecture

has its major impact on several of its areas. Sequence analysis is the primary action in the computation process of biology. There are three types of biological sequences involved such as DNA, RNA (ribose nucleic acid), and protein. DNA is comprised of four monomer bases such as A (adenine), T (thymine), G (gyanine), and C (cytosine). RNA is divided into mRNA (messengerRNA), rRNA (ribosomalRNA), and tRNA (transferRNA).

3 Environmental Sequence Analysis

In the case of DNA, there are sequences which have specific characteristics apart from several extraction techniques for these sequences such as clustering, support vector machine (SVM), neural networks, association, and classification.

The environmental data are highly sensitive data which make major differences in the characteristics, if the single mutant performs its processes on the way to gene prospects. Mutation is the variation that happens in the sequences for their characteristics building. If the mutation process changes the gene activity, then the whole genome processes are varied.

Gene selection can be done in gene expression analysis, in the identification of the respective gene which is connected to the target class or a class provided for finding. In microarray perception, association has the ability to predict or analyze the disease or the output-related data efficiently with their split up.

4 Protein Association

The examination of organic information mining provides a profound comprehension of how the information mining strategies such as grouping, arrangement, affiliation, expectation, and relapse can be produced. This synthesis of clustering is not that easy in practice as it is based on several constraints requiring the separation of objects into different groups to facilitate data mining and further procedures with the purpose of getting meaningful data from the given bunch of data and making it in a better interface to the user ability for proceeding further.

As plant biology paves the way for very smart and interest in finding solutions for every problem which they come across while traveling through time, several methods have already come up with many new interesting algorithms for association problem as well. Again this very special and interesting human nature of dissatisfaction and engineers has a keen interest in innovation things. It motivates to innovate more than one clustering algorithms for clustering which was invented by different engineers and researchers developed into numerous software such as BioWeka and Weka.

In the case of each observation for association methodology, objects inside are the same between themselves, while they are not the same when compared with other rules formed earlier. As mentioned before, there are different association algorithms to form rules where BioWeka is a tool for comparison. Association algorithms are compared in two ways, namely performance analysis and matrix analysis. Performance measures produce the better action for every newly proposed algorithm in different applications such as medical field, environmental field, and education field. Matrix analysis in plant biology and animal biology for association algorithm produces better results for further sequence analysis. This analysis can be used better for different application based on these analyzed parameters.

5 Results and Discussion

All activities in the animals and plants are because of chemical reactions, which results in different chemical bonds. These bonds constitute the behavior of internal organs. An abnormal chemical reaction may result in a difference in the behavior of organs, which provides a base for analyzing, monitoring the anomalies, and functioning of the human genome. Genome analyses can be performed by two processes, viz.PPI—PPI refers to physical contacts established between two or more proteins as a result of biochemical events and NGS (Fig. 4).

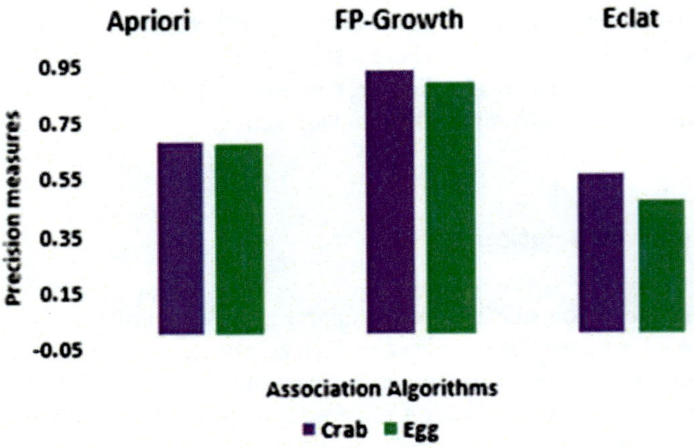

Fig. 4 Matrix-associated information

The main goal of association algorithms in bio-environmental sequence analysis is to identify the similar data with corresponds to a gene, patterns, etc., available in the database. This research work deals with the two types of datasets, viz. first one, the plant biology-related dataset named as Withania somnifera and secondly, the animal biology-related data sets such as crab and egg, which can execute in the wet laboratory with the required water, direct ventilation, and specialized piped utilities for microorganisms identification; then, the merge sets filter used on the datasets for PPI factor followed by association methodology prospects. The second one is data downloaded from the biological databases, such as NCBI, PDB, which is available online with lakhs of datasets incorporated into it. These datasets are used for cross-validation of the proposed system (Fig. 5).

6 Conclusion

The problem addressed in this research work is to effectively identify the learning prospects given any biological data or database, thereby determining the microorganisms in the samples and protein structure available in the sequence produced. Prior to the process of identifying the group, the huge volume of amino acids in the dataset got reduced which are highly correlated and redundant in nature. Moreover, the computational complexity of the pattern identification algorithm got reduced.

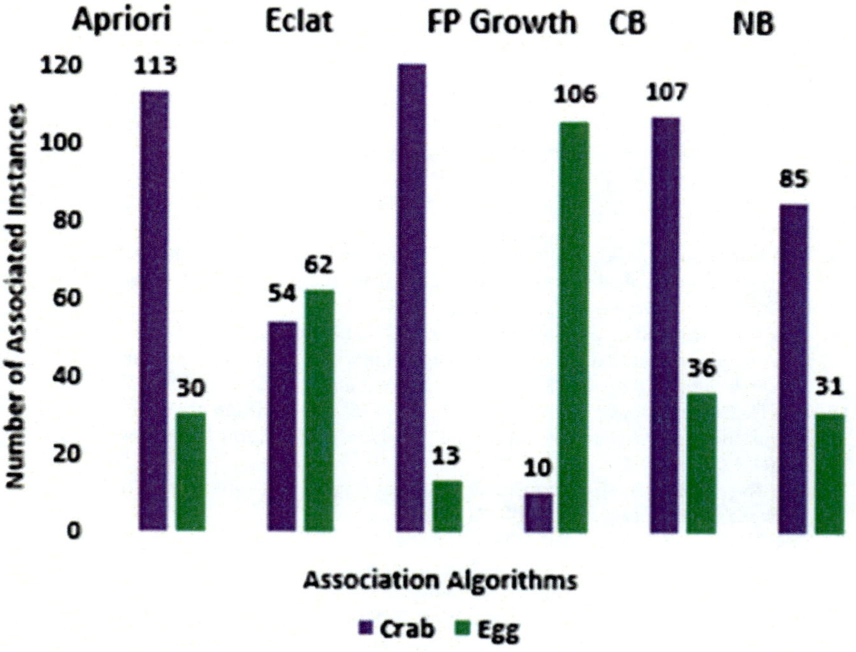

Fig. 5 Performance measures

References

1. Cibulskis, K., McKenna, A., Fennell, T., et al. (2011). ContEst: Estimating cross-contamination of human samples in next-generation sequencing data. *Bioinformatics, 27,* 2601–2602.
2. Vignesh, U., & Parvathi, R. (2018). 3D visualization and cluster analysis of unstructured protein sequences using ARCSA with a file conversion approach. *Journal of Supercomputing.* https://doi.org/10.1007/s11227-018-2319-4.
3. Vignesh, U., & Parvathi, R. (2017). Clustering on structured proteins with filtering instances on Bioweka. *Journal of Engineering Science and Technology, 12*(3), 820–833.
4. Vignesh, U., & Parvathi, R. (2017). Bioweka classification on egg and crab protein structured sequence with filter instances. *Journal of Advanced Research in Dynamical and Control Systems, 9,* 17.
5. Vignesh, U., & Parvathi, R. (2016). Next generation sequencing data analysis software and methods: A survey. *International Journal of Control Theory and Applications, 9*(52), 1–15.
6. Vignesh, U., & Parvathi, R. (2017). Biological data analysis and visualization: A survey—modern technologies for big data classification and clustering. https://doi.org/10.4018/978-1-5225-2805-0.ch010.
7. Ahmad, N., Elhassan, H. M., Dafallaa, I., BurhanurRehman, M., et al. (2019). Comparative study on load balancing algorithm for multiprocessor interconnection networks. *IJATCSE, 8*(3), 410–414.
8. Manoj, B., Sasikanth, K. V. K., Subbarao, M. V., et al. (2018). Analysis of data science with the use of big data. *IJATCSE, 7*(6), 87–90.
9. Siva Nageswara Rao, G., & Srinivasu, S. V. N. (2017). Hybrid approach for task scheduling in heterogeneous cloud based systems. *Journal of Advanced Research in Dynamical and Control Systems,* 9(Special Issue 14), 618–625.

10. Siva Nageswara Rao, G., & Jayanth, T. V. M., Shabzan, S., & Supritha, B. (2017). Airport luggage tracking system using RFID and GSM. *Journal of Advanced Research in Dynamical and Control Systems, 9*, 748–757.
11. Siva Nageswara Rao, G., & Srinivasu, S. (2016). Task scheduling for real time applications using mean-difference round robin (MDRR) algorithm with dynamic time slice (MDDRWDTS). *International Journal of Pharmacy and Technology, 8*(3), 16082–16088.
12. Siva Nageswara Rao, G. (2016). Real time scheduling for dynamic process execution using round robin switching. *Journal of Theoretical and Applied Information Technology, 89*(1) 60–66.
13. Siva Nageswara Rao, G., Srinivasu, S. V. N., Srinivasu, N., &Naga Raju, O. (2015). A new proposed Dynamic dual processor based CPU scheduling algorithm. *Journal of Theoretical and Applied Information Technology, 73*(2), 226–231.
14. Siva Nageswara Rao, G., Srinivasu, S. V. N., Srinivasu, N., & Ramakoteswara Rao, G. (2015). Enhanced precedence scheduling algorithm with dynamic time quantum (EPSADTQ). *Research Journal of Applied Sciences, Engineering and Technology, 10*(8), 938–941.
15. Siva Nageswara Rao, G., Krishna, C. V. P., Rao, K. R. (2014). Multi objective particle swarm optimization for software cost estimation. *Advances in Intelligent Systems and Computing, 248*(I), 125–132.
16. Siva Nageswara Rao, G., Manojkumar, B., Jaya Raj, R., & Sharma, A. (2018). IOT based garbage management system, *10*(4), 31–36.
17. Siva Nageswara Rao, G., Kiran Babu, M., Rao, S. (2019). Efficient scheduling measures for improving real time systems using prediction process. *Journal of Computational and Theoretical Nano Science, 16*(5–6), 1876–1880.
18. Venkateswara Rao, C., Siva Nageswara Rao, G. (2019) An effective research on data mining techniques for intrusion detection & learning classes. *International Journal of Innovative Technology and Exploring Engineering, 8*(11 Special Issue), 845–849.
19. Yaswanth, K. B., Manasa, L., Chetan, B. V., Siva Nageswara Rao, G. (2019). Modern vehicle tracking and monitoring system using embedded technology. *International Journal of Recent Technology and Engineering, 8*(1), 1849–1851.
20. Anusha, A., Guptha, A., Sivanageswar Rao, G., Tenali, R. K. (2019). A model for smart agriculture using IOT. *International Journal of Innovative Technology and Exploring Engineering, 8*(6), 1656–1659.

Neural Network-Based Random-Valued Impulsive Noise Suppression Scheme

Punyaban Patel and Bibekananda Jena

Abstract Denoising of digital images is one of the important pre-processing stages in image processing. In this paper, a new efficient back-propagation neural network-based impulse noise suppression technique is presented. The proposed technique works in two stages, i.e., impulse noise detection and noise filtering. In noise-detection stage, a feed-forward neural network is trained by back-propagation training algorithm to classify the noisy and non-noisy pixels in the test image. The input patterns for training are selected carefully to make the detector more robust for impulse detection. An adaptive median filtering is employed in the second stage for restoring the value of the noisy pixels identified in the first stage keeping the non-noisy ones unaltered. The restoration ability of the proposed method is evaluated using different standard images at different noise levels. It has been observed after several experiments on standard images that the performances of the proposed techniques are superior in terms of both visual quality and quantitative measurement.

Keywords Salt and pepper noise (SPN) · Random-valued impulse noise (RVIN) · Mean square error (MSE) · Peak signal-to-noise ratio (PSNR) · Image quality index (IQI) · Back-propagation neural network (BPNN)

1 Introduction

Images are often contaminated by impulse noise in the procedures of image acquisition and transmission which sometimes create difficulty in taking correct decision. Hence, the denoising technique is very much necessary for all image-processing-related area [1–3]. There are two variations of the impulse noise as per the noise

P. Patel (✉)
Department of Computer Science & Engineering, CMR Technical Campus, Kandlakoya, Medchel, Hyderabad, India
e-mail: punyaban@gmail.com

B. Jena
Department of ECE, Anil Neerukonda Institute of Technology & Sciences, Visakhapatnam, Andhra Pradesh, India
e-mail: bibekananda.jena@gmail.com

S. C. Satapathy et al. (eds.), *Intelligent System Design*, Advances in Intelligent Systems and Computing 1171, https://doi.org/10.1007/978-981-15-5400-1_72

distribution: salt and pepper noise and random-valued impulse noise. Salt and pepper noises can be detected and filtered by most of the reported method efficiently. But their performances are not satisfactory when dealing with random-valued impulse noise (RVIN). Figure 1 shows how an image corrupted by RVIN is different from an image corrupted by salt and pepper noise. It is clearly observed that in salt and pepper noise, the pixels of the original image are replaced by a value 0 or 255, whereas in RVIN, the corrupted pixels can have any value in the available dynamic range, i.e., in between 0 and 255 for 8-bit images. Hence, a single threshold value will not be enough to mitigate the effect of such noise. The selection of the line to separate the noisy and noise-free pixels needs to be chosen carefully. Preferably, the selection process should dynamically adapt to its environment and produce an optimal value. The artificial neural network can be used to classify noises in images [4, 5].

The simple median filter [1, 2] and its modifications [6, 7] were once the standard nonlinear filter for suppressing impulse noise. But the limitation is, all the pixels are participated in the filtering process irrespective of they are noisy or not and results which destroy fine details and yield blotches in the filtered images. A solution to this problem is to distinguish between non-noisy and noisy pixels before filtering, and only the noise candidates are allowed to be filtered. Many authors have proposed several techniques to eliminate impulse like pixel without affecting uncorrupted pixels and preserve the fine details of the test images. One of such schemes is novel adaptive impulse detection using center-weighted median filter (ACWMF) proposed in [8]. The impulse detector takes decision depending upon the difference of the output produced by CWM filter with varied center weight and current test pixel. But the filter fails to work if the noise density exceeds 20%. In [9], Cronjevic proposed

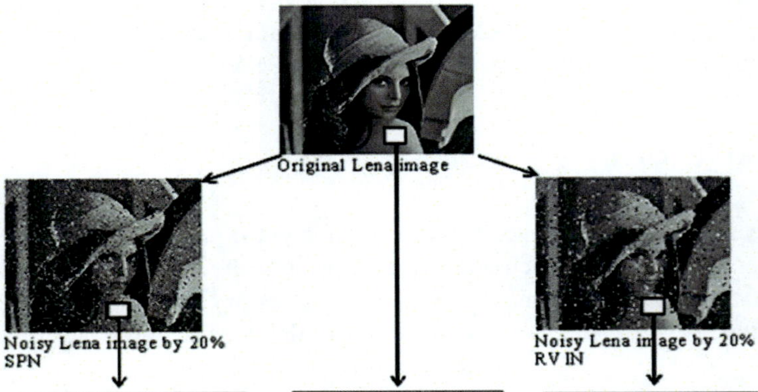

84	86	84	0	85
86	255	85	84	0
85	86	85	0	85
87	87	255	84	84
255	84	86	86	87

84	86	84	86	85
86	87	85	84	86
85	86	85	85	85
87	87	85	84	84
86	84	86	86	87

84	86	84	0	85
86	44	85	84	86
85	86	85	50	85
87	87	85	84	167
255	84	86	128	87

Fig. 1 Original and noisy Lena image with SPN and RVIN

a new technique which is an extension of the switching technique as discussed in [10] known as adaptive MAD-based filter (ADMAD). In this case, the local variance estimated by the median of the absolute deviations (MAD) defines the threshold used for identifying corrupted pixel which changes from pixel to pixel. An adaptive switching median (ASWMF) filter is proposed in [11] where the threshold selection depends upon the weighted mean and standard deviation of the sliding window. It automatically calculates the required threshold for each and every pixels of the image. A new efficient algorithm called An Efficient Detail-Preserving Approach for Removing Impulse Noise in images (ATBMF) is proposed in [12] to remove corrupted pixels from noisy images without disturbing the fine details of the image. In this case, the difference between the current text pixel and the alpha-trimmed means of that pixel is taken as a measure to distinguish between nosy and noise-free pixel elements. In [13], Tao Chenet proposed multi-state median (MSM) filter whose output is adaptively switched among those of a group of center-weighted median filters that have different center weights. As a result, the MSM filter becomes an adaptive CWM filter with a space-varying center weight which relies on local signal statistics. Chung-Chia Kang presented a modified switching median filter (MSWMF) [14] by adding one more another noise detector to improve the capability of impulse noise removal. The impulse noise detector used here is established based on the rank-order arrangement of the pixels in the sliding window. The new impulse detection concept called advanced impulse detection based on pixel-wise MAD (PWMAD) is proposed in [15]. MAD is used to estimate the presence of image details. An iterative pixel-wise modification of MAD (PWMAD) provides reliable removal of arbitrarily distributed impulse noise. In [16], Tao Chen et al. proposed a new median filter, called tri-state median (TSM) filter. This filter is the combination of identity filter, standard median filter, center-weighted median filter, an impulse noise detector, and a switching logic. The outputs of all the filters are taken as the inputs of the noise detector. The noise detector compares these outputs with the test pixel to go for a tri-state decision. The switching logic uses a threshold value to restore the pixel value if it is corrupted, i.e., the text pixel value remains unchanged or replaced with the output of SM filter or CWM filter depending on this threshold value.

Recently, a novel algorithm called standard deviation for obtaining the optimal direction in the removal of impulse noise has been proposed [17] based on finding the normalized distance between the tested pixel, and the pixels present in the optimal direction, known as normalized distance in optimal direction (NDOD). Edge pixels can be preserved if the accurate or optimal direction of the edge is determined. The recent detection scheme for random-valued impulse noise is based on the robust outlyingness ratio (ROR) [18]. This parameter is able to say about a pixel how much it is impulse like. But this scheme provides poor results for low-noise-density images.

In this paper, we propose a neural network detection-based filter (NNDF) to combat RVIN in images. The NNDF scheme aims at efficient classification of pixels as noisy or noise free. The input parameters to the neural network are derived from two different statistical parameters, namely robust outlyingness ratio (ROR) [18] and normalized difference in optimal direction (NDOD), a similarity measures discussed in [17]. Subsequently, the identified noisy pixels are filtered with an adaptive median filter.

The overall paper is organized as follows. Section 1 deals with introduction. Section 2 describes the proposed denoising work. Section 3 discusses the simulation and results. Finally, Sect. 4 provides the concluding remarks.

2 Proposed NNDF Filter

The proposed technique has two stages for impulse noise suppression. The first stage is a noise detector used for classifying the non-noisy pixels and the pixels-corrupted random-valued impulse noise in the image. A feed-forward neural network is trained with a set of known training parameters in this stage. The trained network produces a binary noise map of the corrupted image at the output. *A* value 0 at certain location in the map indicates that the network detect a pixel in the same location of the input image as a noisy pixels.

Similarly, a value 1 in the binary map classifies the pixel present in the same location of the input image as nonnoisy. Once the detection stage is over, with the help of generated binary map, an adaptive median filter restores the value of the noisy pixels by its non-noisy neighbors keeping the value of non-noisy pixels unaltered.

a. Neural network-based impulse noise detection

In this paper, a three-layer NN with two hidden layers and a decision maker as shown in Fig. 2 is proposed for impulse noise detection. The neural network consists of two nodes corresponding to the two known training parameters called as robust outlyingness ratio (ROR) and a normalized difference in optimal direction (NDOD)

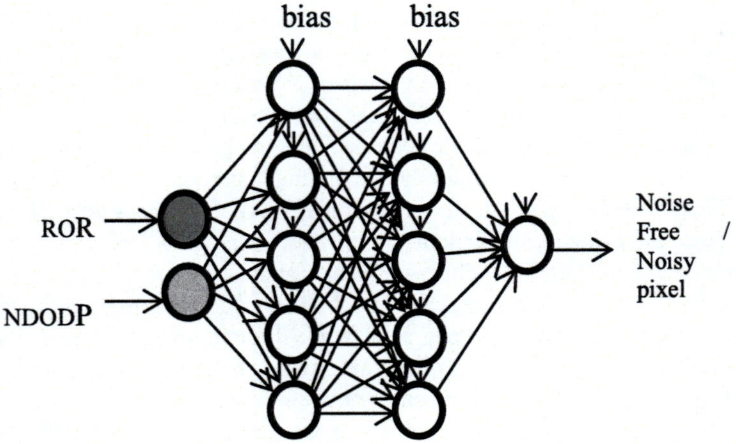

Fig. 2 BPNN structure for noise detection

derived from a 5 × 5 window. The second and third layers are the hidden layers that consist of five neurons each. There is only one node at output to distinguish between "noisy" or "non-noisy," pixels and the log sigmoid function is used as the activation function. The output of the neural network is fed to a decision maker. The decision maker uses a threshold 0.5 to rounds off the output of the neural network. The two input training parameters of the network are discussed as follows:

i. Robust Outlyingness Ratio (ROR) [18]

This is a newly proposed parameter for detection of RVIN noise in images. ROR measures how much each pixel gets affected by impulse. The computation of ROR involves the following steps.

Step 1: Consider a 5 × 5 window, W with center pixel as the test pixel y

Step 2: Compute Med(y) as the median of the window

Step 3: Compute the median of absolute deviation, MAD as,

$$\mathrm{MAD}(y) = \mathrm{Med}\,\{|x - \mathrm{Med}(y)|, x \in W\}$$

Step 4: Compute MADN(y) empirically as,

$$\mathrm{MADN}(y) = \mathrm{MAD}(y)/0.6457$$

Step 5: Compute $\mathrm{ROR}(y) = \left|\frac{(y-\mathrm{Med}(y))}{\mathrm{MADN}(y)}\right|$

ii. Normalized difference in optimal direction (NDOD) [17]

NDOD is calculated in an optimal direction of a moving window in the input image to identify whether the tested pixel is noisy or not. At any point in the image, a moving window has four possible directions as denoted by D_1, D_2, D_3, and D_4 as shown in Fig. 3. The most similar pixels of a moving window are present in the optimal direction of the window. Hence, with the help of optimal direction, we can identify the edge pixels conveniently. Also the pixels which have insignificant deviation from the pixels in optimal direction can be exempted from filtering. The computation of NDOD is given in the following steps.

Step 1: Consider a 5 × 5 window and let the first pixel be denoted as x_{ij}. The entire pixels in the moving window, except the central pixel $x_{i+(k+1)/2,j+(k+1)/2,}$ for $k = 5$ are divided into four directions $D_d, d = 1 : 4$ as shown in Fig. 3. The co-ordinates of the pixels in each direction are as follows:

$$D_1 = \{(0,0), (1,1), (3,3), (4,4)\}, \ldots, D_2 = \{(0,4), (1,3), (3,1), (4,0)\}$$

$$D_3 = \{(0,2), (1,2), (3,2), (4,2)\}, \ldots, D_4 = \{(2,0), (2,1), (2,3), (2,4)\}$$

Step 2: The optimal direction D^{op} is obtained by finding the vector among D_d having smallest standard deviation.

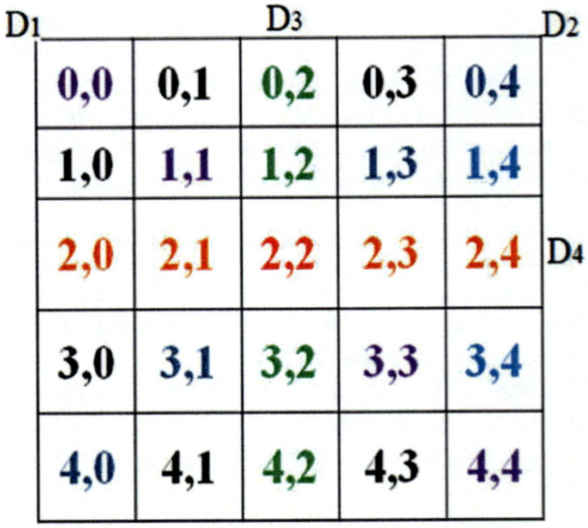

Fig. 3 A 5 × 5 moving window

$$D^{op} = \arg\min_{D_d}\{\sigma_{D_d}\}$$

Here, σ_{D_d} is denoted as standard deviation of D_d.

Step 3: The NDOD for the central pixel $x_{i+(k+1)/2,\,j+(k+1)/2}$ is calculated as,

$$\text{NDOD} = \sum_{s=1}^{k} \frac{\left|x_s^{op} - x_{i+(k+1)/2,\,j+(k+1)/2}\right|}{255}$$

where x_s^{op} is the pixel in optimal direction D^{op}.

To make the neural network as an efficient universal impulse detector, it has to be trained perfectly. The training patterns are generated from the standard Lena image of size 512 × 512 corrupted with 20% of RVIN. A total of 1000 noisy pixels and 1000 non-noisy pixels are collected randomly in the image for generating training patterns. For every training pixel, both ROR and NDOD are computed. The target output of the noise-free pixels is taken as 1 and that of the noisy pixels is 0. The proposed 2-5-5-1 neural network is then trained using back-propagation algorithm. The error goal is set to 0 and the training is carried up to at least 100 no of epochs.

b. Noise Filtering

The output of the noise-detection stage is followed by a noise filtering stage which restores the value of identified noisy pixels. The performance of the filtering method purely depends on the performance of the noise detector. An adaptive median filter is adopted which restores the noisy value by finding the median of its uncorrupted neighbors. The size of the processing window of the filter increases continuously till it finds at least one healthy pixel in the window. The steps of the filtering schemes are as follows:

1. Initialize a window of size $W = 3$. Collect the number of healthy neighbor pixels(S) inside the window (using the proposed detector). If $S = 0$, then go to step 2 else go to step 4.
2. Update the window size by, $W = W + 2$.
3. Find S for the new window. If $S = 0$, then jump to step 2, else proceed to next step.
4. Find the median of the set of pixels in "S".
5. Restore the test pixel by the median value.

3 Simulations and Results

The restoration ability of the proposed NNDF scheme is evaluated by extensive simulations on well-known standard 512 × 512 8-bit gray-scale test images. In this paper, the results obtained from the standard Boat and Bridge images are discussed. The input images are corrupted with RVIN from 10% to 40%. The proposed scheme along with the well-performing schemes like ACWMF, ADMAD, ASWMF, ATBMF, MSM, MSWMF, PWMAD, TSM, and SDOOD are applied to the noisy images. Both subjective and objective valuations have been made for each restored images. Peak signal-to-noise ratio (PSNR) [2, 20] and image quality index (IQI) [19, 20] are taken as two parameters to analyze the denoising performances of the proposed method quantitatively and qualitatively, respectively. PSNR is defined as

$$\text{PSNR}(dB) = 10\log_{10}\left(\frac{255^2}{\text{MSE}}\right) \quad (1)$$

$$\text{MSE} = \frac{1}{M \times N}\sum_{i=0}^{M-1}\sum_{j=0}^{N-1}\left(X_{i,j} - \hat{X}_{i,j}\right)^2 \quad (2)$$

where X and $\hat{X}$ represent the original and restored images, respectively.

The distortion in the image may be due to loss of correlation, changes in luminance, or distortion in contrast level. IQI combines the above parameters to represent the degradation level and defined as:

$$\text{IQI} = (1/M)\sum_{j}\text{IQI}_j \quad (3)$$

where, $j = 1, 2, 3, \ldots M$ and

$$\text{IQI}_j = \text{Corr}\left(X_w, \hat{X}_w\right) * \text{Lum}\left(X_w, \hat{X}_w\right) * \text{Cont}\left(X_w, \hat{X}_w\right)$$

for each 8 ×8 local region. M is the total number of such region inside the image.

$$\text{where, Corr}\left(X_w, \hat{X}_w\right) = \sigma_{X\hat{X}}/\sigma_X\sigma_{\hat{X}}$$

$$\text{Lum}\left(X_w, \hat{X}_w\right) = 2\mu_X\mu_{\hat{X}}/\left(\mu_X^2 + \mu_{\hat{X}}^2\right)$$

$$\text{Cont}\left(X_w, \hat{X}_w\right) = 2\sigma_X\sigma_{\hat{X}}/\left(\sigma_X^2 + \sigma_{\hat{X}}^2\right)$$

The maximum value that the IQI attend is 1, which represents that the restored and original images are almost equal

Tables 1 and 2 represent the comparative analysis of peak signal-to-noise ratio and Image Quality Index measures for Bridge image at various RVIN noise levels,

Table 1 Comparative analysis of PSNR(dB) for various filters in Bridge image

Methods	% of RVIN						
	10	15	20	25	30	35	40
ACWMF	33.65	31.49	29.66	27.72	26.10	24.45	22.91
ADMAD	34.21	32.27	30.19	28.33	26.55	24.62	23.01
ASWMF	34.18	32.03	30.49	28.93	27.34	26.26	24.73
ATBMF	33.97	32.05	30.28	28.59	26.74	25.37	23.90
MSM	34.04	32.59	31.24	29.85	28.40	26.90	25.56
MSWMF	32.12	30.95	29.51	27.98	26.33	24.90	23.22
PWMAD	32.56	31.51	29.98	28.30	26.45	24.65	22.91
TSM	33.54	31.94	29.96	27.93	26.16	24.49	22.68
SDOOD	30.72	29.93	29.21	28.34	27.59	26.77	25.77
PROPOSED	33.62	32.15	31.10	30.08	29.07	27.82	26.61

Table 2 Comparative analysis of IQI for various filters in Bridge image

Methods	% of RVIN						
	10	15	20	25	30	35	40
ACWMF	0.934	0.896	0.852	0.791	0.728	0.647	0.576
ADMAD	0.930	0.902	0.863	0.810	0.747	0.666	0.586
ASWMF	0.945	0.914	0.880	0.830	0.778	0.719	0.655
ATBMF	0.936	0.907	0.866	0.813	0.747	0.686	0.618
MSM	0.945	0.927	0.907	0.872	0.830	0.781	0.723
MSWMF	0.918	0.896	0.860	0.811	0.745	0.674	0.592
PWMAD	0.937	0.920	0.885	0.836	0.759	0.665	0.580
TSM	0.931	0.898	0.858	0.792	0.729	0.652	0.564
SDOOD	0.911	0.889	0.867	0.841	0.811	0.776	0.728
PROPOSED	0.938	0.923	0.913	0.896	0.873	0.836	0.785

respectively. It can be seen that ADMAD shows better result when the noise ratio is less than 20%. But for noise level ≥ 20%, the proposed method provides significant improvement over existing techniques. The simulated PSNR and IQI result of Boat image are plotted in Figs. 4 and 5 respectively. It is clear from the graphs that the performance of the proposed method is better than others. Along with the numerical value, the restored Bridge images of different method have been given in Fig. 6. Each filter processes the same Bridge image corrupted with 40% RVIN. Compared with the original image, the proposed scheme shows a superior restoration performance.

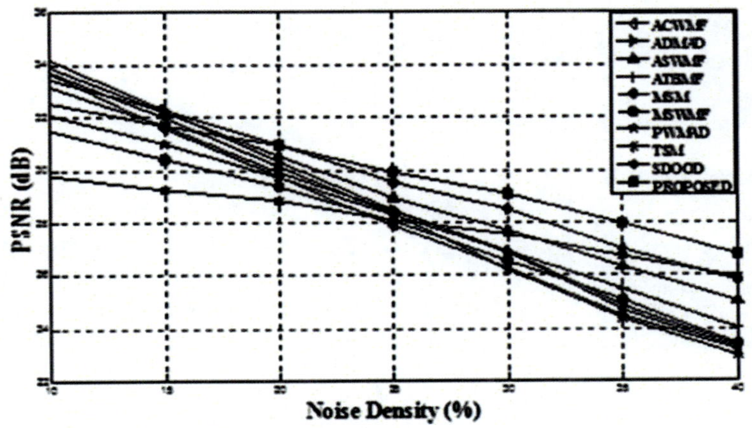

Fig. 4 Comparative analysis of PSNR at various noise densities of Boat image

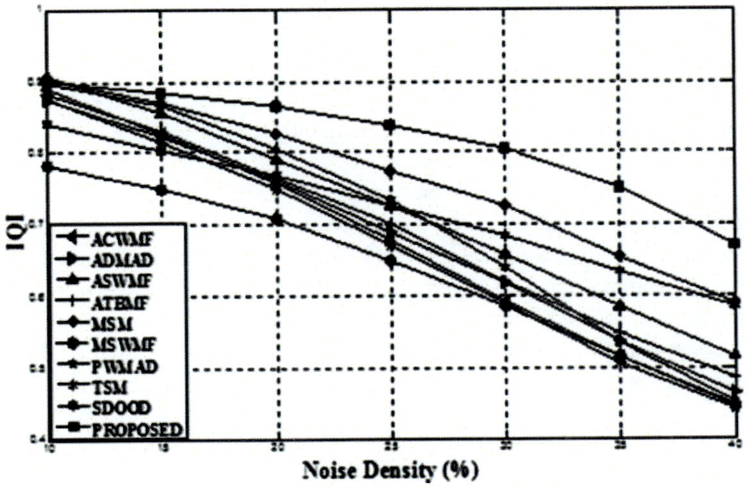

Fig. 5 Comparative analysis of IQI at various noise densities of Boat image

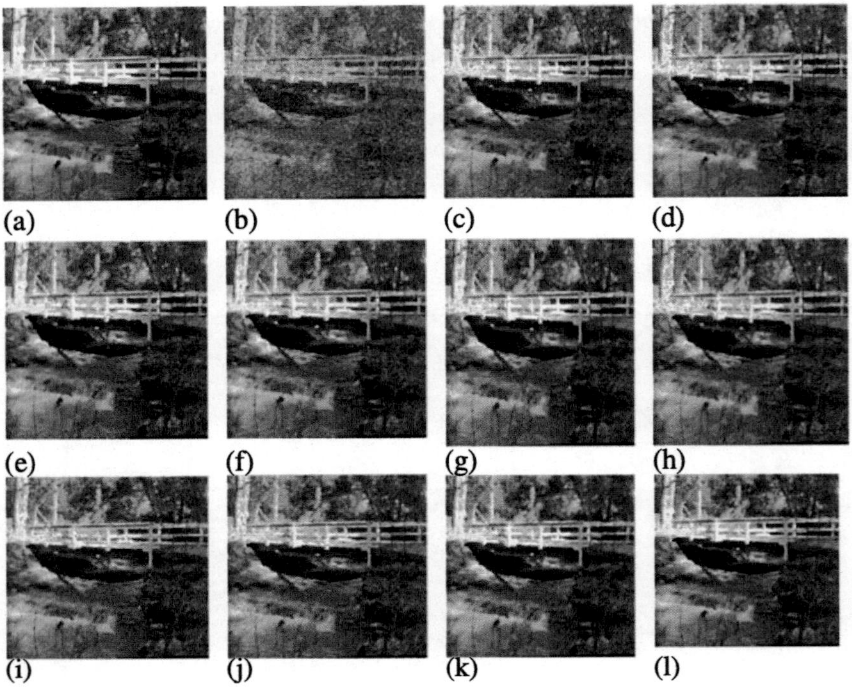

Fig. 6 Restoration results of different methods included in comparison for test image Bridge with 40% of random-valued impulse noise. **a** Original image. **b** Noisy image. **c** ACWMF. **d** ADMAD **e** ASWMF **f** ATBMF **g** MSM. **h** MSWMF **i** PWMAD **j** TSM **k** SDOOD **l** Proposed method

4 Conclusion

The proposed NNDF algorithm recovers images corrupted with RVIN effectively. The filter initially detects the location of corrupted pixels by proposed neural network model followed by the adaptive median filtering operation. The two features used as the input to the neural network make the accuracy of detection high enough and generalized for which the network structure neither required to be trained differently for different noise ratios or different images. Filtering is applied to the identified noisy pixels only keeping the uncorrupted pixels as it is. Both qualitatively and quantitatively measures are used for evaluation of the proposed technique. The comparative results show that the proposed technique gives superior result than other existing technique not only suppression of impulse noise but also preservation of images details.

References

1. Gonzalez, R. C., & Woods, R. E. (2007). *Digital image processing*. Upper Saddle River, New Jersey: Pearson Education.
2. Chanda, B., & Dutta Majumder, D. (2002). *Digital image processing and analysis* (1st ed.). India: Prentice-Hall of India.
3. Pratt, W. K. (1991). *Digital image processing*. New York: Wiley-Interscience.
4. Santhanam, T., & Radhika, S. (2011). Probabilistic neural network—a better solution for noise classification. *Journal of Theoretical and Applied Information Technology, 27*(1), 39–42.
5. Kumar, S. (2004). *Neural networks: A classroom approach* (p. 169). Tata Mc-Graw Hill Publications.
6. Brownrigg, D. R. K. (1984, August). The Weighted Median Filter. *Communications ACM, 27*, 807–818.
7. Ko, S. J., & Lee, Y. H. (1991). Center weighted median filters and their applications to image enhancement. *IEEE Transactions on Circuits and Systems, 38*(9), 984–993.
8. Chen, T., & Wu, H. R. (2001). Adaptive impulse detection using center-weighted median filters. *IEEE Signal Processing Letters, 8*(1), 1–3.
9. Crnojevic, V. (2005, September). Impulse noise filter with adaptive MAD-based threshold. In *IEEE International Conference on Image Processing 2005* (Vol. 3, pp. III-337). IEEE.
10. Sun, T., & Neuvo, Y. (1994). Detail-preserving median based filters in image processing. *Pattern Recognition Letters, 15*, 341–347.
11. Akkoul, S., Ledee, R., Leconge, R., & Harba, R. (2009, September). A new detector for switching median filter. In *2009 Proceedings of 6th International Symposium on Image and Signal Processing and Analysis* (pp. 770–775). IEEE.
12. Luo, W. (2006). An efficient detail-preserving approach for removing impulse noise in images. *IEEE Signal Processing Letters, 13*(7), 413–416.
13. Chen, T., & Wu, H. R. (2001). Space variant median filters for the restoration of impulse noise corrupted images. *IEEE transactions on circuits and systems II: analog and digital signal processing, 48*(8), 784–789.
14. Kang, C.-C., & Wang, W.-J. (2009). Modified switching median filter one more noise detector For impulse noise removal. *AEU-International Journal of Electronics and Communications, 63*, 998–1004.
15. Crnojevic, V., Senk, V., & Trpovski, Z. (2004). Advanced impulse detection based on pixel-wise MAD. *IEEE Signal Processing Letters, 11*(7), 589–592.
16. Chen, T., Ma, K. K., & Chen, L. H. (1999). Tri-state median filter for image denoising. *IEEE Transactions on Image processing, 8*(12), 1834–1838.
17. Awad, A. S. (2011). Standard deviation for obtaining the optimal direction in the removal of impulse noise. *IEEE Signal Processing Letters, 18*(7), 407–410.
18. Xiong, B., & Yin, Z. (2011). A universal denoising framework with a new impulse detector and nonlocal means. *IEEE Transactions on Image Processing, 21*(4), 1663–1675.
19. Wang, Z., & Bovik, A. C. (2002). A universal image quality index. *IEEE Signal Processing Letters, 9*(3), 81–84.
20. Nair, M. S., & Raju, G. (2012). A new fuzzy-based decision algorithm for high-density impulse noise removal. *Signal, Image and Video Processing, 6*(4), 579–595.

Recognition of Change in Facial Expressions From Video Sequence Using Incenter-Circumcenter Pair Gradient Signature

Md Nasir, Paramartha Dutta, and Avishek Nandi

Abstract Facial expression recognition system provides us enormous scope for studying characteristics of different human facial expression. Basic expressions (anger, disgust, fear, happiness, sadness, and surprise) can be distinguished from one another with minimum overlapping features between them. A human facial emotion can be represented uniquely if we are able to capture such features having high discrimination power. In this paper, we propose a method for recognizing facial expression by using salient landmark-induced gradient signature derived from the line segment joining circumcenter and incenter pair of a triangle obtained by joining relevant landmark points. We use appearance-based models to detect essential landmark points on face. We are considering only those landmark points situated at eyes, lips, eyebrows, and nose to get geometric feature. This landmark-based geometric facial feature has been used to describe changes in emotion from neutral through basic expression. We examine our proposed method on Chon-Kanade (CK+), MMI, and MUG benchmark image sequence databases and compare its performance under different motion image databases. In our experiment, dynamic behavior of every individual expression has been recognized with impressive recognition rate of accuracy.

Keywords Landmarks · Gradient signature · Motion images · FER · MLP

Md Nasir (✉) · P. Dutta · A. Nandi
Department of Computer and System Sciences, Visva-Bharati University, Santiniketan, West Bengal 731235, India
e-mail: nasir.vb@gmail.com

P. Dutta
e-mail: paramartha.dutta@gmail.com

A. Nandi
e-mail: avisheknandi10@gmail.com

S. C. Satapathy et al. (eds.), *Intelligent System Design*, Advances in Intelligent Systems and Computing 1171, https://doi.org/10.1007/978-981-15-5400-1_73

1 Introduction

Facial expression analysis from video frames is very challenging and promising task for identification of dynamic changes in human emotion [2]. Facial expressions recognition system [9] has attracted substantial endeavor among researchers in recent days mainly because it has several applications, such as human computer interactions, multimedia, surveillance, and treatment of patients in mentally underdeveloped. In the study of Mehrabian [10], it is acknowledged that when a person is interacting with others through their emotion, 55% of the message are conveyed through facial expression alone, vocal cues provide 38%, and the remaining 7% via verbal cues. This also has motivated researchers to evaluate expression by exploring major parts of face. Ekman and Frisen [7] identified six basic facial expressions, viz anger, disgust, fear, happiness, sadness, and surprise. They earmarked some indications to extract geometric features from crucial parts of face, viz eyes, eyebrows, mouth, lips, etc. Features extracted from image sequences can be mapped into different action units or expression categories. Active appearance model (AAM) [5] is a combination of both shape and texture models. It is very useful to find out the geometric location of landmarks on a face image. In this paper, we propose a gradient signature-based facial expression recognition system using multilayer perceptron (MLP) to classify given image sequence data. Appearance-based method is employed here to retrieve all 68 landmark points on entire face region. Key landmark selection among them is very important to achieve higher performance of our proposed recognition system. In our work, we only take into account six landmark points on mouth region, three landmark points on eyebrows region, three landmark points on nose region, and four landmark points on eyes region. These landmarks are used to generate a triangle shape. For all possible triangles we consider two points namely the circumcenter and incenter from which the gradient of straight line is formed by joining those two points. This gradient information is taken in a specific order, thereby inducing gradient signature which is fed into MLP to classify different types of emotional changes.

2 Method

In this article, we intend to develop a facial expression recognition system extracting features based on geometric signature. We are trying to capture more prominent features from image sequence data that can provide high-level representation of every image frame [16]. Facial expression recognition (FER) system [6] contain following basic components landmarks detection, feature extraction, and classification.

2.1 Landmarks Detection

AAM fitting model is applied on every frame in image sequence corresponding to each expression to sense landmark points. We got geometric position of 68 landmark points along horizontal-axis and vertical-axis on face image plane for every frame in a sequence. These points are sufficient to cover total face region. In [13], it is observed that some landmarks play significant role due to displacement of facial components during dynamic changes of the emotion, and some landmarks have no effective influences in discriminating one expression from another. We have taken only 23 landmark points among them into the account to explore vital face regions. Vital regions are eyebrows, eyes, nose, and mouth [12]. We have already discussed in introduction that how 23 points has been considered from major portions of human face. It has been found that these informative points are very sensible and play important role in analyzing facial expression.

2.2 Feature Extraction

We considered those finest landmark points into a single array and found all possible combinations of three landmark points. We have identified triangles from each combination of points through connecting them. These different types of triangles have been used to find out prominent geometric features which are fed into facial expression recognition system.

2.2.1 Computation of Gradient Signature

We considered landmarks as $L = [l_1, l_2, l_3, \ldots, l_{23}]$ on a face image as shown in Fig. 1a. For every possible combination containing three landmark points, we have formed triangle $T_n = [(l_i, l_j, l_k)|i < j < k]$ where $n = 1, 2, 3, \ldots, m$ and $m = {}^{23}C_3$ are number of triangles in one image frame as shown in Fig. 1b . After generating all triangles, we have computed positions $C_i(\alpha_1, \beta_1)$ and $I_i(\alpha_2, \beta_2)$ of circumcenter and incenter, respectively, for every triangle T_i as shown in Fig. 1c where (α, β) is geometric position corresponding to horizontal-axis and vertical-axis, respectively.

Finally, we computed gradient g_i of these two center points C and I by using following formulas

$$g_i = fn(C_i, I_i) \quad (1)$$

where

$$fn(C_i, I_i) = \frac{\beta_2 - \beta_1}{\alpha_2 - \alpha_1}$$

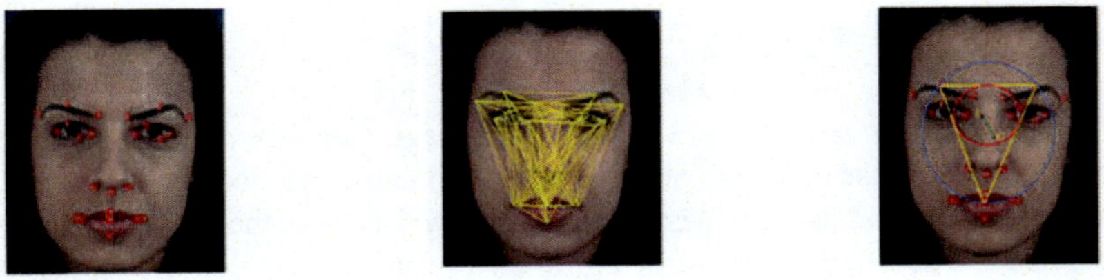

(a) landmarks (b) Traingles (c) circumcenter-incenter

Fig. 1 Gradient formation from landmarks

In this way, gradient signature matrix γ is calculated by given formula

$$\gamma = \sum_{f=1}^{k} \sum_{t=1}^{m} g_{ft} \quad (2)$$

k is total number of frames in a sequence, and m is total number of triangle in each frame.

2.3 *MLP for Classification*

Different types of classifiers are conducted for recognition of human facial expression and those are observed in [6]. Among them, we have selected multilayer perceptron for classification task of human emotion. Basically, it is an artificial neural network containing one input layer, one hidden layer, and one output layer. We have considered gradient feature as feature vector for each frame in a sequence, and by concatenating those feature vectors into single vector that representing one, image sequence is fed into input layer of MLP. We used ten hidden layer units $u = 1, 2, 3, 4, \ldots, 10$ which computes outputs of their activation function $tanh$ [3] by taking input signal coming from input layer and passes its output signals to the output layer. After computing final output, y_k it is compared with target output t_k and generating the error. Network uses scaled conjugate gradient backpropagation as training function and trained by minimizing mean square error (MSE) [4] $\epsilon_k = \frac{1}{2}\sum(t_k - y_k)^2$ for classification. Network stopped the training process when error is going down below the threshold value $\theta > \epsilon_k$, and network performance is evaluated by considering learning rate η. During the training period, weight is adjusted by depending on backpropagation error. We have trained our network $fifty$ times to achieve best performance in recognition.

Algorithm 1: Training Algorithm

Input: A $m \times n$ human face image sequence
Output: Recognition of emotion with maximum accuracy
1. AAM fitting model is used to detect landmarks on face image frame
2. Crucial landmarks extracted from major parts of face
3. Formulate possible triangle from landmarks
4. Find circumcenter and incenter points for every triangle
5. Calculate gradient signature as feature vector for each frame in sequence
6. Concatenate all feature vectors into single vector
7. Single vector for all sequences are fed into MLP
8. Distinguish the sequences into expressions

3 Experiment and Results

Our gradient signature-based method is illustrated with benchmark image sequence databases Cohn-Kanade (CK+) [8], MMI [15], and MUG [1]. Cohn-Kanade database contains 327 image sequences with seven emotion labels and corresponding 68 landmarks. Emotions are anger, contempt, disgust, fear, happiness, sadness, and surprise. MMI database contains 201 image sequences with six basic expressions. We have generated landmarks from image sequences by using AAM model. MUG databases containing 801 image sequences, and here also, landmarks are generated. In both databases, MMI and MUG, contempt emotion label is missing. We have divided our all image sequence databases into three groups as 70% of total sequences are taken for training, 15% are considered for validation, and 15% are used for testing purpose. We are considering ten frames of every sequence into the account for experimental purpose, and dynamic behavior of each individual expression category over the frames is evaluated. Figure 2 shows that how anger expression is changing over the

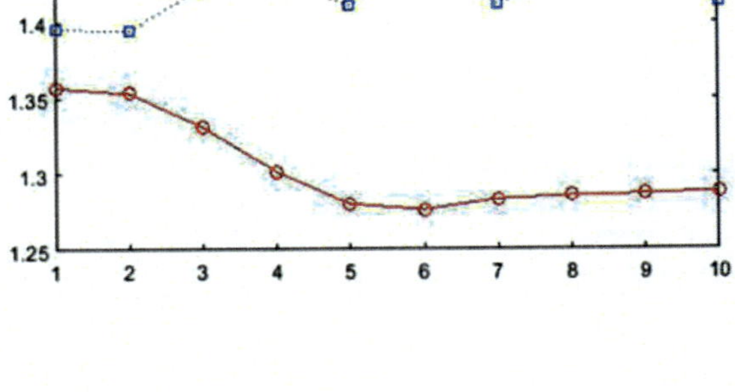

Fig. 2 Gradient values versus frames

Table 1 Confusion matrix on CK+ database

	AN	CON	DI	FE	HA	SA	SU
AN	43	0	2	0	0	0	0
CON	0	18	0	0	0	0	0
DI	1	0	57	0	1	0	0
FE	2	0	0	22	1	0	0
HA	0	0	0	0	69	0	0
SA	1	0	1	0	0	26	0
SU	1	0	0	0	0	0	82

frames under different image sequence databases captured by gradient signature-based features. Details of experimental results for CK+, MMI, and MUG databases are described below.

3.1 Results on CK+ Database

Our proposed gradient feature-based technique has been performed on CK+ sequence database with reasonable recognition rate to identify sequences into specific classes. Table 1 shows the confusion matrix of CK+ database which is obtained by using MLP classifier. Seven basic expressions are denoted by following symbols in the confusion matrix, anger (AN), contempt (CON), disgust(DI), fear (FE), happiness (HA), sadness (SA), and surprise (SU). Anger is identified in 43 sequences but misclassified with 2 disgust sequences. Similarly, contempt and happiness is classified into 18 sequences and 69 sequences, respectively, they are recognized perfectly. Disgust is identified into 57 sequences, but it is misclassified with 1 anger and 1 happiness. Fear is recognized into 22 sequences but misclassified with 2 anger and 1 happiness. Sadness is confused with 1 anger sequence and mapped into 26 sequences properly. Surprise is classified with 82 sequences, but it is confused with 1 anger. All sequences are mapped into their specific class with average accuracy of 96.94%.

3.2 Results on MMI and MUG Database

Performance with MMI database is very challenging to us to recognize sequences into different class [11]. On the other hand, performance label of our method on MUG database is showing very high and giving promising results in recognition comparing with other two benchmark Databases. Table 2 shows the confusion matrices of MMI and MUG databases. Average recognition rate on MMI is shown 94.05%, and in MUG database, our method achieved an average recognition rate 98.5%.

Table 2 Confusion matrix on MMI and MUG databases

	AN	DI	FE	HA	SA	SU
(a) Confusion matrix on MMI database						
AN	30	0	1	0	0	0
DI	2	29	0	1	0	0
FE	3	0	24	0	1	0
HA	1	1	0	40	0	0
SA	1	0	0	0	27	0
SU	0	0	0	0	1	40
(b) Confusion matrix on MUG database						
AN	146	1	0	2	0	0
DI	0	114	0	0	0	3
FE	0	0	149	0	0	1
HA	4	1	0	102	0	0
SA	0	0	0	0	133	0
SU	0	0	0	0	0	145

4 Conclusions

This paper focused on the effective feature extraction [14] method to signify human emotion and extricate different expressions from one another by using gradient signature induced from landmarks on face image. Furthermore, it has clarified our proposed method by comparing results obtained from various sequence databases CK+, MMI, and MUG. Figure 3 shows the comparisons among training, validation, testing,

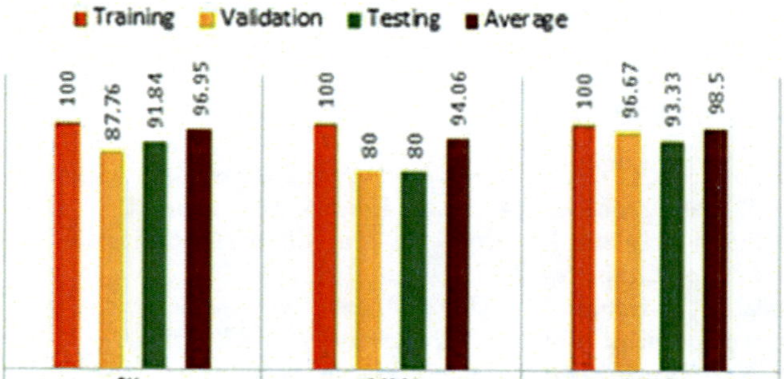

	CK+	MMI	MUG
Training	100	100	100
Validation	87.76	80	96.67
Testing	91.84	80	93.33
Average	96.95	94.06	98.5

Fig. 3 Comparison in different recognition results

and average recognition performances of the proposed method on three benchmark databases. The experimental outcomes encourage us to compare further our method with the state-of-the-art method.

Acknowledgements The authors like to avail this opportunity to express their gratitude to Dr. A. Delopoulos for providing MUG database and Prof. Maja Pantic for providing MMI database free for carrying out this work. The authors would also like to thank the Department of Computer and System Sciences, Visva-Bharati, Santiniketan, for the infrastructure support. The authors gratefully acknowledge the support of DST-INSPIRE Fellowship (INSPIRE Reg. no. IF160285, Ref. No.: DST/INSPIRE Fellowship/[IF160285]) for pursuing Doctoral Research in Department of Science and Technology, Ministry of Science and Technology, Government of India.

References

1. Aifanti, N., Papachristou, C., & Delopoulos, A. (2010). The mug facial expression database. In *11th International Workshop on Image Analysis for Multimedia Interactive Services WIAMIS 10* (pp. 1–4). New York: IEEE.
2. Akagi, Y., & Kawasaki, H. (2014). A method of micro facial expression recognition based on dense facial motion data.
3. Barman, A., & Dutta, P. (2019). Facial expression recognition using distance and texture signature relevant features. *Applied Soft Computing*.
4. Chakrabarti, D., & Dutta, D. (2013). Facial expression recognition using eigenspaces. *Procedia Technology, 10*, 755–761.
5. Cootes, T. F., Edwards, G. J., & Taylor, C. J. (2001). Active appearance models. *IEEE Transactions on Pattern Analysis & Machine Intelligence, 6*, 681–685.
6. Ding, Y., Zhao, Q., Li, B., & Yuan, X. (2017). Facial expression recognition from image sequence based on LBP and Taylor expansion. *IEEE Access, 5*, 19409–19419.
7. Ekman, P., & Friesen, W. V. (1971). Constants across cultures in the face and emotion. *Journal of Personality and Social Psychology, 17*(2), 124.
8. Lucey, P., Cohn, J. F., Kanade, T., Saragih, J., Ambadar, Z., & Matthews, I. (2010). The extended Cohn-Kanade dataset (ck+): A complete dataset for action unit and emotion-specified expression. In *2010 IEEE Computer Society Conference on Computer Vision and Pattern Recognition-Workshops* (pp. 94–101). New York: IEEE.
9. Majumder, A., Behera, L., & Subramanian, V. K. (2018). Automatic facial expression recognition system using deep network-based data fusion. *IEEE Transactions on Cybernetics, 48*(1), 103–114.
10. Mehrabian, A. (2007). *Nonverbal communication*. New Brunswick.
11. Neoh, S. C., Zhang, L., Mistry, K., Hossain, M. A., Lim, C. P., Aslam, N., et al. (2015). Intelligent facial emotion recognition using a layered encoding cascade optimization model. *Applied Soft Computing, 34*, 72–93.
12. Samal, A., & Iyengar, P. A. (1992). Automatic recognition and analysis of human faces and facial expressions: A survey. *Pattern Recognition, 25*(1), 65–77.
13. Tie, Y., & Guan, L. (2013). Automatic landmark point detection and tracking for human facial expressions. *EURASIP Journal on Image and Video Processing, 2013*(1), 8.
14. Uddin, M. Z., Hassan, M. M., Almogren, A., Alamri, A., Alrubaian, M., & Fortino, G. (2017). Facial expression recognition utilizing local direction-based robust features and deep belief network. *IEEE Access, 5*, 4525–4536.

15. Valstar, M., & Pantic, M. (2010). Induced disgust, happiness and surprise: An addition to the mmi facial expression database. In *Proceedings of the 3rd International Workshop on EMOTION (satellite of LREC): Corpora for Research on Emotion and Affect* (p. 65).
16. Zhang, L., & Tjondronegoro, D. (2011). Facial expression recognition using facial movement features. *IEEE Transactions on Affective Computing*, 2(4), 219–229.

A Novel Fusion of Deep Learning and Android Application for Real-Time Mango Fruits Disease Detection

Vani Ashok and D. S. Vinod

Abstract The use of computer vision in agriculture has become popular due to the fact that it provides necessary information in real time. The recent advances in computer vision made possible by deep learning combined with increasing infiltration of smartphones have paved way for smartphone-assisted crop disease diagnosis. The proposed system uses a pre-trained model of deep convolutional neural network (CNN) architecture, to retrain on images of mango fruits with the goal of identifying diseases (or absence thereof) and also on images that the model had not seen before. The system retrains the model on the principle of transfer learning (TL) using a dataset of diseased and healthy mango images acquired under controlled conditions. The system also incorporates functionality of listing out the causes and certain common symptoms of the diseases. Moreover, the proposed system embeds the prediction model within the smartphone application so as to completely eliminate the necessity of internet access for mango disease detection, i.e., the system achieves the disease diagnosis offline. The proposed approach of using TL on deep CNN for mango fruits disease detection and classification has yielded a high level of accuracy. The maximum achieved training accuracy is 98.6%, and maximum validation accuracy is 96.4%.

Keywords Deep learning · Transfer learning · Convolutional neural network · MobileNets · Validation

V. Ashok (✉)
Computer Science and Engineering, Sri Jayachamarajendra College of Engineering, Mysuru, Karnataka, India
e-mail: vanisj@sjce.ac.in

D. S. Vinod
Information Science and Engineering, Sri Jayachamarajendra College of Engineering, Mysuru, Karnataka, India
e-mail: dsvinod@daad-alumni.de

S. C. Satapathy et al. (eds.), *Intelligent System Design*, Advances in Intelligent Systems and Computing 1171, https://doi.org/10.1007/978-981-15-5400-1_74

1 Introduction

Fruit disease identification is one of the most basic and important activities in agriculture. In most cases, it is performed manually, either visually or by microscopy. The problem with visual assessment is that, being a subjective task, it is time-consuming, laborious, and error-prone. The vast majority of automatic disease detection methods proposed so far rely on digital images, which allow the use of very fast techniques. However, some factors associated with the image taken and the disease to be identified have some negative impact on the automatic detection methods making them error-prone and hence pose a challenge to the accuracy of these methods [1].

The traditional computer vision based crop disease detection techniques have employed traditional handcrafted feature based methods which require complex preprocessing, features extraction, classification, and validation. Various image processing methods such as back-propagation neural network (BPNN), k-nearest neighbor (kNN), Mahalanobis distance classifier, and support vector machine (SVM) have been used to identify and classify fungal disease symptoms on four types of horticulture products; fruits, vegetables, cereals, and commercial crops, with accuracies ranging from 76.6% to 94.08% [2]. A smart farming tool was proposed to identify diseases in pomegranate by extracting color, morphological features and then classifying the fruits using SVM [3]. SVM classifier was also used to perform grading of tomatoes into infected and uninfected classes with an accuracy of 92% based on color and texture features [4]. Automated disease detection models have been proposed for apple and mango fruits using probabilistic neural network and artificial neural network [5, 6]. The models were able to achieve disease detection in apple and mango fruits with an accuracy of 86.52% and 90.2%, respectively.

Even though deep learning is lot more popular than any other machine learning algorithms, it is less associated in the field of fruit classification [7]. CNN architecture with SVM classifier at the last layer was used for image classification in MNSIT and fashion MNSIT dataset with 99.04% and 90.72% accuracy, respectively [8]. Comparisons between various deep-meta-architectures such as faster region-based CNN (faster R-CNN), single shot detector (SSD), and region-based fully convolutional network (R-FCN) have been performed to demonstrate how a tool built using deep learning can successfully recognize nine different categories of diseases and pests in tomatoes [9].

The performance of traditional computer vision techniques depends heavily on image classification based preprocessing techniques, extraction of hand-engineered features, and a series of other complex methodologies which are time consuming and require high computational power [10]. The popularity of deep learning in computer vision field has given way for the design of systems that provide end-to-end supervised training and testing even for image classification problems with a very large number of classes, beating the traditional approaches of using hand-engineered features by a substantial margin in standard benchmarks [11].

In this context, we have taken up the challenge to automate the mango fruit disease detection using CNN and deep learning algorithms. In our approach, we focus on the identification and recognition of some common diseases that affect mango fruits. Mangoes being the fruits with high export value, a real-time, non-destructive, cost-effective and a simple quality assessment tool is the need of the hour for the growers and agronomists to identify the diseases in the cultivation fields. Mangoes are affected by various disorders at various stages of growth [12]. These disorders may be inherent or induced due to post-harvest treatments. Since the core application of the proposed work lies in helping farmers in rural areas to identify the diseases that affect mangoes, the proposed system intends to provide a simple solution by integrating the disease detection process with an android application, so as to increase its performance and to make it work even in regions of low internet connectivity.

1.1 Convolutional Neural Network (CNN/ConvNet)

A simple ConvNet is a sequence of layers, and every layer of a ConvNet transforms one volume of activations to another through a differentiable function. Three main types of layers are used to build ConvNet architectures: convolutional layer (CONV), pooling layer (POOL), and fully connected layer (FC) (same as the one seen in regular Neural Networks). These three layers are stacked upon one another in a particular fashion to form a full ConvNet architecture.

Some layers of ConvNet require parameters, and others don't. For example, each CONV/FC layer is specified by weights and the outputs are submitted to an activation function. On the other hand, the RELU/POOL layers do not require any parameter learning. The parameters in the CONV/FC layers will be trained with gradient descent so that the class scores that the ConvNet computes are consistent with the labels in the training set for each image.

1.2 Transfer Learning and MobileNet

Transfer learning or inductive transfer is a research problem in machine learning that focuses on storing knowledge gained while solving one problem and applying it to a different but related problem [13]. MobileNets are a new family of convolutional neural networks that are extremely small and computationally efficient [14]. Generally, when an image has to be classified, that image is sent to a Web service, classified on a remote server, and the result is sent back to smartphone. This can be avoided by using light weight platforms and learning models specifically designed for mobile devices in recent years. The MobileNets are developed in such a way so as to shrink the network complexity required for computer vision tasks and perform accurately even with limited computational power available within a smartphone.

In the proposed model, we present a system that uses TL technique for retraining a deep learning CNN model, MobileNetV1. Later, this retrained model is used to detect and diagnose mango fruit diseases through a smartphone application. The android application developed can be used as a smart crop assistant on smart phones, i.e., crop advisory application for farmers, extension workers, and agronomists across the world. It can diagnose fruit diseases, pests, and nutrient deficiencies affecting crops. It is also smart enough to suggest some common cures that can be applied as soon as the disease is detected. This smartphone application provides a practical real-time fruit disease detection solution for the fruit images captured by different camera devices irrespective of various resolutions, illumination conditions, background and object size.

The proposed system consists of different modules which cannot be implemented as a whole because they are structured under different languages and different architectures. Hence, the TL module of the system is implemented on TensorFlow using python and the disease detection module of the system is implemented as an android application using Java on Android. A database consisting of disease names, causes, and symptoms of the disease are stored on a SQLite database within the android application. A graph file, consisting of the TL CNN model with trained weights and biases, and a labels file, which consists of all the class labels of the fruit dataset, are saved on to the disk from the TL module. After successful training and saving of the output files, these are used by the android application which hosts the disease detection module to identify and classify the mango fruit disease. With the help of the integrated database, other details like symptoms, agent which causes the disease are fetched and provided to the user, if a disease is detected in the fruit.

2 Materials and Methods

All experiments in this study were conducted on computer system with Intel Xeon® CPU E3-1240 V3@3.40 GHz, 16 GB of DDR3 RAM, and NVIDIA Quadro K600 4 GB DDR3 GPU.

The proposed model for automated mango fruit disease detection was implemented using deep learning CNN model, MobileNetV1, from Google TensorFlow. It uses TL technique for retraining a deep CNN. Later, this retrained model is used to detect and diagnose mango fruit diseases through a smartphone application. The overall system follows four-step schema, as shown in Fig. 1.

2.1 *Mango Image Dataset*

For the proposed work, an image dataset was prepared with the 'Alphonso' cultivar of mango. The mangoes were picked from a grove near Mysuru, Karnataka, from two batches of harvests, and based on the opinion of experts, they were divided into *Non-Defective* and *Defective* classes. Defective class consists of mangoes with physical and physiological defects induced from pre and post-harvest methods as well as diseased fruits, whereas non-defective class contains normal-sized, perfectly

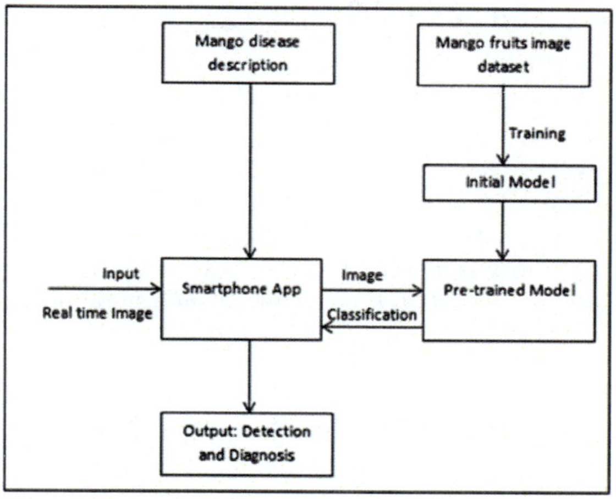

Fig. 1 Deep learning model integrated with mobile application

ripened, disease-free fruits. A third class of C*hilling Injured* mangoes was created by storing a batch of mango fruits at 4–5 °C for about 25 days. In total, the dataset contains defective, non-defective, and chilling injury classes of mango fruits with 621, 840, and 669 samples, respectively. The whole of the image dataset is split into training, validation, testing, and post-retrain random testing dataset with a ratio of ~79:10:10: ~1. The sample fruit images in the three classes are shown in Fig. 2.

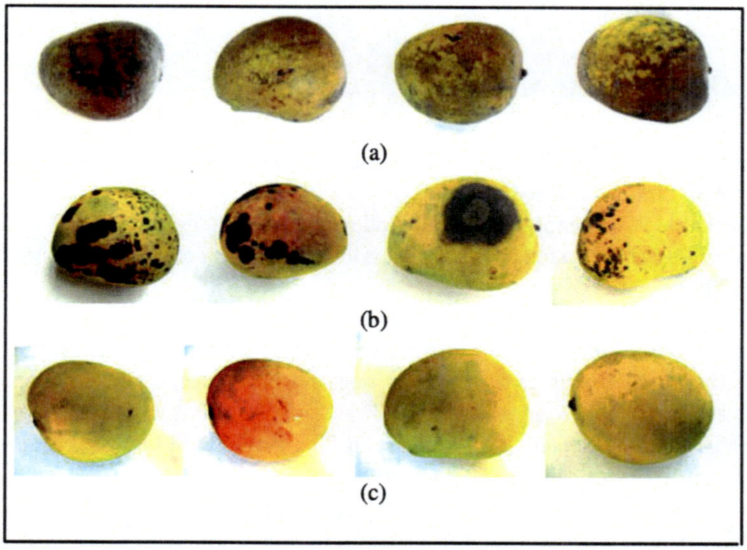

Fig. 2 Sample fruit images for each of the three categorical classes. **a** Chilling injured **b** Defective **c** Non-defective [6]

2.2 Initial Model Training

Training the deep learning neural network is the actual learning step of the algorithm. Ideally, all the available data in the form of images are provided to the system in the training phase. To evaluate the model and to have a feedback about the performance, a separate validation set which is not a part of the training phase is used. After completion of the training phase, it is necessary to estimate how well the system will perform with input data that was not used during the training phase. For this purpose, another separate part of the dataset called testing dataset is used.

2.3 Pre-trained Model

In practice, deep learning networks are not usually trained from scratch. This is because it is relatively rare to have a dataset of sufficient size that is required for complex tasks such as image classification. Instead, it is common to pre-train a network on a very large dataset and then use it as an initialization.

In the proposed system, MobileNet model which is pre-trained on the *ImageNet* dataset containing 1 million images spread over 1000 classes is used and it is retrained for the new classes. Once the training process with batch size of 100 is done, the model is tested for some random mango images. Then, the pre-trained model is frozen and deployed onto a smartphone with the help of TensorFlow Mobile.

2.4 Deploying the Model on a Smartphone

- Freeze the model: A frozen version of the retrained model is created by serializing the graph into a protobuf (.pb file).
- Check nodes: The input and output nodes of the frozen graph are used as starting and ending points in the process of classification. The node names are used within the TF ImageClassifier class.
- Model Optimization: Some of the operations within the model which are adjacent tend to normalize and de-normalize a value leaving it in the original state. Removing such adjacent operations does no harm to the performance of the system.
- Gradle build inclusion: The gradle build files available in Google's TensorFlow are to be included within the application.
- API creation: A custom classification API has to be created in the application, which reads the model and processes an image to predict a class label for the image.
- Compilation and Run: After all the above mentioned operations are carried out, the application is compiled and installed on a smartphone.

3 Results and Discussion

The first thing to focus on when analyzing how well a prediction model works is to look at its accuracy, be it while training, validation, or testing. The proposed approach of using TL on deep CNN for mango fruit disease detection and classification has yielded a high accuracy score and low model loss score. The graph shown in Fig. 3 which is plotted with the help of TensorFlow utility functions and visualized with help of TensorBoard, depicts training (orange line) and validation (blue line) accuracies over a course of 10,000 steps. The maximum achieved training accuracy is 98.6%, and maximum validation accuracy is 96.4%.

With neural networks apart from accuracy, the other important performance measurement in prediction or classification is the cost or loss which is to be minimized. Loss can be simply thought of as the difference between the ground truth of an instance and the value predicted by the model. The model has been successful in minimizing the cross entropy loss by using the gradient descent optimizer. The minimum achieved train loss (orange line) value is 0.0575 and that of validation is 0.1912. The graph shown in Fig. 4 depicts the loss value decreasing over a course of 10,000 steps.

The real-time disease detection capability of the proposed model is evaluated on mango fruit images captured in real-time and also with pre-captured images. The snapshot of the smartphone application shown in Fig. 5 is one of the instances for the image captured in real-time. More than 95% of the images were classified correctly into their respective classes with significant accuracy, and the details of the identified disease were displayed in an interface. The snapshot shown in Fig. 6 is an instance

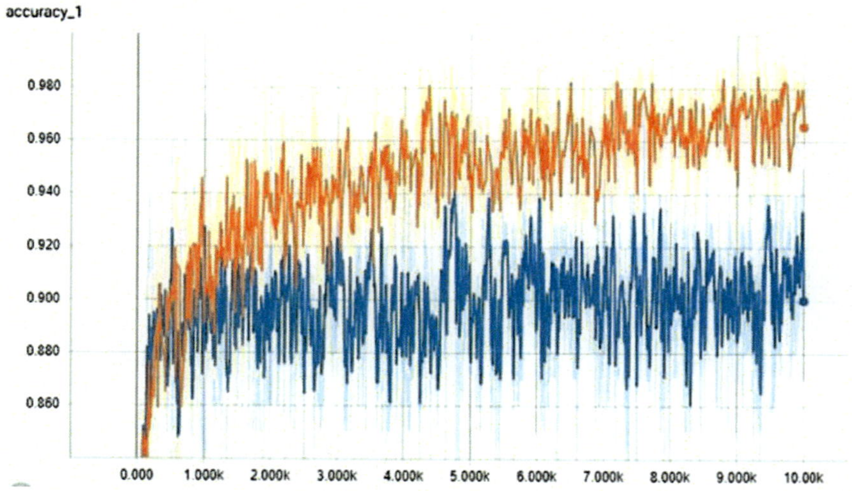

Fig. 3 Accuracy of MobileNetV1 model as a function of iterations

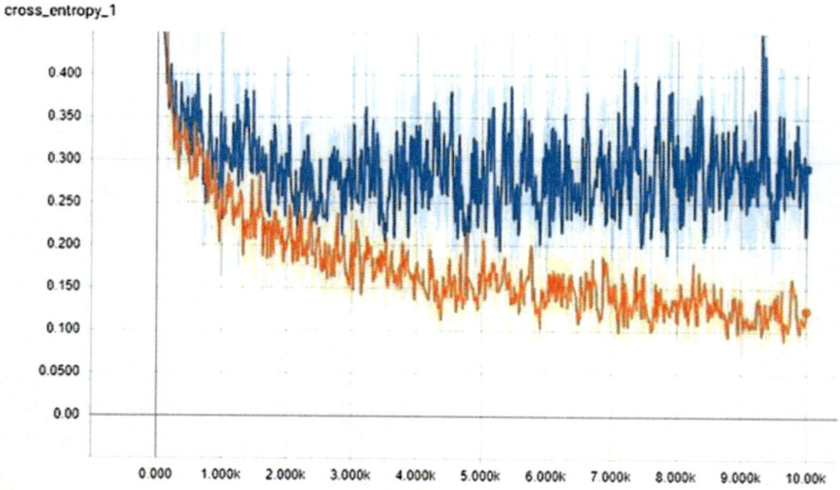

Fig. 4 Loss of MobileNetV1 model as a function of iterations

Fig. 5 Screenshot of real-time image test

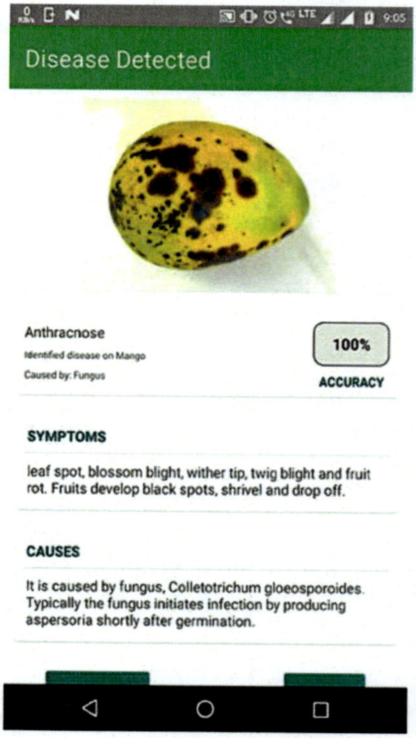

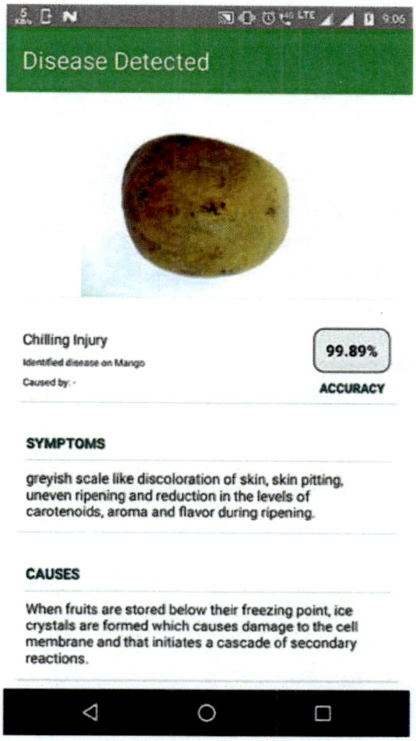

Fig. 6 Screenshot of precaptured image test

where the application is tested with pre-captured mango fruit images with chilling injury [12].

In instances where there are more than 2 classes with unequal number of samples in each class, the performance of the learning algorithm can be summarized by a confusion matrix. The confusion matrix for the proposed model is abridged in Table 1. The primary diagonal elements of the matrix indicate True Positives i.e., correct predictions and the remaining elements indicate False Positives or mis-predictions.

Table 1 Confusion matrix of the proposed model

Class		Prediction			Total
		Chilling injured	Defective	Non-defective	
Ground truth	Chilling injured	659	10	0	669
	Defective	6	601	14	621
	Non-defective	0	10	830	840

4 Conclusions

Using the deep convolutional neural network architecture and with the help of TL technique, we trained the proposed model on images of mango fruits with the goal of classifying and identifying the diseases on images that the model had not seen before with an accuracy of 96.4%. Importantly, while the training of the model takes a lot of time (14–15 minutes on a high performance GPU cluster computer), the classification itself is very fast (less than a second on a CPU) and thus was easily implemented on a smartphone. The proposed system embeds the prediction model within the smartphone application and completely eliminates the necessity of internet access for mango disease detection. It can be extended to incorporate several new features like geo-tagging and remote alerts. Our overall aim is to decentralize artificial intelligence and make use of it for disease detection of global goods like fruits.

Declaration We have taken permission from competent authorities to use the images/data as given in the paper. In case of any dispute in the future, we shall be wholly responsible.

References

1. Barbedo, J. G. A. (2016). A review on the main challenges in automatic plant disease identification based on visible range images. *Biosystems Engineering, 144,* 52–60.
2. Pujari, J. D., Yakkundimath, R., & Byadgi, A. S. (2015). Image processing based detection of fungal diseases in plants. *Procedia Computer Science, 46,* 1802–1808.
3. Bhange, M., & Hingoliwala, H. A. (2015). Smart farming: Pomegranate disease detection using image processing. *Procedia Computer Science, 58,* 280–288.
4. Semary, N. A., Tharwat, A., Elhariri, E., & Hassanien, A. E. (2015). Fruit-based tomato grading system using features fusion and support vector machine. In *Intelligent Systems' 2014* (pp. 401–410). Cham: Springer.
5. Ashok, V., & Vinod, D. S. (2014) Automatic quality evaluation of fruits using probabilistic neural network approach. In *2014 International Conference on Contemporary Computing and Informatics (IC3I)* (pp. 308–311). IEEE.
6. Ashok, V., & Vinod, D. S. (2016, August). A comparative study of feature extraction methods in defect classification of mangoes using neural network. In *2016 Second International Conference on Cognitive Computing and Information Processing (CCIP)* (pp. 1–6). IEEE.
7. Naik, S., & Patel, B. (2017). Machine vision based fruit classification and grading-a review. *International Journal of Computer Applications, 170*(9), 22–34.
8. Agarap, A. F. (2017). An architecture combining convolutional neural network (CNN) and support vector machine (SVM) for image classification. *arXiv preprint* arXiv:1712.03541.
9. Fuentes, A., Yoon, S., Kim, S., & Park, D. (2017). A robust deep-learning-based detector for real-time tomato plant diseases and pests recognition. *Sensors, 17*(9), 2022.
10. Singh, A., Ganapathysubramanian, B., Singh, A. K., & Sarkar, S. (2016). Machine learning for high-throughput stress phenotyping in plants. *Trends in Plant Science, 21*(2), 110–124.
11. Krizhevsky, A., Sutskever, I., & Hinton, G. E. (2012). Imagenet classification with deep convolutional neural networks. In *Advances in neural information processing systems* (pp. 1097–1105).
12. Jha, S. N., Narsaiah, K., Sharma, A. D., Singh, M., Bansal, S., & Kumar, R. (2010). Quality parameters of mango and potential of non-destructive techniques for their measurement—a review. *Journal of Food Science and Technology, 47*(1), 1–14.

13. Torrey, L., & Shavlik, J. (2010). Transfer learning. *Handbook of research on machine learning applications and trends: algorithms, methods, and techniques* (pp. 242–264). IGI Global.
14. Howard, A. G., Zhu, M., Chen, B., Kalenichenko, D., Wang, W., Weyand, T., … & Mobilenets, H. A. (2017). Efficient convolutional neural networks for mobile vision applications. *arXiv preprint* arXiv:1704.04861.

Time Series Forecasting to Predict Pollutants of Air, Water and Noise Using Deep Learning Methods

Nimit Jain, Siddharth Singh, Naman Datta, and Suma Dawn

Abstract This paper analyzes the long-term air, water and noise pollution monitoring data using autoregressive integrated moving averages (ARIMA) modeling and an artificial neural network called long short-term memory network (LSTM). Different features of air, water and noise are predicted using these modeling techniques. This is known as time series forecasting. The accuracies of both the modeling techniques are compared with each other applied on the same datasets. We compare both the modeling techniques, and the models are evaluated with metric root-mean-squared error (RMSE) which is also called the objective function or loss function in ML lingo. The algorithm, equipped with prominent accuracy, is used to predict the future values of different features of air, water and noise. These predicted values are then compared with the health statistics, globally defined by the World Health Organization (WHO), to determine a fit and healthy life of an average human being. The harmful effects, possible disease threats and their precautions are also listed if the predicted values do not lie in the risk-free zone defined by the WHO. Most of the air, water and noise dataset is provided by the Central Pollution Control Board, Govt of India (CPCB) online. We will be using Python to execute this project.

Keywords Pollution · Time series · LSTM · ARIMA · RMSE

N. Jain · S. Singh (✉) · N. Datta · S. Dawn
Department of CSE/IT, Jaypee Institute of Information Technology, Noida, India
e-mail: siddharthprogamer@gmail.com

N. Jain
e-mail: j.nimit99@gmail.com

N. Datta
e-mail: dattanaman213@gmail.com

S. Dawn
e-mail: suma.dawn@gmail.com

S. C. Satapathy et al. (eds.), *Intelligent System Design*, Advances in Intelligent Systems and Computing 1171, https://doi.org/10.1007/978-981-15-5400-1_75

1 Introduction

A polluted environment is truly democratic; it cuts across all caste and class barriers. As the population increases, our country faces different forms of pollution which create challenging environment concerns and abrupt climate changes. India has a huge population and with it a huge pollution level too. Pollution in India includes air pollution, water pollution, noise pollution, garbage pollution, industrial pollution, etc. Due to these factors, the natural environment is severely damaged which majorly affects the health of people living in a polluted surrounding. The health impacts of air, water and noise pollution are extremely severe, sometimes beyond the understanding of the common man. Responsible citizens should take an active interest in keeping their own environment clean, for themselves and all the future generations to come. Air pollution represents the largest environmental risk in India. Over 80% of the population is exposed to annual levels of outdoor fine particulate matter that are above the specified levels in the World Health Organization guidelines. This exposition accounted for a lot of premature deaths [1].

In 2010, India's central pollution control board did a research in this field. The world quality PM10 targets set by WHO are violated by most Indian cities. This is not ideal for a country moving toward being a developed nation. Noise pollution in India can be owed to the increase in the number vehicles, population, and the consequent road congestion and traffic. The noise levels on weekdays and weekends, tested in different settings, were way above the permissible standards. Continued exposure to such levels can harm the auditory system of the people. A noise-induced hearing loss is also a serious problem that should not be ignored. Problems like hypertension, auditory morbidity and increased blood pressure due to noise are easily preventable with just control measures as simple as banning pressure horns and some vegetation plantation. Water pollution is one of the major issues facing the country right now. Virtually all human activities produce some kind of environmental disturbance that contaminates surrounding waters [2].

The air monitoring data analyzed in this present study is reported from Central Pollution Control board (CPCB) air monitoring station situated in a commercial zone of Delhi. These stations have air monitoring equipment which consists of distinct types of sensors to measure the different components that are present in the air. These sensors work on the principle of measuring the attenuation of infrared radiation in the air. The sensors consist of an infrared radiation source, a light-water pipe and the infrared detector with the appropriate filter. Under the current measurement of air quality, 8 parameters are calculated for air quality measurement, namely the pollutants being PM10, PM2.5, NO_2, SO_2, CO, O_3, NH_3 and Pb in which one of PM10 or PM2.5 parameter is mandatory. There are six air quality categories, namely good, satisfactory, moderately polluted, poor, very poor and severe; major part of the water monitoring data analyzed in this research was acquired from the CPCB water monitoring databases. Water monitoring data is inefficiently uploaded on the CPCB site and is not checked nor updated regularly, thus causing some anomalies in the dataset. These anomalies are present in the form of high spikes in the data, wrong/out of the range values which hinder the functioning of the modeling techniques applied in this study. Therefore, anomaly detection using Gaussian distribution was used while preprocessing the data to remove such irregular values. This positively shaped our data for the algorithm, thus increasing the accuracy while predicting the future values of the selected features.

The noise monitoring data analyzed in this research is reported from the CPCB site which provides the noise data in a systematic and a regular manner. It provides data in both automatic and manual monitoring types. The noise monitoring data used for this study is reported from a noise monitoring station situated in a commercial zone in Delhi. The daily noise level is acquired from an operating remote terminal sound-level meter consisting of an outdoor microphone which is then communicated to a central station through an integrated GPRS modem. Noise monitoring data for this study is acquired in an hourly format for 3 months (August to October) of 2018, value being in decibels (dB).

For this study, we chose water data from the year 2014–2015 on weekly basis having features, namely pH, conductivity, bio-chemical oxygen demand (BOD), fecal coliform, total coliform and nitrate (N). Further, we also chose five parameters, namely PM10, PM2.5, NO_2, CO, O_3 and SO_2, in an hourly order fashion for 3 months (June to August) of 2018. Missing values of certain features were replaced by the mean value of that particular feature for the whole dataset. For air, we picked up parameters PM2.5 and PM10. For different values of PM2.5 and PM10, different health quality measures and some precautionary steps are mentioned. Similarly, for water we took prime features: BOD and pH to determine the health quality of the person exposed to such conditions. For noise, standard decibel (dB) levels that are approved by the WHO are used to determine the health quality of people exposed to different noise levels.

1.1 Related Works

As the technology advancements are growing immensely day-by-day, they are also constantly evolving all the industries together with it. In public health and environmental health, technology is increasingly playing a role in almost all processes, from patient registration to data monitoring, from weather reports to predicting of environment quality. Machine learning is one such branch that is now being researched and is being incorporated in the field of health care.

For an instance, a multivariate time series model was designed by MaryzehGhassemi et al. [3] for assessment of severity of illness and forecasting ICU with sparse and heterogeneous clinical data. They evaluated the use of multivariate time series models with the multi-task Gaussian process (MTGP) models using noisy, incomplete, sparse, heterogeneous and unevenly sampled clinical data including physiological signals and clinical notes.

Another take in this field was done by Mooney et al. [4]. They used big data technologies in the field of public health. They considered common public health research and practical uses of big data, including surveillance, hypothesis-generating research and causal inference, while exploring the role that machine learning may play in each use. Finally, they made suggestions regarding structuring teams and training to succeed in working with big data in research and practice.

Much work has also been contributed toward environmental care with the help of machine learning. Sarkar et al. [6] worked on similar grounds by evaluating the river water quality using artificial neural network. They used basic back-propagation technique in studying the water quality of the river Yamuna, one of the largest perennial rivers in India. Water data for this study was also collected from Central Water Commission, monthly data from 1990 to 1996 having features like flow discharge, temperature, pH, bio-chemical oxygen demand (BOD) and dissolved oxygen (DO). Comparison of model was based on RMSE. Similarly, Garg [5] worked on applications of auto-regressive integrated moving average (ARIMA) approach in time series prediction of traffic noise pollution. Several measures of accuracy were applied to ascertain the performance of the ARIMA models so developed such as stationary R2, R2, root-mean-square error (RMSE), mean absolute percentage error (MAPE) and normalized BIC (Bayesian information criterion).

2 Proposed Methodology

2.1 Time Series Forecasting

Time series analysis and forecasting is the field of understanding certain trends in data points collected over a period of time. Data points in a time series data are equally spaced over time. Analysis leads to extraction of meaningful characteristics about the data and helps in predicting possible future values of time series. Time series analysis has been around since the ancient ages and has earlier been used in astronomy to study the movements of planets and the sun. With the incessantly changing world, now time series has wider applications in the field of stock market analysis, business modeling and many more.

Traditionally, time series forecasting has been performed using various statistical models like autoregressive integrated moving average (ARIMA). Recently, deep learning models like recurrent neural networks (RNN) and long short-term memory (LSTM) have emerged and have gained popularity among researchers.

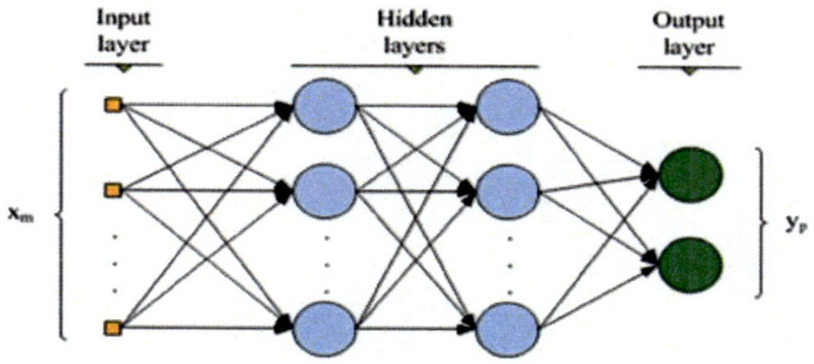

Fig. 1 Feed-forward neural network having one input layer, one or more hidden layers and one output layer. Input layer has x_m nodes, and output layer has y_p nodes

2.2 *LSTM*

Long short-term memory abbreviated as LSTM is a special case of RNN. It was first introduced by Hochreiter and Schmidhuber [8] to address the vanishing and exploding gradient problems faced by conventional recurrent neural network but gained popularity only few years back due to recent advancement in powerful processors.

2.2.1 Multi-layer Feed-Forward Neural Network

Neural network consists of layers of interconnected nodes, called neurons. Each interconnection has a specific weight, and every network has one input layer, one or more hidden layers and one output layer. The simplest form of a neural network is a feed-forward neural network in which information flows only in one direction (Fig. 1).

2.2.2 Recurrent Neural Networks

In order to handle sequential, recurrent neural network is widely used. It is able to memorize parts of input and use them to make accurate predictions. These networks are used in speech recognition, translation, and many more (Fig. 2).

2.2.3 Long Short-Term Memory

Training conventional RNNs over large datasets may lead to vanishing or exploding gradient problems. These problems limit the capability of RNNs for long-range predictions. LSTM has been designed to overcome these problems and has proven to be far better in performance than RNNs. LSTM is composed of memory units. It has a property of selectively remembering patterns for long duration of time. Memory

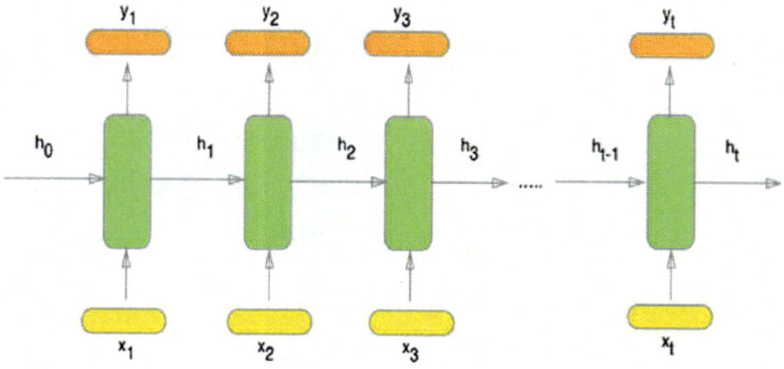

Fig. 2 x_1,x_2, x_3,..., x_t represents the inputs, y_1, y_2, y_3,..., y_t represents the outputs and h_0, h_1, h_2,... h_{t-1} holds the information for the previous inputs

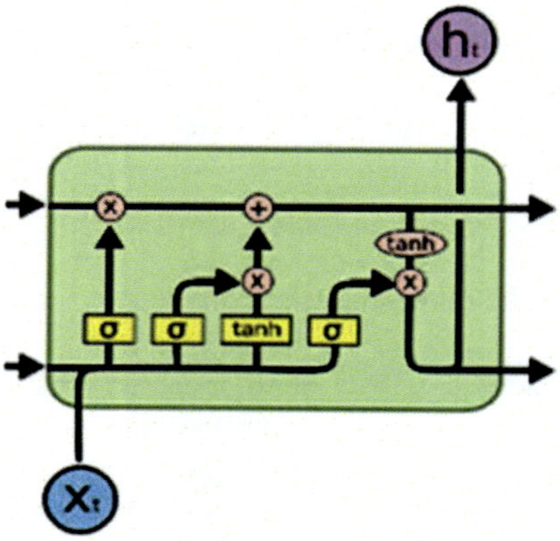

Fig. 3 Structure of a LSTM cell

units consist of three types of gates, namely input gate, forget gate and output gate (Fig. 3).

2.3 ARIMA

ARIMA also referred to as Box–Jenkins model stands for autoregressive integrated moving average and is a more generalized model of autoregressive moving average (ARMA) that incorporates both autoregressive and moving average parameters together. In ARIMA models, a non-stationary time series is made stationary by applying finite differencing of the data points.

The parameters in ARIMA (p, d, q) model denotes-

- p is the number of lag observations (i.e., lag order)
- d is the number of non-seasonal differences needed for stationary (i.e., degree of differencing applied in the time series)
- q is the number of lagged forecast errors (i.e., order of moving average).

2.4 Assessment Metric

The root-mean-square error (RMSE) is the standard deviation of the residuals (prediction errors) and is frequently used in forecasting, regression analysis to verify experimental results. It measures the difference or residuals between actual and predicted values. To find the RMSE, follow the steps:

a. Squaring the residuals
b. Finding the average of the squares of the residuals
c. Taking the square root of the result.

The formula for RMSE is as follows:

$$\text{RMSE} = \sqrt{\frac{\sum_{i=0}^{N}(x_i - p_i)^2}{N}}$$

where N is the total number of observations, x_i is the actual value, and p_i is the predicted value.

The methodology follows the given work flow model as shown in Fig. 4.

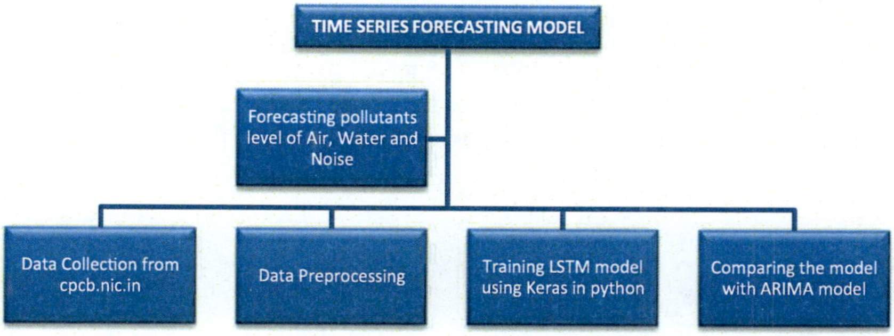

Fig. 4 Proposed model for time series forecasting

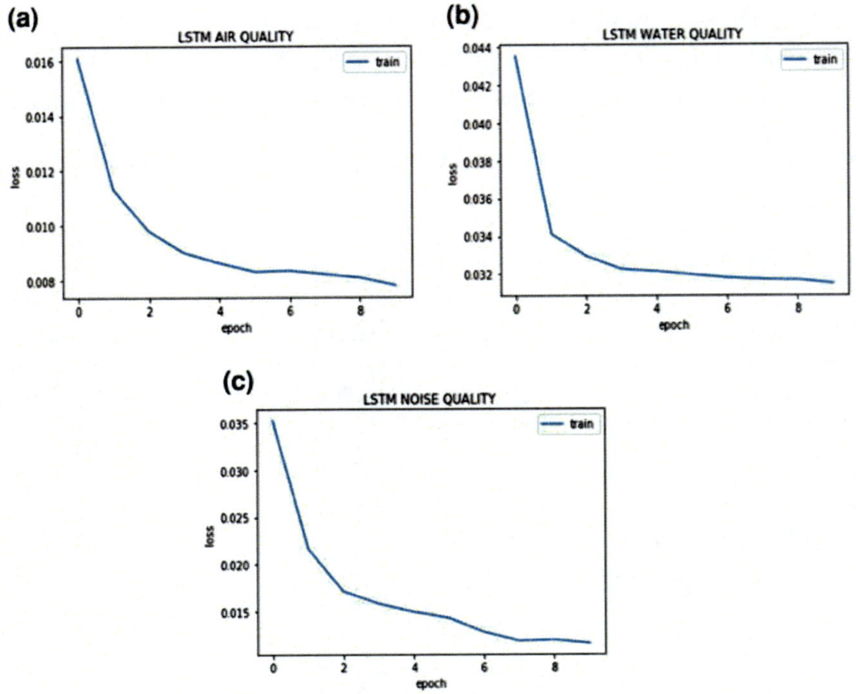

Fig. 5 **a** Training loss of LSTM model on air quality time series w.r.t number of epochs, **b** Training loss of LSTM model on water quality time series w.r.t number of epochs, **c** Training loss of LSTM model on Noise Quality time series w.r.t number of epochs

3 Results

The root-mean-squared error using LSTM model on air quality, water quality and noise quality data are 28.418, 167.758 and 2.157, respectively. RMSE of univariate noise quality data using ARIMA model is 2.430 which is certainly higher than that of LSTM model (Figs. 5, 6 and 7).

4 Conclusions

The goal of the presented work was to evaluate the effectiveness of LSTM for time series analysis for different real-time pollution data, namely air, water and noise and compare the results obtained by traditional models like ARIMA. The LSTM model shows greater reduction in error over ARIMA as seen in the results above.

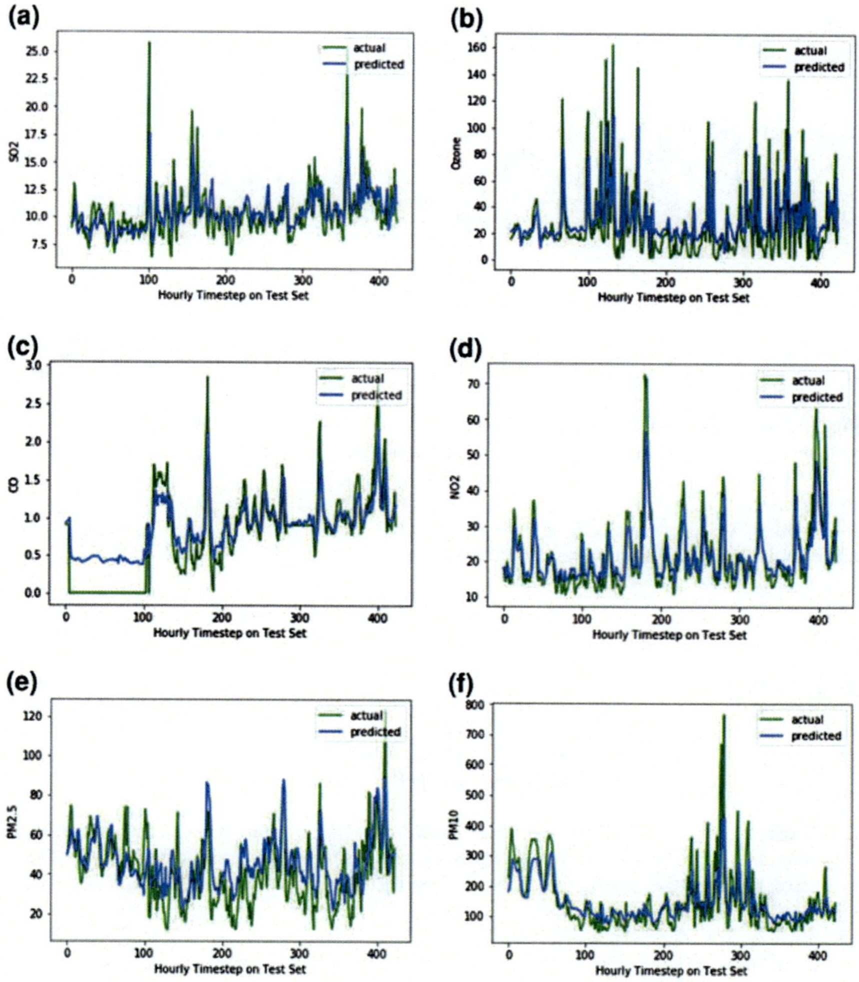

Fig. 6 **a–f** Actual and predicted value on test set using LSTM model

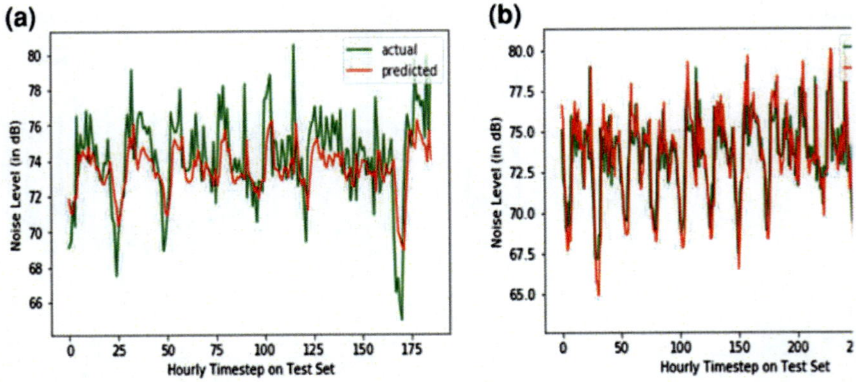

Fig. 7 **a** Actual and predicted value on noise test set using LSTM model (RMSE = 2.157), **b** Actual and predicted value on noise test set using ARIMA model (RMSE = 2.43)

Declaration We have taken permission from competent authorities to use the images/data as given in the paper. In case of any dispute in the future, we shall be wholly responsible.

References

1. https://www.who.int/news-room/fact-sheets/detail/ambient-(outdoor)-air-quality-and-health.
2. http://www.environmentalpollution.in/water-pollution/six-main-sources-of-water-pollution/142.
3. Ghassemi, M., Pimentel, M. A., Naumann, T., Brennan, T., Clifton, D. A., Szolovits, P., & Feng, M. (2015, February). A multivariate timeseries modeling approach to severity of illness assessment and forecasting in ICU with sparse, heterogeneous clinical data. In *Twenty-Ninth AAAI Conference on Artificial Intelligence*.
4. Mooney, S. J., & Pejaver, V. (2018). Big data in public health: terminology, machine learning, and privacy. *Annual Review of Public Health, 39*, 95–112.
5. Garg, N., Soni, K., Saxena, T. K., & Maji, S. (2015). Applications of Autoregressive integrated moving average (ARIMA) approach in time-series prediction of traffic noise pollution. *Noise Control Engineering Journal, 63*(2), 182–194.
6. Sarkar, A., & Pandey, P. (2015). River water quality modelling using artificial neural network technique. *Aquatic Procedia, 4*, 1070–1077.
7. Zhu, D., Cai, C., Yang, T., & Zhou, X. (2018). A machine learning approach for air quality prediction: Model regularization and optimization. *Big Data and Cognitive Computing, 2*(1), 5.
8. Hochreiter, S., & Schmidhuber, J. (1997). Long short-term memory. *Neural Computation, 9*(8), 1735–1780.
9. Curtis, L., Rea, W., Smith-Willis, P., Fenyves, E., & Pan, Y. (2006). Adverse health effects of outdoor air pollutants. *Environment International, 32,* 815–830.
10. Sakizadeh, M. (2016). Artificial intelligence for the prediction of water quality index in groundwater systems. *Modeling Earth Systems and Environment, 2*(1), 8.
11. Oprea, M., & Iliadis, L. (2011). An artificial intelligence-based environment quality analysis system. In *Engineering applications of neural networks* (pp. 499–508). Springer, Berlin, Heidelberg.

MAFCA: Mobility-Aware Fuzzy Clustering Algorithm for Wireless Sensor Networks

Deepika Agrawal and Sudhakar Pandey

Abstract Sensor nodes in wireless sensor networks are equipped with limited resources like storage capacity and battery. Presently, available clustering protocols assume that the sensor nodes are static. If mobility is introduced into the network, then most of the clustering protocol performance degrades. Because of the mobility, topology changes frequently. It increases the network's energy consumption resulting in less network lifetime. Thus, it is beneficial to deal with mobility at the time of the cluster-heads election. In this paper, an algorithm is proposed known as Mobility-Aware Fuzzy Clustering Algorithm (MAFCA). It is used to elect cluster-heads while keeping the utilization of the energy to a minimum. The MAFCA considers mobility factor for the election of cluster-heads. The MAFCA is compared with well-known clustering protocol, LEACH-Mobile and mobile-based clustering (MBC). The experimental result demonstrates that the MAFCA performs better in terms of network lifetime and energy consumption in both the cases.

Keywords Cluster-head · Clustering · Fuzzy logic · Mobility · Mobile wireless sensor network

1 Introduction

There is a current advancement in micro-electro-mechanical systems (MEMS) technology, wireless communication, and digital electronics that have empowered the development of nodes that are low in cost, power, and are little size. Communication between sensor nodes is done using radio frequency (RF) [1]. Sensor nodes have restricted capability, and it is only capable of gathering data from a limited area within its range. A collection of sensor nodes that work collaboratively forms a network known as wireless sensor network (WSN). Sensor nodes exhaust energy

D. Agrawal (✉) · S. Pandey
Department of IT, National Institute of Technology Raipur, Raipur, Chhattisgarh, India
e-mail: deepika721@gmail.com

S. Pandey
e-mail: spandey96@gmail.com

S. C. Satapathy et al. (eds.), *Intelligent System Design*, Advances in Intelligent Systems and Computing 1171, https://doi.org/10.1007/978-981-15-5400-1_76

in sensing, receiving, processing, and transmitting data. Sensor nodes are resource constrained, particularly in the battery. Once nodes are deployed, these batteries cannot be replaced. Often, nodes are left unattended. Hence, minimization of the energy consumption is necessary for the successful working of network operation [2].

Most of the algorithms believe static sensor nodes having a fixed position once they deployed [1]. But, there are some applications like search and rescue, wildlife tracking, habitat monitoring, etc., which need mobility among sensor nodes [2]. Based on the application, nodes, as well as sink, can be mobile. Mobility in the network affects the network operation, particularly in energy consumption [3–5].

The wireless sensor network can be partitioned into small groups, known as clusters to make the network energy efficient [2]. To lessen the utilization of energy by the nodes as well as the network, clustering is one of the techniques [2, 6, 7]. In this technique, cluster-heads (CHs) are chosen, and nearby nodes attach to their relevant cluster-head. After the cluster formation, all the nodes transmit their collected information to their cluster-head. Cluster-head then aggregates all the received data from the cluster members and transmits to the sink. In clustering, all nodes need not directly transmit all their data to the sink, and instead, they transmit their information to the cluster-heads and then cluster-head to the sink. In this way, energy consumption is reduced significantly. There is no need for central organization. It promotes local decision [7, 8].

The use of fuzzy logic [9, 10] benefitted many protocols. Real-time decisions regarding environment are made by fuzzy logic. Different environmental parameters are merged as per predefined rules, and a decision is being made. This is the important application of fuzzy logic. WSNs generally use fuzzy logic for incorporating different clustering parameters to elect CHs. Some overhead is associated with the cluster-head election that is profoundly lessened by utilizing fuzzy logic.

As mobility is an important factor in the lifespan of the network of wireless sensors. During CHs selection, it needs to be considered. In this paper, a fuzzy logic-based clustering algorithm for mobile sensor nodes is introduced to make an improvement in prolonging the WSN lifetime. It is a distributed algorithm which elects CHs via energy and mobility. To make wise decisions, Mobility-Aware Fuzzy Clustering Algorithm (MAFCA) uses the fuzzy if-then rules for calculating the chances of each node to become a CH.

2 Related Works

In this section, various routing protocols proposed so far have been reviewed. The key features of the popular clustering algorithms are explained below.

2.1 Hierarchical Routing Protocols Based on Clustering with Static Nodes

Low-energy adaptive clustering hierarchy (LEACH) [11] is a hierarchical routing protocol. Here, each sensor node has an equal chance to become a cluster-head (CH). It is divided into two phases, namely a setup phase and steady-state phase. Cluster formation is done in the first phase while actual data transmission takes place in the second phase. The cluster-head is chosen as per threshold. SEP [12] is an enhanced version of LEACH protocol. In this protocol, nodes are classified as normal nodes and advanced nodes. Nodes are different with respect to their initial energy. In SEP, nodes that have high initial energy get more frequently elected as CHs.

2.2 Hierarchical Routing Protocols Based on Clustering with Node Mobility

LEACH-Mobile [7] uses a complete probabilistic model for CH selection. It is an improved version of LEACH. In this, when the mobile nodes move, they declare the membership of the cluster. A TDMA schedule is assigned to each node. This protocol confirms that a mobile node communicates with a specific cluster-head within the assigned time schedule. However, the amount of energy dissipation is increased in LEACH-Mobile as compared to LEACH. It also does not take mobility factor into consideration for cluster-heads selection. In mobility-based clustering (MBC) [13], the CH is chosen based on the residual energy and mobility. LEACH-Mobile-enhanced protocol [14] is the enhanced version of LEACH-Mobile protocol. In this protocol, remoteness is used as a factor for CH selection. Remoteness is considered as mobility metric for CH selection. The next cluster-head will be the node having least mobility factor, and the energy level of that node is not below the threshold. Mobility and energy-conscious clustering protocol (MECP) is a mobility-aware and energy-efficient protocol which uses information about the node's movement as well as residual energy during the formation of the network.

2.3 Fuzzy Logic-Based Clustering Protocol

Cluster-head election mechanism using fuzzy logic (CHEF) uses a fuzzy logic approach for CHs election. It uses two parameters, i.e., energy and local distance, to calculate a chance for each node to become CH. Nodes having the highest chance will become CH [15]. A fuzzy logic-based control system is used to elect CH through three linguistic variables, i.e., the distance between the nodes close to the cluster center, total nodes in a cluster, and the communication traffic [16]. The mobility of base station is also taken as a parameter for the election of super cluster-head

(SCH) [17] to extend the network lifetime. Another protocol [18] is proposed for cluster-head selection which uses fuzzy logic with unequal clustering.

Although these related works show improvement over one another, still no algorithm is found which adapts mobility among nodes in the sensor networks and reduces the overall energy consumption in the network. The main aim of MAFCA is to increase the network lifetime where network comprises of mobile sensor nodes. MAFCA delays the first node death and last node death which in turn increases the network lifetime.

3 MAFCA Protocol

In all the protocols mentioned, each of the protocol considered either the static nodes or the sink mobility after deployment. But, some applications like wildlife tracking and medical-related applications require sensor nodes mobility. If the nodes are mobile, then for CH selection, mobility needs to be considered. Mobility is an important metric; since in CH selection, if the node having high mobility will become CH, then after advertisement message, it may possible that CH node relocates from its location to a new location. Hence, cluster members who want to join that CH cannot join. This leads to the wastage of energy. The common clustering protocol for mobile WSNs is LEACH-Mobile and MBC. But, the energy dissipated by all the nodes in LEACH-Mobile is high, and mobility is not considered in CH election. In MBC, mobility factor is considered in the probabilistic formula.

In MAFCA, energy and mobility are considered as a metric for the election of CH, and fuzzy logic is employed for CH selection. Fuzzy logic yields the best result when there are lots of uncertainties. Hence, this approach is used.

MAFCA is also based on a probabilistic model which is used for electing preliminary CHs. Though, it has just elected tentative CHs by depending on this model. CH election is done based on the chances calculated, and the node with the maximum chance among neighbors is then elected as a CH.

In every clustering round, a number between 0 and 1 is randomly chosen. If the chosen random number is smaller than the threshold function, then that node is selected as a preliminary CH.

The chance for each node changes dynamically as it is based on energy and mobility. As the mobile nodes move, their location changes which result in a change in values.

Energy: This parameter denotes the energy of a node. The higher the energy, the higher the chances to become CH.

Mobility: It denotes the speed of a node. The higher the mobility of a node, lessen the chances to become CH.

Network lifetime largely depends on the energy consumption. Hence, the energy of the node takes into consideration as one of the parameters for the election of cluster-head. Mobility implies that the distance between the nodes and base station

Table 1 Fuzzy rules and values of chance

S. no.	Energy	Mobility	Chance
1.	Low	High	Very weak
2.	Low	Medium	Weak
3.	Low	Low	Lower medium
4.	Medium	High	Weak
5.	Medium	Medium	Lower medium
6.	Medium	Low	Medium
7.	High	High	Higher medium
8.	High	Medium	Strong
9.	High	Low	Very strong

increases and decreases with respect to speed and direction of mobile nodes when nodes move in a specific direction.

The main objective of MAFCA is to discover an energy-efficient cluster-head based on the parameters defined formerly, so that the network lifetime is prolonged. Mamdani algorithm is used to realize the fuzzy logic inference. It is the most widely used methods. The input fuzzy linguistic variables are energy and mobility. 9 (3^2) rules are present in the rule base. There is a single output fuzzy variable, namely chance, which determines the chance of each node to become a CH. Table 1 listed the rules used in the fuzzy rule base. After finding preliminary CHs, the competition for final CH begins. A chance value is calculated for each preliminary CH. Each preliminary CH transmits a message to compete with other preliminary CHs within its range. If a preliminary CH receives a message from another preliminary CH and the value of chance of the sending node is greater than the chance value of receiving node, then the receiving node backs out from the competition and waits for the announcement. If a particular preliminary CH has the highest chance among the preliminary CH which it receives a message from, then it becomes a CH. This protocol guarantees that no other CHs exist in the range of a particular CH, and in this way, consumption of the network's energy is balanced. After the election is completed, the nodes that are not elected as a CH join to the closest CH by transmitting a join message. The cluster-head creates a TDMA schedule for each cluster members after the cluster formation. In their respective TDMA schedule only, cluster members can transmit data to the CH.

4 Simulation Results

4.1 Parameters Setting

For evaluating the performance of our proposed protocol MAFCA and comparing it with LEACH-Mobile [7] and MBC [13], experiments have been carried out in

Table 2 Simulation parameters

Parameter	Values
Network boundaries	100 m * 100 m
Location of the sink	(50, 50)
Number of deployed Nodes	100, 200
Mobility model	Random way point
Election probability value of Cluster-heads (P)	0.1
Number of rounds	3500
Initial energy per node (E_0)	0.5 J
ε_{mp}	0.0010 pJ/bit/m^4
$E_{elec} = E_{bit}$	50 nJ/bit
E_{fs}	10 pJ/bit/m^2
E_{DA}	5 nJ/bit
V_{min}	0 m/s
V_{max}	3 m/s

MATLAB. The results are evaluated in a squared sensed area having a length of the sides. Nodes are randomly scattered in the sensing area. Mobility is given to all the nodes across the network. Two scenarios are created that are:

Scenario 1: The number of deployed sensor nodes is 100 nodes.
Scenario 2: The number of deployed sensor nodes is 200 nodes.

The simulation parameters are listed in Table 2.

4.2 Network Lifetime Results

The network lifetime is calculated as the time at which the first node is dead, as we can observe from Fig. 1a, b that the number of dead sensor nodes with respect to the number of rounds is higher in MAFCA in both the scenarios.

Also, from Table 3, we see that the MAFCA outperforms LEACH-Mobile and MBC concerning stability period. Stability period is the period between the start of the network and first node's death. Network lifetime is also high in MAFCA as the death of the last node takes time as compared to LEACH-Mobile and MBC.

For scenario 1 with 100 nodes, the death of the first node and the last node's death are prolonged by 92.16% and 7.99%, respectively, comparing with LEACH-Mobile. And with MBC, the death of the first node and the last node death are prolonged by 42.41% and 5.75%, respectively.

For scenario 2 with 200 nodes, the death of the first node and the last node's death are prolonged by 86.31% and 35.46%, respectively, comparing with LEACH-Mobile. The death of the first node and the last node's death are prolonged by 40.15

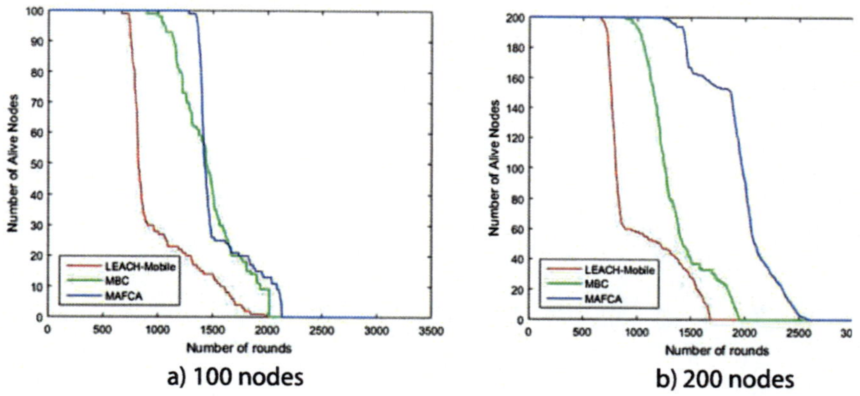

Fig. 1 Number of dead nodes versus number of rounds

Table 3 Simulation results

S. no.	Number of nodes	Protocol name	First node death	Last node death
1.	100	LEACH-Mobile	664	1977
2.		MBC	896	2019
3.		**MAFCA**	**1276**	**2135**
4.	200	LEACH-Mobile	665	1906
5.		MBC	884	1955
6.		**MAFCA**	**1239**	**2582**

and 32.07% when compared with MBC. As the total number of nodes increases, the number of dead nodes gets lesser in MAFCA as compared to LEACH-Mobile and MBC.

4.3 Total Remaining Energy in the Network

Figure 2a, b shows the graph of the total remaining energy on the network with the total number of rounds. The figure shows a good performance on energy efficiency. It is confirmed that the MAFCA has better energy utilization as compared to the LEACH-Mobile and MBC in both the scenarios.

In LEACH-Mobile, nodes having less energy may become CH as it is based on the probabilistic model. Hence, energy consumption in LEACH-Mobile is high. Also, nodes having high mobility may become CH as it does not take mobility into consideration for CH selection. In MBC, mobility is taken into consideration, but it is based on the probabilistic model. In MAFCA, mobility, energy, and centrality take into consideration. Here, the chance of preliminary CHs is calculated by using

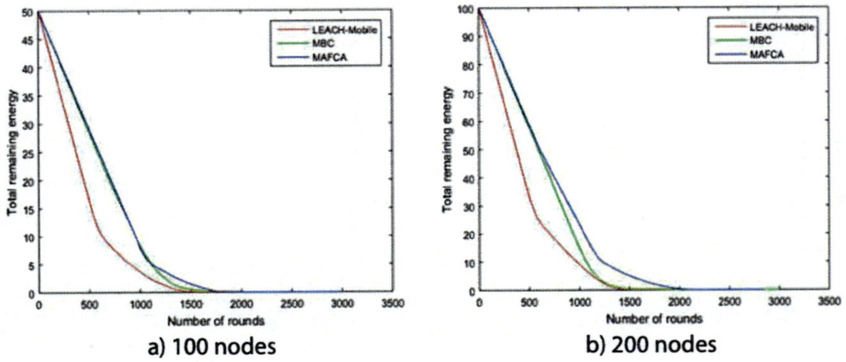

Fig. 2 Total remaining energy versus number of rounds

fuzzy approach. MAFCA is based on probabilistic model only for the selection of preliminary CHs. Hence, energy consumption of the network is less resulting in an increase of network lifetime.

5 Conclusion

Applications of WSNs are usually spread over large areas. Also, they have limited resources like memory and battery. So, there is a requirement of protocols which can manage the WSNs in a better way. The main challenge of the protocol design is energy efficiency and network lifetime due to the limited resources. If sensor nodes are mobile, then energy consumption across the network is high. Through this paper, an attempt has been made to make the clustering protocol for mobile wireless sensor network as energy efficient as possible. MAFCA takes mobility as well as energy into consideration for cluster-heads election. In terms of total remaining energy and the network lifetime, MAFCA performs better as compared to LEACH-Mobile and MBC. For making decisions related to clustering, local information is used primarily, and hence, the proposed protocol, i.e., MAFCA, experiences little overhead in terms of messages.

References

1. Akyildiz, I. F., Su, W., Sankarasubramaniam, Y., & Cayirci, E. (2002). Wireless sensor networks: A survey. *Computer Networks, 38*(4), 393–422.
2. Younis, O., Krunz, M., & Ramasubramanian, S. (2006). Node clustering in wireless sensor networks: Recent developments and deployment challenges. *IEEE Network, 20*(3), 20–25.
3. Howard, A., Matarić, M. J., & Sukhatme, G. S. (2002). An incremental self-deployment algorithm for mobile sensor networks. *Autonomous Robots, 13*(2), 113–126.

4. Shah, R. C., Roy, S., Jain, S., & Brunette, W. (2003). Data mules: Modeling and analysis of a three-tier architecture for sparse sensor networks. *Ad Hoc Networks, 1*(2), 215–233.
5. Ye, F., Luo, H., Cheng, J., Lu, S., & Zhang, L. (2002). A two-tier data dissemination model for large-scale wireless sensor networks. In *Proceedings of the 8th Annual International Conference on Mobile Computing and Networking* (pp. 148–159). ACM.
6. Al-Karaki, J. N., & Kamal, A. E. (2004). Routing techniques in wireless sensor networks: A survey. *IEEE Wireless Communications, 11*(6), 6–28.
7. Awwad, S. A. B., Ng, C. K., Noordin, N. K., & Rasid, M. F. A. (2011). Cluster-based routing protocol for mobile nodes in wireless sensor network. *Wireless Personal Communications, 61*(2), 251–281.
8. Lotfinezhad, M., & Liang, B. (2004). Effect of partially correlated data on clustering in wireless sensor networks. *Sensor and Ad Hoc Communications, and Networks (IEEE SECON), 2004*, 172–181.
9. Gupta, I., Riordan, D., & Sampalli, S. (2005). Cluster-head election using fuzzy logic for wireless sensor networks. In *3rd Annual Communication Networks and Services Research Conference (CNSR'05)* (pp. 255–260). IEEE.
10. Ran, G., Zhang, H., & Gong, S. (2010). Improving on LEACH protocol of wireless sensor networks using fuzzy logic. *Journal of Information & Computational Science, 7*(3), 767–775.
11. Heinzelman, W. R., Chandrakasan, A., & Balakrishnan, H. (2000). Energy-efficient communication protocol for wireless microsensor networks. In: Proceedings of the 33rd Annual Hawaii International Conference on System Sciences (10p). IEEE.
12. Smaragdakis, G., Matta, I., Bestavros, A. (2004). *SEP: A stable election protocol for clustered heterogeneous wireless sensor networks*. Boston University Computer Science Department.
13. Deng, S., Li, J., & Shen, L. (2011). Mobility-based clustering protocol for wireless sensor networks with mobile nodes. *IET Wireless Sensor Systems, 1*, 39–47.
14. Kumar, G. S., Paul, M. V. Vinu, & Jacob, K. P. (2008). Mobility metric based LEACH-Mobile Protocol. In *Proceedings of International Conference Advanced Computing Communication Chennai, India* (pp. 248–253), Dec 2008.
15. Kim, J.-M., Park, S.-H., Han, Y.-J., & Chung, T.-M. (2008). CHEF: cluster head election mechanism using fuzzy logic in wireless sensor networks. In 10th International Conference on Advanced Communication Technology (ICACT 2008) (Vol. 1, pp. 654–659).
16. Khan, F. U. S., Shah, I. A., Jan, S., Imran, K., & Mehmood, M. A. (2015). Fuzzy logic based cluster head selection for homogeneous wireless sensor networks. In 2015 International Conference on Open Source Systems & Technologies (ICOSST) (pp. 41–45). IEEE.
17. Nayak, P., & Devulapalli, A. (2016). A fuzzy logic-based clustering algorithm for WSN to extend the network lifetime. *IEEE Sensors Journal, 16*(1), 137–144.
18. Agrawal, D., & Pandey, S. (2018). FUCA: Fuzzy-based unequal clustering algorithm to prolong the lifetime of wireless sensor networks. *International Journal of Communication Systems, 31*(2), e3448.
19. Kim, D.-S., & Chung, Y.-J. (2006). Self-organization routing protocol supporting mobile nodes for wireless sensor network. In *First International Multi-Symposiums on Computer and Computational Sciences (IMSCCS'06)* (Vol. 2, pp. 622–626). IEEE.
20. Torghabeh, N., Abazari, T., Mohammad Reza Akbarzadeh, M., & Mohammad Hossein, Y. (2010). Cluster head selection using a two-level fuzzy logic in wireless sensor networks. In *2nd International Conference on Computer Engineering and Technology (ICCET)* (Vol. 2, pp. V2-357–V2-361). IEEE.
21. Singh, A., & Singh, A. K.: Mobility-and energy-conscious clustering protocol for wireless networks. In *Proceedings of the International Congress on Information and Communication Technology* (pp. 365–374). Springer, Singapore.

Multiple Hand Gestures for Cursor Movement Using Convolution Neural Networks

G. Santoshi, Pritee Parwekar, G. Gowri Pushpa, and T. Kranthi

Abstract Developing a multi-hand gesture to create a magical mouse movement by which we can tab, swipe, zoom-in, zoom-out by using one or more fingers. The machine learning technique can be used to perform all the operations as a state of art. The objective of the proposed algorithm is to use the hand gesture in front of the built-in webcam in the computers or the cellphones which can control the mouse movement. A multilayer convolution neural network is proposed in this paper which considers the pixel-level segmentation to identify the hand region, in the second stage, it identifies the number of fingers used, and in the last stage, the movement detection of the fingers to identify the operations is to be performed.

Keywords Multi-hand gesture · Multilayer convolution neural networks · Segmentation · Lucas–Kanade

1 Introduction

Ever since the dawn of time, man has worked tirelessly to make life better. With the invention of computers and the Internet, the way of information transfer has become extremely easy. But there is always been one common way of interacting with these machines. The human being specially kids stay closer to the machines and physically connected to interact with it. Applying a strong gesture recognition technique [1] can change the order techniques and can create a new trend. In simple words, gesture recognition is the science of interpreting human gestures as input

G. Santoshi (✉) · P. Parwekar · G. Gowri Pushpa · T. Kranthi
ANITS Engineering College, Visakhapatnam, India
e-mail: gsanthoshi.cse@anits.edu.in

P. Parwekar
e-mail: pritee2000@gmail.com

G. Gowri Pushpa
e-mail: gowripushpa.cse@anits.edu.in

T. Kranthi
e-mail: tkranthi.cse@anits.edu.in

S. C. Satapathy et al. (eds.), *Intelligent System Design*, Advances in Intelligent Systems and Computing 1171, https://doi.org/10.1007/978-981-15-5400-1_77

commands using mathematical algorithms [2]. If we unleash the whole power of latest ground-breaking technologies, then hardware input devices like mice and keyboard can actually become redundant. Hand gesture recognition [3] is based on image processing concept. In recent year, there is a lot of research on gesture recognition using Kinect sensor [4] on using HD camera, but camera and Kinect sensors are costlier. We focus on reducing cost and improve robustness of the proposed system using simple Web camera.

Gesture recognition [3] is the process of understanding and interpreting meaningful movements of the hands, arms, face, or sometimes head. Gestures are a major form of human communication [5]. Gestures are found to be an easy and important way to interact with computers, as the users are already familiar of how we communicate. Today, the industries working on different applications are to make interactions more easy, natural, and convenient without wearing any extra device. The goal of gesture recognition research is to create a dynamic system which can identify specific human gestures and convey the messages to the devices. The ultimate aim of human–computer interaction is to make computers understand human body language and extract as much data as possible from just a mere look through camera. This paper proposes a process of extracting data from hand gestures and performs specific tasks related to the gestures.

The interest of the paper is to learn the direct interaction of humans with the consumer electronic devices built in the computer or laptop. This takes the user experience to a whole new level. The paper consists of related work in the first section, followed by the proposed work in detail. The experiment result shows the detail explanation of the dataset comparison, proposed two-level CNN comparison, and the working experiences of the proposed work. Finally, the paper is concluded with the future extension of the paper.

2 Related Work

Interactive presentation systems use advanced human–computer interaction (HCI) techniques to provide convenient and user-friendly interface for controlling presentation displays, for example, page up/down controls in a slideshow. Compared with traditional mouse and keyboard control, the presentation experience is significantly improved with these techniques. Hand gesture has wide-ranging applications. In this study, it is applied to an interactive presentation system to create an easy-to-understand interaction interface. When compared with the vision-based hand gesture recognition and the glove-based hand gestures [6], the glove uses a flex-based sensor, it is costly, and the paper [7] shows that vision-based dataset is more accurate compared to glove-based dataset. The gestures are naturally used by many people and especially for interaction with deaf people. In India, the classical dance [5] forms have lots of meaning in the mudras that they use in the dance. The yoga practices [8] with different gesture have different meaning and can solve many health issues.

The hand gesture is the best way of replacing mouse [7] for human–computer interaction (HCI) which can be applied to the human–robot interaction as mouse is not suitable for the interaction. The paper aimed at investigating and developing a computer control (CC) system using hand gestures. All laptops and mobiles today are equipped with webcams, which have recently been used insecurity applications utilizing face recognition [9]. In order to harness the full potential of a webcam, it can be used for vision-based CC, which would effectively eliminate the need for a computer mouse or mouse pad. The usage of a webcam can also be extended to other HCI application such as a sign language database or motion controller. Over the past decades, there have been significant advancements in HCI technologies for gaming purposes, such as the Microsoft Kinect and Nintendo Wii [10].

A gesture is a non-vocal and non-verbal communication which provides a message through the actions. Gestures include movement of the hands, face, or other parts of the body. Gestures are natural way of human interaction in daily life. Motion sensors can be used for the identification of small or big gesture in the hand [1]. It may include anything from the pointing of a finger to a roundhouse kick or a nod of the head to a pinch or wave of the hand gestures are classified into two types. Online gesture is used to control the machine as the actions are performed, whereas offline gesture do not show the real world time changes [11]. Depending on the type of interaction, the gestures can be used required.

Due to human interaction system, the need in research and development of new gesture control products has increased with different gesture functionalities. Gestures can be of various forms from any motion, such as face [12], hands [1], and body [11]. Mostly, they are classified as hand and arm gestures recognition for hand poses and arm movements, sign languages of deaf peoples, etc. Head and face gestures [10] are for identifying the facial expressions, face recognition, and head shaking. Body gestures [13] are used to identify the direction of person, yoga poster prediction, medical rehabilitation, etc. In the early days, the development stages of gesture control technology were most on hand gestures and head gestures. Direct control via head gesture and hand posture is immediate, but limited in the number of choices as they use special gloves, headsets with micro-controllers to detect the gestures. For example, Wii and Kinect can recognize a player's body gestures, while leap motion [7] can recognize the fine movement of all ten fingers. There are many other areas of applications for gesture control technology as in automotive, entertainment sector, consumer electronics, healthcare, and education sectors.

For the CNN implementation, [7] proposed a method of low resolution network (LRN) and high resolution network (HRN) together to perform better than the LRN and HRN individually. There are many color space used for skin detection such as RGB [14], Normalized RGB [15], HSV [16]. After comparison, the HSV color space was identified as the better among all the above-mentioned algorithms [17]. CAMSHIFT method used by [18] is used for the hand tracking method. Tracking algorithm [19] creates an ellipse around the region of interest (hand contour) and tracks the center of the ellipse over successive frame. The template matching, convexity defects, and Hough circle transform were proposed, but this algorithm complexity is less but needs a clear background or hand gloves for better output. The recent works are using CNN [13] with deep learning. The segmentation is used for fine images to not to get the noise, and then, the feature extraction is used to classify the images.

3 Proposed Work

The convolution neural network (CNN) is deep learning technique which can be trained using small set of dataset. The dataset can be rotated and resized to make the set large and used for training. The webcam can capture images of small size which has to be preprocessed and trained to identify only the fingers of the hand as shown in Fig. 1. After identifying, the ROI is using a three-layer CNN for segmentation. The output image is converted to a 1D array as input to the next layer of CNN which processes the data, identifies the shape of the image, and decides the number of fingers

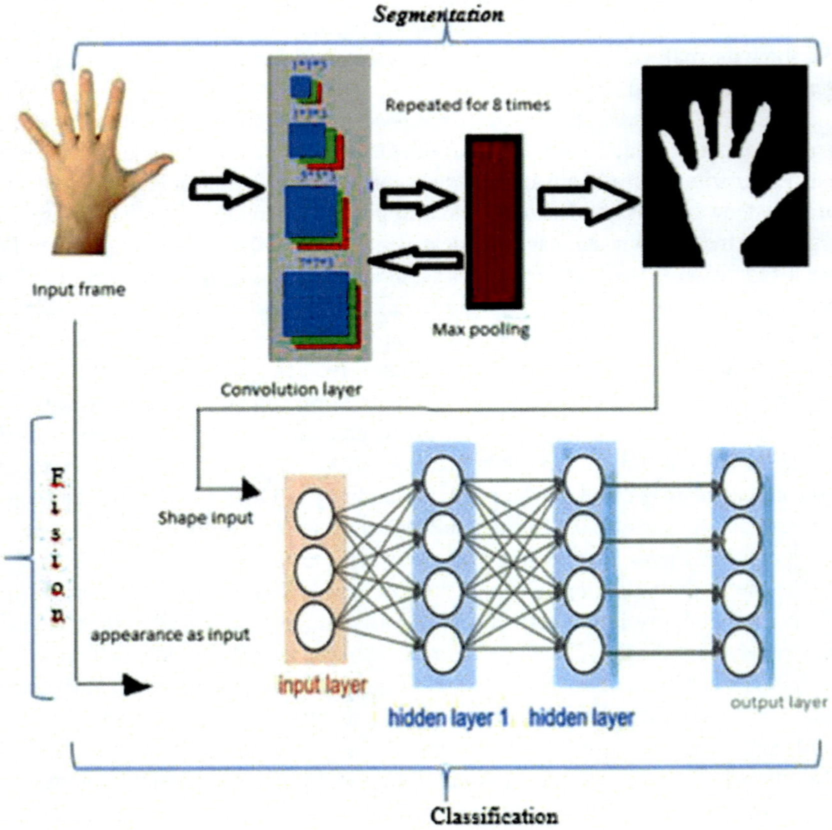

Fig. 1 Proposed CNN architecture

used. The process is continued for every second to get the change in the distance and direction of the finger to detect various operations to perform. The proposed architecture is subdivided into five sections. Section 3.1 consists of selection of dataset, and in Sect. 3.2, segmentation of image using shape-based and appearance-based CNN is explained in brief. The process is repeated after every 1 s, if continuously same gesture is repeated for two times in the frames with some momentum in distance or direction, and Sect. 3.4 explains the operations to perform after calculation of the motion and distance in order to implement the cursor movement.

3.1 Dataset

Under the challenging condition, the webcam can accept a dataset of 115×250 pixels resolution as the distance from the camera which is a challenging job. The webcam has less intensity and resolution under all these conditions, and the images dataset recorded with Microsoft Kinect device is taken into consideration for training the dataset. The dataset consists of ten different gestures of five different people, each persons' ten gestures with 30 repeats and a total of 1500 samples for training. The training dataset contains with noise in the background and also the non-noisy dataset. The dataset is divided into four different classes such that it can identify all the features correctly, and the other gestures are considered as the negative set. The leap motion hand gesture dataset [4] is used for training. The dataset contains all the gestures required for the proposed work and contains all the variations in the gestures which help in training the dataset more efficiently. Only four classes are taken into consideration for the work, and the rest is considered as the negative dataset which are given as input to CNN which further can identify the number of fingers present in the image using convex hull algorithm to draw a polygon over hand, and convexity defects is to identify specific gestures. This is used for cursor movement, and CNN gives the finger count as two fingers, *Class 2* is used for sliding up and down; it is also used for clicking, and the finger count is considered as three fingers, the finger count is two for the *Class 3*, but it is used for zooming in and zoom-out; but Class 1 and class 3 are differentiated using the CNN. Finally, the *Class 4* gives a finger count of five which is used for double click. All the classes with different forms are given in Table 1.

3.2 Segmentation Using Convolution Neural Networks

The segmentation is used for identifying the feature of the image using convolution neural networks. There are three versions of fully convolution neural (FCN), we implement eight-layer FCN, and each layer is divided into two internal group of components, convolution layer and max pooling.

Table 1 Different types of class to be trained and tested

Class 1—cursor move	Class 2—click	Class 3—zoom	Class 4—double click
			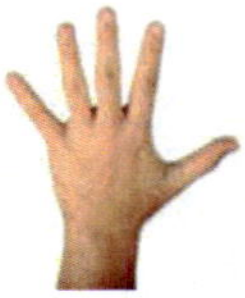

3.2.1 Convolution Layer

The input is an RGB image having three channels. An image of $M \times M$ is taken where M is the height and width of the image. The convolutional kernel of different sizes is used in different layer to get a fined segmentation. As in the figure, 3 * 3, 5 * 5, and 7 * 7 with three channels are taken to extract the color feature of each pixel. These kernels are used to extract color feature of the central pixels and the neighboring pixels. The value of these convolutional kernels is fixed. Fig. 2 shows the example of calculation of the values of padding for a convolutional kernel of 3*3 matrix.

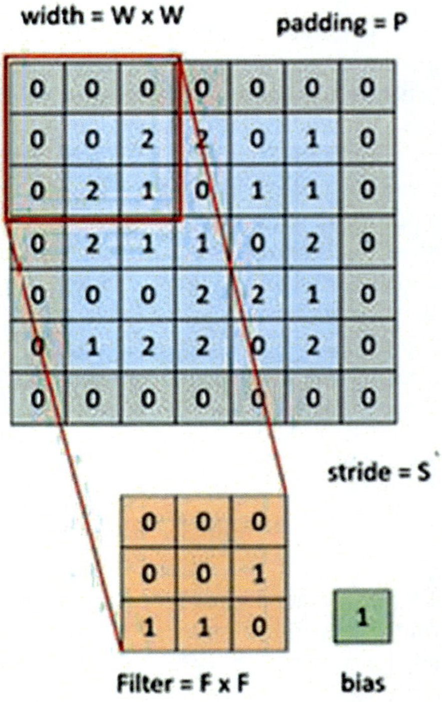

Fig. 2 Padding of the 1D image pixels with a 3 * 3 convolutional Kernel

Fig. 3 Max pooling applied for a matrix of 2 * 2 matrix

3.2.2 Max Pooling

Pooling technique is used to reduce the size of the feature vector by selecting the maximum value of the matrix of 2 * 2 size sub-matrix which in turn supports in reducing the spatial size as well as the computation time. In Fig. 3, we have a matrix of 4 * 4 which has been reduced to 2 * 2 matrix.

3.3 Recognition Using CNN

The hand gesture is predicted using a combination of two CNNs. One network explores the shape-based and the other explores the appearance-based information. Each CNN is subdivided with the rest two operations of fully connected layers dropouts combined with ReLUs which are the hidden layers and fully connected layer at the end and generate output of **four classes**. In this phase, the tensors are used for creating the layers for a FCN, output and the number of classes to classify.

3.3.1 ReLUs

Rectified linear units activation function acts as the transfer function to the CNN. The nonlinear properties of the decision function increase the overall network without affecting the receptive fields of the convolution layer. As compared with the TensorFlow, the Adam optimization algorithm is a proof to be not an optimal solution compared to stochastic gradient descent (SGD) with momentum. To get the optimized output, we would like to propose the SGD as the activation function along with the standard cross-entropy loss for backpropagation and error detection.

3.3.2 Fusion of CNN

By using the SoftMax classifier, we map s^{rec} to a probability distribution over four classes of the image which is classified for both the CNN classifiers. The fusion of CNN is done by using Eq. 1.

$$F^{\text{sum}} = f^S + f^A, \tag{1}$$

$f^S \in R^d$ (shape stream) and $f^A \in R^d$ (appearance stream) be the outputs of the last fully connected layers, and F^{sum} is minimized and trained into SoftMax classifier for joint supervised learning.

3.4 Operations Performed

Using the fusion CNN for feature detection and finger count is identified, then Lucas–Kanade [20] is applied if there is a movement of the same gesture of the algorithm which predicts the distance and the direction of the movement to perform different operation. To implement the movement of the mouse, the CAMSHIFT method is used which helps in cursor movement using the CNN algorithm. The final algorithm gives the detailed implementation of the proposed work stepwise.

Step 1: The algorithm captures the frame for every 1 s which is given as input to CNN for classification.

Step 2: The finger count is identified using convex hull, if finger count == 2 for both image, then check the frame for every 2 ms and repeat the step 1 and 2 until having same condition.

- 2.1 Apply Lucas–Kanade [20], and the frames are checked for the movement
- 2.2 As the hand moves, the same way the mouse is also moved

Step 3: If finger count == 2 in first frame and count == 3 in second frame, then click operation is performed or select the folder is done using the cursor.

Step 4: if finger count == 3 in both frame, then compare the frame for 1 s

Step 5: Apply Lucas–Kanade for movement of the frames

Step 6: If moved right, then swipe right, and if moved left, then swipe right

Step 7: If moved up, then scroll up or vice versa

Step 8: If finger count == 3 in first frame and count == 5 in second frame, then the operation double click takes place.

Step 9: After CNN, the finger count is identified as 1 which is a part of the class. The next frame is changed to 2 count as given in Table 2, and in sl no. 5, the operation of zoom-in and zoom-out is performed.

Step 10: Otherwise do no operation.

4 Experimental Results and Discussions

The python open-source deep learning library with a TensorFlow backend is used to implement the proposed method. As explained in Sect. 3.3.1, we used cross-entropy

Table 2 Operation to be performed on the motion, direction, and change in finger count detection

Sl. No.	First image	Second image	Operation	Sl. No.	First image	Second image	Operation
1			To click	4			To scroll up or down
2			To move the mouse	5			To zoom-in or zoom-out
3		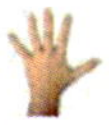	Double click	6			Swipe between pages

loss as the cost functions and applied the SGD optimization algorithm to minimize it. In this section, we aim to prove that the proposed method used to find the cursor movement is one of the best possible ways. The leap gesture dataset is compared with the other standard dataset in Sect. 4.1. The proposed algorithm fusion CNN is compared with other CNN in Sect. 4.2, and the last section shows the experimental results with proposed algorithm.

4.1 *Experimental Results of the Dataset*

We evaluated the leap gesture dataset which contains 1500 RGB images of five individuals performing 30 different hand gestures with ten different resolutions, collected under uncontrolled backgrounds without any face–hand occlusion. The dataset is categorized to five classes, four as required gesture and the extra is the false of negative image set.

The leap gesture dataset's F-score is compared with the standard datasets OUHANDS [21], HGR [22], and LGR [11] datasets using precision and recall [23] as follows:

$$\text{Precision} = \text{tp} + \text{fp}/\text{tp} \tag{2}$$

$$\text{Recall} = \text{tp}/\text{tp} + \text{fn} \tag{3}$$

$$F\text{-score} = 2 \times \text{Precision} \times \text{Recall}/(\text{Precision} + \text{Recall}) \tag{4}$$

where tp = true positive, fp = false positive, and false negative is fn.

Table 3 *F*-score proposed leap gestures dataset, LRG [11], HRG [22], and OUHANDS [21] dataset for hand gesture recognition

Method	*F*-score
HGR	0.935
LGR	0.908
OUHANDS	0.9562
Proposed method	0.966

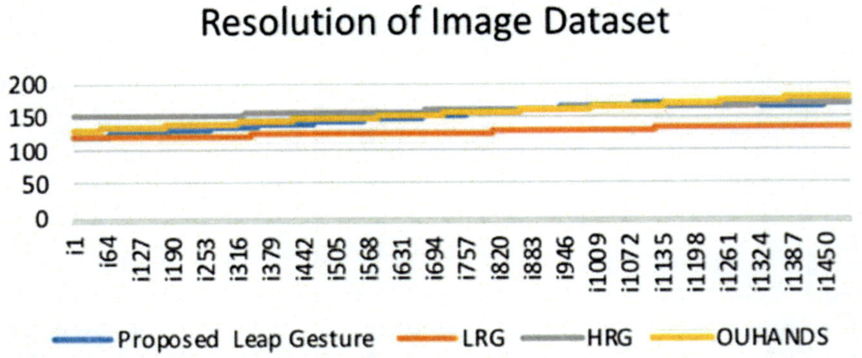

Fig. 4 Different resolutions considered for proposed, LRG [11], HRG [22], and OUHANDS [21] dataset for hand gesture recognition

The prediction in Table 3 shows that OUHANDS [21] and leap gesture recognition have almost same *F*-score, but leap gesture is applicable for the proposed method as the dataset contains the noisy images which help in training the CNN in more efficient way. The used database also contains the data having low resolution as well as high resolution images as shown in Fig. 4 which contributes to recognize the images generated by the built-in camera as required for the cursor movement.

4.2 Experimental Results of Fusion CNN

The confusion matrix Table 5 gives the resultant classification of the classes explained in Sect. 3.1 using CNN which clearly shows that the fusion of CNN gives approximately accurate results. All the classes are classified using two-level CNN or fusion CNN which is 90% approximately correctly predicted. Further for more clarification, the two-level CNN(fusion CNN algorithm), when compared for the complexity and the accuracy of the fusion, gives more accurate results as given in Table 4. Identify the hand gesture.

We can see in the next case by using CNN fusion to train the confusion matrix though it takes time to train, but the accuracy for the prediction is more. We can see Table 4 which shows that the accuracy increases to 5%.

Table 4 Confusion matrix of the classification

	Class 1	Class 2	Class 3	Class 4
Class 1	95	0	5	0
Class 2	6	78	11	5
Class 3	6	4	90	0
Class 4	0	0	3	97

4.3 Experimental Result of Cursor Movement

In this paper, six operations are performed by taking hand gestures as input through Web camera. The operations include cursor movement, sliding, scrolling, and clicking as a part of replacing traditional hardware mouse. The experiments are conducted using the built-in Web camera, In each second, 24 frames are taken as input and stitched as a video. Jupyter is a open-source software which is included with OpenCV to implement the algorithm.

Cursor movement is one of the most used operations performed by mouse. Using gesture count one, the cursor on the screen can be moved as shown in Fig. 5a. The

Table 5 Performance of fusion CNN for the leap gesture dataset for hand gesture recognition

	Testing accuracy (%)	Training dataset (%)	Computing time (ms)
CNN-shape based [19]	85.66	97	15
CNN-appearance based [24]	83.9	90	7
Fusion CNN (proposed model)	88.7	87	16

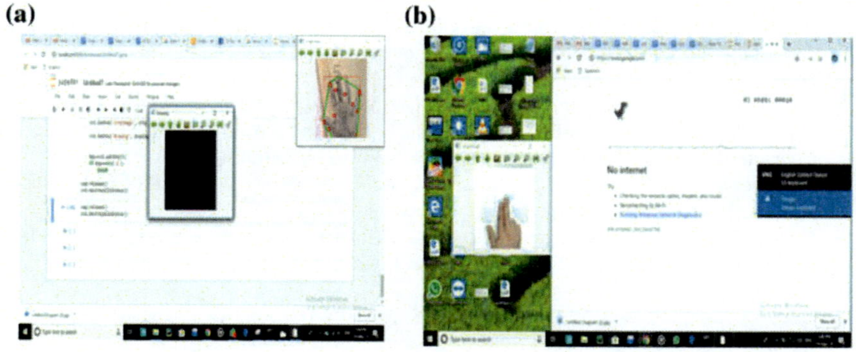

Fig. 5 Experimental result shows the finger count predicted by CNN and running an application using hand gesture

image clearly shows how the finger count is recognized as value five which is used for the double click on the display screen.

Fig. 5b shows how the gesture is used to scroll and play the Google Chrome offline game and shows how the dragon game is implemented, with gesture count three, the dragon character in the game jumps to cross the hurdles by the count three by swiping up and down.

5 Conclusion

The proposed algorithm uses all the properties of the fully connected networks with multilayer approach. The best algorithm practices are followed to implement the algorithm. A two-level CNN is required for implementing the gesture recognition, one for the shape detection and the other for the appearance which helps to remove the background noise and identify the hand exactly. As execution of the two-level CNN take lot of time, to reduce the time complexity, the feature extraction for both the neural network is merged into one and the common feature is given as input to both the classifier. To reduce the conflict between the classifiers, again the output of both the CNNs after recurrent is fused together, and the classification is applied to identify the gesture. As this algorithm needs to identify the finest segmentation and the movements in the fingers, these outputs are used for implementation of the operations of the mouse movement as explained in the expected output. The implementation of this algorithm can best utilize the webcam for all the mouse operation without physical contact of the human with the computer or the mobile.

References

1. Matilainen, M., Sangi, P., Holappa, J., & Silvén, O. (2016). OUHANDS database for hand detection and pose recognition. In 6th *International Conference on Image Processing Theory Tools and Applications* (pp. 1–5). IEEE.
2. Molchanov, P., Gupta, S., Kim, K., & Kautz, J. (2015). Hand gesture recognition with 3D convolutional neural networks: NVIDIA, Santa Clara, California, USA. In *The IEEE Conference on Computer Vision and Pattern Recognition (CVPR) Workshops* (pp. 1–7).
3. Dadashzadeh, A., Targhi, A. T., Tahmasbi, M., & Mirmehdi, M. (2018). HGR-net: A fusion network for hand gesture segmentation and recognition.
4. Marin, G., Dominio, F., & Zanuttigh, P. (2014, October). Hand gesture recognition with leap motion and kinect devices. In *2014 IEEE International Conference on Image Processing (ICIP)* (pp. 1565–1569). IEEE.
5. Liao, Y., Xiong, P., Min, W., Min, W., & Lu, J. (2019). Dynamic sign language recognition based on video sequence with BLSTM-3D residual networks. In *IEEE Access*, March 4, 2019.
6. Molchanov, P., Gupta, S., Kim, K., & Kautz, J. (2015). Hand gesture recognization with 3D convolutional neural networks. In *The IEEE Conference on Computer Vision and Pattern Recognition (CVPR) Workshops* (pp. 1–7).

7. Chanu, O. R., Pillai, A., Sinha, S., & Das, P. (2017). Comparative study for vision based and data based hand gesture recognition techniques. In *2017 International Conference on Intelligent Communication and Computational Techniques (ICCT)* (pp. 26–31).
8. Belgioioso, G., Cenedese, A., Cirillo, G. I., Fraccaroli, F., & Susto, G. A. (2015). A machine learning based approach for gesture recognition from inertial measurements. In *53rd IEEE Conference on Decision and Control.*
9. Ohn-Bar, E., & Trivedi, M. M. (2015). A comparative study of color and depth features for hand gesture recognition in naturalistic driving settings. In *2015 IEEE Intelligent Vehicles Symposium (IV)* (pp. 845–850).
10. Zhang, C., Liu, J., Yu, F., Wan, S., Han, Y., Wang, J., & Wang, G. (2018). Segmentation model based on convolutional neural networks for extracting vegetation from Gaofen-2 images, August 2, 2018.
11. Santoshi, G., & Mishra, S. R. (2015). Pedestrian with direction detection using the combination of decision tree learning and SVM, emerging ICT for bridging the future, In Proceedings of the 49th Annual Convention of the Computer Society of India (CSI) (vol. 1, pp. 249–255).
12. Siswanto, A. R. S., Nugroho, A. S., & Galinium, M. (2014). Implementation of face recognition algorithm for biometrics based time attendance system. In *2014 International Conference on ICT For Smart Society (ICISS).*
13. Roy, K., Mohanty, A., & Sahay, R. R. (2017). Deep learning based hand detection in cluttered environment using skin segmentation. In *The IEEE International Conference on Computer Vision (ICCV)* (pp. 640–649).
14. Brand, J., & Mason, J. (2000). A comparative assessment of three approaches to pixellevel human skin-detection. In *Proc. of the International Conference on Pattern Recognition*, (vol. 1, pp. 1056–1059.
15. Brown, D., Craw, I., & Lewthwaite, J. (2001). A som based approach to skin detection with application in real time systems. In *Proc. of the British Machine Vision Conference.*
16. Hasan, M. M., & Mishra, P. K. (2010). HSV brightness factor matching for gesture recognition system. *International Journal of Image Processing (IJIP), 4*(5), 456–467.
17. Zarit, B. D., Super, B. J., & Quek, F. K. H. (1999). Comparison of five color models in skin pixel classification. In *ICCV'99 Int'l Workshop on recognition, analysis and tracking of faces and gestures in Real-Time systems*, (pp. 58–63).
18. Shrivastava, R. (2013). A hidden Markov model based dynamic hand gesture recognition system using OpenCV, Advance Computing Conference (IACC), 2013 IEEE Anne-Marie Burns.
19. Yang, C., Han, D. K., & Ko, H. (2017). Continuous hand gesture recognition based on trajectory shape information pattern recognition letters.
20. Kamakula, K., Sharmila Rani, J., Santhosi, G., & Gowri Pushpa, G. (2016). Tracking and speed estimation of moving vehicle for traffic surveillance system. In S. C. Satapathy et al. (Eds.), *Proceedings of the Second International Conference on Computer and Communication Technologies, Advances in Intelligent Systems and Computing*. Berlin: Springer.
21. Ameur, S., Khalifa, A. B., & Bouhlel, M. S. (2016). A comprehensive leap motion database for hand gesture recognition. In *2016 7th International Conference on Sciences of Electronics, Technologies of Information and Telecommunications (SETIT).*
22. HGR1. http://sun.aei.polsl.pl/mkawulok/gestures/.
23. Powers, D. M. (2011). Evaluation: From precision, recall and F-measure to ROC, informedness, markedness and correlation.
24. Deng, X., Zhang, Y., Yang, S., Tan, P., Chang, L., Yuan, Y., & Wang, H. Joint hand detection and rotation estimation using CNN. *IEEE Transactions on Image Processing, 27*(4), 1888–1900.

Secure Anti-piracy System

Junaid Khan, Akshatha Shenoy, K. M. Bhavana, Megha S. Savalgi, and Surekha Borra

Abstract This paper proposes a secure movie distribution and playing system which can avoid piracy in theaters. In the proposed system, an IR LED based anti-piracy screen is set up for discouraging illegal recording of movies. The optimization of power is achieved through progressive activation and deactivation of IR LEDs. Copyright protection of movie file is provided using invisible watermarking, and confidentiality and security are provided by encrypting the movie file as well as by enabling the decryption based on the authorized location verification of theaters. The system is also designed to allow only the authorized personnel to operate the system upon receiving the one-time password (OTP) from the owner. The experimental results proved that the developed system is secure and discourages the illegal video recording by creating artificial degradation in the recoded video. The system also helps to resolve rightful ownership via invisible watermarking.

Keywords Piracy · Watermarking · Encryption · Global positioning system (GPS) · Security · Copyright protection

J. Khan · A. Shenoy · K. M. Bhavana · M. S. Savalgi · S. Borra (✉)
Department of ECE, K.S. Institute of Technology, Bengaluru, Karnataka 560062, India
e-mail: borrasurekha@gmail.com

J. Khan
e-mail: junaidkhan4143@gmail.com

A. Shenoy
e-mail: akshathashenoy97@gmail.com

K. M. Bhavana
e-mail: bhavanamanjunath05@gmail.com

M. S. Savalgi
e-mail: megha.savalgi@gmail.com

S. C. Satapathy et al. (eds.), *Intelligent System Design*, Advances in Intelligent Systems and Computing 1171, https://doi.org/10.1007/978-981-15-5400-1_78

1 Introduction

Online piracy and illegal redistribution of movies has become challenging worldwide with the wide availability of inexpensive access to the broadband. Illegal capturing of the high-quality images and videos with the help of devices such as smart phones, cameras, and camcorders, uploading them on the cloud, and making the users to download it for free and watch as per their convenience have become a major issue these days. The entertainment industry is an important economic contributor, but with the increase in piracy, the industry is losing the economic value that it holds. A study says that 30% of UK population watches pirated movies which results £500 m loss for the film industry. Further, the piracy starts majorly at the cinema itself, followed by video on demand Web sites. The film makers, the producers, independent creators, the distributors, exhibitors, and actor's loss out the royalties, get demotivated to make sequels, remakes, and prequels, and lose self-confidence. Lower grade artists lose the jobs and benefits such as welfare and pension.

Indian film industry is the largest entertainment sector in the world. While China ranks first and the US ranks second, India ranks number fifth globally for piracy. According to a report by Irdeto, a global solutions provider in digital platform security and media and entertainment, the Indian media and entertainment industry loses $2.8 billion of its annual revenue to piracy. In India, cities such as Delhi, Mumbai, and Bangalore are renowned for piracy activities [1]. The piracy being quick, inexpensive, and easy encourages many people to follow it. But, under Digital Economy Act, educating the public and pirates, prosecuting the pirates for copyright infringement, helps to combat the piracy issues to some extent. In India, the Copyright Act, 1957, protects all type of music, songs, videos, literary, and artistic work. The Stop Online Piracy Act (STOP) is also one such act that restricts access to sites that host pirated content. The main aim of STOP is to crack down on copyright infringement [2].

A deep technical knowledge of state-of-the-art cybercrime methodologies is required for proper detection, identification, and tracking of unauthorized copies and associated pirates. Gathering of evidences for sending notices to discourage piracy, protect copyrights, and investments and to generate revenue from copyright violators for infringing the content is required. Hence, there is a necessity to develop a system to overcome piracy. The objective of the research is (i) to develop and implement a system which can avoid piracy in theaters, (ii) to degrade the visual quality of pirated video using IR LED screen, (iii) to make the entire theater system secure, and (iv) to optimize power. Section 2 presents literature survey on existing anti-piracy systems. Section 3 presents the proposed anti-piracy system. Section 4 concludes the paper.

2 Literature Survey

There have been many systems explored in order to overcome piracy problems. One of it was to project scanning beam light which disturbs the camera sensors from illegal video capturing. In another method, extra frames were added to cause flickering and distortion in the recorded video. Apart from this, there are other methods such as temporal psychovisual modulation (TPVM) and spatiotemporal modulation of light beam (STM).

Abhishek et al. [3] proposed a prototype system in order to overcome camcorder piracy by using the concept of projection of interference signals like IR radiations. The infrared transmitters are placed on the screen in order to degrade the visual quality of the recorded signal. For security purposes, radio frequency identification (RFID) tags are provided to each person who buys tickets. Due to these cards, person identification will be stored in separate systems. This information is used to tack the person who pirates the movie. Bourdan et al. [4] proposed a system in which the concept of spatiotemporal modulation of light has been used to produce temporal artifacts in the recorded movie. Nakashima et al. [5] model has strength to detect the recording position of the movie pirate using the soundtrack system. The experimental results showed a mean estimation error of 0.44 m with respect to the recording positions in a movie theater. Gao et al. [6] proposed a system which uses temporal psychovisual modulation [TPVM] to destroy recorded movie contents as well as to achieve visual transparency to the human eye. Veeraraghavan et al. [7] proposed a system in which infrared blasters produce IR rays which are easily perceivable by digital cameras. These blasters are inbuilt within the projector circuit, which gives out IR light beams along with the visible rays of the movie projection. It is directly connected to the microcontroller which helps in varying the radiation frequency and wavelength characteristics. This is done to avoid the use of infrared filters by a person while capturing the film using a camera. Karthik et al. [8] came up with a system that uses white LEDs to cause piracy. This system consists of a microcontroller that receives authentication from the owner in the form of a card that is being compared using a comparator. If the authentication is correct, then the signal is sent to microcontroller, activates LED transmitter section, and degrades the quality of the image that is being recorded in an illegal manner. The system makes use of a VLC transmitter and receiver section where constant current is required for illuminating the LEDs. Yamada et al. [9] demonstrated a system in which the camcorder comes at the center of the screen surrounded by IR cut filters. The region which consists of camcorder is seen as a red spot in the system because of the IR light that is being reflected from the cut filter and is being tracked by IR camcorder. Feng et al. [10] describes a system that uses 3D screen, laser, camera, cradle head, Arduino Uno, and LC glass. The job of cradle head is to detect the mobile phones when the system is working. The lasers used for setup emits light to the mobile camera and causes disturbances that make it uneasy to record the movie. The main principle here is the saturation laser attack and the temporal psychovisual modulation.

Video watermarking watermarks each movie with a unique watermark that identifies digital pirates when they illegally re-distribute multimedia content. The transform domain schemes are computationally complex but robust for signal processing attacks and are the right choice for resolving ownership issues [11–13]. Chen et al. [14] proposed a dual watermarking scheme in transform domain, where the algorithm uses contourlet transform to obtain high- and low-frequency coefficients. High-frequency sub-band is embedded with chaotic watermark sequence. To increase imperceptibility, highest energy coefficients are chosen. The histogram of the low-frequency coefficient is obtained, and the difference between neighboring areas is modified to embed the second watermark. Affine scale invariant feature transform (ASIFT) is used to test if the video is affected by geometric attacks, and if so, the video is corrected before the watermark extraction phase.

Mareen et al. [15] presented a framework where the watermark is embedded by modifying only a few coding decisions of the video encoder. As the coding decision is changed only for a particular region, the corresponding pixels in those regions differ only slightly from when optimal coding is used. The watermark is then a group of distributed implicit distortions. However, the proposed framework is suitable only for HEVC standard and is not compatible for other standards. Ali and Ghrera [16] proposed a split watermark embedding algorithm (SWEA) where the watermark is spilt into 64 smaller parts before being inserted into different I-frames of the input video. Since the watermark is inserted in parts, guessing the pattern of the inserted watermark is difficult. The technique is robust but has a tradeoff with the size of the watermark. Ahuja and Bedi [17] presented a semi-blind MPEG-2-based video watermarking framework. The binary watermark is inserted into the candidate I-frames. The luminance component (Y) is divided into 8×8 blocks followed by DCT application and quantization. Mareen et al. [18] proposed a low complexity and robust architecture that can perform high-quality watermarking along with fast compression rate using fewer resources.

3 Proposed Anti-piracy System

A watermark which consists of the producer's logo and theater information is embedded in the movie file to be secured. This file is then encrypted using Arnolds scrambling algorithm using a key to result a noisy looking video file. The encrypted video file is sent to the theater by the producer. The producer of the movie embeds the copyright information in the form of watermark image in the movie file, which can be extracted later as a proof of ownership. As soon as the system is turned ON, the respected owner of the theater receives a message regarding the same. The theater owner communicates the one-time password (OTP) with the theater operator to be entered using a keypad. The key that is entered is displayed on the LCD for visual verification. If the entered OTP matches, then the normal video or the movie starts playing, and at the same time, IR LEDs system turns ON. If in case the OTP/GPS location does not match, the movie does not get played. Simultaneously, a message is sent to the theater owner indicating the unauthorized access. The flowchart of the proposed anti-piracy system is shown in Fig. 1.

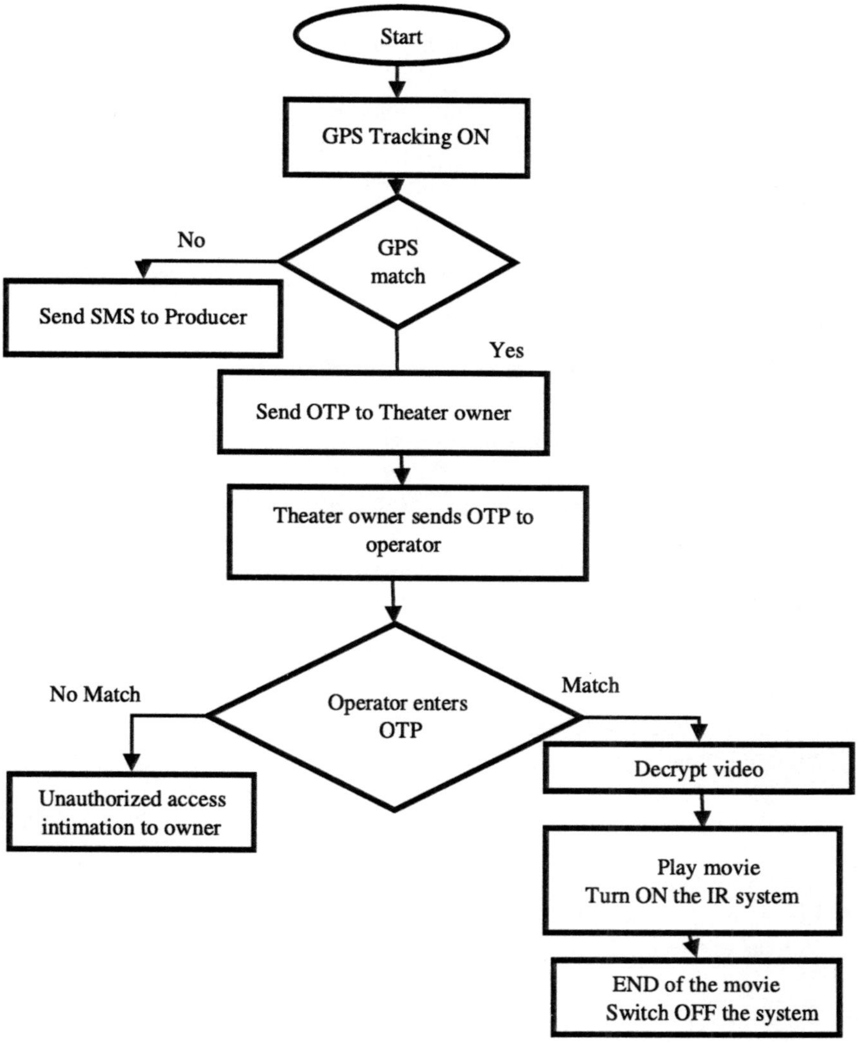

Fig. 1 Flowchart

3.1 Watermark Embedding

Input: Video file, producer secret key, logo to be watermarked.
Output: Watermarked video file.
Step 1: Break the original video which is broken into number of frames. Convert color space from RGB to YUV.
Step 2: Apply first level RDWT on Y frame to obtain wavelet coefficients of sub-band such as LL, LH, HL, and HH.
Step 3: For each frame, insert the watermark into LL sub-band of video frame using the equation below:

$$\mathrm{MLL}_f = \mathrm{LL}_f + \alpha \times W \tag{1}$$

where MLL_f is modified LL sub-band of video frame, LL_f is original LL sub-band of video frame, W is a watermark image, and α is a gain factor.
Step 4: Apply first level inverse RDWT on modified LL sub-band with unmodified sub-bands to get modified Y frames.
Step 5: Convert YUV to RGB to get watermarked frame. Finally, combine all the frames to get watermarked video stream.

3.2 Watermark Extraction

Input: Watermarked video file and extraction key.
Output: Extracted watermark.
Step 1: Break the original video into number of frames. Convert from RGB to YUV color space.
Step 2: Apply first level RDWT on Y frame to get its wavelet coefficients.
Step 3: Break the watermarked video which is broken into number of frames and convert to YUV space.
Step 4: Apply first level RDWT on Y frame to get its wavelet coefficients sub-bands: LLW, LHW, HLW, and HHW.
Step 5: Extract the watermark from each of the watermarked video frame using below equation.

$$\mathrm{EW} = \left(\mathrm{LLW}_f - \mathrm{LL}_f\right)/\alpha \tag{2}$$

where LLW_f is modified LL sub-band of watermarked video frame, LL_f is original LL sub-band of video frame, EW is an extracted watermark image, and α is a gain factor.

3.3 Arnold's Scrambling and Transform

A digital image can be represented as $Z = f(x, y)$ where $x, y \in \{0, 1, 2, 3 \ldots N-1\}$, where N represents order of the digital image. The plain image can be changed into cipher image using Arnold mapping function which changes a point (x, y) to another point (x', y')

$$X' = (x + y) \mod N \tag{3}$$

$$Y' = (x + 2y) \mod N \tag{4}$$

The Arnold transform is lossless invertible, lossless, area preserving, periodic, and scrambling technique that can be used to encrypt and decrypt frames of watermarked video.

The LEDs are placed all over the screen in a matrix form. A specific moving IR LED display pattern indicating the theater name, owner, GPS coordinates, location, and time is projected continuously from behind the screen to meet range of applications including copyright protection, broadcast monitoring, copy detection and control, and video authentication. The IR LED scrolling display pattern behind the screen acts as fingerprint, helps verify the authenticity, and traces the source of illegal copy. The IR LEDs emit infrared light, i.e., in the range of 700 nm to 1 mm. The infrared radiation is invisible and harmless to the human eyes, whereas it has profound effect on the video recorded using camcorders or mobile phones. The recorded video thus resulted is highly distorted, unpleasant to watch, and in turn helps to prevent piracy in theaters.

4 Results

Firstly, the movie file is made secure at the producer level, i.e., by passing it through watermarking and encryption algorithms for the purpose of resolving copyright ownership in the future and to provide high confidentiality. One of the movie file considered for experiments is shown in Fig. 2a. The watermark chosen for experiments is shown in Fig. 2b. The watermark is composed of both producer and owner information. The movie file is then embedded with the watermark using producer's secret key to give the embedded file as shown in Fig. 2c. The watermarked video file is further encrypted with the help of secret key to result a cipher video file which looks as in Fig. 2d. Both the embedding and encryption processes are done at the producer end. The encrypted movie file was decrypted by the theater owner as and when it is to be played using a secret key provided to him by the producer. The decrypted movie file using correct secret key is shown in Fig. 2e, and the extracted watermark is shown in Fig. 2f. The IR LEDs based anti-piracy screen degrades the image quality (as shown in Fig. 3b) when the attackers try to capture the movie using camcorders. In this system, the IR LEDs are placed behind the screen in a matrix

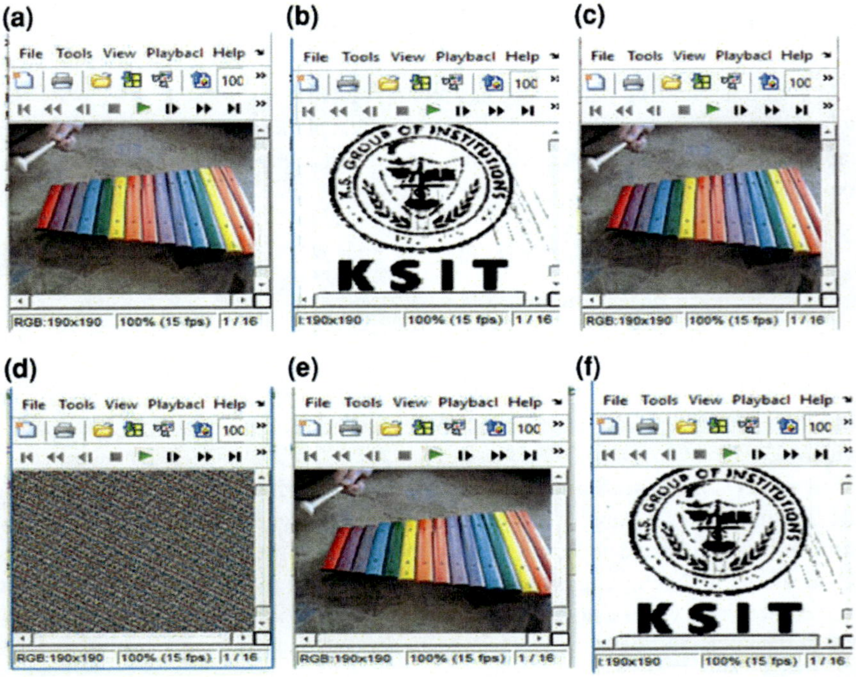

Fig. 2 **a** Movie frame, **b** watermark image, **c** watermarked movie frame, **d** encrypted movie file, **e** decrypted movie file, and **f** extracted watermark

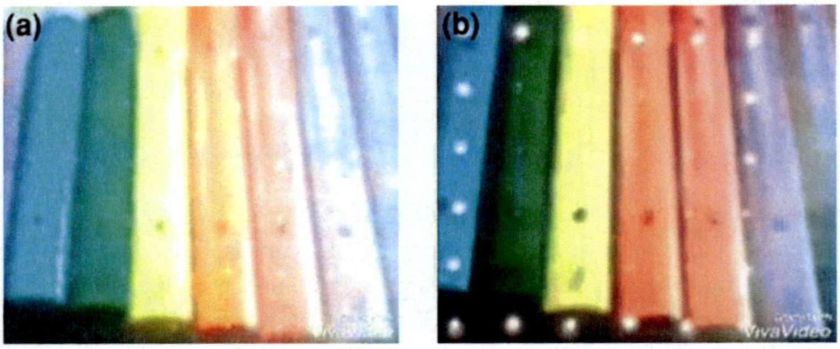

Fig. 3 **a** Original movie frame viewed by audience and **b** distorted movie frames recorded by camera

arrangement and are switched ON/OFF in a controlled manner progressively so as to reduce power consumption. The average PSNR calculated from the image frames of original and degraded movie is 15 dB in the experiments, which indicate poor visual quality of frames in pirated movies. The average PSNR for a degraded movie of 1 h 30 min durations is calculated and is obtained as 19 dB.

Referenes

1. https://www.theguardian.com/film/2014/jul/17/digital-piracy-film-online-counterfeit-dvds.
2. http://www.businessworld.in/article/Online-Piracy-Things-You-Need-To-Know/06-07-2018-154163/.
3. Abhishek, K. A., Chetan, G., Deepak, M. S., Akash, M., & Rohith, M. (2018). Camcorder piracy-RFID based anti-piracy screen. *International Journal of Scientific Research & Development, 6*(3), 1743–1746.
4. Bourdon, P., Thiebaud, S., & Doyen, D. (2008). A theoretical analysis of spatial/temporal modulation-based systems for prevention of illegal recordings in movie theaters. In *International Society for Optics and Photonics, in Security, Forensics, Steganography, and Watermarking of Multimedia Contents X* (Vol. 6819, p. 68190U).
5. Nakashima, Y., Tachibana, R., & Babaguchi, N. (2009). Watermarked movie soundtrack finds the position of the camcorder in a theater. *IEEE Transactions on Multimedia, 11*(3), 443–454.
6. Gao, Z, Zhai, G., Wu, X., Min, X., & Zhi, C. (2014). DLP based anti-piracy display system. In 2014 IEEE Visual Communications and Image Processing Conference (pp. 145–148).
7. Veeraraghavan, A. K., Shreyas Ramachandran, S., & Kaviarasan, V. (2017). A survey on reduction of movie piracy using automated infrared system. *International Journal of Innovative Research in Computer and Communication Engineering, 5*(11), 16633–16637.
8. Karthik, S. (2014). Anti-piracy screen using VLC. *International Journal of Engineering Trends and Technology, 35*(3), 122–124.
9. Yamada, T., Gohshi, S., & Echizen, I. (2012). Enhancement of method for preventing illegal recording of movies to enable it to detect cameras with attached infrared-cut filter. In *2012 IEEE International Conference on Acoustics, Speech and Signal Processing (ICASSP)* (pp. 1825–1828).
10. Lan, F., Zhai, G., Gao, Z., & Yang, X. (2016). Live demonstration: Screen piracy protection using saturation laser attack and TPVM. *In 2016 IEEE International Symposium on Circuits and Systems (ISCAS)* (pp. 2376–2376).
11. Thanki, R., & Borra, S. (2018). A color image steganography in hybrid FRT–DWT domain. *Journal of Information Security and Applications, 40*, 92–102.
12. Thanki, R., Dwivedi, V., Borisagar, K., & Borra, S. (2017). A watermarking algorithm for multiple watermarks protection using RDWT-SVD and compressive sensing. *Informatica: An International Journal of Computing and Informatics, 41*(4), 479–493. ISSN: 0350-5596.
13. Borra, S., Lakshmi, H. R., Dey, N., Ashour, A. S., & Shi, F. (2017). Digital image watermarking tools: State-of-the-art. *Frontiers in Artificial Intelligence and Applications, 296*, 450–459.
14. Chen, S., Li, Z., Cheng, X., & Gao, Q. (2017). Video dual watermarking algorithm against geometric attack based on ASIFT and contourlet transform. In *17th International Conference on Communication Technology (ICCT)* (pp. 1757–1762), October 2017.
15. Mareen, H., De Praeter, J., Van Wallendael, G., & Lambert, P. (2018). A novel video watermarking approach based on implicit distortions. In *2018 IEEE International Conference on Consumer Electronics (ICCE)* (pp. 1–2), January 2018.
16. Ahuja, R., & Bedi, S. S. (2017). Video watermarking scheme based on candidates i-frames for copyright protection. *Indonesian Journal of Electrical Engineering and Computer Science, 5*(2), 391–400.
17. Ali, J., & Ghrera, S. P. (2017). A secure method of copyright protection for digital videos using split watermark embedding algorithm. In *Fourth International Conference on Image Information Processing (ICIIP)* (pp. 1–5), December 2017.
18. Mareen, H., De Praeter, J., Van Wallendael, G., & Lambert, P. (2019). A scalable architecture for uncompressed-domain watermarked videos. *IEEE Transactions on Information Forensics and Security, 6*, 1432–1444.

Analyzing a Cattle Health Monitoring System Using IoT and Its Challenges in Smart Agriculture

Bharat Singh Thakur and Jitendra Sheetlani

Abstract The animal farming is an important component of the global income. Added financial gain will be noticed because of this category of mechanization, like the capability to determine the health issue of animals thereby and early precautions before spreading of its. Nowadays, lots of advance technical strategies in actual operations are produced by engineers as well as researchers. Among the advance specialized development streams is farm automation. Cattle farming is an area which demands understanding and interest on demands of every livestock vary depend on the variety of reproduct of it as well as geographical location. The main problem in the milk business includes absence of information accessibility among farmers, financial problems as well as lack of understanding regarding the health as well as well-being of the cattle. Cattle wellness monitoring system is latest research subject in farm automation. The IoT-based environment and cattle health monitoring technique monitor different livestock health parameters including body temperature, heartbeat, place of creatures as well as green variables like heat, humidity. The involvement of IoT would further elaborate the market as the info is going to be distributed with the whole public.

Keywords Cattle health · Smart agriculture · IoT · Monitoring system

1 Introduction

Animal farming takes a crucial part in the present time and is just about the majority attractive sectors because milk and animal products have an enormous demand around the globe. The quickly growing animal farming industry demands the utilization of more info on an everyday basis. It is vital for the growers to have real-time info about different reproducts of cattle as well as the habits of theirs. Several of the

B. S. Thakur (✉) · J. Sheetlani
Satya Sai University of Technology and Medical Science, Sehore, Bhopal, Madhya Pradesh, India
e-mail: bharattechs@gmail.com

J. Sheetlani
e-mail: dr.jsheetlani@gmail.com

S. C. Satapathy et al. (eds.), *Intelligent System Design*, Advances in Intelligent Systems and Computing 1171, https://doi.org/10.1007/978-981-15-5400-1_79

problems experienced as of today are not enough information regarding the feeding practices of various breeds of livestock, insufficient understanding regarding the sicknesses as well as absence of preventive healthcare procedures [1].

During last 2 decades, an essential requirement of farm automation being researched for wellness monitoring system. In this paper, researcher has focused on monitoring the wellness of cattle by making use of noninvasive, affordable sensors technological innovation which detects abrupt change in body parameter such as temperature and blood pressure. The parameter that's considered by sensors are entry by utilizing wireless engineering collect information use for early detection of disease this particular things will develops by utilizing IoT.

The primary goal of introducing brand-new solutions for this area is obtaining better output and income in a quick duration and establishing the husbandry in the simplest way with the newest IoT system which can help farm owner lessen the hard work as well as male power they have to commit and may also draw in far more individuals to invest into this particular company. In case cattle are not welfare for and treated very well, then the odds that the yield of the animal is impacted are significantly improved. The production area must be handled effectively. The production area needs to be cleaned on a regular basis, and also the animals need to be given regularly. This can alternatively lead to the greater lives and also the cattle giving far more yields [2]

2 Literature Review

Ramnath et al. [3] explained the IoT-based monitoring as well as localization for special packages. The experts proved that IoT-based methods are additional bendy compared to some other localization methods as well as the structures of theirs are undoubtedly more correct on monitoring and localization. The paper pointed out that big unfold deployment of IoT is an organic choice for localization as well as monitoring plans and the status of it is on the rise. Nevertheless, safety-associated elements are count of situation for all those packages [3].

Ilapakurti and Vuppalapati [4] described a novel method to cope with taking care of the important needs of dairy undertaking. We unequivocally acknowledge as legitimate with which the associated dairy will deliver colossal operational efficiencies, price budgetary money, and also substantial bits of expertise to tackle dairy cultivation creature's relevant important problems. Associated milk, drastically, is a record empowered smart unit which encourages much better command of dairy workouts. Sometime, connected milk provides anticipating bits of information that is useful for your time opportunity to milk functional command while in transit to much better method to protect dairy product from unexpected condition connected variants from the majority, dairy cow's health and crises [4].

Prasanna Venkatesan and Jegadeesan [5] have demonstrated concerning technology of wireless sensor network to the developments. Farm owner may support creation with this homestead course of action and make a database with information accumulated from the farm and environmental variable-based devices [5].

Lukonge et al. [6] have given ideas about some other cattle observing techniques and furthermore the advantages of utilizing remote sensor systems for checking steers center body temperature not withstanding place for the homesteads together with their disadvantages and advantages.

3 Methodology

Figure 1 gives a fundamental idea about the proposed method. The health is monitored by the temperature sensor, humidity sensor, and heart rate monitor. The temperature sensor is liable for checking the heat of the animals so that virtually any type of unusual temperature readings or anomalies is quickly monitored. The heart rate sensor is liable; this type of checking is necessary for the health price of breed; therefore, virtually in situation of abnormal heart rate or maybe irregularities perchance administered. Precisely necessary information may possibly be applied to anticipate the feelings on the pet in such a manner, thereby supporting the caretaker to quickly communicate to the creatures. The heart rate could be seen on a routine based to assess whether the breed is good as well as allows the owners to take measures that are essential to make certain the great health of the livestock.

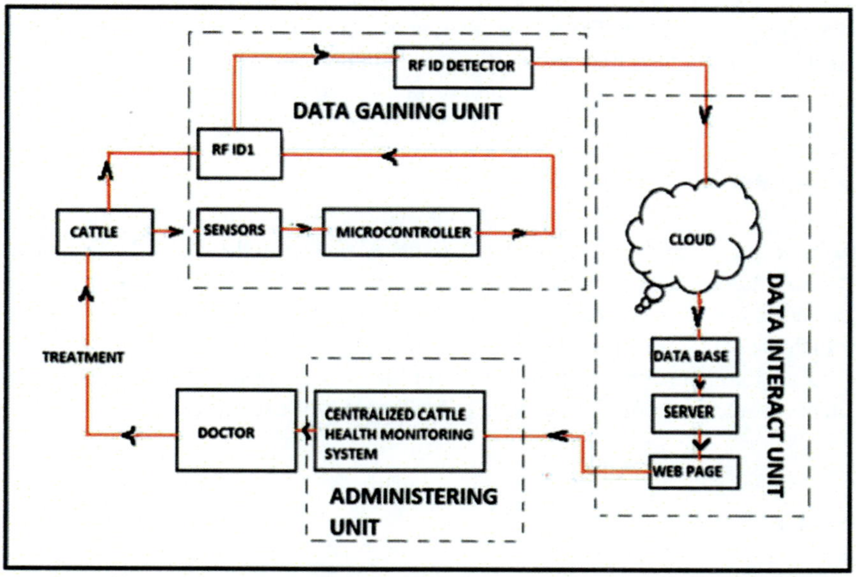

Fig. 1 Architecture of the proposed system

The data has been measured and stored by different sensors over a short time period based on the requirement of farm owner. Collected info delivered to a main processing unit (CPU) that is going to process the information to provide significant info therefore is examined by the pros. Information is stored and displayed on the Web site that will be seen by the administrator. This information is secure since it is entirely uploaded in the Web site. This info is published into the global server; therefore, it may able to commonly display the farmers or local people can use this info whereby they could improve their yield and efficiency by utilizing the information which has already been established to deliver results that are good.

Data gaining unit comprises different sorts of biomedical sensors as body temperature sensor, pulse sensor, dampness sensor, and heartbeat rate recognition sensor that are interfacing with microcontroller. The data gaining units acquire data and furthermore help make it open for any data overseeing just as data cooperate unit (cloud, server, and furthermore site page).

The sensors are used for fundamental and general programmed estimation of numerous therapeutic variables. Such kind of health sensors are introduced on the dairy cattle body that always sees the body issues of the bovines as temperature, heartbeat rate, and so forth and furthermore take paper inside the sort of electric signal. These signals are then set close by a standard point of confinement of typical qualities built up as the beginning spot in data managing unit.

If the managing unit has seen irregular changes or huge changes in certain steers, they can reach to closer creature care doctor, and if a few causes specialist would not offered in closer emergency clinic, after that utilizing IoT overseeing unit controlling individual can convey the diagram to the specialist. So by watching these specific diagrams, doctor can tell about the creature health data and furthermore appropriate treatment that to be actualized on steers without doctor during crisis (Fig. 2).

4 Experimental Results

Subsequent to interfacing the majority of the receptors on the steers' body are used for body temperature and furthermore heart pulse chart. Utilizing this specific chart, caretaker can monitor the fitness of the cattle animals and also identify diseases which the cows' are having? This is shown in Figs. 3 and 4.

The communication device is utilized to inform the farmer with message when there is some type of radical change in the fitness of the livestock.

5 Challenges

- Sensors poor quality: Either caliber of the receptors utilized is not satisfactory; the data gathered with it would not be dependable.

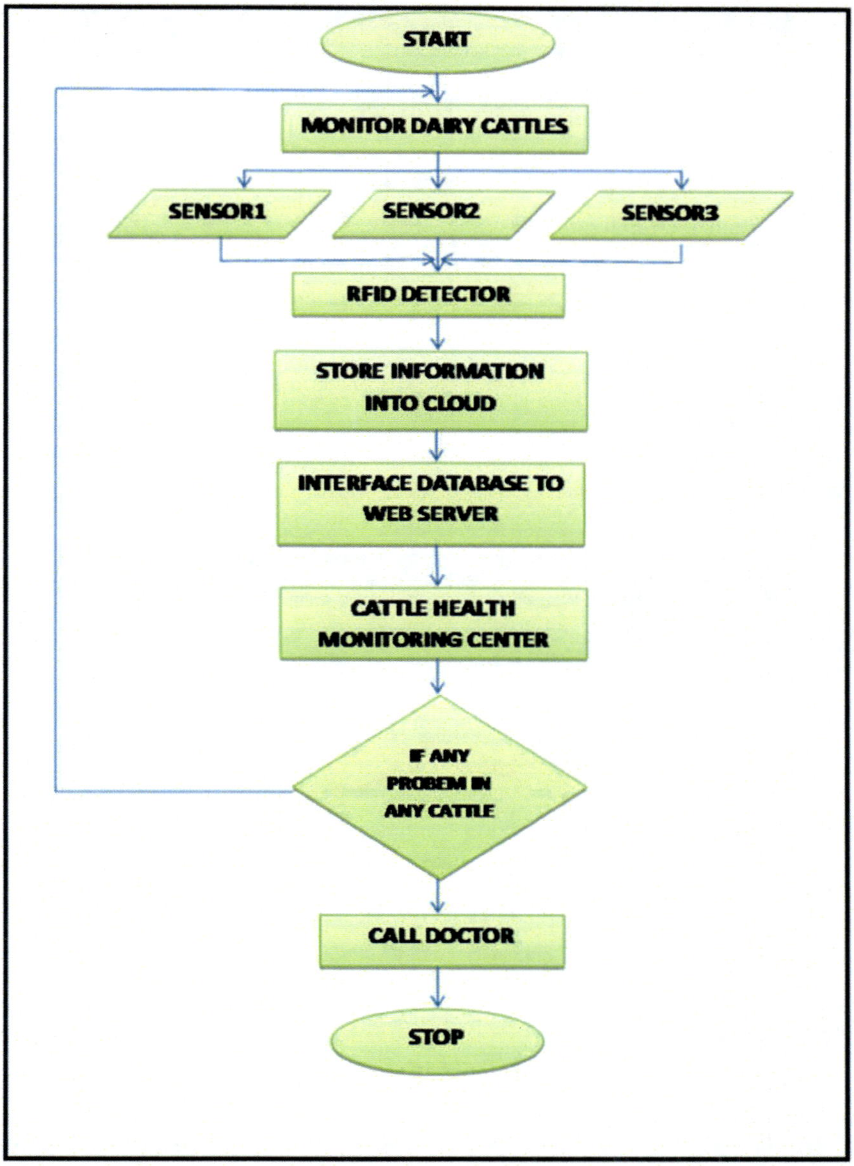

Fig. 2 Flowchart

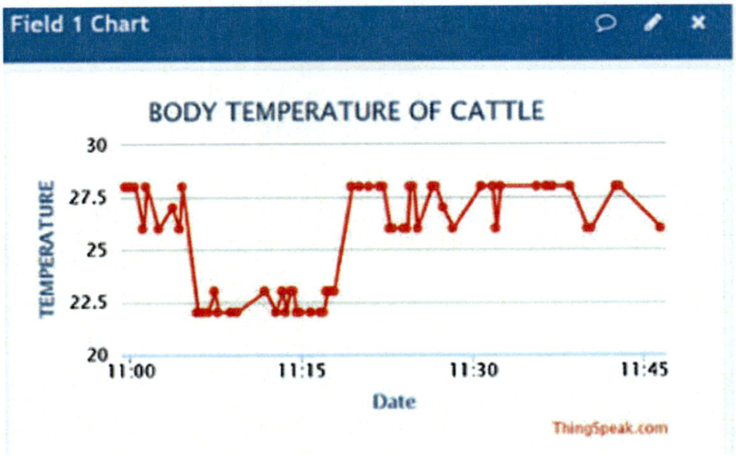

Fig. 3 Body temperature of cattle

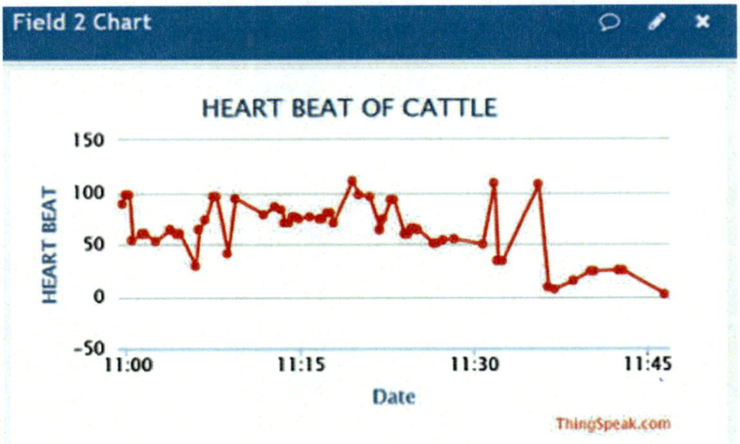

Fig. 4 Heart beat of cattle

- Relevant only for vast scale animal farming: The obtained strategy is fitting just for substantial scale steers cultivating as doing this specific framework for little scale cows cultivating will prompt the sustenance wastage of assets.
- Starting price is truly expansive: The establishment amount of the strategy is amazingly expensive as it uses plenty of detectors and circuit.
- Unusual conditions: Because investigation of data supports in protecting from issues, surprising mishaps cannot be anticipated.

- Partial Automation: Though the devices could be computerized which will bring down labor, total mechanization cannot make sure as despite everything it requires hand help and control.

6 Conclusion

This paper reviews natural checking and health of domesticated animals and furthermore so as to screen spot of the animals. I observe that using of IoT in animal health and condition observing item set on ranches to collect production environmental scenario that will consequently encourage ranchers in checking the creatures from outside the homestead. The proposed procedure actualizes driving edge innovation into this specific market in this way even novices can work pleasantly. The gadget will adequately actualize detecting devices to screen the everyday errands of the cows to assemble facts on standard premise. This fact is cautiously broke down making utilization of the program, and afterward required measures are used to deal with the dairy animals. Using IoT-based technique, it is doable for cultivators to foresee the conduct of creatures in a couple of conditions and perceive sicknesses the dairy cattle could get in the primary stages itself; accordingly; the turmoil might be distinguished just as keeping the spread of it to different animals in the ranch. The program appropriately actualizes most current innovation for draining the milking the cows.

References

1. Helwatkar, A., Riordan, D. (2014). Sensor technology for animal health monitoring. In *The 8th International Conference on Sensing Technology, Liverpool, UK.*
2. Meenakshi, M., & Kharde, S. S. (2016). Advance Cattle health monitoring system using Arduino and IOT. *The International Journal of Advanced Research in Electrical, Electronics and Instrumentation Engineering*. ISSN: 2278-8875.
3. Ramnath, S., Javali, A., Narang, B., Mishra, P., & Routray, S. K. (2017). IoT based localization and tracking. In *International Conference on IoT and Application (ICIOT)* (pp. 1–4).
4. Ilapakurti, A., & Vuppalapati, C. (2015). Building an IoT framework for connected dairy. In *2015 IEEE First International Conference on Big Data Computing Service and Applications (Big Data. Service)* (pp. 275–285).
5. Jegadeesan, S., & Prasanna Venkatesan, G. K. D. (2016). Smart cow health monitoring, farm environmental monitoring and control system using wireless sensor networks. *International Journal of Advanced Engineering Technology, Tamilnadu, VII*(I).
6. Lukonge, A. B., Kaijage, S., & Sinde, R. S. (2014). Review of cattle monitoring system using wireless network. *International Journal of Engineering and Computer Science, 3*(5). ISSN: 2319-7242.

Image Compression and Reconstruction Using Encoder–Decoder Convolutional Neural Network

Ajinkya Prabhu, Sarthak Chowdhary, Swathi Jamjala Narayanan, and Boominathan Perumal

Abstract Image compression has always been an important aspect in computer vision problems where image transmission is involved. Hence, many different image compression techniques have been developed over time. Image compression algorithms using neural network have already been implemented but is quite complex and difficult to run on conventional hardware. Most neural compression algorithms are used in general to input encoding and reconstruction, but the proposed algorithm takes advantage of over-fitting on the train dataset to improve reconstruction accuracy. In this paper, we propose a convolution neural network for lossy image compression and another neural network for the reconstruction of the image which uses encoder–decoder architecture. This kind of architecture enables us to process a large number of images in parallel, thus decreasing the overall time taken to compress a large number of images. This algorithm is easy to run on highly parallel hardware architectures, and it can process large number of images at a time. This has a good application in data centers with a large amount of computing resources and parallel hardware.

Keywords Image compression · Computer vision · Encoder-decoder architecture · Convolutional neural network

A. Prabhu · S. Chowdhary · S. J. Narayanan (✉) · B. Perumal
School of Computer Science and Engineering, Vellore Institute of Technology, Vellore, Tamil Nadu, India
e-mail: swathi.jns@gmail.com

A. Prabhu
e-mail: ajinkyamilind.prabhu2016@vitstudent.ac.in

S. Chowdhary
e-mail: sarthak.chowdhary2016@vitstudent.ac.in

B. Perumal
e-mail: boomi051281@gmail.com

S. C. Satapathy et al. (eds.), *Intelligent System Design*, Advances in Intelligent Systems and Computing 1171, https://doi.org/10.1007/978-981-15-5400-1_80

1 Introduction

Image compression is a compression technique which is used to compress digital images. Compression of image can reduce storage and transmission cost of the image. There are two main types of image compression: lossy and lossless [1]. Lossless compression is used where the decompressed image is to be as identical as possible with the original image, loss of data is minimum, and deviations from the original data are not favorable at all. Lossless data compression is typically used for text documents, source code, executable programs, and some images. Lossless audio compressions are used for archiving music and mass production purposes [1]. Most common lossless image compression is PNG. The term 'lossless' is used since no there is no loss of data during its processing. Most commonly, the lossless compression is achieved by generating a statistical model for the input data, and then this model is used to map input data to bit sequences so that probable data will produce a shorter output that is the compressed data. Various bit encoding systems are used to achieve lossless data compression. Various bit encoding systems are used in lossless data compression. Lossy data compression is also called irreversible compression since some of the data is lost while compressing, which makes it almost impossible to recreate an exact copy of the original image. Most common example of lossy image compression is JPEG. A lossy compression technique is proved to be good if the size of the file is reduced significantly till some sort of degradation is observed by the user. Major performance considerations of a lossy compression technique include: compression ratio, signal-to-noise ratio, and speed of encoding and decoding [1, 2]. The lossy compression on image and video looks for redundant pixel information to discard it and achieve compression by reducing its size frame by frame. This compression technique is used effectively when there is a need to work on the compressed data rather than the original ones that are not in use or unavailable.

Lossless data compression maintains the quality of the data, but it does not reduce the size of the file as much reduced as in lossy data compression does. The decision on the image compression format or the technique that is to be used depends on data that is being compressed. In this paper, the experiment is conducted on MNIST dataset and Fashion MNIST [3] for image compression using convolutional neural network architecture. This paper is organized as follows. In Sect. 2, literature survey is presented. Section 3 describes the problem. Section 4 describes the methodology proposed. In Sect. 5, computational experimental results are presented followed by conclusion, future work, and references.

2 Literature Survey

Image compression has been an interesting area of research in the past literature as well. In the year 2002, Masalmah and Ortiz [4] constructed a multilayer neural network for image compression. The network was trained by back propagation with different learning algorithms, namely one step secant, Newton's method, gradient descent, and adaptive gradient descent learning algorithms.

Charles et al. [5] developed an adaptive method for image compression which is based on complexity level of the image and applied a modified Levenberg–Marquardt algorithm. Anjana and Shreeja [6] used a modified version of Levenberg–Marquardt algorithm for training of neural network. Compression and decompression of image were performed separately, and both required normalization in order to be identified by the neural network. Cavigelli et al. [7] developed a 12-layer deep convolutional neural network which uses compression artifact suppression on JPEG images, trained the neural network using multi-scale loss function, and achieved very good results. Chen et al. [8] developed a hybrid block-based image compression method. A convolutional network that incorporates dense connection and identity skip connection was proposed to enhance the performance of in-loop filtering; then, a CNN-based method was employed to predict probability distribution of syntax elements and reduce the volume of symbols to be encoded by the block-based encoder. The goal of the network was to restore the information lost in the encoding process, where the input and the target are highly similar. Liu et al. [9] proposed a solution which effectively reduces the quantization error based on the frequency component analysis and rectified quantization table, and further increases the compress rate without any accuracy degradation. It works quite differently from the human vision system-inspired JPEG compression. Cheng et al. [10] proposed techniques based on band-pass filtering to increase the flexibility and scalability of deep neural networks for image reconstruction and demonstrated the architecture for reconstruction of subsampled datasets of MRI scans. CNNs were used for image reconstruction and tried to overcome the obstacles which limit the use of CNN in general image reconstruction. On the other hand, though logistic regressions were used for binary classification, still they were incorporated in final layers in deep neural networks and ReLU activation function was used as it provides a solution for the vanishing gradient problem [11]. Convolutional neural networks have achieved state-of-the-art accuracy in image classification problems as shown by Krizhevsky et al. [12]. De-convolution has also been used in multiple computer vision tasks such as semantic segmentation [13]. Implementing neural networks has also been simplified by the use of frameworks such as PyTorch [14].

3 Problem Description

Image compression is performed mainly to achieve minimal storage requirement. Current image compression techniques suffer in achieving good compression results after a certain compression ratio. In this paper, we focus on the research problem of lossy compression to achieve high compression ratio with minimal noise. We propose to use compression and reconstruction technique by taking advantage of convolutional neural networks which have shown state-of-the-art performance in image processing and computer vision problems.

4 Methodology

In our proposed work, image compression and reconstruction are achieved using convolutional neural network. To convert an image into vectors and again to convert it back to image, we have used two convolution neural networks: one for encoding, i.e., for compression, and another one for decoding i.e., reconstruction of original image. The first CNN takes the input image and applies 'same 2d convolution' and '2d max pooling' operations in 3 different layers. Hence, the convolution and max pooling operations are applied 3 times. The number of input and output channels for convolution in each layer is different, with first layer, second layer, and third layer having input and output channels as '1 and 64', '64 and 128', '128 and 256', respectively. The layers have the same kernel size of 3 for both convolution and max pooling. Each convolution operation is applied with padding set as '1'. In each layer, rectified linear unit is used as the activation function. This finally gives us a 256 vector of the image. This vector is given as input to the second CNN which also has 3 layers and in each layer; we apply 2d transpose convolution and 2d max unpooling operation (Fig. 1).

The number of input and output channels used for 2d transpose convolution in these layers namely fourth, fifth, and sixth layer are '256 and 128', '128 and 64', '64 and 1', respectively. The padding is set as 1 for each transpose convolution operation, and the kernel size for both the transpose convolution and max unspooling operation is set as 3. The activation function used in these layers is rectified linear unit, and the output is obtained from the second CNN of the compressed image which has been reconstructed from the 256-dimensional vector. Mean square function is used as the loss function for the proposed model. Adaptive moment optimizer is used as an optimization algorithm. The training of the model is conducted by passing the same target objects as input and output in order to accurately reconstruct the input image from the compressed lower-dimensional space. In order to down sample the image, we apply 2d convolution and 2d max pooling in 3 different layers of CNN and compress it. Rectified linear unit activation function is chosen for the CNN since it is easy to compute, it converges quickly, and it is sparsely activated. It also avoids the vanishing gradient problem encountered with sigmoid activation. Then

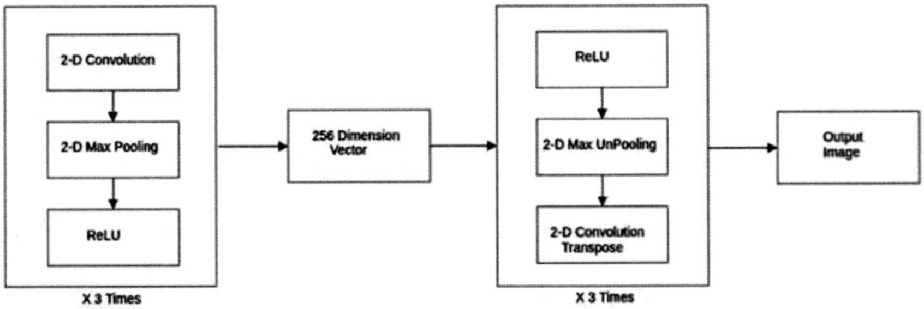

Fig. 1 Proposed CNN architecture for image compression and reconstruction

toward the reconstruction of the original image from the compressed image, another CNN is used in which 3 more layers are used to upscale the image by using the 2d convolution-transpose which applies a 2D transposed convolution operator over the image and 2d max unspooling to compute a partial inverse of MaxPool2d. This methodology creates a kind of bottleneck network for the proposed problem to be solved.

5 Computational Experimental Results

As explained in Sect. 4, we use a convolutional encoder and convolutional decoder model. The training is set up such that the encoder first compresses and then decoder decodes or reconstructs the original image. The loss function (mean square error) is used in order to back propagate and to correct the weights such that the error between the reconstructed image and original input image is minimized. For carrying out the training process, we used an Nvidia Titan X pascal with 12 GB Video Memory. Batch size was set to 128, and adaptive moment optimization algorithm was used to train the network. The framework used to create the whole model was pytorch which has Nvidia CUDA support for parallel computation. Datasets used to produce results were MNIST and Fashion MNIST. We obtained a compression ratio of around 6 with a minor addition of noise in the reconstructed image. This algorithm also takes advantage of hardware accelerators like GPUs. This enables us to compress large amount of images very quickly. The proposed method would be quite effective for a large number of images and takes advantage of over-fitting of the model. Though the compression technique is lossy, still artifacts are preserved in the reconstructed images. This makes it suboptimal in cases where image clarity is required but in cases where image clarity is not a priority for example in archive storage, then our proposed technique is quite powerful. Also we observe that we obtained appreciable accuracy for compression at just one epoch itself. We obtained results for 2 datasets one the MNIST handwritten digits dataset and two the Fashion MNIST dataset. Some sample images after 2–4 epochs are shown in Figs. 2, 3, 4, 5, 6, 7, 8, 9 and 10. The results obtained are run on a Titan XP with 12 GB video Ram and each epoch takes around a minute for batch size of 128. Of course, a smaller batch size will require longer to train but since the network runs quickly it is feasible on slower hardware. In the light of recent literature the proposed model performance is quick in terms of training and has low memory overhead when compared to architecture proposed by Toderici et al. [15] which is very difficult to train and memory intensive due to their use of recurrent units. The loss values achieved for the results in this paper is given in Table 1.

Fig. 2 Fashion MNIST at 5 epochs

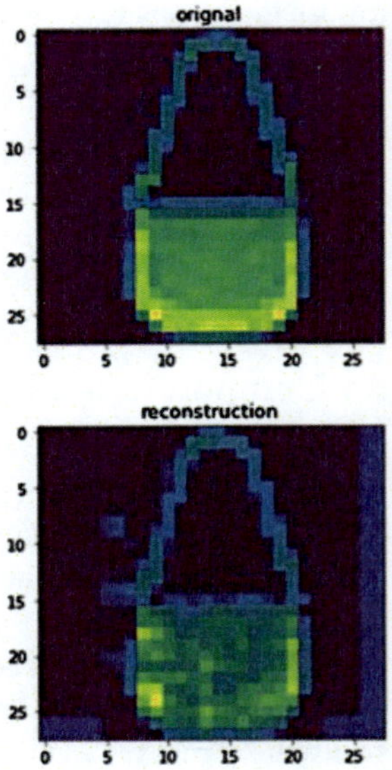

Fig. 3 Fashion MNIST at 1 epoch

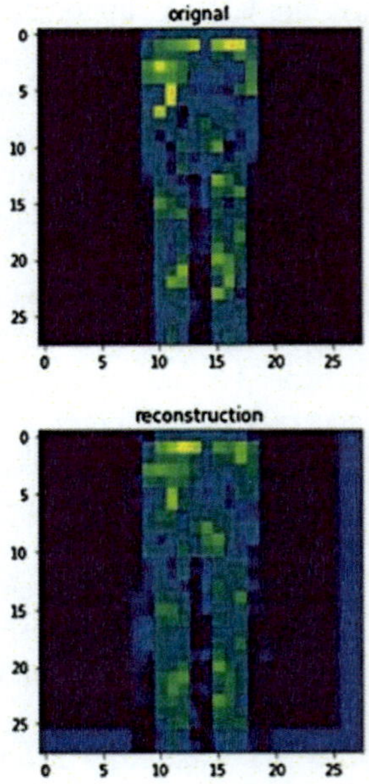

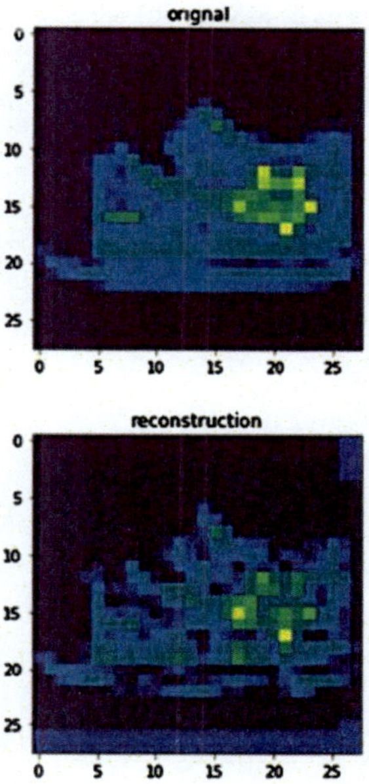

Fig. 4 Fashion MNIST at 2 epochs

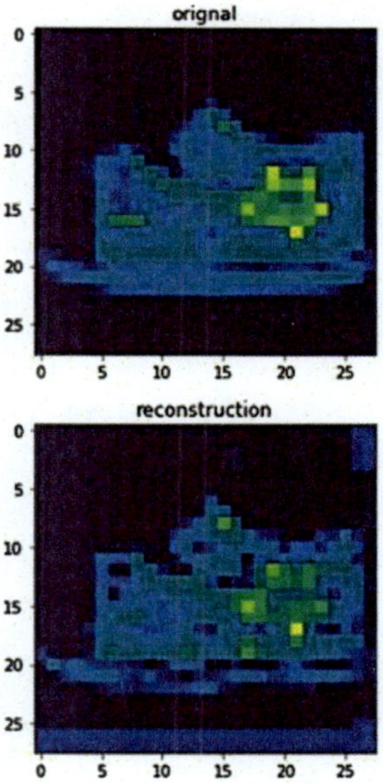

Fig. 5 fashion MNIST at 4 epochs

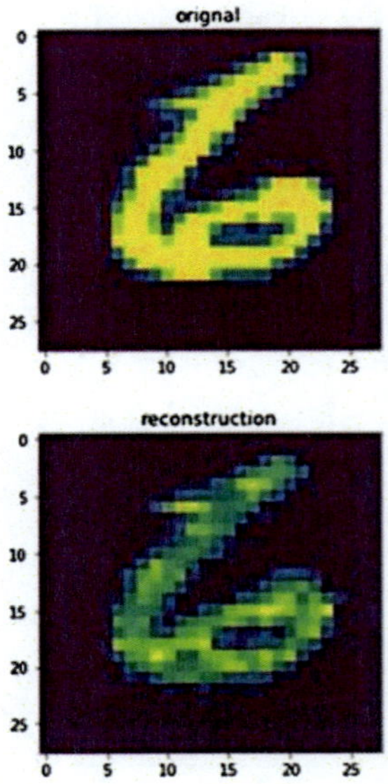

Fig. 6 MNIST at 1 epoch

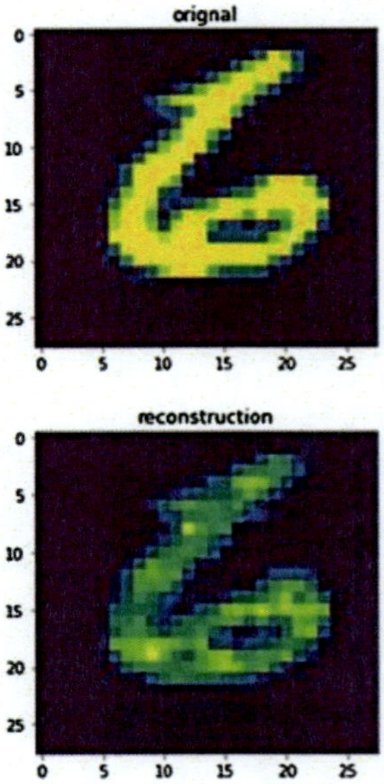

Fig. 7 MNIST at 2 epochs

6 Conclusion and Future Work

Image compression is important research field in problems related to computer vision. The images generally take a lot of space and conventional compression techniques suffer in achieving high compression ratio. To overcome this issue, we proposed

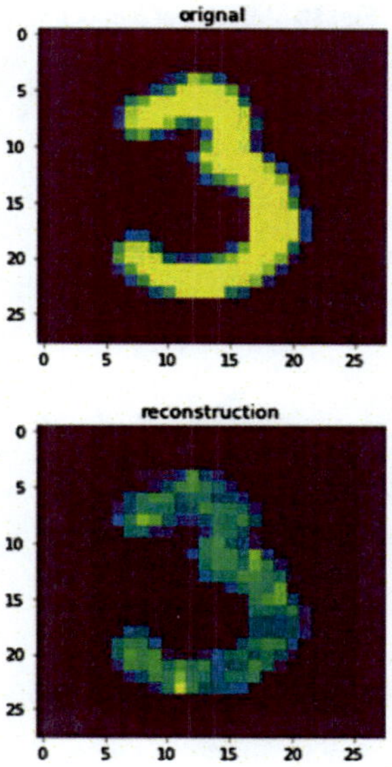

Fig. 8 MNIST at 1 epoch

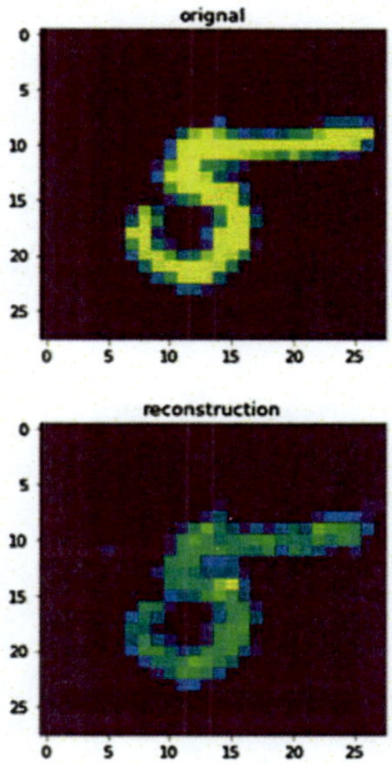

Fig. 9 MNIST at 2 epoch

encoder- decoder CNN architecture which achieved quite a good compression ratio with a minimal amount of training and is quite easy to run on even lower specification hardware. Its high compression ratio makes it ideal for cases where image artifacts are not of concern. The proposed model would also be able to compress a large number of images at a time by employing a parallel algorithm which processes multiple images

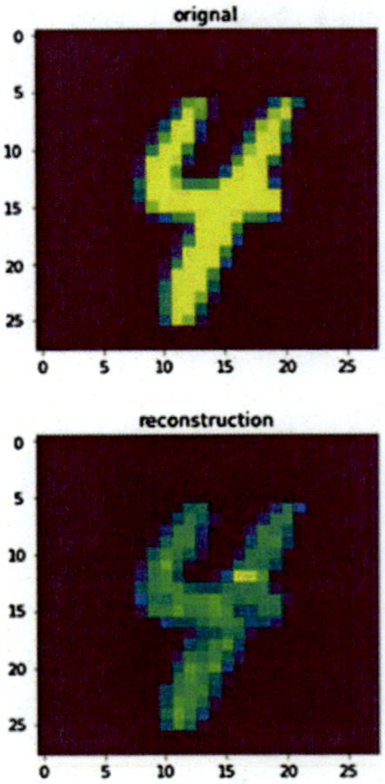

Fig. 10 MNIST at 3 epochs

Table 1 Loss values for the images considered

Fig No.	Loss value
2	50.361
3	24.22
4	22.37
5	20.51
6	138.70
7	70.39
8	22.136
9	12.068
10	8.382

at the same time. Since the proposed model uses GPU acceleration, larger throughput can be achieved, compression of even larger images would be possible and obtaining higher quality reconstruction would be highly feasible. Future work would be to test this model architecture with larger datasets and more complex images with color and a larger number of layers which would only increase reconstruction accuracy.

Acknowledgements The authors thank VIT for providing 'VIT SEED GRANT' for carrying out this research work. We gratefully acknowledge the support of NVIDIA Corporation with the donation of the Titan Xp GPU used for this research on Image Compression and reconstruction using Convolution Neural Networks.

References

1. Vijayvargiya, G., Silakari, S., & Pandey, R. (2013). A survey: Various techniques of image compression. *International Journal of Computer Science and Information Security, 11*(10), 51–55.
2. Mathur, G., Mathur, R., & Mathur, M. R. (2014). A comparative study of various lossy image compression techniques. *ETRASCT, 14*, 165–169.
3. LeCun, Y., Bottou, L., Bengio, Y., & Haffner, P. (1998). Gradient-based learning applied to document recognition. *Proceedings of the IEEE, 86*(11), 2278–324.
4. Masalmah, Y. M., & Ortiz, D. J. (2002). *Image compression using neural networks*. Puerto Rico: Electrical and Computer Engineering Department, University of Puerto Rico.
5. Charles, P. K., Khan, H., Kumar, C. R., Nikhita, N., Roy, S., Harish, V., et al. (2011). Artificial neural network based image compression using Levenberg-Marquardt Algorithm. *International Journal of Modern Engineering Research (IJMER), 1*(2), 482–489.
6. Anjana, B., & Shreeja, R. (2012). Image compression: An artificial neural network approach. *International Journal of Computational Engineering Research, 2*(8), 53–58.
7. Cavigelli, L., Hager, P., & Benini, L. (2017). CAS-CNN: a deep convolutional neural network for image compression artifact suppression. In *International Joint Conference on Neural Networks (IJCNN)* (pp. 752–759).
8. Chen, Z., Li, Y., Liu, F., Liu, Z., Pan, X., Sun, W., Wang, Y., et al. (2018). CNN-optimized image compression with uncertainty based resource allocation. In *Proceedings of the IEEE Conference on Computer Vision and Pattern Recognition (CVPR) Workshops* (pp. 2559–2562).
9. Liu, Z., Liu, T., Wen, W., Jiang, L., Xu, J., Wang, Y., & Quan, G. (2018). DeepN-JPEG: A deep neural network favorable JPEG-based image compression framework In *Proceedings of the 55th Annual Design Automation Conference ACM*.
10. Cheng, J. Y., Chen, F., Alley, M. T., Pauly, J. M., & Vasanawala, S. S. (2018). *Highly scalable image reconstruction using deep neural networks with bandpass filtering* arXiv:1805.03300v2.
11. Nair V, and Hinton GE: Rectified Linear Units Improve Restricted Boltzmann Machines. In: *Proceedings of the 27th international conference on machine learning (ICML-10):* 807–814. (2010).
12. Krizhevsky, A., Sutskever, I., & Hinton, G. E. (2012). ImageNet classification with deep convolutional neural networks. In *Advances in neural information processing systems* (pp. 1097–1105).
13. Noh, H., Hong, S., & Han, B. (2015). Learning deconvolution network for semantic segmentation. In *Proceedings of the IEEE International Conference on Computer Vision* (pp. 1520–1528).
14. Paszke, A., Gross, S., Chintala, S., Chanan, G., Yang, E., DeVito, Z., Lin, Z., et al. (2017). Automatic differentiation in PyTorch. In *Proceedings of 31st Conference on Neural Information Processing Systems.*
15. Toderici, G., Vincent, D., Johnston, N., Jin Hwang, S., Minnen, D., Shor, J., & Covell, M. (2017). Full resolution image compression with recurrent neural networks. In *Proceedings of the IEEE Conference on Computer Vision and Pattern Recognition* (pp. 5306–531).

Enhancement of Synthetic-Aperture Radar (SAR) Images Based on Dynamic Unsharp Masking

Ankita Bishnu, Vikrant Bhateja, and Ankita Rai

Abstract Synthetic-aperture radar (SAR) images are speckle corrupted at the initial stage of their acquisition. For this reason, despeckling of the images is required to preserve the finer details without losing them. But, quite often it was seen that even after despeckling, there persists residual noise at higher noise levels and obscured or weak edges. Therefore, to overcome these problems, in this paper, a method of enhancement using dynamic unsharp masking has been proposed for various noise levels. It uses the concept of variable weight factor which is calculated for a single pixel of the image. Dynamic unsharp masking controls the enhancement of the images adaptively such that there is no overshooting in images. The sharpening competency of the proposed method is then scrutinized and compared with conventional unsharp masking using an evaluation parameter called measure of enhancement (EME).

Keywords SAR images · Speckle · Enhancement · Conventional unsharp masking · Dynamic unsharp masking

1 Introduction

Synthetic-aperture radar is a radar type capable of capturing two-dimensional and three-dimensional imageries of earth's various geographical landscapes. While acquiring images of landscapes, there is an interference phenomenon of the reflecting signals which occurs at the synthetic transducer aperture. These backscattered signals

A. Bishnu · V. Bhateja (✉) · A. Rai
Department of Electronics and Communication Engineering, Shri Ramswaroop Memorial Group of Professional Colleges (SRMGPC), Tiwari Ganj, Faizabad Road, Lucknow, Uttar Pradesh 226028, India
e-mail: bhateja.vikrant@gmail.com

A. Bishnu
e-mail: ankita.bishnu5@gmail.com

A. Rai
e-mail: ankita.rai3131@gmail.com

Dr. A.P.J. Abdul, Kalam Technical University, Lucknow, Uttar Pradesh, India

S. C. Satapathy et al. (eds.), *Intelligent System Design*, Advances in Intelligent Systems and Computing 1171, https://doi.org/10.1007/978-981-15-5400-1_81

when added coherently results in the formation of black and white dots in the images [1]. Therefore, speckle noise is also called granular noise and which can also be modeled as multiplicative noise [2]. Speckle noise predominantly reduces the image representation, image data extraction, and classification. Therefore, SAR imaging is followed by an important step called despeckling which eventually eliminates the speckle noise from the images. But, there are three major challenges faced in the despeckling process of SAR images. The first being the despeckling of the images makes them blurred out mostly at the boundaries or edges [3]. Second, the contours and even some portions of the homogeneous regions become weak as a result information derivation becomes quite a demanding task [3]. Third challenge is that there prevails residual noise even after denoising in high noise variance [4]. The challenges of despeckling are overcome by the enhancement method which increases the contrast enhancement of the images as well as removes the residual noise. In the past few decades, various enhancement techniques had been devised to enhance the features of the images as well as for highlighting the finer details of the edges. Some of the prominent enhancement methods include histogram equalization (HE) [5] and contrast-limited adaptive histogram equalization (CLAHE) [5], which is an advanced version of HE. While HE achieves the image contrast through nearly uniform gray level output, CLAHE processes on small patches of the image. Another contrast enhancement approach which is an alternate of HE is range limited bi-histogram equalization (RLBHE) [6]. Fuzzy sets (FSs) theory [7] and multi-scale transformation [8–10] are some of the widely used solutions applied in the recent years. One of the most appreciated and extensively used methods in the field of enhancement is that of unsharp masking (UM) [11–13]. The purpose of UM is to improve the contour details by adding high-frequency data to the original captured image based on a constant gain factor. In this paper, an improved method of conventional UM has been proposed for enhancing SAR images. The improved version is based on dynamic UM (DUM) where instead of a constant gain factor, an adaptive weight factor is used. The following paper is categorized in the following sections: Sect. 2 contains a brief description of dynamic unsharp masking. Section 3 discusses the proposed enhancement method using (DUM). Section 4 contains the analysis of results and discussions which is followed by conclusion of the overall work stated in Sect. 5.

2 Contrast Enhancement of SAR Images

Satellite borne SAR images have the potential to view broad expanses of earth's landforms in day as well as night time independent of prevalent weather conditions. However, these images lack in meaningful mapping of image data at first sight especially for an inexperienced observer [14]. For consistent maintaining of image perception mainly for study and research purposes, edge detection and enhancement are essential in the post-processing of SAR images. The target is to highlight major useful image information and to segregate the differences between various features, in order to improve the image interpretation, classification, and analysis [15]. The visual appearance of SAR images is vividly emphasized by their high-frequency components to particularly enhance the edge information. The classical (conventional) unsharp masking is widely applied technique in enhancement domain. In the UM technique, a high-passed scaled version of the original input image is first multiplied by a constant gain factor, and then, the resultant edge-enhanced image is added to the original input image itself to give a sharpened (enhanced) image [16]. UM can be expressed by the following formula:

$$U = G + \lambda(G - G \cdot H_{\mathrm{LP}}) \tag{1}$$

where G and U depicts the raw and enhanced image, respectively, λ is a constant gain factor, and H_{LP} denotes a high-pass filter. But, the UM technique has two major shortcomings in the image processing. First, the operation of linear high-pass filter on the image makes it prone to noise sensitivity. Second, when a constant gain factor is multiplied to the entire image, some regions of the image gets over enhanced than other portions that fail to display high image dynamics. Therefore, some sort of overshoot artifacts occurs in the final enhanced image [17]. To overcome the mentioned limitations of UM, an improved variation of UM which is the technique of dynamic unsharp masking (DUM) has been proposed, where an adaptive weight factor has been used to adaptively enhance the image as and where required. The DUM is broadly described in the following section.

3 Proposed Design Methodology

In the proposed enhancement method of DUM, unlike CUM which uses a constant gain factor to increase the high-frequency components throughout the image, DUM uses an adaptive weight factor $h(i, j)$ for individual pixel (i, j) of the image matrix. This adaptive factor is a replacement over the constant gain (G) of CUM for the whole image [18]. The DUM can be expressed by the following formula [18]:

$$U = G + h(i, j)(G - G \cdot H_{\mathrm{LP}}) \tag{2}$$

where U is the sharpened image, G is the raw input, and $h(i, j)$ is the adaptive weight factor for each pixel (i, j) where subscripts i and j are the number of rows and number of columns, respectively, of the image matrix. $h(i, j)$ can be expressed by the following formula [18]:

$$h(i, j) = \alpha \frac{\text{mean}}{\mu(i, j) + \log(\sigma(i, j))} \tag{3}$$

where mean is the gray average for the entire original input image,μ denotes the mean and variance, respectively, of the window taken which is run serially all over the image to process all the image pixels, and α is a constant weight factor used in DUM. α is an important parameter of DUM as it maintains the finer details and to some extent reduces overshooting effects. The logarithm of the variance is taken to ensure enhancement response when the variance is large. Step by step algorithm of the proposed methodology is formulated below. The image quality assessment of the enhanced image is done using an absolute index called EME [19].

BEGIN

Step 1: *Read*: Despeckled SAR image (G)
Step 2: *Process*: High-pass filter on the despeckled SAR image (G)
Step 3: *Compute*: $h(i, j)$ Adaptive weight factor for each pixel using Eq. (3)
Step 4: *Process*: Multiply $h(i, j)$ to the edge image (E) to obtain the edge-enhanced image (E′)
Step 5: *Process*: Add E′ to G
Step 6: *Output*: Enhanced SAR image (U)

END

4 Results and Discussions

For testing the performance of the proposed enhancement method, images are acquired from the dataset of RADARSAT-2 image gallery of geospatial-international resources of MDA Corporation, Canada. The image that has been used for the experiment is a RADARSAT-2 image of Lake Claire, Alberta. The northern notch of Lake Claire which is situated in Wood Buffalo National Park in north of Alberta, Canada, can be visualized in this image. This image is captured from the wide ultra-fine RADARSAT-2 imager. Wide ultra-fine devices from RADARSAT-2 ensure high-resolution imagery (3 m) over a landscape size of 50 km $\times$ 50 km. The area contains the Peace River cutting across the top of the image [20]. The parameters which have been fixed for the experiment are the window size, which is set to 5, and the typical value of α for the experiment **which** is set to 0.3. The proposed DUM technique is applied on the SAR images which have been filtered using local statistics-based adaptive filtering [21]. The despeckled images are then processed further to emphasize the sharpness of the details which got blurred by the despeckling method. The results are shown for three different noise levels which are (i) low noise variance, (ii) medium noise variance, and (iii) high noise variance. The low noise variance is 0.01, medium noise variance is 0.05, while high noise variance is 0.1. The images in results are shown for low and medium noise variance. The enhanced (sharpened) images of all the noise levels are shown in Fig. 1f, g.

Performance metric like EME [19] of the original image, noisy images, filtered images [3], sharpened images by CUM, and sharpened images by DUM are given in Table 1, and it has been seen that the DUM performs reasonably well in all types of speckled images, i.e., low noise level, medium noise level, and high noise level.

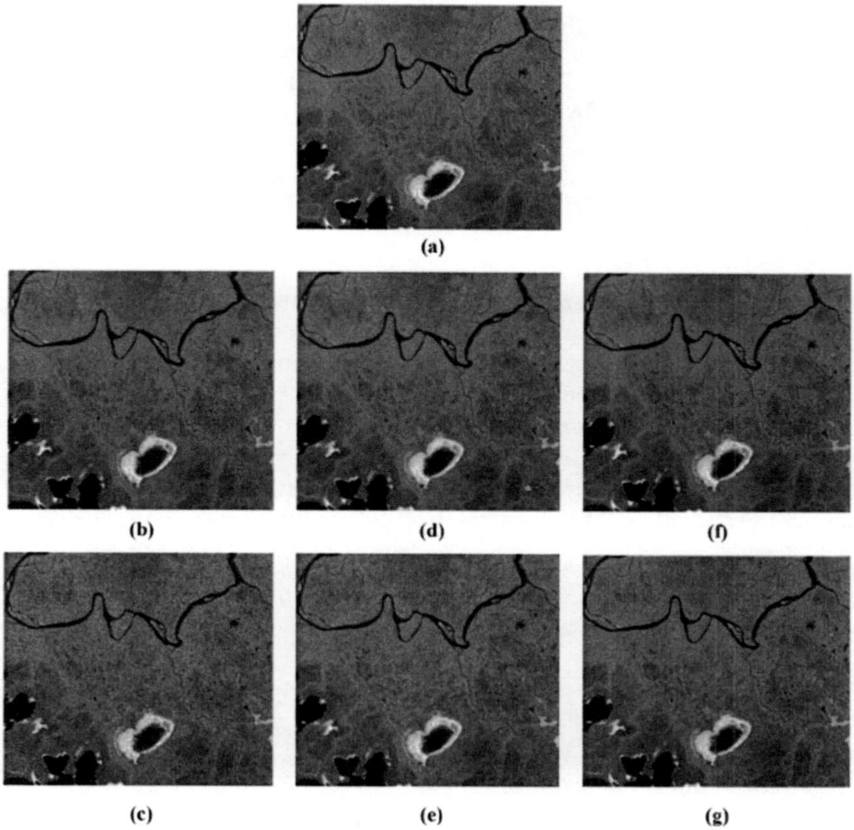

Fig. 1 Simulation results of the proposed DUM. a Original SAR image. b Image with low speckle variance 0.01. c Images with medium speckle variance 0.05. d, e Filtered images of respective noisy images. f, g Enhanced (sharpened) images of respective despeckled images

Table 1 Simulation results of EME enhanced SAR images

Noise variance	Speckled image	Despeckled image	CUM	DUM
0.01	10.8375	11.2413	12.1029	15.3266
0.05	8.3266	10.8203	14.1175	19.8007
0.1	6.7261	8.7502	16.6554	23.3423
Original image		14.5439		

Higher the EME values, greater is the enhancement (sharpening) of the images. It can be seen that in Fig. 1b, for low speckle variance, i.e., in already noiseless image, speckle is almost completely removed from the image and the enhanced image by DUM in Fig. 1f which is almost comparable to the original image (a). However, in medium noise variance, the residual noise has been considerably removed, and edges and features have become more prominent as can be seen in Fig. 1g. The EME [19] values in Table 1 display that performance of DUM technique is appreciable even in higher speckle content. Thus, it is evident that DUM is much more effective in enhancement domain of SAR images than CUM.

5 Conclusion

This paper presented a technique based on DUM for image enhancement. The technique employs dynamic weight factor which changes adaptively according to the individual pixel value of the image. This promotes adaptive enhancement of details where certain regions get enhanced selectively thereby preventing the overshooting and overbrightening phenomenon and also suppresses the minimal residual noise that pertains after despeckling of SAR images. Experimental results shown in this paper demonstrate that the proposed technique works effectively at high noise level as compared to the CUM approach in terms of clarity in image visualization and interpretation.

References

1. Bhateja, V., Gupta, A., & Tripathi, A. (2013). Despeckling of SAR images in contourlet domain using a new adaptive thresholding. In *IEEE 3rd International on Advance Computing Conference (IACC)* (pp. 1257–1261).
2. Bhateja, V., Tripathi, A., & Gupta, A. (2014). *Recent advances in intelligent informatics* (pp. 23–26). Switzerland: Springer International Publishing.
3. Aja-Fernández, S., & Alberola-López, C. (2006). On the estimation of the coefficient of variation for anisotropic diffusion speckle filterin. *IEEE Transactions on Image Processing, 15*(9), 2694–2701.
4. Yu, Y., & Action, S. T. (2002). Speckle reducing anisotropic filtering. IEEE *Transactions on Image Processing, 11*, 1260–1270.
5. Jain, A., Singh, S., & Bhateja, V. (2013). A robust approach for denoising and enhancement of mammographic images contaminated with high density impulse noise. *International Journal for Convergence Computing, 1*(1), 38–48.
6. Bishnu, A., Rai, A., & Bhateja, V. (2018). Despeckling and enhancement techniques for synthetic aperture radar (SAR) images: a technical review. In *2nd International Conference on Computing, Communication and Control Technology (IC4T)*, (Vol. 2(10), pp. 37–41), October 2018.
7. Das, S., & Kundu, M. K. (2013). A neuro-fuzzy approach for medical image fusion. *IEEE Transactions on Biomedical Engineering, 60*(12), 3347–3353.
8. Dippel, S., Stahl, M., Wiemker, R., & Blaffert, T. (2002). Multiscale contrast enhancement for radiographies: Laplacian pyramid versus fast wavelet transform. *IEEE Transactions on Medical Imaging, 21*(4), 343–353.
9. Liu, X., Tang, J., Xiong, S., Feng, Z., & Wang, Z. (2009). A multiscale contrast enhancement algorithm for breast cancer detection using laplacian pyramid. In International Conference on Information and Automation (pp. 1167–1171).
10. Russo, F. (1998). Recent advances in fuzzy techniques for image enhancement. *IEEE Transactions on Instrumentation and Measurement, 47*(6), 1428–1434.

11. Deng, H., Deng, W., Sun, X., Liu, M., Ye, C., & Zhou, X. (2017). Mammogram enhancement using intuitionistic fuzzy sets. *IEEE Transactions on Biomedical Engineering, 64*(8), 1803–1814.
12. Stark, J. A. (2000). Adaptive image contrast enhancement using generalizations of histogram equalization. *IEEE Transactions on Image Processing, 9*(5), 889–896.
13. Bai, X., Zhou, F., & Xue, B. (2012). Image enhancement using multi scale image features extracted by top-hat transform. *Optics & Laser Technology, 44*(2), 328–336.
14. Alonso, M. T., López-Martínez, C., Mallorquí, & Salembier, P. (2011). Edge enhancement algorithm based on the wavelet transform for automatic edge detection in SAR images. *IEEE Transactions on Geoscience and Remote Sensing, 49*(1).
15. Kaplan, N. H. (2018). Remote sensing image enhancement using hazy image model. *Optik—International Journal for Light and Electron Optics, 155*, 139–148.
16. Gupta, R., & Bhateja, V. (2012). A new unsharp masking algorithm for mammography usng non- linear enahncement function. In *Proceedings of International Conference on Information System Design and Intelligent Applications, Vishakhapatnam, India* (pp. 113–114), January 2012.
17. Polesel, A., Ramponi, G., & Mathew, V. J. (2000). Image enhancement via adaptive unsharp masking. *IEEE Transactions on Image Processing, 9*(3), 505–510.
18. Duan, X., Mei, Y., Wu, S., Ling, Q., Qin, G., Ma, J., Chen, J., et al. (2018). A multiscale contrast enhancement for mammogram using dynamic unsharp masking in laplacian pyramid. *IEEE Transactions on Radiation and Plasma Medical Sciences*, 1–9.
19. Banerjee, J., Ray, R., Vadali, S. R. K., Shome, S. N., & Nandy, S. (2016). Real-time underwater image enhancement: An improved approach for imaging with AUV-150. *Sadhana, 41*(2), 225–236.
20. *RADARSAT-2 Images Database*. Available on: https://mdacorporation.com/geospatial/international/resources/image-gallery.
21. Rai, A., Bishnu, A., & Bhateja, V. (2019). Despeckling of synthetic aperture radar (SAR) images using local statistics based adaptive filtering. In *3rd International Conference on Intelligent Computing and Communication (ICICC)* (Vol. 3, No. 6), June 2019.

Cough Sound Analysis in Transform Domain for Classification of Respiratory Disorders

Ahmad Taquee and Vikrant Bhateja

Abstract Cough is hypothesized as the major symptom for all the respiratory disorders. There are not many methods and tools which analyses cough signals for the diagnosis of the respiratory disorders. In the proposed work, the cough signals are analysed in transform domain and classified the cough signals into different respiratory disorders like bronchitis, asthma, etc. Initially, the cough signals are taken from the database for the analysis, further to reduce the effect of noise from the cough signals, thresholding is applied. Furthermore, feature extraction is performed in transform domain and a set of features like zero crossing rate (ZCR), log energy (LogE), Mel-frequency cepstral coefficients (MFCCs) and kurtosis (K) are extracted from cough sound signals. These features are act as input to the support vector machine (SVM) classifier and its associated classifiers for the classification or prediction of the respiratory disorders. The overall execution of the classifier is analysed on the basis of the accuracy achieved by the classifier; the proposed analysis indicates that the best accuracy is achieved by the cubic SVM which is 89.3%.

Keywords Cough signals · Respiratory disorders · SVM · Transform domain · Thresholding

1 Introduction

Respiratory disorders are the major cause of childhood deaths. In 2011, respiratory disorders kill 1.3 million children under five years old [1]. Around 75% of these are centred in poor resource territory and Africa territory, generally pretending the poor

A. Taquee · V. Bhateja (✉)
Department of Electronics and Communication Engineering, Shri Ramswaroop Memorial Group of Professional Colleges (SRMGPC), Tiwari Ganj, Faizabad Road, Lucknow, Uttar Pradesh 226028, India
e-mail: bhateja.vikrant@gmail.com

A. Taquee
e-mail: taquee.ahmad7@gmail.com

Dr. A.P.J. Abdul, Kalam Technical University, Lucknow, Uttar Pradesh, India

S. C. Satapathy et al. (eds.), *Intelligent System Design*, Advances in Intelligent Systems and Computing 1171, https://doi.org/10.1007/978-981-15-5400-1_82

population. Only 65% of the deaths could have been controlled with the right time [2]. So, immediate diagnosis and proper treatment are necessary to prevent these deaths. The approach is mainly centred on automated cough sound signals analysis for the effective diagnosis of respiratory disorders. It has been hypothesized that the cough sound signals are the common symptom of almost all types of respiratory disorders. Cough sound signals are the main symptom of respiratory disorders, but cough sound signals have never been used in diagnosing respiratory disorders mathematically [3]. Cough sound signals have major information in lower respiratory tract as reported in [4]. The major cause of using cough sound signals processing is to help in making decisions for solving respiratory disorders [5]. As the applications of cough sound signals differ, different methods have been proposed and these methods are in time, frequency or in time–frequency based. When the cough sound signals are analysed in time domain, it shows only information contained in time–amplitude plane and unable to view the information contain in frequency–amplitude plane. When the cough signals are analysed using Fourier transform (FT), then, the cough signals cannot describe and explain information contained in time domain [6]. So, the time–frequency analysis of cough sound signals gives the better information in time as well as in frequency domain [6]. In previous few decades, different methods like short-time Fourier transform (STFT), wavelet transform (WT), Weiner distribution function (WDF), etc., have been used. The resolution of the signals in STFT is limited either in frequency–amplitude plane or in time–amplitude plane. The WDF has better result in time domain and frequency domain [7, 8]. The major advantages of using WDF are the resolution is twice as large. The WDF has better clarity so it is easy to analyse and interprete the cough signals [9]. The following equation shows WDF of a continuous signal $s(t)$

$$W(t,\omega) = \frac{1}{2\pi}\int_{-\infty}^{+\infty} s^*\left(t - \frac{\pi}{2}\right)s\left(t + \frac{\pi}{2}\right)\mathrm{e}^{-j\omega\tau}\mathrm{d}\tau \tag{1}$$

where ω is frequency, t is the time, * is the complex conjugate and τ is the shift in time, respectively. A wavelet is mathematical function used in signal processing and can be used to extract features of the signals [10]. To analyse the frequency properties of the cough signals in time–amplitude function $x(t)$ at any moment of time t, WT is an effective method [11, 12]. Let scale of variable, a, and a function of time, t, then WT can be represented as $W(a, t)$. In previous few decades, multiple methods have been introduced by authors for the classification or prediction of respiratory disorders in different domains and are followed by thresholding, pre-processing methods and different types of features extraction. The characteristics of cough signals are analysed in both time–amplitude and frequency–amplitude plane, and thus cough signals can be differentiated into different respiratory disorders. ZCR, autoregressive model and MFCCs are used for the extraction of different features of cough signals to differentiate into respiratory disorders [13]. The classification or prediction of cough signals using different neural network such as artificial neural network (ANN), decision tree and SVM have been proposed in previous decades [14]. Chatrzarrin et al.

proposed that the spectral features of the cough signals and peaks of the energy envelop have been used to differentiate dry and wet cough signals [15]. Murata et al. suggested a method in which the frequency of the cough signals is used to differentiate wet and dry cough signals [16]. Hashimoto et al. distinguish between the two cough signals by calculating the root mean square (RMS) and also measures the rheostatical characteristics of airway mucus [17]. The paper is categorized into different sections which are: Sect. 2 gives the description of the wavelet decomposition process and describes a thresholding-based technique for the preprocessing of the cough signals. Section 3 contains the features extraction in transform domain. Section 4 contains the result and discussion of respiratory disorders using SVM classifier. Section 5 concludes the overall analysis of the proposed technique.

2 Proposed Pre-processing Methodology

The cough sound signals are acquired from International Conference on Biomedical and Health Informatics, 2017, Respiratory Sound Database [18]. At the time of acquisition of cough signals, the cough signals are contaminated with different types of noises. So, wavelet transform is used for the decomposition of the cough signals. The cough signals are decomposed in time–frequency plane [3]. The following equation shows the translated and scaled types of wavelets:

$$\psi_{s,\tau} = \frac{1}{\sqrt{|s|}}\psi\left(\frac{t-\tau}{s}\right) \tag{2}$$

The mother wavelet $\psi(t)$ of a signal $x(t)$ is represented as:

$$T(s,\tau) = \int_{-\infty}^{\infty} x(t)\psi^*\left(\frac{t-\tau}{s}\right)\mathrm{d}t \tag{3}$$

where $\psi(t)$ is the mother wavelet, s is the scale, τ is the translation factor and the symbol * represents the complex conjugate of ψ [3]. The additive white Gaussian noise (AWGN) and residual noise are present in the acquired cough sound signals and these are not suppressed by using multiple filters. Therefore, cough signals are analysed using discrete wavelet transform (DWT) into different subbands at different frequencies. The DWT can be expressed as:

$$\psi_{m,n}(t) = 2^{-m/2}\psi\left(2^{-m}t - n\right) \tag{4}$$

where m and n is integers [3]. Now, the DWT is employed on cough signals at multiple levels, and it gets decomposed into approximation coefficients $a[n]$ and detail coefficients $d[n]$. The $d[n]$ contains low frequency noise and the $a[n]$ contain high frequency noise. A thresholding-based technique has been used in the presented

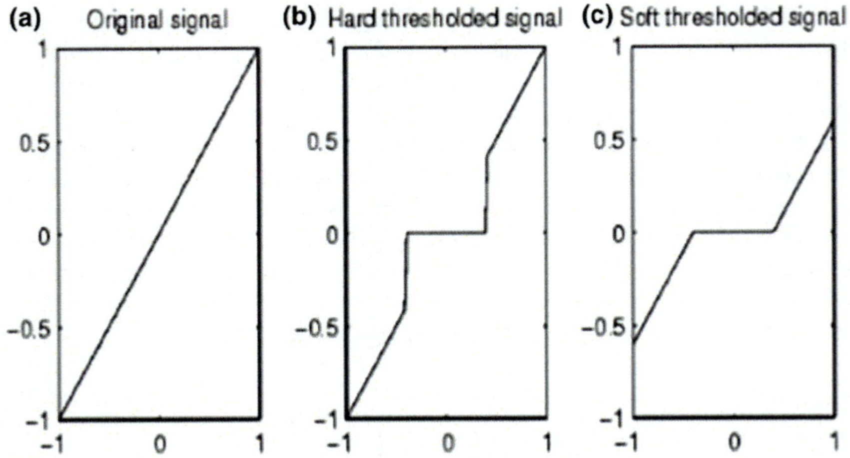

Fig. 1 **a** Original signal, **b** hard threshold signal, **c** soft threshold signal [19]

work. There are mainly two types of thresholding techniques which are soft thresholding (s_j) and hard thresholding technique (H_j) [19]. The following Fig. 1b shows the hard thresholding and Fig. 1c shows the soft thresholding of a signal [19].

The equations given below show the soft thresholding and hard thresholding of signal [20]:

$$s_j = \begin{cases} [\text{sign}(C_j)(|C_j| - \lambda)], & |C_j| > \lambda \\ 0, & |C_j| < \lambda \end{cases} \tag{5}$$

$$H_j = \begin{cases} C_j, & |C_j| > \lambda \\ 0, & |C_j| < \lambda \end{cases} \tag{6}$$

where C_j denotes the coefficients and λ denotes the threshold value. The threshold value λ has been selected using the formula given in [21]:

$$\lambda = \sigma\sqrt{2\log_2 N} \tag{7}$$

where N is the length of the cough signals and σ is standard deviation. The value of standard deviation given in [21]:

$$\sigma = \frac{\text{MAD}(|d[n]|)}{0.6745} \tag{8}$$

where MAD represents median absolute deviation. The calculated value of λ from Eq. 7 is applied $d[n]$ and $a[n]$ of cough signals. Now, signal-to-noise ratio (*SNR*) is used to measure the quality of cough sound signals. The following Eq. 9 is used for the computation of *SNR* [3].

$$\text{SNR}(dB) = 10\log_{10}\frac{\sum_{i=1}^{N}(x_{dn}[n])^2}{\sum_{i=1}^{N}(x[n] - x_{dn}[n])^2} \tag{9}$$

where $x_{dn}[n]$ represents denoised cough signals and $x[n]$ shows original cough signals.

3 Features Extraction in Transform Domain

Features extraction is a method to extract more useful information from the cough signals. In this analysis, the motive of features extraction is not only to extract more information of the signals but also to extract more dominant features present in the cough signals by removing redundant information such as noise. Now, the classification is followed by features extraction process. The overall performance of the classifier depends on the extracted features and the preprocessed cough signals. In this analysis, the transform domain features are extracted [4]:

- ZCR
- MFCCs
- Kurtosis (K)
- Log Energy (LogE).

3.1 ZCR

ZCR has especially been used for the differentiation between unvoiced and voiced signals. Now, the cough signal is separated into small length and then ZCR is calculated. In other words, ZCR can be defined as the number of sign changes in a frame divided by total length of cough signal [4].

3.2 MFCCs

MFCCs is an automatic speech recognition system to extract features so that it identifies the main parts of the cough signals which are good for identifying respiratory problems and suppressing all other parts which carries information like emotion, noises, etc. [4].

3.3 Kurtosis

Kurtosis (K) feature is calculated in transform domain. It shows the peaks of cough signals. It is the fourth-order central moment of the cough signals. It is the probability density distribution of the cough signals [4].

3.4 Log Energy

The log energy feature is calculated in transform domain. It is the energy of each subsegment of the cough sound signals [4].

4 Results and Discussions

The noise is suppressed using hard thresholding on the wavelet coefficients. After that, cough signals are divided into small parts and each part has the duration of 30 ms. Now, the transform features are extracted from each segment of the cough signals. The extracted features in transform domain are the input of the classifier. The classifier(s) used in this analysis are SVM classifier and their associated families. The classifier is used to classify or predict the respiratory disorders. The transform domain features have the size of 374 $\times$ 46 which includes one column of the labels. The testing feature size is 94 $\times$ 45 and there is no column of the labels. Now, the model is trained using classification learner application which is available in MATLAB v9.5 (2018b). The accuracy of the classifier is around 89.3% and the confusion matrix is computed using the classification learner application. False Negative Rate (FNR) and True Positive Rate (TPR) are also calculated using this app and TPR is varied from 65% to 100% and FNR is varied from 1% to 35% for different respiratory disorders, respectively (Fig. 2).

5 Conclusion

The above analysis of cough signals is done in transform domain using wavelet transform followed by features extraction in transform domain and classification of respiratory disorder. From the above analysis, the obtained results are shown in Fig. 1. The performance of the set-up has been verified with the computation of *SNR*, accuracy and confusion matrix in the above analysis. The inferences of the above analysis are given below:

- Cough signals are decomposed at multiple levels using DWT by applying Symlet 4 (sym4) wavelet family.

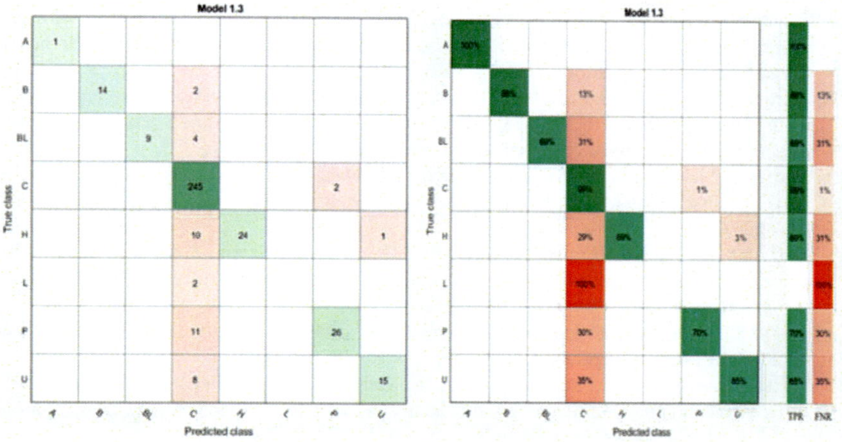

Fig. 2 **a** Confusion matrix, **b** TPR & FNR

- Now, cough signals have been preprocessed using hard thresholding.
- Features such as ZCR, MFCCs, K, etc., have been extracted from the wavelet coefficients.
- The performance of the Cubic SVM classifier is obtained around 89.3%.

References

1. Walker, C. L. F., Rudan, I., Liu, L., Nair, H., Theodoratou, E., Bhutta, Z. A., et al. (2013). Global burden of childhood pneumonia and diarrhoea. *The Lancet, 381,* 1405–1416.
2. Jones, G., Steketee, R. W., Black, R. E., Bhutta, Z. A., & Morris, S. S. (2003). How many child deaths can we prevent this year? *The Lancet, 362,* 65–71.
3. Taquee, A., Bhateja, V., Srivastava, A., & Shankar, A. (2018). Combination of wavelets and hard thresholding for analysis of cough signals. In *Proceedings of 2nd International Conference on Smart Trends in Systems, Security and Sustainability* (WorldS4) (pp. 266—270).
4. Abeyratne, U. R., Swarnkar, V., Setyati, A., & Triasih, R. (2013). Cough sound analysis can rapidly diagnose childhood pneumonia. *Annals of Biomedical Engineering, 41*(11), 2448–2462.
5. Sejdic´, E., Djurovic´, I., & Jiang, J. (2009). Time-frequency feature representation using energy concentration: An overview of recent advances. *Digital Signal Processing, 19*(1), 153–183.
6. Szmajda, M., & Mroczka, J. (2011). Comparison of Gabor-Wigner transform and SPWVD as tools of harmonic computation. *The International Conference on Renewable Energies and Power Quality, 1*(9), 386–391.
7. Rakovic´, P., Sejdić, E., Stanković, L. J., & Jiang, J. (2006). Time-frequency signal processing approaches with applications to heart sound analysis. *Computers in Cardiology, 33,* 197–200.
8. Dong, Y., & Cui, Y. (2012). Analysis of a new joint time-frequency distribution of suppressing cross-term. *Research Journal of Applied Sciences, Engineering and Technology, 4,* 1580–1584.
9. Pries, D., & Georgopoulos, V. (1999). Wigner distribution representation and analysis of audio signals: An illustrated tutorial review. *Audio Engineering Society, 47,* 104–1053.

10. Shankar, A., Bhateja, V., Srivastava, A., & Taquee, A. (2018). Continuous wavelets for pre-processing and analysis of cough signals. In: Proceedings of 3rd International Conference on Smart Computing and Informatics (pp. 1–8).
11. Manjunath, A., & Ravikumar, H. (2011). Comparison of discrete wavelet transform (DWT), lifting wavelet transform (LWT) stationary wavelet transform (SWT) and S-transform in power quality analysis. *European Journal of Scientific Research, 4,* 569–576.
12. Sushama, M., Ram, G., & Laxmi, A. (2010). Detection of Power quality disturbances using wavelet transform. *International Journal of the Computer, the Internet and Management, 18,* 61–66.
13. Hashemi, A., Arabalibiek, H., & Agin, K. (2011). Classification of wheeze sounds using wavelets and neural networks. In *Proceedings of International Conference on Biomedical Engineering and Technology* (pp. 127–131).
14. Barry, S., Dane, A., Morice, A., & Walmsley, A. (2006). The Automatic recognition and counting of cough. *Cough Journal, 2,* 1–9.
15. Chatrzarrin, H., Arcelus, A., Goubran, R., & Knoefel, F. (2011). Feature Extraction for the differentiation of dry and wet cough sounds. In *Proceedings of IEEE International Symposium on Medical Measurements and Applications* (pp. 162–166).
16. Murata, A., Taniguchi, Y., Hashimoto, Y., Takasaki, Y., & Kudoh, S. (1998). Discrimination of productive and non-productive cough by sound analysis. *Internal Medicine, 37,* 732–735.
17. Hashimoto, Y., Murata, A., Mikami, M., Nakamura, S., Yamanaka, E., & Kudoh, S. (2003). Inuence of the rheological properties of airways mucus on cough sound generation. *Respiratory, 8,* 45–51.
18. Rocha, B. M., Filos D., Mendes, L., Vogiatzis, I., Perantoni, E., Kaimakamis, E., Natsiavas, P., et al. (2017). A respiratory sound database for the development of automated classification. In *Precision Medicine Powered by pHealth and Connected Health* (pp. 33–37), November 2017.
19. Srivastava, A., Bhateja, V., Shankar, A., & Taquee, A. (2018). On analysis of suitable wavelet family for processing of cough signals. In Proceedings of 7th International Conference on FICTA (pp. 1–8).
20. Akkar, H. A. R., Hadi, W. A. H., & Al-Dosari, I. H. (2018). A squared-Chebyshev wavelet thresholding based 1D signal compression. In *Defence technology* (pp. 1–6), August 2018.
21. Donoho, D. L. (1995). De-noising by soft-thresholding. *IEEE Transactions on Information Theory, 41*(3), 613–627.

Factor's Persuading 'Online Shopping' Behaviour in Mauritius: Towards Structural Equation Modelling

Vani Ramesh, Vishal Chndre Jaunky, and Christopher Lafleur

Abstract Recent years, the idea of shopping online has become a necessity than a luxury. The singularities of online shopping are budding considerably. The players of this inclination show that there is still a huge possible demand for these shopping trends. There are many contributing factors that are influencing these behaviours. This study intends to scrutinize the contributing factors that are persuading the customers and consumers to opt for online shopping in Mauritius. The sample size is from the respondents having exposure to online shopping or e-stores. The primary data collected through a survey with the help of well-designed questionnaire (5-point Likert scale), with SERVQUAL, CDMM and OSAM models. The data is analyzed and to test the hypothesis and model fit using AMOS. The findings and conclusion of this study understand that there is a significant influence of personality traits (SERVQUAL), internet knowledge, purchase intention and experience, followed by post-purchase satisfaction (OSAM, CDMM). The contribution of this study for corporate academics and existing literature is very stimulating for further research, being first and pioneer in Mauritius.

Keywords Trend · Information technology · Online shopping · Loyalty · Personality

V. Ramesh (✉)
REVA University, Bengaluru, India
e-mail: sarada889@yahoo.in

V. C. Jaunky
LTU, Luleå, Sweden
e-mail: vishal.jaunky@ltu.se

C. Lafleur
University of Mauritius, Moka, Mauritius
e-mail: chrislafleur1234@yahoo.com

S. C. Satapathy et al. (eds.), *Intelligent System Design*, Advances in Intelligent Systems and Computing 1171, https://doi.org/10.1007/978-981-15-5400-1_83

1 Introduction

The term 'Online Shopping' (e-shopping) in 'Digital Age' is an 'i-Generation' or 'i-Gen', 'PostMillennials', where *technology has permitted* electronic commerce (e-commerce) to have business dealings and providing information via 'computer networks' [1]. More than 50% of the business communications are now tangled in or supporter of three elementary 'e-commerce applications, i.e. e-commerce business-to-consumer, business-to-business, and consumer-to-consumer' [2].

As rightly said by Lee et al. [3], online shopping behaviour, online purchase intention, perceived usefulness and subjective norm have an important direct affiliation on repurchase. The advances in technology have stimulated, e-shopping convention is no longer restricted to schmoozing for advertising, also, for sales and marketing led to the growth of online shopping with comfort. Further, 50% of Mauritians are habituated to internet and social networking sites [4]. Bearing in mind these changes, the present study antedates to examine the model that will govern the factors persuading consumer online shopping choices. The outcome of this study is anticipated to be beneficial in serving e-commerce site owners and business entities associated with e-commerce and e-trade in Mauritius.

1.1 Theories Adopted

1.1.1 The Consumer Decision-Making Model (CDMM)

'Nicosia Model' (CDMM), projected by Nicosia [5], exhibits how consumer purchase is a very complex process. Right from need recognition to post-purchase behaviour, the model demonstrates the online shopping behaviour [6]. Considering a positive point, the model aims at inspiring the users, guide the marketers in designing marketing strategies at each stage of the online shopping decision-making process. Also, the model deliberates the element of customers' devotion in searching the information, assessing different alternatives that are available until they choose the end product/service to mollify their expectations.

1.1.2 Online Shopping Acceptance Model (OSAM)

According to OSAM, though a positive relationship exists among users of internet and e-shopping intention [7], in some of the recent studies, the relationship was not found to be significant [8]. This study adopts and examines this model with the help of empirical evidence, whether it has been accepted and what is the acceptance level in Mauritian context.

1.1.3 The Technology Acceptance Model (TAM)

TAM is adopted to test the reasoned action [9], planned behaviour of online customers [10]. Since the technology acceptance model is used to understand user behaviour, the same has been considered to test the online shopping behaviour of Mauritians [11].

1.1.4 Freud's theory of motivation

People behaviour is very unpredictable, and this theory is tested to understand the online consumer behaviour of Mauritians [12].

2 Based on the Review of Previous Studies, the Following Hypothesis was Set for the Present Study

Since the determination is to discover and to test relevant factors influencing 'Online Shopping Technology' adoption and 'Online Shopping Satisfaction' of Mauritians. The study also extends in evaluating the factors influencing motivating in accepting online shopping which leads to 'e-loyalty' and 'e-trust'.

Hypothesis I: Personality traits significantly influence online shopping intentions in Mauritius.
Hypothesis II: Online shopping orientation has significant positive impact on online shopping intension in Mauritius.
Hypothesis III: Information available has significant positive impact on online shopping motivation.
Hypothesis IV: Online shopping satisfaction has significant influence on e-loyalty of Mauritian consumers.

2.1 Methodology

Demographic, psychological, normative beliefs, purchase decision, satisfaction, e-trust and e-loyalty are tested. The questionnaire was constructed codified into six sections are distributed to online shoppers across the island, administered as a drop-off survey. The response rate was 87.3% (443 out of 500). The data is annualized, 'Path analysis', 'multiple regressions' are computed. The results show that the Mauritian e-consumers/e-customers are really fascinated to accept online shopping technology.

3 Literature Review

Online buying intentions are directly influenced by trust and consumer attitudes [13]. With the practical direction, expediency, worth, wide product choice, and earnings also have robust [14]. As rightly whispered by Shergill and Chen [15] in his investigative factor analysis initiate that the 'site design' [16], 'Reliability of online sites', 'Service rendered by the sites', and 'safety of online sites' are leading in influencing online consumer perceptions. Similarly, Heijden et al. [17] identified online purchase intentions with the help of 'Technology_Oriented Perspective', 'Trust_Oriented Perspective'.

3.1 Evolution of Online Shopping

During late 1981, Thomson Holidays invented online shopping to release a first product, the world's first direct Business-to-Business (B2B), where the buyers are connected interactively with the seller's electronic device directly without third party's intervention. There is a theoretical evidence, in the 1980s, the first B2C concept was introduced by the pilots to pay for the social security funding and used to investigate the potential for IT to reduce social disadvantage by offering in the home services to vulnerable and physically challenged citizens. Amazon started eBay bidding in the year 1995, 2015 is given increasingly boost for online shopping with mobile shopping 'staggering $1.55 trillion' in America, later expanded across the globe. China that avowals the major economy, worth $562.66 billion.

3.1.1 Evolution of online shopping-Mauritius

Mauritius has witnessed remarkable growth in information technology (IT) based on business activities. So far, there is no evidence in the literature that any study has focussed on examining the factors that are influencing the adoption of online shopping by the customers and consumers. Moreover, from the point of manufacturers and sellers, this innovation has hiked the turnovers with huge margins, since there are no much overhead costs, of showroom rents, labour wages and others. The present study looks into possible motivating factors for Mauritian context and swotted that there are over 280,000 online customers in Mauritius buying products and services [18]. We are in a new era where every visitor can enjoy a product or service, 'lepoint.mu' or 'tantebazar.com'. Making a 'bazaar' in two plays, Secured payments? Are some of the active online shopping portals that are playing a very vital role in Mauritius? E-secure with SBML, MCB offers secured online transactions. Winners launched online purchases during 2012 also one of the online shopping innovations in Mauritius.

4 Analysis and Discussions

To test the acceptability and reliability of the questionnaire, preliminary test was done with the sample size of 50 respondents from different parts of the country, and have goodness of fit at 0.80 [19]. Accordingly, the Cronbach's Alpha is computed, where the results are within the acceptable range (0.824 and 0.912) [20] to proceed with the study. The total sample collected for the study is 432 (300 female and 132 male). Goodness of fit indices was generated and modified the model with repeated computation to improve the results as per the acceptance range for the study [21]. The female respondents are more in number, it is presumed that the female in Mauritius is more dependent and comfortable to do online shopping.

The respondents are from diverse family structures, different economic background, educational qualifications, regions, types of products purchased and online shopping portals used.

4.1 Exploratory Factor Analysis (EFA)

With the sample size of 432, the factor analysis is extracted [22] (Tables 1, 2 and 3; Fig. 1).

Regression weights are positive and moderate values. Squared multiple correlation R^2 values test reliability is also positive and low. As a final conclusion, the model considered is moderately acceptable. Analysis of variance indicated that overall satisfaction is dependent on the set of independent variables such as connectivity, accessibility and details (Table 4; Figs. 2 and 3).

The factors are divided into personality traits, purchase decision, satisfaction and e-loyalty. The hypothesis is sustained by *p*-value (***).

5 Discussion and Conclusion

The findings showed the hypothesis were supported to the maximum extent of online buyers behaviours towards online shopping in assessment for satisfaction, purchase decision, e-trust and e-loyalty [6]. Consequence of purchase intention towards online

Table 1 KMO and Bartlett's test

Kaiser–Meyer–Olkin measure of sampling adequacy		0.868
Bartlett's test of sphericity	Approx. Chi-square	132.621
	Df	84
	Sig.	0.000

Source Authors computation

Table 2 Model summary of multiple regression

Model	R	R square	Adjusted R square	Std. error of the estimate	R square change	Change statistics			
						F Change	$df1$	$df2$	Sig. F change
1	0.484[a]	0.234	0.217	0.56427	0.234	13.576	3	133	0.000

Source Authors Computation
[a]Predictors: (Constant), Personality Traits, Shopping Decision, Satisfaction and e-loyalty

Table 3 Measurement model fit indices' for 1st output modified CFA

Absolute fit indicators						Incremental fit indicators					
Model fit index	Chi-square (χ^2)	df	*p*-value of χ^2	CMIN/Df ($\chi 2/df$)	RMSEA	CFI	GFI	AGFI	NFI	TLI	$\Delta\chi^2$
Acceptable value	Small		<0.05	<5	<0.05 is good, <0.08 is acceptable	>0.95 is great, >0.7 is tolerable	Same	Same	>0.90 is great, > 0.7 is tolerable	>0.95 is great, >0.7 is tolerable	
First output	132.62	84	0	2.47	0.65	0.934	0.81	0.77	0.843	0.918	
Modified	82.95	57	0.004	2.17	0.058	0.960	0.832	0.795	0.887	0.946	117.9

Source Primary data collected from questionnaire
Source Authors computation

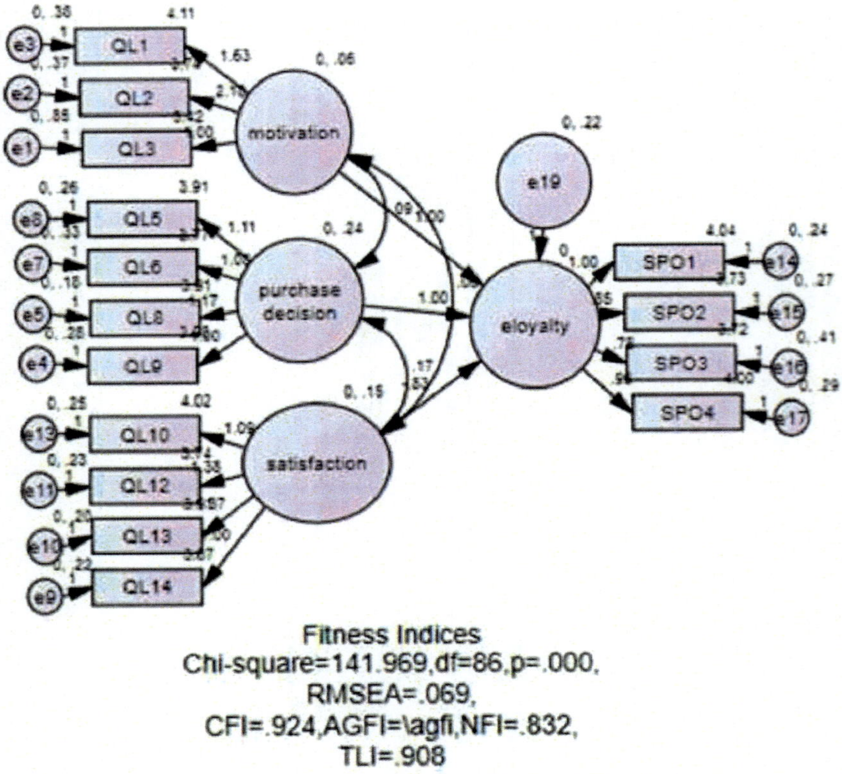

Fig. 1 Structural equation modelling with modified CFA. *Source* Authors computation

Table 4 Coefficients (dependent variable: Overall_Satisfaction)

Model		Unstandardized coefficients		Standardized coefficients	*T*	Sig.
		B	Std. error	Beta		
1	(Constant)	1.640	0.389		4.214	0.000
	Personality Traits	0.097	0.092	0.092	1.059	0.292
	Purchase Decision	0.419	0.107	0.407	3.918	0.000
	e-loyalty	0.046	0.118	0.040	0.391	0.696

Source Authors computation

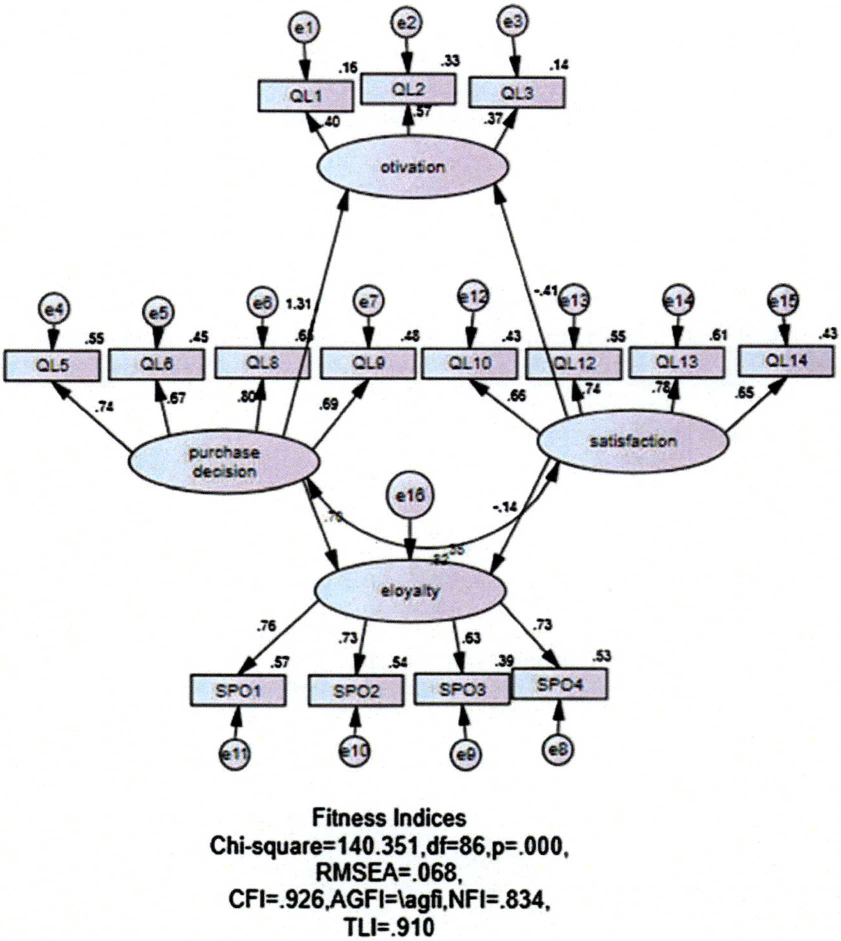

Fig. 2 SEM model modification. *Source* Authors computation

shopping by the Mauritian consumers and customers is reliable as per previous studies [23]. The findings evidenced that Mauritians purchase intention is influenced by personality traits, socio-economic, cultural and law. Though there is a scarcity of resources and infrastructure, still Mauritians are trying to cope up with the technological advances across the globe, by updating themselves. The present study has revealed that there is an amplified illuminating power of Mauritian online shopping intention and also provides guidelines for further research to go in-depth to test more dimensions. Thus it is suggested for further study.

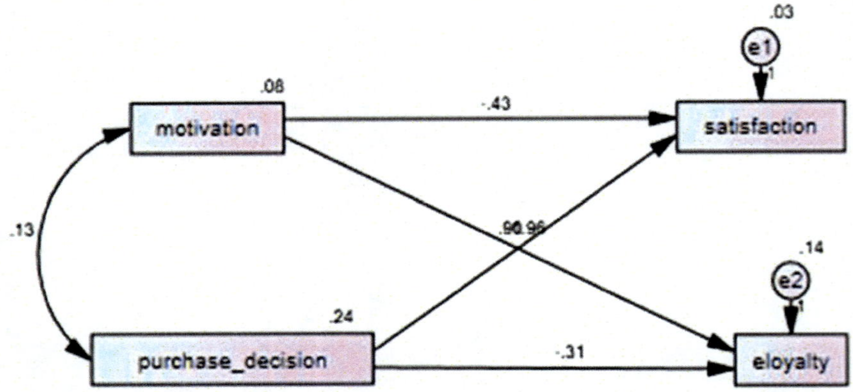

Fig. 3 Multiple regression modified SEM. *Source* Authors computation

References

1. Satapathy, S. C., Bhateja, V., & Das, S. (2016). Smart computing and informatics. In *Proceedings of the First International Conference on SCI* (Vol. 1). O'Brien, J. A., & Marakas, G. M. (2006). *Management information systems* (7th ed.). New York: McGraw Hill. Vandenberg, R. J., & Scarpello, V. (1994). A longitudinal assessment of the determinant relationship between employee commitments to the occupation and the organization. *Journal of Organizational Behavior, 15*(6), 535–547. O'Brien, J. A. (2005). *Introduction to information systems* (12th ed.). New York: McGraw Hill.
2. Law, J. (2018). 11 Street Officially launches in Malaysia. Retrieved June 2018, from hardwarezone http://www.hardwarezone.com.my/tech-news.
3. Lee, C. H., Eze, U. C., & Ndubisi, N. O. (2011). Analyzing key determinants of online repurchase intentions. *Asia Pacific Journal of Marketing and, 23*(2), 200–221
4. Clemes, M. D., Gan, C., & Zhang, J. (2014). An empirical analysis of online shopping adoption in Beijing, China. *Journal of Retailing and Consumer Services, 21*(3), 364–375.
5. Nicosia, F. (1976). Consumer behavior toward sociology of consumption. *Journal of Consumer Research, 5,* 121–133.
6. Bhasin, H. (2010). *Factors affecting consumer buying behavior*. www.marketing91.com. Accessed January 26, 2017.
7. Bhatnagar, A., Misra, S., & Rao, H. R. (2000). On risk, convenience, and internet shopping behavior. *Communications of the ACM, 43*(11), 98–105.
8. Cho, J. (2004). Likelihood to abort an online transaction: Influences from cognitive evaluations, attitudes, and behavioralvariables. *Information & Management, 41,* 827–838.
9. Ajzen, I., & Fishbein, M. (1980). Understanding attitudes and predicting social behavior.
10. Jamil, N. A., Mat, N. K. (2017). To investigate the drivers of online purchasing behavioral in Malaysia based on the theory of planned behavior (TPB): A structural equation modeling (SEM) approach. In *International Coference on Management* (pp. 453–460)
11. Pavlou, P. A., Fygenson, M. (2016). Understanding and predicting electronic commerce adoption: An extension of the theory of planned behavior. *MIS Quarterly, 30*(1), 115–143.
12. Kotler, P., & Armstrong, G. (2005). Atmospherics as a marketing tool. *Journal of Retailing & Consumer Services, 49*(4), 48–64.
13. Delafrooz, N., Paim, L. H., & Khatibi, A. (2011). Understanding consumer's internet purchase intention in Malaysia.

14. Kwek, C. L., Tan, H. P., & Lau, T. C. (2010). The effects of shopping orientations, online trust and prior online purchase experience toward customers' online purchase intention. *International Business Research, 3*(3), 63–76.
15. Shergill, G. S., & Chen, Z. (2005). Web-based shopping: consumers' attitudes towards online shopping in New Zealand. *Journal of Electronic Commerce Research, 6*(2), 79–94.
16. Satapathy, S. C., Bhateja, V., Raju, K. S., & Janakiramaiah, B. (Eds.) Computer communication, networking and internet security. In *Proceedings of Ic3t 2016* (vol. 5). Berlin: Springer.
17. Heijden, H. V., Verhagen, T., & Creemers, M. (2003). Understanding online purchase intentions: Contribution from technology and trust.
18. Clemes, M. D., Gan, C., & Zhang, J. (2014). An empirical analysis of online shopping adoption in Beijing, China. *Journal of Retailing and Consumer Services, 21*(3), 364–375.
19. Sekaran, U. (2003). *Research methods for business: A skill building approach* (4th ed.). New Jersey: Wiley.
20. Nunnally, J. C. (1978). *Psychometric theory*. Michigan: McGraw-Hill.
21. Agresti, A., & Finlay, B. (1997). *Statistical methods for the social sciences*. New Jersey: Prentice Hall.
22. Field, A. (2005). *Discovering statistics using SPSS*. London: Sage Publications.
23. He, D., Lu, Y., & Zhou, D. (2009). Empirical study of consumers' purchase intentions in C2C. *Electronic Commerce, 13*(3), 287–292.

'Customer Satisfaction', Loyalty and 'Adoption' of E-Banking Technology in Mauritius

Vani Ramesh, Vishal Chndre Jaunky, Randhir Roopchund, and Heemlesh Sigh Oodit

Abstract Small Island, Mauritius is adopting drastic changes with reference to technology. The prime focus of extant study is exploratory in nature and trying to examine the relationship between the factors that are inspiring Mauritian consumers and customers to go with technology and technological advances that are user friendly. E-banking services are one among them. Also, try to test which of the dimensions has the strongest potential in influencing 'e-customer'. This study focuses mainly on perception of the customer, e-banking adoption motif, 'e-banking satisfaction' and 'quality' of 'e-banking services' in Mauritius. Primary data is collected with the help of drop-off survey and were statistically analysed with the help of Structural Equation Modelling (SEM) and ordered probit and logit regressions. The empirical evidences of this study reveals, reliability, responsiveness, secured transaction, comfort, efficiency, dependence and ease-of-use has a striking impact on the 'customer satisfaction' and e-loyalty. With these findings, for Mauritian content, 'electronic banking' would be indispensable 'banking service' that can be, provided, well executed to improve 'e-customer satisfaction' and 'e-customer loyalty' to sustain goodwill of the e-banking customers. Relative importance and service quality of e-banking technology at Mauritius would support the banking industry to motivate the customers.

Keywords E-customer · Goodwill · Ease of use · Significance · Efficiency

V. Ramesh (✉) · V. C. Jaunky · R. Roopchund · H. S. Oodit
REVA University, Bengaluru, India
e-mail: drvanisarada@gmail.com

V. C. Jaunky
e-mail: vishal.jaunky@ltu.sc

R. Roopchund
e-mail: rroopchund@udm.ac.mu

H. S. Oodit
e-mail: hoodit@gmail.com

S. C. Satapathy et al. (eds.), *Intelligent System Design*, Advances in Intelligent Systems and Computing 1171, https://doi.org/10.1007/978-981-15-5400-1_84

1 Introduction

Information Technology has been very successful in construction of innumerable characteristics of human life easier for recent societies [1]. Customer Satisfaction is one of the main responsibilities of any business or service. Providing quality service, creating a trust among the customers and captivating their satisfaction and loyalty is a fundamental duty of any business or service industry. Same as applicable for 'Electronic-Banking Services' (e-services) [2].

'e-banking' services are basically count on the information exchange among the bank employees and bank customers, through technology adoption rather traditional modes of 'head-on' interaction [3]. Mauritian banking sector has recognized 'e-banking' technology and tangled enthusiastically in improving its services and business operations constantly. Since then, quality of service is available 24/7, with secured transaction, customer's retention and customer satisfaction is very much practical. Also, the cost effective of their transactions with reference to time, energy and value addition added to strengthen their e-loyalty.

There are ample number of studies that have investigated on e-banking technology adoption and service quality. Considering few of them, those are directly related to the present study, which are measured for 'internet-banking service quality' and 'internet-banking customer satisfaction', the study being first and prime most in Mauritius tries to identify the gap in literature. Similarly, also the aims at filling the gap in existing literature on the area of research, by addressing the 'e-banking adoption', testing eperically the considering all relevant determinants for a small and developing Island, like Mauritius. To our knowledge, so far there is no evidence of such study that has been undertaken earlier. Perhaps, the core findings of the study highlights, importance of 'Internet-Banking Service Quality' on 'Internet-Banking Customer Satisfaction', and reliability, quality of service and perception of 'e-banking customers'.

The structure of the paper is organized as, introduction followed by literature review, methodology, analysis and findings, interpretation and discussion, finally conclusion which recaps the study and presents relevant limitation and commendations for imminent research.

2 Literture Review

2.1 e-Banking

The first ATM (Automated Teller Machine) was invented by Shephered-Barron, a Scottish, on June 27th, 1967 at Barclays bank in Enfiled, North London. The first ATM was installed on July 20th, 2017 at Mauritius Commercial Bank Ltd (MCB) by BIRGER. Though it is recent phenomenon in the country, recently it has been taken momentum in diversification skills and experiencing a rapid economic growth.

Mauritian banking sector plays a key role in contributing for GDP, providing more facilities for the transactions through internet and encouraging the flow of foreign currency.

As on now, the largest networking banks such as, Mauritius Commercial Bank (MCB), Hong Kong Shanghai Bank (HSBC) and the State Bank of Mauritius Ltd are offering the internet banking services. The main services include, inter account transfer, fund transfer to credit card account, transfers to other accounts, SWIFT payments, foreign transfers etc. These e-banking services are spread among the customers so rapidly due to technological advances [4]. According to Lustsik [5], the banking services with the help of technology assistance has numerous option for providing good quality services. Due to drastic changes in lifestyle and vast exposure to global networks, the consumers and customers likes to operate their banking transactions at their convenience to save time and energy in cost effective way. Figure 1 summarizes the most important definitions of 'Internet-banking' (e-banking) services.

Shaikh and Karjaluot [6] has thrown some insights on e-banking with reference to mobile applications. Followed by Tieto [7], and Arcand et al. [8], defining e-banking as a technology which is user friendly and unique in offering customers a quality service.

Mauritius is still developing on updating the technology, yet inadequate to accessibility of internet and mobile technology comparatively, may be due to deliberate progress of Information Technology and infrastructural facilities that are needed for the safe and smooth functioning of e-commerce and e-services. Having considering these hurdle, this study outlines the concept as the e-banking adoption, satisfaction of customer and e-loyalty of e-customer.

Researcher	Definition
Jane *et al.* (2004)	Provide banking services to customers via the Internet starting from the ATM service and the end of the transfer of funds electronically
Riedl *et al.* (2009)	Provide banking services to customers through the Internet and self-service centers.
Ahmad and al Zubi (2011)	The bank gives the opportunity to its customers to gain access to their accounts and the implementation of their transactions and purchase products directly via the Internet.
Ala' Eddin *et al.* (2011)	A portal through which customers can use all kinds of banking services, from paying bills and end of the financial investments.
Asli (2011)	System that enables customers to access their accounts and obtain general information about the services offered by the bank via the Internet.
Jazani *et al.* (2012)	Banking services to provide a high degree of convenience that enables customers to access their accounts at time and place that suits them.
Aliyu (2012)	using technology by the bank by to connect with customers, and presentation of financial services from its path.
Salhieh *et al.* (2013)	Provide banking services to customers and allow them to carry out financial operations of (withdrawals and deposits, transfer funds, pay bills, request loans, make sure the account balances .... and others) 24 hours through the Internet.

Fig. 1 Definition of banking. *Source* Authors Computation

2.2 The Concept—'Customer Satisfaction'

The concept, 'Customer satisfaction', in broader terms, the process of concluding the purchase with postpurchase phenomena, right from the attitude, purchase intention, repeat purchase and loyalty to the brand [9]. As rightly pointed out by Churchill and Surprenanat [10], the post purchase satisfaction that the customer derives out of the the product or service shows a vital part in creating brand doppelgänger and brand loyalty. Generally, sense of gratification ascends once customers starts comparing their insight about the product or service, that they have actually perceived and developed the taste and preference towards it, with same attitude and expectation fro them [11]. Many definitions define 'customer satisfaction' in different ways, but the generally accepted and understood by a lay man, in simple terms is given by Tse and Wilton [12]. According to their definition, "The term, customer satisfaction is a understood as response to the assessment of the apparent variation among expectations and concluding consequence of subsequent consumption". Kotler and Keller [13], describes the satisfaction as "A response on the post purchase valuation and expectations from prior-purchasing stage".

There are some diverse opinion from some studies. Few researchers pragmatic about the effect on the experience in purchasing and consuming phase. This can also have a significant consequence on the customers decisions towards satisfaction [14]. So, the term 'customer satisfaction' can be defined as a is a sensation of liking or disliking that the consumer/customer after utilising the product or service of their anticipation [15].

Concluding, for the purpose of study, it remains tacit, concept 'customer satisfaction' is a perception and insolence of 'e-customer'/'e-consumer' on 'e-banking technology' and quality of 'e-banking service' at Mauritius.

2.3 Affiliation of Internet-Banking (e-Banking) Quality Service with e-Customer Satisfaction

Basic intention of this study is to understand relationship among 'Quality' of e-banking technology, 'services' and 'satisfaction' of customer. There are number of variables that are tested by many studies to examine, which are important motifs behind approval, acceptance of e-banking technology. Table 2 summarises some of components that are tested by different studies.

Referring to Parasuraman, et al. [16], the "service quality of e-banking" and "satisfaction of e-customer" seems very robustic. During 2018, Asiyanbi and Ishola [9], in their emperical evidence, points out, "The unit of satisfaction of customers in the banking sector surges while consuming services of e-banking". According to our view, and the relevant literature review reveals that, the important dimensions to test the e-customer satisfaction and loyalty can be well situated to cluster further down as, an 'efficiency', 'reliability', 'privacy', 'security', 'ease of use', 'responsiveness' and 'good communication' [17].

3 Methodology

Extensive literature review recommends that, secured transaction, reliability, privacy, comfort, 'responsiveness' and 'communication' networks remain very significant dimensions need to be tested for the country. The concept, 'customer satisfaction', 'e-loyalty' with 'e-banking service quality' and 'satisfaction' is tested with the help of empirical evidence in Mauritian context. A survey was conducted at the major commercial points of the country to collect the feed back with the help of well strucuted questionnaire, consisting '5-point' likert scale [18]. Convenience sampling techniques is used to data collection. The reliant variable is 'customer satisfaction', loyalty, adoption of e-banking in Mauritius. The data is analyzed with the help of AMOS (21) for structural equation modelling and soft ware STATA for ordered logit and probit regressions.

The general hypothesis of the study projected a constructive and substantial rapport among service quality of e-banking, e-customer satisfaction and loyalty, as explained below:

- H (1): The 'Efficiency' of 'Internet-banking services' at Maurititus has constructive outcome on 'e-customer satisfaction'.
- H (2): The 'Reliability' of 'Internet-banking services' at Mauritius has constructive outcome on 'customer satisfaction'.
- H (3): The 'secured transaction' of 'Internet-banking services' at Mauritius has positive effect on satisfaction of customer.
- H (4): The 'responsiveness' and 'communication' of 'e-banking' services at Mauritius has constructive outcome on 'customer satisfaction'.

4 Analysis and Findings

4.1 Descriptive Statistics

The sample is normally distributed among male respondents and female respondents across different locations of Mauritius. male respondents constitute 62% and female respondents constitute 38%. Most of the respondents are below middle age, between 25 and 40 years of age group, indicating these groups are more technology usable and having affirmative attitude towards 'e-banking services'. Majority of them are working professionals, which shows the technology is very convenient ant comfortable for them to have safe transactions 24/7. Also, majority of them are earning relatively good pay. About 46% of the respondents are earning more than 30,000 MRU. Tables 1 and 2 explains the logit and probit regressions along with e-satisfaction and e-loyalty.

Linear Modelling Technique ('Ordinary Least Squares', OLS) is computed for testing the variables of e-banking technology. The findings of this analysis demonstrates that, there is an encouraging influence of e-satisfaction of consumer which

Table 1 Descriptive analysis of variables

Variables	Ordered Logit	Outcome 1	Outcome 2	Outcome 3	Outcome 4	Outcome 5
Age	−0.0241412 (−0.0267292)	0.000219 (−0.0002831)	0.0011146 (−0.0012498)	0.0017443 (−0.0019476)	−0.0000622 (−0.0004046)	−0.0030157 (−0.0033113)
Gender: (Baseline Female)	−0.3451426 (−0.3703387)	0.0031304 (−0.0037981)	0.0159359 (−0.0174679)	0.0249374 (−0.0269976)	−0.0008887 (−0.0057248)	−0.043115 (−0.0461423)
Male	−1.089145 (0.6390763)*	0.0098783 (−0.0081134)	0.0502881 (−0.032009)	0.0786934 (0.047082)*	−0.0028043 (−0.017925)	−0.1360555 (0.0788037)*
Educational achievement (Baseline Secondary)	0.874405 (−0.6095082)	0.0079307 (−0.0072847)	0.0403731 (−0.0297689)	0.0631779 (−0.0447184)	−0.0022514 (−0.0144572)	−0.1092303 (−0.0751946)
Diploma	−0.1626591 (−0.6347552)	−0.0026655 (−0.0035072)	0.0075103 (−0.0293919)	0.0117525 (−0.0459039)	−0.0004188 (−0.0031178)	−0.0203193 (−0.0793048)
Undergraduate						
Others						
Residential Area (Baseline Rural)	0.2938893 (−0.3367441	0.0050501 (0.0068247)*	−0.0135695 (0.0158642)	−0.0212342 (0.0243755)	0.0007567 (0.0048892)	0.0367125 (0.04193)
Urban						
Monthly Income (Baseline Above Rs 30,000)	−0.5568011 (0.6743996)	0.010865 (−0.0081415)	0.0257087 (0.031715)	0.0402302 (0.0490637	−0.0014336 (0.0093253)	−0.0695553 (0.0839034)

(continued)

Table 1 (continued)

Variables	Ordered Logit	Outcome 1	Outcome 2	Outcome 3	Outcome 4	Outcome 5
Below Rs 10,000	-1.197936 (0.56877)*	0.0085974 (0.0065394)	0.0553112 (0.0296822)**	0.0865538 (0.0425089)*	−0.0030844 (0.0196998)	−0.1496456 (0.069832)
Rs 10,000-Rs 20,000	−0.9479106 (0.48105)*	0.0008642 (0.0011761)	0.043767 (−0.0251075)	0.0684888 (0.0353908)*	−0.0024407 (0.0155643)	−0.1184125 (0.0594076)*
Rs 20,000–Rs 30,000	−0.0952824 (0.1218743)	0.0026659 (0.003989)	0.0043994 (0.0057626)	0.0068844 (0.0088914)	−0.0002453 (0.0015997)	−0.0119026 (0.015191)
Family size	+					
Civil Status (Baseline Single)	−0.2939363 (0.4235663)	−0.1459896 (8.116549)	0.0135716 (0.0198478)	0.0212376 (0.0308718)	−0.0007568 (0.0048959)	−0.0367184 (0.0530136)
Married	16.09623 (894.8483)	−0.002118 (0.0190263)	−−0.7431962 (41.31742)	−1.162992 (64.65523)	0.0414442 (2.319066)	2.010733 (111.784)
Divorce						
Others	0.2335176 (2.088246)	−0.0000519 (0.0029729)	−0.010782 (0.0964102)	−0.0168722 (0.1508215)	0.0006013 (0.006504)	0.0291709 (0.2608949)
e-Trust: Reliability	0.0057199 (0.3278752)	−0.0052278 (0.0042055)	−0.0002641 (0.0151402)	−0.0004133 (0.0236886)	0.0000147 (0.000847)	0.0007145 (0.0409595)
Efficiency	0.5763975 (0.3316944)	-0.0106083 (0.0070772)	-0.0266135 (0.0163688)*	-0.0416461 (0.0248006)*	0.0014841 (0.0094878)	0.0720033 (0.0408177)002A
Comfort	1.169632 (0.4259147)	−0.0025635 (0.003289)	−0.0540043 (0.0232697)*	−0.0845088 (0.0325553)*	0.0030115 (0.0190801)	0.1461099 (0.0529318)*
Confidence	0.2826459 (0.3221561) +	0.0016727 (0.0021545)	−0.0130503 (0.0150883)	−0.0204219 (0.023455)	0.0007278 (0.0046709)	0.035308 (0.0402463)
Secure transaction	0.1844275 (0.2127307)	−0.0045389 (0.0036447)	0.0085154 (0.01012)	0.0133253(0.0154903	−0.0004749 (0.003109)	−0.0230386 (0.0263928)

(continued)

Table 1 (continued)

Variables	Ordered Logit	Outcome 1	Outcome 2	Outcome 3	Outcome 4	Outcome 5
Dependence	0.5004434 (0.2778108) *	0.0221065 (0.0141747)	−0.0231065 (0.0141747	−0.0361582 (0.0207767)	0.0012885 (0.0082942)	0.0625151 (0.0339362)*
/cut1	0.5712443 (1.806576)					
/cut2l	3.13675 (1.74242)					
/cut3l	4.602404 (1.745298)					
/cut4l	8.03364 (1.805758)					

Source Authors own computation

Table 2 Various factors compiled from author sources

Variables	Mean	Standard Deviation	Minimum	Maximum
Dependent variable: Consumer satisfaction with internet banking in Mauritius	332.24865	0.8423639	1	5
Age	0.951351	9.192554	18	63
Gender				
Male				
Female				
Residential area	0.389189	0.4888895	0	1
Urban				
Rural				
Educational achievement				
Primary	0.6108108	0.4888895	0	1
Secondary				
Diploma	0.189189	0.392721	0	1
Undergraduate	0.459459	0.499706	0	1
Others	0.248648	0.499706	0	1
Income				
Level 1: Below Rs 10,000	0.102702	0.304394	0	1
Level 2: Rs 10,000–Rs 20,000	0.183783	0.388358	0	1
Level 3: Rs 20,000–Rs 30,000	0.194594	0.396962	0	1
Level 4: Above Rs 30000	3.627027	1.357877	0	8
Family Size	0.421621	0.495158	0	1
Marital status				
Single	0.016216	0.126648	0	1
Married	0.005405	0.073521	0	1
Divorced	3.88468	0.672016	1	5
Others	3.81982	0.723128	1	5
e-trust	3.68108	0.65406	1	5
Reliability	3.76576	0.760056	1	5
Efficiency	3.11351	0.905497	1	8.75
Comfort	3.33693	0.79551	1	5
Confidence				
Secure transaction				
Dependence				

Source Authors own computation

is a dependent variable, by the independent varibales, such as quality of e-banking services, reliability, secured transaction, responsiveness and efficiency. The results shows that, most of the variables are significant at 1%, 5 and 10% which is impacting the e-satisfaction and e-loyalty. Also, it is observed that pseudo R^2 is 0–1, which is quite acceptable for the study.

4.2 *Structural Equation Modelling (SEM)- Measurement Model*

SEM, a measurement model is used to evaluate the relationship among the latent construct and dependent variables. Factor Analysis (FA) and Confirmatory Factor Analysis (CFA) is computed, finally path diagram is drawn. This has been demonstrated in a path diagram, Fig. 2.

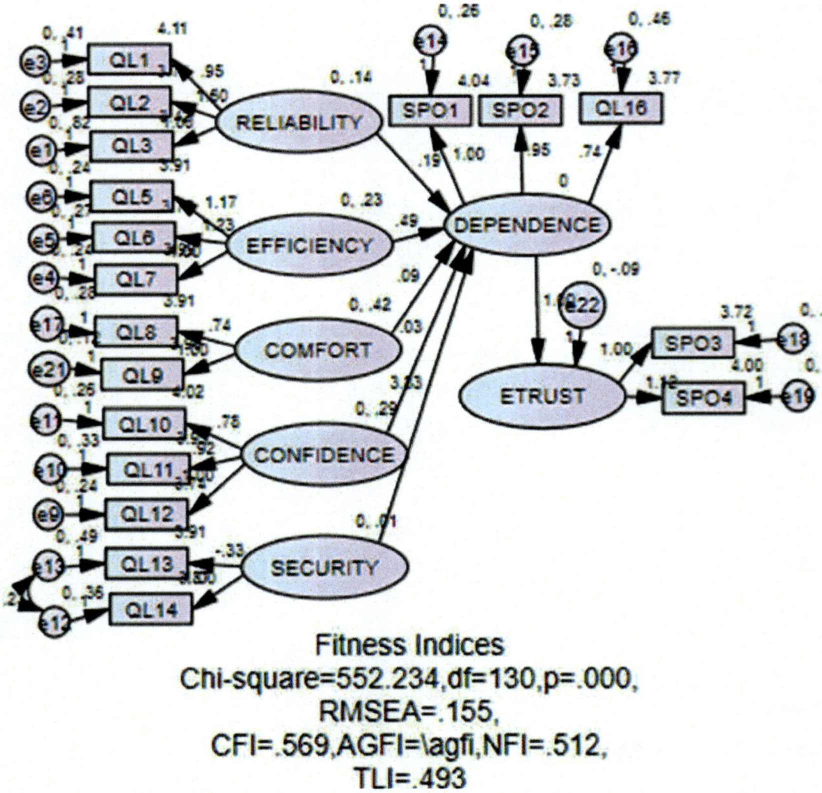

Fig. 2 Path diagram- E-satisfaction and e-loyalty *Source* Author's computation

Table 3 Modelfit indices (*t*-test and *p* value)

Statement on environmental perceptions	Mean	SD	*t* value	*P* value
Dependency	3.90	1.07	16.776	<0.001**
Security	4.10	1.03	21.291	<0.001**
Confidence	3.99	1.14	17.472	<0.001**
Comfort	3.90	1.21	14.878	<0.001**
Reliability	3.80	1.20	13.878	<0.001**

Note **denotes significant at 1% level
Source Author's computation

The measurement model shows standard model fit indices and found adequate. These are explained in a Table 3.

There are exogenous variables (quality of e-banking services, reliability, secured transaction, efficiency) and one endogenous Moreover, *t*-test for specified value of all the independent and dependent variables are at significant level which is displayed in Table 3. There fore, and e-satisfaction of the customer is acceptable.

5 Interpretation and Discussion

The study identifies the 'service quality' has significant rapport on e-customer satisfaction and e-loyalty. Among all, reliability, ease of use, secured transaction are the sturdiest dimensions of quality of e-banking services at Mauritius. This is evidenced and supported by many studies earlier [19]. The study also identifies there is a significant effect on the e-loyalty of Mauritian customers with reference to secured transactions and good communication. Finally, the dimensions that are selected to test empirically for the Mauritian context has the noteworthy effect on the 'e-banking adoption', e-cutomer satisfaction and e-loyalty, as supported by earlier studies in the literature.

6 Conclusion

The present research intendeds to test the effect of 'e-banking' adoption and 'service quality' on customer sataisfaction (e-satisfaction) and customer loyalty (e-loyalty). Referring to earlier studies in other countries (ref. literature section), no study has been done at Mauritian context so far. The study adopted quantitative approach and gathered the feedback from the respondents with the help of well structured questionnaire. The data were analysed and interpreted with the help of both STATA (ordered logit and probit regressions of e-trust and e-loyalty) and AMOS (SEM and path diagram). Study findings recommends that the set hypotheses for this research is satisfactorily reinforced for empirical evidence (primary data). The key impact of this study is, all the selected variables for testing, such reliability, secured transaction, ease of use, service quality, good communication are the main interpreter of e-satisfaction and e-loyalty.

Though the e-banking services are taking a shape and trying to stabilize, comparatively, it is still developing and need to upgrade the technology and infrastuctre in Mauritius. The country need to upgrade the infrastructural facilities for the banks to perform better and provide quality service. One among them may be a good governing with reference to strict law and security for the customer to have more trust on e-technology.

The study recommends for further study, based on the limitations encountered would be, most importantly, since the perception and understanding differs from person to person, deep investigation need to be done in testing socio-cultural, economic, and environmental contexts of different parts of the country carefully.

Acknowledgements Sincere thanks gratitude goes to Dr. M. Dhanamjaya, Registrar, REVA University, Dr. Ramesh, Director Planning, REVA University and Dr. S. Ramesh, Dean and Director, MCC.

References

1. Rust, R. T., & Oliver, R. W. (1994). The death of advertising. *Journal of Advertising, 23*(4), 71–77.
2. Stone, G., & Joseph, M. (2003). An empirical evaluation of US bank customer perceptions of the impact of technology on service delivery in the banking sector. *International Journal of Retail & Distribution Management, 31,* 190–202.
3. Lakhtaria, K. I., et al. (2011). The impact of the new Web 2.0 technologies in communication, development, and revolutions of societies. *Journal of Advances in Information Technology, 2,* 204–216.
4. Khadem, P., & Mousavi, S. (2013). Effects of self-service technology on customer value and customer readiness: The case of banking industry. *Management Science Letters, 3,* 2107–2112.
5. Lustsik, O. (2004, November). Can E-Banking services be profit-able? (University of Tartu Economics and Business Adminis-tration Working Paper No. 30-2004). https://doi.org/10.2139/ssrn.612762.
6. Shaikh, A. A., & Karjaluoto, H. (2015). Mobile banking adoption. A literature review. *Telematics and Informatics, 32*(1), 129–142.
7. Tieto. (2016). *History of banking.* http://www.tieto.com/sites/default/files/atoms/files/26nov_history_of_banking-reddad_v3.pdf [accessed:217-02-28].
8. Arcand, M., Prom Tep, S., Brun, I., & Rajaobelina, L. (2017). Mobile banking service quality and customer relationships. *International Journal of Bank Marketing, 35*(7), 1068–1089.
9. Asiyanbi, H., & Ishola, A. (2018). E-Banking services impact and customer satisfaction in selected bank branches in Ibadan metropolis, Oyo state, Nigeria. *Accounting, 4*(4), 153–160.
10. Churchill, G. A., & Surprenant, C. (1982). An investigation into the determinants of customer satisfaction. *Journal of Marketing Research, 19,* 491–504.
11. Oliver, R. L. (1981). Measurment and evaluation of satisfaction processes in retail settings. *Journal of Retailing, 57*(3), 25–48.
12. Tse, D. K., & Wilton, P. C. (1988). Models of consumer satisfaction formation: An extension. *Journal of Marketing Research, 25,* 204–212.

13. Kotler, P., & Keller, K. (2011). *Marketing management* (14th ed.). New York, NY: Pearson.
14. Koschate, N., et al. (2006). The role of cognition and affect in the formation of customer satisfaction: A dynamic perspective. *Journal of Marketing, 70*(3), 21–31.
15. Keller, K. L., & Lehmann, D. R. (2006). Brands and branding: Research findings and future priorities. *Marketing Science, 25,* 740–759.
16. Parasuraman, A., et al. (2002). SERVQUAL: A multiple-item scale for measuring consumer perceptions of service quality. *Retailing: Critical Concepts, 64*(1), 140–161.
17. Tan, M., & Teo, T. S. H. (2017). Factors influencing the adoption of intenet banking. *Journal of the AIS, 1*(1).
18. Hunain, M., et al. (2016). The impact of E-Banking on customer satisfaction: Evidence from banking sector of Pakistan. *Journal of Business Administration Research, 5*(2), 27–40.
19. Vanijikovan. (2017). Customer acceptance of online banking: an extension of the technology acceptance model. *Internet Research, 14*(3), 224–235